Introducing Biological Rhythms

INTRODUCING BIOLOGICAL RHYTHMS

A Primer on the Temporal Organization of Life, with Implications for Health, Society, Reproduction and the Natural Environment

Willard L. Koukkari and Robert B. Sothern

College of Biological Sciences
University of Minnesota
St. Paul, MN, USA

Willard L. Koukkari
Plant Biology
University of Minnesota
250 Bio Sciences Building
1445 Gortner Avenue
St. Paul, MN 55108
USA
koukk001@umn.edu

Robert B. Sothern
Plant Biology & Laboratory Medicine
  and Pathology
University of Minnesota
250 Bio Sciences Building
1445 Gortner Avenue
St. Paul, MN 55108
USA
sothe001@umn.edu

Library of Congress Control Number: 2005926926

ISBN-10: 1-4020-3691-4
ISBN-13: 978-1-4020-3691-0

Printed on acid-free paper.

© 2006 Springer Science+Business Media, Inc.

All rights reserved. This work may not be translated or copied in whole or in part without the written permission of the publisher (Springer Science+Business Media, Inc., 233 Spring Street, New York, NY 10013, USA), except for brief excerpts in connection with reviews or scholarly analysis. Use in connection with any form of information storage and retrieval, electronic adaptation, computer software, or by similar or dissimilar methodology now known or hereafter developed is forbidden.
The use in this publication of trade names, trademarks, service marks, and similar terms, even if the are not identified as such, is not to be taken as an expression of opinion as to whether or not they are subject to proprietary rights.

Printed in the United States of America.   (SPI/SBA)

9 8 7 6 5 4 3 2 1

springer.com

571.77 KOU 2006

Koukkari, Willard L.
Introducing biological
rhythms : a primer on the

"Be ruled by time, the wisest counselor of all."
—Plutarch (~46–120 AD)
Greek essayist and biographer

DEDICATION

To my family, colleagues and friends
and all who endured the many inquiries and requests
and otherwise offered their patience, understanding,
advice and support during the creation of this text (*RBS*).

Preface

Reading this book will not necessarily make you a better student of life. It will, however, help you to see, to hear, and to observe "that" which would otherwise escape your attention. In the present context, "that" represents the rhythms of life, which occur within us and around us.

This book has been written for students and others, including teachers in the life sciences, clinical personnel, biomedical researchers, and the general public, who are interested in the topic of biological rhythms. It is not a comprehensive treatise, but a "primer" that serves to introduce individuals to the important and fascinating topic regarding the rhythmic changes that affect all living organisms. During the more than three decades that we have been engaged in research and teaching the subject of biological rhythms, our students have depleted the available supply of different books[1] that provided supplemental reading material for these courses. Discussions with students, colleagues, and others often led to the question *"Where can I find out more about this topic?"* Faced with such a situation, we embarked upon the journey of writing this book, first putting fingers to the keyboard in the fall of 1998. Since biological rhythms are a fundamental property of all life and encompass a wide range of frequencies, from seconds to a century or more, we chose to write a book that is broad in scope, covers all major groups of organisms, and discusses rhythms with periods that span the ultradian, circadian, and infradian domains.

In many educational institutions, a biological rhythms course serves as an elective for satisfying a science requirement, as a seminar or for continuing education credit. Even for those who are outside the academic disciplines, the book serves as a primer that introduces the major characteristics of biological rhythms, cites scientific results and references, and discusses the implications and applications of rhythms. Essays provide in-depth historic and other background information that relate to the main topics.

---

[1] E. Bünning (1958) *The Physiological Clock.* Berlin: Springer-Verlag.

Moore-Ede MC, Sulzman FM, Fuller CA. (1982). *The Clocks That Time Us. Physiology of the Circadian Timing System.* Cambridge: Harvard University Press.

We trust that the following pages will serve as an easy to understand book that will enlighten and involve the novice, while being sufficiently technical and detailed for the experienced professional. The book includes a unique chapter on self-monitoring of one's own rhythms (autorhythmometry), which has been one of the more fascinating topics for our students in helping them to develop a greater understanding of the characteristics and role of biological rhythms in the temporal organization of their own lives.

Rather than focus on a few specialized areas, our aim has been to provide a book that broadly covers the field of biological rhythms, without the pretext of being completely comprehensive. For this reason, we may have overlooked certain topics or missed citing some key references. If we did not delve deeply enough into an area, we trust that the material presented will sufficiently intrigue the reader to seek out additional information.

New information on molecular clocks, genes, and pathways continues to accumulate at a rapid pace, as do findings from chronotherapy (timing of medications), investigations into sleep and its disorders, the clinical relevance of melatonin, and the effects of excess light or unusual lighting schedules on the health and survival of organisms, including humans. Names of authors and key words in the references cited provide a starting point for the search for additional information or clarification on topics of further interest.

Finally, while we have tried to be as conscientious as possible in citing and writing about topics and facts contained herein, an error now and then might have occurred. In spite of the eyes of many, any such errors that slipped by our attention are entirely our responsibility. We certainly welcome comments from the readers on any aspect of this book.

June 2005

Willard L. Koukkari
Robert B. Sothern

Acknowledgments

We wish to thank both old and new associates with whom we have collaborated and learned from over several decades. In addition, we are indebted to more than a hundred students who read and commented upon the many early drafts of various chapters as part of their biological rhythms coursework or independent study. We especially thank Van D. Gooch, Mary Hoff, and Tom Soulen for providing invaluable comments and suggestions to early versions of this book. We thank Kathleen Ball for checking facts in the final version against the references.

We gratefully acknowledge our many colleagues, co-workers, faculty, staff, students, and friends who provided invaluable assistance with several aspects of this book, including comments, data, drawings, photos, references, and reviews. These include (in alphabetical order):

Martina Abramovich, Mary Albury-Noyse, Kathy Allen, Tracy Anderson, Leonard Banaszak, Frank Barnwell, Evelyn Borchert, Robert Brambl, Ris Charvat, Anita Cholewa, Philip Clausen, Germaine Córnelissen-Guillaume, Jayna Ditty, Jay Dunlap, Ann Dunnigan, Mark Ebbers, Linda Eels, Roger Eggen, John Erwin, Jim Fuchs, David Gartner, Steve Gantt, Florence Gleason, Stuart Goldstein, William Gray, Francis Guillaume, Franz Halberg, David Hansen, Leon Hardman, Geoffrey Harms, Lynn Hartweck, Odette Holter, David Houghton, Roberta Humphries, Anders Johansson, Eugene Kanabrocki, Daniel Kaplan, John Kettinger, Marcia Koukkari, Mark Koukkari, Kristin Kramer, Mitch Lazar, Don Luce, Scott Madill, Julie Martinez, Krishona Martinson, C. Robertson McClung, David McGlaughlin, Robert McKinnell, Bernard Millet, Cathy Moore-Arcand, Kris Morgan, Mary Mortenson, David Mottet, Dwight Nelson, Sally Noll, Karen Oberhauser, Adeola Okusami, Neil Olszewski, Sandra Orcutt, James Orf, Dana Peterson, Richard Philips, Brian Piasecki, Nora Plesofsky, Korise Rasmusson, Anke Reinders, Tony Sanderfoot, Mark Sanders, Bill Schmidt, Janet Schottle, Michael Simmons, Carolyn Silflow, Gail Slover-Arcari, Pete Snustad, Jo Strand, Régine Terracol, John Tester, James Thompson, Edgar Wagner, John Ward, Chad Walstrom, Thomas Waters, George Weiblen, Andi Weydahl, and Michael Young.

To all of those listed above go our profound and heartfelt thanks for their unstinted and cheerfully rendered help!

Contents

Preface vii
Acknowledgments ix

1 The Study of Biological Rhythms
Introduction 1
A Time for Everything 4
Three Rhythm Domains 6
Implications of Body Clocks 11
Essay 1.1: Accidents and Catastrophes 11
Chronobiology: An Integrating Discipline 12
Chapters in This Book 14
Take-Home Message 15

2 General Features of Rhythms: Terminology and Characteristics
Introduction 19
Details of a Rhythm 20
    Period and Frequency 21
    Circadian Domain 21
    Essay 2.1: Use of the term "Circadian" 23
    Other Domains 25
    Essay 2.2: Time on Earth as We Know It 26
    Amplitude 29
    Phase 29
Genetics and Inheritance 29
    Period 30
    Amplitude and Phase 30
Primary Circadian Clocks 32
    The Brain 33
    SCN Identification 34
    SCN as Synchronizer 36
    Retinal Melanopsin 36
    Retinohypothalamic Tract 37

Characteristics of the Period 38
   Stability and Free-Running 38
   Frequency Multiplication/Demultiplication 39
   Light Quality 40
   Temperature 41
   Chemicals 42
Characteristics of the Amplitude 42
   Damping 43
Characteristics of the Phase 44
   Synchronizers 44
   Phase-Shifting by Light 44
   Phase Shifts by Chemicals or Temperature 45
   Phase–Response Curve 46
Masking 47
Need for a Cyclic Environment 48
In Darwin's Footsteps 49
   Exercise 2.1: An Ultradian Experiment: 52
      Circumnutation
   Exercise 2.2: A Circadian Experiment: 54
      Leaf Movements
   Mechanisms 56
   Take-Home Message 56

3 Physical and Biological Time

Introduction 66
Rotations and Revolutions of the Earth and Moon 67
   The Day 68
   The Year 69
   The Month 71
Clocks and Calendars: Ancient Times 72
   The Month and Year 72
   The Day 77
   The Week 78
Clocks and Calendars: Middle Ages to Now 79
   Longitude and Clocks 80
   Springs to Atoms 82
   Essay 3.1: Wristwatches 83
   Time Zones 84
   Daylight Saving Time 85
   Recording Date and Time 85
   Recording Biological Time 86
The 24-h Biological Clock Concept 87
   Early Studies 87
   Considerations 89

| Ultradian and Infradian Clocks 89
           Endogenous vs. Exogenous 90
           Comparison with Manufactured Clocks 90
       Evolution of the Clock 92
           Molecular Building Blocks 93
           Geological History and Rhythmic Components 95
           Adaptation to Avoid Harmful Light 99
           Ancestral Traits and Convergent Evolution 101
       Take-Home Message 102

4 Photoperiodism
       Introduction 107
       Daylight and Seasons 107
       Photoperiodism: The Process 111
       Response Types 113
           Critical Daylength 114
       Diversity of Responses 115
           Early Studies 116
           Latitude 116
       Light and Photoreceptive Regions 117
           Extraretinal Photoreceptors 118
           Pigments 119
           Cryptochromes 121
           Spectra 121
       Rhythmic Association 123
           Endogenous Oscillators 123
           Bünning's Hypothesis 125
           Phase–Response Curves 125
           Ultradian Cycles 125
           Circannual Cycles 126
           Bird Migration 126
           Deer Antlers 128
           Vernalization 128
       Photoperiodism and Humans 129
           Birth Patterns 129
           Indoor vs. Outdoor Light 130
           Disorders 130
       Take-Home Message 132

5 Biological Oscillators and Timers: Models and Mechanisms
       Introduction 138
       Approaches to Models and Mechanisms 139
       Mechanical Models 141
           Pendulum 142

xiv Contents

    Hourglass 142
    External Coincidence 143
    Hands of a Clock 144
Mathematical Models 145
    Differential Equations 148
    Limit Cycles and Topography 149
    Chaos 153
    Spatiotemporal Systems 153
Biochemical and Metabolic Models 155
    Chemical Systems 155
    Biochemical Systems 155
    Essay 5.1: Selected Biochemical Notes 156
    Glycolytic Oscillations 157
    Nucleotides and Enzymes 158
Membrane Models 158
    Essay 5.2: Membranes and the Phospholipid Bilayer 159
    Lipids and Proteins 163
    Transport and Feedback 165
Molecular Models 167
    Essay 5.3: From Genes to Proteins and Mutants 167
    Genes and Nomenclature 171
    Clock Mutations 173
    Circadian: System and Clock 173
    Transcription/Translation Feedback Loops 174
    Light 177
    Temperature 178
    Five Circadian Clocks 179
    Neurospora Circadian Clock 179
    Advantages 179
    Overt Rhythms 179
    Genetic Highlights 181
    Feedback Loops and Components 181
    Drosophila Circadian Clock 181
    Advantages 182
    Overt Rhythms 183
    Genetic Highlights 183
    Feedback Loops and Components 183
    Mammalian Circadian Clock 184
    Advantages 184
    Overt Rhythms 184
    Genetic Highlights 184
    Feedback Loops and Components 186
    Arabidopsis Circadian Clock 187
    Advantages 187
    Overt Rhythms 187

Genetic Highlights 187
Feedback Loops and Components 187
Cyanobacteria Circadian Clock 189
Advantages 190
Overt Rhythms 191
Genetic Highlights 191
Feedback Loops and Components 191
Models in Perspective 192
Generalized Schematic Model for 192
    Biological Rhythms
Take-Home Message... 193

6 Tidal and Lunar Rhythms

Introduction.. 207
Moon and Light 208
Moon and Tides 210
    High and Low Tides... 210
    Spring and Neap Tides... 214
    Earth Tide... 214
    Essay 6.1: Earth Tides.. 214
Marine Organisms.. 216
    Circatidal Rhythms 216
    Crab Activity.. 217
    Circadian vs. Circatidal 218
    Other Organisms 219
    Reproduction.. 220
    Color Change 224
Terrestrial Organisms 226
    The Menstrual Cycle.. 227
    Atmospheric Tides 227
    Insects 228
Plants.. 228
Lunar/Tidal Clock Hypotheses 229
    Circadian vs. Circalunidian 229
    Interacting Oscillators?.. 230
Take-Home Message... 231

7 Sexuality and Reproduction

Introduction.. 237
    Essay 7.1: Parasexuality 238
Nuclear Division and Genetics 240
    Essay 7.2: Mitosis, Meiosis, and 240
        the Punnett Square
Sex and Reproduction: The Difference...................................... 250

Essay 7.3: Artificial Hybridization and 251
How Sex Produces both "Lunch"
and an Embryo
Asexual Reproduction 255
Essay 7.4: An Abbreviated Life History of 256
Neurospora crassa
Courtship and Mating 257
    Photoperiodism and Sexuality 259
    Diet 259
    Flowers 260
Rhythmic Phases of Sexual Behavior in Humans 261
    Activity 261
    Disease 261
The Menstrual Cycle 262
    Duration and Phase 262
    Essay 7.5: Brief Physiology of Menstrual Cycle 262
    Events and Phases
    Social Synchronization 268
    Sexual Activity and Birth Control 268
Primary and Secondary Sex-Related Rhythms in Men 272
    Ultradian and Circadian Cycles 272
    Infradian Cycles 273
    Essay 7.6: 17th Century Notes of Monthly 273
    Rhythms in Males
    Body Weight 274
    Grip Strength 274
    Cutaneous Pain 277
    Hormones 277
    Essay 7.7: More About Infradians 277
    in Male Hormones
    Emotions 278
    Facial Sebum 279
    Beard Growth and Body Hair 280
    Sexual Activity 282
Take-Home Message 284

8 Natural Resources and Agriculture
Introduction 293
Photoperiodism 294
    Essay 8.1: Photoperiodism as a Basic 294
    Principle of Biology and Its Applications
Thermoperiodism and Temperature Cycles 297
    Vernalization 298
    Temperature Compensation 300
Migration 300

Birds.. 301
Butterflies... 302
Pest Management and Agents of Stress 304
Herbicides 304
Pest Control 306
Plant Responses to Injury 307
Plant Diseases 308
Production of Produce 309
Fisheries and Aquaculture 310
Weather Patterns and Agriculture 313
Gardens 313
Outdoor Hobbies 317
Birding 317
Fishing 318
Essay 8.2: Fly-fishing for Trout 318
Rural and Urban Development 322
Telemetry Tracking Systems 324
Muskrats 324
Squirrels and Foxes 324
Hare 324
Ruffed Grouse 326
The Outdoor Laboratory 327
Temporal Agroecosystems 327
Light Pollution 328
Aquatic Animals 328
Vertical Migration 329
Drift 330
Trout 330
Turtles 331
Insects 331
Birds 332
Take-Home Message 332

9 Veterinary Medicine

Introduction 341
Body Temperature and Activity 342
Diurnal vs. Nocturnal 343
Timing of Food 344
Masking 346
Environment 346
Cattle 348
Dogs and Cats 348
Poultry 349
Hematology and Urology 349
Sampling Blood 349

xviii   Contents

    Multiple Rhythms 349
    Peak Times 351
    Collecting Urine 351
    Excretion Rates 353
    Urinary Rhythms 354
    Interpreting a Sample 354
A Primary Circadian Oscillator 355
    The Suprachiasmatic Nucleus 356
    The Pineal Gland 356
Diseases, Pests, and Stress 357
    Parasites 357
    Bacterial Infections 358
    Seasonal Diseases 358
    Flies 359
    Fleas 360
Reproduction and Photoperiodism 360
    Photoperiod 361
    Melatonin 361
    Domestic Fowl 362
    Sheep 363
    Horses 364
    Pigs and Goats 365
    Artificial Insemination 365
    Semen Quality and Season 365
Implications 366
Take-Home Message 366

10 Society

Introduction 376
Past and Present 377
    The Natural Day vs. 24/7 377
    Time Schedules 378
Social Synchronization 380
    Essay 10.1: Social Synchrony in 380
        Animals
    Circadian Events 381
    Ultradian and Infradian Events 384
Aggression and Violence 384
Night and Shiftwork 385
    Time for Sleep 385
    Problems with Shiftwork 386
    Adjusting to Shiftwork 387
The Global Workplace 388
    Communication 388
    Work Schedules and Outsourcing 389

Sports and Performance 389
   Body Temperature and Performance Variables 391
   A Time to Train or Win 391
   Jet Lag and Professional Sports 392
   Allowing for Jet Lag 392
Travel on the Earth's Surface 393
   Driver Fatigue and Vehicle Accidents 394
   Alcohol, Driving, and Fatigue 395
   The Post-Lunch Dip 396
   Animal Activity and Vehicle Accidents 396
Travel Above the Earth's Surface 396
   Jet Lag 397
   Life in Space 398
Travel Beneath the Seas 399
Mealtimes and Health 401
   Essay 10.2: Preclinical Meal-Timing Studies 402
   Changes in Body Weight and Rhythms 404
   Food: What, How Much, and When 408
Light Pollution 408
   Effects on Melatonin Production 409
   Effects on Clinical Health 410
   Better Lighting Practices 411
Pseudoscience: Birthdate-Based Biorhythms 412
   Essay 10.3: Development of the Biorhythm 412
      "Theory"
   Lack of Scientific Support 413
   Rigidity vs. Elasticity of Infradian Periods 414
   An Oversimplification of Rhythms 414
Take-Home Message 415

11 Clinical Medicine
Introduction 426
Circadian Rhythms in Health 437
   Essay 11.1: Adjusting Urinary Concentrations 438
      for Volume and Time
Overview of Rhythms in Body Systems 430
   What and When is Normal? 433
   Time-Specified Normal Limits 438
Circadian Rhythms in Symptoms and Disease 441
   Birth and Death 442
   Cardiovascular Disease 443
Circannual Rhythms in Health 445
Circannual Rhythms in Symptoms and Disease 446
   Cardiovascular Disease 448
   The Coagulation System 449

xx   Contents

    Cholesterol 450
    Respiratory Illness 452
    Mental Disorders 452
    Seasonal Affective Disorder 452
The Menstrual Cycle 453
    Disorders 454
    Medical Procedures 455
    Male Cycles 455
Melatonin and Human Health 456
    Darkness and Melatonin 456
    Sexuality 458
    Immune Function 459
    Light, Melatonin and Cancer 460
    Light Leaks at Night 461
When to Sample? 462
    Diagnosing Normal Levels 463
    Diagnosing Infectious Agents 464
    Diagnosing Abnormal Levels 467
    Using Rhythm Characteristics in Diagnosis 467
Hours of Changing Resistance 470
    Early Pre-Clinical Findings 471
    Time-Related Responses to 471
        Anti-Cancer Drugs
    Stage of Rhythm vs. Time of Day 473
    Varying Positive or Negative Effects 473
Timing Treatment: Chronotherapy 475
    Three Times a Day? 476
    Constant Dosing 476
    Rhythm-Dependent Effects of Some Drugs 478
    Administering Chronotherapy 478
Examples of Applied Chronotherapy 480
    Asthma 481
    Cancer–Animal Studies 482
    Cancer–Human Trials 484
Cellular Clocks and Chronotherapy 486
    Time-Indicating Genes 487
    Molecular Machinery Underlies Physiology 487
Marker Rhythms 488
The Medical Community and the Concept of Timing 489
Take-Home Message 491

12 Autorhythmometry

Introduction 526
Measuring Your Own Body Rhythms 527
    School Children 528

Adults 529
　　Performance 529
　　Ultradian Rhythms 529
　　Self-help Health Care 529
　　Monitoring Symptoms 530
Body Temperature 530
　　Internal Marker Rhythm 531
　　Measurement Site 532
　　What and When is Normal? 533
Blood Pressure 534
　　Monitoring Hypertension 534
　　Ambulatory Monitoring 535
Morningness–Eveningness 535
　　Questionnaires 535
　　Morningness vs. Life Factors 536
　　Endogenous Disposition 536
　　Body Temperature Phase 536
　　Cognitive Tasks 537
When and How Long to Measure? 537
　　Self-Measurements During Travel 537
　　Self-Measurements During Isolation 540
　　Essay 12.1: Self-Measurements in 540
　　　　"Aschoff's Bunker"
　　Long Self-Measurement Series 544
What can be Self-Measured? 545
Equipment 545
　　Internal or External Body Temperature 546
　　Temperature Devices 547
　　Blood Pressure Devices 550
　　Automatic Devices 550
　　Other Equipment 552
　　Saliva, Urine, and Blood 555
Procedures for Self-Measurements 555
　　Keeping Records 555
　　Sampling Sequence 555
Looking at the Data 556
　　Making Graphs 556
　　Testing for a Time-Effect 557
Take-Home Message 558
Appendix
　　Item 1: Sample sheet for recording oral 568
　　　　temperature and other functions
　　Item 2: Sample sheet for recording and 569
　　　　graphing temperature, pulse and
　　　　blood pressure

xxii   Contents

Item 3: Detailed instructions for performing 570
  self-measurements
Item 4: Random Number Adding Speed Test 572
  sample pages
Item 5: Random Number Memory Test 574
  sample pages

13 Chronobiometry: Analyzing for Rhythms
Introduction.. 577
Data Collection 578
  Number of Timepoints 579
  How Long? 580
  Sampling Often Enough 580
  Aliasing 580
  Decision Making 580
Data Preparation 582
  Graphs and Visual Inspection 582
  Editing or Transforming Data 583
  Essay 13.1: Standard Deviation and Error 584
  Normalizing Data 584
  Partitioning Data Spans 585
Statistical Detection of Time Effects 586
  Using Two Timepoints 586
  Using Three or More Timepoints 586
Statistical Detection of Rhythms 588
  Analyzing Time-Series by Standard Methods 588
  Limitations of Standard Methods 588
  Analyzing Time-Series by Curve Fitting 589
  The Least-Squares Technique 590
  The Best-Fitting Curve 591
  Statistical Significance 592
  Complex Waveforms 592
  Rhythm Parameter Comparisons 593
  Lack of Rhythm Detection 593
Descriptive Rhythm Parameters 595
  The Cosinor Illustrated 596
Example of a Cosinor Program 596
Take-Home Message 600

Author Index **603**
Subject Index **649**

1
The Study of Biological Rhythms

"Dost thou love life? Then do not squander time,
for that's the stuff life is made of."
—Benjamin Franklin (1706–1790)
American Statesman & Philosopher

Introduction

How could you tell if a person is alive or dead? Chances are that your answer will depend upon the presence or absence of a rhythm, the beating of the heart. Biological rhythms are inherent to life itself and can be detected by all the senses. We can see them, hear them, feel them, smell them, and we may even taste evidence of them (Figure 1.1). Perhaps the sense of time itself could be considered a sixth sense. In many ways we can sense how long a time has elapsed since some occurrence was last noted, as well as the time of day, time of month, and time of year from cues all around us (Hering, 1940; Binkley, 1990). Life moves in synchrony to the beat of clocks and calendars, some outside the body and some within the very cells of all living things. Rhythms are among the common strands from which the web of life itself is spun.

By definition, a *rhythm* is a change that is repeated with a similar pattern. Humans, like all other organisms that inhabit this Earth, have a rhythmic order underlying life. Actually, change, not constancy, is the norm for life and the rhythmic timing of change makes predictability a reality. For example, throughout each 24-h day, leaves change their orientation (Figure 1.2), human body temperature rises and falls (Figure 1.3), fungi time their sporulation (Figure 1.4), and activity levels fluctuate (Figure 1.5). These rhythmic changes of life represent only a small segment of an enormous network of biological rhythms, passed on from one generation to the next.

All known variables of life, be they levels of potassium ions in a cell, stages of sleep, or the opening and closing of flowers, have either directly or indirectly been found to display rhythms. Furthermore, adaptations of organisms for survival relative to geophysical cycles, such as the solar day, seasons, and tides, attest to the evolution of the genetic aspects of certain types of rhythmic timing. The objective

4   1. The Study of Biological Rhythms

FIGURE 1.5. Comparison of activity in animals and position of leaves in plants over 24 h. Diurnal animals, such as humans and birds, are active during the day, while nocturnal animals, such as hamsters and rodents, are at rest (*top panel*). At night, the reverse happens and while the diurnal animals are at rest, the nocturnal animals, such as the hamster and rat, are active (*bottom panel*). Also, the orientation of leaves changes from the horizontal during the day to the vertical at night.

of this chapter is to briefly introduce the topic of biological rhythms by emphasizing the commonality, presence, and importance of the subject in nearly every facet of life, as well as its status as an integrating discipline of biology.

A Time for Everything

The rhythmic nature of life influences the very existence of organisms, commencing before conception and extending beyond death. Rhythms may be the most ubiquitous, yet overlooked, phenomena of life (Luce, 1970). They are such an integral part of life that the absence or perturbation of specific oscillations (e.g., brain waves and heart beats) in humans and other animals is used in the practice of medicine to distinguish between life and death, as well as between illness and good health (Figures 1.6 and 1.7). How organisms respond to certain chemicals, be they drugs or house dust, insecticides or herbicides, or any of a long list of agents is often time dependent (Table 1.1), with effects more pronounced at certain times than others. In some cases, what may be beneficial at one time may be noneffective or lethal at another. Even the perception of pain, be it disease-related (e.g., from headache, arthritis) or experimentally induced (e.g., by heat or cold or electrical stimuli), often shows a daily rhythm (Labrecque, 1992).

Pincus DJ, Humeston TR, Martin RJ. (1997) Further studies on the chronotherapy of asthma with inhaled steroids: the effect of dosage timing on drug efficacy. *J Allergy Clin Immunol* 100(6 Pt 1): 771–774.

Presser HB. (1974) Temporal data relating to the human menstrual cycle. In: *Biorhythms and Human Reproduction*. Ferin M, Halberg F, Richert RM, Vande Wiele R, eds. New York: Wiley, pp. 145–160.

Reinberg A, Zagula-Mally Z, Ghata J, Halberg F. (1969) Circadian reactivity rhythm of human skin to house dust, penicillin, and histamine. *J Allergy* 44(5): 292–306.

Reinberg A, Clench J, Aymard N, Galliot M, Bourdon R, Gervais P, Abulker C, Dupont J. (1975) [Circadian variations of the effects of ethanol and of blood ethanol values in the healthy adult man. Chronopharmacological study] [French]. *J Physiol (Paris)* 70(4): 435–456.

Reinberg A, Reinberg MA. (1977) Circadian changes in the duration of local anesthetic agents. *Naunyn-Schmiedebergs Arch Pharmacol* 297: 149–152.

Reinberg A, Smolensky MH. (1985) Chronobiologic considerations of the Bhopal methyl isocyanate disaster. *Chronobiol Intl* 2(1): 61–62.

Reinberg A, Pauchet F, Ruff F, Gervais A, Smolensky MH, Levi F, Gervais P, Chaouat D, Abella ML, Zidani R. (1987) Comparison of once-daily evening versus morning sustained-release theophylline dosing for nocturnal asthma. *Chronobiol Intl* 4(3): 409–419.

Scheving LE, Vedral D, Pauly JA. (1968) Circadian susceptibility rhythm in rats to pentobarbital sodium. *Anat Rec* 160(4): 741–750.

Smiley AM. (1990) The Hinton train disaster. *Accid Anal Prev* 22(5): 443–455.

Smolensky M, Halberg F, Sargent F. (1972) Chronobiology of the life sequence. In: *Advances in Climatic Physiology*. Ito S, Ogata K, Yohimura H, eds. Tokyo: Igaku Shoin, pp. 281–318.

Spruyt E, Verbelen J-P, De Greef JA. (1987) Expression of circaseptan and circannual rhythmicity in the imbibition of dry stored bean seeds. *Plant Physiol* 84: 707–710.

Spruyt E, Verbelen J-P, De Greef JA. (1988) Ultradian and circannual rhythmicity in germination of *Phaseolus* seeds. *J Plant Physiol* 132: 234–238.

Sundararaj BI, Vasal S, Halberg F. (1982) Circannual rhythmic ovarian recrudescence in the catfish, *Heteropneustes fossilis* (Bloch). In: *Toward Chronopharmacology*. Takahashi R, Halberg F, Walker C, eds. New York: Pergamon, pp. 319–337.

Svanes C, Sothern RB, Sørbye H. (1998) Rhythmic patterns in incidence of peptic ulcer perforation over 5.5 decades in Norway. *Chronobiol Intl* 15(3): 241–264.

Willich SN, Levy D, Rocco MB, Tofler GH, Stone PH, Muller JE. (1987) Circadian variation in the incidence of sudden cardiac death in the Framingham Heart Study population. *Amer J Cardiol* 60: 801–806.

Jores A. (1938) First Conf. Ronneby, Sweden, August 13–14, 1937. *Dtsch Med Wochenschr* 64(21/28): 737–989.
Jores A. (1975) The origins of chronobiology: an historical outline. *Chronobiologia* 2(2): 155–159.
Kalmus H. (1974) The foundation meeting of the international society for biological rhythms. *Chronobiologia* 1: 118–124.
Kettlewell PS, Sothern RB, Koukkari WL. (1999) U.K. wheat quality and economic value are dependent on the North Atlantic Oscillation. *J Cereal Science* 29: 205–209.
Kleitman N. (1963) *Sleep and Wakefulness*, 2nd edn, Chicago, IL: University of Chicago Press.
Koukkari WL. (1974) Rhythmic movements of *Albizzia julibrissin* pinnules. In: *Chronobiology*. Scheving LE, Halberg F, Pauly JE, eds. Tokyo: Igaku Shoin, pp. 676–678.
Koukkari WL, Johnson MA. (1979) Oscillations of leaves of *Abutilon theophrasti* (velvetleaf) and their sensitivity to bentazon in relation to low and high humidity. *Physiol Plant* 47: 158–162.
Koukkari WL, Duke SH, Bonzon MV. (1985) Circadian rhythms and their relationships to ultradian and high frequency oscillations. In: *Les Mecanismes de l'Irritabilité et du Fonctionnment des Rythmes chez les Végétaux, 1977–1983*. Grippin H, Wagner E, eds. Genève: Université de Genève, pp. 106–126.
Kripke DF. (1972) An ultradian biological rhythm associated with perceptual deprivation and REM sleep. *Psychosom Med* 34(3): 221–234.
Kyriacou CP, Hall JC. (1980) Circadian rhythm mutations in *Drosophila melanogaster* affect short-term fluctuations in the male's courtship song. *Proc Natl Acad Sci USA* 77(11): 6729–6733.
Labrecque G. (1992) Inflammatory reaction and disease states. In: *Biological Rhythms in Clinical and Laboratory Medicine*. Touitou Y, Haus E, eds. Berlin: Springer-Verlag, pp. 483–492.
Lavie P. (1979) Ultradian rhythms in alertness—a pupillometric study. *Biol Psychol* 9(1): 49–62.
Lloyd D, Edwards SW, Fry JC. (1982) Temperature-compensated oscillations in respiration and cellular protein content in synchronous cultures of *Acanthamoeba castellanii*. *Proc Natl Acad Sci (USA)* 79(12): 3785–3788.
Lovett-Douse JW, Payne WD, Podnieks I. (1978) An ultradian rhythm of reaction time measurements in man. *Neuropsychobiology* 4(2): 93–98.
Luce G. (1970) *Biological Rhythms in Psychiatry and Medicine*. Washington, DC: Natl Inst Mental Health, US Dept. Health, Education and Welfare, 183 pp.
Ludwig H, Hinze E, Junges W. (1982) Endogene Rhythmen des Keimverhaltens der Samen von Kartoffeln, insbesondere von *Solanum acaule*. *Seed Sci Technol* 10: 77–86.
Marshall J. (1977) Diurnal variation in the occurrence of strokes. *Stroke* 8(2): 230–231.
Martinson KB, Sothern RB, Koukkari WL, Durgan BR, Gunsolus JL. (2002) Circadian response of annual weeds to Glyphosate and Glufosinate. *Chronobiol Intl* 19(2): 405–422. [*Erratum: Chronobiol Intl* 2002; 19(4): 805–806]
Millet B, Melin D, Bonnet B, Ibrahim CA, Mercier J. (1984) Rhythmic circumnutation movement of the shoots in *Phaseolus vulgaris* L. *Chronobiol Intl* 1(1): 11–19.
Moore-Ede M. (1993) *The Twenty-Four Hour Society: Understanding Human Limits in a World That Never Stops*. Reading, MA: Addison-Wesley, 230 pp.
Pengelley ET, Fisher KC. (1963) The effect of temperature and photoperiod on the yearly hibernating behavior of captive golden-mantled ground squirrels (*Citellus lateralis testcorum*). *Can J Zool* 41: 1103–1120.

DeCoursey PJ. (1960) Phase control of activity in a rodent. In: *Biological Clocks. Cold Spring Harbor Symposia on Quantitative Biology*, Vol 25. New York: Long Island Biol Assoc, pp. 49–55.

DeVecchi A, Halberg F, Sothern RB, Cantaluppi A, Ponticelli C. (1981) Circaseptan rhythmic aspects of rejection in treated patients with kidney transplant. In: *Chronopharmacology and Chronotherapeutics*. Walker CA, Winget CM, Soliman KFA, eds. Tallahassee: Florida A&M University Foundation, pp. 339–353.

Drake DJ, Evans JW. (1978) Cortisol secretion pattern during prolonged ACTH infusion in dexamethasone treated mares. *J Interdiscipl Cycle Res* 9: 89–96.

Eesa N, Cutkomp LK, Cornélissen G, Halberg F. (1987) Circadian change in Dichloros lethality (LD50) in the cockroach in LD 14:10 and continuous red light. In: *Advances in Chronobiology, Part A*. Pauly JE, Scheving LE, eds. New York: Alan R. Liss, pp. 265–279.

Ehret CF. (1980) On circadian cybernetics, and the innate and genetic nature of circadian rhythms. In: *Chronobiology: Principles and Applications to Shifts in Schedules*. Scheving LE, Halberg F, eds. Alphen aan den Rijn: Sijthoff & Noordhoff, pp. 109–125.

Gervais P, Reinberg A, Gervais C, Smolensky MH, De France O. (1977) Twenty-four-hour rhythm in the bronchial hyperreactivity to house dust in asthmatics. *J Allergy Clin Immunol* 59(3): 207–213.

Goss RJ. (1969) Photoperiodic control of antler cycles in deer: I. Phase shift and frequency changes. *J Exp Zool* 170: 311–324.

Guillaume FM, Koukkari WL. (1987) Two types of high frequency oscillations in *Glycine max* (L.) Merr. In: *Advances in Chronobiology, Part A*. Pauly JE, Scheving LE, eds. New York: Alan R. Liss, pp. 47–58.

Gullion GW. (1982) Forest wildlife interactions. In: *Introduction to Forest Science*. Young RA, ed. New York: Wiley, pp. 379–407.

Gullion GW. (1985) Ruffed grouse research at the University of Minnesota Cloquet Forestry Center. *Minn Dept Nat Res Wildl Res Unit 1985 Report*, pp. 40–49.

Gwinner E. (1977) Circannual rhythms in bird migration. *Annu Rev Ecol Syst* 8: 381–405.

Gwinner E. (2003) Circannual rhythms in birds. *Curr Opin Neurobiol* 13(6): 770–778.

Halberg Fcn, Halberg F, Sothern RB, Pearse JS, Pearse VB, Shankaraiah K, Giese AC. (1987) Consistent synchronization and circaseptennian (about 7-yearly) modulation of circannual gonadal index rhythm of two marine invertebrates. In: *Advances in Chronobiology—Part A*. Pauly JE, Scheving LE, eds. New York: Alan R. Liss, pp. 225–238.

Haus E, Halberg F, Scheving LE, Pauly JE, Cardoso S, Kühl JFW, Sothern RB, Shiotsuka RN, Hwang DS. (1972) Increased tolerance of leukemic mice to arabinosyl cytosine with schedule adjusted to circadian system. *Science* 177(43): 80–82.

Henson CA, Duke SH, Koukkari WL. (1986) Rhythmic oscillations in starch concentration and activities of amylolytic enzymes and invertase in *Medicago sativa* nodules. *Plant Cell Physiol* 27: 233–242.

Hering DW. (1940) The time concept and time sense among cultured and uncultured peoples. In: *Time and Its Mysteries, Series II*. New York: New York University Press, pp. 3–39.

Holaday JW, Martinez HM, Natelson BH. (1977) Synchronized ultradian cortisol rhythms in monkeys: persistence during corticotropin infusion. *Science* 198(4312): 56–58.

Hoppenstaedt FC, Keller JB. (1976) Synchronization of periodical cicada emergences. *Science* 194(4262): 335–337.

Janzen DH. (1976) Why bamboos wait so long to flower. *Annu Rev Ecol Syst* 7: 347–391.

Johnsson A. (1973) Oscillatory transpiration and water uptake of Avena plants: I. Preliminary observations. *Physiol Plant* 28: 40–50.

organisms (be they plant or animal) to given durations and times of light and darkness over the seasons, is included as a chapter. Chapters that deal with tidal and lunar rhythms, sexuality and reproduction, natural resources and agriculture, society, and clinical and veterinary medicine, describe from a practical perspective how knowledge of biological rhythms applies to, or affects, nearly all aspects of life.

Take-Home Message

Biological rhythms are common phenomena of life and are found in all major groups of organisms, but are often overlooked. The length of time required to repeat a rhythmic cycle is called the period, a characteristic that has been used to categorize rhythms into three major groups: circadian (20–28 h), ultradian (<20 h), and infradian (>28 h). Circadian rhythms are usually synchronized by cyclic changes in light and darkness and/or temperature. When isolated from such environmental cues, the rhythm continues (free-runs), but usually with a period slightly longer or shorter than precisely 24.0 h. Some typical examples of biological variables for rhythms include (a) *circadian*: body temperature in humans and leaf movements of plants; (b) *ultradian*: brain waves of humans and twining of movements of bean shoots; and (c) *infradian*: the menstrual cycle of human females and the annual germination of certain seeds.

References

Aschoff J, Giedke H, Pöppel E, Wever R. (1972) The influence of sleep-interruption and of sleep-deprivation on circadian rhythms in human performance. In: *Aspects of Human efficiency. Diurnal Rhythm and Loss of Sleep*. Colquhoun WP, ed. London: English University Press, pp. 133–150.
Aserinsky E, Kleitman N. (1953) Regularly occurring periods of eye motility, and concomitant phenomena, during sleep. *Science* 118(3062): 273–274.
Balzer I, Neuhaus-Steinmetz U, Quentin E, van Wüllen M, Hardeland R. (1989) Concomitance of circadian and circa-4-hour ultradian rhythms in *Euglena gracilis. J Interdiscipl Cycle Res* 20: 15–24.
Binkley S. (1990) *The Clockwork Sparrow*. Englewood Cliffs, NJ: Prentice-Hall, 262 pp.
Bitman J, Lefcourt A, Wood DL, Stroud B. (1984) Circadian and ultradian temperature rhythms of lactating dairy cows. *J Dairy Sci* 67(5): 1014–1023.
Bruguerolle B, Prat M, Douylliez C, Dorfman P. (1988) Are there circadian and circannual variations in acute toxicity of phenobarbital in mice? *Fundam Clin Pharmacol* 2(4): 301–304.
Bünning E, Müssle L. (1951) Der Verlauf der endogen Jahresrhythmik in Samen unter dem Einfluss verschiedenartiger Aussenfaktoren. *Z Naturforsch* 6b: 108–112.
Chance B, Estabrook RW, Ghosh A. (1964) Damped sinusoidal oscillations of cytoplasmic reduced pyridine nucleotide in yeast cells. *Proc Natl Acad Sci (USA)* 51: 1244–1251.
Chiba Y, Cutkomp LK, Halberg F. (1973) Circaseptan (7-day) oviposition rhythm and growth of Spring Tail, *Folsomia Candida* (Collembola: Isotomidae). *J Interdiscipl Cycle Res* 4: 59–66.

Implications of Body Clocks

We have become a "clock driven" society, one that arranges time according to the demands of a commercial or industrial complex often fostered by profit and/or leisure, rather than arranging time so that our internal body clocks are in synchrony with the natural environment of this "clockwork Earth." Accidents, catastrophes, and illnesses are inevitable when the time cycle of society does not heed the biological rules that underlie the rhythms of humans or other organisms.[3] For example, the reason that many traffic accidents occur during night and early morning hours is not only due to the difficulty in seeing in reduced light, but also to a decline in the alertness of the driver who is trying to overcome the physiological urge to sleep. Alertness is but one of the many performance variables that are under rhythmic control (Figure 1.10). It may be no coincidence that some of the major industrial catastrophes of our time were not related to weather, but were associated with erratic work–rest schedules and fatigue (see Essay 1.1).

Essay 1.1 (by RBS): Accidents and Catastrophes

Four major industrial accidents that gained worldwide attention were each attributed to human error due to shift-work and its accompanying fatigue. The methyl isocyanate chemical accident in Bhopal, India, nuclear accidents at 3-Mile Island in the USA and in Chernobyl in the Ukraine, and the oil spill from the tanker Exxon Valdez in Alaska, all occurred at night (Reinberg & Smolensky, 1985; Moore-Ede, 1993). The nuclear accidents resulting in the release of radioactive gases occurred at 3-Mile Island at 04:00 h on March 28, 1979 and at 01:23 h in Chernobyl on April 26, 1986. The Exxon Valdez accident resulting in an environmentally catastrophic oil spill, occurred at 00:04 h on March 24, 1989, while the gas leak in Bhopal occurred at 00:56 h on December 3, 1984. All of these accidents occurred after midnight and were determined to be the result of operator error due to poor work–rest schedules, monotony, and fatigue. Interestingly, while hundreds of villagers and thousands of cattle in Bhopal died as they slept when the gas cloud swept over them, night-shift workers in the chemical plant, and nocturnally active rats observed scurrying around dead bodies all survived with little, if any, ill-effects from the toxic gas. Presumably, due to internal circadian time (stage) and not the external clock time, they were less susceptible to the toxic effects of the gas during the activity portion of their circadian sleep–wake cycle: rats are naturally active during the night, while the workers were in the active portion of their shifted circadian cycle.

Another less well-known, but major disaster occurred in western Canada on February 8, 1986 at 08:41 h, when a freight train going 94 km/h collided head-on with a passenger train going 78 km/h, killing 23 people and causing over $30 million in damages (Smiley, 1990). Weather was not a factor, nor was there any hardware error; a red light was ignored by the engineer of the freight train and no brake application was made. The night before the accident, the freight engineer slept only 3.5 h, while the brakeman slept only 5 h before their train departed at 06:25 AM. Both had worked irregular schedules during

---

[3] This statement does not refer to the popular pseudoscience notion that good days and bad are based upon birth date (see discussion on Biorhythm Theory in Chapter 10).

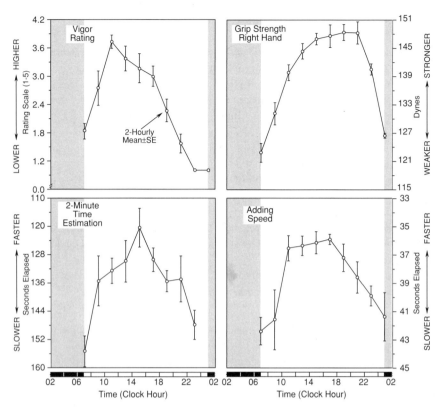

FIGURE 1.10. Circadian patterns in some performance tests by a young man (RBS, 23 years of age) doing self-measurements every 2–4 h during waking-only for 2 weeks reveal that mental and physical performance were at their best in the middle of the daily activity span. Two-hour averages computed from a total of 90 measurements. Dark bar and shaded area indicate times of sleep when no data were obtained.

the previous month, with the engineer finishing a 6-h run at 23:30 h the night before. It was concluded that the number one factor underlying this catastrophe was human error brought on by the interaction between 24-h biologic rhythms and lengthy, irregular work–rest schedules, and consequent negative effects on alertness and performance of the freight crew.

The broad and interesting topic of the applications and implications of rhythms in society as concerns both humans and organisms in their environment will be discussed in Chapter 10 on Society.

Chronobiology: An Integrating Discipline

The study of *biological rhythms*, known as *chronobiology*, is an integrating discipline that has been ranked parallel with the more classical disciplines of

TABLE 1.4. A comparison among the four disciplines of biology relative to duration and episodic designation (cf. Ehret, 1980).

Discipline	Period or duration	Episodic designation
Biological rhythms	Less than a second to years	Existence
Development	Seconds to years	Life span
Genetics	Minutes to years	Dynasty
Evolution	Hours to years	Age

development, *genetics*, and *evolution* (Ehret, 1980). All four of these disciplines span the structural levels of organization from molecules to ecosystems. A comparison among the four disciplines in regard to duration and episodic designation is presented in Table 1.4. The inclusion of chronobiology in formulating questions and hypotheses that lend themselves to experimentation now provides a broad unifying approach that extends from single-celled organisms to higher plants and animals. This can be seen by the continuing annual increase in the number of published scientific papers, which are accruing in the thousands per year since the late 1960s, and can be found in an online literature search[4] that uses circadian and other rhythm domain terms as key words (Figure 1.11).

Historically, the interdisciplinary nature of chronobiology was evident even in the composition of the small group of seven individuals who met in August 1937 in Sweden to create the first international organization focusing on the study of biological rhythms.[5] Of the three review papers presented at this conference that were of a more theoretical nature, one was based upon plant rhythms (by Anthonia Kleinhoonte from Holland), the second focused on animal rhythms (by Hans Kalmus from England), and the third dealt with human rhythms (by Arthur Jores from Germany) (Jores, 1938, 1975) (cf. Kalmus, 1974).[6]

---

[4] The online search engine PubMed (http://www.ncbi.nlm.nih.gov/PubMed/) was used for this literature search of articles indexed at the National Center for Biotechnology Information of the National Library of Medicine. There are hundreds of additional papers on rhythms in books, meeting proceedings, and journals, some of which can be found with other online search engines (e.g., Medline, BIOSIS, AGRICOLA).

[5] The next meeting of the International Society for the Study of Biological Rhythms was held with 12 participants in Holland just before World War II. The third meeting took place in Hamburg, Germany in 1949, with 50 participants. This organization changed its name to the International Society for Chronobiology at its first meeting in the USA in 1971 and included several hundred participants. There are now numerous societies studying rhythmic phenomena, including the American Association of Medical Chronobiology and Chronotherapeutics; the European Pineal Society; the European Society for Chronobiology; Groupe d'Etude des Rythmes Biologiques; Societa Italiana di Cronobiologia; the Japanese Society for Chronobiology, the Society for Light Treatment and Biological Rhythms; the Society for Research on Biological Rhythms; the Mediterranean Society for Chronobiology; among others.

[6] For historical notes see *Bulletin du Groupe d'Etude des Rythmes Biologiques* (1989) 21(1) (1er trimestre); and Cambrosio A, Keating P. (1983) *Social Studies of Science* 13: 323–353.

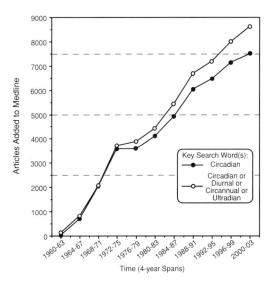

FIGURE 1.11. The number of articles on biological rhythms found by an online search engine (PubMed) using the key words circadian, diurnal, circannual, and ultradian, now number in the thousands over spans of 4 years. There are hundreds of additional papers on rhythms in books, meeting proceedings, and journals, some of which can be found with other online search engines (e.g., Medline, BIOSIS, AGRICOLA, etc.).

Chapters in This Book

In this book, chapters have been prepared around main topics on biological rhythms. However, in some cases there is considerable overlap of material among chapters, which is unavoidable due to the ubiquity of the topic. Biological rhythms have a number of unique characteristics and terms that will be introduced in a separate chapter, and referred to throughout the book. Some of these characteristics, such as the stability of the period of circadian rhythms and how their phase[7] can be shifted, have been incorporated into models to better understand the mechanisms of timing (see Chapter 5 on Models) and how they contribute to the temporal organization of life. Clocks, calendars, and units of time that arose from cycles in the natural environment are described and compared with internal biological clocks in Chapter 3 on Time.

The rhythmic interactions between living things and the things that make them live are complex. Some are easily observed and/or measured, whereas others can be examined only indirectly by specific assays, images, or mathematical interpretations. How the rhythmic nature of life is best observed and monitored, as well as how data can be analyzed for rhythm characteristics, represent chapter topics (autorhythmometry, analyzing data) that help provide a basic understanding of biological rhythms. Because there is often a direct relationship between the internal rhythms of an organism and the external rhythms of the environment, the topic of photoperiodism, which focuses upon the response of

---

[7] Any point on a rhythmic pattern relative to a fixed reference point can be a phase. The time of a peak or trough relative to local midnight are two commonly used phases. In many laboratory studies, the zero point coincides with the start of the light span (e.g., circadian time), rather than midnight.

TABLE 1.3. Examples of variables illustrating a range of rhythmic periods in the infradian domain.

Time	Period	Variable	Organism	Reference
Week	7 days	Oviposition (egg laying)	Spring Tail (*Folsomia candida*)	Chiba et al., 1973
	7 days	Organ transplant rejection	Human (*Homo sapiens*)	DeVecchi et al., 1981
	7 days	Imbibition of seeds	Bean (*Phaseolus vulgaris*)	Spruyt et al., 1987
Month	27–34 days	Menstrual cycle	Woman (*Homo sapiens*)	Presser, 1974
	6 months	Ulcer perforation	Human (*Homo sapiens*)	Svanes et al., 1998
Year	1 year	Seed germination	*Digitalis lutea*, *Chrysanthemum corymbosum*, and *Gratiola officinalis*	Bünning & Müssle, 1951
			Wild potato (*Solanum acaule*)	Ludwig et al., 1982
			Pole bean (*Phaseolus vulgaris*)	Spruyt et al., 1988
	1 year	Antler replacement	Sika deer (*Cervus nippon*)	Goss, 1969
	1 year	Migration	Willow warbler (and others) (*Phylloscopus trochilus*)	Gwinner, 1977
	1 year	Hibernation	Golden-mantled ground squirrel (*Citellus lateralis*)	Pengelley & Fisher, 1977
	1 year	Ovarian weight	Yamuna River catfish (*Heteropneustes fossilis*)	Sundararaj et al., 1982
	1 year and 7 years	Gonadal weight	Purple sea urchin (*Strongylocentrotus purpuratus*)	Halberg et al., 1987
	8 years	Alpha-amylase activity	Wheat (*Triticum* spp.)	Kettlewell et al., 1999
	8–10 years	Population	Ruffed Grouse (*Bonasa umbellus*)	Gullion, 1982, 1985
	13 or 17 years	Emergence	Periodical Cicada (*Magicicada* spp.)	Hoppensteadt & Keller, 1976
	100–120 years	Flowering	Chinese bamboo[a] (*Phyllostachys bamusoides*)	Janzen, 1976

[a]While the range between successive synchronized reproductions by seed is from 3 to 120 years, most bamboo species have shorter, yet still very long times between seeding (15–60 years).

TABLE 1.2. Examples of variables illustrating a range of rhythmic periods in the ultradian and circadian domains.

Time	Period	Variable	Organism	Reference
Seconds	<1 s	EEG activity (delta frequency)	Human (*Homo sapiens*)	Krippke, 1972
	<1 s	ECG (depolarization of heart ventricles)	Human (*Homo sapiens*)	see Figure 1.5
	30–45 s	NADH levels	Yeast (*Saccharomyces carlsbergensis*)	Chance et al., 1964
Minutes	1 min	Courtship song (male)	Fruit fly (*Drosophila melanogaster*)	Kyriacou & Hall, 1980
	2–4 min	Leaflet movements	Telegraph plant (*Desmodium gyrans*)	Koukkari et al., 1985
	4–15 min	Reaction time	Human (*Homo sapiens*)	Lovett-Douse et al., 1978
	15 min	Cortisol secretion	Horse (mares) (*Equus caballus*)	Drake & Evans, 1978
	30 min	Transpiration	Oat (*Avena sativa*)	Johnsson, 1973
	45–62 min	Leaflet movements	Soybean (*Glycine max*)	Guillaume & Koukkari, 1987
	76 min	Cellular protein content	*Acanthamoeba castellanii*	Lloyd et al., 1982
	75–100 min	Pupillary motility	Human (*Homo sapiens*)	Lavie, 1979
	85–90 min	Cortisol secretion	Monkey (*Macaque mulatta*)	Holaday et al., 1977
	90 min	Deep body and udder temperatures	Holstein cow (*Bos taurus*)	Bitman et al., 1984
	90–100 min	REM-NREM sleep	Human (*Homo sapiens*)	Aserinsky & Kleitman, 1953
	100 min	Shoot movements (circumnutation)	Pole bean (*Phaseolus vulgaris*)	Millet et al., 1984
Hour	4 h	Enzyme activity	*Euglena* (*Euglena gracilis*)	Balzer et al., 1989
	12 h	Amylase activity	Alfalfa (*Medicago sativa*)	Henson et al., 1986
Day	24 h	Body temperature	Human (*Homo sapiens*)	Aschoff et al., 1972
	24 h	Sleep–wakefulness	Human (*Homo sapiens*)	Kleitman, 1963
	24 h	Leaf movements	Albizzia (*Albizzia julibrissin*)	Koukkari, 1974
	24 h	Activity	Flying squirrel (*Galucomys volans*)	DeCoursey, 1960

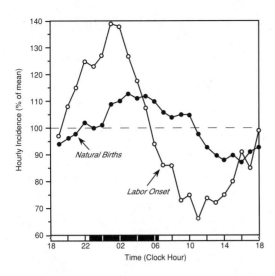

FIGURE 1.8. Comparison of the circadian (*circa* = about, *dies* = day) waveform for the onset of labor and birth in humans, illustrating a natural sequence of events beginning with a peak in the onset of labor early in the night (dark bar) followed by a peak in natural births early in the morning. Hourly incidence of onset of >200,000 spontaneous labors and >2,000,000 natural births (redrawn from Smolensky et al., 1972).

examples illustrating the diversity in periods and types of variables are presented in Table 1.2 (ultradian and circadian domains) and Table 1.3 (infradian domain). Multiyear cycles, such as the emergence of the periodical cicada (*Magicicada* spp.) every 13 years in the south and midwestern USA or every 17 years in the northeastern USA, the 8- to 10-year population cycles of the ruffed grouse (*Bonasa umbellus*) in Minnesota, and the 15- to 120-year cycles in flowering of various species of bamboo, are among the spectacular rhythmic events found in nature that are little understood.

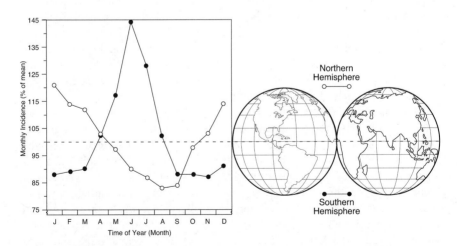

FIGURE 1.9. Monthly cardiac mortality shows circannual rhythm in the northern and southern hemispheres. A greater proportion of overall annual deaths from heart disease occurs in the winter for each hemisphere (redrawn from Smolensky et al., 1972).

TABLE 1.1. Examples of time-dependent responses of organisms to various chemicals.

Chemical	Organism	Comments	Reference
Ethanol	Human	Blood levels greater at 07:00 h; poorer performance after 19:00 and 23:00 h	Reinberg et al., 1975
Histamine	Human	Greater reaction of lungs or skin near midnight	Reinberg et al., 1969
House dust	Human	Bronchial hyperreactivity greater during nighttime than daytime	Gervais et al., 1977
Lidocaine (anesthetic)	Human	Duration of anesthesia in teeth and skin longer in mid-afternoon	Reinberg & Reinberg, 1977
Theophylline (asthmatic drug)	Human	Sustained-release dose at 20:00 h better than 08:00 h in controlling nocturnal dip in lung function	Reinberg et al., 1987
Corticosteroids	Human	Once-daily dosing of inhaled steroid between 15:00 and 17:30 h optimal for asthma control	Pincus et al., 1997
Pentobarbital (anesthetic)	Rat	Duration of anesthesia longer in early dark (activity)	Scheving et al., 1968
Pentobarbital (anesthetic)	Mouse	Maximal mortality in early dark (activity)	Bruguerolle et al., 1988
Arabinosyl cytosine (anticancer drug)	Mouse	Increased tolerance in mid-light (resting)	Haus et al., 1972
Dichlorvos (insecticide)	Cockroach	Greatest toxicity in early dark	Eesa et al., 1987
Bentazon (herbicide)	Velvetleaf	Less injury near middle of the light span	Koukkari & Johnson, 1979
Glyphosate and glufosinate (herbicides)	Weeds	Maximum control at midday	Martinson et al., 2002

continue with a period close to 24.0 h when the organism is isolated from external cues. Thus, periods of about (*circa*) a day (*dies*) are present when organisms are isolated from environmental 24-h cycles, such as the alternating light and darkness of the solar day, and/or changes in temperature. This "free-running" period is an important characteristic of a circadian rhythm. In addition, the period is relatively consistent over a range of temperatures (temperature compensation). These and other characteristics are discussed more extensively in Chapter 2 on General Features of Rhythms.

While the circadian rhythms represent the dominant cycle that has been studied relative to the activities of humans and other organisms, cycles having periods shorter or longer than circadian, such as 90-min or seasonal cycles, are also important. Biological cycles that have periods less than 20 h are called *ultradian rhythms*, while cycles with periods longer than 28 h are called *infradian rhythms*. Depending upon the variable, infradian periods can be measured in weeks, months, years (circannual), and longer. Collectively, these three rhythmic domains comprise a network or web of rhythmic oscillations that in many ways can be likened to the various chemical pathways that perform different functions, but occur simultaneously within the same organelle or cell.

The range of periods for biological rhythms is broad, extending from cycles that are measured in milliseconds to cycles that are over 100 years in length. Some

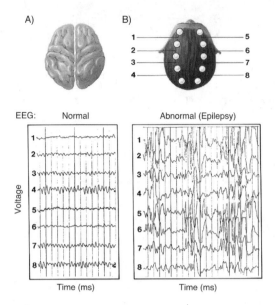

FIGURE 1.7. Diagram illustrating dorsal views of the brain (A) and scalp showing the position of electrodes (B), and two electroencephalograms (EEG). *Left*: Normal brain EEG measuring electrical potential difference with respect to time shows posterior "alpha rhythm." *Right*: Abnormal EEG in a young patient with epilepsy shows abnormal high amplitude discharges during a seizure. EEG panels provided by Mark Koukkari.

The concept of timing in relation to life and death is not of recent origin, as can be noted in the biblical book of Ecclesiastes 3:2–3:

To everything there is a season and a time to every purpose under heaven: a time to be born and a time to die; a time to plant and a time to pluck up that which is planted; a time to kill and a time to heal; a time to break down and a time to build up ... .

The rhythmic nature of birth[1] (Figure 1.8), morbidity (presence of disease), and mortality[2] are widely documented in the scientific literature. While these are but a few of the many examples of daily changes, there are seasonal differences as well, which show an annual rhythm in these same events (Figure 1.9). Even the common things that we are aware of, such as the migration of birds (Gwinner, 1977, 2003), or the less common events, such as the germination of seeds in certain species, display an annual rhythm (Bünning & Mussle, 1951).

Three Rhythm Domains

Much of the early work on biological rhythms focused on cycles in which the *period* was 24 h. These rhythms are referred to as *circadian rhythms* because they reoccur with a period of 24 h during usual light–dark conditions, and can

---

[1] Results from a classic study which summarized rhythms in human natality based upon the hourly incidence of 207,918 spontaneous labors and 2,082,453 natural births (Smolensky et al., 1972) showed that the onset of labor and birth under usual circumstances was most frequent between midnight and 06:00 h.

[2] Sudden cardiac death (Willich et al., 1987) and cerebral infarction (stroke) or hemorrhage (Marshall, 1977) are more prevalent between midnight and 06:00 h.

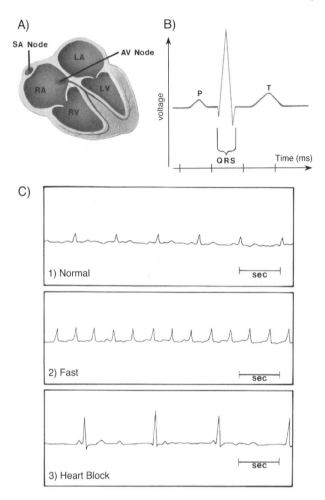

FIGURE 1.6. Diagram illustrating (*A*) the human heart and the specialized cardiac conduction system composed of the sinus node (SA), atrioventricular node (AV), right and left atrial (RA and LA) chambers, and the right and left ventricles (RV and LV). A small electrical current propagates from the sinus node through the atrial muscle to the AV node, which, after a short delay (usually 0.12 s), is transmitted to the ventricles. The electrical events cause both atrial and ventricular contraction. To record the electrical signals from the heart, electrodes are placed on the surface of the body at specific locations (e.g., right arm and left leg). The signals collected in this manner are amplified, and recorded on charts. This recording is called an electrocardiogram (ECG). (*B*) A record of an electrical cycle from a single lead (pair of electrodes). P = atrial electrical activity (depolarization), QRS complex defines the depolarization of ventricles; and T wave signifies repolarization of ventricles. (*C*) Three panels of ECGs (provided by Ann Dunnigan) illustrating their use in the diagnosis of normal as well as abnormal rhythms. *Panel 1*: Normal sinus rhythm rate (rate 75 beats per minute [bpm]); *Panel 2*: abnormally fast regular rhythm at rate of >175 bpm; *Panel 3*: Abnormally slow rhythm where atria do not communicate with ventricles, called complete heart block. In *Panel 3*, sinus node rate ("P") is normal (75 bpm), while ventricular rate is abnormal (40 bpm).

2
General Features of Rhythms: Terminology and Characteristics

"What is time, then? If nobody asks me, I know;
if I have to explain it to someone who has asked me,
I do not know."
—St. Augustine (354–430), Christian philosopher & church father

Introduction

In order to best understand the characteristics of biological rhythms, like the prerequisites for most disciplines in biology, one must be familiar with the use of certain terms that are used. In the case of biological rhythms, the discipline is known as *chronobiology*, a term introduced in the mid-1960s to unify the study of temporal characteristics of biological phenomena (Halberg, 1969; Cambrosio & Keating, 1983). Some of the more common terms used in chronobiology to describe rhythms are presented in a glossary in Table 2.1.

Because words such as rhythm, period, phase, and amplitude are so basic, the first objective of this chapter is to define or describe how these terms apply to all rhythms, be they biological, physical, or strictly mathematical. They are often expressed in numerical units, such as time or angular degrees, and provide a way to categorize biological rhythms (e.g., *Is a cycle completed in approximately 90 min, 1 day, or 1 year?*). We also discuss the use of the term *circadian* and how the length of the day has changed over the history of the Earth.

The second objective is to introduce the genetic basis of biological rhythms and discuss the suprachiasmatic nucleus (SCN) in the brain as a master circadian clock and the role of melanopsin in retinal ganglia in perceiving light for synchronization to alterations in external light and dark.

The third objective is to discuss the characteristics that distinguish biological rhythms from other types of rhythms. Included here will be discussions on three categories or domains of rhythms (ultradian, circadian, and infradian) and responses of their amplitude (damping) and/or phase (phase-shifting) to changes in the environment. Environmental factors that superficially alter or enhance endogenous characteristics (e.g., the waveform) of a rhythm, known as *masking*, will also

2. General Features of Rhythms: Terminology and Characteristics

TABLE 2.1. A glossary of some terminology used in chronobiology.

Term	Definition or description
Acrophase	Peak of a mathematical curve fit to data. Units (time or degrees) expressed as the lag from a reference point (e.g., midnight, L-onset)
Aliasing	Detection of a false period that is longer than the underlying true period due to long time intervals between successive measurements
Amplitude	Distance from rhythmic mean to the peak or to the trough of a mathematical model (e.g., cosine) used to approximate a rhythm
Biological Rhythm	Changes in a biological variable that recur with a similar pattern and systematic interval (period)
CC	Constant environmental conditions
Circadian Time	Time that spans the circadian period in relation to the LD regimen under synchronized conditions; zero time (00:00 h) usually corresponds to the start of the light span
Cycle	A recurring pattern
Damping	Decrease in amplitude of a rhythm over time
DD	Continuous darkness
Domains	See Table 2.2 (e.g., circadian, ultradian, and infradian)
Entrainment	Coupling of the period and phase of a biological rhythm (e.g., circadian) with another cycle (e.g., 24 h solar day)
Free-running	Desynchronization of the period of a biological rhythm from the period of a known environmental synchronizer (e.g., LD); status of a rhythm under constant conditions (absence of synchronizers)
Frequency (f)	The number of cycles per unit of time; the reciprocal of the period ($f = 1/\tau$)
LD	Light span followed by a Dark span (e.g., LD 12:12 = 12 h of L followed by 12 h D)
LL	Continuous illumination
Masking	Superficial change of rhythm characteristics (e.g., amplitude) by external environmental conditions
Period (τ)	Duration of one complete rhythmic cycle
Phase	Indicates the location in time for a value within a rhythm
Phase–shift	A change in the timing of the phase a rhythm to occur earlier or later
Synchronizer (Zeitgeber)	Environmental signal or cycle that entrains a biological rhythm (e.g., sets the phase and/or period)

be discussed, as will the need of a cyclic environment (e.g., light alternating with darkness) for the normal development and synchronization of various organisms.

A final objective is to encourage you to observe firsthand some rhythms in the movements of stems and leaves of a bean plant. Darwin observed and described the rhythmic movements of plants and so can you! In some simple experiments, you can easily measure both ultradian and circadian rhythms of a bean plant and obtain firsthand knowledge about a period, amplitude, and phase.

Details of a Rhythm

Many individuals first encounter the term *rhythm* in the context of music, the repetition or recurrence of note values or beats (meter). In this book, a *rhythm* is defined as a change that is repeated with a similar pattern, probability, and period.

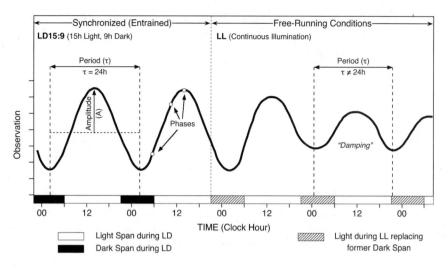

FIGURE 2.1. Terminology used for rhythms under synchronized and free-running conditions. A reduction in amplitude during free-running conditions, such as continuous illumination (LL), is a process called damping. Units of time (e.g., hours) are often labeled along the horizontal X axis (abscissa) and units of measure (e.g., grams) along the Y axis (ordinate).

If the variable that changes is *biological* and *endogenous* (driven from within, as opposed to simply a response to the environment), the observed oscillation can be objectively referred to as a *biological rhythm*. When biological variables display a rhythm,[1] such as the movements of leaves on a plant or the changes in body temperature of a human, three features are often used to describe the rhythm. These characteristic features are the *period, amplitude,* and *phase* (Figure 2.1). Each can be expressed in numerical units, thus providing a means to quantify the rhythm.

Period and Frequency

The *period* (τ) represents the time required to complete a cycle. It may range from milliseconds to more than a century. *Frequency* (f) is the number of cycles in a unit of time and is the reciprocal of the period ($f = 1/\tau$). For example, a leaflet that moves up and down four times each hour displays a frequency of four cycles per hour, which is the same as a period of 15 min. Throughout this book we will usually refer to the period.

Circadian Domain

Biological rhythms are usually distinguished or grouped in reference to the length of their period (Table 2.2 and Figure 2.2). The daily movements of the leaves of a

---

[1] Sometimes the term *overt rhythm* is used to describe changes in the variable being observed (e.g., running activity, body temperature, etc.).

2. General Features of Rhythms: Terminology and Characteristics

TABLE 2.2. Examples of domains used to categorize biological rhythms according to period.[a]

Domain	Period
Ultradian	<20 h
Circadian	20–28 h
Infradian	>28 h
Circaseptan	7 ± 3 days
Circatrigintan	30 ± 5 days
Circasemiannual	~6 months
Circannual	1 year ± 2 months
Infrannual	>1 year
Semitidal	12.4 h
Tidal	24.8 h
Lunar	29.53 days

[a] Ultradian, circadian, and infradian represent the three broad domains often used in categorizing rhythms. For example, a circannual rhythm is also an infradian rhythm.

bean plant and the daily changes in the body temperature of humans are classic examples of variables that display rhythms close to 24 h, and are classified as *circadian* rhythms.[2]

The reason for emphasizing the word *circa* is the fact that although the period is 24 h under "normal" synchronizing environmental conditions of alterations between day and night of light and/or temperature, it is only approximately 24 h when organisms are isolated from these external cycles. For instance, the body

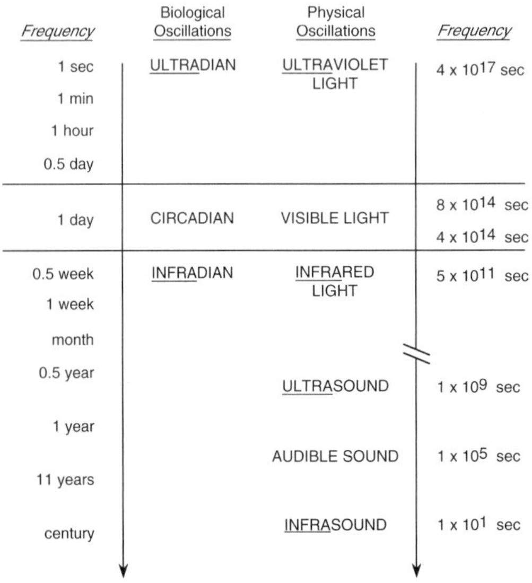

Frequency	Biological Oscillations	Physical Oscillations	Frequency
1 sec	ULTRADIAN	ULTRAVIOLET LIGHT	4 x 10^{17} sec
1 min			
1 hour			
0.5 day			
1 day	CIRCADIAN	VISIBLE LIGHT	8 x 10^{14} sec
			4 x 10^{14} sec
0.5 week	INFRADIAN	INFRARED LIGHT	5 x 10^{11} sec
1 week			
month			
0.5 year			
		ULTRASOUND	1 x 10^9 sec
1 year			
		AUDIBLE SOUND	1 x 10^5 sec
11 years			
		INFRASOUND	1 x 10^1 sec
century			

FIGURE 2.2. Spectrum of biological oscillations compared with the electromagnetic spectrum from light to sound. *Note:* The terminology for "shorter" begins with "ultra" and for "longer" begins with "infra" for the two spectra.

---

[2] The word *diel* (pronounced like the word *dial*) is sometimes used, especially in the studies of aquatic animals, to indicate a 24-h daily rhythm.

temperature rhythm of a human has been shown to have a period closer to 25 h when isolated from the usual 24-h cycles of light and darkness and other known environmental oscillations or time cues that synchronize the rhythm to 24 h (Wever, 1979).

Even under synchronized conditions, however, the prefix "circa" can be applied to biological rhythms since there is invariably some margin of error due to internal and external phenomena that do not always recur at identical sequences at identical intervals (Halberg & Cornélissen, 2001). There is sometimes hesitation to use the word *circadian* in describing 24-h patterns under synchronized conditions, although the real question should be whether the underlying cause of the observed constrained circadian rhythm is *endogenous* or *exogenous* (see Essay 2.1).

Essay 2.1 (by RBS): Use of the term "Circadian"

In the field of biological rhythms, as in any other discipline, confusion may arise from discrepancies in the use of terminology. For example, the term *daily*, which is sometimes used to describe a 24-h variation or cycle, can also be used to indicate day-to-day variations. Terms such as *diurnal rhythm* or *nocturnal rhythm* are used to indicate a part of the 24-h period (daytime vs. nighttime) when something peaks or occurs (e.g., peak activity, sleeping times, patterns in epileptic seizures). By the same token, "24-h" or "diel" may (incorrectly) imply that a rhythm is very precise in its period length (e.g., exactly 24 h), when in fact the amplitude and phase may (and usually does) alter from day to day due to many interacting intrinsic and extrinsic factors. Under 24-h synchronized conditions, the differing degrees of interaction among these factors bring about an observed pattern that may only appear to be 24 h while it is constrained by the synchronizer(s).

Origin. To standardize the terminology used when discussing rhythms with periods close to 24 h, Franz Halberg (b. 1919) introduced the term *circadian* in 1959. His original definition follows:

The term "circadian" was derived from "circa" (about) and "dies" (day); it may serve to imply that certain physiologic periods are close to 24 hours, if not of exactly that length. Herein, "circadian" might be applied to all "24-hour" rhythms, whether or not their periods, individually or on the average, are different from 24 hours, longer or shorter, by a few minutes or hours. "Circadian" thus would apply to the period of rhythms under several conditions. It would describe: 1. rhythms that are frequency synchronized with "acceptable" environmental schedules (24-hour periodic or other) as well as 2. rhythms that are "free-running" from the local time scale, with periods slightly yet consistently different from 24 hours (e.g., in relatively constant environments). (Halberg, 1959, p. 235)

The following definition of "circadian" was adopted in 1977 by the International Committee on Nomenclature of the International Society for Chronobiology:

Circadian: relating to biologic variations or rhythms with a frequency of 1 cycle in 24 ± 4 h; circa (about, approximately) and dies (day or 24 h). Note: term describes rhythms with an about 24-h cycle length, whether they are frequency-synchronized with (acceptable) or are desynchronized or free-running from the local environmental time scale, with periods slightly yet consistently different from 24-h, ... (Halberg et al., 1977)

This latter definition is used as a guideline in the journal *Chronobiology International* when using the word *circadian* (M. H. Smolensky, Editor in Chief, personal communication).

Use. Those who delve into the literature on biological rhythms will soon sense that the term "circadian" is not always used in the same way. Among those who study biological rhythms, views may differ as to whether a free-running ("circa") period is a prerequisite for a rhythm to be called "circadian." In other words, should a daily (24-h) period of any variable be called circadian when the variable has not been studied under free-running (e.g., constant) environmental conditions and there is no evidence that the period differs slightly from 24 h? In the report of the Dahlem Workshop on the molecular basis of circadian rhythms held in 1975 in Berlin, several authors emphasized the need to show free-running as a prerequisite for using the word *circadian*. These included the following:

Circadian rhythm may be considered as those biological rhythms which under constant permissive conditions continue to exhibit oscillations with a period of about (= circa) one day (= diem). . . . The period of an entrained circadian rhythm matches that of the zeitgeber[3]. . . . (Hastings et al., 1976)

Unfortunately the term "circadian," originally coined to indicate that in constant conditions the rhythm persists but may deviate from exactly 24 hours, is now used rather indiscriminately for almost any kind of daily rhythm, regardless of whether or not its endogenous nature has been established. This may be due to the tacit assumption that, since so many daily cyclic processes have been shown to have an endogenous component, this holds for all daily cycles. (Hoffman, 1976)

A circadian rhythm is an oscillation in a biochemical, physiological, or behavioral function which under conditions in nature has a period of exactly 24 hours, in phase with the environmental light and darkness, but which continues to oscillate under constant but permissive conditions of light and temperature with a period of approximately but usually not exactly 24 hours. (Sweeney, 1976)

One of the authors (WLK) of this book is more conservative (cf. Bünning, 1973) in the use of the term, emphasizing the free-running (*circa*) characteristic of circadian rhythms. The other author (RBS) is more impartial and agrees with the original definition of the word that circadian can infer a rhythm that is synchronized to an environmental 24-h schedule without relying upon a prerequisite description of a free-running endogenous about-24-h component. Since the term circadian refers to the domain of about-daily rhythms (just as ultradian refers to rhythms that are shorter and infradian refers to rhythms with periods longer than circadian), a recurring oscillation with a 24-h period found under 24-h synchronized conditions can be described as a circadian rhythm without implying its endogenous or exogenous nature.

While endogenicity proves the built-in basis of circadian periodicity, it has been pointed out that synchronization is essential for the normal functioning of an organism (Went, 1974). Thus, under entrained conditions, circadian rhythms may show larger amplitudes and more precise peaks that are useful to the organism and the observer. So as not to confuse the reader, and in accord with the norm of papers published in various scientific journals, the sense of a free-running period usually appears throughout the

---

[3] Zeitgeber (German for "time-giver") is often used synonymously for the word *synchronizer*.

book, except in Chapter 11 on Clinical Medicine and in cases where some clinical situations are discussed (cf. Smolensky & Lamberg, 2000, p. 26). We (WLK and RBS) feel that this criterion is fully justified for the latter, since in many, if not in most, medical papers, the term circadian is used whatever the causative mechanism, be it directly dependent on daily changes in the environment (exogenous) or independent of all external variations (endogenous) (e.g., Moore Ede, 1973; Arendt et al., 1989). In fact, more recently, the use of the term circadian was justified as follows:

A process that consistently repeats itself every 24 h is said to have a daily (or nycthemeral) rhythm. If this rhythm persists with approximately the same period in the absence of external time cues (and, therefore, is endogenously generated), it is called a circadian rhythm. Naturally, an endogenously generated rhythm may be referred to as a circadian rhythm not only when the animal is isolated from external time cues (the free-running state), but also when the external time cue is present and controls the period and phase of the rhythm (the entrained state). (Refinetti & Menaker, 1992)

In the absence of external oscillations of light and darkness and/or temperature changes, the range of the circadian period depends upon the species and is usually between 20- and 28 h. However, since exceptions have been observed, this range need not be viewed always as a strict rule and could be expanded to include periods between 14 and 34 h (Hassnaoui et al., 1998). For example, eclosion and locomotor activity rhythms in the clock mutant per^s fruit fly have periods of 19 h (Knopka & Benzer, 1971), and the short-day flowering plant *Chenopodium rubrum* displays a light-sensitive 30-h rhythm in flowering when continuous light is interrupted by a single dark span of increasing lengths (Cumming, 1969; King, 1975).

Other Domains

Rhythms having periods *shorter* than the circadian range (usually <20 h) are referred to as *ultradian* rhythms (Table 2.3), while those having periods greater than the circadian range (usually >28 h) are called *infradian* rhythms (Table 2.4).

TABLE 2.3. Examples of variables having ultradian rhythms.

Organism	Variable	Period	Reference
Nematode (*Caenorhabditis elegans*)	Defecation	45 s	Liu & Thomas, 1994
Yeast (*Saccharomyces carlsbergensis*)	Glycolysis	1.8 min	Pye, 1969
Yeast (*Saccharomyces cerevisiae*)	Respiration	48 min	Lloyd et al., 2002
Oat (*Avena sativa*)	Transpiration	30 min	Johnsson, 1973
Bean (*Phaseolus vulgaris*)	Shoot movement	100 min	Millet et al., 1984
Cat (*Felix domesticus*)	Catecholamine release	~60 min	Lanzinger et al., 1989
Golden-mantled ground squirrel (during hibernation) (*Spermophilus lateralis*)	Sleep-wake	6 h	Larkin et al., 2002
Human (*Homo sapien*)	Core temperature	64 ± 8 min	Lindsley et al., 1999
Human (*Homo sapien*)	EEG during sleep	80–120 min	Armitage et al., 1999
Human (*Homo sapien*)	Insulin	6–10 min and 140 min	Schmitz et al., 2002

TABLE 2.4. Selected examples of variables having infradian rhythms.

Organism	Variable	Period	Reference
Bean (*Phaseolus vulgaris*)	Seed imbibition	~7 days	Spruyt et al., 1987
Insect (*Folsomia candida*)	Egg-laying (25°C)	~7 days	Cutkomp et al., 1984
Human	Menstrual cycle	~25–30 days	Presser, 1974
Bird (*Junco hyemalis*)	Migration	~1 year	Wolfson, 1959
Bamboo (*Phyllostachys bamusoides*)	Seeding cycle	~120 years	Janzen, 1976

Various schemes have been proposed for dividing the ultradian rhythms further into subgroups, such as those having ultrahigh frequency rhythms <1 min, periods <30 min, about 90 min (50–130 min), 3–4 h, and 12 h. Subdivisions proposed for the infradian rhythms, such as periods of about a week (*circaseptan*), month (*circatrigintan*), year (*circannual*), and even longer, will be discussed in later chapters.[4] Other rhythmic domains include *tidal* (24.8 h), *semi-tidal* (12.4 h), and *lunar* (29.53 days) periods. The lengths of the circadian and circannual periods are in synchrony with the current cycles experienced on Earth (Essay 2.2).

Essay 2.2 (by RBS): Time on Earth as We Know It

Today, when we discuss *circadian* and *circannual* rhythms, we refer to periods of 24.0 h and 365.25 d, respectively. However, according to the geologic time scale that accounts for the history of Earth, these numbers are but a temporary phenomenon. Since its formation, the duration of time for Earth to orbit the Sun may have changed very little, but because of the change in the speed of its rotation, the length of the day has changed dramatically. With the use of radioactive techniques for dating mineral elements in rocks, such as using K^{40} that has a half-life of 1.3 billion years,[5] it has been possible to estimate the age of Earth to be approximately 4.5 billion years[6] (Hartmann & Miller, 1991). Yet Earth is relatively young[7] when compared with the formation of the Milky Way Galaxy (ca. 14 billion years ago).

---

[4] The term *chronome* has been suggested for modeling the complex time-dependent structure in an organism that encompasses not only built-in rhythms with a variety of periods, but also the influence of non-photic external cycles, chaos, and trends (Halberg et al., 2001).

[5] C^{14} with a half-life of 5,730 years is used for dating shorter time spans, such as time since death of a living organism.

[6] This is inconceivably old and a far cry from the 6,000 years that was thought to be the age of the Earth from detailed study of scriptures by English Archbishop James Ussher (1581–1656), who claimed that the Earth was created on October 26, 4004 BC.

[7] Geologists and astronomers now agree that when the Earth was about 50 million years old, a giant planetesimal collided with the Earth, forcing a huge amount of debris into space. Most of this debris fell back to the Earth, but some of it sprayed into an orbit around the Earth where most of it eventually coalesced into our moon (Hartmann & Miller, 1991). One of the benefits of the Apollo lunar landings between 1969 and 1972 was that astronauts brought back rocks that were determined to have the same composition as those on the Earth and could be dated as far back as 4.1 billion years. Meteorites that have been found on the Earth have been dated back 4.55 billion years.

TABLE 2.5. Due to the gravitational effect of the Moon since its formation 4.45 billion years ago, the Earth's day has been slowly becoming longer.[a]

Time scale (approx.)	Duration of one rotation of the earth (h)
4.5 billion years ago	6.0
4.4 billion years ago	10.0
4.0 billion years ago	13.5
900 million years ago	18.17
400 million years ago	<22.0
245 million years ago	22.75
100 million years ago	23.5
Today	24.0
225 million years from now	25.0

[a]As the Moon has slowly moved away from Earth, its effect on the Earth's daylength has also been slowing down. One second is added to our day every 62,500 years. Fossil traces of true cells from bacteria have been dated to ~3.5 billion years ago, and the first oxygen-dependent life appeared ~1.7 billion years ago (Hartmann & Miller, 1991; Barnett, 1998).

Based upon calculations that indicate the effects of the Moon on the rotation of Earth today, it appears that 4.5 billion years ago the length of a day on Earth was 6.0 h (Hartmann & Miller, 1991). Due to the gravitational effect of the newly formed Moon, the day would have slowed down to about 10.0 h a hundred million years later (Table 2.5). Since then, the rotation of the Earth has continued to slow down due to the gravitational pull of the Moon, but this process has also been slowing down as the moon's spiral away from the Earth has slowed and exerts less of an effect. Based upon observations of coral rings, the day was about 22.0 h 400 million years ago. During the reign of the dinosaurs 165 million years ago, the day was 22 h 45 min long. As the Moon continues to recede from Earth at the rate of about two inches (5 cm) per year, about one second is added to our day every 62,500 years, which means that in another 225 million years, a day on Earth will last 25.0 h and a year will last just over 350 of those days (Barnett, 1998).

Thus, the timescale of Earth's history compared to our time of existence is almost beyond comprehension. Using a scale of one meter, geologic time is compared with the length of the day and year and the development of life on Earth in Figure 2.3. This compressed time scale is almost unfathomable to the human mind. For example, while the dinosaurs may have existed for more than 165 million years, they have been extinct for nearly 100 million years. On such a time scale, where 1 nanometer (nm) equals 4.5 years, a single DNA double helix strand 2 nm in diameter would equal 9 years and a strand of protein 6–10 nm in diameter would equal 27–90 years, a range approximating human life spans. These distances/diameters are so small that they can only be seen with an electron microscope. To put Earth's time scale in another perspective, a human hair that is 0.003 inches (76.2 μm) in diameter would equal 342,900 years. If this hair was sliced into 171 equal slices and one slice was laid on the end of the meter stick at the year zero, it would represent the last 2,000 years!

The word *circadian*, while it means about a day, does not denote a fixed length and has not always meant near 24.0 h, as we assume today. Life on Earth today experiences a 24.0-h solar day, which has been the day length for only the last 2% of Earth's history. Therefore, the vast majority of life forms on Earth experienced shorter days during their

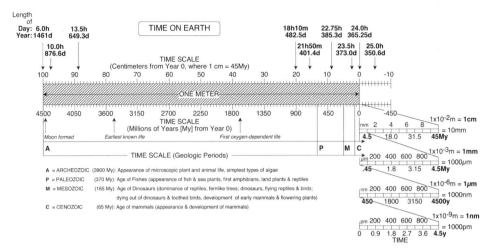

FIGURE 2.3. A time scale for the history of the Earth, where one meter equals the length of time from Earth's formation 4.5 billion years ago until the year zero. The last 5 million years that prehistoric humans and their ancestors walked on the Earth are represented by about the last millimeter of the Cenozoic period. The last 2,000 years of human history beginning with year zero would be represented by a line about 450 nm thick, a width too thin to be reproduced here, while a human living 75 years would be represented by a line only 16.7 nm thick, a line visible only with an electron microscope. Due to gravitational effects on Earth's rotation by the Moon, the length of 1 day on the Earth has been slowing down from the approximate 6.0 h when the Earth was first formed to 24.0 h today. The length of the year has also slowed down from more than 1,000 days to 365.25 days today.

existence (e.g., even 100 million years ago, the day was only 23.5 h long). Since we can calculate that there must have been well over a trillion days and nights since the appearance of the first simplest life forms 3.5 billion years ago, it's not surprising that clock genes developed to help organisms adapt to and measure the regular alterations between light & dark and warmth & cold. While it was long thought that circadian rhythms were exclusive to eukaryotes (organisms having nucleated cells), daily rhythms in nitrogen fixation have been demonstrated in prokaryotes (cells without nuclei) such as cyanobacteria (blue-green algae). This suggests that these photosynthetic organisms were the first inventors of a biological clock, since they appear in fossil records dating 3.5–3.8 billion years ago (Johnson et al., 1998). In turn, clock genes may have translocated from prokaryotic organelles to the nuclei of eukaryotic organisms, thus providing higher plants and organisms with a biological clock. Furthermore, additional circadian clocks may have evolved independently several times during evolution. This temporal organization to life undoubtedly helped a species survive by anticipating the predictable changes in the environment throughout the day and year. In most instances, these molecular aspects of biological clocks must also have evolved as the Earth slowed down and the days lengthened (see Table 3.5 in Chapter 3 on Time).

As a practical concern, humans and their endogenous biological clocks leaving Earth for the Moon or any other planetary body will have to consider the length of their daily schedules not only during the journey, but also while living on a planet where the day length may be shorter or longer than 24.0 h. For example, since Mars has a day length of 24.66 h and a year nearly double that on Earth (687 days), will humans traveling there have their watches slowed down to 61.65 min/h to show that 24.0 h have elapsed between sunrises, thus staying in harmony with the Martian day length, or will they stay on Earth time and thus watch the sunrise 40 min later each day? Similarly, the year on Mars is about 1.88 times as long (686.973 days) as it is on Earth; thereby providing 22.6 "Earth months" of 30 days each that will have to be redivided into meaningful seasons on Mars (e.g., 3 month spans resulting in 7.5 seasons, or four seasons each consisting of 5–6 months).

Amplitude

The magnitude of the variable from the baseline (e.g., usually the middle value[8] or mean) to the peak or trough of the mathematical model (e.g., a cosine) used to describe a rhythm is called the *amplitude*. An amplitude must be detected in order to indicate the presence of a rhythm. The value of the amplitude can sometimes be difficult to interpret, since it depends upon the strength of the biological signal and sensitivity of the monitoring system. The amplitude for a biological variable is recorded in standard units, such as mg, cm, beats/min, or mmHg. However, these units may need to be normalized or manipulated in some manner, such as expressing each value as a percent of overall mean, when comparisons are to be made among rhythms for other variables.

Phase

Another characteristic of a rhythm is the *phase*. Generally, a phase can be viewed as any instantaneous repeatable state of a cycle (e.g., think about the phases of the Moon). The *peak* (highest point) is a definable point of a cycle that could be called a phase, as is the *trough* (lowest point), or any point on the cycle that occurs between the peak and trough.

When a cosine curve has been used to quantify rhythm characteristics, a special term called the *acrophase* is used to designate the distance in time of the peak (= acro) phase of the mathematical curve from an arbitrary reference point (e.g., 0° can be midnight [00:00 h], time of midsleep, time of light onset or awakening, etc.). Similarly, the lowest point on the fitted cosine is called the *bathyphase*.

Genetics and Inheritance

The possibility that rhythms are inherited was suggested from the results of early studies with bean plants *(Phaseolus)* in which the F_1 generation obtained from a

---

[8] The middle value of a cosine is called the *mesor* and represents a rhythm-adjusted mean (see footnote 13 in Chapter 13 on Analyzing for Rhythms).

crossing of two parents with different circadian periods in their free-running leaf movement rhythms displayed the normal Mendelian distribution (Bünning, 1932, 1935; see Chapter 7 on Sexuality and Reproduction for more detailed explanation). Subsequent studies have utilized mutants[9] of various organisms, including *Arabidopsis* (plant) (Somers & Kay, 1998), *Chlamydomonas* (alga) (Bruce, 1974), *Synechococcus* (cyanobacterium) (Kondo et al., 1994), *Neurospora* (fungus) (Feldman & Hoyle, 1973), *Drosophila* (insect) (Konopka & Benzer, 1971), and the *Mesocricetus* (hamster) (Ralph & Menaker, 1988).

Period

Results from studies with numerous and diverse species, especially genetically altered mutants, have demonstrated that the period of a biological rhythm is an inherited trait. Organisms maintained for generations under standardized environmental synchronizing schedules or subjected to abnormal environmental cycles exhibit a circadian periodicity that persists with great similarity from generation to generation. Some experiments have even involved maintaining many successive generations under *noncyclic* conditions (e.g., in LL, DD, constant temperature) with the same result: the rhythm persists with the same approximate period and pattern from generation to generation.

Studies with fruit flies (*Drosophila*) superbly illustrate the genetic nature of the circadian and ultradian periods. In a classic study published in the early 1970s (Konopka & Benzer, 1971), mutants were isolated that had circadian periods[10] differing from normal flies (wild type) in *eclosion* (when the adult emerges from its pupal case) and in their locomotor activity. One of the isolated mutants had short periods (per^s = 19 h), another had long periods (per^l = 28 h), and a third did not seem to have a circadian rhythm (per^o) (Figure 2.4). All three rhythm mutations appear to involve the same functional gene, located on the polytene X chromosome (Figure 2.5).

The *per* gene also affects high-frequency ultradian periods in a similar manner (Kyriacou & Hall, 1980). The average 55-s period of the high-frequency courtship song of the male (*Drosophila*) (Figure 2.6) is lengthened to 82 s in per^l, shortened to 41 s in per^s, and is arrhythmic in per^o. Evidence also exists for a genetic basis for an ultradian clock controlling the about 45-s defecation rhythm in the nematode *C. elegans* (Iwasaki et al., 1995).

Amplitude and Phase

While phase and amplitude are also inherited characteristics, they may be somewhat independent of each other. For example, it has been possible to select for

---

[9] Organisms expressing certain abnormal proteins due to changes that have occurred in their DNA are called mutants.

[10] The abbreviations used for the per genes is as follows: per^s = s for short period, per^l = l for long period, and per^o = 0 for no circadian rhythm (arrhythmic).

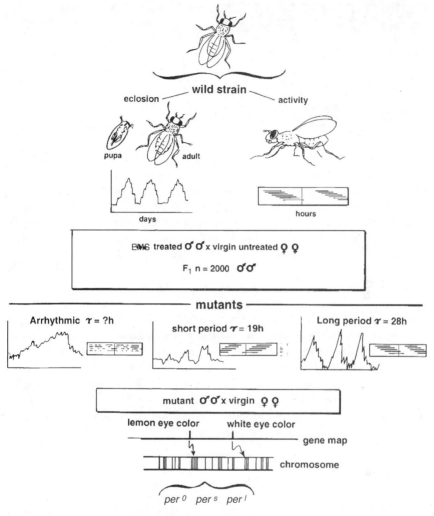

FIGURE 2.4. Illustration of the classic study in which X-linked mutations that affected rhythmic behavior in *Drosophila* eclosion and activity were discovered (Konopka & Benzer, 1971). Wild-type males were mutagenized with the chemical ethylmethane sulfonate (EMS) and mated to virgin females with attached-X chromosomes. Of the 2,000 stocks that were screened, three types of mutants were obtained that had circadian periods differing from normal flies (wild type) in eclosion and in their locomotor activity. One of the isolated mutants had short periods (per^s = 19 h), another had long periods (per^l = 28 h), and a third did not seem to have a circadian rhythm (per^o). These mutations were alleles of the same functional gene on the X chromosome. This gene was subsequently named *period* (*per*). The location based on the gene map is between lemon eye color and white eye color.

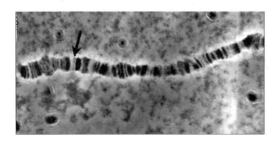

FIGURE 2.5. Photograph of a portion of the X chromosome of *Drosophila melanogaster* showing the approximate location of the *period* (*per*) gene (*arrow*). Photograph by Korise Rasmusson, Genetics & Cell Biology, University of Minnesota.

strains of fruit flies that show 4 h differences in their phase of eclosion, but have similar circadian periods (Pittendrigh, 1967). Although there are mutants that have been identified on the basis of their amplitude, it is possible that the variable selected or monitored (e.g., leaf movement) as an expression of the amplitude is in reality an expression of an anatomical or physiological change, rather than a change in a temporal component or an endogenous oscillator. The situation can be further complicated by environmental factors, such as humidity or temperature, which can affect the amplitude. When statistical models are used for analyses of rhythm data, the error for the amplitude is often much larger than it is for the phase.

Much of our knowledge of the genetics and inheritance of circadian rhythms, as well as some of our knowledge of ultradian rhythms, comes from the results of experiments specifically designed to focus on the genetic aspects of these rhythms. In the case of infradian rhythms, evidence for the inheritance of periodicity comes more from association or implication than from the actual identification and characterization of specific genes. Infradian rhythms are part of our temporal organization, and in the case of some human rhythms (e.g., menstrual cycle), can start, free-run, and stop as a function of age or maturation.

Primary Circadian Clocks

Insects and mammals are but two of the many large groups of organisms in which rhythms have been studied. These animals have organs, appendages, and a nervous system that includes a brain. Because of the large number of rhythmic

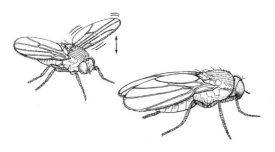

FIGURE 2.6. A male fruit fly (*left*) performs its courtship song for a female (*right*) by approaching her from behind and extending and waving one of its wings. This courtship song repeats as a high-frequency ultradian rhythm with a period of approximately 60 s.

variables (overt rhythms) that are known for insects and mammals, even for individual cells of tissues, it seems appropriate to ask: *is there a master or primary circadian clock, and if so, where is it located?* The answers are *"yes"* and *"a primary oscillator resides in the brain."*

The Brain

In some insects (moths and fruit flies) it has been possible to excise and transfer the brain of one insect into that of another. The technique is rather unique, since the donor brain need not be inserted into the head, but rather can be placed in the abdomen (Truman & Riddiford 1970; Truman 1972). In nature, the rhythm of eclosion in a giant silk moth, *Hyalophora cecropia* (Figure 2.7, *left*), peaks in the morning, while in the Chinese oak silk moth, *Antheraea pernyi* (Figure 2.7, *right*), eclosion peaks in late afternoon (Truman & Riddiford, 1970). When the brain of *H. cecropia* in a diapausing pupa is removed and transferred into the abdomen of an *A. pernyi* pupa that has had its brain removed (debrained), eclosion occurs in phase with that of *H. cecropia* (Truman & Riddiford, 1970). Similarly, the donor brain of *A. pernyi* will determine the phase of eclosion in the recipient *H. cecropia* (Figure 2.8). Thus, an eclosion hormone produced in the brain provides the signal for eclosion (Truman, 1972).

With regard to *Drosophila,* brains have been dissected one to two days *after* eclosion from short-period and arrhythmic mutants and implanted into the abdomen of host flies (Handler & Konopka, 1979). Results from these studies show that brains isolated from short-period mutants can produce short-period (e.g., 16–18 h) activity rhythms in arrhythmic hosts.

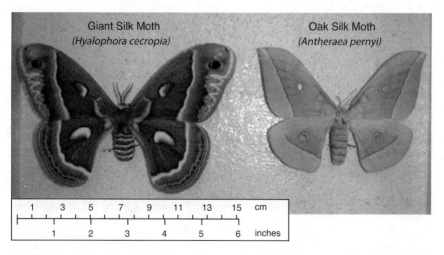

FIGURE 2.7. Adult moths of the two species used in neuroendocrine studies in which the brain was identified as necessary for 24 h timing of eclosion. Specimens provided by Philip Clausen, Insect Museum, Dept of Entomology, University of Minnesota.

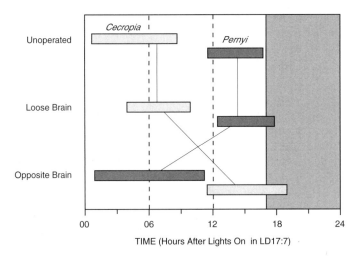

FIGURE 2.8. Time spans of majority emergence of Cecropia (*Hyalophora cecropia*) and Pernyi (*Antheraea pernyi*) moths provide evidence for clock control in the brain. Under intact, unoperated conditions, Cecropia emerges early in the light span, while Pernyi emerges later in the light span. Following transplant of the brain of a diapausing pupa to its abdomen (loose brain), the same emergence pattern for each species persisted. However, when the brain of Cecropia in a diapausing pupa is removed and transferred into the abdomen of a Pernyi pupa that has had its brain removed (debrained), eclosion occurs in phase with that of Cecropia. Similarly, the donor brain of Pernyi will determine the phase of eclosion in the recipient Cecropia (adapted from Figure 2 in Truman & Riddiford, 1970).

SCN Identification

In the search for the central biological clock in mammals, extensive studies were carried out from 1957 to the mid-1960s by Curt Paul Richter (1894–1988), a psychobiologist at the Johns Hopkins Medical School and an early pioneer in biological rhythms. Results from studies using over 900 rats identified the *hypothalamus*[11] of the brain as the location of a primary oscillator (Richter, 1965). The effects due to destroying or removing parts of the brain and glands were studied by the use of charts to record running and resting activity of individual rats that had been blinded. Surgical ablation (destruction or removal) of various parts of the brain, as well as the removal of various glands, such as the pineal[12] or the adrenals, had no effect upon the period of the free-running activity rhythm (Figure 2.9, *left*). However, following a series of hypothalamic lesions, a general area was identified in the hypothalamus, which when destroyed, resulted in a loss of the circadian

---

[11] The control of internal body temperature, like the control of many other functions in mammals, such as blood pressure, reproduction, sexuality, and the production of neurohormones, resides in the hypothalamus.

[12] See Chapter 9 on Veterinary Medicine for discussion of the pineal as a primary oscillator.

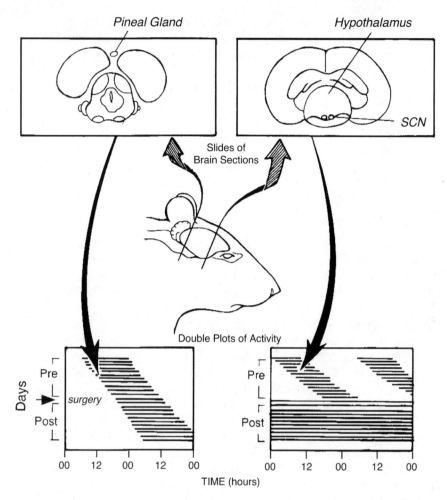

FIGURE 2.9. Spontaneous circadian-running activity in blinded rats before and after various brain surgeries led to the identification of the hypothalamus as a region of the brain containing a master clock (Richter, 1965). Pinealectomy did not affect the clock, as illustrated in the double plot of activity (*left*), while hypothalamic lesions resulted in complete loss of any circadian pattern (*right*).

rhythms for activity (Figure 2.9, *right*), as well as for drinking and feeding (Richter, 1967). Anatomically, this area is located above the optic chiasm where the two optic nerves cross and includes two clumps of nuclei, each containing about 10,000 neurons (cf. Rietveld, 1992). The name for this anatomical area is the *suprachiasmatic nucleus* (SCN).[13]

---

[13] The derivation of the name SCN helps describe its location: *supra (L.)* = above; *chiasm or chiasma (Gr.)* = two lines crossed; and *nucleus (L.)* = a nut or kernel.

A loss of the circadian rhythms of locomotor activity, drinking (Stephan & Zucker, 1972) and adrenal corticosterone (Moore & Eichler, 1972) when the SCN is destroyed by lesion, helped to identify the SCN as a master clock. Further support comes from experiments with mutants and tissue transplants. Here, the transplantation of SCN tissue from a mutant strain of hamster that had a short circadian period in running wheel activity restored the circadian rhythm in arrhythmic hamsters whose own nuclei had been ablated (Ralph et al., 1990). Likewise in mice, the transplantation of wild-type SCN tissue grafts into the hypothalamus restored locomotor activity rhythms in genetically arrhythmic mice, with characteristics of the circadian periodicity determined by the grafted donor SCN (Sujino et al., 2003).

SCN as Synchronizer

As stated earlier, oscillations occur throughout the body (cf. Hastings et al., 2003),[14] but to delineate them from those of the SCN, they are referred to as peripheral oscillations, and the tissues where they occur, as peripheral tissues. Peripheral tissues are capable of self-sustained circadian oscillations, although the properties (period and phase) of the oscillations are tissue specific (Yoo et al., 2004). Peripheral oscillations may differ from those of the SCN in a number of ways (Lowrey & Takahashi, 2004), including the phase that can be delayed by about 4 h from the SCN (cf. Balsalobre, 2002). For example, there is a difference in the phase for the abundance of mPer RNA levels in the SCN and the eyes (Zylka et al., 1998). Thus, the SCN does not necessarily drive the other oscillators, but serves as a synchronizer (cf. Yoo et al., 2004). Regions of the SCN may also transform and emit signals differently to synchronize downstream targets in diurnal and nocturnal animals (cf. Smale et al., 2003; Schwartz et al., 2004). The SCN of the brain can perhaps be likened to the conductor of an orchestra who coordinates the musical output (rhythms) of all the individual members to be in harmony as a single body that plays a circadian symphony. However, even without input from a conductor (SCN), most of the players (peripheral tissues) have their own musical score (clock) and are capable of producing their own rhythm, albeit with some damping.

Retinal Melanopsin

The daily light–dark (LD) cycle serves as a major synchronizer of circadian rhythms, which means that organisms must have photoreceptors. In mammals, the photoreceptive regions that receive the light signals reside in the eyes.[15] What

---

[14] The detailed review by Hastings et al. (2003) includes a number of excellent diagrams or illustration of circadian timing in the brain and other parts of the body.

[15] Lower vertebrates, such as birds, possess extraocular circadian photoreceptors (Menaker et al., 1970; Underwood & Menaker, 1990) and are discussed in Chapters 4 and 9 on Photoperiodism and Veterinary Medicine, respectively.

makes this statement to be more than an intuitive conjecture is the fact that mammals, such as mutant mice lacking rods and cones (blind)[16] can, nevertheless, be entrained by LD cycles (Foster et al., 1991; Freedman et al., 1999; Lupi et al., 1999). However, removal of the eyes from mice or hamsters, abolishes all responses to light, indicating that there is something in the eye necessary for photoreception (Freedman et al., 1999; Yamazaki et al., 1999). In other words, the retinal photoreceptor system for circadian synchronization must include something more than rods, cones, and their opsin-based visual pigment, rhodopsin.[17]

That "something" was initially identified from light sensitive structures[18] in the skin of frogs (*Xenopus laevis*) and named *melanopsin* (Provencio et al., 1998). Melanopsin is an opsin-like protein and because its molecular message (mRNA) was also detected in the SCN, it was quickly proposed to be a mechanism by which circadian rhythms can be entrained (Provencio et al., 1998). Soon melanopsin was cloned from human tissue and found to be expressed (mRNA) in retinal ganglion cells (RGCs), but not in the opsin-containing cells that initiate vision (Provencio et al., 2002).

Non-image-forming photoreception mediated by melanopsin is now recognized as a major component in the synchronization of circadian clocks, as well as in a number of other systems (Berson et al., 2002; Gooley et al., 2003; Hattar et al., 2002, 2003; Rollag et al., 2003). Some of the features reported for circadian entrainment of mammals, such as the sensitivity and kinetics of light (Nelson & Takahashi, 1991, 1999), are also evident in photosensitive ganglion cells (Berson, 2003). When the melanopsin gene is ablated, photosensitivity is also eliminated (Lucas et al., 2003). Mice created in the laboratory that lack melanopsin and are coupled with disabled rod and cone photo-transduction mechanisms (i.e., triple-knockout mice) do not entrain to LD cycles (Hattar et al., 2003). These results suggest that the two systems, melanopsin-associated and the classical rod/cone, work together to provide photic input (Hattar et al., 2003; Lucas et al., 2003). While other photoreceptive pigments do occur in mammals, including the cryptochromes of the circadian clock itself (Chapter 5), melanopsin is the prime pigment in the RGCs leading to the SCN (cf. Brown & Robinson, 2004).

Retinohypothalamic Tract

For decades it has been known that photic (light) information is conveyed from the eyes to the hypothalamus via a retinohypothalamic tract (RHT), where its

---

[16] The capacity of blind humans to entrain to a LD cycle is discussed in Chapter 11 on Clinical Medicine.

[17] Rhodopsin in a molecule that absorbs light and is present in the eyes of animals. The molecule consists of a protein named *opsin* and an 11-*cis* retinal chromophore (light-absorbing group).

[18] These photosensitive structures are called dermal melanophores. They contain cellular pigment granules that are reorganized in response to light and thus produce a change in color.

neural projections terminate in the bilaterally paired SCN (Hendrickson et al., 1972; Moore & Lenn, 1972). It is now known that the RGCs of the RHT, which are selectively expressed for melanopsin, project to the SCN (Gooley et al., 2001; Hannibal et al., 2002; Hattar et al., 2002) and primary RHT neurotransmitters, which mediate the circadian effects of light, may be the amino acid, glutamate (Ebling, 1996; Meijer & Schwartz, 2003) and the polypeptide, PACAP[19] (Hannibal & Fahrenkrug, 2004). Thus, in the intact organism, photic information from both the rod/cone and melanopsin systems is received by the SCN and converted (transduced) into neural and hormonal output signals that affect various rhythms of the animal (cf. Lowrey & Takahashi, 2004).

Characteristics of the Period

Although the circadian period can be described as being relatively stable, it may be altered slightly by light intensity, quality, and/or duration, by temperature, and by certain chemicals. For the most part, these effects are small, but they have been useful in developing models, formulating hypotheses, and designing experiments to better understand the physiology of biological rhythms.

Stability and Free-Running

The most distinguishing characteristic of a circadian rhythm is the *stability* of the period. So important is the stability of the period that it has been referred to as "*the ranking cornerstone of the circadian edifice*" (Ehret, 1980).

Organisms in their natural environment of the solar day generally display 24-h periods in metabolism, development, and behavior. However, when organisms are isolated from such external cycles, the period will differ slightly and become about (*circa*) 24 h. Under such constant conditions (CC), which customarily consist of either continuous illumination (LL) or darkness (DD) and constant temperature, the period will usually become established somewhere between 20 and 28 h. This period, which is known as the *free-running* period (Table 2.6), remains relatively stable with regard to the period length and reflects that organism's genetically determined circadian period.

Before the pattern of the cycle becomes stable under free-running conditions, there might be some unstable transitional cycles called *transients*.[20] Because of the transient cycles that occur between the stable periods established under synchronizing LD conditions and those in CC, it may be necessary to monitor a number of cycles in order to determine the "true" free-running period. Transients that persist for a long duration have been called "*aftereffects*" (cf. Moore-Ede

---

[19] PACAP = pituitary adenylate cyclase-activating polypeptide.

[20] Transients can also occur during the first days after changes in environmental LD entrained conditions, such as happens during jet-lag or shift-work.

TABLE 2.6. Comparison of the free-running circadian period in some organisms during continuous light (LL) or dark (DD).

Organism	Rhythm in	Length of period in:		Reference
		LL	DD	
Plant (*Coleus*)	Leaf movement	Longer	Shorter	Halaban, 1968
Rodent (mouse)	Activity	Longer	Shorter	Aschoff, 1955; Pittendrigh & Bruce, 1957
Bird (Bullfinch)	Activity	Shorter	Longer	Aschoff, 1953
Fruit fly (*Drosophila*)	Emergence	Shorter	Longer	Bruce & Pittendrigh, 1956
Aquatic plant (*Lemna*)	Enzyme activity	Present	Absent	Gordon & Koukkari, 1978
Fungus (*Pilobolus*)	Sporulation	Absent	Present	Schmidle, 1951; Uebelmesser, 1954
Aquatic insect (*Brachycentrus*)	Activity	Shorter	Longer	Koukkari et al., 1999

et al., 1982) and are typically found in cases where organisms (e.g., mice) have been subjected to unusual cycles prior to constant conditions.

Also during free-running in constant conditions (e.g., LL or DD), a phenomenon known as *splitting* can occur, wherein the single-circadian activity rhythm (e.g., of a hamster) can "split" into two components with different frequencies. These two frequencies persist for some time until they eventually share a common period, but often in antiphase to each other (cf. Pittendrigh, 1981). A different type of splitting has been observed in the human rest/activity and body temperature rhythms during isolation, wherein these rhythms initially share a common frequency, but eventually free-run with periods that can differ significantly from each other (Wever, 1979).[21]

Frequency Multiplication/Demultiplication

In some instances, a frequency, which is the inverse of the period, occurs as a multiple or division harmonically related to an originally prominent frequency. For example, if the number of cycles increases from one to two over a similar time span, *frequency multiplication* has occurred, such that the new cycles are shorter than the original. Conversely, when the number of cycles decreases from two to one over a similar time span, *frequency demultiplication* has occurred, indicating that the new period is longer than the original. These higher or shorter frequencies correspond to integer submultiples (e.g., 1/2, 1/3, 1/4, etc.) or multiples (e.g., x2, x3, x4, etc.) of the original frequency.

In addition, sometimes a high-frequency LD cycle (e.g., 12 h, 8 h, 6 h) may entrain a circadian rhythm. Thus, entrained by frequency demultiplication is said

---

[21] This form of splitting was observed in the body temperature and sleep/wake rhythms in one of us (RBS) upon completion of a 3-week stay in isolation in Aschoff's Bunker (see Essay 12.1 in Chapter 12 on Autorhythmometry).

to have occurred when a rhythm becomes entrained to an environmental cycle that is a subharmonic of it's innate frequency (e.g., a 24-h rhythm is entrained by a 12-h LD cycle) (cf. Bruce, 1960). For example, in *Neurospora*, intact WT strains could be entrained by demultiplication when exposing them to warm/cold temperature cycles of 4.5 h/4.5 h, 5 h/5 h, or 6 h/6 h, which yielded conidiation rhythms of ~18, ~20, and ~24 h, respectively, and were within their circadian range on entrainment (Pregueiro et al., 2005).

Light Quality

The circadian period might be *different* depending on whether the constant condition is continuous illumination (LL) or continuous darkness (DD). The period of a given rhythm might be longer or shorter under LL than DD, or rhythmicity might be detected under only LL or DD, but not under both.

Particular wavelengths of light can also influence the free-running period of some variables, although the effect may differ among species. For example, under continuous red light (610–690 nm), the period of the leaf movement rhythm for a bean plant *(Phaseolus coccineus)* is longer than in DD (Lörcher, 1958), while the period of the leaf movement rhythm for *Coleus blumei* is shorter than in DD (Halaban, 1969).

With some animals, such as the *nocturnal* (night-active) mouse *(Mus musculus)* and the *diurnal* (day-active) bird (a starling: *Sturnus vulgaris),* there is a linear relationship between the period of the circadian rhythm and the logarithm of *light intensity* (Aschoff, 1960). More specifically, increasing light intensity during LL often *shortens* the circadian period of *day-active* (diurnal) animals and *lengthens* that of *night-active* (nocturnal) animals. This phenomenon, sometimes referred to as one of *Aschoff's Rules*, is illustrated in Figure 2.10. Many animals follow this rule.

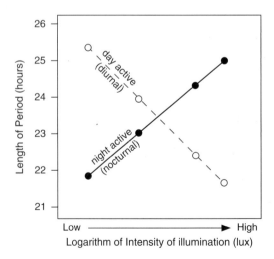

FIGURE 2.10. A generalized diagrammatic illustration of Aschoff's rule (Aschoff, 1960) showing that when light intensity increases, the free-running period shortens for diurnal animals (day-active, *open circles*) and lengthens for nocturnal (night-active, *closed circles*) animals.

Temperature

The period of a circadian rhythm has the unique characteristic of being relatively temperature independent or, more likely, *temperature compensated*. In other words, the effect of temperature on the period of circadian rhythms is surprisingly slight. From a biochemical perspective this is unusual since the rate of most metabolic processes increases with increasing temperature. Due to greater kinetic energy, the rate of chemical reactions often at least doubles for each 10°C rise in temperature.

To express the effect of temperature on the rate of the reaction, or in the case of a rhythm, on the change in period or frequency, biologists refer to a value called the Q_{10}. The Q_{10} temperature coefficient (quotient) expresses the rate or frequency at one temperature divided by the rate or frequency at a temperature of 10°C less. Within the range of environmental temperatures at which an organism remains physiologically viable, the Q_{10} for most chemical reactions is approximately 2 to 3. However, circadian rhythms usually have a Q_{10} within the range of about 0.8–1.2 (cf. Sweeney & Hastings, 1960 for examples). This observation holds even for mammals, in which the effects of temperature might seem unsubstantial because of their near-constant body temperature. Nonetheless, circadian rhythms with a period of approximately 24 h can still be found in hibernating animals in which body temperature is lowered (Wollnik & Schmidt, 1995) and in tissues isolated from mammals (Andrews & Folk, 1964).

Most ultradian rhythms are dramatically affected by temperature, although some are temperature compensated when the temperature is in a physiologic range. Included in the latter are ultradian rhythms in unicellular (eukaryotic) organisms with periods ranging from 1 min to 5 h, where temperature compensation of the period is as effective as that usually associated with circadian clocks (Kippert, 1997). For example, a period of about 76 min remained approximately the same at temperatures from 20°C to 30°C for the levels of total protein and the respiration rate in the soil amoeba *Acanthamoeba castellanii* (Lloyd et al., 1982). Likewise, the about 55-s courtship song, which male *Drosophila melanogaster* flies produce by vibrating their wings, is temperature compensated, remaining nearly unchanged for temperatures ranging from 16°C to 35°C (Kyriacou & Hall, 1980). This is also the case for the nematode *C. elegans*, where an averaged 45 s defecation rhythm remains nearly constant over a temperature range of 19°C to 30°C (but is lengthened at 13°C and 16°C) (Liu & Thomas, 1994).

Ultradian rhythms in higher organisms characteristically exhibit shorter periods with increasing temperature. For example, the periods of horizontal shoot movements (*circumnutations*) of the small plant, *Arabidopsis* (Schuster & Engelmann, 1997), and garden bean *(Phaseolus)* shoot tips (Millet & Badot, 1996) have a Q_{10} of about 2. Perhaps it is safe to conclude that ultradian rhythms usually exhibit shorter periods with increasing temperature, although this view has not been fully addressed in cases where ultradian oscillations have been observed as part of an overall circadian pattern of a single variable (e.g., ultradian movements of a soybean leaflet that also displays circadian movements) (cf. Guillaume & Koukkari, 1985).

Compared to ultradian rhythms, less is known about the effects of temperature on infradian rhythms. An example of temperature compensation for an infradian rhythm was shown in the ~40-week pupation rhythm of the varied carpet beetle *Anthrenus verbasci*. Interestingly, temperature compensation has been observed under DD (Blake, 1959) and LD 12:12 (Nisimura & Numata, 2001) when temperatures were between 17.5°C and 27.5°C.

Chemicals

The effects of some chemicals have been shown to slightly alter the period of circadian rhythms. Leading this list are lithium ions, D_2O (heavy water), and ethanol (ethyl alcohol), which have been shown to lengthen the free-running circadian period in both plants and animals (cf. Engelmann & Schrempf, 1980). Alkaloids, such as theobromine, theophylline, and caffeine, can lengthen the circadian period (cf. Bünning, 1973), while lithium ions have been shown to lengthen some ultradian periods in plants such as *Phaseolus* and *Desmodium* (Millet et al., 1984; Chen et al., 1997).

Except for the effects of certain birth control pills on the circatrigintan (~30 days) rhythm of the female menstrual cycle, very little work, if any, has been done to identify a chemical that changes the period of an infradian rhythm. The presence of a mammary tumor has been shown to lengthen the period of the estrus cycle of a mouse, possibly by mediating sex and stress hormones (Ratajczak et al., 1986). In the latter case, the cycle length returned to normal after the tumor was surgically removed.

Characteristics of the Amplitude

The mathematical analysis of data for determining the presence or absence of a rhythm, as well as for describing rhythm characteristics, involves the use of statistical tests (see Chapter 13 on Analyzing for Rhythms). The amplitude is an important characteristic in the detection of rhythms. For example, an amplitude of zero would indicate the absence of an oscillation or rhythm at the period being tested. This would occur if the data representing the variable (1) did not change during the course of time, and thus appeared as a flat line, (2) fluctuated randomly, or (3) oscillated with a period other than that of the mathematical model being fitted.

Although a single-component (e.g., 24 h) cosine curve may be an adequate model for defining an amplitude, rhythmic patterns can vary and may need more complex cosine models involving the combination of two or more periods (e.g., 24 h & 12 h or 24 h & 12 h & 8 h, etc.) to properly describe them. A pattern could be as simple as a series of spikes, which is typical of some pulsatile hormone secretions in humans (e.g., insulin), or it may be a complex series of short-period oscillations superimposed upon the regular pattern of a longer cycle. In some studies, such as in the case of the bioluminescence of an alga species, zero is used as the value for the baseline, and the amplitude is defined as the magnitude from zero to the peak.

In general, the larger the amplitude, the more prominent a rhythm in the data will appear to the naked eye. However, the magnitude of an amplitude does not always indicate whether a rhythm is strong or weak. While a large amplitude may *subjectively* seem to indicate a stronger rhythm, such as that found in the hormones cortisol and melatonin, it does not indicate that the rhythm is any more or less *biologically significant* than a small-amplitude rhythm, such as that observed in body temperature. In other cases, a rhythm may appear weaker when the amplitude is reduced during certain situations (e.g., during illness or phase-shifting) when compared with the amplitude found during a normal (e.g., healthy, synchronized) situation. The important point is that a rhythm is *objectively* validated to be *statistically significant* by having an amplitude significantly different from zero.

Damping

When organisms are synchronized under LD conditions (e.g., 16-h L followed by 8-h D) and then transferred to LL or DD, the amplitude of the circadian rhythm may become smaller for each subsequent cycle and the rhythm might thus appear to overtly fade out. This characteristic is called *damping* (Figure 2.1; see also Figure 13.1 in Chapter 13 on Analyzing for Rhythms), and although the rhythm might no longer be observed, it does not necessarily indicate that the organism or cell no longer has circadian oscillations (Bünning, 1973). This phenomenon could be due to the oscillation mechanism itself damping. Another possibility is that a variety of cells or individual oscillators with slightly different periods become desynchronized from each other, giving the overall appearance of damping.

A rhythm that dampens to the extent that it can no longer be detected under one type of constant condition can sometimes be reinstated and detected again when the organism or material is transferred to another type of constant condition (e.g., from LL to DD). Sometimes only a given duration (e.g., 12 h, 18 h) of the opposite condition, such as providing a dark span during LL (Wasserman, 1959) or changing the light intensity in LL, will restore the pattern or amplitude (see Figure 2.11). For example, changes in light intensity can either inhibit or initiate a detectable carbon dioxide emission rhythm of the leaves of the succulent plant *Bryophyllum* (Wilkins, 1960).

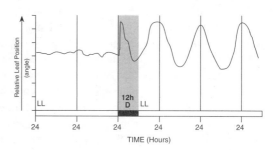

FIGURE 2.11. A generalized diagrammatic illustration of the initiation of a circadian rhythm in continuous light (LL) by a single span of darkness (D) shown for the leaf movement of a plant.

Characteristics of the Phase

The phases of a rhythm are determined by the schedule of environmental synchronizer(s). Entrainment to the daily periodicity in the external environment (usually the LD schedule) is an essential characteristic of circadian rhythms, whereby stable phase relationships between the endogenous circadian system and exogenous synchronizers become established. Under fixed LD schedules, the phase of a rhythm is repeated at about the same time every day. A change in the environmental cycle will cause a shift in the cyclic pattern of a biological variable so that the phase comes earlier or later compared to the phase of the rhythm prior to the change in the environment.

Synchronizers

We, like most other groups of organisms, live in an environment where light and darkness, as well as changes in temperature, follows the cyclic pattern of the solar day. These external 24-h cycles of light and temperature are primary *synchronizers* (or *zeitgebers*) for the phases of circadian rhythms. While LD and temperature cycles are the dominant environmental agents for synchronizing (*entraining*) the phases of circadian rhythms, other possible synchronizers include the timing of water or humidity (cf. Bünning, 1973), feeding (Nelson et al., 1975), certain chemicals, and, in the case of animals, factors in the social environment, such as sound for a bird (Menaker & Eskin, 1966) and a cricket (Marques & Hoenen, 1999). In addition, the phases of circadian clocks in peripheral tissues (liver, kidney, and heart) of mice are entrainable by the cortisol rhythm (Balsalobre et al., 2000). The more natural gradual changes in light intensity that occur at dawn and dusk (called LD-twilight), as compared to abrupt LD transitions (called LD-rectangular), increase the strength of the LD synchronizer in rodents (e.g., hamsters) (Boulos et al., 1996, 2002), as well as in humans (Wehr et al., 2001).

Phase-Shifting by Light

In order to look at relationships between the external environment and endogenous rhythms, numerous studies using rodents, humans, and other organisms have been performed to study the behavior and adaptation of circadian rhythms (e.g., in whole body activity, temperature, etc.), following phase advances or delays of the LD schedule. A major characteristic of circadian rhythms, as well as of many ultradian and infradian rhythms, is a labile phase. In other words, the phase can be moved (*phase-shifted*) to occur earlier or later. A change in the phase during one or more cycles in reference to previous cycles is called *phase-shifting*. For example, the peak in the circadian body temperature rhythm is in the afternoon for most humans that are 24 h synchronized with sleep at night. However, when a person living in central North America crosses seven time zones by traveling quickly to Central Europe by jet airliner, the afternoon peak in his or her body temperature will, after about 5–7 days of transients, have advanced by about 7 h to occur again

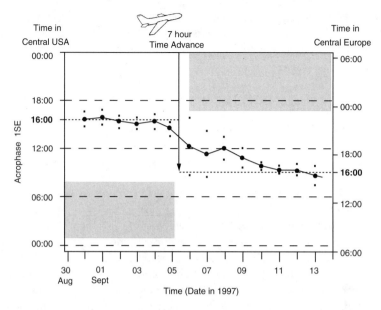

FIGURE 2.12. Phase-shifting of the circadian peak (acrophase) in oral temperature of a healthy man (RBS, 51 years of age) traveling from east to west (St. Paul, Minnesota to Paris, France). *Note*: After several days, the acrophase advanced by 7 h to again occur at 16:00 h local time, which represents the time change between central USA and central Europe (Sothern, unpublished).

during the afternoon at the new destination (Figure 2.12). Stated more simply, the phase is shifted from its location in the original time zone so that the rhythm is similarly synchronized to a new time zone.

Phase Shifts by Chemicals or Temperature

Peaks of rhythms can sometimes be moved (shifted) due to the effect of certain chemicals or changes in environmental temperature. Some of the well-known chemicals that have been shown to affect the circadian clock in various organisms include caffeine (Mayer & Scherer, 1975), theophylline (Ehret et al., 1975; Mayer et al., 1975), and ethanol (Enright, 1971; Bünning & Moser, 1973; Sweeney, 1974). Protein synthesis inhibitors can also reset the circadian clock (Taylor et al., 1982; Takahashi & Turek, 1987; Inouye et al., 1988; Shinohara & Oka, 1994; Murakami et al., 1995). Glucocorticoids have also induced phase shifts in circadian gene expression in peripheral tissues, but not in the SCN, of mice (Balsalobre et al., 2000).

With regard to temperature, an increase in 24-h water temperature from 20°C to 28°C induced an 8-h shift in the peak of the daily serum prolactin rhythm in the gulf killi fish (*Fundulus grandis*) with respect to the same LD 12:12 photoperiod (Spieler et al., 1978). Even brief exposure to a higher or lower temperature may shift a phase. For example, when *Phaseolus* plants were maintained in LL at a constant temperature of 20°C, a 4-h span of 28°C during their lowest position of the leaves delayed the following same phase in the next cycle (cf. Bünning, 1973). Such changes in the sensitivity, which may cause the phase to occur earlier or later depending upon the time within a cycle when the agent (stimulus) is applied, is referred to as a *phase response*.

Phase–Response Curve

When the phase is *advanced* so that the event occurs earlier, the shift is *positive* and indicated by "+" symbol. When the phase is *delayed* so that the event occurs later, the shift is *negative* and indicated by "−" symbol. The reason for emphasizing phase-shifting is the fact that the phase of a free-running circadian rhythm may be either advanced or delayed in reference to *when* a synchronizing agent is introduced. Only a brief exposure (*pulse*) might be necessary to produce the shift. The extent and direction of the shift as a function of the time of the pulse can be graphically illustrated by a *phase–response curve* (Figure 2.13).

Most studies on phase-shifting and phase–response curves have centered on circadian rhythms. How applicable this characteristic is to rhythms with longer or shorter cycles is still in question. We do know, however, that phase-shifting can take place in the infradian female menstrual cycle of humans and the ultradian glycolytic cycle of yeast. In the first case, the menstrual cycle can be reset due to various causes, such as use of the pill, social synchronization, or disease

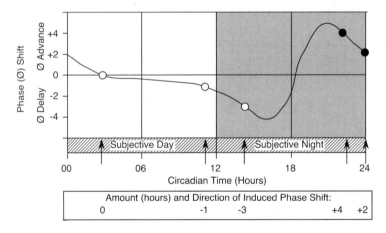

FIGURE 2.13. A generalized phase–response curve (PRC) in continuous darkness (DD). A light pulse (e.g., 5 min of bright light) given at a specific time (*arrows*) during the subjective times of previous light or dark spans results in either an advance (+) or delay (−) of phase, depending on the time the pulse is given. Phase-shift points resulting in a delay are shown as *open circles*. A relatively small phase shift of up to 6 h is called a Type 1 PRC, whereas Type 0 PRCs show larger phase shifts (Johnson, 1999). Stimuli that induce a PRC include pulses of light, temperature, or chemicals. *Note:* The original PRC graphs showed an advance downward and a delay upward (cf. Pavlidis, 1973). The current convention for PRC charts is to show the opposite (advance upward, delay downward).

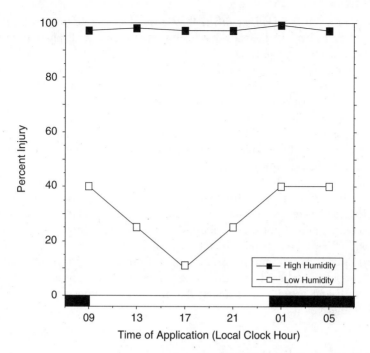

FIGURE 2.14. High humidity (*closed boxes*) masks the daily rhythm in the response of *Abutilon theophrasti* leaves (% injury) to the herbicide bentazon applied at the same concentration at different times during the light and dark spans during low humidity (*open boxes*) (dark span = dark bar) (modified from Koukkari & Johnson, 1979).

(see section on the Menstrual Cycle in Chapter 7 on Sexuality and Reproduction). In the latter case, the phase of the short-period NADH[22] biochemical oscillations (e.g., ~5 min) can be delayed when ADP is added during the trough, advanced when ADP is added during the rising of the curve, or remain unchanged when ADP is added at the peak (Pye, 1971).

Masking

Characteristics of a rhythm may be superficially altered by the environment, a process called *masking*. For example, a rhythm in the response of the weed velvetleaf (*Abutilon theophrasti*) to a herbicide, bentazon, has a larger amplitude when the humidity is low than when the humidity is high (Koukkari & Johnson, 1979) (Figure 2.14). In this instance, the amplitude of the rhythm is greatly reduced by a high relative humidity, which masks (hides) the large amplitude. In

---

[22] NADH is the reduced form of nicotinamide adenine dinucleotide and ADP is the molecule, adenosine disphosphate. Both occur naturally in living cells (see Essay 5.1 in Chapter 5 on Models and Mechanisms).

the winter, a red fox (*Vulpes fulva*) that normally hunts during the night for rodents located beneath the snow has been shown to change its phase of hunting activity to capturing squirrels during daylight hours when the snow becomes covered with ice (Tester, personal communication; Tester & Figala, 1990).

In humans, the body temperature will respond to exogenous activities and meals, resulting in either small or large changes being superimposed on the endogenous circadian pattern. Subjecting an individual to a constant routine consisting of bedrest and equally spaced, frequent same-size meals has been shown to "unmask" and thereby "purify" the waveform of the endogenous circadian pattern in body temperature (Minors & Waterhouse, 1989).

Meal timing has also been shown to alter the circadian patterns in food-filtering behavior of caddisflies *(Brachycentrus occidentalis)*. During their aquatic larval stage, these insects are nocturnal feeders under *ad libitum* (upon desire) feeding conditions. In a laboratory setting, when food was supplied to the caddisflies in the middle of the day or at 8-h intervals, the underlying natural circadian rhythmic pattern, while still present, was masked, since the circadian patterns altered accordingly so that an increase in filter-feeding coincided with and actually anticipated the next presentation of food (Gallepp, 1976; Sothern et al., 1998).

Need for a Cyclic Environment

The need of many organisms for a favorable cyclic environment and the consequences of constant or unusual environmental conditions (e.g., short or long days, light interruptions at night) are often ignored by humans, even in the study of biological rhythms. The pathological consequences resulting from constant conditions greatly impinge upon the critical evaluation of the previous history (aftereffects) as a characteristic of free-running periods.

In addition, the pathological effects of constant conditions may lead to abnormal development, disease, and even death. This has been dramatically illustrated in a number of plant studies (Bünning, 1932; Hillman, 1956; Went, 1960). The green leaves of tomato plants (*Solanum lycopersicum*) remain healthy when maintained under 24-h cycles of LD or temperature, but become yellow (chlorotic) and brown (necrotic) when plants are maintained under LL and constant temperature (Figure 2.15). However, by providing a 24-h temperature cycle to these plants during LL, the injury associated with the constant light is eliminated (Hillman, 1956).

With regard to mammals, there have been several reports on the deleterious effects from unusual light exposures. For example, exposing the social vole (*Microtus socialis*) to brief spans of light during the long nights of winter in Israel resulted in death, which may be related to their thermoregulation (Haim et al., 2005). In rats and mice, tumor growth was enhanced when they were kept in continuous light (LL) or long days (Waldrop et al., 1989; Sanchez, 1993; Dauchy et al., 1997). Mammary gland development, reflecting early sexual maturity, was accelerated in rats kept in LL from birth (Anderson et al., 2000). Several articles have also discussed the

FIGURE 2.15. Leaves of tomatoes (*Solanum lycopersicum*) from plants maintained in our lab under either LD (*left*) or continuous light (LL) (*right*). Pathological effects, such as the yellowing of leaves (chlorosis), can be attributed to the noncyclic photoperiod (LL) and illustrate the importance of a cyclic environment for normal development of these plants (Koukkari, unpublished).

influence of night-time light as a possible risk factor for human breast cancer (Stevens et al., 1992; Waalen, 1993; Stevens & Davis, 1996) (see full discussion of light, melatonin and cancer in Chapter 11 on Clinical Medicine).

In Darwin's Footsteps

Many, if not most individuals, are aware of Charles Darwin and his studies of evolution and natural selection, which he published in *The Origin of Species* (Darwin, 1859). Fewer, however, are aware of his studies of circumnutations, tropisms,[23]

---

[23] Tropisms are growth movements of plants directed toward or away from a stimulus, such as a shoot bending toward light (phototropism).

and the movements of leaves and stems, which he and his son published in *The Power of Movements in Plants* (Darwin & Darwin, 1897). From the perspective of chronobiology, Charles Darwin and his son, Francis, also possessed insight into the genetic basis of rhythms: ". . . we may conclude that the periodicity of their movements is to a certain extent inherited" (Darwin & Darwin, 1897).[24]

In this chapter we have described the characteristics of biological rhythms. Here we introduce some "hands on" learning activities that should help you better understand and appreciate some of these characteristics, by following the old adage: *"Those who hear, forget. Those who see, remember. Those who do, understand."*[25]

Not only are these activities fun to do, but also the only item required that is not usually found or easily constructed in a "home environment," is a pole bean (*Phaseolus vulgaris*).[26] The pole bean is ideal because of its two well-defined overt rhythms and the ease by which this classic organism can be maintained and studied[27] by almost anyone. In pole beans these two rhythms are the ultradian circumnutation movements of the stem or shoot (Figure 2.16) and the circadian movements of the leaves (Figure 2.17).

Like Darwin, but following a slightly different procedure,[28] you too should be able to estimate the period and pattern of the circumnutation and leaf movement rhythms. The procedures are relatively simple and can be performed by one person, or possibly, by a group of two or three individuals with each person being responsible for a given number of measurements (cf. Koukkari, 1994) (see Exercises 2.1 and 2.2).

---

[24] The first edition of their book was published in 1880. This quote refers to leaf movements and appears on p. 413 in the 1897 edition of their book (D. Appleton & Co., New York).

[25] One of us (WLK) often refers to this quote when teaching. Its original source may have been the undated Chinese proverb: *I hear and I forget. I see and I remember. I do and I understand.*

[26] Seeds of pole bean are sown in sterilized potting soil contained in pots or cups, all of which can be purchased commercially. Usually two to four seeds are sown in each container, covered with 2 to 3 cm of soil, watered, and cared for in the manner prescribed for most houseplants. When the pair of foliage leaves appear above the cotyledons and begin to expand (~10 days after sowing) the number of plants per container should be reduced to one. The extra plants can be repotted (one per pot) and also used in the study.

[27] Years ago, one could borrow a mouse from an animal research facility to demonstrate running wheel locomotor activity rhythms, or purchase a minnow from a local bait store to study Q_{10}, without requesting prior written approval from an institutional committee that must oversee the use of animals. These rules are now necessary. For a short teaching exercise, crickets can be substituted for mice, and pole beans for minnows.

[28] For example, a small thin glass filament was attached with shellac to the part of the plant to be observed and the plant covered by a horizontal glass plate (top) and a vertical glass plate (side). Dots corresponding to the location of the observed part of the plant were made on the glass and later copied on tracing paper (Darwin & Darwin, 1897).

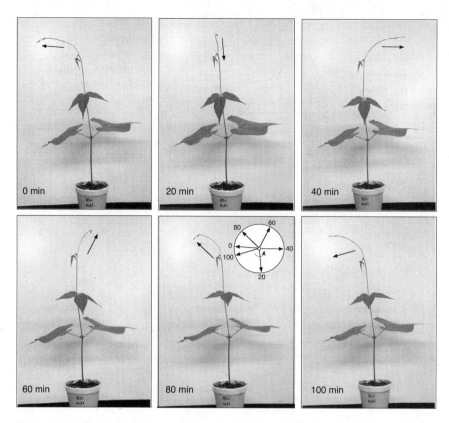

FIGURE 2.16. A series of photographs every 20 min for 100 min showing the ultradian rhythm in circumnutation of a 17-day-old pole bean *(Phaseolus vulgaris,* cv. Kentucky Wonder). Note the elongated apical shoot and its arched (bending) position as it moves in a counterclockwise direction. Time elapsed from start listed in each panel and direction of shoot noted by an arrow in each panel and in the circle in lower middle panel. Photos by R. Sothern.

FIGURE 2.17. Day (up) and night (down) positions of the leaves of a young bean plant.

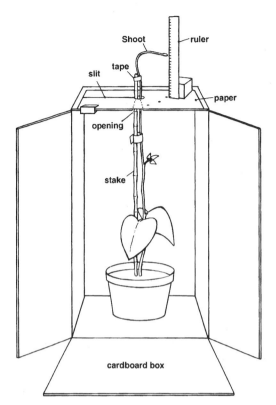

FIGURE 2.18. A simple setup for monitoring circumnutation movements of a pole bean. The ruler can be attached to a small wooden block for stability and can also be used to measure the vertical distance of the shoot tip from the surface of the plate (paper) if also monitoring vertical movements.

Exercise 2.1: An Ultradian Experiment: Circumnutation

A 2- to 3-week-old pole bean that has an elongated apical shoot is ideal for this exercise (Figure 2.16). The plant can be placed in an open box with the top (side) of the box serving as a plate (Figure 2.18), or as an alternative setup, a wood or plastic plate may be supported and positioned above the plant (Figure 2.19a). In either case, the plate should be level and have a 25-mm hole in the center. It will be easier to position a shoot into the hole if you cut an opening that is 5–10 mm wide and extends from the middle of one edge to the center hole. Prior to placing the plant and plate into position, the shoot must be secured to the tip of a stake, preferably with a tape fastened loosely at a location that is about 8 cm below the shoot apex.[29] The stake should protrude about 8–10 cm above the plate so that the plate will not interfere with the vertical (up and down) movements of the shoot apex.

Obtain a square sheet of paper and cut a straight slit leading from the middle of one edge to the center. Because the dimensions of the paper need to be large enough

---

[29] If the shoot is very long, lower portions of the stem may also be secured to the stake. Actually, the stem could be allowed to bulge out from the stake in order to have the plant fit better into the box. Generally, this will not interfere with the movements of the shoot tip.

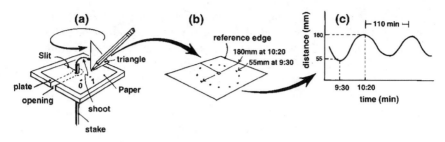

FIGURE 2.19. Ultradian shoot movements of a pole bean: (*a*) monitoring, (*b*) transcribing, and (*c*) graphing horizontal circumnutation movements of a bean shoot.

to extend beyond the tip of the shoot when it is moving (see Figures 2.18 and 2.19a), two to four smaller sheets of paper can be secured with a tape and trimmed to make one large sheet. Next, cut out a circular hole in the center of the paper (about 30 mm), which is slightly larger in diameter than the hole in the center of the plate. Slide the sheet onto the plate and position the hole around the stem, taking care not to cut the stem while aligning the two holes and taping or clamping the sheet to the plate. In order to have a solid surface to write on, the slit in the paper should not be placed above the horizontal opening that was cut out of the plate.

To monitor the horizontal circumnutation movements, place the measuring device (e.g., ruler or triangle) on the plate and align the edge of the measuring device with the tip of the shoot (almost touching). Next, place a dot on the sheet where the bottom of the measuring device meets the paper and label the dot as 0 (zero). This procedure is illustrated in Figure 2.19a. On another sheet of paper or in a notebook, record the time that the dot was made and the number representing the dot (0).[30] Ten minutes later, repeat the procedure, but label the second dot as 1, and again record both the time and corresponding number (e.g., 1). The process of measuring and recording (monitoring) the location of the shoot tip at 10-min intervals should be continued for the next 3–5 h. If necessary, the interval between measurements could be lengthened to 15 or 20 min.

After the monitoring has been completed, carefully remove the sheet of paper from the plate[31] and connect the series of dots to see better the horizontal pattern of circumnutation. Select one of the four edges (margins) of the paper to serve as a reference edge (it does not matter which edge) and use a ruler to measure the perpendicular distance (mm) between the reference edge and each of the points (Figure 2.19b). Next, prepare a graph to illustrate the rhythm (Figure 2.19c). Because each dot is 10 min apart, you should be able to count the number of dots until they come full circle (back to the 0 mark) and estimate the period in minutes.

---

[30] One could measure the vertical movements along with the horizontal movements by recording the distance in millimeters from the shoot tip to the surface of the paper.

[31] If this is a group project and all persons are not present when the final measurement is taken, photocopies of the entire pattern of dots can be made and distributed to all members of the group.

By viewing the graph, the amplitude can be estimated by calculating half the distance between the lowest and highest distance of the mm marks. Using the same data points, another graph could be prepared that includes measurements from two different reference edges to show a different phase relationship (but the same period and amplitude). The frequency can also be calculated and the relationship between period and frequency noted. The frequency of the circumnutation rhythm can be used to determine the Q_{10} of the rhythm, although this usually will require two facilities (rooms) or controlled environments where two different temperatures can be maintained during the course of measurements.

Often in a university setting, finding the time or space to introduce the topic of biological rhythms into a course that may already be overloaded is difficult. However, the addition of this exercise to a number of our biology courses has not led to the elimination of other material, mostly, because the exercise involves only a few minutes of actual work by any one student and can occur when the focus is on another topic[32] (Koukkari, 1994).

Exercise 2.2: A Circadian Experiment: Leaf Movements

In a research laboratory, leaf movements are often monitored continuously by electronic or video methods. For a didactic exercise that could be performed simultaneously by a large number of people in a variety of different locations (e.g., laboratory, home, dormitory room, closet, etc.), a simple inexpensive measuring card is more than adequate (Figure 2.20a). The position of any simple leaf or leaflet at any point in time can be determined by placing the measuring card over the midrib and reading the degrees indicated by a string attached to a metal nut.

If only one plant is available, leaf movements may be monitored for the first day or two when the plant is maintained in a LD-synchronized environment. The LD environment chosen will depend on whatever is available, such as the "natural" solar day viewed from a window, beneath a fluorescent lamp connected to a timer, or better yet, a controlled environmental chamber programmed to a LD 12:12 cycle. In order to determine if the leaf movement rhythm continues in the absence of a LD-synchronized environment, the plant should be transferred to a light-tight enclosed location (e.g., closet), where a fluorescent lamp could be safely placed to provide continuous illumination (LL)[33] for a few days. If two

---

[32] Even in a lecture session or during a seminar, a pole bean with a paper collar (plate) attached with tape to the top of the pot provides an impressive means to demonstrate a circumnutation rhythm. In such cases, I have selected a different student to be responsible for each time point and dot. Actually, monitoring and the placement of dots (time points) may commence before class. Most people are not aware that the tip of a plant will move so rapidly and display a rhythm.

[33] A low watt fluorescent lamp, similar to the type used for reading, may be a good choice. Incandescent lamps usually produce too much heat, especially in a small-enclosed area, and should not be used without a means for temperature control. In fact, providing a constant temperature during both LD and LL would be desirable.

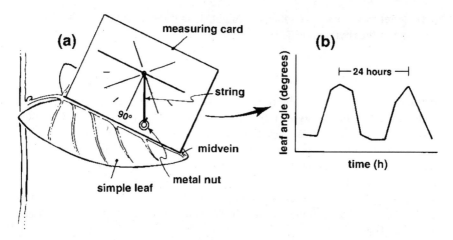

FIGURE 2.20. Circadian leaf movements of a pole bean: (*a*) use of an angle card to measure the position (angle) of the blade of a leaf relative to the horizontal or vertical axis, as determined by a plumb line (metal nut suspended by a string); (*b*) leaf angles (*y* axis) plotted for each time point (*x* axis).

plants are available, one could be maintained in LD and the other in LL, thus reducing the total duration of the study.

The interval between successive measurements can be arranged to best fit the academic or social schedule of the person who does the monitoring, although one of the measurements should be conducted when the L span begins, another near mid-L, and another measurement conducted when the L span is about to end. Equally spaced 3- to 4-h intervals between measurements are desirable and intervals longer than 6 h should be avoided. However, all measurements scheduled to occur during times of D, except for the final time point, could be omitted.[34] For this last measurement in the LD series, a light can be turned on, the angle of the leaf measured immediately, and the monitoring procedure terminated. In some "less-controlled facilities," such as a living room in a house, there may be "stray light" or a fixed "night light" that will enable a person to monitor leaf positions during "darkness."

The data collected should be plotted as illustrated in Figure 2.20b and the results analyzed to estimate the period, amplitude, and phase. Select the time of usual L onset as the reference point for Circadian Time (CT = 0) of the leaf movement rhythm under both synchronized (LD) and free-running (LL) conditions. In LL, the time span that corresponds to the D span that was provided during entrainment

---

[34] *"It is troublesome to observe the movements of leaves in the middle of the night, but this was done in a few cases"* (Darwin & Darwin, 1897). *Note*: Plant physiologists often use dark rooms having special "green safe lights."

(LD) is referred to as the subjective night. It may be possible also to observe damping of the rhythm during LL.

Mechanisms

Although the objective of the preceding experiments was to help you obtain a better understanding of some of the characteristics of rhythms, you may wonder about the mechanisms that cause these rhythmic movements. Briefly, the rhythmic circumnutation movements of the shoot appears to be due to a change in turgor that spirals around the twining apical section and produces a rhythm of growth in the epidermal cells of the bending zone (Millet et al., 1984, 1988). The shoot tip of *Phaseolus* circumnutates in a counterclockwise direction, but in hops (*Humulus lupulus*) the apex of the twining vine circumnutates in a clockwise direction.

Plants such as the garden bean (*Phaseolus*), soybean (*Glycine*), and velvetleaf (*Abutilon*) have leaves that move up and down due to rhythmic turgor changes in specialized regions, called pulvini (singular = pulvinus), joint-like thickenings near the base of leaf blades, which act like cushioned hinges. The blade of the bean leaf moves as a result of turgor changes in cells located on opposite sides of the pulvinus. These turgor or osmotic changes involve membranes and the movements (flux) or ratios of ions, such as K^+ (Kiyosawa & Tanaka, 1976). Perhaps the very mention of these factors brings back memories of "osmosis experiments" that you may have performed in a general biology class, and which now apply to the mechanisms of leaf movements.[35] Finally, the physiology of shoot and leaf rhythms is much more complex than described here. Additional information on the role of leaf and flower movements in the study of rhythms is addressed further in the Chapter 3 on Time. The molecular basis of the circadian clock, as well as the circadian system, is discussed in the Chapter 5 on Models and Mechanisms.

Take-Home Message

There is a genetically determined, endogenous basis for biological rhythms in all living organisms. Periods are grouped into three main categories: ultradian (<20 h), circadian (20–28 h), and infradian (>28 h). Rhythms are characterized by their period, amplitude, and phase, which are stable under synchronized conditions, but their pattern may be altered by changes in the environment or during disease. A cyclic environment, such as light alternating with darkness, is generally needed for normal development and health. The SCN and pineal are examples of primary circadian clocks located in the brain.

---

[35] Not all plant species that display leaf movements have pulvini. Some depend upon unequal growth rates and possibly plant hormones (cf. Yin, 1941).

References

Anderson LE, Morris JE, Sasser LB, Stevens RG. (2000) Effect of constant light on DMBA mammary tumorigenesis in rats. *Cancer Lett* 148(2): 121–126.
Andrews RV, Folk GE. (1964) Circadian metabolic patterns in cultured hamster adrenals. *Comp Biochem Physiol* 11: 393–409.
Arendt J, Minors DS, Waterhouse JM, eds. (1989) *Biological Rhythms in Clinical Practice*. London: Wright, 299 pp.
Armitage R, Hoffmann RF, Rush AJ. (1999) Biological rhythm disturbance in depression: temporal coherence of ultradian sleep EEG rhythms. *Psychol Med* 29(6): 1435–1448.
Aschoff J. (1953) Aktivitätsperiodik bei Gimpeln unter naturlichen und kunstlichen belichtungsverhaltnissen. *Z Vergl Physiol* 35: 159–166.
Aschoff J. (1955) Exogene und endogene Komponente der 24-Stunden-Periodik bei Tier und Mensch. *Naturwiss* 42: 569–575.
Aschoff J. (1960) Exogenous and endogenous components in circadian rhythms. In: *Biological Clocks. Cold Spring Harbor Symposia on Quantitative Biology*, Vol. 25. New York: the Biological Laboratory, pp. 11–28.
Balsalobre A, Brown SA, Marcacci L, Tronche F, Kellendonk C, Reichardt HM, Schutz G, Schibler U. (2000) Resetting of circadian time in peripheral tissues by glucocorticoid signaling. *Science* 289(5488): 2344–2347.
Balsalobre A. (2002) Clock genes in mammalian peripheral tissues. *Cell Tissue Res* 309(1):193–199 (Review).
Barnett JE. (1998) *Time's Pendulum: The Quest to Capture Time from Sundials to Atomic Clocks*. New York: Plenum, 340 pp.
Berson DM, Dunn FA, Takao M. (2002) Phototransduction by retinal ganglion cells that set the circadian clock. *Science* 295(5557): 1070–1073.
Berson DM. (2003) Strange vision: ganglion cells as circadian photoreceptors. *Trends Neurosci* 26(6): 314–320.
Blake G. (1959) Control of diapause by an "internal clock" in *Anthrenus verbasci* (L.) (Col., Dermestidae). *Nature (London)* 183: 126–127.
Boulos Z, Macchi M, Terman M. (1996) Twilight transitions promote circadian entrainment to lengthening light–dark cycles. *Amer J Physiol* 271(3 Pt 2): R813–818.
Boulos Z, Macchi MM, Terman M. (2002) Twilights widen the range of photic entrainment in hamsters. *J Biol Rhythms* 17(4): 353–363.
Brown RL, Robinson PR. (2004) Melanopsin—shedding light on the elusive circadian photopigment. *Chronobiol Intl* 21(2): 189–204.
Bruce VG, Pittendrigh CS. (1956) Temperature independence in a unicellular "clock." *Proc Natl Acad Sci USA* 42: 676–682.
Bruce VG. (1960) Environmental entrainment of circadian rhythms. In: *Biological Clocks. Cold Spring Harbor Symposia on Quantitative Biology*, Vol. 25. New York: the Biological Laboratory, pp. 29–48.
Bruce VG. (1974) Recombinants between clock mutants of *Chlamydomonas reinhardi*. *Genetics* 77: 221–230.
Bünning E. (1932) Über die Erblichkeit der tagesperiodizität bei den Phaseolus-Blättern. *Jahrbücher Wiss Bot* 77: 283–320.
Bünning E. (1935) Zur Kenntnis der erblichen Tagesperiodizität bei den Primärblattern von *Phaseolus multiforus*. *Jahrbücher Wiss Bot* 81: 411–418.
Bünning E. (1973) *The Physiological Clock*, 3rd edn. (revised). Berlin: Springer-Verlag, 258 pp.

Bünning E, Moser I. (1973) Light-induced phase shifts of circadian leaf movements of *Phaseolus*. *Proc Natl Acad Sci USA* 69: 2732–2733.

Cambrosio A, Keating P. (1983) The disciplinary stake: the case of chronobiology. *Soc Stud Sci* 13(3): 323–353.

Chen J-P, Eichelmann C, Engelmann W. (1997) Substances interfering with phosphatidyl inositol signaling pathway affect ultradian rhythm of *Desmodium motorium*. *J Biosci* 22: 1–12.

Cumming BG. (1969) *Chenopodium rubrum* L. and related species. In: *The Induction of Flowering. Some Case Histories*. Evans LT, ed. New York: Cornell University Press, pp. 156–185.

Cutkomp LK, Halberg F, Cornelissen G. (1984) Temperature effect on infradian oviposition rhythms in the Springtail *Folsomia candida (Willem)*. In: *Chronobiology 1982–1983*. Haus E, Kabat H, eds. Basel: S. Karger. pp. 1–9.

Darwin C. (1859) *The Origin of Species*. London: John Murray, 488 pp.

Darwin C, Darwin F. (1897) *The Power of Movement in Plants*. New York: D. Appleton & Co., 592 pp.

Dauchy RT, Sauer LA, Blask DE, Vaughan GM. (1997) Light contamination during the dark phase in "photoperiodically-controlled" animal rooms: effect on tumor growth and metabolism in rats. *Lab Anim Sci* 47(5): 511–518.

Ebling FJ. (1996) The role of glutamate in the photic regulation of the suprachiasmatic nucleus. *Prog Neurobiol* 50(2–3): 109–132 (Review).

Ehret CF, Potter VR, Dobra KW. (1975) Chronotypic action of theophylline and of pentobarbital as circadian zeitgebers in the rat. *Science* 188: 1212–1215.

Ehret CF. (1980) On circadian cybernetics, and the innate and genetic nature of circadian rhythms. In: *Chronobiology: Principles and Applications to Shifts in Schedules*. Scheving LE & Halberg F, eds. Alphen aan den Rijn: Sijthoff & Noordhoff, pp. 109–125.

Engelmann W, Schrempf M. (1980) Membrane models for circadian rhythms. In: *Photochemical and Photobiological Reviews*, Vol. 5. Smith KC, ed. New York: Plenum, pp. 49–85.

Enright JT. (1971) The internal clock of drunken isopods. *Z Vergl Physiol* 75: 332–346.

Feldman JF, Hoyle MN. (1973) Isolation of circadian clock mutants of *Neurospora crassa*. *Genetics* 75: 605–613.

Foster RG, Provencio I, Hudson D, Fiske S, DeGrip W, Menaker M. (1991) Circadian photoreception in the retinally degenerate mouse (Rd/Rd). *J Comp Physiol* 169(1): 39–50.

Freedman MS, Lucas RJ, Soni B, von Schantz M, Muñoz M, David-Gray, Z, Foster R. (1999) Regulation of mammalian circadian behavior by non-rod, non-cone, ocular photoreceptors. *Science* 284(5413): 502–504.

Gallepp G. (1976) Temperature as a cue for the periodicity in feeding of *Brachycentrus occidentalis* (Insecta: Trichoptera). *Animal Behav* 24(1): 7–10.

Gooley JJ, Lu J, Chou TC, Scammell TE, Saper CB. (2001) Melanopsin in cells of origin of the retinohypothalamic tract. *Nature Neurosci* 4(12): 1165.

Gooley JJ, Lu J, Fischer D, Saper CB. (2003) A broad role for melanopsin in nonvisual photoreception. *J Neurosci* 23(18): 7093–7106.

Gordon WR, Koukkari WL. (1978) Circadian rhythmicity in the activities of phenylalanine ammonia-lyase from *Lemna perpusilla* and *Spirodela polyrhiza*. *Plant Physiol* 62: 612–615.

Guillaume FM, Koukkari WL. (1985) Two types of high frequency oscillations in *Glycine max* (L.) Merr. In: *Advances in Chronobiology, Part A. Prog in Clin & Biol Res*, Vol 227A. Pauly JE, Scheving LE, eds. New York: Alan R. Liss, Inc., pp. 47–57.

Haim A, Shanas U, Zubidad Ael S, Scantelbury M. (2005) Seasonality and seasons out of time—the thermoregulatory effects of light interference. *Chronobiol Intl* 22(1): 59–66.

Halaban R. (1968) The circadian rhythm of leaf movement of *Coleus blumei x C. frederici*, a short day plant. I. Under constant light conditions. *Plant Physiol* 43: 1883–188.

Halaban R. (1969) Effects of light quality on the circadian rhythm of leaf movement of a short-day plant. *Plant Physiol* 44: 973–977.

Halberg F. (1959) Physiologic 24-hour periodicity; general and procedural considerations with reference to the adrenal cycle. *Z für Vitamin, Hormon u Fermentforsch* 10(3/4): 225–296.

Halberg F. (1969) Chronobiology. *Annu Rev Physiol* 31: 675–725 (Review).

Halberg F, Carandente F, Cornelissen G, Katinas GS. (1977) Glossary of Chronobiology. *Chronobiologia* 4(Suppl 1): 1–189.

Halberg F, Cornélissen G. (2001) Chronobiology: rhythms, clocks, chaos, aging, and other trends. In: *Encyclopedia of Aging*, 3rd edn. Maddox GL, ed. New York: Springer, pp. 196–201.

Halberg F, Cornélissen G, Otsuka K, Schwartzkopff O, Halberg J, Bakken EE. (2001) Chronomics. *Biomed Pharmacother* 55: 153–190.

Handler AM, Konopka RJ. (1979) Transplantation of a circadian pacemaker in *Drosophila*. *Nature* 279(5710): 236–238.

Hannibal J, Hindersson P, Knudsen SM, Georg B, Fahrenkrug J. (2002) The photopigment melanopsin is exclusively present in pituitary adenylate cyclase-activating polypeptide-containing retinal ganglion cells of the retinohypothalamic tract. *J Neuroscience* 22(RC191): 1–7.

Hannibal J, Fahrenkrug J. (2004) Target areas innervated by PACAP-immunoreactive retinal ganglion cells. *Cell Tissue Res* 316(1): 99–113.

Hartmann WK, Miller R. (1991) *The History of Earth. An Illustrated Chronicle of an Evolving Planet*. New York: Workman, 260 pp.

Hassnaoui M, Pupier R, Attia J, Blanc M, Beauchaud M, Buisson B. (1998) Some tools to analyze changes of rhythms in biological time series. *Biol Rhythm Res* 29(4): 353–366.

Hastings JW, Aschoff JWL, Bünning E, Edmunds LN, Hoffmann K, Pittendrigh CS, Winfree AT. (1976) Basic feature group report. In: *The Molecular Basis of Circadian Rhythms*. Hastings JW, Schweiger H-G, eds. Berlin: Abakon, pp. 49–62.

Hastings MH, Reddy AB, Maywood ES. (2003) A clockwork web: circadian timing in brain and periphery, in health and disease. *Nat Rev Neurosci* 4(8): 649–661.

Hattar S, Liao HW, Takao M, Berson DM, Yau KW. (2002) Melanopsin-containing retinal ganglion cells: architecture, projections, and intrinsic photosensitivity. *Science* 295(5557): 1065–1070.

Hattar S, Lucas RJ, Mrosovsky N, Thompson S, Douglas RH, Hankins MW, Lem J, Hofmann F, Foster RG, Yau K-W. (2003) Melanopsin and rod-cone photoreceptive systems account for all major accessory visual functions in mice. *Nature* 424(6944): 76–81.

Hendrickson AE, Wagoner N, Cowan WM. (1972) An autoradiographic and electron microscopic study of retino-hypothalamic connections. *Z Zellforsch Mikrosk Anat* 135(1): 1–26.

Hillman WS. (1956) Injury of tomato plants by continuous light and unfavorable photoperiodic cycles. *Planta* 114: 119–129.

Hoffmann K. (1976) The adaptive significance of biological rhythms corresponding to geophysical cycles. In: *The Molecular Basis of Circadian Rhythms*. Hastings JW, Schweiger H-G, eds. Berlin: Abakon, pp. 63–75.

Inouye ST, Takahashi JS, Wollnik F, Turek FW. (1988) Inhibitor of protein synthesis phase shifts a circadian pacemaker in mammalian SCN. *Amer J Physiol* 255(6 Pt 2): R1055–1058.

Iwasaki K, Liu DW, Thomas JH. (1995) Genes that control a temperature-compensated ultradian clock in *Caenorhabditis elegans*. *Proc Natl Acad Sci USA* 92(22): 10317–10321.

Janzen DH. (1976) Why bamboos wait so long to flower. *Annu Rev Ecol Syst* 7: 347–391.

Johnson CH, Knight M, Trewavas A, Kondo T. (1998) A clockwork green: circadian programs in photosynthetic organisms. In: *Biological Rhythms and Photoperiodism in Plants*. Lumsden PJ, Millar AJ, eds. Oxford: BIOS, pp. 1–34.

Johnson CK. (1999) Forty years of PRCs—What have we learned? *Chronobiol Intl* 16(6): 711–743.

Johnsson A. (1973) Oscillatory transpiration and water uptake of Avena plants. I. Preliminary observations. *Physiol Plant* 28: 40–50.

King RW. (1975) Multiple circadian rhythms regulate photoperiodic flowering responses in *Chenopodium rubrum*. *Can J Bot* 53: 2631–2638.

Kippert F. (1997) The ultradian clocks of eukaryotic microbes: timekeeping devices displaying a homeostasis of the period. *Chronobiol Int* 14(5): 469–479 (Review).

Kiyosawa K, Tanaka H. (1976) Change in potassium distribution in a *Phaseolus* pulvinus during circadian movement of the leaf. *Plant Cell Physiol* 17: 289–298.

Kondo T, Tsinoremas NF, Golden SS, Johnson CH, Kutsuna S, Ishiura M. (1994) Circadian clock mutants of cyanobacteria. *Science* 266(5188): 1233–1236.

Konopka RJ, Benzer S. (1971) Clock mutants of *Drosophila melanogaster*. *Proc Natl Acad Sci USA* 68: 2112–2116.

Koukkari WL, Johnson MA (1979) Oscillations of leaves of *Abutilon theophrasti* (velvetleaf) and their sensitivity to bentazon in relation to low and high humidity. *Physiol Plant* 47: 158–162.

Koukkari WL. (1994) Movement of a bean shoot: an introduction to chronobiology. *Chronobiol Intl* 11(2): 85–93.

Koukkari WL, Parks TW, Sothern RB. (1999) Individual circadian rhythms in filtering behavior of Trichoptera during synchronized and constant lighting conditions (abstract 435). *NABS* 16(1): 218.

Kyriacou CP, Hall JC. (1980) Circadian rhythm mutations in *Drosophila melanogaster* affect short-term fluctuation in the male's courtship song. *Proc Natl Acad Sci USA* 77: 6729–6733.

Lanzinger I, Kobilanski C, Philippu A. (1989) Pattern of catecholamine release in the nucleus tractus solitarii of the cat. *Naunyn Schmiedebergs Arch Pharmacol* 339(3): 298–301.

Larkin JE, Franken P, Heller HC. (2002) Loss of circadian organization of sleep and wakefulness during hibernation. *Amer J Physiol Regul Integr Comp Physiol* 282(4): R1086–1095.

Lindsley G, Dowse HB, Burgoon PW, Kolka MA, Stephenson LA. (1999) A persistent circhoral ultradian rhythm is identified in human core temperature. *Chronobiol Intl* 16(1): 69–78.

Liu DW, Thomas JH. (1994) Regulation of a periodic motor program in *C. elegans*. *J Neurosci* 14(4): 1953–1962.

Lloyd D, Edwards SW, Fry JC. (1982) Temperature-compensated oscillations in respiration and cellular protein content in synchronous cultures of *Acanthamoeba castellanaii*. *Proc Natl Acad Sci USA* 79: 3785–3788.

Lloyd D, Salgado LE, Turner MP, Suller MT, Murray D. (2002) Cycles of mitochondrial energization driven by the ultradian clock in a continuous culture of *Saccharomyces cerevisiae*. *Microbiol* 148(Pt 11): 3715–3724.

Lörcher L. (1958) Die Wirkung verschiedener Lichtqualitäten auf die endogene Tagesrhythmik von *Phaseolus*. *Z Bot* 46: 209–241.

Lowrey PL, Takahashi JS. (2004) Mammalian circadian biology: elucidating genome-wide levels of temporal organization. *Annu Rev Genomics Hum Genet* 5: 407–741.

Lucas RJ, Hattar S, Takao M, Berson DM, Foster RG, Yau KW. (2003) Diminished pupillary light reflex at high irradiances in melanopsin-knockout mice. *Science* 299(5604): 245–247.

Lupi D, Cooper HM, Froehlich A, Standford L, McCall MA, Foster RG. (1999) Transgenic ablation of rod photoreceptors alters the circadian phenotype of mice. *Neuroscience* 89(2): 363–74.

Marques M, Hoenen MM. (1999) Altered circadian patterns in a cave insect: signs of temporal adaptation? (Abstract 78). In: *Proc Intl Cong Chronobiol*, Aug. 29–Sep. 1, 1999, Washington, DC, p. 75.

Mayer W, Gruner R, Strubel H. (1975) Period-lengthening and phase-shifting of the circadian rhythm of *Phaseolus coccineus* L. by theophylline. *Planta* 125: 141–148.

Mayer W, Scherer I. (1975) Phase shifting effect of caffeine in the circadian rhythm of *Phaseolus coccineus* L. *Z Naturforsch* 30: 855–856.

Meijer JH, Schwartz WJ. (2003) In search of the pathways for light-induced pacemaker resetting in the suprachiasmatic nucleus. *J Biol Rhythms* 18(3): 235–249 (Review).

Menaker M, Eskin A. (1966) Entrainment of circadian rhythms by sound in *Passer domesticus*. *Science* 154: 1579–1581.

Menaker M, Roberts R, Elliott J, Underwood H. (1970) Extraretinal light perception in the sparrow, III: The eyes do not participate in photoperiodic photoreception. *Proc Natl Acad Sci USA* 67(1): 320–325.

Millet B, Melin D, Bonnet B, Assad C, Mercier J. (1984) Rhythmic circumnutation movement of the shoots in *Phaseolus vulgaris* L. *Chronobiol Intl* 1: 11–19.

Millet, B, Melin D, Badot P-M. (1988) Circumnutation in *Phaseolus vulgaris*. I. Growth, osmotic potential and cell ultrastructure in the free-moving part of the shoot. *Physiol Plantarum* 72: 133–138.

Millet B, Badot PM. (1996) The revolving movement mechanism in *Phaseolus*: new approaches to old questions. In: *Vistas on Biorhythmicity*. Greppin H, Degli Agosti R, Bonzon M, eds. Geneva: University of Geneva, pp. 77–98.

Minors DS, Waterhouse JM, eds. (1989) Masking and biological rhythms (special issue). *Chronobiol Intl* 6(1): 1–102.

Moore RY, Eichler VB. (1972) Loss of a circadian adrenal corticosterone rhythm following suprachiasmatic lesions in the rat. *Brain Res* 42(1): 201–206.

Moore RY, Lenn NJ. (1972) A retinohypothalamic projection in the rat. *J Comp Neurol* 146(1): 1–14.

Moore-Ede MC. (1973) Circadian rhythms of drug effectiveness and toxicity. *Clin Pharm Therapeut* 14(6): 925–935.

Moore-Ede MC, Sulzman FM, Fuller CA. (1982) *The Clocks That Time Us. Physiology of the Circadian Timing System*. Cambridge: Harvard University Press, 448 pp.

Murakami N, Nishi R, Katayama T, Nasu T. (1995) Inhibitor of protein synthesis phase-shifts the circadian oscillator and inhibits the light induced-phase shift of the melatonin rhythm in pigeon pineal cells. *Brain Res* 693(1–2): 1–7.

Nelson W, Scheving L, Halberg F. (1975) Circadian rhythms in mice fed a single daily meal at different stages of lighting regimen. *J Nutr* 105(2): 171–184.

Nelson DE, Takahashi JS. (1991) Sensitivity and integration in a visual pathway for circadian entrainment in the hamster (*Mesocricetus auratus*). *J Physiol* 439: 115–145.

Nelson DE, Takahashi JS. (1999) Integration and saturation within the circadian photic entrainment pathway of hamsters. *Amer J Physiol* 277(5 Pt 2): R1351–R1361.

Nisimura T, Numata H. (2001) Endogenous timing mechanism controlling the circannual pupation rhythm of the varied carpet beetle *Anthrenus verbasci*. *J Comp Physiol* 187(6): 433–440.

Pavlidis T. (1973) Phase shifts and phase response curves. In: *Biological Oscillators: Their Mathematical Analysis*. New York: Academic Press, pp. 49–70.

Pittendrigh CS, Bruce VG. (1957) An oscillator model for biological clocks. In: *Rhythmic and Synthetic Processes of Growth*. Rudnick D, ed. Princeton, NJ: Princeton University Press, pp. 75–109.

Pittendrigh CS. (1967) Circadian systems. I. The driving oscillation and its assay in *Drosophila psuedoobscura*. *Proc Natl Acad Sci USA* 58: 1762–1767.

Pittendrigh CS. (1981) Circadian systems: general perspective. In: *Handbook of Bahvioral Neurobiology, Vol. 4: Biological Rhythms*. Aschoff J, ed. New York: Plenum, pp. 57–80.

Pregueiro AM, Price-Lloyd N, Bell-Pedersen D, Heintzen C, Loros JJ, Dunlap JC. (2005) Assignment of an essential role for the *Neurospora frequency* gene in circadian entrainment to temperature cycles. *Proc Natl Acad Sci USA* 102(6): 2210–2215.

Presser HB. (1974) Temporal data relating to the human menstrual cycle. In: *Biorhythms and Human Reproduction*. Ferin M, Halberg F, Richert RM, Vande Wiele R, eds. New York: John Wiley and Sons, Inc, pp. 145–160.

Provencio I, Jiang G, De Grip WJ, Hayes WP, Rollag MD. (1998) Melanopsin: An opsin in melanophores, brain, and eye. *Proc Natl Acad Sci USA* 95(1): 340–345.

Provencio I, Rodriguez IR, Jiang G, Hayes WP, Moreira EF, Rollag MD. (2000) A novel human opsin in the inner retina. *J Neurosci* 20(2): 600–605.

Pye EK. (1969) Biochemical mechanisms underlying the metabolic oscillations in yeast. *Can J Botany* 47: 271–285.

Pye EK. (1971) Periodicities in intermediary metabolism. In: *Biochronometry*. Menaker M, ed. Washington, DC: National Academy of Sciences, pp. 623–636.

Ralph MR, Foster RG, Davis FC, Menaker M. (1990) Transplanted suprachiasmatic nucleus determines circadian period. *Science* 247(4945): 975–978.

Ralph MR. Menaker M. (1988) A mutation of the circadian system in golden hamsters. *Science* 241(4870): 1225–1227.

Ratajczak HV, Sothern RB, Hrushesky W. (1986) Single cosinor analysis of vaginal smear cell types quantifies mouse estrous cycle and its alteration by mammary adenocarcinoma. In: *Ann Rev Chronopharm*, Vol 3. Reinberg A, Smolensky M, Labrecque G, eds. New York: Pergamon Press, pp. 223–226.

Ratajczak HV, Sothern RB, Hrushesky WJM. (1988) Estrous influence on surgical cure of a mouse breast cancer. *J Exp Med* 168: 73-83.

Refinetti R, Menaker M. (1992) The circadian rhythm of body temperature. *Physiol & Behav* 51(3): 613–637.

Richter CP. (1965) *Biological Clocks in Medicine and Psychiatry*. Springfield, IL: CC Thomas, 108 pp.

Richter CP. (1967) Sleep and activity: their relation to the 24-hour clock. *Res Publ Assoc Res Nerv Ment Dis* 45: 8–29.

Rietveld WJ. (1992) The suprachiasmatic nucleus and other pacemakers. In: *Biological Clocks. Mechanisms and Applications*. Touitou Y, ed. Amsterdam: Elsevier, pp. 55–64.
Rollag MD, Berson DM, Provencio I. (2003) Melanopsin, ganglion-cell photoreceptors, and mammalian photoentrainment. *J Biol Rhythms* 18(3): 227–234.
Sanchez de la Peña S. (1993) The feedsideward of cephalo-adrenal immune interactions. *Chronobiologia* 20(1–2): 1–52.
Schmidle A. (1951) Die Tagesperiodizität der asexuellen Reproduktion von *Pilobolus sphaerosporus*. *Arch Mikrobiol* 16: 80–100.
Schmitz O, Brock B, Hollingdal M, Juhl CB, Porksen N. (2002) High-frequency insulin pulsatility and type 2 diabetes: from physiology and pathophysiology to clinical pharmacology. *Diabetes Metab* 28(6 Suppl): 4S14–20 (Review).
Schuster J, Engelmann W. (1997) Circumnutations of *Arabidopsis thaliana* seedlings. *Biol Rhythm Res* 28(4): 422–440.
Schwartz MD, Nunez AA, Smale L. (2004) Differences in the suprachiasmatic nucleus and lower subparaventricular zone of diurnal and nocturnal rodents. *Neuroscience* 127(1): 13–23.
Shinohara K, Oka T. (1994) Protein synthesis inhibitor phase shifts vasopressin rhythms in long-term suprachiasmatic cultures. *Neuroreport* 5(16): 2201–2204.
Smale L, Lee T, Nunez AA. (2003) Mammalian diurnality: some facts and gaps. *J Biol Rhythms* 18(5): 356–366 (Review).
Smolensky M, Lamberg L. (2000) *The Body Clock Guide to Better Health*. New York: Henry Holt & Co., 428 pp.
Somers DE, Kay SA. (1998) Genetic approaches to the analysis of circadian rhythms in plants. In: *Biological Rhythms and Photoperiodism in Plants*. Lumsden PJ, Millar AJ, eds. Oxford: BIOS, pp. 81–98.
Sothern RB, Hermida RC, Nelson R, Mojón A, Koukkari WL. (1998) Reanalysis of filter-feeding behavior of Caddisfly (*Brachycentrus*) larvae reveals masking and circadian rhythmicity. *Chronobiol Intl* 15(6): 595–606.
Spieler RE, Meier AH, Noeske TA. (1978) Temperature-induced phase shift of daily rhythm of serum prolactin in gulf killifish. *Nature* 271(5644): 469–470.
Spruyt E, Verbelen J-P, DeGreef JA. (1987) Expression of circaseptan and circannual rhythmicity in the imbibition of dry stored bean seeds. *Plant Physiol* 84: 707–710.
Stephan FK, Zucker I. (1972) Circadian rhythms in drinking behavior and locomotor activity of rats are eliminated by hypothalamic lesions. *Proc Natl Acad Sci USA* 69(6): 1583–1586.
Stevens RG, Davis S, Thomas DB, Anderson LE, Wilson BW. (1992) Electric power, pineal function, and the risk of breast cancer. *FASEB J* 6(3): 853–860.
Stevens RG, Davis S. (1996) The melatonin hypothesis: electric power and breast cancer. *Environ Health Perspect* 104(Suppl. 1): 135–140.
Sujino M, Masumoto K, Yamaguchi S, van der Horst GT, Okamura H, Inouye SI. (2003) Suprachiasmatic nucleus grafts restore circadian behavioral rhythms of genetically arrhythmic mice. *Curr Biol* 13(8): 664–668.
Sweeney BM, Hastings JW. (1960) Effects of temperature upon diurnal rhythms. In: *Biological Clocks. Cold Spring Harbor Symposia on Quantitative Biology*, Vol. 25. New York: the Biological Laboratory, pp. 87–104.
Sweeney BM. (1974) The potassium content of *Gonyaulax polyhedra* and phase changes in the circadian rhythm of stimulated bioluminescence by short exposure to ethanol and valinomycin. *Plant Physiol* 53: 337–342.

Sweeney BM. (1976) Circadian rhythms, definition and general characterization. In: *The Molecular Basis of Circadian Rhythms*. Hastings JW, Schweiger H-G, eds. Berlin: Abakon, pp. 77–83.

Takahashi JS, Turek FW. (1987) Anisomycin, an inhibitor of protein synthesis, perturbs the phase of a mammalian circadian pacemaker. *Brain Res* 405(1): 199–203.

Taylor WR, Dunlap JC, Hastings JW. (1982) Inhibitors of protein synthesis on 80S ribosomes phase shift the Gonyaulax clock. *J Exp Biol* 97: 121–136.

Tester JR, Figala J. (1990) Effects of biological and environmental factors on activity rhythms of wild animals. In: *Chronobiology: Its Role in Clinical Medicine, General Biology and Agriculture, Part B, Progress in Clinical and Biological Research*, Vol. 341B. Hayes DK, Pauly JE, Reiter RJ, eds. New York: Wiley, pp. 809–819.

Truman JW, Riddiford LM. (1970) Neuroendocrine control of ecdysis in silkmoths. *Science* 167: 1624–1626.

Truman JW. (1972) Physiology of insect rhythms II. The silkmoth brain as the location of the biological clock controlling eclosion. *J Comp Physiol* 81: 99–114.

Uebelmesser ER. (1954) Über den endonomen Rhythmus der Sporangientrager bildung von *Pilobolus*. [The endogenous daily rhythm of conidiospore formation of Pilobolus.] *Arch Mikrobiol* 20(1): 1–33.

Underwood H, Menaker M. (1970) Photoperiodically significant photoreception in sparrows: Is the retina involved? *Science* 167: 298–301.

Waalen J. (1993) Nighttime light studied as possible breast cancer risk. *J Natl Cancer Inst* 85(21): 1712–1713.

Waldrop RD, Saydjari R, Rubin NH, Rayford PL, Townsend CM Jr., Thompson JC. (1989) Photoperiod influences the growth of colon cancer in mice. *Life Sci* 45(8): 737–744.

Wasserman L. (1959) Die Auslösung endogen-tagesperiodischer Vorgänge bei Pflanzen durch einmalige Reize. *Planta* 53: 647–669.

Wehr TA, Aeschbach D, Duncan WC Jr. (2001) Evidence for a biological dawn and dusk in the human circadian timing system. *J Physiol* 535(Pt 3): 937–951.

Went FW. (1960) Photo- and thermoperiodic effects in plant growth. In: *Biological Clocks. Cold Spring Harbor Symposia on Quantitative Biology*, Vol. 25. New York: the Biological Laboratory, pp. 221–230.

Went FW. (1974) Reflections and observations. *Annu Rev Plant Physiol* 25: 1–26.

Wever RA. (1979) *The Circadian System of Man. Results of Experiments Under Temporal Isolation*. New York: Springer-Verlag, 276 pp.

Wilkins MB. (1960) The effect of light upon plant rhythms. In: *Biological Clocks. Cold Spring Harbor Symposia on Quantitative Biology*, Vol. 25. New York: the Biological Laboratory, pp. 115–129.

Wolfson A. (1959) The role of light and darkness in the regulation of spring migration and reproductive cycles in birds. In: *Photoperiodism and Related Phenomena in Plants and Animals*. Withrow RB, ed. Washington: American Association for the Advancement of Science Publication 55, pp. 679–716.

Wollnik F, Schmidt B. (1995) Seasonal and daily rhythms of body temperature in the European hamster (*Cricetus cricetus*) under semi-natural conditions. *J Comp Physiol* 165(3): 171–182.

Yamazaki S, Goto M, Menaker M. (1999) No evidence for extraocular photoreceptors in the circadian system of the Syrian hamster. *J Biol Rhythms* 14(3): 197–201.

Yin HC. (1941) Studies on the nyctinastic movement of the leaves of *Carica papaya*. *Amer J Bot* 28: 250–261.

Yoo SH, Yamazaki S, Lowrey PL, Shimomura K, Ko CH, Buhr ED, Siepka SM, Hong HK, Oh WJ, Yoo OJ, Menaker M, Takahashi JS. (2004) PERIOD2::LUCIFERASE real-time reporting of circadian dynamics reveals persistent circadian oscillations in mouse peripheral tissues. *Proc Natl Acad Sci USA* 101(15): 5339–5346.

Zylka MJ, Shearman LP, Weaver DR, Reppert SM. (1998) Three *period* homologs in mammals: differential light responses in the suprachiasmatic circadian clock and oscillating transcripts outside of brain. *Neuron* 20(6): 1103–1110.

3
Physical and Biological Time

"Time has no divisions to mark its passage, there is never a thunderstorm or blare of trumpets to announce the beginning of a new month or year. Even when a new century begins it is we mortals who ring bells and fire off pistols."
—Thomas Mann (1887–1955), German author

Introduction

The recording of time has been a hallmark of civilization and culture, deeply rooted in the activities of individuals and society. We often take for granted the units of time and their quantification, but to more fully appreciate their structure and how they were acquired, we must turn to astronomy and to the fascinating history that led to the discovery and development of clocks.

Actually there are two types or groups of clocks, each with its own historical path. One group includes the mechanical, electrical, and atomic clocks that are used in society to determine the time of day or year. Its technology is based upon physical components, either natural (e.g., sand, water, stars, radiation) or more likely, engineered and manufactured (e.g., springs and gears). The other type of clock is biological, and its history of development is called evolution. Because most of the units of time were developed and described first for physical clocks, and later carried over and applied to biological clocks, they will be discussed first.

The first objective of this chapter is to help you develop an appreciation of how the rotations and revolutions of the Earth and Moon helped to give rise to many of our units of time. These movements account for the natural synchronizers of circadian and circannual rhythms (by the Earth) and tidal rhythms (by the Moon), and in the case of the rotations and revolutions of the Earth, the changes in duration and timing of light and dark spans involved in photoperiodism. The second objective is to assist you in obtaining a greater appreciation of the units of time by becoming familiar with some of the events or benchmarks that are associated with the development of clocks or timepieces.

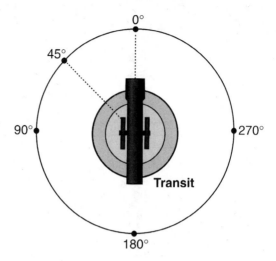

FIGURE 3.1. A circle with a diagram of a transit in the center and showing sightings of 0° and 45°. A transit is an instrument that can be used for sighting straight lines and measuring angles. Degrees are presented here in a counterclockwise direction because the rotation of the Earth and its revolution around the Sun occur in a counterclockwise direction.

Along with the discussion of clocks and calendars, the third objective is to provide some background information about topics such as longitude, meridian, time zones, and Daylight Saving Time (DST) so that you may better understand the resynchronization of rhythms, especially as they relate to transmeridian flights. The fourth objective is to discuss the methods of reporting dates and times and its importance in everyday life and in laboratory studies. A fifth objective is to introduce the concept of biological clocks and some of the analogies that exist between internal (endogenous) biological rhythms and external (exogenous) clocks. Finally, the evolution of the biological clock throughout the history of the Earth is discussed.

Rotations and Revolutions of the Earth and Moon

The 24-h cycle of light alternating with darkness that occurs each day is the dominant natural synchronizer of circadian rhythms. One common notion is that the period of 24 h represents the time required by the Earth to complete one full rotation. However, the Earth makes one complete rotation on its axis in about 23 h and 56 min.[1] Based upon this fact, we must now ask: what happens to the extra four minutes, or is there something drastically wrong with our external clocks? To find the answer, we turn to basic astronomy and elementary geometry (Figure 3.1). From the study of astronomy we have learned that the Earth revolves around the

---

[1] The precise duration for one complete rotation is closer to 23 h, 56 min, and 4.09 s (Time & Frequency Division, National Institute of Standards and Technology (NIST), US Department of Commerce, Boulder, CO. Complete info online at: http://www.physics.nist.gov/GenInt/contents.html).

68   3. Physical and Biological Time

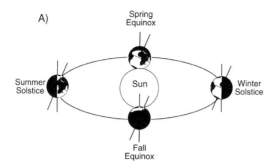

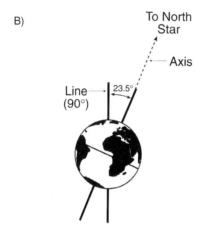

FIGURE 3.2. The ecliptic plane formed by the Earth as it revolves around the Sun *(A)* and 23.5° tilt of Earth relative to a vertical line drawn perpendicular to the ecliptic plane *(B)*.

Sun[2] in an elliptical plane while rotating on its axis (Figure 3.2). Both the rotation of the Earth and its revolution around the Sun occur in a counterclockwise direction.

The Day

To more easily resolve this apparent difference between the 24-h clock of society and the duration of about 23 h and 56 min for the Earth to make one complete rotation of 360°, let us make two assumptions (see Figure 3.3). First, suppose that

---

[2] Today, we accept this as a factual statement, but up to the early 1600s and even into the 1700s, the Earth was considered by many to be at the center and the Sun revolved around it. Individuals, such as Heracleides (ca. 388–315 BC), Arlistarchus (310–230 BC), Copernicus (1473–1543), and Kepler (1571–1630) are included among those who claimed that the Earth moved around the Sun, but it took centuries or decades before most people would come to the same conclusion, or for many to be allowed to accept the truth. The latter represents a case where the implications of religion were strongly woven into a time-related matter. Perhaps the most classic example occurred in 1633, when the Italian astronomer and physicist Galileo Galilei (1564–1642), fearing for his life, was brought before a church inquisition in Rome and forced to recant his view that the Earth moved around the Sun.

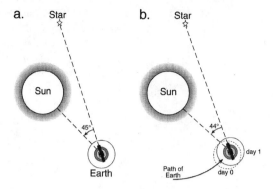

FIGURE 3.3. A transit sighting a far away star that will serve as a reference (0°) and assuming that the position of the Sun from Earth at that point in time is located 45° in a counterclockwise direction from 0° *(a)*. After a span of about 23 h and 56 min the Earth has made one complete rotation on its axis *(b)*. The reference star is again at 0°, but the angle for the position of the Sun is only 44°. The one additional degree of rotation (1° + 44° = 45°) by the Earth requires an additional 4 min.

a surveyor on the Earth positions a special astronomical transit to locate a faraway star for a reference and sets the dial of the transit at 0°. Second, let us assume that the angle for the position of the Sun is 45° from the 0° setting. After a span of about 23 h and 56 min, the Earth has made one complete rotation on its axis and the transit dial again reads 0° for the reference star. However, the angle for the position of the Sun is only 44°. The original 45° position for the Sun will not be reached until one more degree of rotation has occurred. This will require about four more minutes. The reason for the lag is quite simple. Not only is the Earth rotating, it is revolving around the Sun. As a result, the location of the Earth relative to the Sun has changed by about one degree. The span of time of 23 h and 56 min is called the *sidereal day* (= 360°), while our clocks are based upon the *solar day* (360° + 1° = 361°) of 24 h.

The Year

The seasons of the year, which in nature provide time cues via photoperiodism, also are caused by the movements of the Earth. Many people know that the seasons are based upon the revolution of the Earth around the Sun. However, the warmth and long daylight hours of summer are not the result of the Earth being closer to the Sun in summer and further in winter. In fact, the Earth is closest to the Sun in January and furthest in July and these distances cannot account for phase differences among seasons that exist between the Northern and Southern Hemispheres.

To better understand what accounts for the seasons, let us first consider a sphere (Earth) and a unilateral source of light (Sun). One will note that one-half of the sphere is illuminated, while the other half remains in the shade or darkness

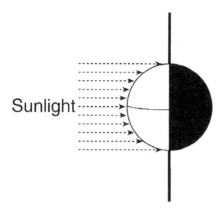

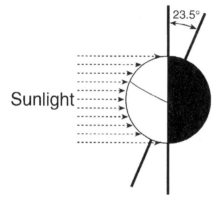

FIGURE 3.4. When a unilateral source of light shines of a sphere, one-half of the sphere is illuminated while the other half remains in the shade or darkness, regardless of how the sphere is tilted. The axis upon which the Earth rotates is tilted by 23.5° and points toward the North Star (Polaris) as the Earth revolves around the Sun. Consequently, during the course of the year, there is a seasonal difference in the duration of daylight and darkness for the two hemispheres.

(Figure 3.4). Even if the axis of the sphere is tilted, the same conditions prevail: one-half of the sphere is in light and one-half remains in darkness. The Earth revolves around the Sun in an elliptical orbit on a plane called the *ecliptic*. It takes slightly longer than 365 days to complete one full cycle. The axis upon which the Earth rotates, however, is tilted by 23.5° to a line perpendicular to its orbit (Figure 3.2B) and continues to point toward the North Star as the Earth revolves around the Sun. In other words, the axis maintains the same tilt of 23.5° as the Earth revolves around the Sun (Figure 3.2). Together, the fixed tilt and the position of the Earth relative to the Sun account for the seasons (Figure 3.2A).

The summer solstice,[3] which occurs near June 21, is when the duration of light (and the most direct sunlight) is the longest in the Northern Hemisphere

---

[3] The annual occurrence of the summer solstice was one of the time points of the year that was indicated by the design of some of the structures erected by ancient civilizations. Examples include Stonehenge in England and the Sun Dagger in New Mexico (USA). The latter is a spiral that was carved on the vertical face of a butte (small isolated mountain with steep sides). At noon, and only during the summer solstice, sunlight passes between some stones in front of the spiral and produces an image in the shape of a dagger, which goes through the center of the spiral (cf. Bennett et al., 1999).

FIGURE 3.5. Various phases of the Moon observed from the Earth as it orbits the Earth. From one phase to the same phase again is about 29.5 days, the lunar month.

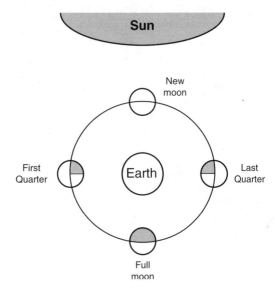

and shortest (least direct sunlight) in the Southern Hemisphere (Figure 3.4B). The opposite is true for the winter solstice near December 22. During the spring (vernal) equinox near March 21 and the fall (autumnal) equinox near September 23, the Northern and Southern Hemispheres receive the same duration of light. Our calendar is based upon the *tropical year*, derived from the duration of time from one spring equinox to the next at the equator (365 days, 5 h, 49 min).

The Month

Tidal rhythms, which will be discussed in a separate chapter, are controlled in large measure by the movements of the Moon. The Moon revolves around the Earth, completing one full revolution (360°) in about 27.25 days, the *sidereal month*.[4] Again, due to the revolution of the Earth around the Sun, which is about 30° per month (360° divided by 12 months = 30°), the duration of time from one new moon phase to the next is about 29.5 days, the *lunar month*[5] (Figures 3.5 and 3.6). The moon and its phases not only cause and influence the tides and biologic phenomena, but, as we will discuss in the following section, they also serve as units of time and the basis for a number of calendars.

---

[4] We always see the same side of the Moon since it takes the same amount of time for it to turn once around on its axis as it takes to revolve around the Earth (known as synchronous rotation).

[5] The lunar month was shorter millions of years ago. Since the Moon is distancing itself from the Earth by 1.5 inches (3.8 cm) a year, it is estimated that a billion years ago the lunar month lasted 6.5 h when it was nearer to Earth.

72   3. Physical and Biological Time

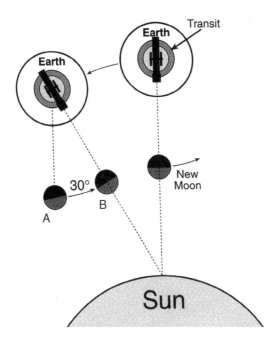

FIGURE 3.6. The concept for this illustration is similar to the one illustrated in Figure 3.3, except that emphasis is upon the Moon. As shown by assumed sightings with a surveyor's transit, the Moon completes one full revolution (360°) around the Earth in about 27.25 days, but the phase for the new Moon is not reached until about 2.25 additional days have passed (lunar month = 29.5 days). On average, the Earth travels about 30° per month while the Moon completes one rotation (360° + 30° = 390°).

Clocks and Calendars: Ancient Times

The length of a rhythmic cycle, which we call the *period*, is measured in units of time. For early primitive people, it was necessary not only to predict the best and safest times to gather food, seek shelter, etc., but also their relationship with the "spiritual world" was often a rhythm-dependent activity. Except for a more sophisticated technology, the same can be said for much of our society in the 21st century.

Time, as it relates to the invention and discovery of clocks, is a fascinating topic. The consciousness of time occurred relatively late in human history, but developed rapidly and became paramount when commerce, business, and travel began to flourish (Hering, 1940). To facilitate the discussion of clocks, a map with illustrations (Figure 3.7) and supplementary comments (Table 3.1) have been included for you to examine as you read and study the narrative material that follows. Many different clocks and calendars based upon astronomical events were devised throughout history in order to enable humans to reliably figure out the "time" of day or year (Figure 3.8, Table 3.1).

The Month and Year

Perhaps the earliest means of measuring the passing of time between recurring events was based upon the appearance, disappearance, or phases of astronomical bodies, such as the Sun, Moon, and certain stars. Because the average female

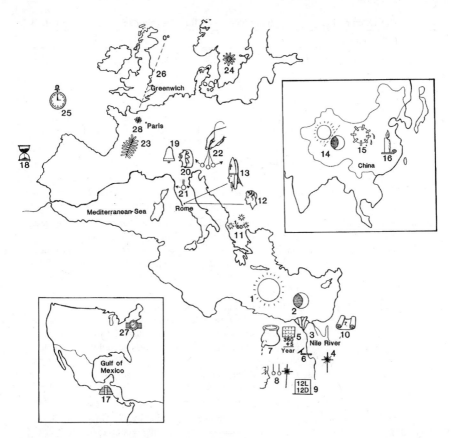

FIGURE 3.7. A map with illustrations representing various time measuring devices, locations, and events that relates to clocks and time (see Table 3.1 for comments related to numbering).

menstrual cycle is about 29.5 ± 3 days (cf. Cutler, 1980; Cutler et al., 1986), a series of spans between full moons (29.53 days each) may have provided primitive women with a time measure for the length of a pregnancy and thus may have been one of the oldest units of time (Breasted, 1936). However, a series of 12 moons does not equal a year and by not relying upon the Moon, the ancient Egyptians obtained the first close estimate of the year being about 365 days. They based their predictions upon the seasonal flooding of the Nile River and the accompanying annual reappearance of the Dog Star in Canis Major, which we call Sirius, in the eastern sky next to the Sun at sunrise.[6] The date for the introduction of this early calendar of 365 days in a year has been traced back to 4236 BC (cf. Breasted, 1936).

---

[6] The flooding of the Nile could vary, but not the rising of Sirius. In this region of Egypt, Sirius rises about 4 min earlier each day and for a given duration it is not visible (during the light hours), only to reappear again on a day that was marked as the beginning of the year (cf. Breasted, 1936).

74   3. Physical and Biological Time

TABLE 3.1. A legend corresponding to numbers 1–27 for illustrations presented in Figure 3.7.

#	Illustration	Comments
1	Sun	Sunrise, sunset, and midday have served time points for various groups of people throughout history (e.g., Romans divided the day into two parts: ante meridiem (AM), before midday, and post meridiem (PM), after midday
2	Moon	Lunar cycles mark certain occasions and events in primitive and modern societies (e.g., to set times of certain religious events)
3	Rivers	Floods have been used to mark seasons and the approximate time of year (see Egyptian Calendar #5)
4	Stars and planets	Appearance of Sirius in the eastern sky near sunrise used by ancient Egyptians to develop their calendar of 365 days. Also stars used to measure time during the night (see # 8)
5	Ancient Egyptian Calendar	Flooding of the Nile River near the time when Sirius, the brightest star in the sky, first appears in the eastern horizon before sunrise after not having been seen for several months, provided the basis for their calendar of 12 months of 30 days each, plus 5 days to equal a year of 365 days. Calendar traced back to about 4236 BC (Breasted, 1936)
6	Sundial or shadow clock	Used by the Egyptians to measure the time of day by the Sun's shadow (ca. 1500 BC). Some older versions were shaped like an inverted T-square that was positioned toward the east in the morning and rotated at noon to face west, in order to divide the sunlit day into 10 parts, plus 2 parts for a twilight hour in the morning and evening. Obelisks were built as early as 3500 BC and the moving shadow of the Sun was used to divide the day into 2 parts centered at noon (Ward, 1961)
7	Water flow devices	Used by the Egyptians as early as 1500 BC, the Greeks (ca. 400 BC) and others to measure various durations of time. A water clock (Greek *clepsydra*, "water thief") could be used at night, since it depended upon the amount of water that moved from one container operation to another
8	Merkhet	Device developed by ancient Egyptians at least by 600 BC consisting of a pair of merkhets (plumb lines) lined up with a star to measure time during the night (Whitrow, 1988)
9	Egyptian 12 part day and 12 part night	"Hours" varied in length depending upon the season of the year
10	Week	The origin of the 7-day week is not clear, but was observed as part of ancient Jewish life (Deuteronomy 5:13–14)
11	Hellenistic astronomers	Credited with the division of each of the 24 h into 60 min each. The Hellenistic period was ~323 BC to ~31 BC
12	Julian calendar	Adopted by Julius Caesar on January 1, 45 BC and consisted of 365 1/4 days in a year; 1/4 day achieved by adding 1 day every 4 years
13	Gregorian calendar	Instituted by Pope Gregory XIII in 1582 AD. In order to correct for the error in days accumulated by the Julian Calendar
14	Old Chinese calendar	Astronomical calendars ca. 2637 BC used both the Sun and phases of the Moon and are still used for determining festival dates
15	Water-wheel clock	A Chinese astronomical type of mechanical clock (tower) that used a water-wheel escapement has been traced to about 1090 AD (cf. Ward, 1961)

Clocks and Calendars: Ancient Times 75

TABLE 3.1. (*Continued*)

#	Illustration	Comments
16	Incense sticks and candle clocks	Incense sticks can be traced back to the Sung Dynasty in China (960–1279), where a different aroma was produced every hour as the incense burned at a constant rate. Each hour could be *smelled*. Candles that burned at a known rate were marked with horizontal lines to count the hours (credited to the Saxon king Alfred the Great [849–899]) (cf. Whitrow, 1988; Barnett, 1998)
17	Mayan calendar	Calendar had a solar year of 365 days, as perhaps illustrated by the number of steps in one of their pyramids (four stairways of 91 steps each = 364 steps plus the top platform). Accuracy in astronomy and in the dating events were remarkable
18	Sandglasses	Often called hourglasses, were better than water devices for measuring short intervals and were used by explorers (e.g., Christopher Columbus)
19	Bell	Early clocks did not have a dial, but rather a sound was produced by a weight striking a bell
20	Mechanical clock	Mechanical clocks may have been invented about 1300 AD (e.g., perhaps after 1271 AD, see Thorndike, 1941)
21	Pendulum clocks	In about 1581 AD, Galileo discovered some of the characteristics of the pendulum in regard to time. Credit for actually building the first pendulum clock is given to Christiaan Huygens, ca. 1656
22	Pulsilogium	A pendulum-type device for measuring pulse rate developed by Galileo and used by Santorio Santorio (1561–1636) in the early practice of medicine (cf. Major, 1954; Talbott, 1970)
23	Biological clock	The term may have been first introduced in the mid-1950s relative to rhythms, but the persistence of a rhythm (e.g., leaf movements) in the absence of the daily light/dark cycle was observed by de Mairan and reported by Marchant in 1729
24	Flower clock	Developed by Linnaeus (1707–1778) to indicate the time of day based upon when flowers or an inflorescence (cluster of flowers) of a species would be open or closed
25	Marine watch	Invented by John Harrison (in 1759 or 1773) for use on vessels at sea and served to assist navigators in determining longitude. (cf. Quill, 1966)
26	Prime meridian	An established imaginary north-south line that passes through Greenwich, England designated as 0° longitude and became the basis for time around the world in 1884. This location serves as an international standard for time around the world and is known a Greenwich Mean Time (GMT) or Universal Time (UT)
27	Quartz crystal watches	Windup clocks were replaced by electronically run, ultra-precise quartz crystal watches in the 1960s and remain the basis for millions of watches and clocks in calculators and computers today
28	Atomic clock	Most accurate clock, based on electromagnetic radiation of a cesium-133 atom. Today time is determined by counting seconds, with a second defined as the length of time for 9,192,631,770 cycles of a cesium-133 atom at zero magnetic field

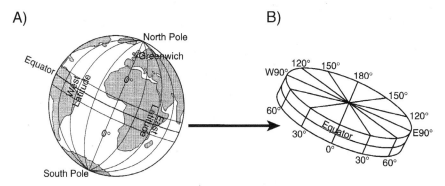

FIGURE 3.8. Globe showing east and west longitudes *(A)*, and a disk of 360° showing longitude divisions *(B)*.

Prehistoric astronomical observations were made outside of the Mediterranean region as well. For example, archaeologists have determined that Stonehenge in England, consisting of large standing stones set in a circle, was originally built in about 3100 BC, and updated in ~2100 BC and 1550 BC (Barnett, 1998). This structure and hundreds of analogous structures throughout Britain, all of which are oriented with the Sun and Moon, may have served as calendars to indicate the time of the month and year anchored by the summer solstice.[7]

The Egyptian calendar was adopted and modified to include 365.25 days by Julius Caesar in 45 BC. Because the year is a little shorter than 365.25 days, the date of the spring equinox was occurring earlier in the "Julian" calendar as the centuries progressed. In 1582 on the advice of a special commission, Pope Gregory XIII modified the Julian calendar by 10 days. He ordered that the day after October 4, 1582 be called October 15, 1582,[8] thus restoring the spring equinox to again occur on March 21 in 1583,[9] The Gregorian calendar, which it became known as and which has been adopted worldwide, includes special rules for leap years as follows: a day is added to February in years divisible by 4, except in years divisible by 100, but a leap day is added to a year divisible by 400 (e.g., the year

---

[7] Centuries later, other civilizations were still using sundial-like structures to determine the time of year. This was evident in South America in the 16th century, where the Inca used stone structures, such as the Intihuatana at Machu Pichu in the Andes Mountains of Peru, as a precise indicator of the winter solstice. During the 15th–16th centuries, cathedrals in Italy were constructed so that light projecting through holes in a great window would move along a line (called a *meridiana*) on the floor. The church building itself thus served as a solar observatory and helped to precisely determine time of year (cf. Van Helden, 1999).

[8] This particular interval was chosen for religious reasons (e.g., fewer ecclesiastical days) (cf. Whitrow, 1988). The 10-day correction had various social repercussions, such as salary issues, changes in the age of individuals, and ushered in a number of political and regional conflicts. The Gregorian calendar was not accepted until 1752 by the American colonies and Great Britain and is still not accepted by some Orthodox Christians.

[9] The spring equinox occurs every fourth year on March 21 (cf. Bennett et al., 1999).

TABLE 3.2. English names for months of the year originated from Roman names.[a]

Name	Origin of name
January	The god Janus
February	The purification festival, Februa
March	The god Mars
April	The goddess Aphrodite or Latin *aperire*, to open
May	The goddess Maia
June	The goddess Juno
July	Quintilis, the 5th month, from the word *quintus*, fifth, renamed for Julius Caesar in 45 BC
August	Sextilis, the 6th month, from the word *sextus*, sixth, renamed for Emperor Augustus in 4 BC
September	The 7th month, from the word *septem*, seven
October	The 8th month, from the word *octo*, eight
November	The 9th month, from the word *novem*, nine
December	The 10th month, from the word *decem*, ten

[a] Originally, the Roman calendar started on March 1 and consisted of only 10 months, with an unnamed winter period. Sometime before 673 BC, February and January, in that order, were added as months between December and March. In 450 BC, February was moved to its current position in the calendar (Tøndering, 2000).

1900 was not a leap year, but 2000 was). The end result is that the Gregorian calendar year is almost precisely the same as the length of the tropical year. Our modern calendar is still based upon names and numbers of days assigned by the Romans during the time of Julius Caesar in 45 BC (Table 3.2).

The Day

Devices were also implemented to subdivide the spans of light and dark into day and night. The invention of the shadow clock (sundial), the water clock (*clepsydra*),[10] and the use of a 24-h day can also be attributed to the ancient Egyptians. In turn, these devices and procedures were adopted and modified by others. For example, for the actual division of equal length hours, and subdivisions of 60 parts for minutes, one needs to turn to antiquity and credit the Hellenistic astronomers[11] who modified the Egyptian hours with the Babylonian sexagesimal (base 60) measurement system for minutes (cf. Neugebauer, 1957). Our unit of time known as the second is a further subdivision of the minute into 60 equal parts.

---

[10] Clepsydra was the term used by the Greeks and derived in part from their word, *kleptein* = to steal + *hydor* = water (e.g., "water thief"). In a water clock, water drained from one container at a fixed rate, which in turn filled another container or moved more elaborate mechanisms, e.g., balls or pellets falling to produce a ringing sound (cf. De Solla Price, 1964; Hill, 1976). A sundial, water clock, and other clocks can be seen in motion at the following or similar website(s): http://www.brittanica.com/clockworks.

[11] The duration of the Hellenistic period was from approximately 323 BC (when Alexander the Great died) to 31 BC (Adriani, 1963).

Candles with gradation for hours and incense that produced different aromas, both of which burned at known rates, were used in England ~980 AD and in China in ~1000 AD to mark the passing of time. Although less accurate, these were useful indoors, during the night or on a cloudy day. Water clocks were still popular in Germany during the early 1200s, but because the fluidity of water was influenced by temperature, the sandclock began to replace it. Sandclocks, which today are often referred to as "hourglasses," may have been an improvement, but they too were somewhat restrictive and cumbersome. One of their main uses was on board ships, where they were used to measure durations of activities and events, such as the speed of travel (knots).[12]

The Week

Having introduced some of the devices and units for measuring time within a day and over a year, we turn our attention to the week. Our lives are usually organized along a weekly pattern of orienting ourselves with a 5- or 6-day workweek that usually begins on Monday and ends on Friday or Saturday, with the weekend reserved for leisure, religious, or other activities.[13] While there were some attempts by various civilizations to use weeks of different lengths,[14] it is likely that the 7-day week was being followed by the Jews during the time of Moses (ca. 14th century BC), based upon their belief that God worked for six days creating the world and rested on the 7th day. As noted in Table 3.1, item #10, the recurrence of the Lord's day every seventh day was already a part of the Jewish life.[15] The familiarity of Jewish Christians with the sabbatical week and Gentile Christians with the planetary week may have set the stage for the 7-day weekly cycle that was noted already in the first century (79 AD) and was in common use throughout the Roman Empire by the end of the third and beginning of the fourth century (cf. Colson, 1926).

The astrological 7-day week is essentially a Hellenistic invention, with the contemporary names for the days of the week and their origin depending upon nationality, language, and religion (cf. Zerubavel, 1985). Names for the days of the week have been derived from various sources, including the names of planets, names of

---

[12] A log tethered to a line having equally spaced knots along its length would be tossed into the water in the direction of the stern from the front end of the ship and sailors would count the number of knots along the rope that was passed as the ship moved forward during a given interval of time, which was measured by a sand glass (e.g., 0.5 min).

[13] Zerubavel (1985) noted that some civilizations with complex divinitory systems and/or market economies cherished regularity. He called this a particular mentality of "*homo rhythmicus*" man.

[14] The Mayan calendar included both a 13-day and a 20-day week. Between 1793 and January 1806, the French Revolutionary calendar divided each month into 10-day weeks, while the Soviet Union used a 5-day week in 1929–1930 and a 6-day week from September 1931 until June 26, 1940. Interestingly, a 5-day week fits the solar year the best, since 365 days/5 = exactly 73 weeks.

[15] Deuteronomy 5: 13–14; Exodus 20: 8–11.

TABLE 3.3. English names for days of the week and origin.

Day	"Planet"[a]	Latin	Comments
Sunday	(Sun)	*Solis*	From the Latin planetary name for the Sun
Monday	(Moon)	*Lunae*	From the Latin planetary name for the Moon
Tuesday	Mars	*Martis*	From Tiu, the Anglo-Saxon name for Tyr, the Norse god of war
Wednesday	Mercury	*Mercurii*	From Odin, or Wodin, the Norse supreme deity
Thursday	Jupiter	*Jovis*	Honoring Thor, the Norse god of thunder
Friday	Venus	*Veneris*	Honoring Frigg, the wife of Odin, who represented love and beauty
Saturday	Saturn	*Saturni*	From the Latin planetary name for Saturn

[a]The Romans based names for days of the week on the Sun, Moon, and the then five known planets. Romance languages, such as French, Spanish, and Italian, still use this Roman system, while English uses names derived from Norse legend for four of the days (Tuesday–Friday).

mythical characters, or a combination of both, or they have been numbered without specific names (Table 3.3). In Romance languages such as French, Italian, and Spanish, the days of the week are based upon the names for the Sun, Moon, and the five planets known to the Romans (Mars, Mercury, Jupiter, Venus, and Saturn). Some names in English were changed to reflect Anglo-Saxon gods. In Hebrew, the first six days are designated only by number (e.g., Yom Ri-shon = first day, Yom Sha-ynee = second day, etc.) and only the seventh has a name, Yom Shabbat. The Quakers omit names entirely and only use numbers.

Most Christians have designated Sunday as their day of worship and rest, whereas this day is Saturday for Jews and Friday for Muslims. Most of today's printed calendars show Sunday as the first day of the week, although there are some notable exceptions (e.g., some European calendars, work and school schedules show Monday as the first day of the week). The International Organization for Standardization has officially set Monday as the first day of the week.

While many historians may be of the view that we do not know the origin of the week, one should not go as far as those who claim that there are no natural cycles of seven days. Evidence for presence of weekly (circaseptan) rhythms has been found at the cellular, individual, and population levels representing such diverse organisms as algae, vascular plants, insects, and humans (Schweiger et al., 1986; Spruyt et al., 1987; Halberg et al., 1990; Cornelissen et al., 1996).[16]

Clocks and Calendars: Middle Ages to Now

During the Middle Ages (ca. 500–1500 AD), various types and styles of sundials, water clocks, and sandglasses were developed, but their technology was still based upon the principles from ancient Egypt. The development of the mechanical clock,

---

[16] See Chapter 2 on Rhythm Characteristics for examples of circaseptan rhythms. It has been suggested that 7-day cycles in energy output from the Sun (e.g., sunspots, solar flux, etc.) may act as an environmental signal for certain 7-day biological rhythms in life on the Earth (Cornelissen et al., 1996).

which probably occurred near the end of the 13th century in Europe (Thorndike, 1941; Whitrow, 1988), was strongly influenced by the church (North, 1975), and its need to set and adhere to times designated for prayers and other monastic activities.[17]

Early versions of mechanical clocks did not have a dial. Instead, time was indicated by the sound produced by a bell. The clocks were large and usually housed in a tower in order to serve as a public clock, hence the word "bell tower."[18] In fact, our word for clock is derived from European words for bell, such as *clocca* (Latin), *glocke* (German), and *cloche* (French). These early clocks were powered by a falling weight that caused a drum to rotate as the rope, which suspended the weight, unwound. An oscillating escapement mechanism controlled the fall of the weight by engaging and disengaging cogs on a wheel that would turn, thus providing a timing mechanism.[19]

Longitude and Clocks

The development of clocks that could be kept in homes or onboard ships at sea had far reaching effects, not only in establishing schedules for various activities, but also for oceanic travel and navigation. In addition, one of the driving forces that led to the development of timepieces that had even greater precision was the need for navigators of vessels at sea, and out of sight of land, to locate longitude.[20]

---

[17] Churches were the centers of education in Europe and elsewhere. They controlled much of the power, had skilled craftsmen, and knowledge of engineering.

[18] In the 1800s, there were time balls in many cities around the world, each connected by telegraph to an observatory, which would drop at noon each day as a public signal of midday (12:00 h) to ships in harbors and rivers and also to townspeople. People could stop at noon and reset their watches by it. This relic of the past can still be seen on New Year's Eve in Times Square in New York when a ball drops at midnight (cf. Barnett, 1998).

[19] The mechanism of these early mechanical clocks consisted of a horizontal bar with a balancing weight on each end and secured to a vertical spindle that contained two flat flanges (pallets) spaced apart and set at different angles to engage alternately the teeth at opposite positions of a wheel. The wheel, which is called the escape wheel, was powered to rotate by the unwinding of a rope from around a drum, which was caused by the pull of a weight suspended from the other end of the rope. The design of the teeth on the wheel was such that when the flanges on the spindle engage a tooth, the weight was prevented from falling. The wheel also caused the spindle to be rotated back and forth, first in one direction and than in the other direction (e.g., clockwise and then counterclockwise). The back and forth oscillations of the horizontal bar, to which the spindle and flanges were secured, allow for a controlled escapement (one tooth at a time with a brief pause) (Ward, 1961).

[20] The British Government offered a prize in 1714 (today's equivalent of several million dollars) for a practical method that could be used at sea to determine longitude. John Harrison (1693–1776) solved the problem by developing a number of precision timepieces that minimized the effect of changes in temperature and surpassed the requirements for precision. His spring and balance wheel-driven marine chronometer kept time on a rolling ship to within 1/5 of a second a day, nearly as well as a pendulum clock on land. On a voyage to Barbados in 1765, the mean error for one of his instruments was only 39.2 s (about 1/2 of a degree of longitude) (Quill, 1966).

FIGURE 3.9. Assume that the person shown is standing somewhere in the Northeastern part of the United States of America, facing directly South (notice the shadows and the compass, which has been corrected to read true North and South). It is noon (mean solar time = 12:00 h) because the Sun is just passing the person's meridian, an imaginary line that goes from his horizon due south, through his zenith directly overhead, and continues to his horizon due north. The watch that he is wearing has been set on Greenwich Mean Time (GMT) and now reads 17:00 h. What is his longitude? (You should try and answer this question before you look it up in footnote 26).

But how can a clock that is very precise help tell where you are relative to longitude, especially if one has not been able to maintain an accurate record of speed, direction, and distance?

To answer this question, we start with the knowledge that a hemisphere is half of a sphere (Gk. *hemi* = one-half), and although we often view the Earth as having a Northern Hemisphere and the Southern Hemisphere, map makers also view the Earth as having hemispheres of east and west longitude (see Figure 3.8). These two hemispheres have been divided along two imaginary lines of longitude, which run north and south and are separated from each other by 180°.[21] By international

---

[21] Because the Sun and the path of the Sun are circular, it is possible that the "o" superscript, which is used to designates degrees (e.g., 360°), could be a primitive hieroglyph for the Sun (Boorstin, 1985).

agreement in 1884, zero degrees (0°) longitude passes through Greenwich, England[22] and is referred to as the prime meridian.[23]

Longitude is an angular distance marked in degrees,[24] starting with zero and moving both east and west (Figure 3.8). They meet at longitude 180°, the *International Date Line*, which in reality has been drawn to zigzag in certain places in an attempt to avoid a nation having two different dates on the calendar for the same day.[25] The distance between lines of longitude is widest at the equator and becomes narrower as the lines approach and finally meet at both the North and South Poles. Because of the counterclockwise rotation of the Earth, the path of light from the Sun moves at a rate of about 15°/h from east to west. For this reason, the mean solar time at any one location is either ahead or behind the solar time at any other location that is east or west of it.

Based upon the background information presented in the preceding paragraphs, the reader should now be able to answer the question as to how a clock can help tell where you are relative to longitude. First, if you know the time at 0° longitude (Greenwich Mean Time, GMT, also known as Universal Time, UT), and second, the mean solar time where you are, it is possible to calculate longitude from the difference between the two, as shown by the example in Figure 3.9.[26]

Most navigation today relies upon a global positioning system (GPS) that receives signals from satellites. In fact, small hand-held GPS devices that display longitude, latitude, and GMT are readily available to the general public and can be used for a variety of purposes, such as hiking, finding a specific location for a picnic, or to meet someone, or even the distance to the next hole when playing golf.

Springs to Atoms

Spring- and pendulum-driven clocks came into use ~1500 AD, both of which improved the precision of time keeping. These spring-powered clocks allowed

---

[22] The great pyramid of Egypt, Jerusalem, and the Bering Straights were also suggested as possible places for the prime meridian, but Greenwich had a high quality observatory (cf. Barnett, 1998).

[23] A meridian is a half-circle, a line extending from the north pole to the south pole and passing over one's zenith (a point 90° overhead).

[24] Each degree of longitude is divided into units called minutes (1 degree = 60 min) and each minute is divided into units called seconds (1 min = 60 s). *Note of caution:* Longitude is expressed in degrees, and in this case, the divisions of minutes and seconds are parts of a degree, not units of time.

[25] This involves places such as Siberia, the Aleutian Islands, and the Fiji Islands. A new day begins on the western side of the international date line (180th meridian), and as the Earth rotates counterclockwise, the new date moves westward and arrives 12 h later at the prime meridian (0°). A 24-h difference in time exists between 180° west longitude and 180° east longitude.

[26] In this example, there are 5 h between local noon (12:00) and GMT (17:00), so the longitude would be $5 \times 30°/h = 150°W$. The procedure illustrated in the figure is more complicated than it appears, because it may be difficult to determine when it is exactly noon and where true south is. Today, one is able to purchase a GPS device that receives signals from satellites and displays very precise longitude and latitude readings.

people who could afford them to bring clocks into the home for the first time. Even though these clocks slowed down as the mainspring unwound, they were precursors to the more accurate timepieces to come. Knowing the time of day, as well as measuring spans of time, has become an integral component of our society. Today, nearly every adult person in the more technologically advanced nations has a watch (see Essay 3.1).

Essay 3.1 (by RBS): Wristwatches

The realization that every individual should have an accurate indication of the time became a practical concern and one of the social consequences of World War I (1914–1917) (cf. Whitrow, 1988). A watch strapped to a wrist was more convenient on the battlefield when using weapons or synchronizing events among widely separated individuals. Prior to this time, most watches were designed to be carried in a pocket, while watches worn on the wrist were mostly decorative ornaments.

Both the pocket watch and the wristwatch were mechanical. They needed to be wound and reset regularly, since even some of the better ones could be off by one or more seconds per day. In the early 1920s, W.H. Shortt, a railway engineer, devised a clock that included a "master" clock having a pendulum of steel and nickel that was relatively temperature-independent, and a subsidiary "slave" clock[27] (Whitrow, 1988). These two "clocks" attained an accuracy to 10 sec/year. Even greater precision in timing devices (to within 0.002 sec/day) was achieved by clocks that contained a quartz crystal that would vibrate (oscillate) at a known frequency (32,768 Hz) when stimulated by an electric field.[28]

Individuals who need or want to know the "exact time" rely upon an atomic clock. This clock, which is based on electromagnetic radiation of a cesium-133 atom, is the standard for counting a second, the international unit of time. The atomic clock is so accurate that it will stay within one second of true time for 6 million years. Except for certain highly technical scientific work (e.g., computers, global positioning satellites, space exploration) that depends upon this and the next generation of atomic clocks,[29] most individuals do not need such accuracy, although many homes in the United States and elsewhere have devices,

---

[27] The use of a "slave clock" with a "master clock" as a means for obtaining greater precision appears also as an analogy in one of the models for a more precise biological clock (see Chapter 5 on Models and Mechanisms).

[28] In 1928, W.A. Marrison, a scientist at Bell Laboratories in the USA, invented this clock that in the 1960s became the basis for the millions of clocks in both analog and digital wristwatches, calculators, and computers.

[29] The "atomic fountain," is 5–10 times more accurate than the current atomic clock and is accurate to within 1 sec in 60 million years. The ultimate atomic clock that will produce hyper-accurate time may be the "mercury ion optical clock," that will monitor ions rather than atoms and may be 1,000 times more accurate than today's cesium-133 atomic clock. This clock may be especially useful in communications, navigation, and tracking, especially between satellites and spacecraft. These clocks are capable of measuring a femtosecond (10^{-15} s), while future atomic clocks are likely to be able to measure an attosecond (10^{-18} s) or zeptosecond (10^{-21} s) (Diddams et al., 2004).

such as computers and clocks, by which they receive precise time from an atomic clock.[30]

Official local clock times and dates, which are acknowledged and used throughout the world, have now been established by national and international agreements or laws. There are also international standards for writing dates and times.

Time Zones

The times for sunrise and sunset change with longitude, but external watches and clocks are synchronized to a standard time determined by the laws passed by federal governing bodies. How did this happen? In the United States alone there were over 300 different "Sun" times in use during the early 1800s,[31] and due to the increased rapidity of transportation by train, there was much confusion associated with the use of local time derived from the position of the Sun at each station.[32] At precisely noon on November 18, 1883, the USA was divided by the railroads into four time zones centered on the 75th, 90th, 105th, and 120th meridians, with all clocks in a given zone set the same.[33] Although there was some resentment by those who felt it was not natural, the "railroad times" were officially legalized nationally in 1918 during the debate for Daylight Saving Time (Barnett, 1998).

The whole Earth was divided into 24 time zones in 1884, each 15 degrees of longitude apart, thereby creating a temporal map of the Earth, with Greenwich England being the location of the Prime Meridian. In most cases, the new time zones were first accepted by the railroads and later adopted as local time by individual countries (e.g., France adopted its current time zone in 1911, while Liberia adopted their time zone only in 1972). The boundaries of these time zones have been influenced by social, political, economic, and other reasons, and do not strictly adhere to straight north–south longitudinal lines. For example, England would have two time zones if zero longitude was used, but it has chosen to have a single time zone. Generally, each time zone is one hour earlier than the time zone east of it and one hour later than the time zone west of it. The International Date

---

[30] Today in the USA, you can synchronize your computer's clock or wristwatch to the nearest second by logging onto the internet at www.time.gov and selecting your time zone (this site is provided by the National Institute of Standards and Technology in Boulder, CO and the U.S. Naval Observatory). A link at this site (www.bsdi.com) can connect you to an international time website, as well as dozens of sites related to time services.

[31] These time zones differed in minutes, as well as in hours (Barnett, 1998).

[32] This also led to discrepancies in documenting times of events, such as differences in the reported time of day of the Custer Massacre at the Battle of the Little Big Horn in Montana on June 25, 1876. Some watches which were found and had stopped during the battle showed different times, probably since they were not on local time, but still set to local time of the owner's origin (e.g., St. Louis, Boston, or California), while reports by Sioux Indians cited time according to the position of the Sun.

[33] Telegraph lines transmitted GMT to all major cities in order to reset all clocks correctly to the new zone's time.

Line is an exception, since there is a 24-h difference between 180° west longitude and 180° east longitude.

Daylight Saving Time

In order to conserve fuel, Daylight Saving Time (DST) was originally used by several countries during WWI, and also by the USA year-round in 1942–1945 during WWII. From 1946–1966, individual states and regions were free to choose when DST began and ended, which led to some confusion in telecommunications, train, and airplane schedules. Legislation enacted in 1986 (US Time Code Section 260) established uniform DST dates across the whole USA to begin on the last Sunday in April and end the last Sunday in October, but in 1986, the onset of DST in the USA was changed from the last Sunday to the first Sunday in April.[34]

Although we have grown accustomed to DST (or "summertime" as it is known in Europe and many other countries), many equatorial and tropical countries do not use DST since there is so little change in daylight hours throughout the year. DST is used between October and March in the Southern Hemisphere. In addition, some countries (e.g., Russia, with 11 time zones) observe DST year round, thereby effectively placing each time zone 15° to the east. However, Russia still adds an extra hour in the summer, making each time zone 2 h ahead of standard time.

Most of us also might be aware of a "mini jet lag" twice each year due to the onset and cessation of DST in the spring and fall, respectively. For example, in the USA, we lose 1 h on the first Sunday of April when 1 AM becomes 2 AM and we need to get up an hour earlier according to the new setting on an alarm clock. Conversely, when DST ends on the last Sunday in October, we "gain" an extra hour when the clocks are reset back from 2 AM to 1 AM, but often wake up an hour earlier than necessary for a day or two since we are accustomed to DST. While there may be complaints about having to reset our many mechanical clocks (e.g., wristwatches, wall clocks, VCRs, computers, TVs, etc.), we also need to reset our biological clocks by an hour in order to adjust to new sleeping schedules after we "spring forward" and "fall back."

Recording Date and Time

Even though we now have clocks and calendars that are relatively precise, with established reference points for the beginning of a new day, a new month, a new year, and a new time zone, there are various, sometimes confusing ways of recording and reporting dates and/or times. For example, different countries use different customs to note the date as follows: a day-month-year format (25/7/2004 or 25-7-2004) is used in most countries, while a month-day-year format (7/25/2004

---

[34] Some areas in the USA have still not adopted DST (e.g., parts of Arizona and Indiana). Also, much of Africa does not observe DST, nor does, Japan, China, Indonesia, or the Indian subcontinent.

or 7-25-2004) is more commonly used in the USA. This can sometimes become confusing when a date is written 07/05/2004 – is it July 5 or May 7? The International Standard requires a year-month-day format (20040725 or 2004-07-25), that is especially useful in computers. The year is often annotated with BC ("Before Christ"), AD ("Anno Domini" = in the year of our Lord), or BCE (Before the Common or Christian Era) and CE (Common or Christian Era).

With regard to time of day, most of the world use 24-h "military" time by reporting hours within a day as hours after midnight (e.g., 06:00 h, 13:00 h, 18:00 h, etc.), while the USA divides the day into two 12-hour parts, with the first 12 hours after midnight annotated with AM (*ante meridiem* = before midday) and the next 12 hours annotated with PM (*post meridiem* = after midday).[35] Thus, when it is 13:00 h or 18:00 h in a European household, an American would say it was 1 PM or 6 PM, respectively. It is also sometimes necessary to report local time appended with the local time zone and whether or not daylight saving time is in effect (e.g., 09:00 CST or CDT for Central Standard Time or Central Daylight Time in the USA). With modern telecommunications, especially email, the hour and minute of a transaction are usually coded in relation to Universal Time (GMT).

Recording Biological Time

While time zones have been established by social laws, the rhythms of organisms have evolved to be in synchrony with physical laws, be they natural, such as the solar day, or a programmed regimen maintained in a controlled environmental chamber or building. Therefore, to study and analyze the characteristics of rhythms, such as phase, period, and amplitude, one needs to be familiar with how times are recorded and standardized.

The use of local clock hour and minute can be misleading when it does not coincide with the experimental conditions within the laboratory (e.g., one lab may turn lights on at 6 AM and off at 6 PM, another at 8 AM and off at 8 PM, etc.), when different photoperiods are used (e.g., 12L:12D, 8L:16D, etc.), or when data from nocturnal organisms vs. diurnal organisms are reported or compared (Sothern, 1995). One common solution to avoid the ambiguities of local time is to report time as hours after some external reference point other than local midnight, such as light-onset, dark-onset, mid-light, mid-dark, restricted food access onset, or activity onset. In circadian studies, it is common to find time coded in reference to lighting onset, such as CT (Circadian Time), ZT (Zeitgeber Time), or HALO (Hours After Lights On).

An advantage of such time scales is to assure that results are interpreted in relation to internal, biologic time (stage of rhythm) and not external, solar time (time of day). For example, when local midnight is used as the time reference point, a

---

[35] The Romans used A.M. and P.M. when they divided their day into two parts, with the dividing line being when the Sun crossed the meridian immediately overhead (Boorstin, 1983).

peak in bone marrow DNA S-phase synthesis is found at 13:04 h in diurnally active humans and at 23:08 h in nocturnally active mice, seemingly a 10-h difference according to an external clock. However, when activity onset is used as a reference point, the peaks occur at 06:04 h and 05:08 h, respectively, only 1 h apart, indicating that bone marrow is most prolific at the same circadian *stage* of both species (e.g., during mid-activity) (Sothern, 1995).

The 24-h Biological Clock Concept

Biological rhythms having periods of about 24 h were initially referred to as diurnal rhythms until the terminology of *circadian* was introduced by F. Halberg in 1959 (Halberg, 1959), and their general characteristics summarized in the classic papers of the 1960 Cold Spring Harbor Symposium (e.g., Pittendrigh, 1960) (see Essay 2.1 in Chapter 2 on Characteristics). The term *biological clock* had been introduced about 5 years earlier[36] and referred to about 24-h periods that later became known as circadian. Originally, the concept of the biological clock excluded ultradian and infradian periods, which had not yet been introduced as domains.

Early Studies

To more fully understand the clock concept as it applies to circadian rhythms, we turn first to early studies of flowers and bees. One of the early scientific reports that associated a rhythmic occurrence with a given phase of the day has been credited to Linnaeus (1707–1778) and his floral clock in Sweden that indicated times when a flower or inflorescence (cluster of flowers) of a given species would open or close throughout the day (Figure 3.10) (cf. Kerner von Marilaun, 1895).[37] A comment made to Linnaeus by one of his correspondents during the 18th century that his floral clock would replace mechanical clocks, and thus caused some concern to clockmakers of his day, has turned out to be more amusing (or anecdotal) than practical.[38]

Results from early studies with bees showed that they could be trained to come to a source of food at any standardized time of day, providing that the intervals were about 24 h, but not 19 h, apart (Beling, 1929). Furthermore, it was shown that bees which were entrained in France to seek their food between the hours of 20:15 and 22:15 h in a closed room and then transported overnight to New York,

---

[36] Frank A. Brown of Northwestern University published the term in the April issue of *Scientific American* (Brown, 1954). The use of the term "clock system" appeared later that same year in the *Proc Natl Acad Sci USA* (Pittendrigh, 1954).

[37] A reproduction of the "Flower Clock" designed by Linneaus can be found on p. 12 in *The Clocks That Time Us* (Moore-Ede et al., 1982) or on p. 3 in *Biological Rhythms and Medicine* (Reinberg & Smolensky, 1983).

[38] "*– put all the watchmakers in Sweden out of business*" (Blunt, 1971).

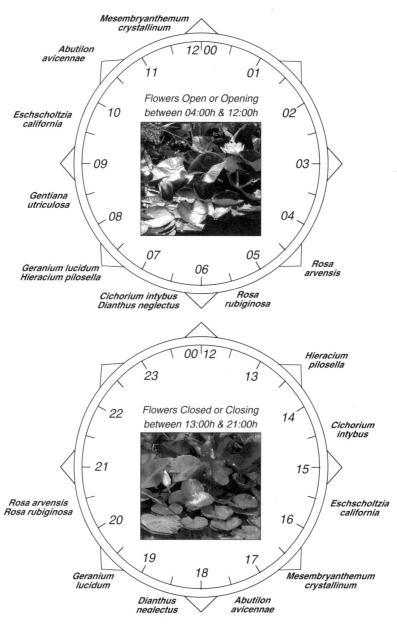

FIGURE 3.10. Two 12-h flower clocks indicating times of the day relative to Sun-related clock hours when the same flowers of certain species open or close at Innsbrück, 47°N (based on Tables in Kerner von Marilaun, 1895). According to Karl Linnaeus, who designed his "Flower-Clock" in Uppsala, Sweden (60°N) in 1745, a trained botanist could estimate the time of day in the summer by noting which flowers of selected plants were open and/or closed. Insets = Example of Water Lily flowers that open between 06:00 h and 07:00 h and close between 16:00 h and 17:00 h. Photos by R. Sothern.

responded as if they were still in France (e.g., they would seek food in the new laboratory between 14:15 and 16:15 h in New York when it was between 20:15 and 22:15 h in France) (Renner, 1955). These results indicated that there were no external time cues, such as cosmic rays that were telling the bees when food would be present, but rather it was due to the presence of an internal circadian timing mechanism.[39] The important point is that a physiological system functioning in a manner analogous to a clock provides bees, birds, plants, and other organisms with "information" that identifies a certain time point relative to another time point 24 h apart, such as dawn (also see Pittendrigh & Bruce, 1957).

Considerations

The "biological clock" concept has both advantages and disadvantages. One of the major benefits of using the word "clock" is that it provides a concept that is easily visualized as a functional entity. From the perspective of physiology, the concept lends itself well to the formulation and testing of hypotheses in basic research. Results from such research have produced a number of models on how autonomous oscillators could function. The concept also implies that the clock is a device to measure time (Pittendrigh & Bruce, 1957; Pittendrigh, 1981). Here too, the functional connotations help foster exciting and important research. This is such an important area of study that an entire chapter in this book has been devoted to the topic of models and mechanisms related to biological clocks (Chapter 5). A highly simplified diagram of the features of a circadian system with a central clock is illustrated in Figure 5.23 in Chapter 5 on Models and Mechanisms.

Carried to the extreme, however, such a "device" can become too inclusive, divisive, and misleading. The overuse of analogies between biological clocks and mechanistic models is a temptation far too common, even for the authors of this book. Two drawbacks of the biological clock concept occur when (1) it implies that a single entity or master clock is likely to be situated in a specific anatomical location and limited to one group of organisms (e.g., the suprachiasmatic nucleus of the mammalian brain); and (2) when it is confined to circadian rhythms, thus excluding ultradian and infradian rhythms that are also part of the network of frequencies that comprise the temporal organization of life.

Ultradian and Infradian Clocks

Results from studies as diverse as the courtship song of fruit flies (*Drosophila*), defecation behavior of nematodes (*Caenorhabditis elegans*), and membrane potentials of the mammalian heart, clearly attest to the presence of various clock-like features of ultradian rhythms (cf. Iwasaki & Thomas, 1997). Ultradian and infradian rhythms encompass a wide range of periods, and very likely, great diversity in the mechanisms that generate the frequency. As with circadian rhythms, there is a tendency to focus on one variable, such as pulse, and refer to it as the

---

[39] This ability was called "Zeitgedächtnis" (time memory) by August Forel in 1910.

"biological clock," the *"chronometer of humans"* (Talbott, 1970). In the context of this book, which addresses biological rhythms in all three domains (ultradian, circadian, and infradian), biological clocks are viewed to be self-sustained oscillators that produce biological rhythms in the absence of external periodic sources (Haus & Touitou, 1994).

Endogenous vs. Exogenous

The first known scientific paper that demonstrated the possible presence of an endogenous 24-h biological timing mechanism was published in L'Academie Royale des Sciences in the early 18th century. This brief communication by M. Marchand described the botanical observations of the French astronomer J. De Mairan (1678–1771), who had observed that in the absence of sunlight, the leaves of a sensitive plant would be open during the usual daylight hours and closed at night (De Mairan, 1729). In other words, a daily rhythm persisted in the absence of the natural environmental synchronizer of alternating light and darkness. Results from studies that followed showed that it was not the higher or lower environmental temperatures throughout the day that caused the rhythms of leaf movements in constant darkness[40] (Duhamel Du Monceau, 1759; Zinn, 1759) and that the "circadian clock"[41] would free-run with a period that is close to, but not exactly, 24 hours (De Candolle, 1832; Figure 3.11).

Comparison with Manufactured Clocks

During the development of a more precise mechanical timepiece in the first quarter of the 18th century, two clock-centered issues emerged related to circadian rhythms. These two issues were the temperature compensation of clocks (see Chapter 2 on General Features of Rhythms) and locating longitude (this chapter). Not only is temperature compensation a necessary feature for an exogenous marine clock if it is to maintain accuracy, it also is one of the major characteristics of a circadian rhythm (cf. Pittendrigh, 1960). Also, the phases of circadian rhythms in nature are synchronized according to the duration and timing of L and D at the prevailing longitude and latitude.

The relationship between endogenous (driven internally) and exogenous (driven externally) rhythms is much more than an analogy between models of biological clocks and manufactured clocks. Biological rhythms have been used to quantify external oscillations and these oscillations have been used to quantify endogenous rhythms. One classic example involves the human pulse and the pendulum. Supposedly, when Galileo (1564–1642) observed and noted that the time for the swing of the altar lamp varied with the length of the pendulum, not the

---

[40] However, temperature that cycles over 24 h during constant darkness *can entrain* the plant rhythm to 24 h.

[41] The terms, "circadian" and "clock" were introduced much later and not used by these investigators.

A) Leaflets open and close during DD *(de Mairan, 1729)*

B) Leaflets open and close during DD & constant temperature *(Duhamel Du Monceau, 1759)*

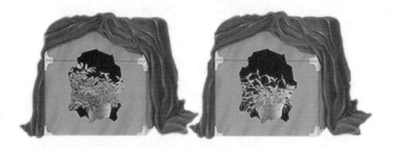

C) Leaflets open and close during LL *(De Candolle, 1832)*

FIGURE 3.11. A diagrammatic representation of the studies of de Mairan *(A)*, Duhamel Du Monceau *(B)*, and De Candolle *(C)* with sensitive plants (e.g., *Mimosa*) that illustrates results of their observations. Opening and closing of leaflets of plants maintained in DD *(A, B)* continued to display a circadian rhythm in the absence of sunlight *(A)*, as well as at constant temperature *(B)*. When movements were monitored under continuous illumination (LL) *(C)*, the rhythm free-ran with a period of 22–23 h.

FIGURE 3.12. Galileo Galilei (1564–1642) used his pulse as a human clock to time the swinging of a pendulum in the baptistry of the Cathedral of Pisa (building in the foreground). He observed and noted that the time for the swing of the altar lamp varied with the length of the pendulum and not the width of the swing. The significance of this relationship is strongly emphasized in the history of medicine, where the pulse is called the biological clock, the chronometer of human beings. Photo by R. Sothern.

width of the swing, he used his pulse (an ultradian biological clock) to quantify the swinging motions (Figure 3.12).[42] Later, in the early practice of medicine, physicians such as Santorius (1561–1636) used a simple pendulum device, known as the pulselogium, which consisted of a string with a weight at one end, to quantify the pulse rate (rhythm) of their patients. The significance of this relationship is emphasized in the history of medicine, where the pulse has been called a biological clock, the chronometer of human beings.[43]

Evolution of the Clock

Earlier in this chapter we turned to astronomy and the history of human civilization to learn how the physical clocks of our society were developed. Now we turn to geology and the history of life, which is called *evolution*, to learn how the biological clock may have developed. Evolution is a biological discipline that focuses on the genetic changes that have occurred over time in members of a population. Except for those changes that occur over a span of a few generations (microevolution), time is measured often in millions of years (mya = million years ago), starting with the present and going back.

---

[42] Galileo Galilei may have made this observation while in the baptistry of the Cathedral of Pisa (shown in forefront of Figure 3.12) (cf. Major, 1954), although the actual construction of the first pendulum clock is often attributed to the research of the Dutch scientist Christiaan Huygens (1629–1695).

[43] *"Just as the foundations of modern physics rest on the accurate measurement of time, the chronometer of humans is the biological clock, the pulse."* (Talbott, 1970)

Events and conditions experienced by the Earth changed dramatically over the eons[44] and differed greatly from what we experience today. Some of these changes, such as fluctuations in sea levels and temperature, are relatively easy to envision, while others, such as the change in the length of time that Earth takes to complete a rotation on its axis, are more difficult to envision. Today, the period to complete one rotation is about 23 h 56 min and much slower than the period of 6 h or less, which may have been present when biological molecules and processes evolved (cf. Lathe, 2004; Tauber et al., 2004). Although the period of rotation has changed greatly, the period of Earth's revolution around the Sun has remained the same as it is today (about 365.25 days) (Tauber et al., 2004).

Studies designed to examine past life on the Earth became easier and more precise near the middle of the 20th century, due in large part to techniques that became available for measuring levels of radioactivity from naturally occurring isotopes present in fossils and rocks. Much of what we have learned about the history of life comes from fossils, the preserved remains and impressions of past organisms in the materials that formed rocks. To trace the evolutionary development of rhythmic timing systems and biological clock components, we will rely upon fossil records, the geological history of Earth, and comparisons of *molecular sequences* among various taxa.[45]

The timing systems in both prokaryotes (cyanobacteria) and eukaryotes (fungi, plants, and animals) have a number of key molecular components. These include, among others, clock genes, cycling proteins, and photoreceptors. Photoreceptors are part of an *input pathway* that receives and transmits signals from external environmental synchronizers to a *central clock* that generates rhythmicity. In turn, an *output pathway* transmits temporal signals from the clock to biological variables that oscillate. These oscillations are the overt rhythms of variables such as activity cycles, leaf movements, spore formation, and enzyme activity. Sometimes these variables are called the "hands of the clock."

Molecular Building Blocks

Nucleic acids, *proteins*, and *photoreceptors*, as well as the molecules from which they are made, provide the building blocks or components of timing systems (Table 3.4). A *gene* is an information-carrying unit (DNA) of heredity, and by a process called *transcription*, genetic information from DNA is transferred through the base-pairing (synthesis) of the correct sequence of nucleotides to produce RNA (ribonucleic acid). Next, with the help of messenger RNA (mRNA) and other components, amino acids are assembled into a polypeptide (protein), a process known as *translation*. In other words, genetic information is expressed in

---

[44] The geological history of the Earth is divided into eons. The most recent is called the Phaneozoic eon, which started about 600 mya and it is further subdivided into three eras (Paleozoic, Mesozoic, and Cenozoic—see Figure 2.3). Subdivisions of eras are called periods.

[45] Taxa (plural of taxon), as used here, refers to groups of organisms such as various species, genera, or families.

TABLE 3.4. Examples of molecules referred to when discussing the evolution of biological timing systems.

Molecules	Description	Examples cited	Comments
Pyrimidines	Nitrogen base-type molecules with one ring and present in nucleic acids (Figure 5.21)	DNA has cytosine and thymine; RNA has cytosine and uracil	Two pyrimidines joined together form a dimer; caused by UV-B radiation and is harmful
Purines	Nitrogen base-type molecules with two rings and present in nucleic acids (Figure 5.21)	Both DNA and RNA have adenine and guanine.	Flow of information is from DNA to RNA to Protein
Genes	Information-carrying units (DNA) of heredity that are expressed as proteins (Essay 5.3)	*kaiA*, *kaiB*, and *kaiC*; *cry*	In cyanobacteria; In plants and animals
Proteins	Polymers of amino acids and have many functions (e.g., enzymes)	a) BMAL1 b) WC-1 c) KaiA, KaiB, KaiC	a) in mammalian clock; b) in *Neurospora* clock; c) in cyanobacteria clock

an organism as a protein. Models of the central clock for eukaryotic organisms include transcription/translation loops with positive and negative feedback that produce delays and generate rhythmicity (Figure 3.13).

Proteins are composed of chains of *amino acids*. Because each of the 20 different amino acids found in proteins can be used numerous times in a given chain, there are a large number of possible arrangements or sequences that could specify a particular protein. By analyzing and comparing sequences, one can determine the presence of absence of relationships among different species, mutants, and macromolecules (Figure 3.14). Studying and comparing *sequences* are important steps in the study of evolution, and in the genetic dissection of biological clocks and timing systems.

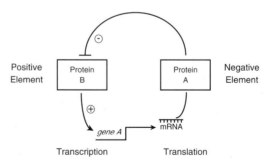

FIGURE 3.13. Positive and negative elements in the transcription/translation feedback loop of a circadian central clock. Transcription of *gene A* yields, via translation (mRNA), the Protein A (negative element), which blocks (negative feedback) the activity of Protein B (positive element) that functions to activate the clock *gene A*. For names of specific genes and proteins, see Chapter 5.

FIGURE 3.14. Two amino acid sequences before *(A)* and after *(B)* insertion of "gap" between "val" and "gly" to align and compare homologous amino acids. Six amino acids are shown: valine (val), leucine (leu), glycine (gly), phenylalanine (phe), serine (ser), and argine (arg) See Figure 5.18 (in Models, Chapter 5) for molecular structures.

Geological History and Rhythmic Components

The early atmosphere of Earth is considered to have been a *reducing environment*, one without the presence of free oxygen (O_2). The oxygen was bound with hydrogen (H_2) to form water (H_2O) or with materials in the Earth's crust to form oxides, such as carbon dioxide and iron oxides. Astronomical events that produced the Moon and fast terrestrial rotations likely served as a setting where early biological molecules could have been produced.

Results from experiments conducted in the 1950s showed that under conditions that then were thought to simulate those of primitive Earth, amino acids, such as glycine and alanine, and simple acids, such as acetic acid, could be produced (Miller, 1953; Miller & Urey, 1959). It is now realized that these assumptions were not completely correct. Some scientists contend that the early building blocks or biological molecules may have been deposited on Earth from meteorites and comets originating elsewhere in the galaxy. Nevertheless, the making of large molecules from subunits of smaller molecules (polymerization) by dehydration (removal of water molecules) and phosphorylation (addition of phosphate groups) could lead to the production of proteins, polysaccharides (long chains of sugars), and nucleic acids. Perhaps the first pre-RNA molecules may have arisen in this early environment. Because the Moon was closer to the Earth during its early history and the rotations of the Earth were much faster, Earth would have experienced rapid cycles of diluting (flooding) and concentrating (drying) of solutes, thus producing salt-dependent association/dissociation conditions. Under such conditions, the replication and amplification of DNA-like polymers could have occurred in a manner somewhat conceptually analogous with the well-known polymerase chain reaction[46] (Lathe, 2004).

---

[46] Polymerase chain reaction (PCR) involves the enzymatic amplification of DNA *in vitro* (Mullis et al., 1986). PCR is a cyclic process of heating and cooling, plus the addition of nucleotides and an enzyme called DNA polymerase. Each cycle requires only a few minutes and the cycles can be repeated over and over again. The amount of DNA is doubled with each cycle.

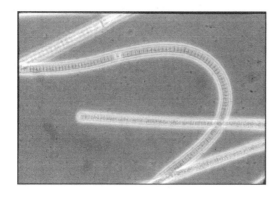

FIGURE 3.15. *Oscillatoria*, one of the common present day genera of cyanobacteria. Photo courtesy of Professor David McLaughlin, Department of Plant Biology, University of Minnesota.

Having discussed briefly how the molecular components could have been produced, although speculative, we can now focus on the specific clock components that evolved from such molecules. A logical place to start this evolutionary journey is with prokaryotic organisms known as *cyanobacteria*. They are among the oldest inhabitants of the Earth. Microfossils of cyanobacterium-like microorganisms, which were discovered in well-preserved sedimentary and volcanic rocks deposited during the early Archean Eon, confirm their presence as early as 3,465 million years ago (Schopf, 1993). Some of these fossil species from Western Australia are indistinguishable from modern species of cyanobacteria that are members of the genus *Oscillatoria* (Figure 3.15).

The central circadian clock of cyanobacteria, at least in the most extensively studied species, *Synechococcus elongates* PCC 7924, includes proteins (KaiA, KaiB, KaiC) that are encoded by a cluster of genes: *kaiA*, *kaiB*, and *kaiC* (Figure 3.16). The origin and evolutionary patterns of these circadian clock genes in prokaryotes have been reconstructed by using available sequences of DNA and protein found in databases[47] and the phylogenetic analysis of these genes (Dvornyk et al., 2003).

The evolutionary history or lineage (Phylogeny) of decent can be diagramed in the form of a tree, a *phylogenetic tree* (Figure 3.17). A molecule known as 16S rRNA[48] is a good candidate for identifying major branches since it evolved slowly. Phylogenetic trees of 16S rRNA and *kaiC* homologs of prokaryotes indicate that the *kaiC genes* occur also in Proteobacteria and Archaea[49] (Dvornyk et al., 2003). Evolutionarily, *kaiC* is the oldest of the three clock genes, followed by *kaiB*. The *kaiA* gene is the youngest and likely evolved only in cyanobacteria (Table 3.5).

---

[47] Sequence data of prokaryotic genomes obtained from public depositories.

[48] The upper case S refers to svedbergs (a lower case) and represents sedimentation coefficient values. Svedberg (1884–1971) designed and built the first ultracentrifuge (cf. Freifelder, 1982). rRNA = RNA molecules in the ribosome.

[49] Archaea is the name of another domain of prokaryotes, often used in a three-domains system where the prokaryotes are placed in two separate domains. In such a system, protista, fungi, plants, and animals represent kingdoms in the domain Eukarya.

Evolution of the Clock 97

FIGURE 3.16. Three Kai proteins (A, B, C) comprise the central oscillator of the cyanobacteria circadian clock (see Figure 5.29 for more detail), here superimposed on a mechanical watch for emphasis.

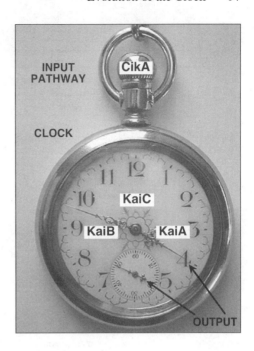

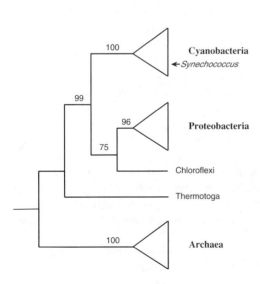

FIGURE 3.17. Phylogenetic tree based upon 16S rRNA genes of prokaryotes. Three major lineages shown (Cyanobacteria, Proteobacteria, Archaea). *Synechococcus* belongs to the cyanobacteria lineage. Bootstrap resampling was used to estimate support for clade (a group of organisms that share features inherited from a common ancestor). Bootstrapping is an analytical method that seeks to evaluate the strength of support for phylogenetic relationships by resampling the data with replacement (i.e., a small set of labeled data is used to induce a classifier in a larger set of unlabeled data). Bootstrap values ranging from 50% to 100% are shown above the branches, with higher values indicating stronger support (based on Figure 1 in Dvornyk et al., 2003).

TABLE 3.5. Biological events in the geological history of Earth that relate to rhythmic timing system of life (see text for references).

Time (mya)[a]	Biological events	Geological notes
2–500	Duplication of clock genes; a number of mass extinctions occur; many new species of animals and plants evolve.	Post-Cambrian = Ordovician Period (500 mya) leading to Quaternary Period (2 mya); massive glaciations and changes or shifts in temperature
544–560	Presence of animals possibly linked to arthropods, annelids mollusks, and jellyfish (*Ediacaran fauna*)	Vendian period, the last period of the Proterozoic Eon
1,000	*KaiA* and *kaiABC* cluster in cyanobacteria evolved	O_2 levels much lower than current levels
1,900–3,500	Organisms have ultradian periods; O_2-producing cyanobacteria out-competed early-evolved photosynthetic bacteria; cyanobacteria *kaiB* gene evolved.	Rotation period of Earth still short, <20 h; reducing geochemical environment replaced by oxidizing environment
540	Clock genes present; animal phyla present.	Cambrian Period; rotation period of Earth perhaps 22 h, close to present
3,500–3,800	Cyanobacteria *kaiC* gene evolved	O_2 appears in atmosphere and combines with H_2 and other materials
3,465	Filamentous microbial prokaryotes, including cyanobacteria are present	Archean Eon; O_2 production evolved
3,900–4,000	About the time that life on Earth originated; self-replicating pre-RNA	Depending upon the reference, rotation period of Earth 4–10 h

[a] In depicting events over evolutionary time, the most recent appear at the top and the oldest at the bottom of the table. Time is presented in mya (million years ago).

The evolution of the prokaryotic clock may have occurred in parallel with the geological history of Earth. If the primitive clock was an ultradian oscillator, which it most likely was, the periods of early cyanobacteria would have been in synchrony with the phase and short-period rotations of the Earth (cf. Tauber et al., 2004). For both organisms and the Earth, the periods would have been less than 20 h.[50] In evolutionary terms, today's ultradian clock could have been yesterday's circadian clock.

The cyanobacteria that evolved during each geological period of the Earth would have been the ones with biological clocks that best synchronized physiological and metabolic process to occur at optimal phases of the prevailing environmental cycle. Within a population, those with the best clock for the given geological period should have an adaptive advantage that will enable them to out number or defeat those with a clock that ran too fast (shorter period) or too slow (longer period). The adaptive value of this type of biological clock has been

---

[50] Ultradian cycles of 2–6 h from tides may have been influential on replication of early biomolecules on primitive Earth (Lathe, 2004).

TABLE 3.6. Winners in cyanobacteria competition[a] in two different environmental [LD cycles, each with equal amounts (hours) of Light (L) and Darkness (D)].

Competitors	LD15:15	LD11:11
WT vs. SPM	WT	SPM
WT vs. LPM	LPM	WT

[a]Growth rates of WT = wild-type cells; SPM = short free-running period (22 h) *kaiB* mutant; LPM = long free-running period (30 h) *kaiB* mutant (based upon Woelfle et al., 2004).

superbly demonstrated in the laboratory with various strains or mutants of cyanobacteria having different free-running periods (FRP) (Johnson et al., 1998; Ouyang et al., 1998; Woelfle et al., 2004). Strains with either short FRPs (22 h) or long FRPs (30 h) have been "set-up" to compete against wild-type cells in cyclic LD environments, in which the durations of light (L) and darkness (D) were equal, but with the periods (complete cycles) being either 22 h or 30 h. When the strains (mutants) were grown separately under the same conditions as used in the experiments, there did not appear to be any differences in growth rates. However, in mixed cultures of wild type (WT) with either short-period mutants (SPM) or with long-period mutants (LPM), there were differences. Growth rates were enhanced when the clock oscillated with a period that was similar to the environmental cycle (Table 3.6) (Woelfle et al., 2004).

Cyanobacteria rely upon light for photosynthesis, a process in which O_2 is "split" from water. These oxygen-producing organisms changed the physical atmosphere of Earth from a reducing environment to an oxidizing environment. In the upper atmosphere, chemical reactions of O_2, which is a product of photosynthesis, and ultraviolet light (UV) radiation from the Sun, produced ozone (O_3) to form an ultraviolet-absorbing stratosphere. UV causes damage to DNA and organisms need to be protected from its harmful rays.[51] The presence of UV and the amounts that reach our environment have contributed to the evolution of photolyases (repair mechanisms), photoreceptors, and circadian systems.

Adaptation to Avoid Harmful Light

Light is required by a large number of biological processes,[52] including, although in no means limited to, photosynthesis, vision, and circadian synchronization. Natural solar radiation reaching Earth includes wavelengths that are both shorter

---

[51] Because UV from the Sun acts upon certain lipids of the body to produce vitamin D (calciferol) in the skin, some longer wavelengths of UV are helpful.

[52] Included in this list would be photomorphogenesis (the development of stems and leaves), photoperiodism (Chapter 4), phototropism (structures growing towards light), and phototaxis (the movement of an organism or part of an organism toward light).

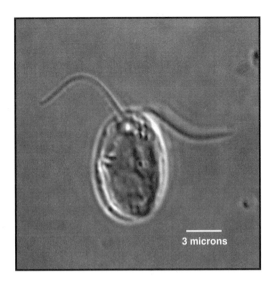

FIGURE 3.18. *Chlamydomonas reinhardtii*, a motile unicellular freshwater green alga with two equal flagella, is widely used in molecular studies. For scale, white bar = 3 micrometers. Photo courtesy of Professor Carolyn Silflow and Brian P. Piasecki, Department of Plant Biology, University of Minnesota.

(UV)[53] and longer (infrared) than those of the visible spectrum, which we see as the colors of the rainbow or the light passing through a glass prism. While the ozone layer does provide protection for present-day organisms from much of the harmful UV (UV-C) radiation, some, such as UV-B does reach Earth. As organisms evolved, a biological timing system that enabled them to escape the harmful effect of UV would have been conserved, even in present day species.

Avoiding sunlight as much as possible would be one method for escaping the harmful effects of UV, but for organisms that depend upon light from the visible portion of the spectrum for photosynthesis, this poses a problem. One strategy would be to time those events that are most vulnerable to the deleterious effect of UV, such as DNA replication and cell division, to occur at night. This hypothesis has been tested in experiments with the unicellular alga *Chlamydomonas reinhardtii* (Figure 3.18). *Chlamydomonas* cells were found to display a circadian rhythm of survival from UV-C radiation (Nikaido & Johnson, 2000). The most sensitive phases occurred during late day and the beginning of the night, which also corresponded to the times of nuclear division.

An interesting relationship may exist that links the coevolution of blue-light photoreception and circadian timing, an evolutionary process involving UV avoidance and the photolyase enzymes that repair DNA damage. *Photolyases* are found in bacteria and higher organisms. They are activated by blue light. Closely related to the photolyases are the *cryptochromes* (CRYs), blue-light photoreceptors, which were first discovered in plants (Ahmad & Cashmore, 1993), and now known to be present in insects, humans, and bacteria (Cashmore, 2003). In plants, cryptochromes are part of the circadian system, whereas in fruit flies,

---

[53] UV wavelengths are often divided into three regions: UV-A 400–320 nm; UV-B 320–290 nm; UV-C 290–100 nm.

cryptochrome may actually be a component of the central clock (Emery et al., 1998). In mammals, CRY protein may function as a negative regulator of the circadian clock (Shearman et al., 2000).

Possibly, the UV component of sunlight contributed to the selective pressure for the evolution of cryptochrome and that some of the early species of animals (metazoans) in the ocean would descend to greater depths to avoid daylight (Gehring & Rosbash, 2003). Direct measurements of radiation from the Sun and sky to depths in the ocean, where "midwater" fauna are found, lend support to this hypothesis. The vertical migrations of an animal community, which in the ocean is called a sonic-scattering layer, has been shown to compare favorably with the descent and ascent of a constant level of blue light (Boden & Kampa, 1967). The peak in the irradiance spectrum during midday at the depth where the fauna layer is located ranges between 470 and 480 nm. A peak in the blue regions of the spectrum at a greater depth would be expected, since wavelengths of blue light penetrate to a greater depth in water than do longer wavelengths, such as red light (e.g., 660 nm). The fauna in this "blue zone" seek to live within a relatively narrow range of intensity.[54]

Being able to differentiate between UV and blue light is important for survival. For example, the water flea, *Daphnia magna,* reacts differently to visible light in the range of 420–600 nm than to UV light (260–380 nm) (Storz & Paul, 1998). They will swim away from UV light (negative phototaxis), but toward (positive phototaxis) blue light. A vertical migration response of *Daphnia* to UV occurs under full spectrum solar radiation, which has been demonstrated by either allowing or blocking natural UV radiation (Leech & Williams, 2001). The actual range in depth of descent in lakes, however, can be influenced by predation, as well as by UV avoidance (Boeing et al., 2004).

As observed in a number of present day aquatic organisms by their avoidance of UV light and dependence upon blue light, and in terrestrial organisms having blue-light photoreceptors that perform many functions in circadian clock systems, coevolution appears inevitable. An extensive duplication of the *Cry* genes has occurred and with it a variety of modulations of function (Tauber et al., 2004). This could apply to other components of biological clocks and timing systems. Often it is difficult to distinguish between ancestral traits and derived traits, since they may diverge to perform different functions.

Ancestral Traits and Convergent Evolution

One often asked question is: *do present day biological clocks have a mechanism that is derived from a common ancestral mechanism?* The answer is very likely, no, not even for circadian clocks. Both the prokaryotic cyanobacteria clock and the eukaryotic clocks of fungi, plants, and animals perform a similar function as

---

[54] Light intensities for the animals in this scattering layer are about 10^8–10^9 quanta/cm^2/s at 474 nm and within the range required for many photochemical reactions (Boden & Kampa, 1967).

an oscillator, a conserved trait. However, the self-regulating feedback loop in cyanobacteria, unlike those of eukaryotes, does not require specific clock gene promoters (Xu et al., 2003) and it can keep time independent of *de novo* processes of transcription and translation (Tomita et al., 2005). When different structures, such as heterologous promoters vs. specific gene promoters, perform a similar function, they have evolved by a process called *convergent evolution*.[55] The same can be said when comparing the clocks of *Neurospora* and *Drosophila*. Their genes are different, but their functions in transcriptional/translational feedback loops are similar.

Some support for a common ancestral clock for plant, fungal, and metazoan proteins is the presence of a particular protein–protein interaction domain[56] found in molecular clock systems (cf. Young & Kay, 2001). Perhaps these domains could have evolved to help coordinate transcription and/or metabolic activities leading to development in a time-delayed manner to be in harmony with favorable environmental phases. There are a limited number of suggestions that certain clock proteins, such as BMAL1 in mammals and WC-1 in fungi, have a common evolutionary relationship (Lee et al., 2000). CRY proteins and their orthologs,[57] which are present in the circadian timing system of both plants and animals, represent the earliest evolved proteins that have shown extensive duplication (Tauber et al., 2004). Phylogenetic analyses of the photolyase-blue-light photoreceptor family indicate that the gene duplication of their ancestral proteins occurred a number of times before the divergence between eubacteria and eukaryotes (Kanai et al., 1997).

Take-Home Message

The rotations and revolutions of the Earth account for the natural synchronization of circadian and circannual rhythms, as well as for the changes in duration and timing of light and dark spans in photoperiodism. Two types or groups of clocks are paramount in the study of biological rhythms, each with a fascinating history of discovery. One group includes the mechanical, electrical, and atomic clocks of society. Its technology is based upon physical components, either natural (e.g., sand, water, stars, radiation) or manufactured (e.g., springs and gears). The other type of clock is biological, present in the cells of organisms and having a mechanism that is based upon feedback loops, genes, and cycling proteins. The evolutionary development of the biological clock can be traced back to the early forms of cyanobacteria.

---

[55] A more common example would be the difference between the wings of a bird and an insect. Both are used for flying, but their structures are entirely different.

[56] This refers to what is known as a PAS domain and is described in Chapter 5 on Models.

[57] An *ortholog* is any gene in two or more species that can be traced to a common ancestor, but which evolved in a different way.

References

Adriani A. (1963) *Hellenistic Art. Encyclopedia of World Art,* Vol. VII. New York: McGraw-Hill, pp. 283–392.

Ahmad M, Cashmore AR. (1993) *HY4* gene of *A. thaliana* encodes a protein with characteristics of a blue-light photoreceptor. *Nature* 366(6451): 162–166.

Barnett JE. (1998) *Time's Pendulum: The Quest to Capture Time from Sundials to Atomic Clocks.* New York: Plenum, 340 pp.

Beling I. (1929) Über das Zeitgedächtnis der Bienen. *Z Vergl Physiol* 9: 259–338.

Bennett J, Donahue M, Schneider N, Voit M. (1999) *The Cosmic Perspective.* Menlo Park, CA: Addison-Wesley Longman, Inc., 698 pp.

Blunt W. (1971) *The Compleat Naturalist. A Life of Linnaeus.* New York: The Viking Press, Inc., 256 pp.

Boden BP, Kampa EM. (1967) The influence of natural light on the vertical migrations of an animal community at sea. *Symp Zool Soc Lond* 19: 15–26.

Boeing WJ, Leech DM, Williamson CE, Cooke S, Torres L. (2004) Damaging UV radiation and invertebrate predation: conflicting selective pressures for zooplankton vertical distribution in the water column of low DOC lakes. *Oecologia* 138(4): 603–612.

Boorstin DJ. (1985) *The Discoverers.* New York: Random House, 745 pp.

Breasted JH. (1936) The beginnings of time-measurement and the origins of our calendar. In: *Time and Its Mysteries Series I.* New York: New York University Press, pp. 59–94.

Brown FA. (1954) Biological clocks and the fiddler crab. *Scientific Amer* 190(4): 34–37.

Cashmore AR. (2003) Cryptochromes: enabling plants and animals to determine circadian time. *Cell* 114(5): 537–543 (Review).

Colson FH. (1926) *The Week. An Essay on the Origin & Development of the Seven-Day Cycle.* Cambridge: Cambridge University Press, 126 pp.

Cornélissen G, Halberg F, Wendt HW, Bingham C, Sothern RB, Haus E, Kleitman E, Kleitman N, Revilla MA, Revilla M Jr, Breus TK, Pimenov K, Grigoriev AE, Mitish MD, Yatsyk GV, Syutkina EV. (1996) Resonance of about-weekly rhythm in human heart rate with solar activity change. *Biologia (Bratisl)* 51(6): 749–756.

Cutler WB. (1980) Lunar and menstrual phase locking. *Amer J Obstet Gynecol* 137(7): 834–839.

Cutler WB, Petri G, Krieger A, Huggins GR, Garcia CR, Lawley HJ. (1986) Human axillary secretions influence women's menstrual cycles: the role of donor extract from men. *Horm Behav* 20(4): 463–473.

De Candolle AP. (1832) *Physiologie Végétale,* Vol. 2. Paris: Bechet Jeune.

De Mairan J. (1729) Observation Botanique. *Histoire de l'Academie Royale des Sciences,* pp. 35–36.

De Solla Price DJ. (1964) Mechanical water clocks of the 14th century in Fez, Morocco. *Proc Tenth Intl Cong Hist of Science.* Paris: Hermann, pp. 599–602.

Diddams SA, Bergquist JC, Jefferts SR, Oates CW. (2004) Standards of time and frequency at the outset of the 21st century. *Science* 306(5700): 1318–1324.

Duhamel Du Monceau HL. (1759) *La Physique Des Arbres.* Seconde Partie. Paris: HL Guerin and LF Delatour.

Dvornyk V, Vinogradova O, Nevo E. (2003) Origin and evolution of circadian clock genes in prokaryotes. *Proc Natl Acad Sci USA* 100(5): 2495–2500.

Emery P, So WV, Kaneko M, Hall JC, Rosbash M. (1998) CRY, a *Drosophila* clock and light-regulated cryptochrome, is a major contributor to circadian rhythm resetting and photosensitivity. *Cell* 95(5): 669–679.

Forel A. (1910) *Das Sinnesleben der Insekten* (Semon M., trans.). München: Ernst Reinhardt Verlag.

Freifelder D. (1982) *Physical Biochemistry*, 2nd edn. New York: WH Freeman, 761 pp.

Gehring W, Rosbash M. (2003) The coevolution of blue-light photoreception and circadian rhythms. *J Mol Evol* 57 (Suppl 1): S286–289 (Review).

Halberg F, Halberg E, Barnum CP, Bittner JJ. (1959) Physiologic 24-hour periodicity in human beings and mice, the lighting regimen and daily routine. In: *Photoperiodism and Related Phenomena in Plants and Animals*, Whitrow RB, ed. Publication No. 55 of the Amer Assoc Adv Sci, Washington, DC, pp. 803–878.

Halberg F, Marques N, Cornélissen G, Bingham C, Sánchez de la Peña S, Halberg J, Marques M, Jinyi W, Halberg E. (1990) Circaseptan biologic time structure. *Acta Entomol Bohemoslov* 87: 1–29.

Haus E, Touitou Y. (1994) Principles of clinical chronobiology. In: *Biologic Rhythms in Clinical and Laboratory Medicine*. Touitou Y, Haus E, eds. Berlin: Springer-Verlag, pp. 6–34.

Hering DW. (1940) The time concept and time sense among cultured and uncultured peoples. In: *Time and Its Mysteries, Series II*. New York: New York University Press, pp. 3–39.

Hill DR. (1976) *On the Construction of Water-Clocks*. London: Turner & Devereux, 46 pp.

Iwasaki K, Thomas JH. (1997) Genetics in rhythm. *Trends in Genetics* 13(3): 111–115.

Johnson CH, Knight M, Trewavas A, Kondo T. (1998) A clockwork green: circadian programs in photosynthetic organisms. In: *Biological Rhythms and Photoperiodism in Plants*. Lumsden PJ, Millar AJ, eds. Oxford: BIOS Scientific, pp. 1–34.

Kanai S, Kikuno R, Toh H, Ryo H, Todo T. (1997) Molecular evolution of the photolyase-blue-light photoreceptor family. *J Mol Evol* 45(5): 535–548.

Kerner von Marilaun A. (1895) *The Natural History of Plants, their Forms, Growth, Reproduction, and Distribution*. New York: H. Holt, pp. 216–217.

Lathe R. (2004) Fast tidal cycling and the origin of life. *Icarus* 168: 18–22.

Lee K, Loros JJ, Dunlap JC. (2000) Interconnected feedback loops in the *Neurospora* circadian system. *Science* 289(5476): 107–110.

Leech DM, Williamson CE. (2001) *In situ* exposure to ultraviolet radiation alters the depth distribution of *Daphnia*. *Limnol Oceanogr* (Suppl): 416–420.

Major RH. (1954) *A History of Medicine*, Vol 1. Springfield, IL: Charles C. Thomas, p. 486.

Miller SL. (1953) A production of amino acids under possible primitive earth conditions. *Science* 117(3046): 528–529.

Miller SL, Urey HC. (1959) Organic compound synthesis on the primitive earth. *Science* 130(3370): 245–251.

Moore-Ede MC, Sulzman FM, Fuller CA. (1982) *The Clocks That Time Us. Physiology of the Circadian Timing System*. Cambridge, MA: Harvard University Press, 448 pp.

Mullis K, Faloona F, Scharf S, Saiki R, Horn G, Erlich H. (1986) Specific enzymatic amplification of DNA in vitro: the polymerase chain reaction. *Cold Spring Harbor Quant Biol* 51: 263–273.

Neugebauer O. (1957) *The Exact Sciences In Antiquity*, 2nd edn. Providence, RI: Brown University Press, 240 pp.

Nikaido SS, Johnson CH. (2000) Daily and circadian variation in survival from ultraviolet radiation in *Chlamydomonas reinhardtii*. *Photochem Photobiol* 71(6): 758–765.

North JD. (1975) Monasticism and the first mechanical clocks. In: *Study of time II: Proc 2nd Conf Intl Soc for the Study of Time*. Lake Yamanaka, Japan. Berlin: Springer-Verlag, pp. 381–398.

Ouyang Y, Andersson CR, Kondo T, Golden SS, Johnson CH. (1998) Resonating circadian clocks enhance fitness in cyanobacteria. *Proc Natl Acad Sci USA* 95(15): 8660–8664.

Pittendrigh CS. (1954) On temperature independence in the clock system controlling emergence time in *Drosophila. Proc Natl Acad Sci USA* 40: 1018–1029.

Pittendrigh CS, Bruce VG. (1957) V. An oscillator model for biological clocks. In: *Rhythmic and Synthetic Processes in Growth.* Rudnick D, ed. Princeton: Princeton University Press, pp. 75–109.

Pittendrigh CS. (1960) Circadian rhythms and the circadian organization of living systems. *Cold Spring Harbor Symp Quant Biol* 25: 159–182.

Pittendrigh CS. (1981) Circadian Systems: General Perspective. In: *Handbook of Behavioral Biology. Vol 4: Biological Rhythms.* Aschoff J, ed. New York: Plenum Press, pp. 57–80.

Quill H. (1966) *John Harrison. The Man Who Found Longitude.* London: John Baker Publication Ltd.

Reinberg A, Smolensky M. (1983) *Biological Rhythms and Medicine. Cellular, Metabolic, Physiopathologic, and Pharmacologic Aspects.* New York: Springer-Verlag, 305 pp.

Renner M. (1955) Ein transozeanversuch zum zeitsinn der honigbiene. *Naturwissenschaften* 42: 540–541.

Schopf JW. (1993) Microfossils of the Early Archean Apex chert: new evidence of the antiquity of life. *Science* 260: 640–646.

Schweiger HG, Berger S, Kretschmer H, Morler H, Halberg E, Sothern RB, Halberg F. (1986) Evidence for a circaseptan and a circasemiseptan growth response to light/dark cycle shifts in nucleated and enucleated cells of *Acetabularia*, respectively. *Proc Natl Acad Sci USA* 83: 8619–8623.

Shearman LP, Sriram S, Weaver DR, Maywood ES, Chaves I, Zheng B, Kume K, Lee CC, van der Horst GT, Hastings MH, Reppert SM. (2000) Interacting molecular loops in the mammalian circadian clock. *Science* 288(5468): 1013–1019.

Sothern RB. (1995) Time of day versus internal circadian timing references. *J Infus Chemother* 5(1): 24–30.

Spruyt E, Verbelen J-P, De Greef JA. (1987) Expression of circaseptan and circannual rhythmicity in the imbibition of dry stored bean seeds. *Plant Physiol* 84: 707–710.

Storz UC, Paul RJ. (1998) Phototaxis in water fleas (*Daphnia magna*) is differently influenced by visible and UV light. *J Comp Physiol A* 183: 709–717.

Talbott JH. (1970) "Santorio Santorio (1561–1636)," *A Biographical History of Medicine: Excerpts and Essays on the Men and Their Work.* New York: Grune & Stratton, pp. 87–89.

Tauber E, Last KS, Olive PJ, Kyriacou CP. (2004) Clock gene evolution and functional divergence. *J Biol Rhythms* 19(5): 445–458.

Thorndike L. (1941) Invention of the mechanical clock about 1271 A.D. *Speculum* 16: 242–243.

Tomita J, Nakajima M, Kondo T, Iwasaki H. (2005) No transcription-translation feedback in circadian rhythm of KaiC phosphorylation. *Science* 307(5707): 251–254.

Tøndering C. (2000) Frequently asked questions about calendars. http://www.tondering.dk/claus/calendar.html, 53 pp.

Van Helden A. (1999) Cathedrals as astronomical instruments. *Science* 286: 2279–2280.

Ward FAB. (1961) How timekeeping mechanisms became accurate. *The Chartered Mechanical Engineer*, pp. 604–609 (also p. 615).

Whitrow GJ. (1988) *Time in History: The Evolution of Our General Awareness of Time and Temporal Perspective.* Oxford: Oxford University Press, 217 pp.

Woelfle MA, Ouyang Y, Phanvijhitsiri K, Johnson CH. (2004) The adaptive value of circadian clocks: an experimental assessment in cyanobacteria. *Curr Biol* 14(16): 1481–1486.

Xu Y, Mori T, Johnson CH. (2003) Cyanobacterial circadian clockwork: roles of KaiA, KaiB and the kaiBC promoter in regulating KaiC. *EMBO J* 22(9): 2117–2126.

Young MW, Kay SA. (2001) Time zones: a comparative genetics of circadian clocks. *Rev Genet* 2(9): 702–715 (Review).

Zerubavel E. (1985) *The Seven Day Circle: The History and Meaning of the Week.* New York: Free Press, 206 pp.

Zinn JG. (1759) [On the sleep of plants.] [German]. *Hamburgisches Magazin* 22: 40–50.

4
Photoperiodism

> *"Time is very bankrupt and owes more than he's worth to season.*
> *Nay, he's a thief too: have you not heard men say,*
> *that Time comes stealing on by night and day?"*
> —William Shakespeare (1564–1616), British dramatist, author

Introduction

In rural settings of decades past, it was common to see and experience the wonders of biological transformations associated with the seasons of the year. As the hours of daylight changed with the seasons, so too did life in the countryside. Animals gave birth, hibernation ended or started, some species left the region and others arrived, flowers, leaves, and fruit appeared and disappeared—almost everything was in a constant change, with the same sequence of events recurring again the following year (Figure 4.1). Today, the trends of urbanization and the utilization of electrical power and rapid transportation shield most of us from witnessing much of the seasonal biological diversity found in nature. Fresh flowers, fruits, and meats that were once associated with a particular season can now be found in the market throughout the year (Figure 4.2). Some are imported from regions where the season is different, while others are produced locally through the technical manipulations of a light/dark environment within special structures, such as greenhouses (Figure 4.3) or controlled environment chambers.

Daylight and Seasons

Even though we need not depend exclusively upon the seasons of our own environment for the biological products of a season, most of us are aware of seasonal differences in the hours of daylight or darkness, especially the difference between summer and winter (Figure 4.4). The further a location is from the equator (increasing latitude), the greater the extent of change between night and day, even to the point of 24 h of daylight in the summer (the midnight Sun) and 24 h of

FIGURE 4.1. Seasonal changes in development and/or activity that take place from spring (*left*) to fall (*right*) in a rural setting.

darkness in mid-winter above the Arctic Circle (Figure 4.5). Included in these 24-h spans are the durations of less-intense visible light that occur before and after the Sun is at the horizon (dawn and dusk), a time known as "civil twilight." Thus, there are still several hours of dim light in the arctic region during the middle of the day in winter. Subtle changes in light intensity are also found at the equator throughout the year (e.g., changes in the weather). Changes in the lengths of day and night are precise and predictable, unlike that of environmental temperature or precipitation, which can sometimes deviate greatly from the norm for any given time of the year.

FIGURE 4.2. Fresh flowers, fruits, and meats once associated with a particular season are now available year-round due to imports and manipulation of the photoperiod.

FIGURE 4.3. Greenhouse with lights at night in use to provide photoperiods different from the natural environment. Photos by R. Sothern.

Native plants and animals have become adapted to the seasons of their environment by responding to the changes in the lengths of daylight and night in preparation for the climatic changes that are to come (Figure 4.6). These responses, which are associated with the seasons of the year, involve a physiological process known as *photoperiodism* (photo = light, period = duration, ism = process). *Photoperiodism* is a broad topic that is one of the basic principles of biology. The objectives of this chapter are to help you obtain a better knowledge of photoperiodism,

FIGURE 4.4. Two positions of the Earth during its orbit around the Sun result in different lengths of daylight during different seasons.

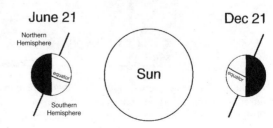

110   4. Photoperiodism

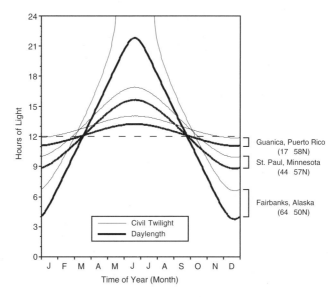

FIGURE 4.5. Amount of light throughout the year (hours of daylength and civil twilight) at three different latitudes.

FIGURE 4.6. Snowshoe hares *(Lepus americanus)* respond to photoperiod by having darker fur color in summer (top) and lighter color in winter (bottom) (photographs courtesy of Professor Richard Phillips, University of Minnesota).

develop an appreciation for its role in nature, and to more fully understand its complex status within the temporal organization of biological rhythms.

Photoperiodism: The Process

In the context of biological rhythms, photoperiodism can be defined as a *response of an organism to the timing and duration of light and dark*. Response to photoperiod often centers on reproduction and survival, such as the production of flowers, the migration of birds, or any one of a number of variables (Table 4.1). Being able to *adapt* and *survive* by *predicting* what is to come, be it either a favorable or unfavorable season, depends upon the organism being able to sense a reliable environmental cue. For this purpose, the recurring seasonal changes in the lengths of daylight and darkness are dependable and precise. For example, a given species of plants will produce flowers if the length of the light span is 15.5 h or

TABLE 4.1. Examples of various organisms and their response to changes in photoperiod. Responses listed do not necessarily apply to all members of the taxa (cf. Hendricks, 1956; Hillman, 1979; Takeda & Skopic, 1997).

Organism	Response	Comments
Algae	Changes in development and morphology	Species of Chlorophyta, Phaeophyta, and Rhodophyta
Bryophytes	Sexual reproduction and vegetative growth	Liverworts (*Marchantia* sp. and *Lunularia* sp.)
Seedless vascular plants	Production of reproductive structures (*sporocarps*)	Water fern (*Salvinia* sp.)
Seed plants (Angiosperms)	Initiation of flowers and tuber formation	Soybean (*Glycine max*) and Jerusalem artichoke (*Helianthus tuberosus*)
Seed plants (Gymnosperms)	Formation of terminal buds (budset)	Norway spruce (*Picea abies*)
Arthropods (Insects)	Diapause (winter resting phase); parthenogenetic production of larvae	Many species; parthenogenetic females in some aphids (e.g., *Megoura* sp.)
Arthropods (Mites)	Diapause	Fruit tree red spider mite (*Metatetranchus ulmi*)
Fishes	Sexual activity; spermatogenesis	Various species (e.g., *Gasterosteus aculeatus*)
Reptiles	Gonadal development, mating, egg-laying, hibernation	Fringe-toed desert lizard (*Uma notata*)
Birds	Molting; mating; laying of eggs; testicle development; migration	Relationship to rhythms has been extensively studied in birds
Mammals	(a) Testicle development (b) Sexual behavior and mating (c) Color change (d) Antler development (e) Production of winter coat	(a) Golden Hamster (*Mesocricetus aureus*) (b) Sheep (*Ovis aries*) (c) Snowshoe Hare (*Lepus americanus*) (d) Sika Deer (*Cervus nippon*) (e) Horse (*Equus equus*)

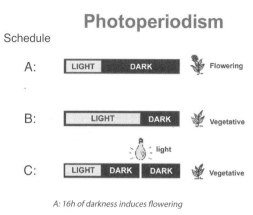

FIGURE 4.7. Importance of timing in photoperiodism illustrated by the effects of three LD schedules on the flowering response of a hypothetical short-day plant (SDP). Note that the total hours of L (8 h) and D (16 h) are similar in schedules A (8 h) and C (7.75 h + 15 min), but the timing of L is different. The responses (vegetative or flowering) are illustrated to show that flowering during A, but not during conditions B and C.

less, but not if it is over 15.5 h. Likewise, for gonadal enlargement in the spring, a certain species of birds may require a light span of 8.5 h or more, but not less. Regardless of the type of response, timing is a *critical feature* of photoperiodism, especially how light and darkness succeed each other (Hillman, 1962).

In nature, the matter of timing is quite simple and warrants little, if any, elaboration. Each daily 24-h cycle consists of a single light span followed by a single dark span, both of which change in duration throughout the year. When one span lengthens, the other span shortens (Figures 4.4 and 4.5). The changes in the spectrum of light (red vs. far-red) during these transitions from one span to the other (light to dark = dusk and dark to light = dawn) may also allow organisms to better discriminate the lengths of day and night due to photoreceptors (e.g., the pigments known as phytochromes). In the laboratory setting, however, both the number and the duration of each span can be manipulated. It has been through the use of various manipulations, such as interrupting the dark span with one or more brief pulses of light, that the important role of timing in photoperiodism has become evident. As illustrated in Figure 4.7, a short-day plant (SDP) will remain vegetative (no flowers produced) if the dark span is interrupted by light, even though all plants receive a total of 8 h of light and 16 h of darkness. What makes the difference? The answer is not the total duration of light or darkness, but the duration and timing of individual light and dark spans (e.g., the single dark span of 16 h was divided into two dark spans of 7.75 h and 8 h separated by a 15-min span of light).[1] Generally, in cases such as the one in Figure 4.7 where short days and long nights are required for floral induction of a SDP, the duration of the uninterrupted dark span is very important. Even in long-day plants (LDPs), which flower in response to long days and short nights, the dark span continues to be important, but since the photoperiodic process

---

[1] One may also question whether moonlight can have a nighttime interrupting effect on plants. Maximum moonlight can occur at different times of the dark span, but is generally not great enough to exert an influence on plant photoperiodism, although a slight response has been reported for a few plants (Von Gaertner & Braunroth, 1935; Kadman-Zahavi & Peiper, 1987).

may be more complicated for LDPs, the dark span may not be as important as it is in SDPs (cf. Thomas & Vince-Prue, 1997).

Beginning after December 22 in the northern hemisphere, the duration of the light span increases and the length of the dark span decreases until the latter part of June, when the sequence reverses, so that the duration of the light span progressively decreases as the duration of the dark span increases (Figure 4.4).[2] Somewhere during this sequence of change, the correct combination of light and dark is reached to induce a particular response in a specific group of organisms (e.g., sexual reproduction in some birds).

Response Types

Photoperiodic requirements of organisms for the induction of a response are referred to as *"response types."* In a classic paper that reported the results of studies on the flowering of plants, two scientists from the U.S. Department of Agriculture introduced the term photoperiodism, as well as the two categories known as *short-day* and *long-day* plants (Garner & Allard, 1920). Of special interest was how a soybean *(Glycine max)* cultivar (variety) known as Biloxi, and a tobacco plant *(Nicotiana tabacum)* called Maryland Mammoth responded to the duration of light and darkness. In the fields near Washington, DC, Biloxi soybeans would flower in September, regardless of whether the seeds were sown in April or July (see Table 4.2). Their observations of Maryland Mammoth were equally astonishing, since these plants tended to remain vegetative when grown in the field during the summer, but flowered when kept in a greenhouse during the winter. However, the plants remained vegetative during winter if the duration of the light span was lengthened with electric lamps. Based upon their requirements for floral induction, both Biloxi soybean and Maryland Mammoth tobacco were designated as short-day plants. Plants such as radish *(Raphenus sativus)*, and hibiscus *(Hibiscus moscheutos* L.),[3] which require a longer light span to flower, were called long-day plants.

TABLE 4.2. Effects of date of sowing and emergence of seedlings on the date when the first blossoms of Biloxi soybeans *(Glycine max)* appeared in a field in Arlington, Virginia, USA during 1919 (Garner & Allard, 1920).

Date Sown	Date emerged	Date first blossom
April 09	May 02	September 04
May 27	June 02	September 04
June 26	July 03	September 18
July 29	August 02	September 29

---

[2] The Sun is farthest north from the equator (summer solstice) on about June 21–22 and farthest south of the equator (winter solstice) on about December 21–22, resulting in the longest and shortest days, respectively.

[3] *Hibiscus moscheutos* L., found in North America and called Rose-mallow, was used by Garner & Allard (1920). *H. syriacus,* which is known as Rose of Sharon and native to E. Asia, is often cited as a "typical" LDP by various authors.

TABLE 4.3. Three basic photoperiodic response types and examples of plants induced to flower under these photoperiodic conditions.

Type	Example
Short-day	Soybean (*Glycine max* L. merr. cv. Biloxi), morning glory (*Ipomoea hederacea*)
Long-day	Ryegrass (*Lolium temulentum* L. Darnel), barley (*Hordeum vulgare*)
Daylength indifferent (day-neutral)	Onion (*Allium cepa*), tomato (*Solanum lycopersicum*)

Critical Daylength

Distinguishing between a short-day and a long-day response is not based upon the absolute length of the light span, but upon a value called the *critical daylength*. For example, a SDP could have a critical daylength of 14 h, while a LDP could have a critical daylength of 9 h. The clue to this seeming paradox lies upon what occurs before and after the critical daylength. Responses that occur when the light span or daylength is *shorter* than the critical daylength are called *short-day*, while responses that occur when the light span is *longer* than the critical daylength are called *long-day*. Because of the importance of the dark span, reference sometimes is made to the critical *nightlength*, rather than the critical daylength.

A third response type or category, called *daylength-indifferent* (DI) or *day-neutral*, is used when there appears to be no relationship between development and photoperiodism. A truly day-neutral plant, which should flower regardless of daylength, is not as common as the short- and long-day plants (cf. Salisbury & Ross, 1992). Exceptions abound in all of these categories, since even some SDPs may not flower if the light span is too short. A similar behavioral response to a short light span relative to reproduction can also be observed in insects and other animals. Collectively, short-day, long-day, and daylength-neutral can be viewed as the three basic photoperiodic response types or categories (Table 4.3).

One should be aware of the fact that a process or activity that has been classified within one of the basic categories is not necessarily limited to that category or photoperiod. The requirement could be modified or changed by conditions other than photoperiod, such as the age of the organism or a change in temperature. Sometimes a given photoperiod merely hastens the process.[4] In fact, an absolute requirement or response, referred to as *qualitative*, may be less common than the *quantitative* (how much) response. In addition to qualitative and quantitative responses, there are other types of responses such as *dual-daylength* or

---

[4] Because of the duration of time required to maintain organisms and obtain results, data are often collected when the organism is in an early stage of development (e.g., dissecting shoot apices to score for the induction of flowering or using young animals).

TABLE 4.4. Additional response types relative to flowering (A) or Diapause (B).[a]
(A) Flowering Response (cf. Salisbury, 1963; Lang, 1965)

Type	Plant
Qualitative response types	
Short-day	Soybean (*Glycine max* cv. Biloxi)
Long-day	Black-eyed Susan (*Rudbeckia hirta*)
Quantitative response types	
Short-day	Redroot Pigweed (*Amaranthus retroflexus*)
Long-day	Arabidopsis (*Arabidopsis thaliana*)
Dual daylength response types	
Long-short-day	Aloe (*Aloe bulbilifera*)
Short-long-day	White clover (*Trifolium repens*)
Other response types	
Intermediate-day	Sugarcane (*Saccharum spontaneum*)
Ambiphotoperiodic (long or short, but flowering inhibited by intermediate)	Tarweed (*Media elegans*)

(B) Diapause Response (cf. Lofts, 1970)

Type	Insect	Comments
Short-day	Silkworm (*Bombyx mori*)	Normal development under SD, but diapause under LD
Long-day	Colorado beetle (*Leptinotarsa decemlineata*)	Normal development under LD, but diapause under SD
Short or long-day	European corn borer (*Ostrinia nubilalis*)	Diapause in midrange (ca. 8–16 h)
Intermediate-day	Peach-fruit moth (*Carposina niponensis*)	Diapause induced by either SD or LD

[a] Not all members of the species are necessarily represented by the examples in these tables. Some species have cultivars or strains in more than one category and the response may be affected by age, stage of development, environment, and how the response is scored.

intermediate-day, as illustrated in the case of the flowering response in plants and diapause in insects[5] (Table 4.4).

Diversity of Responses

In the previous section, Garner and Allard (1920) were acknowledged for introducing the term "photoperiodism" and providing quantitative data (Table 4.2) that were instrumental in establishing photoperiodism as a biological principle. It would be amiss, however, at least from a historical perspective, not to mention the contributions of four other individuals to the early work on photoperiodism.

---

[5] Both flowering and diapause relate to reproduction, with diapause being an adaptive process by which reproduction is delayed in order to prevent high mortality that would occur under adverse climatic conditions, such as low winter temperatures.

Early Studies

In France, Julien Tournois (1914) while studying flowering in hops *(Houblon japonais)* and hemp *(Cannabis sativa)* noticed exceptionally early flowering when these two species were exposed to short light spans and long nights.[6] Georg Klebs (1918) from Germany also became aware of the importance of daylength for flowering to occur in *Sempervivum funkii* plants after they had been exposed to the required light span.[7] Soon after Garner and Allard's classical paper on photoperiodism in plants had been published, results from experiments with insects (sexual forms, bearing wings, and laying eggs) reported by Marcovitch in 1923, and with birds (gonadal growth and migration of the snow bird, *Junco hyemalis*) reported by Rowan in 1926,[8] confirmed the role of photoperiodism in animals. Subsequently, a large number of photoperiodic responses have been recorded for species of algae, bryophytes, angiosperms, gymnosperms, insects, fish, reptiles, birds, and mammals (Table 4.1).

Latitude

Often the role of photoperiodism in the life of a given species or cultivar is not appreciated or realized until an organism is moved to a different latitude. Historically, these differences were noted when groups of people moved north or south, bringing with them their domestic plants and animals. Reproductive cycles of both plants and animals can be affected, occurring either too early or too late for the optimum chance of survival or normal development. To this end, a number of plant and animal breeding programs have focused on producing and selecting organisms that are better suited for different latitudes. In the case of soybeans, the species *Glycine max* is currently grown and harvested from an area that extends from the tropical regions of Central and South America to Scandinavia. However, specific cultivars have been selected from breeding programs for specific latitudes and regions. The soybean cultivar Biloxi, which played an important role in the history of photoperiodism, would be killed by cold temperatures before it could produce flowers in Minnesota (45°N) or Sweden (65°N). Therefore, a cultivar

---

[6] As stated by Tournois (1914), "floraisons progénétiques apparaissent sur de jeunes plantes de Houblon japonais ou de Chanvre lorsque, à partir de la germination, elles sont soumises à un éclairement quotidien de très courte durée." ("Early flowering appeared in young plants of hops (*Houblon japonais*) or hemp (*Cannabis sativa*) when they were subjected to a very short daily irradiation from germination onwards.")

[7] "Beide Arten, die unter den Bedingungen der freien Natur eine ganz bestimmte eng begrenzte Blütezeit haben, lassen sich auf Grund der Kenntnisse der äusseren Blütenbedingungen zu jeder Zeit des Jahres zur Entwicklung der Infloreszenz bringen." ("Both species (*S. albidum and S. funkii*), which under natural conditions have a specific, precisely limited flowering time, can be induced to flower at any time of the year due to the information which we have gained on the conditions for natural flowering.") Klebs also conducted experiments of the effects of light quality (blue, red, far-red) on flowering.

[8] Garner and Allard (1920) noted that it occurred to them that the migration of birds may be in response to the daylength, rather than their instinct or volition.

TABLE 4.5. Various characteristics of plant (*A*) and insect (*B*) development or behavior that may be under photoperiodic control in certain species or types.

(A) Plant (cf. Salisbury & Ross, 1992)	
Germination of seeds	Vegetative reproduction (plantlets)
Shoot elongation	Floral induction
Branching	Floral development
Tillering (sprouting)	Expression of sex
Formation of roots	Seed development
Production of bulbs	Dormancy
Formation of tubers	Winter hardiness

(B) Insect (Masaki, 1978, 1984)	
Egg stage of development	Migratory flight
Larval instars	Mating
Sex ratio of progeny	Oviposition
Diapause	Seasonal polyphenism

known as "Evans" is commonly grown in Minnesota, while another, "Fiskeby," has been adapted for Swedish growing conditions (personal communication[9]).

Various stages of development in plants and animals can be affected by photoperiodism (Table 4.5), as well as the number of inductive cycles necessary to evoke a response. In the case of floral induction, there are numerous examples of SDP and LDP that require only a single-inductive cycle, although there may be more that require several cycles (Table 4.6).

Light and Photoreceptive Regions

In most instances, the *photoresponsive* region of an organism is located some distance away from the *photoreceptive* region. For example, flowers arise from a photoresponsive region known as the *shoot* meristem,[10] but the photoreceptors are in the *leaf*. Some SDPs maintained under long-days will produce flowers when leaves from SDPs maintained under short-days have been grafted onto them. In the case of the purple common perilla *(Perilla crispa)*, a SDP, even excised leaves from long-day grown plants will induce flowering if they are first placed under short-days and then grafted onto plants kept under long-days (Zeevaart, 1958; cf. Zeevaart, 1969). Successful grafts and the subsequent transmission of the

---

[9] Information provided by Professor James H. Orf, Department of Agronomy & Plant Genetics, University of Minnesota.

[10] Meristems are tissues located in specific regions of the plant and are composed of undifferentiated cells that can give rise to differentiation in cells (e.g., parenchyma), which comprise a tissue (e.g., epidermis). The term "shoot" refers to a structure that is comprised of a young stem with leaves. An apical meristem is located at the tip of a shoot or root. A bud is actually a young, undeveloped shoot. Flowers are determinate, reproductive short shoots, anatomically having leaf-like structures plus stamens or carpels (imperfect flowers) or both stamens and carpels (perfect flowers).

TABLE 4.6. Examples of SDP and LDP species induced to flower in response to one or several photoperiodic inductive cycles.[a]

Only a single cycle required	Several cycles required
Cocklebur (SDP) (*Xanthium strumarium*)	Soybean (SDP) (*Glycine max*)
Darnel ryegrass (LDP) (*Lolium temulentum*)	Black henbane strain (LDP) (*Hyoscyamus niger*)

[a]Not all members of the species are necessarily represented by the example. Some species have cultivars or strains in more than one category and the response may be affected by age, stage of development, environment, and how or when flowering is scored.

stimulus to induce flowering have been achieved between species from different families. Transmission of a response to receptor plants from grafted donor leaves has also been shown for the production of tubers of the Jerusalem artichoke *(Helianthus tuberosus)* (Hamner & Long, 1939).

Extraretinal Photoreceptors

It is easy to assume that the eyes are the sole photoreceptors in animals, since they convey information about environmental light conditions to specific regions of the brain. However, the perception of light cycles is usually coupled to a complex system. For example, circadian rhythms in some blinded animals can remain synchronized to an environmental light–dark cycle and free-run in constant darkness. Results from such studies provided one of the early clues to scientists that *extraretinal photoreceptors* in the brain, rather than just the eyes, could participate in photoperiodic timing.

While the circadian rhythm in perching activity of blinded house sparrows *(Passer domesticus)* free-ran in continuous darkness with periods longer than 24 h, a photoperiod of 12 h of light with intensity as low as 0.1 lux entrained the rhythm to 24 h in 50% of blinded house sparrows (and in all sighted birds) (Menaker, 1968a). If the light level was reduced even further to a sub-threshold intensity, the activity rhythm in blinded birds again free-ran with a longer cycle. This rhythm could again become synchronized when feathers from the top of the head were plucked in order to increase the amount of light penetrating the skull to the brain (Menaker et al., 1970), while injection of India ink under the head skin prevented sufficient light to penetrate the brain and the rhythm again free-ran (Menaker, 1968b).

Extraretinal photoreception has also been demonstrated by normal testicular recrudescence (renewed growth) in blinded sparrows exposed to artificial long days (Menaker & Keatts, 1968; Underwood & Menaker, 1970). When sighted birds that had their head feathers plucked and India ink injected under the skin of their heads in order to prevent light from penetrating the brain were exposed to the same lighting regimen, there was very little increase in their testis size, indicating that the eyes do not participate in those photoperiodic aspects of light detection that influence gonadal changes associated with reproduction (Menaker et al., 1970). In addition, extraocular, nonpineal brain photoreceptors have been found to be involved in photoentrainment of amphibians, fish, and reptiles whether or not both eyes and/or the pineal gland[11] have been removed (cf. Foster et al., 1993).

Photoperiodic photoreception in insects and mites represents still another complex situation, due, at least in part, to an extensive phylogeny (species history) that is represented by a large number of species and by various stages of development and metamorphosis. Photoreceptors of these arthropods are probably located in the brain, but in one way or another could involve the compound eyes.[12] In mammals, the eyes appear to be involved, but probably not solely, via the normal pathway of vision, since in addition to the image-forming system that is associated with vision, the mammalian eye has non-image-forming photoreceptors that perceive light. If the light response for photoperiodism is the same or similar to the light response in circadian timing, then results from recent studies further suggest that neither the rods nor the cones at the back of the eye are needed for photoperiodic reception (cf. Freedman et al., 1999).

Pigments

A generalized scheme of what may take place during the process of photoperiodism is illustrated in Figure 4.8. The visible light that is effective in most photoperiodic responses lies within a spectrum that spans from violet to far-red. Certain wavelengths of radiant energy that reach an organism are preferentially absorbed by receptor molecules known as *pigments* (Figure 4.9). Great diversity exists in the molecular structure of pigments, as well as in their associations with proteins, but each has a light-absorbing portion called a *chromophore* (Gr. *chroma* = color, *phoros* = carrying). *Phytochromes*, which constitute a major class of pigments that function in the photoperiodic response of plants (Figure 4.10), have a chromophore that consists of four rings (*tetrapyrrole*).

In higher animals the major visual pigment is *rhodopsin*, a molecule that contains a light absorbing part of a molecule (chromophore) called 11-*cis* retinal and a protein known as opsin. The molecular structure of 11-*cis*-retinal is similar to the molecular structure of vitamin A, which in the photochemistry of vision, is converted to 11-*cis*-retinal. In fact, the stereoisomers of the retinal group of molecules, such as vitamin A, 11-*cis*-retinal, and all-*trans*-retinal, are interconvertible by photochemical reactions (Clayton, 1971) and are functional in the vision of animals.[13] However, these photoreceptive pigments are not necessarily limited to eyes. For example, 11-*cis*-retinal has been found in the pineal of rainbow trout

---

[11] A structure in the brain of animals which, in certain species, may function as a photoreceptor.

[12] In addition to a pair of large compound eyes, adult insects usually have three simple eyes (ocelli), which, at least in a species of crickets, do not perceive the light/dark cycles that entrain certain rhythms (cf. Sokolove & Loher, 1975).

[13] The 11-*cis* designation of retinal refers to carbon number 11 in the chain of carbon atoms that comprise the molecule, as well as to the form (*cis*) that this chromophore displays at the location of carbon number 11.

120   4. Photoperiodism

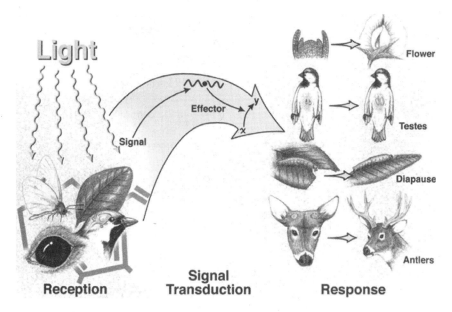

FIGURE 4.8. Diagram illustrating photoperiodic photoreceptive regions (e.g., leaf, brain, eye, and head), signal transduction (sequence from light signal (hυ) to timer (oscillator) to chemical signal ("inactive" X producing active "Y") to an intercellular messenger (e.g., hormone), which activates the responsive or target regions and responsive regions (apical shoot meristem, testis, larva, and head). Note changes in morphology, such as floral induction, testis development, diapause, and antler production.

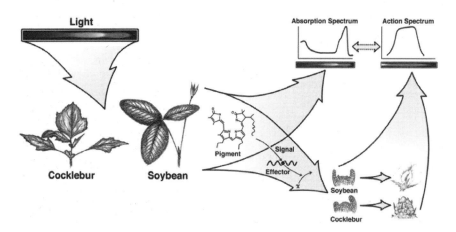

FIGURE 4.9. Diagram illustrating the visible light spectrum and the leaves of cocklebur and soybean that contain phytochrome (a photoreceptive pigment). Certain wavelengths of radiant energy are preferentially absorbed by receptor molecules (pigments) to induce flowering (signal-transduction effector as in Figure 4.8). The absorption spectrum of phytochrome and the action spectrum for regulating photoperiodic floral induction of these two SDPs both

FIGURE 4.10. Transformation of red (Pr) and far-red (Pfr) forms of phytochrome (P) illustrates that phytochrome is photo-reversible.

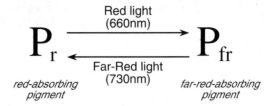

(Salmo gairdneri) (Tabata et al., 1985). Results from studies conducted with an eyeless species of predacious mites (*Amblyseius potentillae*), suggest that photoperiodic photoreception can occur extraretinally, and that *carotenoids* (colored pigments) and vitamin A appear to be essential for photoperiodic induction of diapause (Van Zon et al., 1981; Veerman et al., 1983).

Cryptochromes

Cryptochromes represent another group of pigments that also function in biological timing (discussed in Chapter 5 on Models). These pigments, which are found in plants, are also found in the eyes of mammals and have a role in circadian photoreception (Thresher et al., 1998). Results from research in which a number of locations on the human body have been exposed to light (e.g., the popliteal region directly behind the knee joint) suggest the possible involvement of heme or other pigments associated with various tissues (e.g., blood) in extraretinal phototransduction in humans (Campbell & Murphy, 1998). Examples of pigments and their possible or actual role in photoperiodic responses are listed in Table 4.7. Mechanisms of photoentrainment, however, very likely involve multiple ocular photopigments and reactions. Included here would be *melanopsin*, an essential component in the phototransduction cascade in the retinal photoentrainment pathway that has been identified in retinal ganglia and the pineal and deep brain of higher animals (Foster & Hankins, 2002; Rollag et al., 2003). For additional discussion of melanopsin, see Chapter 2 on Rhythm Features and Chapter 11 on Clinical Medicine.

Spectra

Much of our knowledge of the mechanisms of how light of specific wavelengths regulates photoperiodism did not start with the identification of pigments and their *absorption spectra*. Rather, predictions and subsequent identification of active pigments were often based upon the results obtained from an *action spectrum* (Figure 4.9), after which comparisons were made between the action spectrum

---

show peaks in the red region (the action spectrum is limited to the red region, which therefore has been expanded in the illustration). *Note*: the spectra for the Pr and Pfr forms of phytochrome have slightly different absorption peaks (666 nm vs. 730 nm), but have been combined in this illustration, as has the action spectra for cocklebur and soybean. For the actual absorption spectra of Pr and Pfr, see Vierstra & Quail, 1983. For the action spectrum of photoperiodic control of floral induction of cocklebur and soybean, see Parker et al., 1946, 1949.

TABLE 4.7. Pigments[a] that function in photoperiodism and/or circadian rhythms (phy = phytochrome; cry = cryptochrome; LDP = long-day plants; SDP = short-day plant; L = light; D = dark).

Pigment	Comments
Phytochromes	Red, far-red, photoreversible group of chromoproteins found in plants and designated as phyA, phyB, phyC, phyD, and phyE (phyto = plant; chrome = colored)
PhyA	Functions under low intensity red light and in sensing daylength in LDP; may interact with phyC in SDP; may also interact with cry1
PhyB	Functions as primary high-intensity red light receptor in circadian systems; influences sensitivity to photoperiod for flowering and tuberization under noninductive conditions in LDP and SDP
PhyC	May interact with phyA in SDP and be involved in sensitivity to L during D span
Cryptochromes	Flavin-type blue/UV-A absorbing pigments found in both plants and animals and designated as cry1 and cry2. The name, cryptochrome, is related to its association with nonflowering plants known as cryptogams
Cry1	May mediate signal of high-intensity blue light for control of circadian period; may interact with phyA to transmit low-fluence blue light to timing mechanism
Cry2	May modulate circadian responses (e.g., in mice)
Flavins	Group of yellow pigments; see cryptochromes
Carotenoids	Large group of fat-soluble pigments, which include carotenes and in some instances (e.g., eyeless predacious mites) may be involved in photoperiodic induction
Rhodopsin	Found in retinal rods of animals and could in some form be involved in photoperiodic photoreception of certain organisms
Vitamin A	Functions in vision, along with 11-*cis*-retinal
11-*cis*-retinal	Functions in vision
Melanopsin	Essential component in the phototransduction cascade in the retinal photoentrainment pathway of higher animals
Pterins	Light-absorbing derivative of pteridine that often functions in insects, fishes, and birds. Possibly associated with the role of Cry1 in blue-light responses in plants

[a]For more details, see Veerman et al.,1983; Lumsden, 1991; Salisbury & Ross, 1992; Jackson & Thomas, 1997; Ahmad et al., 1998; Lucas & Foster, 1999; Somers et al., 1998; Thresher et al., 1998; Taiz & Zeiger, 2002; Foster & Hankins, 2002.

and the absorption spectrum. For example, the inhibition of flowering in SDPs by a brief exposure to light during the long night span has an action spectrum that indicates the red region, and in turn, compares favorably with the absorption spectrum of phytochrome.

The series of events following the absorption of light by plants are not completely understood, but they do involve signal transduction, in which messengers are formed and translocated to target sites, where they initiate the response, whether it is the induction of flowering or the migration of animals. Depending upon the organism, the stage of development and the environment, hormones, inhibitors, neurosecretory products, and catalysts are likely candidates in signal transduction, but none have been exclusively identified. In the case of mammals, a specific retinohypothalamic tract provides the pathway by which light reaches the suprachiasmatic nuclei of the brain (Foster, 1998).

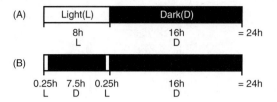

FIGURE 4.11. An example of an 8-h photoperiod (A) and a skeleton 8 h photoperiod (B) created by using two 15-min spans of light.

Rhythmic Association

We preface this section with the following quote from the mid 1970s: "*There is no more confusing literature than that on the relationship between circadian rhythmicity and photoperiodism.*" (Hillman, 1976)

Decades later, this statement still remains applicable, although much has been learned about this important relationship. In some cases, a "dark-period hourglass" model could easily account for the means by which the time is measured in photoperiodism. The beginning of darkness following a light span or light break starts the timing, just as the sand starts to move down to the lower chamber when an hourglass is inverted.

Endogenous Oscillators

Photoperiodic time measurements for a number of variables in the aphid *Megoura* may be best explained by a non-self-sustaining *hourglass model* (Lees, 1966; Hillman, 1973), unlike most photoperiodic variables which are best explained by the functioning of a self-sustaining endogenous circadian oscillator that can be reset or manipulated. Strong support for an endogenous circadian oscillator comes from the results of studies where the total hours of light plus darkness have been manipulated to deviate greatly from 24 h or when skeleton (interrupted) photoperiods have been used (Figure 4.11). By keeping the light span at a desired or given length (e.g., 6 h or 8 h of L), but increasing the dark span in increments to have cycles that are as long as 60 or more hours,[14] photoperiodic responses in both plants (flowering of soybeans) and animals (testicular development in birds) have been shown to depend upon an endogenous cycle or oscillator that is approximately 24 h (Figure 4.12). Other examples from similar type experiments, which indicate the interaction between circadian rhythms and photoperiodism, have been found for an annual herb, *Chenopodium rubrum*, a SDP that requires only one inductive cycle for floral induction and can be maintained in LL prior to and after darkness (Cumming, 1967; King, 1975), the flesh fly, *Sarcophaga argyrostoma* (Saunders, 1973), and the hamster *Mesocricetus aureus* (Stetson et al., 1975).

---

[14] The Nanda-Hamner protocol is used to refer to experiments that include treatments where the light span is set for a given duration (e.g., 6 h) but the dark span is progressively lengthened, resulting in cycles that may span up to 60 or more hours. Experiments involving such treatments were introduced by Nanda & Hamner (1958).

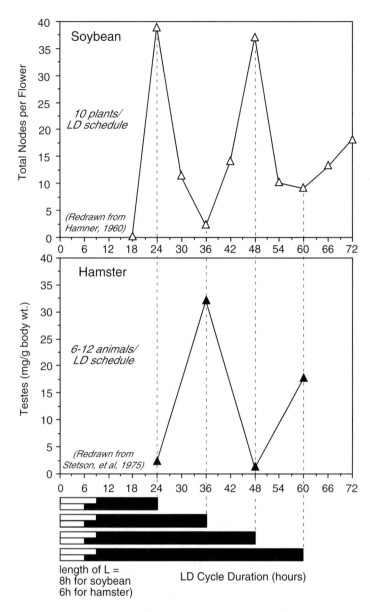

FIGURE 4.12. Circadian rhythm in photosensitivity revealed by effects of different total LD cycle lengths on flowering in Biloxi soybeans (*top*) and testis weight in Golden hamsters (*bottom*). The light span was kept constant (8 h for soybeans or 6 h for hamsters), but the dark span was lengthened in a series of separate experiments to equal the times indicated. A 24-h periodicity is indicated for both the plant study ($n = 10$ plants per LD schedule) and animal study (6–12 animals per LD schedule), since similar results were obtained when the total cycle lengths were integral multiples of 24 h (based on Hamner, 1960 and Stetson et al., 1975).

Relationships, such as those that exist between circadian rhythmicity and photoperiodism in diverse groups of organisms, appear to be straightforward and obvious. The confusion arises, however, when these two time-dependent processes are incorporated into various models, none of which by themselves adequately account for all of the photoperiodic processes or events found in the various taxa (classifications). Three of the more common models of photoperiodism (resonance, external coincidence, internal coincidence), which are based upon an endogenous circadian oscillator or clock, are presented in Table 5.1 (a detailed explanation of how each of them function is presented in Chapter 5 on Models and Mechanisms). It is paramount, however, to note that *phase* is a key component of each model. More specifically, these models focus on phase relationships, as can be illustrated in phase–response curves where exogenous L and D pulses evoke endogenous phase advances and phase delays (cf. Saunders, 1981; Pittendrigh et al., 1984).

Bünning's Hypothesis

In the 1930s, Erwin Bünning (1906–1990), a German botanist and pioneer in biological rhythm research who laid out the basis of photoperiodism for developmental cycles in plants and insects, suggested that the circadian cycle could be divided into two phases, a *photophil* (light-loving) phase and a *scotophil* (dark-loving) phase, and that these phases were important in the photoperiod response of flowering (Bünning, 1936, 1960). In other words, there is a phase in the circadian cycle where the light span is important and another phase where the dark span has a specific role. The suggestion that plants display endogenous phases of light and dark sensitivity, known as "Bünning's Hypothesis," was refined by C.S. Pittendrigh in the mid-1960s (Pittendrigh & Minis, 1964; Pittendrigh, 1966).

Phase–Response Curves

Phase–response curves, which were discussed in Chapter 2 and illustrated in Figure 2.13, tell us something about the entrainment of the rhythm relative to what happens when phases of the rhythm are exposed to light or dark pulses under free-running conditions. The half-cycle that corresponds to the dark span (subjective night) is viewed as being more responsive to light than the half-cycle that corresponds to the light span (subjective day). Phase delays occur when light is given during the early part of the subjective night and phase advances occur when light is given during the latter part of the subjective night (Pittendrigh et al., 1984).

Ultradian Cycles

The relationship between photoperiodism and rhythmicity is not limited to the circadian domain. A timing system that functions in some photoperiodic responses can include frequencies that are higher or lower than circadian. Photoperiodic phenomena displaying rhythms shorter than a day (*ultradian*), especially cycles of 12 h, are common in both plants and animals. An endogenous ultradian rhythm with a period of about 12 h (sometimes referred to as a *semidian* rhythm) has been observed for the flowering response of the Japanese morning glory *(Pharbitis nil)* to far-red light (Heide et al., 1986). The fact that a 12-h cycle for diapause is present in a mutant strain of *Drosophila melanogaster*, which is no longer circadian (Saunders, 1990), provides additional support for the involvement of semidian periodicity in photoperiodism.

Circannual Cycles

Annual cycles of photoperiods that are indicative of the seasons serve as synchronizers of various circannual rhythms (Gwinner, 1977a). The yearly periodicity of a number of variables, especially in birds where it has been studied extensively, persists under seasonally controlled environmental conditions, although under such conditions the period often deviates slightly from precisely one year (Gwinner, 1977b). The endogenous nature of the circannual rhythm in reproductive function has been shown for the gonads of equatorial weaver finches *(Quelea quelea)*, which when kept in a constant LD12:12 environment for 2.5 years, came into full breeding condition and later regressed with the same yearly pattern anticipated for the gonads of birds living in the wild (Lofts, 1964). The role of the photoperiod in synchronizing infradian rhythms has been superbly demonstrated in studies with the European starling (*Sturnus vulgaris*). In the laboratory, circannual rhythms of molt and gonadal size have been synchronized with photoperiodic cycles as short as 2.4 months, although the phase of the rhythm becomes progressively more delayed as the photoperiodic cycle is reduced (Gwinner, 1977b).

Bird Migration

Interactions between photoperiod and rhythms longer than a day (*infradian*) have been extensively studied in bird migration (cf. Gwinner, 1977a). By their very nature, these studies often span many years and may include both field and laboratory experiments. Conceptually, the distance that a bird travels during migration (e.g., flights of hundreds or thousands of miles or kilometers) would appear beyond study in a normal laboratory protocol. In the laboratory, however, a night behavior commonly called "*Zugunruhe*" (migratory restlessness), which indicates migratory readiness, can be measured by the quantity and direction of footprints produced by a bird as it hops either from an ink pad to a piece of paper or on a perch (Figure 4.13*a,b*). During nonmigratory seasons, perch hopping is confined to the light span, while during migratory seasons, the bird remains active throughout the night (Figure 4.13*c,d*). When results obtained from Zugunruhe-type studies under controlled photoperiods were compared to

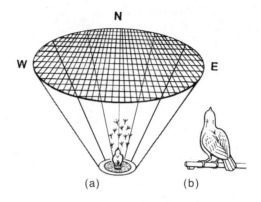
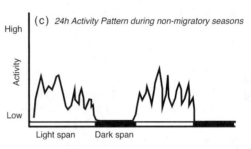
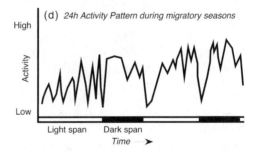

FIGURE 4.13. Migratory restlessness (*Zugunruhe*) of a bird can be monitored in the lab. The direction of footprints on the wall of a conical cage can be noted as the bird hops from an ink pad (*a*) or activity can be recorded electronically via a microswitch under a perch as the bird hops restlessly (*b*). During nonmigratory seasons, perch-hopping activity is confined to the daylight hours (*c*), whereas during a migratory season, the bird remains active throughout the night (*d*).

results of migration times of free-living birds, similar phases of activity have been observed. In the case of willow warblers *(Phylloscopus trochilus)*, which migrate between Africa to Europe, the most intense span of Zugunruhe coincides with the span of highest speed, which occurs when free-living birds cross the Sahara desert (Gwinner, 1977a). An endogenous time-keeping system that mediates photoperiod-induced seasonality, has been described for the migratory Blackheaded Bunting *(Emberiza melanocephala)* (Malik et al., 2004). Thus, a photoperiodic clock responds to changes in light wavelength and light intensity at the right time of day, especially during dawn and dusk, by the development of Zugunruhe-associated symptoms (intense nighttime activity, increases in body mass and testes size).

Deer Antlers

The seasonal photoperiod cycle synchronizes the circannual rhythms of the antler cycle of deer, such as sika *(Cervus nippon)*, white-tailed (*Odocoileus virginianus*), roe (*Capreolus capreolus*), and reindeer (*Rangifer tarandus*).[15] This cycle includes the shedding of antlers in the spring, the development and elongation of new antlers that are covered by a skin layer called velvet in the summer, shedding of velvet in late summer, and finally the presence of the hard, dead, bony antlers for the autumn breeding season. Results achieved from studies where the photoperiod cycle was reversed from the normal outdoor temperature, as well as from studies where the seasonal photoperiod cycle was either decreased to 3 or 4 months or increased to 24 months, clearly show that the seasonal photoperiod cycle, rather than temperature, serves as the synchronizer for the antler cycle (Goss, 1969a; Bubenik et al., 1987; Sempere et al., 1992).

Vernalization

An interesting relationship involving photoperiodism, a low-temperature treatment called *vernalization* (a process that hastens or induces flowering), and a circannual rhythm has been observed for a cultivar of Japanese radish (*Raphanus sativus* L.), a plant that flowers more rapidly under long days (Yoo & Uemoto, 1976). When seeds were removed each month from storage at room temperature and sown, a circannual rhythm was observed for the number of days from germination to the opening of flowers (anthesis). However, if the seeds were vernalized prior to sowing, the number of days from germination to anthesis remained almost constant over the 2-year duration of the study and the circannual rhythm was not observed (Figure 4.14).

Much remains to be learned about the interactions between photoperiodism and rhythms, especially in relation to seasonal photoperiod cycles and circannual rhythms. Commonly, the interactions between photoperiodism and rhythms are studied relative to a single terminal event in the life history of an organism (e.g., floral induction of plants or eclosion of insects) and require a population of organisms. The use of perennial plants and noninvasive practices with animals that can be followed for years may warrant further attention, although current research trends emphasize the use of clock gene mutants (organisms with genetically based circadian rhythms that are longer or shorter than 24 h).

---

[15] Sika deer are native to Japan and have adapted to temperate regions of the world. The study referred to here was conducted in light-tight barns in Rhode Island, USA at approximately 42° N latitude under both natural and controlled light/dark conditions (Goss, 1969a,b). White-tailed deer, which are native to North America, have a similar seasonal antler cycle (Bubenik et al., 1987). Reindeer (or caribou) (*Rangifer tarandus*) is the only species of deer in which both males and females exhibit a yearly antler cycle (Lincoln & Tyler, 1999).

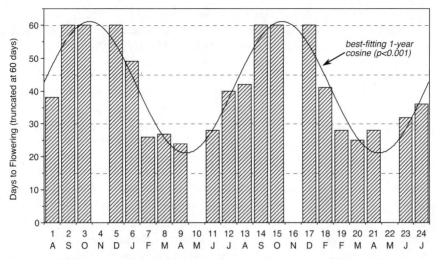

FIGURE 4.14. Annual rhythm in readiness to flowering by a radish (*Raphanus sativus* cv. Waseshijunichi). Seed harvested in July and stored at room temperature (25°C) flower at different rates depending on the months after seed production, indicating an endogenous annual rhythm in readiness to flower (data from Figure 1 in Yoo & Uemoto (1976) redrawn and analyzed using the least-squares fit of a 1-year cosine). Cold treatment (10 days at 5°C) abolishes this rhythm.

Photoperiodism and Humans

When university students are introduced to the topic of photoperiodism, a lingering question often remains unanswered: *"Do human beings respond to photoperiods, and if so, how?"*

Humans have been the subject in numerous experiments that have focused on rhythms, but the use of human subjects in photoperiodic experiments, especially in ones that are of an experimental nature and comparable to those conducted with plants, insects, birds and certain mammals do not appear in the literature. However, a limited number of epidemiologic studies designed to search for statistically significant correlations between variables within a population have included photoperiod as one of the factors.

Birth Patterns

Results from one of the most extensive "worldwide" studies of this nature, which included data on human conceptions, hours of sunlight, temperature and relative humidity from a number of geographical locations, showed a significant correlation between conception rates and photoperiod (Roenneberg & Aschoff, 1990a,b). In locations where changes in daylength are more pronounced (e.g., higher latitudes, see Figure 4.5), the peak in human conception was found to coincide with the vernal equinox.[16] For example, when birth records between 1778 and 1940 were reviewed for Eskimos living in northern Labrador, Canada (52°N), an 80% difference between the peak in March and a trough in June was noted, which agreed with observations recorded by early Arctic explorers (Ehrenkranz, 1983a,b).

In general, photoperiodic control is less rigid in mammals, including humans, than in birds and lower invertebrates, and generally serves to time well-established endogenous rhythms that lead to development of reproductive conditions (Lofts, 1970). The influence of photoperiod on humans appeared to be more dominant prior to 1930, perhaps because humans have since acquired the technology to regulate their physical environment, especially in controlling spans of light, darkness, and temperature. It is possible that the impact of photoperiodism on the lives of most humans who live in a modern society has become less pronounced as a primary synchronizer because individuals in industrialized societies are shielded, at least in part, from the full impact of the natural seasons by secondary synchronizers, including artificial lighting, curtains, scheduled activities, and lifestyles (Van Dongen et al., 1998; Roenneberg, 2004).

Indoor vs. Outdoor Light

There is a large variation in natural light exposure between persons who work indoors and who work outdoors. The amount of light that most of us are exposed to when working in buildings during the day is much less than the amount naturally present outdoors, while at night we are exposed to unnatural light levels due to electric lighting. In one study on 12 healthy men and women in Montreal who wore ambulatory photosensors for a week in the summer and in the winter, it was found that there was no seasonal variation in the amount of time spent under moderate (100–1,000 lux) illumination and more than 50% of their time they were exposed to illumination less than 100 lux, even in summer (Hébert et al., 1999). However, the amount of time spent under bright light (>1,000 lux) was much greater in summer (2 h 37 min) as compared with the winter (26 min), which suggests seasonal differences in overall light exposure. Some domesticated animals and plants also appear to have lost their dependence on photoperiodism, such as the rapid flowering *Brassica* which, because of its short life cycle, is used in biology laboratories (Williams & Hill, 1986).

Disorders

Several human aggressive behaviors or psychological disorders have been found to correlate with the time of year (Table 4.8). In a review of 2,131 acts of hostility, the opening month of wars showed a peak in mid-summer in both hemispheres and it was suggested that the lengthening of the daily photoperiod might serve as an environmental cue for an increase in affective aggressiveness (Schreiber et al., 1991).

---

[16] Date in the spring when the light and dark spans are about equal.

TABLE 4.8. Examples of human behaviors or disorders that have been found to correlate with the time of year and photoperiodism.

Variable	Location of peak(s)	Reference
Alcoholism (hospitalizations)	Spring	Eastwood & Stiasny, 1978
Bipolar and schizoaffective illness, violent episodes	June and December (Israel)	Roitman et al., 1990
Chronic fatigue syndrome	Winter worsening	Terman et al., 1998
Depression (rating)	Winter months	Hansen et al., 1998
Depression (hospitalizations)	Spring and Autumn	Eastwood & Peacocke, 1976
Eating disorders	Higher severity scores in winter	Brewerton et al., 1994
Acts of hostility[a] and individual violent crimes[b]	Summer in N and S hemispheres	Schreiber et al., 1991, 1997
Mania	Summer (England and Wales)	Symonds & Williams, 1976; Hare & Walter, 1978
	Summer (England and Scotland)	Myers & Davies, 1978
	Spring (Greece)	Frangos et al., 1980
	Spring–Summer (New Zealand)	Mulder et al., 1990
Menarche	December–January and August–September	Albright et al, 1990
Mood (normal and clinical populations)	Feeling worst, more sleep, weight gain in Winter	Kasper, et al., 1989; Hardin et al., 1991
Seasonal affective disorder (SAD)	More cases in Winter and at higher latitudes	Rosen et al., 1990; Hardin et al., 1991; Mèrsch et al., 1999a,b
Schizophrenia (hospital admissions)	July–August (England and Wales)	Hare & Walter, 1978
Sexual conception	Increase near Vernal Equinox	Roenneberg & Aschoff, 1990b
Sleep problems	Winter	Hansen et al., 1998
Suicide	Spring months in N and S hemispheres	Lester, 1985, 1997; Preti, 1997; Retamal & Humphreys, 1998

[a]Opening dates of wars in both hemispheres. Onset of wars near the equator showed a near-constant monthly rate throughout the year.
[b]Includes offenses against the human body, sex offenses, forcible rape, and aggravated assault.

Another major example of annual photoperiodism and humans is *seasonal affective disorder,* commonly referred to by its acronym SAD (Rosenthal et al., 1984). SAD is a syndrome characterized by recurrent major depressions at the same time each year, usually just before and during the winter season (Jacobsen et al., 1987), with a full spontaneous remission during the following spring and summer. While many people occasionally have the "winter blues" that can last a few days or weeks and is psychological in origin (e.g., from unrealistic expectations for holidays, loneliness, etc.), SAD is a result of physiological reactions to seasonal changes in photoperiod. Because bright artificial light phototherapy has an antidepressant effect (Rosenthal et al., 1988),[17] SAD is thought to be the result

---

[17] Bright light between 6 AM and 8AM in the morning is more antidepressant than between 7 PM and 9 PM in the evening. Bright light immediately upon awakening is the recommended circadian time for light treatment of SAD patients (Lewy et al., 1998).

of inadequate light reception and/or processing related to decreased sunlight during winter months (Oren et al., 1991). Thus, vulnerability to short photoperiods may be related to depression and winter SAD (Oren et al., 1994). SAD is discussed in more detail in Chapter 11 on Clinical Medicine.

Extending light into the night has also been shown to have a negative impact on circadian rhythms in aquatic animals, insects, birds, and mammals, including humans, and there are now several reports on the association between excessive light and disease. This topic is discussed in a section on light pollution in Chapter 10 on Society.

Take-Home Message

The normal development, reproduction, and activities of organisms are often dependent upon the duration and timing of light and dark spans, which in nature are associated with the seasons. This process is called *photoperiodism*, a complex and diverse process that is a basic principle of biology and is intrinsic to the temporal organization of life.

References

Ahmad M, Jarillo JA, Smirnova O, Cashmore AR. (1998) The CRY1 blue light photoreceptor of *Arabidopsis* interacts with Phytochrome A in vitro. *Mol Cell* 1: 939–948.

Albright DL, Voda AM, Smolensky MH, Hsi BP, Decker M. (1990) Seasonal characteristics of and age at menarche. *Prog Clin Biol Res 341A: Chronobiology: Its Role in Clinical Medicine, General Biology, and Agriculture, Part A*. 341A: 709–720.

Brewerton TD, Krahn DD, Hardin TA, Wehr TA, Rosenthal NE. (1994) Findings from the seasonal pattern assessment questionnaire in patients with eating disorders and control subjects: effects of diagnosis and location. *Psychiatry Res* 52(1): 71–84.

Bubenik GA, Schams D, Coenen G. (1987) The effect of artificial photoperiodicity and antiandrogen treatment on the antler growth and plasma levels of LH, FSH, testosterone, prolactin and alkaline phosphatase in the male white-tailed deer. *Comp Biochem Physiol A* 87(3): 551–559.

Bünning E. (1936) Über die Erblichkeit der Tagesperiodizität bei den Phaseolus-Blättern. *Jarhbücher Wiss Bot* 77: 283–320.

Bünning E. (1960) Circadian rhythms and the time measurement in photoperiodism. In: *Biological Clocks. Cold Spring Harbor Symposia on Quantitative Biology, Vol 25*. New York: Long Island Biol Assoc, pp. 249–256.

Campbell SS, Murphy PJ. (1998) Extraocular circadian phototransduction in humans. *Science* 279: 396–399.

Clayton RK. (1971) Light and Living Matter, Vol. 2: The Biological Part. New York: McGraw-Hill, 243 pp.

Cumming BG. (1967) Circadian rhythmic flowering responses in *Chenopodium rubrum*: effects of glucose and sucrose. *Can J Bot* 45: 21783–2193.

Eastwood MR, Peacocke J. (1976) Seasonal patterns of suicide, depression and electroconvulsive therapy. *Br J Psychiatry* 129: 472–475.

Eastwood MR, Stiasny S. (1978) Psychiatric disorder, hospital admission, and season. *Arch Gen Psychiatry* 35(6): 769–771

Ehrenkranz JRL. (1983a) A gland for all seasons. *Nat Hist* 6: 18–23.

Ehrenkranz JRL. (1983b) Seasonal breeding in humans: birth records of the Labrador Eskimo. *Fert Steril* 40: 485–489.

Foster RG, Garcia-Fernandez JM, Provencio I, De Grip WJ. (1993) Opsin localization and chromophore retinoids identified within the basal brain of the lizard *Anolis carolinensis*. *J Comp Physiol A* 172: 33–45.

Foster RG. (1998) Photoentrainment in the vertebrates: a comparative analysis. In: *Biological Rhythms and Photoperiodism in Plants*. Lumsden PJ, Millar AJ. eds. Oxford: BIOS, pp. 135–149.

Foster RG, Hankins MW. (2002) Non-rod, non-cone photoreception in the vertebrates. *Prog Retin Eye Res* 21(6): 507–527 (Review).

Frangos E, Athanassenas G, Tsitourides S, Psilolignos P, Robos A, Katsanou N, Bulgaris C. (1980) Seasonality of the episodes of recurrent affective psychoses. Possible prophylactic interventions. *J Affect Disord* 2(4): 239–247.

Freedman MS, Lucas RJ, Soni B, von Schantz M, Munoz M, David-Gray Z, Foster R. (1999) Regulation of mammalian circadian behavior by non-rod, non-cone, ocular photoreceptors. *Science* 284: 502–504.

Garner WW, Allard HA. (1920) Effect of the relative length of day and night and other factors of the environment of growth and reproduction in plants. *J Agric Res* 18: 553–606.

Goss RJ. (1969a) Photoperiodic control of antler cycles in deer. I. Phase shift and frequency changes. *J Exp Zool* 170: 311–324.

Goss RJ. (1969b) Photoperiodic control of antler cycles in deer: II. Alterations in amplitude. *J Exp Zool* 171: 223–234.

Gwinner E. (1977a) Circannual migrations in bird migration. *Ann Rev Ecol Syst* 8: 381–405.

Gwinner E. (1977b) Photoperiodic synchronization of circannual rhythms in the European starling *(Sturnis vulgaris)*. *Naturwiss* 64: S44.

Hamner KC, Long EM. (1939) Localization of photoperiodic perception in *Helianthus tuberosus*. *Bot Gazette* 101: 81–90.

Hamner KC. (1960) Photoperiodism and circadian rhythms. *Cold Spring Harbor Symp Quant Biol* 25: 269–277.

Hansen V, Lund E, Smith-Sivertsen T. (1998) Self-reported mental distress under the shifting daylight in the high north. *Psychol Med* 28(2): 447–452.

Hardin TA, Wehr TA, Brewerton T, Kasper S, Berrettini W, Rabkin J, Rosenthal NE. (1991) Evaluation of seasonality in six clinical populations and two normal populations. *J Psychiat Res* 25: 75–87.

Hare EH, Walter SD. (1978) Seasonal variation in admissions of psychiatric patients and its relation to seasonal variation in their births. *J Epidemiol Community Health* 32(1): 47–52.

Hébert M, Dumont M, Paquet J. (1998) Seasonal and diurnal patterns of human illumination under natural conditions. *Chronobiol Intl* 15(1): 59–70.

Heide OM, King RW, Evans LT. (1986) A semidian rhythm in the flowering response of *Pharbitis nil* to far-red light. I. Phasing in relation to the light-off signal. *Plant Physiol* 80: 1020–1024.

Hendricks SB. (1956) Control of growth and reproduction by light and darkness. *Amer Sci* 44(3): 229–247.

Hillman WS. (1962) *The Physiology of Flowering*. New York: Holt, Rinehart & Winston, 164 pp.

Hillman WS. (1973) Non-circadian photoperiodic timing in the aphid *Megoura*. *Nature* 242: 128–129.

Hillman WS. (1976) Biological rhythms and physiological timing. *Ann Rev Plant Physiol* 27: 159–179.

Hillman WS. (1979) *Photoperiodism in Plants and Animals. Carolina Biology Readers No. 107*. Burlington: Carolina Biological Supply Co., 16 pp.

Jackson S, Thomas B. (1997) Photoreceptors and signals in the photoperiodic control of development. *Plant, Cell Environ* 20: 790–795.

Jacobsen FM, Wehr TA, Sack DA, James SP, Rosenthal NE. (1987) Seasonal affective disorder: a review of the syndrome and its public health implications. *Amer J Pub Health* 77(1): 57–60.

Kadman-Zahavi A, Peiper D. (1987) Effects of moonlight on flower induction in *Pharbitis nil*, using a single dark period. *Ann Bot* 6: 621–623.

Kasper S, Wehr TA, Bartko JJ, Gaist PA, Rosenthal NE. (1989) Epidemiological findings of seasonal changes in mood and behavior. A telephone survey of Montgomery County, Maryland. *Arch Gen Psychiatry* 46(9): 823–833.

King RW. (1975) Multiple circadian rhythms regulate photoperiodic flowering responses in *Chenopodium rubrum*. *Can J Bot* 53: 2631–2638.

Klebs G. (1918) Über die Blütenbildung von *Sempervivum*. [On flowering formation in *Sempervivum*]. *Flora (Jena)* 111: 128–151.

Lang A. (1965) Physiology of flower initiation. In: *Encyclopedia of Plant Physiology*, Vol 15. Ruhland W, ed. Berlin: Springer-Verlag, pp. 1380–1535.

Lees AD. (1966) Photoperiodic timing mechanisms in insects. *Nature* 210: 986–989.

Lester D. (1985) Seasonal variation in suicidal deaths by each method. *Psychol Rep* 56(2): 650.

Lester D. (1997) Spring peak in suicides. *Mot Skills* 85(3 Pt 1): 1058.

Lewy AJ, Bauer VK, Cutler NL, Sack RL, Ahmed S, Thomas KH, Blood ML, Jackson JM. (1998) Morning vs evening light treatment of patients with winter depression [see comments in: *Arch Gen Psychiatry* 1998; 55(10): 861-864] *Arch Gen Psychiatry* 55(10): 890–896.

Lincoln GA, Tyler NJ. (1999) Role of oestradiol in the regulation of the seasonal antler cycle in female reindeer, *Rangifer tarandus*. *J Reprod Fertil* 115(1): 167–174.

Lofts B. (1964) Evidence of an autonomous reproductive rhythm in an equatorial bird (*Quelea quelea*). *Nature* 201: 523–524.

Lofts B. (1970) *Animal Photoperiodism*. The Institute of Biology's *Studies in Biology* No. 25. London: Edward Arnold Ltd., 64 pp.

Lucas RJ, Foster RG. (1999) Photoentrainment in mammals: a role for cryptochrome? *J Biol Rhythms* 14(1): 4–10 (Review).

Lumsden PJ. (1991) Circadian rhythms and phytochrome. *Ann Rev Plant Physiol Plant Mol Biol* 42: 351–371.

Malik S, Rani S, Kumar V. (2004) Wavelength dependency of light-induced effects on photoperiodic clock in the migratory blackheaded bunting (*Emberiza melanocephala*). *Chronobiol Intl* 21(3): 367–384.

Marcovitch S. (1923) Plant lice and light exposure. *Science* 58: 537–538.

Masaki S. (1978) Seasonal and latitudinal adaptations in the life cycles of crickets. In: *Evolution of Insect Migration and Diapause*. Dingle H, ed. New York: Springer, pp. 72–100.

Masaki S. (1984) Unity and diversity in insect photoperiodism. In: *Photoperiodic Regulation of Insect and Molluscan Hormones.* Ciba Foundation Symposium 104. London: Pitman, pp. 7–25.
Menaker M. (1968a) Extraretinal light perception in the sparrow: I. Entrainment of the biological clock. *Proc Natl Acad Sci USA* 59: 414–421.
Menaker M. (1968b) Light perception by extra-retinal receptors in the brain of the sparrow. *Proc 76th Ann Conv Amer Psychol Assoc* 3: 299–300.
Menaker M, Keatts H. (1968) Extraretinal light perception in the sparrow: II. Photoperiodic stimulation of testis growth. *Proc Natl Acad Sci USA* 60: 146–151.
Menaker M, Roberts R, Elliott J, Underwood H. (1970) Extraretinal light perception in the sparrow: III: The eyes do not participate in photoperiodic photoreception. *Proc Natl Acad Sci USA* 67(1): 320–325.
Mersch PP, Middendorp HM, Bouhuys AL, Beersma DG, van den Hoofdakker RH. (1999a) The prevalence of seasonal affective disorder in The Netherlands: a prospective and retrospective study of seasonal mood variation in the general population. *Biol Psychiat* 45(8): 1013–1022.
Mersch PP, Middendorp HM, Bouhuys AL, Beersma DG, van den Hoofdakker RH. (1999b) Seasonal affective disorder and latitude: a review of the literature. *J Affect Disord* 53(1): 35–48 (Review).
Mulder RT, Cosgriff JP, Smith AM, Joyce PR. (1990) Seasonality of mania in New Zealand. *Aust N Z J Psychiat* 24(2): 187–190.
Myers DH, Davies P. (1978) The seasonal incidence of mania and its relationship to climatic variables. *Psychol Med* 8(3): 433–440.
Nanda KK, Hamner KC. (1958) Studies on the nature of the endogenous rhythm affecting photoperiodic response of Biloxi soybean. *Bot Gazette* 120: 14–25.
Oren DA, Shannon NJ, Carpenter CJ, Rosenthal NE. (1991) Usage patterns of phototherapy in seasonal affective disorder. *Comprehensive Psychiatry* 32(2): 147–152.
Oren DA, Moul DE, Schwartz PJ, Brown C, Yamada EM, Rosenthal NE. (1994) Exposure to ambient light in patients with winter seasonal affective disorder. *Amer J Psychiat* 151(4): 591–593.
Parker MW, Hendricks SB, Borthwick HA, Scully NJ. (1946) Action spectrum for the photoperiodic control of floral initiation of short-day plants. *Bot Gazette* 108: 1–26.
Parker MW, Hendricks SB, Borthwick HA, Went FW. (1949) Spectral sensitivities for leaf and stem growth of etiolated pea seedlings and their similarities to action spectra for photoperiodism. *Amer J Botany* 36: 194–204.
Pittendrigh CS, Minis DH. (1964) The entrainment of circadian oscillations by light and their role as photoperiod clocks. *Amer Naturalist* 98: 261–294.
Pittendrigh CS. (1966) The circadian oscillation in *Drosophila psuedoobscura* pupae: a model for the photoperiodic clock. *Z Pflanzenphysiol* 54: 275–307.
Pittendrigh CS, Elliott J, Takamura T. (1984) The circadian component in photoperiodic induction. In: *Photoperiodic Regulation of Insect and Molluscan Hormones, Ciba Foundation Symposium 104.* Porter R, Collins GM, eds. London: Pitman, pp. 26–39.
Preti A. (1997) The influence of seasonal change on suicidal behaviour in Italy. *J Affective Disorders* 44(2–3): 123–130.
Retamal P, Humphreys D. (1998) Occurrence of suicide and seasonal variation. *Rev Saude Publica* 32(5): 408–412.
Roenneberg T, Aschoff J. (1990a) Annual rhythm of human reproduction. I: Biology, sociology or both? *J Biol Rhythms* 5(3): 195–216.

Roenneberg T, Aschoff J. (1990b) Annual rhythm of human reproduction. II: Environmental correlations. *J Biol Rhythms* 5(3): 217–239.

Roenneberg T. (2004) The decline in human seasonality. *J Biol Rhythms* 19(3): 193–195.

Roitman G, Orev E, Schreiber G. (1990) Annual rhythms of violence in hospitalized affective patients: correlation with changes in the duration of the daily photoperiod. *Acta Psychiatr Scand* 82: 73–76.

Rollag MD, Berson DM, Provencio I. (2003) Melanopsin, ganglion-cell photoreceptors, and mammalian photoentrainment. *J Biol Rhythms* 18(3): 227–234 (Review).

Rosen LN, Targum SD, Terman M, Bryant MJ, Hoffman H, Kasper SF, Hamovit JR, Docherty JP, Welch B, Rosenthal NE. (1990) Prevalence of seasonal affective disorder at four latitudes. *Psychiatry Res* 31(2): 131–144.

Rosenthal NE, Sack DA, Gillin JC, Lewy AJ, Goodwin FK, Davenport Y, Mueller PS, Newsome DA, Wehr TA. (1984) Seasonal affective disorder. A description of the syndrome and preliminary findings with light therapy. *Arch Gen Psychiatry* 41(1): 2–80.

Rosenthal NE, Sack DA, Skwerer RG, Jacobsen FM, Wehr TA. (1988) Phototherapy for seasonal affective disorder. *J Biol Rhythms* 3(2): 101–120 (Review).

Rowan W. (1926) On photoperiodism, reproductive periodicity, and the annual migrations of birds and certain fishes. *Proc Boston Soc Nat Hist* 38: 147–189.

Salisbury FB. (1963) *The Flowering Process*. New York: Macmillan, 234 pp.

Salisbury FB, Ross CW. (1992) *Plant Physiology*, 4th edn. Belmont: Wadsworth, 682 pp.

Saunders DS. (1973) The photoperiodic clock in the flesh-fly, *Sarcophaga argyrostoma*. *J Insect Physiol* 19(10): 1941–1954.

Saunders DS. (1981) Insect photoperiodism. In: *Handbook of Neurobiology 4. Biological Rhythms*. Aschoff J, ed. New York: Plenum, pp. 411–447.

Saunders DS. (1990) The circadian basis of ovarian diapause regulation in *Drosophila melanogaster*: is the period gene causally involved in photoperiodic time measurement? *J Biol Rhythms* 5(4): 315–331.

Schreiber G, Avissar S, Tzahor Z, Grisaru N. (1991) War rhythms. *Nature* 352: 574–575.

Schreiber G, Avissar S, Tzahor Z, Barak-Glantz I, Grisaru N. (1997) Photoperiodicity and annual rhythms of wars and violent crimes. *Med Hypoth* 48: 89–96.

Sempere AJ, Mauget R, Bubenik GA. (1992) Influence of photoperiod on the seasonal pattern of secretion of luteinizing hormone and testosterone and on the antler cycle in roe deer (*Capreolus capreolus*). *J Reprod Fertil* 95(3): 693–700.

Sokolove PG, Loher W. (1975) Role of eyes, optic lobes, and pars intercerebralis in locomotory and stridulatory circadian rhythms of *Teleogryllus commodus*. *J Insect Physiol* 21(4): 785–799.

Somers DE, Devlin PF, Kay SA. (1998) Phytochromes and cryptochromes in the entrainment of the *Arabidopsis* circadian clock. *Science* 282: 1488–1490.

Stetson MH, Elliott JA, Menaker M. (1975) Photoperiodic regulation of hamster testis: circadian sensitivity to the effects of light. *Biol Reprod* 13(3): 329–339.

Symonds RL, Williams P. (1976) Seasonal variation in the incidence of mania. *Br J Psychiat* 129: 45–48.

Tabata M, Suzuki T, Niwa H. (1985) Chromophores in the extraretinal photoreceptor (pineal organ) of teleosts. *Brain Res* 338(1): 173–176.

Taiz L, Zeiger E. (2002) *Plant Physiology*, 3rd edn. Sunderland, Mass: Sinauer Assoc., 690 pp.

Takeda M, Skopik SD. (1997) Photoperiodic time measurement and related physiological mechanisms in insects and mites. *Ann Rev Entomol* 42: 323–349.

Terman M, Levine SM, Terman JS, Doherty S. (1998) Chronic fatigue syndrome and seasonal affective disorder: comorbidity, diagnostic overlap, and implications for treatment. *Amer J Med* 105(3A): 115S–124S.

Thomas B, Vince-Prue D. (1997) *Photoperiodism in Plants*. San Diego: Academic Press, 428 pp.

Thomas B. (1998) Photoperiodism: an overview. In: *Biological Rhythms and Photoperiodism in Plants*. Lumsden PJ, Millar AJ, eds. Oxford: BIOS, pp. 151–165.

Thresher RJ, Hotz Vitaterna M, Miyamoto Y, Kazantsev A, Hsu DS, Petit C, Selby CP, Dawut L, Smithies O, Takahashi JS, Sancar A. (1998) Role of mouse cryptochrome blue-light photoreceptor in circadian photoresponses. *Science* 282: 1490–1494.

Tournois J. (1914) Études sur la sexualité du Houblon. *Ann Sci Nat (Bot)* 19: 49–191.

Underwood H, Menaker M. (1970) Photoperiodically significant photoreception in sparrows: is the retina involved? *Science* 167: 298–301.

Van Dongen HP, Kerkhof GA, Souverijn JH. (1998) Absence of seasonal variation in the phase of the endogenous circadian rhythm in humans. *Chronobiol Intl* 15(6): 623–632.

Van Zon AQ, Overmeer WPJ, Veerman A. (1981) Carotenoids function in photoperiodic induction of diapause in a predacious mite. *Science* 213: 1131–1133.

Veerman A, Overmeer WPJ, van Zon AQ, de Boer JM, de Waard ER, Hiusman HO. (1983) Vitamin A is essential for photoperiodic induction of diapause in an eyeless mite. *Nature* 302: 248–249.

Vierstra RD, Quail PH. (1983) Photochemistry of 124 kilodalton *Avena* phytochrome *in vitro*. *Plant Physiol* 72: 264–267.

Von Gaertner T, Braunroth E. (1935) [About the influence of moonlight on flowering date of long- and short-day plants.] [German] *Botan Centralblatt Abt* A53: 554–563.

Williams PH, Hill CB. (1986) Rapid-cycling populations of *Brassica*. *Science* 232: 1385–1389.

Yoo KC, Uemoto S. (1976) Studies on the physiology of bolting and flowering in *Raphanus sativus* L: II. Annual rhythm in readiness to flower in Japanese radish, cultivar 'Wase-shijunichi.' *Plant Cell Physiol* 17: 863–865.

Zeevaart JAD. (1958) Flower formation as studied by grafting. *Meded LandbHoogesch Wageningen* 58(3): 1–88.

Zeevaart JAD. (1969) Perrila. In: *The Induction of Flowering. Some Case Histories*. Evens LT, ed. Ithaca, NY: Cornell University Press, pp. 116–155.

5
Biological Oscillators and Timers: Models and Mechanisms

"Time, space, and causality are only metaphors of knowledge, with which we explain things to ourselves."
—Friedrich Nietzsche (1844–1900), German philosopher

Introduction

Within the realm of life, there is a structural organization, a hierarchy of components, extending from subatomic particles (e.g., electrons, protons) to organisms and ecosystems. How these components function (physiology), individually and in networks, depends not only upon the presence of the correct components and where they occur, but *when*. Individual components and sites of action are represented by structural units, while *"when"* represents the temporal organization of life and the measurable unit known as time.

While the word "rhythm" is a descriptive word, "clock" implies a measuring device. Clocks of one type or another quantify time. In the case of manufactured or astronomical clocks, which were discussed in Chapter 3 on Time, the mechanisms by which they operate are known and have been described relatively well. Such is not the case for biological clocks or oscillators, and may well account for the extensive use of the term "model" in the literature that focuses on the mechanisms of biological timing and how rhythms are generated.

Models and mechanisms of biological oscillators, especially those of circadian clocks, have been described and discussed in numerous papers with the assumption often, and rightly so, that the reader is knowledgeable in the more advanced areas of biology. Lower division undergraduate students and others who wish merely to explore the phenomena of biological rhythms are attracted to the subject often before they have taken courses in biochemistry, physiology, genetics, neuroscience, differential equations, and nonlinear systems. While such subjects are excellent prerequisites and highly desirable, a person can learn much without a mastery of them—a premise that will hopefully become evident in this chapter.

The major objectives of this chapter are: (1) to introduce five general categories of models that have been used in studies to elucidate the nature of biological clocks; and (2) to enhance the reader's understanding of mechanisms that may generate and regulate oscillations. The goal is not to list and discuss fully, or even in part, all models, mechanisms, and approaches that have been followed. Rather, it is to help one understand the nature of some of the key components and how they may function in generating rhythms. In some cases, major components, rather than a specific model or system, will be emphasized. This applies especially to membrane models, which are no longer popular and have been replaced by molecular models. For those individuals who are well-versed in the subject of rhythms, we can but echo the words of the Australian botanist and biologist, Rutherford Robertson (1913–2001): "*I crave the patience of my specialist colleagues who do not need the book anyway, unless they use it to improve or correct a concept which I stated imperfectly or too speculatively.*" (Robertson, 1983)

Approaches to Models and Mechanisms

A *model* represents an analogy that helps one to visualize something, often something that has not been directly observed. In the present context of time keeping, the analogy will be of an oscillator or an oscillating system, which in many cases is described best as a clock or pacemaker. *Mechanisms*, on the other hand, focus on the various natural chemical and physical processes that are involved in the process of biological time keeping. Models based upon known mechanisms are important research tools for analyzing characteristics of biological rhythms and in predicting certain patterns, activities, and behavior. They provide clues or guidelines for formulating hypotheses and postulates that lend themselves to experimental protocols, which can be applied to test the accuracy and validity of the model.[1] Some models of oscillating systems are likened to common items, such as a pendulum or a gate that can be visualized easily by most individuals. Others are not as easily visualized, especially those that are based upon biochemical and biophysical processes, mathematical equations, or include a more advanced biological nomenclature.

The strategy that is followed in an experimental protocol is often referred to as an *approach*. Many different approaches have been followed in studying the mechanisms of biological clocks (cf. Edmunds, 1988, 1992). They include: (1) a search for specific anatomical parts or regions at various levels of structural organization; (2) an attempt to trace the steps or pathways from environmental synchronizers to the variable that oscillates; (3) the application of biochemical and biophysical procedures and techniques; (4) the utilization of the tools of molecular biology and genetics; and (5) the formulation of mathematical equations that could describe oscillations relative to the properties of biochemical and

---

[1] Models are commonly used also in the discipline of chronobiology for the statistical analyses of rhythm characteristics (e.g., period, amplitude, and phase) and are addressed in Chapter 13 on Analyzing Rhythmic Data.

140   5. Biological Oscillators and Timers: Models and Mechanisms

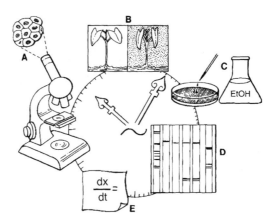

FIGURE 5.1. Some of the approaches that have been used in attempts to determine the nature of biological clocks: (*A*) Microscopic examination of cells and tissues; (*B*) Determining the steps and relations from the synchronizer to the variable that displays the rhythm; (*C*) Studying the effects of chemical and physical agents on molecules or organelles closely associated with the overt rhythm; (*D*) Molecular biology and genetics; and (*E*) Mathematical equations.

biophysical components (Figure 5.1). Not emphasized in this chapter, but illustrated and discussed in other chapters (e.g., Chapters 2 and 9) are certain *pacemakers* or *master oscillators* and their anatomical loci. Included among these "master oscillators" are the suprachiasmatic nuclei (SCN) of the vertebrate hypothalamus and the pineal gland of birds.[2]

Models of oscillators and oscillating systems represent analogies, usually based upon results obtained from studies where a particular approach or various approaches have been followed to understand the mechanisms of biological rhythms and oscillators. Many of the earlier models and their components (e.g., entrainment, gating) that helped guide the research of decades past continue to be in vogue, even as more molecular components are discovered. Groups or categories of models (e.g., mechanical, molecular, etc.), as well as approaches to elucidate the mechanisms of oscillators, are not clearly delineated.

The spectrum of biological rhythms spans the range from fractions of a second to centuries. Much of the emphasis on biological clock models has been upon the circadian domain. However, a number of excellent mathematical and biochemical models have been known since the 1920s for high-frequency ultradian rhythms (e.g., electrical model of the heart by van der Pol & van der Mark, 1928). Likewise, models (or hypotheses) of circannual rhythms, which have a long history of being associated with photoperiodism (e.g., Gwinner, 1977), lend themselves to advances in the genetic analysis of seasonal clocks in animals, such as *Drosophila*

---

[2] The eyes of marine gastropods (*Aplysia* and *Bulla*) could be included also, since studies with these structures contributed greatly to our knowledge of cellular and chemical oscillators (cf. Jacklet, 1988–89).

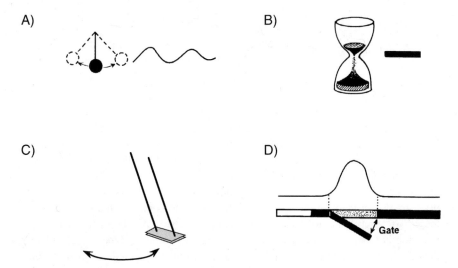

FIGURE 5.2. Some common categories or analogies of mechanical models: (*A*) Pendulum; (*B*) Hourglass; (*C*) Swing; (*D*) Gate.

(Majercak et al., 1999), hamsters (Gorman, 2003; Prendergast et al., 2004), and sheep (Lincoln et al., 2003).

Components and characteristics of mathematical, physical, and biological rhythms, such as phase shifts and period stability, have been used extensively in model building to explain the mechanisms of biological clocks.[3] Even though five categories of models will be delineated in this chapter, a considerable amount of integration exists among them, since some models incorporate features from several or all of these categories.

Mechanical Models

While "mechanical" type models often do not emphasize specific biological molecules, they do help scientists distinguish if the circadian rhythm is generated by an external environmental oscillator (e.g., the solar day), or if it will free-run for more than one cycle in the absence of environmental synchronizers (e.g., a light/dark cycle). Two models that attracted much attention in the 1960s and 1970s, were the *pendulum model* (Figure 5.2A) and *hourglass model* (Figure 5.2B).

---

[3] Excluded from this discussion are the external timing hypothesis and the endogenous timing hypothesis, two opposing mechanisms that were postulated for circadian rhythms (Brown et al., 1970). The existence of an unknown exogenous environmental or cosmic factor (e.g., "X") that controls biological clocks is no longer a major issue or component in most models of circadian clocks (also see discussions by Salisbury & Ross, 1992; Foster & Kreitzman, 2004).

Pendulum

A pendulum-type clock continues to oscillate, while the hourglass clock must be reset after each cycle or the completion of a span (e.g., darkness). Closely related to the pendulum, at least in the way it moves, is a common object known as a *swing* (Figure 5.2C). The comparison of a swing that is pushed at certain times and thereby causing its speed to transitionally change, represents an excellent analogy of how the response of an oscillator depends upon phase (Roennenberg et al., 2003). These responses of a mechanical oscillator (swing) have been graphed as a phase–response curve (PRC) and compared as a model to the PRCs of circadian oscillators. The pendulum model is applicable to a number of variables, including such classical examples as the cyclic pattern of testicular development in birds (Stetson et al., 1975) and floral induction in plants (Hamner, 1960). Illustrations of the rhythmic nature of these two variables are found in Figure 4.12 in Chapter 4 on Photoperiodism.

A number of variations have been proposed for pendulum-type models. One of these variations is the addition of a secondary oscillator, called the *slave oscillator*. Its role is to make the entrainment oscillating systems more precise[4] (cf. Pittendrigh, 1981), be it a protein in the phytochrome response of *Arabidopsis* (Kuno et al., 2003), a photoperiodic oscillator associated with control of diapause in *Drosophila* (Gillanders & Saunders, 1992), or separate slave oscillators associated with the feeding and drinking behavior of rats (Strubbe et al., 1986).

Hourglass

The hourglass analogy represents a clock that must be reset after each cycle or the completion of a span (e.g., darkness), while as mentioned above, a pendulum-type clock continues to oscillate. Classical examples of biological variables that fit the hourglass model are diapause (physiological inactivity) in aphids (*Megoura*) (cf. Lees, 1966; Hillman, 1973) and seed germination in plants (cf. Sweeney, 1974a). A mitotic clock (Harley, 1991) that is based upon shortening in the length of telomeres[5] and determines the number of cell divisions that can occur before a cell can no longer divide, could be another example of an hourglass clock (Rensing et al., 2001). In fact, aging itself, which in the "popular press" is sometimes referred to as the "biological clock," can be seen as an hourglass clock that measures the lifetime of one individual.

The hourglass analogy has also been applied to a number of physiological processes or systems where it is sometimes referred to as either an *hourglass*

---

[4] This feature of the biological model could be analogous to a feature found in a mechanical clock known as the Shortt Free-Pendulum Clock, where more precise timing is accomplished by the addition of a "slave clock" that reduces the interference of the free motion of a pendulum (cf. Whitrow, 1988).

[5] After each round of DNA division and replication, the 3' end is incompletely replicated, resulting in 50–150 base pairs being lost from the ends (telomeres) of chromosomes per cell division in human fibroblasts (Martens et al., 2000). Fibroblasts are cells that give rise to connective tissues.

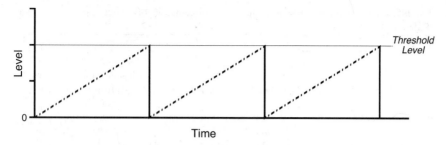

FIGURE 5.3. The Relaxation Oscillator or Integrate and Fire model has been applied to a number of physiological processes or systems. A variable increases up to a threshold level, when there is a sudden relaxation or firing back to the original lower level.

relaxation oscillator (Klotter, 1960) or an *integrate and fire model* (Figure 5.3) (Keener et al., 1981). In these *self-sustained* systems there is an increase in a variable (e.g., amount, activity) up to a threshold level, where when reached, there is a sudden relaxation or firing back to the original lower level. Examples of self-sustained hourglass oscillations of this type can be found in mitosis (Tyson & Kauffman, 1975) and the filling and emptying of the bladder (Glass & Mackey, 1988).

External Coincidence

Research in chronobiology that occurred prior to the early 1970s, especially in the disciplines of plant and animal physiology, centered on studies that combined photoperiodism and circadian rhythms. Here, the pendulum and hourglass models, as well as variations of them, were relevant tools of research. However, the interactions between photoperiodism and rhythms are highly complex, and much of the literature that addresses this relationship can be confusing (Hillman, 1976). Nevertheless, models and mechanisms that focus upon interactions between photoperiodism and circadian rhythms (Table 5.1) continue to be relevant in molecular biology (cf. Lumsden & Millar, 1998).

One of the more popular examples is the *external coincidence model* and its analogy to a gate (Table 5.1). A gate forms a barrier that can open and close, thereby determining whether something enters or remains outside. In the case of photoperiodism, the "external something" could be light (photo) or darkness, which is "let in" when the organism is most receptive (i.e., the gate is open). If two events occur at the same time or phase, such as the light phase of the solar day cycle being present when the inductive or responsive phase of a rhythmic biological variable is also present, they coincide ("coincidence"). However, to more fully understand the nature of this model, we may need to review the status of photoperiodism in nature and then examine how it can be manipulated in a laboratory setting.

Under natural conditions of our solar day, photoperiodism refers to the response of organisms to the relative lengths and timing of light and darkness

TABLE 5.1. Three models based upon the role of circadian involvement in photoperiodism (Pittendrigh, 1972).

Model	Comments
Resonance	In the physical sense, a resonance relates to a large amplitude that is produced by another oscillation with the same period. Here, the amplitude (magnitude) of the photoperiodic response depends upon the circadian resonance with the light cycle
External coincidence	This involves the coincidence of light with a phase in the circadian cycle (e.g., Bünning's Hypothesis)[a] and fits results from many experiments that have been conducted with insects and other organisms
Internal coincidence	Based on the premise that circadian systems include many oscillations and their phase relations may be altered by a change in photoperiod (e.g., one oscillator could be entrained by the first part of the L span and the other by the latter part of the L span or beginning of D

[a]In 1936, Bünning introduced and later modified a view (Bünning, 1936), which has been cited by various authors as a hypothesis, postulate, or even a model. Briefly, the 24-h cycle was viewed originally as being divided into two parts: a "photophil" (light-liking) phase that promotes a developmental change or activity and a "scotophil" (dark-liking) phase that does not promote the change or activity.

(e.g., hours of daylight and hours of night). This process is associated with the seasons of the year and is evident in changes that occur during the development and behavior of many organisms. For example, some plants are induced to produce flowers when the light span is short and the dark span is long, while others may require a long light span and a short dark span to produce flowers. Similarly, in some animals the changes in the lengths of light and dark spans affect the development of reproductive structures, such as the testes (see Chapter 4 on Photoperiodism).

In the laboratory, however, an organism maintained under a photoperiodic regime that in nature would normally induce or inhibit a change in development can be "tricked" not to do so. This is accomplished relatively easily by subjecting the organism to a short pulse (e.g., 15 min) of the nonprevailing span, such as a light pulse during the "normal" dark span (see Chapter 4 on Photoperiodism). However, the external pulse is effective or works best (or only) at a certain time during the 24-h environmental cycle. In other words, there is a specific span of time within the light–dark cycle when an organism responds best or not at all to the signal or pulse (Figure 5.2D). This time slot or span has been likened to a gate, a phase in the cycle that recurs with a circadian periodicity. The phenomenon is known as *gating* and represents a feature incorporated into the external coincidence model (Table 5.1).

Hands of a Clock

Historically, studies of biological rhythms and biological clocks can be traced back to the observations of recurring changes in the movements of structures (e.g., leaves), stages of development (pupa to adult), and behavioral activities (feeding, resting, etc.). These recurring changes of a given variable, such as the position of a leaf or the activity of an insect, were readily visible and represented the "hands of a clock." Mechanical-type models were well suited for such studies.

However, the mechanisms that generate the recurring position or status of the "hands" occur at a molecular level. Today, the "hands of the biological clock" are often viewed or represented as ions, electrons, atoms, and molecules, paralleling our technological society, where the most accurate time is provided by an atomic clock (see Chapter 3 on Time). Nevertheless, the "hands of a clock" is an important analogy that is commonly used to indicate output, the overt rhythm, which is generated by a central clock. In addition, just as removing the hands of a mechanical clock does not change the mechanical mechanisms, inhibiting a biological process to the extent that it is not visible does not mean that the biological clocking mechanism itself is not functioning or driving other biological oscillations.

Mathematical Models

Bridging the gap between mechanical models on the one hand, and the biochemical or molecular models on the other, are the mathematical models. Perhaps here and in many other cases, the term "integrating" would be better than "bridging," since both mathematical and molecular components are present in a single model. The *Goodwin model* (Goodwin, 1965) is a superb example, containing three simple biochemical or molecular components (nuclear messenger, cytoplasmic messenger, and repressor), and three mathematical equations that express the relationships between the three molecular components and how they change in time.[6]

Of the various types of models that are used for studying biological clocks and oscillators, mathematical models are viewed as being the most difficult to understand. A strong background in higher mathematics[7] helps in understanding them, but the lack of such a background need not be viewed as an excuse for ignoring them. At the very least, one should attempt to develop an appreciation for the concepts upon which mathematical models are based and to become aware of some of their applications and implications.

The mathematics that is relevant to modeling biological rhythms is the mathematics of change: calculus. Developed in the 17th century by Newton and Liebnitz, calculus gave mathematicians the ability to relate how quantities, such as the position of an object, change in time. One of the major concepts is that of the *state* of a system: calculus can be used to describe how the state changes in time as a function of the state itself. Abstractly, we can think of the state of a system as a set of coordinates. For example, the state of the spring depicted in Figure 5.4 can be represented by the position and velocity of the mass: the change in the state comes from the acceleration imposed by gravity and the force exerted by the spring

---

[6] For more recent discussions, equations, and illustrations, see Rensing et al., 2001 and Ruoff et al., 1999.

[7] This applies particularly to those who have not been exposed to courses such as calculus, linear algebra, and nonlinear systems.

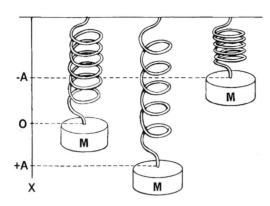

FIGURE 5.4. A simple harmonic oscillator illustrated by the moving mass (*M*) on a spring.

(which is a function of position). So, for the spring, two numbers suffice to describe the state. The change of state can be described as a flow, much like of flowing water: the flow dictates how each point in the water is moved along in time. For a biochemical system, the state may involve more numbers: the concentrations of each of the molecular components.

The state of a rhythmic system can be described by a *phase*, which quantifies what position in the cycle the system is currently at, just as the hour shown by a clock tells us where we are in the daily cycle. When there are interacting rhythms, there are multiple phases, one for each of the rhythmic components. Mathematical models of these rhythms describe the flow. In fact, the word *rhythm* is derived from the Greek *rhein* "to flow." This carries over to a philosophical statement that one cannot step into the same river twice, and a retort that one cannot step into the same river even once.[8] With mathematical oscillators, time, rather than water, is the variable that "flows," and thus, mathematical models allow one to "step" in the same flow more than once.

In the flow of a cycle, any instantaneous point (i.e., position) is a phase, and phases or phase angles are defined by *coordinate systems*. Very likely, most individuals have used coordinates when trying to locate places on a geographical map. In such instances, the coordinates were very likely given by the numbers (or letters) found along the margins of a map.

The mathematical flows used to model rhythms assign a specific direction of change to each state. Figure 5.5 shows a possible flow on a state that consists of only a single number. We can think of this as a fixed wire where the state is represented by the position of a bead on the wire. The flow tells what direction the bead should be moved, as in a simple-minded board game where the flow dictates the rules. It may be obvious in such a system that the bead can never go back and forth; at any point the bead is moving either to the right or the left, so it never returns to any previous position.

---

[8] The source or the gist of stepping into the same river twice has been attributed to the early Greek philosopher, Heraclitus. In regard to the origin of "*stepping into the same river once*," one of us (WLK) recalls reading about this "bafflement" many years ago. Perhaps in a statement attributed to Plato in his dialogue entitled "Cratylus."

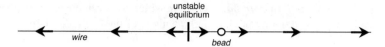

FIGURE 5.5. An example of a possible single-number flow state that consists of only a single number illustrated by a fixed wire where the state is represented by the position of a bead on the wire.

In contrast, the two-dimensional flow shown in Figure 5.6 can easily produce a cycle. The state, like a leaf from a plant flowing in a whirlpool, can return to its previous position. For this reason, throughout this book, data for a biological cycle have been illustrated in a two-dimensional plot, a *Cartesian coordinate system*. In these illustrations, two coordinates, the horizontal X axis (abscissa) and the vertical Y axis (ordinate) locate a point on a plane. The same data can be

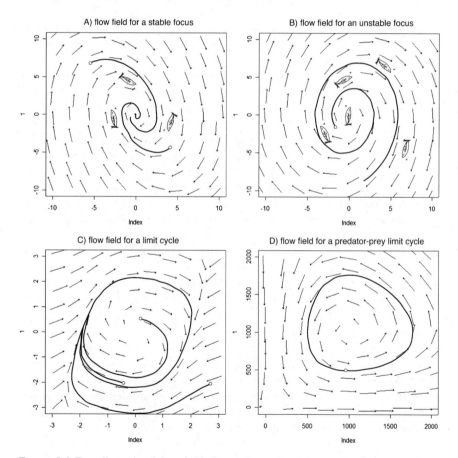

FIGURE 5.6. Two-dimensional flow fields illustrating the path followed by a proverbial leaf (*A*, *B*); a limit cycle (*C*); and a predator–prey limit cycle (*D*) (courtesy of Daniel Kaplan, Macalester College, St. Paul, MN).

illustrated also in a *polar coordinate system* in which a coordinate is defined by its distance from a given point on a line and the angle of the line (see Chapter 13 on Analyzing for Rhythms).

Differential Equations

The mathematical language for describing flows is that of differential equations, and so differential equations are central to most mathematical models of rhythms and cycles. The *derivative* of a quantity expresses the rate with which it changes. For example, imagine a quantity 'x' that changes in time. We can think of this as 'x' being a function of time: $x(t)$. The derivative, denoted dx/dt, is another function, one that tells, at each instant in time, how 'x' is changing. In terms of a graph of $x(t)$ versus t, the derivative is the slope of the graph at any time t. For this reason, calculus is sometimes described as slope finding (Pine, 1983).

A flow is described by relating the derivative of the state to the state itself. Because of this, it follows that one needs to know what these equations contain and reveal. Briefly, differential equations are equations that contain derivatives of variables (cf. Kaplan & Glass, 1995). Perhaps from an over simplistic view, a derivative can be visualized as a special kind of ratio that indicates a *change* in something (x) that corresponds to the change in another (independent) variable as it approaches zero (0). For example, a change in something (x) that occurs over time (t) is the derivative dx/dt and is called a *first-order derivative*. Derivatives are used or appear in the Goodwin oscillator, as well as in models for cardiac oscillations (*van der Pol equations*), predator/prey cycles (*Lotka-Volterra equations*), and in one of the most commonly cited mathematical models of the biological clock, the *limit cycle oscillator*.

Finding the rhythm that results from a specified flow is as simple as dropping a proverbial leaf into the flow and watching the path that it follows (Figure 5.6). This path is called the "trajectory" of the system. In mathematical language, finding the path is called "solving the differential equations." Sometimes, the flow is such that the path ends at a single point: the leaf stays put. This point is called an *attractor*, also called an equilibrium point or fixed point: a state where the flow is such that the state does not change. There are other sorts of attractors as well, which we will discuss below.

For the most part, mathematical models of oscillators can be divided into two types: those based upon *linear differential equations* and those based upon *nonlinear differential equations*.[9] A classical example of a linear differential equation is the spring-mass system, called in physics the *harmonic oscillator*. Linear systems are easily solved using simple mathematical techniques. However, the linear structure imposes severe limitations. For example, a linear system can have only

---

[9] Those who are mathematically inclined and/or have a background in calculus may find it advantageous to refer to references by Andronov et al., (1966) and Kaplan & Glass (1995).

a single-point attractor. Unlike linear systems, nonlinear systems can be difficult to solve using simple mathematical techniques, but they are generally easily solved by the leaf-watching technique, which is usually implemented by computer simulation. However, nonlinear systems can have much more interesting types of attractors, particularly for the purposes of modeling rhythms: limit cycles.

Limit Cycles and Topography

The trajectory of a limit cycle is a closed loop. Often, we are forced to study systems where we cannot trace the state of the system continuously in time as it moves around the loop. Instead, the only information available to us is the timing of certain events. Think of racecars moving around a looped track: so long as we know when they cross a fixed line across the loop, we can effectively monitor the progress of the race. Similarly, in studying biological rhythms, we may only know the timing of certain events: when fireflies flash, whether the leaf is dropping or already tucked into nighttime position, when a person goes to sleep or wakes up. Henri Poincaré introduced the idea of studying a flow by studying the times of specific events: he turned the continuous time representation of dynamics into a discrete-time one, giving us the mathematics of studying periodic events (cf. Poincaré, 1881).

In a topological description, only the relevant parameters specifying the overall behavior are plotted in the phase space and the cumbersome "microinformation" is thrown away (e.g., in the case of the harmonic oscillator, it would be the position vs. the speed). This description is equivalent to positioning a small "frictionless" steel ball on the three-dimensional map of a mountain range and watching the ball roll. Depending on the energy and the direction, the ball may spiral down to a valley, be stuck on a pass, stay in unstable equilibrium on a peak, etc. Supposedly, the ball could be caught on a closed path, retracing "forever" (periodically) the same trajectory. This is a *limit cycle*, and perhaps best defined as "an isolated closed trajectory" (Minorsky, 1962). A limit cycle may either be stable or unstable. In the case of most biological oscillators, a light pulse given during constant darkness will often result in a phase shift, but the original trajectory (period of the cycle) is reestablished as a stable limit cycle.

An excellent example of how a simple physical self-sustained oscillator can generate a stable limit cycle has been demonstrated by a model consisting of a block of wood connected to a fixed location by a spring and placed on a conveyer belt that moves at a constant speed (Kondo & Ishiura, 1999). An apparatus of this type, along with a drawing (solution) helps one to visualize how the stable limit cycle could be generated (Figure 5.7).

A biological system that illustrates the same concept, although not exactly a limit cycle, is the relationship between predators and prey (Figure 5.8). As the population of prey increases, the population of the predator increases because of the availability of the prey. However, the prey population drops as more predators

150   5. Biological Oscillators and Timers: Models and Mechanisms

a)

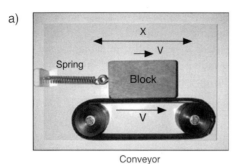

b)

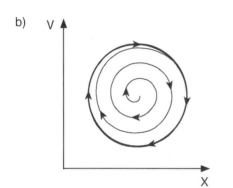

FIGURE 5.7. An apparatus (*A*) illustrating a model of a physical self-sustained oscillator and how the process of a block of wood moving on a conveyer belt could generate a stable limit cycle (*B*) (adapted from Figure 2 in Kondo & Ishiura, 1999).

are produced. Once the prey population is reduced, the predatory population starts declining (dying off) because of the lack of food, allowing the prey population to increase again. The population cycle continues and the two populations regulate (limit) each other in a cyclic manner. In nature, however, a predator–prey limit cycle is more complex, and may be either slow or fast (Rinaldi & Scheffer, 2000) as a result of food availability, increased predation, or abrupt changes in the

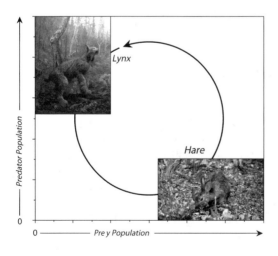

FIGURE 5.8. An idealized predator–prey limit cycle model illustrating the relationship between predators (lynx) and prey (hare). Photos of lynx and snowshoe hare courtesy of James Ford Bell Museum of Natural History and Professor Richard Phillips, University of Minnesota, respectively.

ecosystem and other climatic factors (Scheffer, 1989; Holmgren & Scheffer, 2001; Scheffer et al., 2003). For example, predation by fish can lead to a dramatic collapse in *Daphnia* population rather than a gradual decline, when a critical density of fish is exceeded (Scheffer et al., 2000).

Limit cycles have provided tools by which the rhythmic behavior of a biological variable under various conditions and treatments of light and temperature have been simulated and studied (Johnsson & Karlsson, 1972; Johnsson et al., 1972; Peterson, 1980; Peterson & Saunders, 1980; Engelmann, 1996). When limit cycles are illustrated, they are often presented (drawn or sketched) in a two-dimensional surface known as a *phase plane*. Upon this plane is traced a path called the *trajectory*. At first glance, phase plane illustrations of the trajectory (Figure 5.6B) could appear to be irrelevant to the subject of biological rhythms, except for the fact that from a mathematical perspective they display flow (i.e., the entire pattern of trajectories). Because differential equations that pertain to phase planes are beyond the scope of this book, they have been omitted. Instead, we have provided some diagrams (Figure 5.9) and accompanying terminology (Table 5.2) that could be helpful in understanding, or at least appreciating, equations and descriptions that may appear in the literature.

The diagrams in Figure 5.9 are more symbolic or artistic, rather than actual, especially for the Lorenz attractor[10] (Figure 5.9J), which mathematicians generate

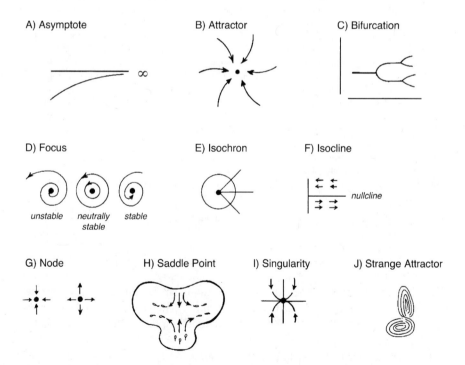

FIGURE 5.9. Illustrations corresponding to flow model terms found in Table 5.2.

TABLE 5.2. Special terms that apply to differential equations and limit cycle models (Letters (A–J) that appear in the second column (Fig) correspond to the illustrations in Figure 5.9).

Term	Fig	Definition and/or Analogies and Comments[a]
Asymptote	A	The asymptote of a curve is the straight line that closely approximates the curve as the curve heads off to infinity
Attractor	B	The set of points (or even a single point) to which the state of the system moves when following a flow. The set of points can be said to "attract" the state
Bifurcation point	C	In many systems, there are parameters that set the behavior of the system, such as the level of friction in a harmonic oscillator. When these parameters are changed, the behavior of the system can also change. Sometimes a small change in the parameters leads to a big change in the behavior. This is called a bifurcation
Focus	D	The equilibrium point in a linear flow field where the flow spirals around the equilibrium. It can be either stable (with the spiral trajectories heading toward the equilibrium) or unstable (with the trajectories heading away)
Isochron	E	Each point on a limit cycle can be assigned a phase, just as the numbers on a clock assign a time-of-day to each point on the clock's cycle. In a flow field, we can also assign a phase to every point in the flow, by following a trajectory starting at that point and seeing where it joins up with the limit cycle. The set of points in the flow field that all join the limit cycle at the same place should all be assigned the same phase. This set of points is called an isochron, for same ("iso") and time ("chron")
Nullcline	F	In a flow field, there may be points where one of the variables, say x, is not changing. At these points, $dx/dt = 0$, that is, the slope of x with respect to time is zero. The set of points where the variable is unchanging is called a nullcline, for ("null") change ("cline"). There will be a separate nullcline for each variable. Where they all intersect, the system has an equilibrium point
Node	G	An equilibrium point in a flow, such as a focus. It can be stable or unstable. See saddle point
Saddle point	H	A kind of node, an equilibrium point in a flow which is stable if approached from some directions and unstable in other directions, like a marble rolling on a horse's saddle
Singularity	I	Another term for an equilibrium point. In an oscillator, a singularity is a point in the flow field where the isochrons come together, just as the lines indicating the time zones come together at the north and south poles on earth. For any limit cycle oscillator, there must be at least one singularity. When pushed to a singularity, the system's oscillation stops. A singularity is an equilibrium point, and can be stable or unstable
Strange attractor	J	An attractor that has a "strange" and complex dimensional orbital structure. The Lorenz oscillator is an excellent example

[a] Assistance of D. Kaplan, Macalester College, St. Paul, MN, gratefully acknowledged.

by numerical approximations on a computer. In accord with two of the views presented near the beginning of this section on mathematical models, the highly complex Lorenz attractor has a "flow," but the nonlinear dynamics or mathematical solutions for the attractor are extremely difficult (cf. Stewart, 2000) and were published only recently (Tucker, 1999). The equations for the attractor show that dif-

---

[10] Equations relating to flow in the atmosphere were studied in the early 1960s by Edward N. Lorenz, an MIT meteorologist.

ferences in initial condition become larger (e.g., two mathematical points moving further apart) with the passing of time. While this feature is associated with chaos, the Lorenz attractor symbolizes order in chaos (cf. Stewart, 2000).

Chaos

Perhaps at this juncture in our discussion, it is best to pause and ask, what is "chaos" and how does it pertain to biological oscillators? Certainly, the double-lobed shape of the Lorenz attractor does not equate readily to what we customarily visualize as the phase plane, whereupon lies a circular circadian cycle. When one examines a time-series (x axis) for a variable that fluctuates (y axis), it is relatively easy to assume the presence of rhythmicity where none exists. Statisticians have demonstrated this with data generated by *randomization*. In the context of everyday speech, randomization is something associated with disorganization and confusion, and referred to as *chaos*. In the context of mathematical models, chaos is more difficult to describe without referring to differential equations, although there are features, such as the presence of an aperiodic state, that help to characterize it (cf. Glass & Mackey, 1988; Kaplan & Glass, 1995). Despite an aperiodic paradigm, a mathematical model has been proposed where the circadian or ultradian oscillator could be a *chaotic attractor* (Lloyd & Lloyd, 1993).

Spatiotemporal Systems

Moving points produce geometric lines or curves in a geometric plane and provide a framework for mathematical models of biological oscillators. Often, they can be described by equations and illustrated topologically. These oscillating waves with different geometries were initially studied in certain chemical systems and in the movements (aggregation) of tiny organisms called "cellular slime molds." In the latter case, single-celled, independently living amoebae of *Dictyostelium discoideum* (Figure 5.10) produce limit cycle induced ultradian pulses (e.g., 1–2 min) of cyclic AMP after the onset of starvation (cf. Gross, 1994). In turn, groups of individual cells stimulated by cyclic AMP make a directional movement for about one minute, then release cyclic AMP themselves, resulting in successive signals inducing concentric waves of inward movement. Aggregating cells thus result in both concentric and spiral waves[11] of amoebae moving toward centers of aggregation (Gerisch, 1976, 1987; Gross et al., 1976). Much of the early work and popularization of spatiotemporal systems can be accredited to the theoretical biologist, Arthur T. Winfree (1942–2002) (e.g., Winfree, 1972, 1987, 2001).

It is tempting to view spatiotemporal systems as purely novel events within the confines of the basic sciences. To overcome such a notion, one need but turn to the

---

[11] High magnification of these concentric waves (produced by cells that emit cyclic AMP spontaneously) can be seen in Figure 2 in Gross, 1994, while the sinusoidal oscillations of spiral chemotactic movement during aggregation (generated by excitations related continuously around loops) can be seen in Figure 3 of the same article.

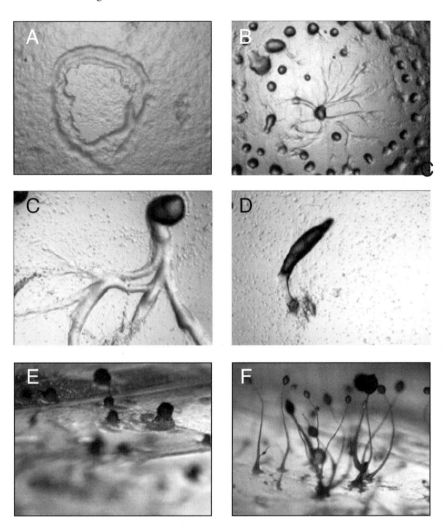

FIGURE 5.10. Life history of the cellular slime mold *Dictyostelium discoideum* from aggregation to differentiation following ultradian pulses of cAMP. Cellular slime molds exist as free-living cells (myxamoebas) found in soils where they feed on bacteria. Starved individual amoebas release rhythmic waves of cAMP that last 1–2 min, causing the myxamoebas to aggregate (*A–C*) to form a slug-like mass (*D*) that moves to a new area and stops. Aggregating cells result in both concentric (microscopic) and spiral (*A*) waves of amoebae moving toward centers of aggregation. Cells in anterior of slug become stalk cells, while posterior cells at the base of this stalk climb to top of the stalk (*E*) and become dormant spores (*F*) ready to germinate in warm and damp conditions. Photos by R. Sothern.

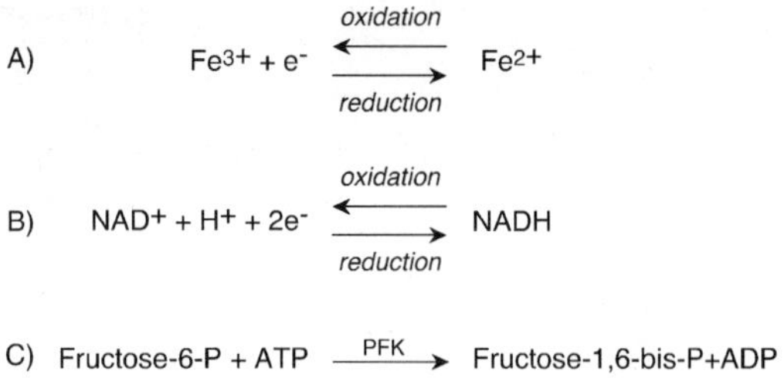

FIGURE 5.11. Examples of oxidation–reduction reactions (A, B) and the reaction catalyzed by phosphofructokinase (PFK) in glycolysis (C).

A) $Fe^{3+} + e^- \underset{reduction}{\overset{oxidation}{\rightleftarrows}} Fe^{2+}$

B) $NAD^+ + H^+ + 2e^- \underset{reduction}{\overset{oxidation}{\rightleftarrows}} NADH$

C) Fructose-6-P + ATP $\xrightarrow{PFK}$ Fructose-1,6-bis-P + ADP

topic of ultradian cardiac rhythms (Chapter 1), where it appears that geometric spiraling waves may have implications in cardiac arrhythmias. More specifically, some of the most serious cardiac arrhythmias, such as ventricular tachycardia and fibrillation, could be due to the behavior of spiral waves in the muscle (Davidenko et al., 1992).

Biochemical and Metabolic Models

A chemical or biochemical model may be relatively simple, focusing on one or two reactions, such as the gain or loss of electrons (Figures 5.11A,B), or it may include a vast network of biochemical pathways and processes.

Chemical Systems

The *Belousov-Zhabotinsky* reaction[12] is a classical example of an oscillation in a chemical system (cf. Hess, 1977). In this model, cerium ions oscillate between the oxidized (Ce^{4+}) and reduced (Ce^{3+}) state in a solution that contains bromate ions and acids. The "hands" of this "chemical clock" can be monitored by changes in optical density (e.g., Busse, 1973) or in color of the solution. A number of different chemical systems are known to oscillate. The observed changes in color can be quite dramatic (Figure 5.12), including those that display a periodic structure in the form of waves (spatiotemporal systems).

For the most part, chemical oscillators are far more complicated than just a single loss or gain of electrons. Often they include numerous reactions and catalytic activity coupled to other reactions in ways that provide feedback. In fact, periodic chemical reactions may be viewed as models for biochemical oscillations (cf. Hess & Boiteux, 1971). These oscillations are useful in formulating hypotheses that can be readily tested in a laboratory.

Biochemical Systems

In cases where the focus is on specific biochemical molecules and processes (e.g., metabolism), the "hands of the clock" may include nucleotides (e.g., NADH, ATP,

---

[12] B.P. Belousov (1893–1970) was a Russian chemist/biophysicist who discovered the oscillation and A.M. Zhabotinsky continued to study it.

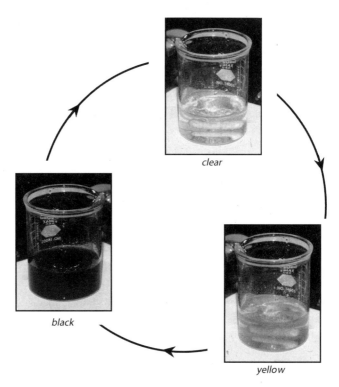

FIGURE 5.12. A series of photographs illustrating an oscillating iodine clock with the "hands" corresponding to a cyclic change in color from clear to yellow to black to clear, etc. At 25°C, this oscillation typically occurs with a period of 10–15 s (from a demonstration used in the laboratory of the authors and based on descriptions in Briggs & Rauscher, 1973).

cAMP), ions (e.g., H^+, K^+, Ca^{++}), enzymes, nucleic acids, or other molecules that can be monitored for a number of cycles. [Individuals who wish to review some basic biochemistry before proceeding further may find it helpful to examine the topics presented in Essay 5.1.]

Essay 5.1 (by WLK): Selected Biochemical Notes

Enzyme Reactions—Enzymes are a special group of proteins that function as *catalysts* by decreasing the energy of activation in chemical reactions (i.e., making it easier for a reaction to occur). The region of the enzyme where it combines temporarily with the substrate is called the *active site*. Certain enzymes have additional sites, known as *allosteric sites* (*allo* = other), where other molecules can "fit." When molecules are temporarily bound to these sites, they can have either a stimulatory or inhibitory effect on the activity of the enzyme. Allosteric enzymes, such as phosphofructokinase (PFK), as well as allosteric effectors (e.g., ATP, ADP, etc.), are important components in a number of biochemical pathways. Phosphofructokinase is one of many *kinase* enzymes, which transfer phosphate groups from ATP to other molecules, including proteins (e.g., protein kinases). Kinases serve as important components in many molecular models of circadian biological clocks.

Oxidation/Reduction—Oxidation–reduction reactions involve the transfer of *electrons* between molecules or ions; frequently—but not always—these electrons are transferred as part of hydrogen atoms. The molecule losing electrons is *oxidized*, and the molecule receiving electrons *reduced*. Oxidation therefore may be defined as a loss of electrons and reduction as a gain of electrons (e.g., Fe^{+++} to Fe^{++}). A nucleotide can serve as an excellent example of a participant in such a reaction. A *nucleotide* molecule consists of three parts: a nitrogen-containing base (e.g. adenine, nicotinamide, riboflavin, thiamine), a 5-carbon sugar (ribose or deoxyribose), and one or more phosphate groups. Sometimes the essential parts of two nucleotides are linked together into a single molecule referred to as a *dinucleotide*. Appropriate sequences of nucleotides linked together form molecules of DNA and RNA (see Essay 5.3: From Genes to Proteins and Mutants).

Individual nucleotides or their derivatives frequently serve also as *coenzymes*, which are relatively low-molecular weight molecules (much smaller than enzymes) that are essential in many metabolic reactions. Some function as group transfer coenzymes. An example would be adenosine triphosphate (ATP), which can transfer one of its three phosphate groups to a metabolic intermediate, thereby phosphorylating it (e.g., ATP + glucose → ADP + glucose-6-phosphate). Again, the process of phosphorylation is an essential component in many molecular models that will be discussed in another section of this chapter. In addition to group transfer coenzymes, there are oxidation–reduction coenzymes that oxidize or reduce metabolic intermediates. An example would be nicotinamide adenine dinucleotide (NAD), which exists in the oxidized (NAD^+) or reduced (NADH) form.

Oxidation–reduction reactions are important also in those aspects of metabolism that relate to energy, since oxidation of any molecule results in the release of some energy from it. It can thus be said that more reduced molecules possess the higher energy content. Therefore, NADH and NADPH are energy-rich molecules.

Glycolytic Oscillations

Models and/or systems that focus on biochemical oscillations and oscillators have been studied extensively (cf. Betz & Sel'kov, 1969; Hess & Boiteux, 1971; Pye, 1971; Chance et al., 1973; Goldbeter & Caplan, 1976; Hess, 1977; Estabrook & Srere, 1981). Glycolysis[13] may well be one of the most common processes studied as a model system, especially in yeast (*Saccharomyces*). Levels of the nucleotide NADH provide the "hands" of the oscillator and can be monitored for many cycles in populations of cells, as well as in cell-free extracts. The mechanism controlling the observed oscillations of NADH includes a number of factors, although an important point of regulation is a reaction catalyzed by the enzyme phosphofructokinase (PFK) (Figure 5.11C).

PFK provides a key control point for the overall regulation of glycolysis, and its activity in turn is influenced by a wide variety of substances. Of most importance

---

[13] Glycolysis (glucose and Gr. *lysis*, loosening) is an energy-producing process involving the breakdown of glucose to pyruvic acid.

in terms of the role of PFK in glycolytic oscillation models is the fact that the enzyme in yeast is activated allosterically by one of its products, ADP, the concentration of which also can (1) affect the rate of other reactions of glycolysis and (2) be itself influenced by still other enzymatic activities in the cell or system. In some ways, this type of feedback is much like that of prey–predator cycle discussed earlier, where populations (animals rather than chemical molecules) increase and decrease in a cyclic manner. It is important to point out that the glycolytic oscillations discussed here are created by "non-natural" conditions in a laboratory and probably are the mere consequences of a feedback-controlled system of glycolysis.

The glycolytic oscillations induced in the laboratory have ultradian periods measured in seconds or minutes. However, these and other high-frequency oscillations could, with the inclusion of certain assumptions, serve as components of models for circadian oscillators. In fact, a simple model that gives rise to circadian oscillations has been proposed, which is based on an *allosteric effect* (e.g., a change in the shape and activity of an enzyme) and the involvement of two enzymes (Pavlidis & Kauzmann, 1969).

Nucleotides and Enzymes

Molecules, such as the nucleotide cAMP (adenosine $3', 5'$-cyclic monophosphate), have been assigned major or specific roles in certain models, such as being a *messenger, coupler*, or as a particular type of *activator*. Levels of cAMP display a free-running rhythm in certain biological systems (Dobra & Ehret, 1977) and have a role in the periodic aggregation of *Dictyostelium* (Gerisch, 1976, 1987; Goldbetter & Caplan, 1976). One biochemical model that includes cAMP, ATP, AMP, adenyl cyclase, and phosphodiesterase, assumes limit-cycle oscillations to occur through the effects of allosteric feedback (Cummings, 1975). The nucleotide ATP, which may be a product or substrate in a biochemical reaction, is often required for membrane transport. Rhythmic changes in enzyme activity are common features in the temporal organization of life (cf. Koukkari & Soulen, 1981; North et al., 1981; Table 1 in Hardeland et al., 2003). Enzymes known as the ATPases and the kinases, which transfer phosphate groups from ATP, are especially important in many of the molecular models that will be discussed elsewhere in this chapter.

Membrane Models

For the most part, membrane models (Table 5.3) of biological clocks have given way to molecular models. Their inclusion or emphasis in the broad scheme of modeling is now primarily from either a historical perspective or the fact that so much of the biology of cells and their temporal organization (e.g., photoreceptors, cell cycles, signaling, transport, etc.) involve membranes. Chemical reactions and metabolic processes, including those that were discussed in the preceding section,

TABLE 5.3. Membrane model mechanisms proposed by various authors (cf. Edmunds, 1988, 1992) (CR = circadian rhythm; AT = active transport).

Author(s), year	Comments and Key Points of the Model
Sweeney, 1974b	CR of photosynthesis continues without nucleus in an alga; transplanted nucleus controls phase of CR; CR generated by feedback; active transport (AT) across organelle membrane dependent upon distribution of "X" molecule being transported; AT stops when critical level reached; even distribution established by passive diffusion; light resets CR by activating transport
Njus et al., 1974	Three major points: role of ions and chemicals (K^+), effects of light (ion gates), and lipids; temperature compensation dependent upon variations in fatty acids; ion concentrations related to passive and AT channels; lateral diffusion of intercalated (embedded) particles in membrane
Wagner et al., 1974	Circadian changes in sensitivity to photostimulation determined through energy-dependent structural changes of membranes to which photoreceptors are bound
Schweiger & Schweiger, 1977	Coupled translation-membrane model: synthesis of essential membrane proteins regulated by feedback
Chernavskii et al., 1977	Mathematical model by which plasma membrane regulates cell cycle
Burgoyne, 1978	Synthesis of a postulated membrane protein that is involved in ion transport and regulated through feedback by the concentration of intracellular monovalent ions; translation of mRNA sensitive to changes in monovalent ion levels
Konopka & Orr, 1980	Establishing and depleting of an ion gradient across a membrane: ion pump establishes gradient during subjective L; gradient depleted through open photosensitive ion channels during subjective D
Satter & Galston, 1981	Turgor regulation involving K^+, Cl^-, other ions; includes H^+ pumps, H^+/sucrose symporters, OH^-/anion antiporters, and H^+ and K^+ diffusion pathways
Chay, 1981	Ion gradients across membrane; active H^+ transport; key enzymes with pH-dependent activity and either translocating H^+ from outside or producing it
Iglesias & Satter, 1983	Coupled H^+/K^+ fluxes and their direction promoted by light and/or darkness

occur within specific areas or compartments of the cell. Within these compartments, molecules, ions, substrates, products, catalysts, etc., often depend upon structures that provide surfaces onto which to bind. The major cellular component that provides such a surface, as well as separating the compartments, is a *membrane* (Essay 5.2).

Essay 5.2 (by WLK): Membranes and the Phospholipid Bilayer

Biological membranes (Latin *membrana* = skin) are found in all living cells. They surround protoplasts, nuclei, vacuoles, mitochondria, plastids, as well as various sac or tube-like structures, such as thylakoids, vesicles, and the endoplasmic reticulum. A membrane that surrounds cellular structures may be either a single bilayer, such as the membranes of vacuoles and vesicles, or a double bilayer, as in the case of the nuclei, mitochondria, and chloroplasts of eukaroytic organisms. Membranes, such as those that

160   5. Biological Oscillators and Timers: Models and Mechanisms

surround the protoplast, are relatively thin. When observed in cross-section under an electron microscope, they may appear as two dark lines separated by a lighter area (Figure 5.13). The general design of a biological membrane is described as a *fluid mosaic* (Singer & Nicolson, 1972), composed primarily of a phospholipid bilayer that is intercalated (embedded) with protein (Figure 5.14).

The word *lipid* is derived from the Greek, *lipos* for fat, which explains why lipids are commonly referred to as fats and oils. Fats are solid at room temperature (20°C), while oils are liquid at room temperature. Both are insoluble in water and have diverse chemical structures. Fats and oils are called triglycerides. They are formed from glycerol, a small molecule with three hydroxyl groups (–OH), and three fatty acids, each with carboxyl group (–COOH) and a hydrocarbon tail (see Figure 5.15).

What makes the lipid a phospholipid is the substitution of a compound containing a phosphate group for one of the three fatty acids (Figures 5.16). This portion of the phospholipid molecule has one or more charged atoms. Because it attracts water molecules, this part is referred to as the *hydrophilic* (*hydro* = water; *phil* = loving) head. The two fatty acid parts of the molecule are *hydrophobic* (*hydro* = water; *phobos* = fearing) and form the interior of the membrane. In a water environment, these hydrophilic and hydrophobic forces cause the membrane to be a bilayer with the two

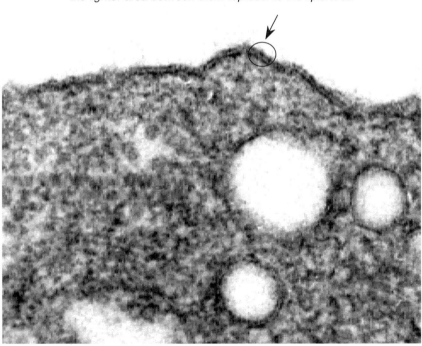

FIGURE 5.13. An electron micrograph of a section of a cell, illustrating a membrane in the circled area (courtesy of Mark Sanders, Program Director, Imaging Center, University of Minnesota).

Membrane Models   161

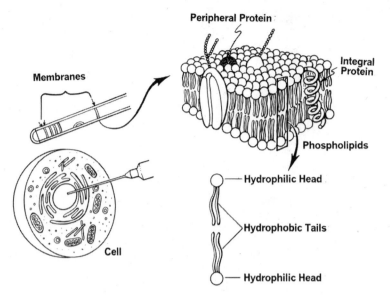

FIGURE 5.14. Diagram illustrating various membranes of cellular components that could be encountered hypothetically during a biopsy of a cell (*left*), a fluid mosaic structure of a membrane (*top right*), and two phospholipids (*lower right*) showing hydrophilic heads and hydrophobic tails.

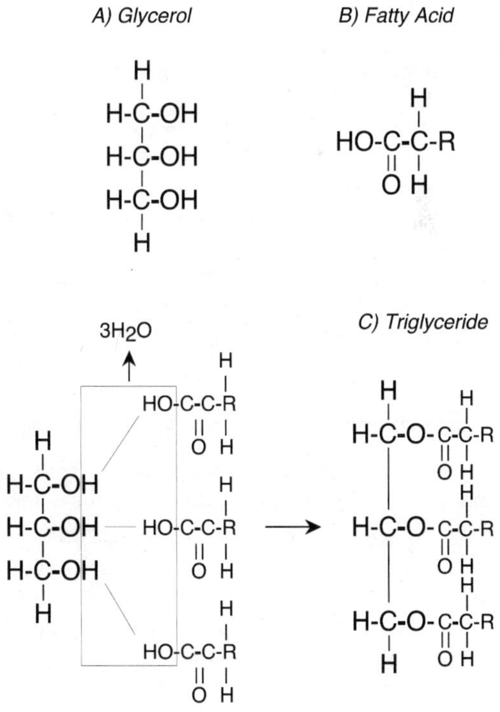

FIGURE 5.15. Structures of glycerol (*A*) and a generic fatty acid (*B*) and how they combine to form a triglyceride (*C*). The letter 'R' represents various organic chains, the presence of which forms specific fatty acids.

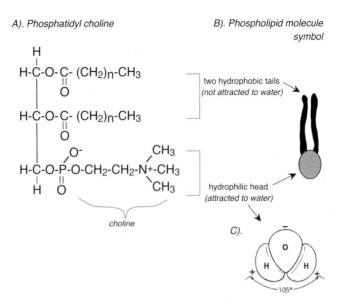

FIGURE 5.16. A phospholipid molecule, phosphatidyl choline (*A*), and the symbol used to illustrate a phospholipid in a membrane (*B*). Water molecule (*C*) illustrates the angle between two hydrogen atoms attracted to the oxygen atom. The attraction of the negative (−) oxygen to the positive (+) hydrogen of another molecule produce a hydrogen bond.

fatty acid tails of one phospholipid facing the fatty acid tails of another phospholipid molecule (Figure 5.14).

Two major groups of proteins are found in membranes, the *peripheral membrane proteins* and the *integral membrane proteins* (Figure 5.14). The peripheral proteins do not penetrate the lipid bilayer, but are attached to the integral membrane proteins or phospholipid molecules by various bonds. Because the environment of the cell on one side of the membrane is likely to differ from what it is on the other side (e.g., with respect to pH), so too are the proteins found on each of the two surfaces.

Integral membrane proteins, unlike the peripheral membrane proteins, can extend into the bilayer, or completely through it. These proteins provide a possible connection from one side of the membrane to the other (Figure 5.14). Some integral membrane proteins move within the phospholipid bilayer, possibly forming gates or channels. Such movements have been incorporated into models of circadian oscillators.

One reason that membrane models were so appealing in the past was the fact that many of the variables that display rhythms could be traced to changes that recurred between or among "biological compartments," especially processes that included intra- and/or inter-cellular transport of ions and molecules from one area to another (Figure 5.13). Not only can the composition and concentration of substances be different on either side of the membrane, but membranes also can create these differences. Depending upon which membranes are involved, they can transport and pump materials, trap light, synthesize molecules, recognize, and censor what may enter or leave, regulate events, and participate in various metabolic reactions.

These regulatory activities have a temporal organization, which made membranes a prime structure for studying the mechanisms of biological clocks (Table 5.3). An understanding of the structure and function of two principle components, lipids and proteins, provides insight, not only as to how a membrane could function as an oscillator or clock, but how these two chemicals can serve as components of other models as well.

Lipids and Proteins

One could question what the structure of lipids has to do with biological oscillators. Perhaps very little, if it were not for the effect that temperature has on the fatty acid composition of cellular membranes on the one hand, and the properties of a circadian oscillator to compensate for changes in temperature on the other. These two features, *fatty acid composition* and *temperature compensation*, have been important components in a number of membrane models, since in order to maintain accuracy (i.e., a stable period), a circadian oscillator, like Harrison's marine mechanical clock (see Chapter 3 on Time), must not be affected significantly by changes in temperature.[14]

How can this be modeled as an oscillator? Ratios in the levels of saturated and unsaturated fatty acids appear to be temperature dependent. Results from studies of the membrane lipid composition of *E. coli* cultured at different temperatures have shown that there is an increase in saturated fatty acids and a decrease in unsaturated fatty acids when the temperature of the culture conditions is elevated (Haest et al., 1969). When the temperature is lowered, the proportion of unsaturated fatty acids increases (Shaw & Ingraham, 1965). Similarly, in the fungus *Neurospora crassa*, there is an increase in the degree of unsaturation of fatty acids at lower temperatures (cf. Lakin-Thomas et al., 1997). Models of a saturated fatty acid and an unsaturated fatty acid are shown in Figure 5.17.

The restructuring of a lipid bilayer to adapt to changes in temperature by desaturating a significant percentage of single bonds in fatty acids to double bonds, or saturating double bonds to form single bonds, must occur during spans that are well within a circadian domain (ca. 24 h). This has been found to be true in both prokaryotic and eukaryotic systems, where the process may occur within three to four hours (cf. Rao, 1967; Okuyama, 1969). How the restructuring of lipids could relate to function is evident in the term "fluid mosaic." Conceptually, the greater the fluidity, the more rapid the movement of substances. Not only is the available space and possible movements between straight (saturated) vs. kinked (unsaturated) chains different, but the thickness of a lipid bilayer may also change due to temperature-related lengthening or shortening of the hydrocarbon chains (cf. Robertson, 1983).

---

[14] A Q_{10} of approximately one (Q_{10} = 0.7–1.3) is a major distinguishing feature of circadian rhythms (see Chapter 2).

164   5. Biological Oscillators and Timers: Models and Mechanisms

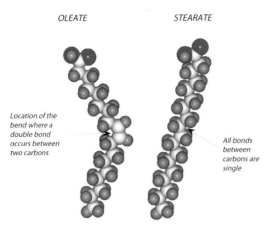

FIGURE 5.17. Model of a saturated fatty acid (stearate), where all bonds between carbon atoms are single, and an unsaturated fatty acid (oleate), where a double bond between two carbons causes the molecule to bend or kink (arrow) (courtesy of Leonard Banaszak and James Thompson, Dept. of Biochem., Mol. Biol., & Biophysics, University of Minnesota).

Fungi, especially mutants of *Neurospora*, have provided excellent systems for studying the physiology of fatty acid composition as it applies to biological rhythms. For example, the "normal development" of one mutant (*bd csp cel*) requires the addition of fatty acids to the medium. The circadian period for sporulation of this mutant is 21.5 h, but it can be lengthened by supplementing the medium with an unsaturated fatty acid (oleic, linoleic, or linolenic acid). This increase in period length can be reversed, however, by the addition of the saturated fatty acid, palmitic acid (Brody & Martins, 1979).

In matters relating to temperature compensation, proteins may well contribute to thermal adaptation (Hazel, 1995). A special group of proteins known as the NADH oxidases (ECTO-NOX proteins) occur on the external surfaces of cells and may have a special role in biological timing. The ECTO-NOX proteins display temperature-compensated ultradian periods that might be ultradian pacemakers of the cellular circadian clock (Morré et al., 2002). Additional experiments are needed to further test the role that specific proteins serve as biochemical ultradian drivers of biological clocks.

Only a slight change in the structure of a protein can affect function and stability. Amino acids (Figure 5.18) are the building blocks of proteins and the substitution of a single amino acid may be sufficient to alter thermal stability. For example, the replacement of threonine by isoleucine in malate dehydrogenase from a thermophilic bacterium modifies catalytic functions and causes a slight decrease in the heat stability of the enzyme (Nishiyama et al., 1986). While high temperatures can denature most proteins, there are proteins that can adapt to function under extremely high temperatures. Prime examples are found among the hyperthermophilic archaebacteria that live in near boiling, harsh acidic environments (e.g., in deep sea volcanic faults) (cf. Somero, 1995).

In addition to adaptation, special proteins, known as *heat shock proteins*, are synthesized when the temperature is suddenly raised, thus allowing for developmental and metabolic processes to continue. Heat shock proteins have been found in many of organisms, such as the fungus *Neurospora* (Plesofsky-Vig & Brambl,

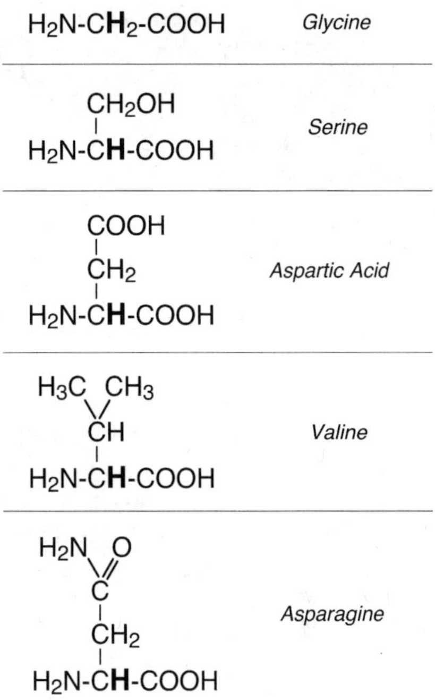

FIGURE 5.18. Diagram illustrating how the structures of four amino acids can be derived from glycine by substituting for the middle hydrogen atom.

H_2N-CH_2-COOH Glycine

$$\begin{array}{c} CH_2OH \\ | \\ H_2N-CH-COOH \end{array}$$ Serine

$$\begin{array}{c} COOH \\ | \\ CH_2 \\ | \\ H_2N-CH-COOH \end{array}$$ Aspartic Acid

$$\begin{array}{c} H_3C CH_3 \\ \diagdown \diagup \\ CH \\ | \\ H_2N-CH-COOH \end{array}$$ Valine

$$\begin{array}{c} H_2N O \\ \diagdown \diagup \\ C \\ | \\ CH_2 \\ | \\ H_2N-CH-COOH \end{array}$$ Asparagine

1985), the fruit fly *Drosophila* (Ashburner & Bonner, 1979), and the soybean *Glycine* (Lin et al., 1984), as well as the mammalian brain, heart, and kidney (Cosgrove & Brown, 1983). Given that the expression of heat shock genes is enhanced following exposure to a higher temperature, their specific role, if any, in circadian periodicity is not clear (cf. Rensing et al., 1997). Resistance of proteins to high temperature in living systems may be further aided by the presence of low-weight protein stabilizers.

Transport and Feedback

An important feature of most membrane models of biological oscillators is a *feedback system*, which could be provided by ion concentration and ion transport channels (cf. Njus et al., 1974). These channels are formed by proteins and have been visualized in drawings (Figure 5.19) as coils, tubes, etc. Ions such as K^+ are transported from one side of a membrane to the other, as well as between cells. In cells and tissues where intercellular transport occurs, ion concentrations contribute to osmotic changes, which in turn underlie the "hands" of many biological rhythms. Classical examples include the ultradian and circadian movements of plant structures, such as the circumnutation of twining shoot tips, the movements of leaves, and the opening and closing of the stomatal pores.

Leaves of legumes, such as *Albizzia* and *Samanea,* have special anatomical structures that function as a hinge (pulvinus), where changes in cellular levels or ratios of K^+ and Ca^{++} between different regions or cells follow a circadian rhythm

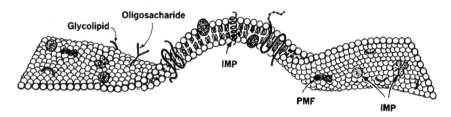

FIGURE 5.19. A biological membrane in the form of a "ribbon-like" phospholipid bilayer that is twisted to show the outer top surface (*left*), side section (*center*), and inner lower surface (*right*). Illustrated are various integral membrane proteins (IMP), peripheral membrane proteins (PMP), a glycolipid chain, and an oligosaccharide chain.

(Figure 5.20). These rhythmic changes in ions and turgor often drive the rhythmic movements of leaves[15] and leaflets. Interestingly, the addition of some ions, such as Li^+, are known to slow down the circadian rhythms in petal movements of flowers (*Kalanchoe blossfeldiana*) and the circadian activity rhythm of a small mammal (a desert rat, *Meriones crassus*) (Engelmann, 1973).

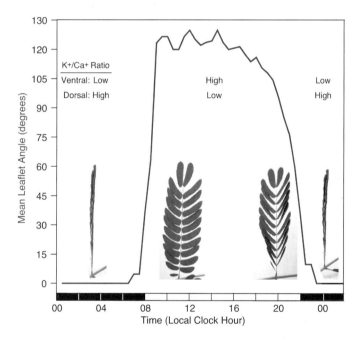

FIGURE 5.20. *Albizzia* pinnae illustrating closed and open leaflets and status of the K^+ and Ca^{++} ratios of the motor cell regions in the ventral and dorsal areas of the pulvinules ("hinge" areas) (cf. Satter & Galston, 1971). Diagram indicating angles in degrees formed by a pair of leaflets on a pinna of a plant maintained under a 24-h LD schedule (14-h L, 10-h D) in our laboratory.

---

[15] Some of the older literature on leaf movements also includes examples where daily rhythmic changes in leaf position are caused by oscillations in rates of growth and/or hormone concentrations (cf. Koukkari & Warde, 1985).

Biological membranes and their components are definitely within the "big picture" of biological clocks and circadian systems, but the central oscillator is best described by molecular models.

Molecular Models

Molecular models of biological clocks have contributed greatly to our understanding of the mechanisms of biological timing. The transcription of mRNA from DNA became a component of molecular models during the late 1960s,[16] but the major breakthrough that contributed to the building of nearly all future models of the biological clock was initiated with the identification of "clock" genes and mutants in the 1970s (Konopka & Benzer, 1971; Feldman & Hoyle, 1973).

At the very core of most molecular models is the basic dogma for molecular biology, where information flows from DNA to RNA to protein (see Essay 5.3). It is here that a number of possible gene-regulated events take place, including the expression of various clock genes that can negatively control their own transcription. In other words, cycling occurs at the mRNA and/or protein level of structural organization (cf. Harmer et al., 2001).

Essay 5.3 (by WLK): From Genes to Proteins and Mutants

Genes are the information-carrying units of *heredity* and are located in the *nucleus* of eukaryotic cells in structures called *chromosomes* (Figure 5.21). They are also found in mitochondria and chloroplasts, but our discussion here will focus on events that take place in the nucleus and on molecular machines called *ribosomes*. The molecular structure of a gene is DNA (deoxyribonucleic acid), a double-stranded helix composed of sugar (deoxyribose), phosphate groups, and four base *nucleotides* called adenine (A), thymine (T), guanine (G), and cytosine (C). The information contained in a gene is expressed in the organism, be it a plant, animal, or any other living thing, as a *protein*.[17] This flow of information from DNA to the synthesis of a particular protein is accomplished in a series of steps that involves the participation of RNA (ribonucleic acid). RNA differs from DNA in that it is usually a single polynucleotide strand, contains ribose instead of deoxyribose, and uracil is substituted for thymine. In this essay we will mention three types of RNA based upon their function. The RNA that has the role of a *messenger* is called *mRNA*, while *tRNA* is involved in the *transfer* of amino acids and *rRNA* makes up part of the *ribosome*. Proteins are composed of chains of *amino acids*, and because there are 20 different amino acids and each can be used many times in a given chain, there are a large number of possible arrangements or sequences, each of

---

[16] The chronon concept was one of the early molecular models for the circadian clock (Ehret & Trucco, 1967). It described how the genetic material in the form of transcribable DNA would be essential for the hour to hour functioning of the circadian clock in eukaryotic cells (Ehret, 1980).

[17] There are some exceptions in this gene to protein process, which are not discussed here.

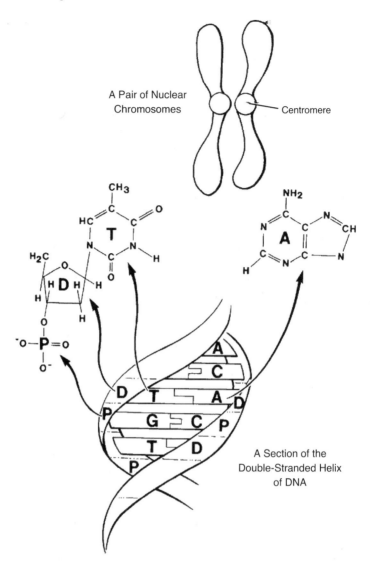

FIGURE 5.21. Diagrams illustrating a pair of nuclear chromosomes and a section of the double-stranded helix of DNA. The double helix is composed of deoxyribose (D), phosphate groups (P), and the four bases: adenine (A), thymine (T), guanine (G), and cytosine (C).

which could specify a particular protein. The exact sequence of amino acids in a protein chain is determined by the sequence of nucleotides in the corresponding part of DNA. These proteins, which are large *polypeptides*, play important roles in the structure and function of cells. Some proteins, usually having names that end in "*ase*," are *enzymes* and function as catalysts. For example, enzymes known as the kinases are important in many molecular models. A kinase catalyzes the transfer of a phosphate group from ATP to another molecule, a process that is called *phosphorylation*.

Molecular Models   169

How does the cell use DNA to make protein? (Figure 5.22). First, in the nucleus and with the help of the enzyme *RNA polymerase*, a region of one of the DNA strands partially unwinds and serves as a template for the synthesis of RNA. By a process called *transcription*, the genetic information from DNA is transferred through the base-pairing (synthesis) of the correct sequence of nucleotides to produce RNA. In this process, A always pairs with T in DNA and U in RNA, while C pairs with G in both. The transcribed section of DNA rewinds back to its two strand helical structure as the RNA moves out. Which strand of DNA is transcribed and where on the strand the RNA polymerase begins

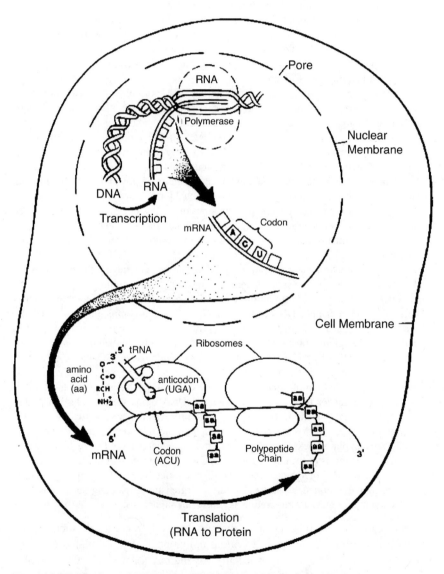

FIGURE 5.22. The flow of information from DNA to the synthesis of protein (polypeptide) and the series of steps involving transcription and translation.

its activity is a complex process that is regulated by *promoters*. The intricacies of this process are beyond the scope of this brief essay, except to state that the promoter is a specific sequence in a DNA strand that functions as the initial binding site for the RNA polymerase; one promoter per gene (eukaryotes) or gene set (prokaryotes). Regulatory proteins bind to the promoter and either inhibit or promote transcription of the gene.

After transcription, mRNA travels out of the nucleus to a ribosome. Here, amino acids are assembled into a *polypeptide* (protein). This process is called *translation* and involves the participation of tRNA, ribosomes, enzymes, and amino acids. Each specific amino acid, which has been produced by metabolic reactions that have occurred elsewhere, becomes associated with a specific tRNA. Enzymes, known as the *aminoacyl-tRNA synthetases*, catalyze the union of specific amino acids to specific tRNAs. These activities occur on the ribosome, where mRNA with the code it received from DNA (codons) and the tRNAs with their amino acids accomplish the translation process of building the polypeptide.

Each sequential set of three nucleotides of the RNA molecule is called a *codon*. Each codon represents the information or code that is necessary to either position a single amino acid into a protein chain, or to start or stop the addition of amino acids to the polypeptide chain. For example, the codons AGU (adenine-guanine-uracil) of mRNA code for the amino acid serine, whereas UGA is a "stop" codon leading to the termination of translation. Interestingly, AUG is a "start" codon and also the code for methionine.

Changes in DNA sequence (mutations) may result in the production of proteins with an altered amino acid sequence. This can occasionally result in a change in the function of the protein and a mutant phenotype. Mutations may be natural or they may be induced by chemical or other environmental agents. If mutations did not occur, there would be no evolution. Organisms that express mutations in one way or another are called *mutants* and are useful scientifically because they can be used to study the function of individual genes. In addition to mutants, organisms produced through mating of dissimilar parents, known as *hybrids*, have been instrumental in studying inherited traits.

Scientists can easily examine the expression of a gene using recombinant DNA technology. For example, DNA copies of mRNA, known as complementary or cDNAs, can be synthesized from mRNA using the enzymes *reverse transcriptase* and *DNA polymerase*. Regions of DNA that control gene transcription, called *promoters*, can be isolated from DNA by the use of various *restriction endonucleases*, which cleave DNA at given locations. The cDNA and the promotor regions can be combined into a single molecule with the help of repair enzymes called *DNA ligases*. After the molecule is attached to a carrier (e.g., a circular bacterial DNA unit called a "plasmid"), it is inserted into an organism. If the events proceed properly, the promoter will lead to the expression of the cDNA sequence. For example, if a cDNA copy is made from firefly luciferase and placed next to a promoter from a gene that is turned on by high temperature, then synthesis of the enzyme luciferase will be induced by high temperature. Its activity can be monitored by the amount of light emitted when the organism is exposed to exogenously applied substrate, luciferin. Organisms possessing such constructs (cDNA and promoter) are referred to as *transgenic* organisms.

Gene regulation can also occur at the post-transcriptional level. Such regulation is often complex, but is an essential component of molecular oscillators, which in one way or another, especially for the circadian clock, must involve *environmental receptors* for internal synchronization and account for *temperature compensation*.

Generally, components of an oscillator should satisfy four basic criteria (Kay & Miller, 1995): (1) they should oscillate in activity and have the same period as the overt rhythm; (2) fixing or holding ("pegging") the component to any constant activity level (peak, trough, etc.) should cause arrhythmicity; (3) rapid changes in the oscillator's activity should produce predictable phase shifts of the clock; and (4) signals (e.g., LD, temperature) that shift the phase of the overt rhythm should change the activity of the oscillator component within one cycle. A key component or starting place to describe most molecular models is the gene, although in reality, a number of genes or their proteins may be necessary for an oscillator to function properly.

Genes and Nomenclature

The molecular structure of a gene is a specific sequence of DNA (deoxyribonucleic acid), and the location it occupies on a chromosome is called the *locus*. By the use of forward genetic screens,[18] it has been possible to obtain "clock" mutants that have altered rhythm characteristics (see Figure 2.8 in Chapter 2 on Characteristics). These differences, such as the period being longer or shorter than the wild type, can be traced back to genetic loci where changes have occurred in the gene. The change in the DNA of a given gene may be as simple as a change in one nucleotide, which can lead to a change in transcription and the subsequent expression of the protein.

Major strides in the genetic dissection of biological clocks began during the early 1970s when the gene *period* (*per*) was discovered in fruit flies (Konopka & Benzer, 1971), and alleles[19] of the gene *frequency* (*frq*) were identified in fungi (Feldman & Hoyle, 1973).[20] Genes that are identified in one species may have homologues in other species, or may even play multiple roles in certain conserved molecular clock mechanisms, such as in *C. elegans*.[21] Italic print, letters, and

---

[18] In forward genetics, mutants of a species are created, often through the use of chemical or physical agents. The resulting mutant population is then screened for a specific phenotype, with the eventual goal to determine the gene(s) responsible for the observed phenotype

[19] For definition and characteristics of alleles, see Chapter 7 on Sexuality and Reproduction.

[20] Two decades later, the discovery of the first two mammalian clock genes, *clock* and *per*, were cited as one of the breakthrough runners-up of the year in 1997 (Science Magazine Editors, 1997). The following year, the discovery that biological clocks use oscillating levels of proteins in feedback loops to keep time and that this mechanism is preserved across species separated by 700 million years of evolution was cited as the first runner-up breakthrough of the year (Science, 1998).

[21] In the nematode *Caenorhabditis elegans, lin-42* resembles *per* in *Drosophila* and *tim-1* resembles *timeless* in the mouse (Jeon et al., 1999), while *kin-20* resembles *timeless* in *Drosophila* (Banerjee et al., 2005). In *Drosophila, per* oscillates with a 24-h periodicity, while in *C. elegans, lin-42* oscillates approximately every 6 h. Also, while there are genes *clk-1, clk-2,* and *clk-3* called "clock genes" in the *C. elegans* model system, they function in a general physiological clock determining development and lifespan, and not in a 24-h clock (Lakowski & Hekimi, 1999). This suggests a conserved molecular clock mechanism that is used for different types of biological timing (developmental and circadian), and thus multiple roles for circadian genes (Banerjee et al., 2005).

superscript are used to indicate these homologues or types. For example, a *per* gene could be written as *dper* (*Drosophila*), *hper1* (*Homo*), *mper* (*Mus*), or *dper^s* (*Drosophila* short period). Unfortunately, rules on nomenclature that would be consistent for all species have not been established, and variations exist in how letters, numbers, superscripts, and subscripts are employed to denote genes. For example, the human clock gene *period* has been denoted as *per*, *Per*, *hper*, or h*Per*. Also, for *Arabidopsis*, uppercase italics indicates the wild-type allele, while lowercase italics indicates a mutant allele (e.g., *TOC1* and *toc1*). Mutant clock genes are followed by a superscript (e.g., *per^s* or *per^l* for short and long mutants of *period* in *Drosophila*). A partial list of names and derivations of genes related to the molecular clock mechanism found in various organisms is provided in Table 5.4.

TABLE 5.4. Explanation of meaning or origin of some gene names related to circadian molecular clocks in one or more organisms and example (reference) where rhythmicity has been described.

Gene Name	Meaning or Origin[a]	Organism[b]	Reference
Bmal	Brain and muscle ARNT-like protein	M	Ikeda & Nomura, 1997
CCA1	Circadian clock associated 1	A	Alabadi et al., 2002
clk	Clock (circadian locomotor output cycles kaput)	M	Vitaterna et al., 1994
cry1	Cryptochrome 1	M	Kume et al., 1999
cyc	Cycle	D	Rutila et al., 1998
Dec1	Differentially expressed in chondrocytes 1	M	Honma et al., 2002
dbt	Doubletime	D	Kloss et al., 1998
ELF	Early flowering	A	Hicks et al., 2001
frq	Frequency	N	McClung et al., 1989
KaiA,B,C	"Cycle" or "rotation" (Japanese)	S	Ishiura et al., 1998
LHY	Late elongated hypocotyl	A	Schaffer et al., 1998
Mop3	Member of PAS superfamily	M	Bunger et al., 2000
per	Period	D	Reddy et al., 1984
Pdp1	PAR domain protein 1	D	Cyran et al., 2003
PRR	Pseudo-response regulator	A	Eriksson et al., 2003
Rev-Erbα	Reverse strand of TRα1 (homolog of c-erbAα)	M	Preitner et al., 2002
RIGUI	Ancient Chinese sundial	M	Sun et al., 1997
sgg	Shaggy	D	Martinek et al., 2001
TIC	Time for coffee	A	Hall et al., 2003
tim	Timeless	D	Myers et al., 1995
TOC1	Timing of CAB-1	A	Millar et al., 1995
vrille	Spiral [French]	D	Blau & Young, 1999
wc	White collar	N	Ballario et al., 1996
ZTL	Zeitlupe (German) = slow motion	A	Somers et al., 2000

[a]ARNT = aryl hydrocarbon receptor nuclear translocator; PAS = PER, ARNT, and SIM (single-minded); PAR = proline and acidic rich; TRα1 = thyroid receptor alpha gene 1; c-erbAα = cellular homolog of viral oncogene v-erbAα; CAB = chlorophyll a/b.
[b]A = *Arabidopsis* (plant); D = *Drosophila* (fruit fly); M = *Mus* (mouse); N = *Neurospora* (fungus); S = *Synechococcus* (cyanobacterium).

Clock Mutations

Much of the progress that has occurred in understanding the molecular mechanisms of biological oscillators can be attributed to the *genetic screening* for clock mutations (cf. Young, 1998). Briefly, a mutation represents a change in the DNA, although it may be quite simple. For example, both the short-*period* (per^s) and the long-*period* (per^l) mutations in *Drosophila* are missense[22] mutations, expressed as single amino acid substitutions in the PER protein (Baylies et al., 1987; Yu et al., 1987). In per^s, a serine is replaced by asparagine (G to A substitution), while in per^l, a valine is replaced by aspartic acid (T to A substitution) (see Figure 5.18). The arrhythmic mutation (per^0) is a nonsense mutation, which creates a premature stop-codon and results in a truncated protein.

Mutants with periods that differ from the normal species type have been identified from natural occurring populations, produced with chemical mutagens (see Figure 2.8 in Chapter 2 on General Features of Rhythms), or obtained by recombinant DNA technology. Transgenic mutants[23] have been of great value in studies that have focused on cyanobacteria (Kondo et al., 1994) and *Arabidopsis* (Millar et al., 1992).

Circadian: System and Clock

The word "model," as it applies to molecular circadian clocks, is used in a variety of contexts. In the scientific literature, the genus or common name of the organism is often added as part of the title to identify a specific model[24] (Table 5.5).

Before proceeding any further in our discussion, it is necessary to distinguish the difference between a circadian system and a circadian clock. Briefly, a *circadian system* consists of three parts: (1) an input pathway, which receives and transmits signals from external environmental synchronizers, to the central clock; (2) the central clock or oscillator that generates rhythmicity; and (3) an output pathway, which transmits temporal signals from the clock to physiological processes. This three-part system will be discussed in greater detail later in this chapter as a generalized schematic model for biological rhythms.

The *central circadian clock* is in itself a defined molecular entity, but instead of gears, springs, cogs, and balance beams that are engineered and arranged to make a mechanical clock function properly, the central circadian clock consists of positive

---

[22] A "missense" mutation occurs when a different amino acid is inserted into the polypeptide chain (protein).

[23] General background information is presented in the last paragraph of Essay 5.3. More specific information appears in the individual sections for models of cyanobacteria and *Arabidopsis*.

[24] This may be confusing, since the name (e.g., genus) remains the same year after year, although more is added to, or changed in, the model. However, taking a cue from the automobile industry, the problem could be resolved if one would cite the year of the reference where it was described.

TABLE 5.5. Names and biological categories applied to molecular circadian clock models.

Model	Category	Comments
Mammalian clock	Class	Theoretically, the Mammalia include all mammals, but "clock" studies are often restricted to hamsters, mice, rats and humans
Fly clock, mouse clock	Common names	Often used, but scientific species names should appear in the paper
The biological clock	Definite article	When used as a definite article, "the" (singular) often appears in the titles of biological clocks model, which in reality are not, as intended or perceived to be, all-inclusive
Circadian clock	Domain	Broad category applied to timers of circadian periodicity
Photoreceptive clock	Function	An example where emphasis is placed upon a physiological function
Drosophila clock, *Neurospora* clock	Genera	The genus (noun) is used as an adjective to indicate the organism, and assumes that the clock is a central oscillator
Cyanobacteria clock	Group	Group with many genera that have chlorophyll a, but term used often in reference to the genus, *Synechococcus*, and not to all cyanobacteria
Central oscillator model	Master timer	Provides a unifying concept in modeling, especially in regard to circadian rhythms

and negative feedback loops where *clock genes*[25] are turned on or off by the cycling proteins that they encode (Table 5.6). The entire process is sequential, with built-in delays and molecular receptors, producing a *self-sustaining* network that has a circadian rhythm.

Because research is progressing so rapidly in the area of molecular biology, models of biological clocks are continually being modified and changed to keep pace with the discovery of new genes and other components. For this reason, we will start with a simple generic model to discuss some general features of a circadian molecular clock (Figure 5.23), followed by an overview of circadian clock models for *Neurospora, Drosophila*, cyanobacteria, mammals, and *Arabidopsis*.

Transcription/Translation Feedback Loops

Molecular models of the circadian clock are based upon *feedback loops* that lead back to transcription. Within these loops, mRNAs and proteins of the clock continue

TABLE 5.6. Three circadian clocks and their positive and negative elements (proteins) in the feedback loop(s).

Clock	Positive Element	Negative Element
Neurospora	WC-1:WC-2 PAS	FRQ
Drosophila	CYC:CLK	PER:TIM
Mammals	BMAL1:CLK	PER and CRY

---

[25] A clock gene is defined as a gene whose protein product is required for the generation and/or maintenance of the circadian clock (cf. Balsalobre, 2002).

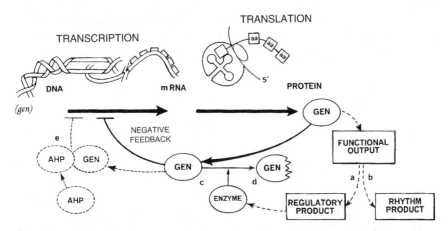

FIGURE 5.23. A simple molecular model of a biological clock showing a hypothetical gene *generic* (*gen*), a hypothetical protein (GEN), and a negative feedback loop. The possibility of a different negative feedback loop (e) is shown by including *a*nother *h*ypothetical protein (AHP), which could be connected with GEN by their PAS protein-dimerization domain. In molecular clock diagrams, negative regulation is denoted by a line terminating with a bar.

to be degraded and synthesized. As is commonly observed, the peak in mRNA levels occurs before the peak in protein (Figure 5.24). Calculations, which have been based upon data for the SCN and cells of rats, indicate that the accumulation of signaling peptides requires that both mRNAs and proteins decay with a *relatively short half-life* (e.g., of about 3 h) (Schibler et al., 2003). Although the individual molecules that comprise feedback loops in bacteria, fungi, plants, and animals may differ, the manner in which they are regulated is much the same (cf. Williams & Sehgal, 2001).

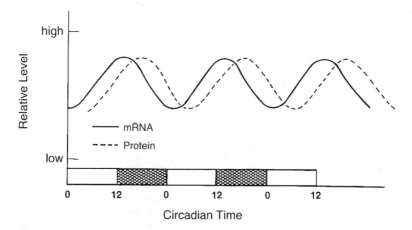

FIGURE 5.24. Circadian expression of mRNA level is shown here to peak about 5 h before the encoded protein level reaches its peak.

Transcriptional factors, which are proteins that bind to regulatory elements in promoters to regulate transcription, represent the positive elements, and the proteins that block or inhibit these factors represent the negative elements. Stated in the form of a general rule or principle, *"the transcription of clock genes give rise to messages whose translation produces clock proteins (negative elements) that serve to block the action of positive elements, which activate the clock genes"* (cf. Dunlap, 1999).

This positive/negative relationship is well illustrated in *Neurospora* where the protein complex, WCC (White Collar Complex), is a positive element that activates the *frequency* gene. In turn, the FRQ protein acts as a negative element in the loop. Usually, negative regulation is illustrated in diagrams by a line that terminates with a bar, instead of an arrow head.

The structure, and thus the function, of negative elements may be altered. For example, in Figure 5.23, if we consider the hypothetical gene *generic* (*gen*), the hypothetical protein (GEN) that is expressed by *gen*, could be modified by the addition of a phosphate group. In fact, *phosphorylation*[26] is known to play an important regulator role in the kinetics (i.e., changes or rate of change) of oscillations (e.g., Liu et al., 2000). Often, this can be attributed to the catalytic activities of *kinases* and their role in transferring phosphate groups to proteins. In the "real world" of circadian clocks, an excellent example of the critical role of phosphorylation can be found in the circadian clock of *Drosophila*. The *per* gene encodes the protein, PER, and the phosphorylation of PER is believed to regulate the stability of PER accumulation rate and thus the timing of the clock (Harms et al., 2004).

A time-dependent feedback loop, which can switch or regulate transcription by causing it to continue or to stop (delay), controls the production of the protein. The three rectangular boxes and four dashed arrows in Figure 5.23 show where both regulatory products (a) and rhythm products (b) could arise from functional outputs. Delays in the feedback loop (III) regulate transcription (I), which is expressed through the mechanism of functional output in rhythm products that oscillate (hands of the clock). In Figure 5.23, the amount of GEN that is available for negative feedback can be affected by action of an enzyme that is influenced by a GEN-controlled regulatory product. Although this model includes only one gene, two proteins, and a feedback loop, most oscillators have more than one feedback loop, plus a number of genes and proteins. Another *hypothetical protein* (AHP) has been inserted into this generic molecular model to illustrate a dimer (two proteins) composed of GEN and AHP. Two proteins can be connected by their *protein-dimerization domain* to form a complex (negative element). A single protein with a PAS dimerization domain[27] is in need of a partner (Dunlap, 1998). This is shown within the dashed line (e) in Figure 5.23. PAS protein–protein

---

[26] Phosphorylation is a process in which a phosphate-containing group is added to a molecule. Three of the many over-simplified analogies that have been made to explain what the phosphate group contributes are: (1) a support that provide stability; (2) a handle that will enhance chemical reactions; and (3) a switch that allows changes in a molecular structure.

[27] PAS domains are protein interaction domains. PAS is an acronym that is named for three proteins in which it has been identified: PER (period); ARNT (aryl hydrocarbon receptor

interaction domains are common to molecular clock systems (Dunlap, 1998; McClung, 2001).

Within the molecular circuitry of feedback loops are found transcriptional regulatory entities or elements. One example is an *E-box* (enhancer), a specific short-DNA-regulatory sequence that promotes the transcription of clock genes.[28] The clock proteins bind E-boxes and activate gene transcription. For example, in the model for interlocked feedback loops in *Drosophila,* CLK and CYC form complexes that bind E-box enhancers to activate the clock genes *per, tim, vri,* and *Pdpd1*. As these clock transcriptional regulators reach high levels, they feed back to control the transcription of their own genes, and thereby a self-sustained rhythm in gene expression is generated (Hardin, 2004). Similar-type binding of BMAL-1/CLK to E-box of mouse *per1* and *per2* genes accelerates their transcription in the mammalian circadian clock (Harms et al., 2004; Okamura, 2004).

Because clock proteins are degraded fairly rapidly, there must be mechanisms to accomplish this process. One example is the 26S *proteasome,* a large complex that recognizes and degrades proteins that are tagged with a small protein called *ubiquitin*.[29] Participating in the identification and operation of the process are ubiquitin-ligases, enzymes that must recognize the substrate, sometimes not until the substrate has been modified by a signal (e.g., light, hormone, etc.). Once the protein is tagged with ubiquin, it is recognized by the proteasome and degraded. In the circadian clock loops of *Arabidopsis*, the degradation of the TOC1 protein is triggered by the addition of ubiquitin to the F-box protein ZTL and the activity of ZTL is negatively regulated by light (Salomé & McClung, 2004). The fly protein TIM may also be degraded through an ubiquitin–proteasome mechanism (Naidoo et al., 1999).

Light

The resetting of the clock by light occurs in various ways, including the induction of clock genes or transcription factors, and the degradation of protein (Crosthwaite et al., 1997; Williams & Sehgal, 2001; Salomé & McClung, 2004). *Phytochromes* (PHYs) and *cryptochromes* (CRYs) are two common photoreceptors.[30] The phytochromes of plants absorb light most efficiently in the red to

---

nuclear translocator), and SIM (single-minded) (cf. Taylor & Zhulin, 1999). A basic helix-loop-helix (bHLH) domain, which functions in DNA binding, can accompany the PAS domain (Crews & Fan, 1999).

[28] A transcriptional factor is a protein that binds to a regulatory element (E-box) and regulates gene expression. The E-box is but one example, others (e.g., D-box, F-box) are included in some molecular models.

[29] Ubiquitin (ubiquitous = being everywhere, widespread) forms small chains on the protein that is to be degraded. This serves to attract larger molecules that do the actual degrading (e.g., an enzyme or proteasome).

[30] These photoreceptors, especially the phytochromes, have been studied extensively in plants where brief exposures of light inhibit the elongation of hypocotyls (photomorphogenesis) and influences flowering (photoperiodism). The hypocotyl is the region located below the cotyledons (e.g., the whitish stem-like part of a bean sprout that has been cultured or grown in darkness).

far-red regions of the spectrum, while the cryptochromes absorb light in the UV-A/blue region (cf. Millar, 2003). The expressions of phytochrome and cryptochrome genes are regulated by the circadian clock (Tóth et al., 2001). Another light-related component of the *Arabidopsis* clock is the gene *CIRCADIAN CLOCK ASSOCIATED 1 (CCA1)*, which is transitionally induced by red and far-red light (i.e., phytochrome) to encode the protein CCA1 (Wang & Tobin, 1998).

The photoreceptive pathway in the cyanobacterium *Synechococcus elongatus* clock includes a bacterial phytochrome family kinase (CikA), which is encoded by the gene *cikA* (*c*ircadian *i*nput *k*inase) (Schmitz et al., 2000). Two other genes, *ldpA* (*light-dependent period*) and *pex* (*period extender*) encode the proteins LdpA and Pex, which may be elements in the input pathway (cf. Ditty et al., 2003; Golden, 2003).

In *Drosophila*, levels of the clock protein TIM are reduced by light, a feature that correlates with resetting of the clock (cf. Williams & Sehgal, 2001). Very likely, the light acts upon the clock through blue-light cryptochrome photoreceptors (Emery et al., 1998). It should be noted, however, that in the absence of a *cry* gene, photic input from the visual system of *Drosophila* can, in some way, still drive the clock (Ashmore & Sehgal, 2003).

Cryptochrome genes are present in mammals, although the photoreceptor system for circadian synchronization in mammals appears to center largely upon *melanopsin*. This pigment is found in retinal ganglion cells, which project from the eyes to the suprachiasmatic nucleus (SCN) of the brain (discussed in Chapter 2).

Neurospora also respond to blue light (Linden et al., 1997), but the photoreceptor is WC-1 (White Collar-1), a protein with a flavin adenine dinucleotide (FAD) chromophore (He et al., 2002). Light/darkness serves as an independent synchronizer, as does temperature, for the circadian rhythm in *Neurospora*. Timing of the light can set or reset the phase of the rhythm. Advances or delays in phase, which are caused by the timing of light pulses, occur in response to the induction of *frq* and its status (i.e., when levels are rising or falling) through the operative mechanisms of the WC-1, WC-2, and WCC system (cf. Crosthwaite et al., 1997; Dunlap et al., 2004).

Temperature

In matters pertaining to synchronization and entrainment, the circadian clock in a variety of organisms is highly sensitive to temperature changes. A difference of 2.5°C or less between a high and low temperature during a cycle is sufficient to entrain the rhythms of some organisms (Sweeney & Hastings, 1960; Bünning, 1973; Francis & Sargent, 1979; Underwood & Calaban, 1987). At the molecular level, recent studies have shown that the expression of *Per1::luc* in supportive tissue (glia cells) of the SCN in the rat brain could be synchronized to a daily 1.5°C temperature cycle (Prolo et al., 2005). Yet, certain mammalian species are not sensitive to even large differences in the range of temperatures over 24 h (cf. Bünning, 1973; Moore-Ede et al., 1982).

In *Neurospora*, the effects of temperature appear to be mediated largely through translational control, in contrast to the effects of light, which primarily affects transcriptional regulation (Dunlap, 1999). As discussed earlier for *Neurospora*, temperature can be a stronger entraining agent than light (Liu et al., 1998). Temperature resets the circadian cycle by its effects upon the operation of the feedback loop and changes in the relationship among components of the clock (Liu et al., 1998).

Individual clock proteins may respond differently to light and heat. For example, dPER is sensitive to heat, but not to light (Sidote et al., 1998). In *Arabidopsis*, two circadian oscillators have been distinguished by their sensitivity to temperature (Michael et al., 2003). One oscillator, which regulates the expression of a certain gene (*CAB2*), responds preferentially to LD cycles. The other oscillator, which regulates the expression of a different gene (*CAT3*), responds preferentially to a temperature cycle.

Five Circadian Clocks

Much of what we know about models of molecular circadian clocks has come from five organismal systems: *Neurospora, Drosophila*, mammals, *Arabidopsis*, and cyanobacteria (Figures 5.25–5.29). For each of the five systems we will discuss briefly: (1) advantages of the system and why it is an attractive model; (2) examples of overt rhythms that are created by the output components of the circadian system; (3) some of the genetic highlights that have been revealed during modeling; and (4) their transcriptional/translational autoregulatory feedback loops and components.

Neurospora Circadian Clock

Neurospora crassa is a fungus, commonly called bread mold (Figure 5.25). The life history of this organism is illustrated in Figure 7.18 in Chapter 7 on Sexuality and Reproduction.

Advantages

The organism is easy to culture on a defined medium and analyze for metabolic and developmental changes or deficiencies. The nuclei of *Neurospora* are haploid, making it easy to detect mutant phenotypes. Today, much of what we take for granted in genetics and molecular biology (e.g., one gene responsible for producing one polypeptide) has an origin in studies conducted with *Neurospora*. Its genome has been sequenced. Because of its genetics and life history, the fungus *N. crassa* has provided one of the best model systems for studying and analyzing the molecular basis of circadian rhythms (cf. Bell-Pedersen et al., 2004; Dunlap & Loros, 2004).

Overt Rhythms

The classical developmental circadian rhythm in *N. crassa* is the patterning of asexual spore formation by hyphae[31] that develop along the surface of a solid-agar

---

[31] Hyphae (singular hypha) are long filament like structures.

180   5. Biological Oscillators and Timers: Models and Mechanisms

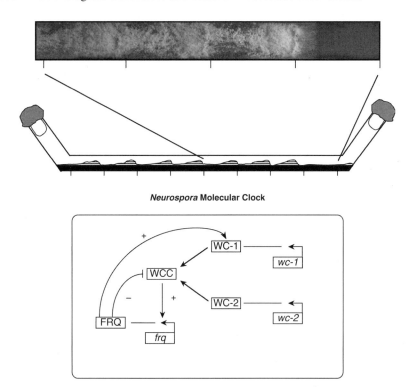

FIGURE 5.25. Simplified model of the *Neurospora* molecular clock. Photo shows top view of development of spore (conidia) formation of *Neurospora crassa* in a glass tube (courtesy of Van D. Gooch, University of Minnesota-Morris) and a side view (drawing) of a growth culture tube showing dense spans of spores (bumps) occurring at about 24-h intervals. The schematic shows an example of a positive element (WCC = white collar complex) and a negative element (FRQ) found in the *Neurospora* molecular circadian clock (adapted from Figure 1B in Dunlap & Loros, 2004).

medium (Pittendrigh et al., 1959; Sargent et al., 1966; Sargent & Briggs, 1967). The mycelium (mass of hyphae) produces dense areas (bands) of spores (conidia) alternating with less-dense areas and few spores. The rhythmic developmental pattern can be monitored in cultures maintained on an agar medium in petri plates or in glass race tubes (e.g., 1 cm diameter, 32 cm long with each end bent at a 45° angle, and closed with a cotton plug; see Figure 5.25). After a desired number of days, the cultures can be analyzed or they can be monitored continually with a time-lapse video system (e.g., Gooch et al., 2004). Molecular rhythms have been monitored by assays from liquid cultures (Loros et al., 1989), as well as with the use of firefly luciferase (Morgan et al., 2003).

Also, one of the major contributions of the system was the initial development of the field of molecular analysis of circadian output. The first systematic screens for clock-controlled genes (ccgs) were carried out in *Neurospora* (Loros et al., 1989), the term "ccg" was coined from this, and the general idea that clocks regulate metabolism through daily regulation of transcription of genes that are not a part of the core feedback loop, came from *Neurospora*. In general, this is now how output is viewed in circadian systems.

Genetic Highlights

The first of many highlights was the initial identification of clock mutant strains of *Neurospora* (Feldman & Hoyle, 1973), followed by the cloning of the clock gene *frequency (frq)* (McClung et al., 1989); six additional clock genes have since been cloned. Other highlights include finding the negative feedback autoregulation of *frq* by the FRQ protein (Aronson et al., 1994), elucidation of the molecular mechanism by which light (Crosthwaite et al., 1995) and temperature steps (Liu et al., 1998) reset circadian clocks, the discovery that a complex of two different proteins having PAS domains is the activator in circadian feedback loops (Crosthwaite et al., 1997), the cloning of the *wc-1* & *wc-2* genes (Ballario et al., 1996; Linden & Macino, 1997) and the proof that WC-1 is the circadian blue-light photoreceptor (Froehlich et al., 2002), the discovery of interlocked feedback loops in circadian systems (Lee et al., 2000), delineating the temporal fate of FRQ (Garceau et al., 1997; Cheng et al., 2001a), and its interactions with WCC (Cheng et al., 2001b; Denault et al., 2001; Merrow et al., 2001; Froehlich et al., 2003). The *frq* gene has also been identified as playing an essential role in endogenous entrainment to temperature cycles (Pregueiro et al., 2005).

Feedback Loops and Components

The central components of the circadian clock of *Neurospora* includes the gene *frequency (frq)*, *frq* mRNA, and the proteins FRQ, WC-1 (White Collar-1), and WC-2 (White Collar-2) (Lee et al., 2000; Dunlap et al., 2004). WC-1 and WC-2 interact by means of their PAS domains and form a WCC dimer (cf. Dunlap, 1998). WCC activates expression of *frq* leading to *frq* mRNA, which in turn encodes FRQ protein that feeds back to block the activation rhythms. In accord with the transcription/translation feedback loops that comprise most molecular models of the clock, the transcription factors (WC-1/WC-2) provide positive elements or drive, while the FRQ proteins that feed back to block the activation serve as the negative elements or drive (Figure 5.25). A second role of FRQ is seen in feeding forward to promote the synthesis of WC-1 from a constant pool of *wc-1* mRNA; this interlocked loop keeps the clock from winding down. The *Neurospora* molecular clock model continues to be modified. Very likely, noncircadian oscillators, such as the FLOs (FRQ-less oscillators) may become established as part of the overall circadian system, coordinated with the FRQ/WCC or other feedback loops (cf. Dunlap & Loros, 2004; Pregueiro et al., 2005).

Drosophila Circadian Clock

Drosophila is the genus name for a number of species that are commonly called fruit flies. They are often seen flying and gathering in areas where there are well-ripened fruits. *Drosophila melanogaster* is the species now used in most clock

182   5. Biological Oscillators and Timers: Models and Mechanisms

studies (Figure 5.26), although another species, *D. pseudoobscura,* had been used earlier for studying circadian rhythms (e.g., Pittendrigh, 1967).

Advantages

The reasons for using the species *D. melanogaster* are many and it is one of the best-described and analyzed molecular systems (Hall, 1995). Fruit flies are small

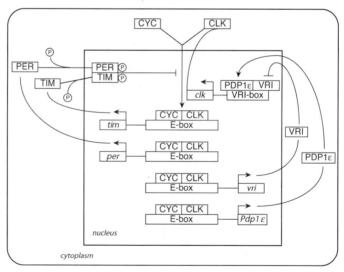

FIGURE 5.26. Simplified model of the *Drosophila* molecular clock. Photos of lateral and dorsal views of an individual fly (courtesy of Michael Simmons, University of Minnesota) and an activity monitor tray (courtesy of Michael Young, Rockefeller University, NY). The schematic shows the proteins PER, TIM, CYC, CLK, and E-box enhancers that activate four genes of the *Drosophila* circadian clock (adapted from Figure 1 in Hardin, 2004).

insects that have a short-generation time and are easily reared or cultured in the laboratory. Because geneticists have used fruit flies for decades, a large amount of genetic information is available; including gene maps, chromosome maps, and more recently, the complete genome sequence.

Overt Rhythms

The two common variables monitored for rhythmic behavior in *D. melanogaster* are *eclosion* and *locomotor activity*. Eclosion is a process in which the adult insect emerges from its pupal case. In nature this event occurs close to dawn. The rhythm is monitored in the laboratory by recording the number of flies that emerge at various times during each day. Locomotor activity is monitored and recorded as to the times when flies are actively moving or resting. Rhythms of molecular clock components, such as the TIM and PER proteins, are monitored from extracts of structures (e.g., heads) obtained at spaced-time intervals and assayed by various laboratory techniques (e.g., Western Blot Analysis).

Genetic Highlights

The isolation of three alleles of *period* (arrhythmic, long period, and short period) on the X chromosome of *D. melanogaster* by Konopka & Benzer (1971) helped usher in the molecular dissection of biological clocks (see Figure 2.18 in Chapter 2 on Characteristics). The effects of this single-gene mutation are seen in the circadian rhythms of eclosion and locomotor activity, as well as in a high-frequency ultradian rhythm (courtship song) (Kyriacou & Hall, 1980). About 14 years after it was discovered, the *per* locus of *Drosophila* was isolated (Bargiello & Young, 1984; Reddy et al., 1984).

Other highlights include the discovery of cycling gene regulation in this system (Hardin et al., 1990), the cloning of the *timeless* (*tim*) gene (Myers et al., 1995), identifying interactions between PER and TIM (Gekakis et al., 1995), and the construction of a model depicting various times during a 24-h clock cycle when the two genes and their proteins might generate circadian cycles in feedback regulation (Sehgal et al., 1995), and a secondary feedback loop involving *vrille*, *Pdp1*, and *dClock* (Cyran et al., 2003).

Feedback Loops and Components

The proteins PER, TIM, CLK (CLOCK), and CYC (CYCLE) are essential components of the molecular clock of *Drosophila*, although very likely, VRI (VRILLE), PDP1ε (PAR DOMAIN PROTEIN 1), and others should be included. CLK and CYC function as transcription factors, which positively regulate the expression of *per* and *tim*, while the PER and TIM proteins negatively regulate (i.e., "shut down") their own synthesis.

After translation has occurred, PER and TIM begin to accumulate in the cytoplasm (Hardin, 2004; Harms et al., 2004). Here, PER is phosphorylated by Doubletime (DBT), a kinase encoded by the *dbt* gene (Kloss et al., 1998, 2001; Price et al., 1998). The destabilization of PER by phosphorylation continues until high enough levels of TIM are reached to bind with PER and protect it from further degradation by DBT. The lag between the time of PER and TIM transcription and the binding of PER with TIM gives rise to a time delay for the clock that is necessary for producing an about 24-h circuit (Stanewsky, 2002; Harms et al., 2004). The entry of the PER–TIM complex into the nucleus appears to be influenced by DBT and other kinases. One of these proteins is the SHAGGY (SGG) kinase that influences phosphorylation and mobility of TIM (Martinek et al., 2001), although it has been stated that the phosphorylation of clock proteins is currently not linked to any particular kinase (Harms et al., 2004).

PER/TIM heterodimers move into the nucleus where they interact with the CLK/CYC heterodimers. Following the degradation of PER and TIM, the heterodimerization of CLK/CYC can occur to bind E-box enhancers to activate the expression of *per, tim, vri, and Pdp1ε* (see Figure 5.26). As is the case with most molecular models, the *Drosophila* model is being repeatedly refined as other components, domains, and genes are discovered.

Mammalian Circadian Clock

Of the approximately 4,000 species of living mammals, the mammalian molecular circadian clock has been modeled almost entirely on what had been discovered in mice (Figure 5.27).[32] On the other hand, biological rhythms and the circadian system have been studied in a large number of wild and domesticated species (see Chapter 9 on Veterinary Medicine), including humans.

Advantages

Human beings are interested in their own biological clocks and for a number of reasons, mice are a good substitute. Inbred strains and mutants are available for experiments in the laboratory. A wealth of genomic and cDNA[33] sequences are readily available in accessible databases (cf. Takahashi et al., 1994; King & Takahashi, 2000).

Overt Rhythms

The rest/activity cycle is the most common behavioral rhythm monitored in mice and other rodents. Generally, running wheels or activity boxes with electronic sensors and recorders are used to measure rhythmic outputs, a classic and well-used technique, even for the screening of mutants (see below).

Genetic Highlights

In 1994, a mutation that lengthens the circadian period of wheel running activity of mice was identified and named *Clock* (*c*ircadian *l*ocomotor *o*utput *c*ycles

---

[32] A clock mutant named *tau,* which shortens the circadian period, is known for the golden hamster (Ralph & Menaker, 1988).

[33] Information of cDNA is presented in last paragraph of Essay 5.3: From Genes to Proteins and Mutants.

Mouse Molecular Clock

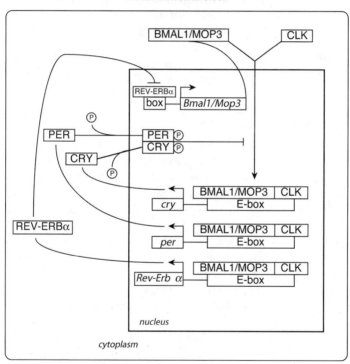

FIGURE 5.27. Simplified model of the mammalian (mouse) molecular clock. Photo of cage and running wheel used to monitor activity rhythms of the mouse (courtesy of Dwight E. Nelson, University of St. Thomas, St. Paul, MN). The schematic shows the proteins PER, CRY, CLK, BMAL1/MOP3, and E-box enhancers that activate three genes of the mouse circadian clock (adapted from Figure 3 in Hardin, 2004). Box indicates elements where transcription factors bind.

*k*aput)³⁴ (Vitaterna et al., 1994). Three years later, the *Clock* gene was cloned (King et al., 1997a,b) and it was shown that long period and loss-of-rhythm phenotypes in *Clock* mutant mice could be completely rescued via BAC (bacterial artificial chromosome) transgenesis (Antoch et al., 1997). A human *Clock* gene similar to the mouse gene was discovered later, and like mouse *Clock*, it is present in the SCN and other tissues (e.g., skeletal muscle) (Steeves et al., 1999). In 1997, the *per* gene in the mouse (m*Per*) and human (h*Per*) was identified (Tei et al., 1997) (each now denoted as *Per1*), as was m*Per2* in the mouse (Shearman et al., 1997). The following year, the CLOCK–BMAL1 heterodimers were shown to drive the positive component of *per* transcription (Gekakis et al., 1998).

The genes, *Mop3* and *Rev-erbα*, are essential components of the mammalian circadian clock. Their encoded protein, MOP3, is a heterodimeric partner of mCLOCK (Bunger et al., 2000) and REV-ERBα affects phase-shifting properties of the clock (Preitner et al., 2002). The expression of *Rev-erbα* serves as a major negative regulator of *Bmal1*, a positive component of the circadian clock, both of which display a rhythm in their mRNA levels that are in antiphase (cf. Yin & Lazar, 2005). Two regulators of the circadian clock, *Dec1* and *Dec2*, were also shown to be expressed in the mouse SCN, with a peak in the subjective day (Honma et al., 2002).

A number of genes now known to be components of the mammalian clock are homologous to the clock genes of *Drosophila*. These include *per* (m*per1*, m*per2*, m*per3*), *tim* (m*tim*), *clk* (m*clk*), and *cyc* (also called *Bmal1* or *Mop3*).

Feedback Loops and Components

The molecular model of the circadian clock continues to become more complex as new components from the broad field of molecular biology *per se* become incorporated into the functional mechanisms of the mammalian clock. In an overly simplistic framework, we can start with the transcription of the genes *per1* and *per2* (Okamura, 2004). In the loop, CLOCK/BMAL1 activates feedback regulation via E-box-mediated transcription (Hardin, 2004). The basic mechanism involving loops and the negative and positive elements is much the same as described in a more generalized manner in the main text of this section, only more complicated due to possible complexes that may be formed (Figure 5.27). For example, the mPER proteins that are translocated into the nucleus form various complexes composed of mCRY1, mCRY2, mPER1, mPER2, mPER3, and mTIM, which suppress the transcription of *mPer1* and *mPer2* by binding to CLOCK and BMAL1 (Okamura, 2004).

In the case of humans, where mutants are not readily available for research, the expression of clock genes has nevertheless been studied. For example, the relative RNA expression of *hClk*, *hTim*, *hPer1*, *hCry1*, and *hBmal1* of oral mucosa and skin from healthy human males collected over a 24-h span, each have a circadian profile similar to that found in the SCN and peripheral tissues of rodents (in relation of usual activity/rest schedules) (Bjarnason et al., 2001). Peaks in expression occur at different times throughout the day, with *per1* peaking in the morning,

---

[34] Note: *kaput* (German) = broken, unable to function or continue.

cry1 during late afternoon, and *bmal1* at night. At least in this initial study, and as in the mouse, the expression of *clk* and *tim* was not rhythmic.

In addition to these tissues, circadian clock genes (*hPer1*, *hPer2*, *hPer3*, *hDec1*) are expressed in a circadian manner in circulating human blood mononuclear cells during "time-free" conditions, with highest levels during times of habitual activity (Boivin et al., 2003). *hPer1* has also been shown to exhibit a 24-h cycle in blood polymorphonuclear cells (Kusanagi et al., 2004), while *hPer1*, *hPer2*, and *hCry2* displayed 24-h cycles in bone marrow CD34+ cells (Tsinkalovsky et al., in preparation).

Arabidopsis Circadian Clock

Arabidopsis thaliana is a small flowering plant (Figure 5.28) belonging to the mustard family, a family of plants that includes the radish (*Raphanus sativus*).

Advantages

Arabidopsis has been extensively used in genetics studies and its small genome has been sequenced. Unlike many of the common garden plants, *Arabidopsis* has a short life history (about 6 weeks from seed to mature plant).

Overt Rhythms

Included among the common overt rhythms studied are leaf movements, transpiration, hypocotyl elongation, and photoperiodic control of flowering. More recently, bioluminescence has been monitored following the insertion of marker genes (see the following section).

Genetic Highlights

Plants have played a major role in the study of circadian rhythms (see Chapter 3 on Time) and display a diverse list of rhythmic variables and periodicities (Koukkari & Warde, 1985). The prior lag of molecular information was caused by the lack of suitable genetic systems and the isolation of mutants with defective circadian rhythms (Carré, 1999). The successful genetic insertion of the firefly luciferase reporter gene into *Arabidopsis* (Millar et al., 1992) during the early 1990s ushered in an approach that has contributed greatly to the development of a circadian molecular clock model for plants. One of the major mutants isolated by this method was *TOC1* (*TIMING OF CAB EXPRESSION 1*), a gene responsible for shortening the period of a number of circadian variables, such as the movements of primary leaves and the transcription of the gene encoding a chlorophyll a/b binding protein (Millar et al., 1995). It is now known that over 450 genes cycle in this model organism (Harmer et al., 2000).

Feedback Loops and Components

Briefly, the molecular model of the *Arabidopsis* clock (Salomé & McClung, 2004) includes three major proteins, CCA1 (CIRCADIAN CLOCK ASSOCIATED 1), LHY (LATE ELONGATED HYPOCOTYL), and TOC1 (TIMING OF CAB EXPRESSION 1). The PHYs (PHYTOCHROMES) and CRYs

188   5. Biological Oscillators and Timers: Models and Mechanisms

(CRYPTOCHROMES) absorb light and induce the expression of CCA1 and LHY (transcription factors). Apparently, CCA1 is phosphorylated, but regulatory activities of CCA1 and LHY and how they are degraded remains unanswered. One of the negative regulatory activities of CCA1 and LHY is on the expression of the *TOC1* gene (Figure 5.28), while ZTL (ZEITLUPE = slow motion [German]),

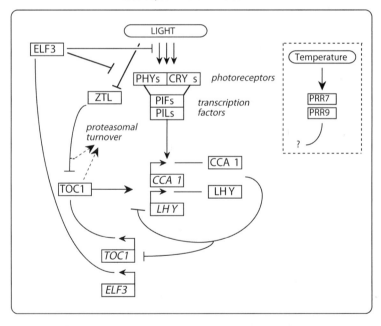

FIGURE 5.28. Simplified model of the *Arabidopsis* molecular clock (adapted from Figure 1 in Salomé & McClung, 2004). Photos of a tray of plants and an individual plant, all about 4 weeks of age. The schematic shows PHYs and CRYs photoreceptors, which induce the expression of *CCA1* and *LHY*. Shown in a dashed diagonal box is the temperature input on PRR7 and PRR9, which may be critical components for the temperature responsiveness of the clock. *Note*: ZTL inhibits TOC1 activity by targeting it for proteasomal turnover.

a protein encoded by *ztl* (Somers et al., 2000) mediates the degradation of TOC1 (Más et al., 2003). The activity of ZTL is negatively regulated by light, with a peak occurring close to dusk (near lights off) and trough near dawn. Its degradation is mediated by the proteasome (Kim et al., 2003). *TOC1* is a member of a small family of *PRR* genes encoding Pseudo Response Regulators, related to the response regulators of two component signaling cascades.

Quite recently, two *TOC1* relatives, *PRR7* and *PRR9*, have been suggested as elements of a second interlocking feedback loop (Farré et al., 2005; Nakamichi et al., 2005; Salomé & McClung, 2005). This second loop has also been implicated in the pathway of temperature input to the clock (Salomé & McClung, 2005). The model also includes the gene, *ELF3 (EARLY FLOWERING 3)*, whose mRNA oscillates (peak after dusk) and its encoded protein ELF3 (Hicks et al., 2001). Pertinent to our discussion is that the LHY mRNA encoded by the wild-type *lhy* gene shows a circadian rhythm with a peak near dawn. A mutant *lhy* gene encodes a constant expression of LHY and thereby disrupts this pattern, indicating that LHY is associated with the central clock (Schäffer et al., 1998). Many components of the clock, as well as their roles, remain to be identified. For example, a mutant named *TIME FOR COFFEE* (*tic*) that stops the clock in the morning (Hall et al., 2003), remains to be characterized. Also, while *PRR5* has been shown to play an antagonistic role to the CCA1 clock component (Fujimoro et al., 2005), it is not certain whether other PRR members are involved with clock function. Finally, the *Arabidopsis* molecular clock, like the other molecular clock models, includes a negative feedback loop and post-transcriptional regulation by phosphorylation, as well as proteasomal degradation.

Cyanobacteria Circadian Clock

Cyanobacteria are prokaryotic organisms and like other bacteria, they lack a true nucleus and membrane-bound organelles, such as mitochondria and chloroplasts. As recently as the mid and late 1980s, a major characteristic of circadian rhythms was that they existed only in eukaryotes (plants, fungi, animals). This long-held view changed with the discovery of circadian rhythms in cyanobacteria for photosynthesis and nitrogen fixation[35] (Grobbellaar et al., 1986; Mitsui et al., 1986, 1987), and a temperature-compensated 24-h rhythm in cell division (Sweeney & Borgese, 1989). The molecular model of the cyanobacteria circadian clock is based upon *Synechococcus elongatus* PCC 7942 (Figure 5.29), a fresh water strain that does not fix nitrogen (cf. Ditty et al., 2003).

---

[35] Nitrogen fixation is a process in which atmospheric nitrogen (N_2) is reduced to ammonia (NH_3). Cyanobacteria carry out photosynthesis (PS II), which produces oxygen. Because nitrogen fixation is inhibited by oxygen, the two processes appear to be incompatible. Yet, this does not appear to be a problem in some *Synechococcus* species where, even under free-running conditions (3d in LL), the two rhythms continue and are out of phase with each other (Mitsui et al., 1986).

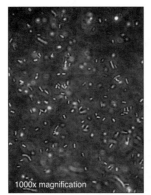
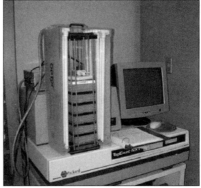
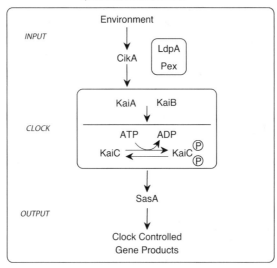

FIGURE 5.29. Simplified model of the Cyanobacteria molecular clock. Photos show a simple phase contrast view (1,000 × magnification) of *Synechococcus elongatus* PCC 7942 and the apparatus used to monitor luciferase activity (courtesy of Jayna L. Ditty, University of St. Thomas, St. Paul, MN). Three Kai proteins (KaiA, KaiB, and KaiC) comprise the basic circadian timekeeping clock of the Cyanobacteria *Synechococcus elongatus* (adapted from Figures 2 and 3 in Golden, 2003).

Advantages

Cyanobacteria are relatively easy to culture on a defined medium, and a number of mutants and genetic tools are available for *S. elongatus*. Actually, the cyanobacteria clock can be viewed as a unique structure called a "periodosome" that assembles and disassembles every day and defines the circadian period (Golden, 2004).

Overt Rhythms

Bioluminescence and protein cycling (Western blot) have become standard techniques for measuring rhythmic outputs (see Genetic Highlights). In earlier studies with different strains, nitrogenase activity (acetylene reduction method) and oxygen exchange (oxygen electrode) were used to monitor the output rhythms of nitrogen fixation and photosynthesis.

Genetic Highlights

Mutants having short-period, long-period, and arrhythmic features have been identified by the use of genetic engineering. The methods are somewhat unique in that a bacterial luciferase gene set (*luxAB*), which is controlled by a photosystem II[36] gene promoter[37] (*psbAI*), was inserted into *S. elongatus*. This procedure enabled scientists to monitor the luciferase expression in chemically mutagenized bioluminescent cells automatically (Kondo et al., 1994). Three genes cloned from *S. elongatus*, *kaiA*, *kaiB*, *kaiC*, are essential for the functioning of the circadian clock (Ishiura et al., 1998; Iwasaki et al., 1999), since disruption of any one of them nullifies circadian rhythms (Iwasaki & Kondo, 2000).

Feedback Loops and Components

The three Kai proteins (KaiA, KaiB, and KaiC) comprise the basic circadian timekeeping mechanism of *S. elongatus* (Figure 5.29) (Golden, 2003). Direct protein–protein association among KaiA, KaiB, and KaiC may be critical in generating circadian rhythms (Iwasaki et al., 1999). It has been suggested that KaiA functions as a positive element, activating *kaiBC* expression (Iwasaki & Kondo, 2000). It also promotes the phosphorylation of KaiC (Xu et al., 2003). The rhythmic expression of KaiC is a critical component of the clock (Xu et al., 2000) and may function as the negative element.

A KaiC-interacting histidine kinase, known as SasA, forms a secondary loop in the molecular clock model that amplifies the circadian rhythm (Iwasaki et al., 2000; Iwasaki & Kondo, 2004). Unlike the transcription/translation processes based upon negative feedback of clock genes in most model systems, the cyanobacteria clock can keep time independent of the *de novo* transcription/translation process (Tomita et al., 2005). Also, the feedback loop in cyanobacteria does not require specific clock gene promoters (Xu et al., 2003). Some features of eukaryotic circadian clocks that are found also in prokaryotic cyanobacteria, include a daily cycle in

---

[36] Two photosystems (PS II and PS I) use light energy in photosynthesis. In PS II, electrons are passed from water to a reaction center called P680. The reaction center for PS I is called P700. Together, PS II and PS I function to transfer the electrons to NADPH (molecule described in Essay 5.1: Biochemical Notes). The 680 and 700 represent nm (nanometers) in the light spectrum where highest efficiency absorption occurs in the reaction center.

[37] Promoters, as well as the procedure used, are introduced in Essay 5.3: From Genes to Proteins and Mutants.

abundance of key components and a rhythmic change in the phosphorylation state of a critical factor (Golden, 2004).

Models in Perspective

In this chapter we have discussed numerous models, mechanisms, and variations thereof. Each model has been developed to explore a given approach in order to better understand biological clocks and timing systems. There are unifying themes, such as mutual interactions among components (Edmunds & Cirillo, 1974), which have been described or reviewed in a number of articles and monographs, and many are cited in this chapter.

Mechanical models maintain their status in the broad field of biological rhythms and continue to be cited in the modern era of molecular biology. Mathematical models, like mechanical models, have both an analytical and a supplemental role in understanding the mechanisms of timing. The dynamic of these two models and how they apply to physical models (e.g., pendulum, harmonic oscillators) and their analyses (limit cycles, bifurcation) provide conceptual utility in understanding how biological oscillations and oscillators may function.

For the most part, membrane models have been replaced by molecular models, especially as generators of the central clock or oscillator. However, in matters relating to input pathways, environmental sensors, output pathways, and overt rhythms of the circadian system, membrane modeling remains an important entity. The same applies to biochemical models, although it is often difficult to distinguish between what is biochemical and what is molecular.

The study of molecular models is where the primary interest and research is currently centered. Molecular models are growing in complexity and with it is coming a better understanding of how biological clocks function. Many of the molecular models, or at least their components, need no longer be viewed strictly as models. Often, they explain and account for the mechanisms that occur at the molecular level and drive the overt rhythms of physiological, developmental, and behavioral variables that provide organisms with a temporal organization.

Generalized Schematic Model for Biological Rhythms

It seems only fitting that we conclude this chapter with a generalized model for the circadian system (Figure 5.30). The system includes environmental receptors (e.g., light-absorbing pigments), biochemical and/or biophysical pathways leading to the clock (primary oscillators), and output pathways leading to the oscillation or overt rhythm. Multiple environmental receptors and input pathways may function individually, interact with each other, or act in combination. For example, some plants need either a cyclic-LD cycle or a cyclic-temperature cycle during LL to be healthy (Hillman, 1956; Went, 1974), indicating that the internal clocking mechanism need not be directly linked with a pigment system (Went,

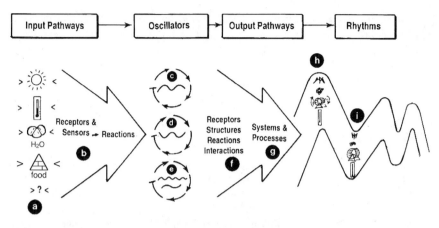

FIGURE 5.30. A simple conceptual model for a circadian system with five possible synchronizers (light–dark, temperature, water, food, and others) in the input pathway (*a*), with receptors, sensors, and reactions (*b*), leading to central oscillators (*c, d, e*) with negative and positive components that generate rhythms. Multiple output pathways with various receptors and reactions (*f*), as well as systems and processes (*g*), lead to the overt rhythms (e.g., leaf movements, protein levels, locomotor activity, and body temperature) with their time-dependent phases shown by peaks (*h*) and troughs (*i*).

1960). Results from these older classical studies with whole intact organisms have been supported by results from molecular studies in *Neurospora* where temperature was found to be the stronger entraining agent than light for the rhythmic levels of a clock protein (Liu et al., 1998).

The generalized model in Figure 5.30 includes two of the more common external synchronizers (light/darkness and high/low temperature), plus other synchronizers that may be more specific to given organisms or systems, such as water or humidity, feed/fast regimens, social cues and noise. These external agents do not "drive" the oscillator(s), but serve as synchronizers. In some cases, the "hands" (e.g., the rhythm) of the oscillator that account for the amplitude may "disappear" (dampen) when the agents of entrainment are absent. In addition, there may be multiple interactions among environmental components, receptors, pathways and oscillators. Central oscillators no doubt have positive and negative feedback components. All oscillators do not generate the same frequency, nor can it be assumed that they are linked together, at least not in the same manner.

Take-Home Message

Many different models and approaches have been used in studying the mechanisms of biological clocks that quantify time. These include mechanical, mathematical, biochemical, membrane, and molecular models. Molecular clocks and various clock genes have been best described so far in five organismal systems:

Neurospora, *Drosophila*, cyanobacteria, *Arabidopsis*, and mammals. A generalized model of a circadian system consists of an input pathway that receives and transmits signals from external environmental synchronizers to the central clock; a central clock or oscillator that generates rhythmicity; and an output pathway that transmits temporal signals from the central clock to biological processes (overt rhythms).

References

Alabadi D, Yanovsky MJ, Mas P, Harmer SL, Kay SA. (2002) Critical role for CCA1 and LHY in maintaining circadian rhythmicity in Arabidopsis. *Curr Biol* 12(9): 757–761.

Andronov AA, Vitt AA, Khaikin SE. (1966) *Theory of Oscillators*. New York: Dover Publication, 815 pp.

Antoch MP, Song EJ, Chang AM, Vitaterna MH, Zhao Y, Wilsbacher LD, Sangoram AM, King DP, Pinto LH, Takahashi JS. (1997) Functional identification of the mouse circadian Clock gene by transgenic BAC rescue. *Cell* 89(4): 655–667.

Aronson BD, Johnson KA, Loros JJ, Dunlap JC. (1994) Negative feedback defining a circadian clock: autoregulation in the clock gene frequency. *Science* 263(5153): 1578–1584.

Ashburner M, Bonner JJ. (1979) The induction of gene activity in *Drosophila* by heat shock. *Cell* 17(2): 241–254.

Ashmore LJ, Sehgal A. (2003) A fly's eye view of circadian entrainment. *J Biol Rhythms* 18(3): 206–216.

Ballario P, Vittorioso P, Magrelli A, Talora C, Cabibbo A, Macino G. (1996) *White collar-1*, central regulator of blue-light responses in *Neurospora*, is a zinc-finger protein. *EMBO J* 15(7): 1650–1657.

Balsalobre A. (2002) Clock genes in mammalian peripheral tissues. *Cell Tissue Res* 309(1): 193–199 (Review).

Banerjee D, Kwok A, Lin SY, Slack FJ. (2005) Developmental timing in *C. elegans* is regulated by *kin-20* and *tim-1*, homologs of core circadian clock genes. *Dev Cell* 8(2): 287–295.

Bargiello TA, Young MW. (1984) Molecular genetics of a biological clock in *Drosophila*. *Proc Natl Acad Sci USA* 81: 2142–2146.

Baylies MK, Bargiello TA, Jackson FR, Young MW. (1987) Changes in abundance or structure of the *per* gene product can alter periodicity of the *Drosophila* clock. *Nature* 326(6111): 390–392.

Bell-Pedersen D, Crosthwaite SK, Lakin-Thomas PL, Merrow M, Økland M. (2004) The *Neurospora* circadian clock: simple or complex? *Phil Trans R Soc Lond* 356: 1697–1709.

Betz A, Sel'kov E. (1969) Control of phosphofructokinase [PFK] activity in conditions simulating those of glycolysing yeast extract. *FEBS Lett* 3(1): 5–9.

Bjarnason GA, Jordan RC, Wood PA, Li Q, Lincoln DW, Sothern RB, Hrushesky WJ, Ben-David Y. (2001) Circadian expression of clock genes in human oral mucosa and skin. *Amer J Path* 158(5): 1793–1801.

Blau J, Young MW. (1999) Cycling *vrille* expression is required for a functional *Drosophila* clock. *Cell* 99(6): 661–671.

Boivin DB, James FO, Wu A, Cho-Park PF, Xiong H, Sun ZS. (2003) Circadian clock genes oscillate in human peripheral blood mononuclear cells. *Blood* 102(12): 4143–4145.

Briggs TS, Rauscher WC. (1973) An oscillating iodine clock. *J Chem Edu* 5: 496.
Brody S, Martins SA. (1979) Circadian rhythms in *Neurospora crassa*: effects of unsaturated fatty acids. *J Bacteriol* 137(2): 912–915.
Brown F, Hastings W, Palmer JD. (1970) *The Biological Clock, Two Views*. New York: Academic Press, 94 pp.
Bunger MK, Wilsbacher LD, Moran SM, Clendenin C, Radcliffe LA, Hogenesch JB, Simon MC, Takahashi JS, Bradfield CA. (2000) *Mop3* is an essential component of the master circadian pacemaker in mammals. *Cell* 103(7): 1009–1017.
Bünning E. (1936) Die endogene Tagesrhythmik als Grundlage der photoperiodischen Reaktion. *Ber Dtsch Bot Ges* 54: 590–607.
Bünning E. (1973) *The Physiological Clock*, 3rd edn. Berlin: Springer-Verlag, 258 pp.
Burgoyne RD. (1978) A model for the molecular basis of circadian rhythms involving monovalent ion-mediated translational control. *FEBS Lett* 94(1): 17–19.
Busse H-G. (1973) Some experiments of a chemical periodic reaction in liquid phase. In: *Biological and Biochemical Oscillators*. Chance B, Pye EK, Ghosh AK, Hess B, eds. New York: Academic Press, pp. 63–69.
Carré IA. (1999) Putative components of the *Arabidopsis* circadian clock. *Biol Rhythm Res* 30 (3): 259–263.
Chabre M. (1987) The G protein connection: is it in the membrane or cytoplasm? *Trends Biochem Sci* 12: 213–215.
Chance B, Pye EK, Ghosh AK, Hess B. (1973) *Biological and Biochemical Oscillators*. New York: Academic Press, 534 pp.
Chay TR. (1981) A model for biological oscillations. *Proc Natl Acad Sci USA* 78(4): 2204–2207.
Cheng P, Yang Y, Heintzen C, Liu Y. (2001a) Coiled-coil domain-mediated FRQ-FRQ interaction is essential for its circadian clock function in *Neurospora*. *EMBO J* 20(1–2): 101–110.
Cheng P, Yang Y, Liu Y. (2001b) Interlocked feedback loops contribute to the robustness of the *Neurospora* circadian clock. *Proc Natl Acad Sci USA* 98(13): 7408–7413.
Chernavskii DS, Palamarchuk EK, Polezhaev AA, Solyanik GI. (1977) A mathematical model of periodic processes in membranes (with application to cell cycle regulation). *Biosystems* 9(4): 187–193.
Cosgrove JW, Brown IR. (1983) Heat shock protein in mammalian brain and other organs after a physiologically relevant increase in body temperature induced by D-lysergic acid diethylamide. *Proc Natl Acad Sci USA* 80(2): 569–573.
Crews ST, Fan C-M. (1999) Remembrance of things PAS: regulation of development by bHLH-PAS proteins. *Curr Opin Genet Dev* 9(5): 580–587.
Crosthwaite SK, Loros JJ, Dunlap JC. (1995) Light-induced resetting of a circadian clock is mediated by a rapid increase in *frequency* transcript. *Cell* 81(7): 1003–1012.
Crosthwaite SK, Dunlap JC, Loros JJ. (1997) Neurospora *WC-1* and *WC-2*: transcription, photoresponses, and the origins of circadian rhythmicity. *Science* 276(5313): 763–769.
Cummings FW. (1975) A biochemical model of the circadian clock. *J Theor Biol* 55(2): 455–470.
Cyran SA, Buchsbaum AM, Reddy KL, Lin M-C, Glossop NR, Hardin PE, Young MW, Storti RV, Blau J. (2003) *vrille*, *Pdp1*, and *dClock* form a second feedback loop in the *Drosophila* circadian clock. *Cell* 112(3): 329–341.
Davidenko JM, Pertsov AV, Salomonsz R, Baxter W, Jalife J. (1992) Stationary and drifting spiral waves of excitation in isolated cardiac muscle. *Nature* 355(6358): 349–351.

Denault DL, Loros JJ, Dunlap JC. (2001) WC-2 mediates WC-1–FRQ interaction within the PAS protein-linked circadian feedback loop of *Neurospora*. *EMBO J* 20(1–2): 109–117.

Ditty JL, Williams SB, Golden SS. (2003) A cyanobacterial circadian timing mechanism. *Rev Genet* 37: 513–543 (Review).

Dobra KW, Ehret CF. (1977) Circadian regulation of glycogen, tyrosine aminotransferase, and several respiratory parameters in solid agar cultures of *Tertahymena pyriformis*. In: *Proc XII Intl Conf Intl Soc Chronobiol*, Washington, DC, 1975. Milano: Il Ponte, pp. 589–594.

Dunlap J. (1998) Circadian rhythms. An end in the beginning. *Science* 280(5369): 1548–1549.

Dunlap JC. (1999) Molecular bases for circadian clocks. *Cell* 96(2): 271–290 (Review).

Dunlap JC, Loros JJ. (2004) The *Neurospora* circadian system. *J Biol Rhythms* 19(5): 414–424.

Dunlap JC, Loros JJ, Denault D, Lee K, Froehlich A, Colot H, Shi M, Pregueiro A. (2004) Genetics and molecular biology of circadian rhythms. In: *The Mycota III. Biochemistry and Molecular Biology*, 2nd edn. Brambl R, Marzluf GA, eds. Berlin: Springer-Verlag, pp. 209–229.

Edmunds LN Jr, Cirillo VP. (1974) On the interplay among cell cycle, biological clock and membrane transport control systems. *Intl J Chronobiol* 2(3): 233–246 (Review).

Edmunds LN Jr. (1988) *Cellular and Molecular Bases of Biological Clocks. Models and Mechanisms for Circadian Timekeeping*. New York: Springer-Verlag, 497 pp.

Edmunds LN Jr. (1992) Cellular and molecular aspects of circadian oscillators: models and mechanisms for biological timekeeping. In: *Biological Rhythms in Clinical and Laboratory Medicine*. Touitou Y, Haus E, eds. Berlin: Springer-Verlag, pp. 35–54.

Ehret CF, Trucco E. (1967) Molecular models for the circadian clock: I. The chronon concept. *J Theor Biol* 15(2): 240–262 (Review).

Ehret CF. (1980) The chronon theory of circadian rhythm control. In: *Chronobiology: Principles and Applications to Shifts in Schedules*. Scheving LE, Halberg F, eds. Alphen aan den Rijn: Sijthoff & Noordhoff, pp. 229–247.

Emery P, So WV, Kaneko M, Hall JC, Rosbash M. (1998) CRY, a *Drosophila* clock and light-regulated cryptochrome, is a major contributor to circadian rhythm resetting and photosensitivity. *Cell* 95(5): 669–679.

Engelmann W. (1973) A slowing down of circadian rhythms by lithium ions. *Z Naturforsch* 28: 733–736.

Engelmann W. (1996) Leaf movement rhythms as hands of biological clocks. In: *Vistas on Biorhythmicity*. Greppin H, Degli Agosti R, Bonzon M, eds. Geneva: University of Geneva Press, pp. 51–76.

Eriksson ME, Hanano S, Southern MM, Hall A, Millar AJ. (2003) Response regulator homologues have complementary, light-dependent functions in the *Arabidopsis* circadian clock. *Planta* 218(1): 159–162.

Estabrook RW, Srere P, eds. (1981) *Current Topics in Cellular Regulation, Vol. 18 –Biological Cycles*. New York: Academic Press, 573 pp.

Farré EM, Harmer SL, Harmon FG, Yonovsky MJ, Kay SA. (2005) Overlapping and distinct roles of *PRR7* and *PRR9* in the *Arabidopsis* circadian clock. *Curr Biol* 15(1): 47–54.

Feldman JF, Hoyle MN. (1973) Isolation of circadian clock mutants of *Neurospora crassa*. *Genetics* 75(4): 605–613.

Foster RG, Kreitzman L. (2004) *Rhythms of Life. The Biological Clocks that Control the Daily Lives of Every Living Thing*. New Haven and London: Yale University Press, 276 pp.

Francis CD, Sargent ML. (1979) Effects of temperature perturbations on circadian conidiation in *Neurospora*. *Plant Physiol* 64(6): 1000–1004.

Froehlich AC, Liu Y, Loros JJ, Dunlap JC. (2002) White Collar-1, a circadian blue light photoreceptor, binding to the *frequency* promoter. *Science* 297(5582): 815–819.

Froehlich AC, Loros JJ, Dunlap JC. (2003) Rhythmic binding of a White Collar containing complex to the *frequency* promoter is inhibited by FREQUENCY. *Proc Natl Acad Sci USA* 100(10): 5914–5919.

Fujimori T, Sato E, Yamashino T, Mizuno T. (2005) PRR5 (pseudo-response regulator 5) plays antagonistic roles to CCA1 (circadian clock-associated 1) in *Arabidopsis thaliana*. *Biosci Biotechnol Biochem* 69(2): 426–430.

Garceau N, Liu Y, Loros JJ, Dunlap JC. (1997) Alternative initiation of translation and time-specific phosphorylation yield multiple forms of the essential clock protein FREQUENCY. *Cell* 89(3): 469–476.

Gekakis N, Saez L, Delahaye-Brown AM, Myers MP, Sehgal A, Young MW, Weitz CJ. (1995) Isolation of *timeless* by PER protein interaction: defective interaction between *timeless* protein and long-period mutant PER^L. *Science* 270(5237): 811–815.

Gekakis N, Staknis D, Nguyen HB, Davis FC, Wilsbacher LD, King DP, Takahashi JS, Weitz CJ. (1998) Role of the CLOCK protein in the mammalian circadian mechanism. *Science* 280(5369): 1564–1569.

Gerisch G. (1976) Extracellular cyclic-amp phosphodiesterase regulation in agar plate cultures of *Dictyostelium discoideum*. *Cell Differ* 5(1): 21–25.

Gerisch G. (1987) Cyclic AMP and other signals controlling cell development and differentiation in *Dictyostelium*. *Ann Rev Biochem* 56: 853–879.

Gillanders SW, Saunders DS. (1992) A coupled pacemaker-slave model for the insect photoperiodic clock: interpretation of ovarian diapause data in *Drosophila melanogaster*. *Biol Cybern* 67(5): 451–459.

Glass L, Mackey MC. (1988) *From Clocks to Chaos, The Rhythms of Life*. Princeton, NJ: Princeton University Press, 248 pp.

Goldbeter A, Caplan SR. (1976) Oscillatory enzymes. *Ann Rev Biophys Bioeng* 5: 449–476 (Review).

Golden SS. (2003) Timekeeping in bacteria: the cyanobacterial circadian clock. *Curr Opin Microbiol* 6(6): 535–540 (Review).

Golden SS. (2004) Meshing the gears of the cyanobacterial circadian clock. *Proc Natl Acad Sci USA* 101(38): 13697–13698.

Gooch VD, Freeman L, Lakin-Thomas PL. (2004) Time-lapse analysis of the circadian rhythms of conidiation and growth rate in *Neurospora*. *J Biol Rhythms* 19(6): 493–503.

Goodwin BC. (1965) Oscillatory behaviour in enzymatic control processes. *Adv Enzyme Regul* 3: 425–438.

Gorman MR. (2003) Melatonin implants disrupt developmental synchrony regulated by flexible interval timers. *J Neuroendocrinol* 15(11): 1084–1094.

Grobbelaar N, Huang TC, Lin HY, Chow TJ. (1986) Dinitrogen-fixing endogenous rhythm in *Synechococcus* RF-1. *FEMS Microbiol Lett* 37: 173–178.

Gross JD, Peacey MJ, Trevan DJ. (1976) Signal emission and signal propagation during early aggregation in *Dictyostelium discoideum*. *J Cell Sci* 22(3): 645–656.

Gross JD. (1994) Developmental decisions in *Dictyostelium discoideum*. *Microbiol Rev* 58(3): 330–351 (Review).

Gwinner E. (1977) Circannual rhythms in bird migration. *Ann Rev Ecol Syst* 8: 381–405.

Haest CWM, De Gier J, van Deenen LLM. (1969) Changes in the chemical and the barrier properties of the membrane lipids of *E. coli* by variation of the temperature of growth. *Chem Phys Lipids* 3(4): 413–417.

Hall JC. (1995) Tripping along the trail to the molecular mechanisms of biological clocks. *Trends Neurosci* 18(5): 230–240.

Hall JC. (2003) Genetics and molecular biology of rhythms in *Drosophila* and other insects. *Adv Genet* 48: 1–280 (Review).

Hall A, Bastow RM, Davis SJ, Hanano S, McWatters HG, Hibberd V, Doyle MR, Sung S, Halliday KJ, Amasino RM, Millar AJ. (2003) The *TIME FOR COFFEE* gene maintains the amplitude and timing of *Arabidopsis* circadian clocks. *Plant Cell* 15(11): 2719–2729.

Hamner KC. (1960) Photoperiodism and circadian rhythms. In: *Biological Clocks. Cold Spring Harbor Symposia on Quantitative Biology*, Vol 25. New York: Long Island Biol Assoc, pp. 269–277.

Hardeland R, Coto-Montes A, Poeggeler B. (2003) Circadian rhythms, oxidative stress, and antioxidative defense mechanisms. *Chronobiol Intl* 20(6): 921–962 (Review).

Hardin PE, Hall JC, Rosbash M. (1990) Feedback of the *Drosophila* period gene product on circadian cycling of its messenger RNA levels. *Nature* 343(6258): 536–540.

Hardin PE. (2004) Transcription regulation within the circadian clock: the E-box and beyond. *J Biol Rhythms* 19(5): 348–360.

Harley CB. (1991) Telomere loss: mitotic clock or genetic time bomb? *Mutat Res* 256(2–6): 271–282.

Harmer SL, Hogenesch JB, Straume M, Chang HS, Han B, Zhu T, Wang X, Kreps JA, Kay SA. (2000) Orchestrated transcription of key pathways in *Arabidopsis* by the circadian clock. *Science* 290(5499): 2110–2113.

Harmer SL, Panda S, Kay SA. (2001) Molecular bases of circadian rhythms. *Ann Rev Cell Dev Biol* 17: 215–253 (Review).

Harms E, Kivimae S, Young MW, Saez L. (2004) Posttranscriptional and posttranslational regulation of clock genes. *J Biol Rhythms* 19(5): 361–373.

Hazel JR. (1995) Thermal adaptation in biological membranes: is homeoviscous adaptation the explanation? *Ann Rev Physiol* 57: 19–42 (Review).

He Q, Cheng P, Yang Y, Wang L, Gardner KH, Liu Y. (2002) White collar-1, a DNA binding transcription factor and a light sensor. *Science* 297(5582): 840–843.

Hess B, Boiteux A. (1971) Oscillatory phenomena in biochemistry. *Ann Rev Biochem* 40: 237–258 (Review).

Hess B. (1977) Oscillating reactions. *Trends Biochem Sci* 2: 193–195.

Hicks KA, Albertson TM, Wagner DR. (2001) *EARLY FLOWERING3* encodes a novel protein that regulates circadian clock function and flowering in Arabidopsis. *Plant Cell* 13(6): 1281–1292.

Hillman WS. (1956) Injury of tomato plants by continuous light and unfavorable photoperiodic cycles. *Amer J Bot* 43: 89–96.

Hillman WS. (1973) Non-circadian photoperiodic timing in the aphid *Megoura*. *Nature* 242: 128–129.

Hillman WS. (1976) Biological rhythms and physiological timing. *Ann Rev Plant Physiol* 27: 159–179.

Holmgren M, Scheffer M. (2001) El Niño as a window of opportunity for the restoration of degraded arid ecosystems. *Ecosystems* 4: 151–159.

Honma S, Kawamoto T, Takagi Y, Fujimoto K, Sato F, Noshiro M, Kato Y, Honma K. (2002) *Dec1* and *Dec2* are regulators of the mammalian molecular clock. *Nature* 419(6909): 841–844.

Iglesias A, Satter RL. (1983) H⁺ fluxes in exercised *Samanea* motor tissue: II. Rhythmic properties. *Plant Physiol* 72: 570–572.

Ikeda M, Nomura M. (1997) cDNA cloning and tissue-specific expression of a novel basic helix-loop-helix/PAS protein (BMAL1) and identification of alternatively spliced variants with alternative translation initiation site usage. *Biochem Biophys Res Commun* 233(1): 258–264.

Ishiura M, Kutsuna S, Aoki S, Iwasaki H, Andersson CR, Tanabe A, Golden SS, Johnson CH, Kondo T. (1998) Expression of a gene cluster *kaiABC* as a circadian feedback process in Cyanobacteria. *Science* 281(5382): 1519–1523.

Iwasaki H, Taniguchi Y, Ishiura M, Kondo T. (1999) Physical interactions among circadian clock proteins KaiA, KaiB and KaiC in cyanobacteria. *EMBO J* 18(5): 1137–1145.

Iwasaki H, Williams SB, Kitayama Y, Ishiura M, Golden SS, Kondo T. (2000) A kaiC-interacting sensory histidine kinase, SasA, necessary to sustain robust circadian oscillation in cyanobacteria. *Cell* 101(2): 223–233.

Iwasaki H, Kondo T. (2000) The current state and problems of circadian clock studies in cyanobacteria. *Plant Cell Physiol* 41(9): 1013–1020 (Review).

Iwasaki H, Kondo T. (2004) Circadian timing mechanism in the prokaryotic clock system of cyanobacteria. *J Biol Rhythms* 19(5): 436–444.

Jacklet JW. (1988–89) Circadian pacemaker neurons: membranes and molecules. *J Physiol (Paris)* 83(3): 164–171.

Jeon M, Gardner HF, Miller EA, Deshler J, Rougvie AE. (1999) Similarity of the *C. elegans* developmental timing protein LIN-42 to circadian rhythm proteins. *Science* 286(5442): 1141–1146.

Johnsson A, Karlsson HG. (1972) A feedback model for biological rhythms: I. Mathematical description and basic properties of the model. *J Theor Biol* 36(1): 153–174.

Johnsson A, Karlsson HG, Engelmann W. (1972) Phase shift effects in the *Kalanchoë* petal rhythm due to two or more light pulses. A theoretical and experimental study. *Physiol Plant* 28: 134–142.

Kaplan D, Glass L. (1995) *Understanding Nonlinear Dynamics*. New York: Springer-Verlag, 420 pp.

Kay SA, Millar AJ. (1995) New models in vogue for circadian clocks. *Cell* 83(3): 361–364 (Review).

Keener JP, Hoppensteadt FC, Rinzel J. (1981) Integrate-and-fire models of nerve membrane response to oscillatory input. *Siam J Appl Math* 41(3): 503–517.

Kim WY, Geng R, Somers DE. (2003) Circadian phase-specific degradation of the F-box protein ZTL is mediated by the proteasome. *Proc Natl Acad Sci USA* 100(8): 4933–4938

King DP, Vitaterna MH, Chang AM, Dove WF, Pinto LH, Turek FW, Takahashi JS. (1997a) The mouse *Clock* mutation behaves as an antimorph and maps within the W^{19H} deletion, distal of Kit. *Genetics* 146(3): 1049–1060.

King DP, Zhao Y, Sangoram AM, Wilsbacher LD, Tanaka M, Antoch MP, Steeves TD, Vitaterna MH, Kornhauser JM, Lowrey PL, Turek FW, Takahashi JS. (1997b) Positional cloning of the mouse circadian clock gene. *Cell* 89(4): 641–653.

King DP, Takahashi JS. (2000) Molecular genetics of circadian rhythms in mammals. *Ann Rev Neurosci* 23: 713–742 (Review).

Kloss B, Price JL, Saez L, Blau J, Rothenfluh A, Wesley CS, Young MW. (1998) The *Drosophila* gene *double-time* encodes a protein closely related to human casein kinase Iε. *Cell* 94(1): 97–107.

Kloss B, Rothenfluh A, Young MW, Saez L. (2001) Phosphorylation of period is influenced by cycling physical associations of double-time, period, and timeless in the *Drosophila* clock. *Neuron* 30(3): 699–706.

Klotter K. (1960) General properties of oscillating systems. In: *Biological Clocks. Cold Spring Harbor Symposia on Quantitative Biology*, Vol. 25. New York: The Biological Laboratory, pp. 185–187.

Kondo T, Tsinoremas NF, Golden SS, Johnson CH, Kutsuna S, Ishiura M. (1994) Circadian clock mutants of Cyanobacteria. *Science* 266(5188): 1233–1236.

Kondo T, Ishiura M. (1999) The circadian clocks of plants and cyanobacteria. *Trends in Plant Science* 4(5): 171–176 (Review).

Konopka RJ, Benzer S. (1971) Clock mutants of *Drosophila melanogaster*. *Proc Nat Acad Sci USA* 68(9): 2112–2116.

Konopka RJ, Orr D. (1980) Effects of a clock mutation on the subjective day—Implications for a membrane model of the *Drosophila* circadian clock. In: *Development and Neurobiology of Drosophila*. Siddiqi O, Babu P, Hall LM, Hall JC, eds. New York: Plenum Press, pp. 409–416.

Koukkari WL, Soulen TK. (1981) Circadian time structure of vascular flowering plants. In: *Neoplasms–Comparative Pathology of Growth in Animals, Plants, and Man*. Kaiser HE, ed. Baltimore: Williams and Wilkins, pp. 175–184.

Koukkari WL, Warde SB. (1985) Rhythms and their relations to hormones. In: *Encyclopedia of Plant Physiology, New Series, Vol. 11: Hormonal Regulation of Development III*. Pharis RP, Reid DM, eds. Berlin: Springer-Verlag, pp. 37–77.

Kume K, Zylka MJ, Sriram S, Shearman LP, Weaver DR, Jin X, Maywood ES, Hastings MH, Reppert SM. (1999) mCRY1 and mCRY2 are essential components of the negative limb of the circadian clock feedback loop. *Cell* 98(2): 193–205.

Kuno N, Møller SG, Shinomura T, Xu X, Chua NH, Furuya M. (2003) The novel MYB protein Early-phytochrome-responsive1 is a component of a slave circadian oscillator in *Arabidopsis*. *Plant Cell* 15(10): 2476–2488.

Kusanagi H, Mishima K, Satoh K, Echizenya M, Katoh T, Shimizu T. (2004) Similar profiles in human period1 gene expression in peripheral mononuclear and polymorphonuclear cells. *Neurosci Lett* 365(2): 124–127.

Kyriacou CP, Hall JC. (1980) Circadian rhythm mutations in *Drosophila melanogaster* affect short-term fluctuation in the male's courtship song. *Proc Nat Acad Sci USA* 77(11): 6729–6733.

Lakin-Thomas PL, Brody S, Coté CG. (1997) Temperature compensation and membrane composition in *Neurospora crassa*. *Chronobiol Intl* 14(5): 445–454.

Lakowski B, Hekimi S. (1996) Determination of life-span in *Caenorhabditis elegans* by four clock genes. *Science* 272(5264): 1010–1013.

Lee K, Loros JJ, Dunlap JC. (2000) Interconnected feedback loops in the *Neurospora* circadian system. *Science* 289(5476): 107–110.

Lees AD. (1966) Photoperiodic timing mechanisms in insects. *Nature* 210(40): 986–989.

Lin CY, Roberts JK, Key JL. (1984) Acquisition of thermotolerance in soybean seedlings: synthesis and accumulation of heat shock proteins and their cellular localization. *Plant Physiol* 74: 152–160.

Lincoln GA, Andersson H, Hazlerigg D. (2003) Clock genes and the long–term regulation of prolactin secretion: evidence for a photoperiod/circannual timer in the pars tuberalis. *J Neuroendocrinol* 15: 390–397.

Linden H, Macino G. (1997) White collar 2, a partner in blue-light signal transduction, controlling expression of light-regulated genes in *Neurospora crassa*. *EMBO J* 16(1): 98–109.

Linden H, Ballario P, Macino G. (1997) Blue light regulation in *Neurospora crassa*. *Fungal Genet Biol* 22(3): 141–150 (Review).
Liu Y, Merrow M, Loros JJ, Dunlap JC. (1998) How temperature changes reset a circadian oscillator. *Science* 281(5378): 825–829.
Liu Y, Loros J, Dunlap JC. (2000) Phosphorylation of the *Neurospora* clock protein FREQUENCY determines its degradation rate and strongly influences the period length of the circadian clock. *Proc Natl Acad Sci USA* 7(1): 234–239.
Lloyd AL, Lloyd D. (1993) Hypothesis: the central oscillator of the circadian clock is a controlled chaotic attractor. *Biosystems* 29(2–3): 77–85.
Lloyd AL, Lloyd D. (1995) Chaos: its significance and detection in biology. *Biol Rhythm Res* 26(2): 233–252.
Loros JJ, Denome SA, Dunlap JC. (1989) Molecular cloning of genes under the control of the circadian clock in *Neurospora*. *Science* 243(4889): 385–388.
Lumsden PJ, Miller AJ. (1998) *Biological Rhythms and Photoperiodism in Plants*. Oxford: BIOS Scientific Publication, 284 pp.
Majercak J, Sidote D, Hardin PE, Edery I. (1999) How a circadian clock adapts to seasonal decreases in temperature and day length. *Neuron* 24(1): 219–230.
Martens UM, Chavez EA, Poon SSS, Schmoor C, Landsdorp PM. (2000) Accumulation of short telomeres in human fibroblasts prior to replicative senescence. *Exp Cell Res* 256(1): 291–299.
Martinek S, Inonog S, Manoukian AS, Young MW. (2001) A role for the segment polarity gene *shaggy*/GSK-3 in the *Drosophila* circadian clock. *Cell* 105(6): 769–779.
Más P, Kim WY, Somers DE, Kay SA. (2003) Targeted degradation of TOC1 by ZTL modulates circadian function in *Arabidopsis thaliana*. *Nature* 426(6966): 567–570.
McClung CR. (2001) Circadian rhythms in plants. *Ann Rev Plant Physiol Plant Mol Biol* 52: 139–162 (Review).
McClung CR, Fox BA, Dunlap JC (1989) The *Neurospora* clock gene *frequency* shares a sequence element with the *Drosophila* clock gene *period*. *Nature* 339(6225): 558–562.
Merrow M, Franchi L, Dragovic Z, Gorl M, Johnson J, Brunner M, Macino G, Roenneberg T. (2001) Circadian regulation of the light input pathway in *Neurospora crassa*. *EMBO J* 20(3): 307–315.
Michael TP, Salomé PA, McClung CR. (2003) Two *Arabidopsis* circadian oscillators can be distinguished by differential temperature sensitivity. *Proc Natl Acad Sci USA* 100(11): 6878–6883.
Millar AJ, Short SR, Chua NH, Kay SA. (1992) A novel circadian phenotype based on firefly luciferase expression in transgenic plants. *Plant Cell* 4(9): 1075–1087.
Millar AJ, Carré IA, Strayer CA, Chua NH, Kay SA. (1995) Circadian clock mutants in *Arabidopsis* identified by luciferase imaging. *Science* 267(5201): 1161–1163.
Millar AJ. (2003) A suite of photoreceptors entrains the plant circadian clock. *J Biol Res* 18 (3): 217–226.
Minorsky N. (1962) *Nonlinear Oscillations*. Princeton, NJ: Van Nostrand, 714 pp.
Mitsui A, Kumazawa S, Takahashi A, Ikemoto H, Cao S, Arai T. (1986) Strategy by which nitrogen-fixing unicellular cyanobacteria grow photoautotrophically. *Nature* 323: 720–722.
Mitsui A, Cao S, Takahashi A, Arai T. (1987) Growth synchrony and cellular parameter of unicellular nitrogen-fixing marine cyanobacterium, *Synechococcus* sp. strain Miami BG 043511 under continuous illumination. *Physiol Plant* 69(1): 1–8.
Moore-Ede MC, Sulzman FM, Fuller CA. (1982) *The Clocks That Time Us. Physiology of the Circadian Timing System*. Cambridge, MA: Harvard University Press, 448 pp.

Morgan LW, Greene AV, Bell-Pedersen D. (2003) Circadian and light-induced expression of luciferase in *Neurospora crassa*. *Fungal Genet Biol* 38(3): 327–332.

Morré DJ, Chueh PJ, Pletcher J, Tang X, Wu LY, Morré DM. (2002) Biochemical basis for the biological clock. *Biochemistry* 41(40): 11941–11945.

Myers MP, Wager-Smith K, Wesley CS, Young MW, Sehgal A. (1995) Positional cloning and sequence analysis of the *Drosophila* clock gene, timeless. *Science* 270(5237): 805–808.

Nakamichi N, Kita M, Ito S, Sato E, Yamashino T, Mizuno T. (2005) The *Arabidopsis* Pseudo-Response Regulators, PRR5 and PRR7, coordinately play essential roles for circadian clock function. *Plant Cell Physiol* 46(4): 609–619).

Naidoo N, Song W, Hunter-Ensor M, Sehgal A. (1999) A role for the proteasome in the light response of the timeless clock protein. *Science* 285(5434): 1737–1741.

Nishiyama M, Matsubara N, Yamamoto K, Iijima S, Uozumi T, Beppu T. (1986) Nucleotide sequence of the malate dehydrogenase gene of *Thermus flavus* and its mutation directing an increase in enzyme activity. *J Biol Chem* 261(30): 14178–14183.

Njus D, Sulzman FM, Hastings JW. (1974) Membrane model for the circadian clock. *Nature* 248(444): 116–120.

North C, Feuers RJ, Scheving LE, Pauly JE, Tsai TH, Casciano DA. (1981) Circadian organization of thirteen liver and six brain enzymes of the mouse. *Amer J Anat* 162(3): 183–199.

Okamura H. (2004) Clock genes in cell clocks: roles, actions, and mysteries. *J Biol Rhythms* 19(5): 388–399 (Review).

Okayama H. (2004) Clock genes in cell clocks: roles, actions, and mysteries. *J Biol Rhythms* 19(5): 388–399.

Okuyama H. (1969) Phospholipid metabolism in *Escherichia coli* after a shift in temperature. *Biochim Biophys Acta* 176(1): 125–134.

Pavlidis T, Kauzmann W. (1969) Toward a quantitative biochemical model for circadian oscillators. *Arch Biochem Biophys* 132: 338–348.

Peterson EL. (1980) A limit cycle interpretation of a mosquito circadian oscillator. *J Theor Biol* 84(2): 281–310.

Peterson EL, Saunders DS. (1980) The circadian eclosion rhythm in *Sarcophaga argyrotoma*: a limit cycle representation of the pacemaker. *J Theor Biol* 86: 265–277.

Pine ES. (1983) *How to Enjoy Calculus*. New Jersey: Steinlitz-Hammacher Co., 155 pp.

Pittendrigh CS, Bruce VG, Rosenzweig NS, Rubin ML. (1959) A biological clock in *Neurospora*. *Nature* 184(4681): 169–170.

Pittendrigh CS. (1967) Circadian systems I. The driving oscillation and its assay in *Drosophila pseudoobscura*. *Proc Natl Acad Sci USA* 58(4): 1762–1767.

Pittendrigh CS. (1981) Circadian organization and the photoperiodic phenomena. In: *Biological Clocks and Reproductive Cycles*. Follett BK, ed. Bristol: John Wright, pp. 1–35.

Pittendrigh CS. (1972) Circadian surfaces and the diversity of possible roles of circadian organization in photoperiodic induction. *Proc Natl Acad Sci USA* 69(9): 2734–2737.

Plesofsky-Vig N, Brambl R. (1985) Heat shock response of *Neurospora crassa*: protein synthesis and induced thermotolerance. *J Bacteriol* 162(3): 1083–1091.

Poincaré H. (1881) Mémoire sur les coubes définies par une équation différentielle. *J de Math Pur Appl, 3d ser*, 7: 375–422.

Pregueiro AM, Price-Lloyd N, Bell-Pedersen D, Heintzen C, Loros JJ, Dunlap JC. (2005) Assignment of an essential role for the *Neurospora frequency* gene in circadian entrainment to temperature cycles. *Proc Natl Acad Sci USA* 102(6): 2210–2215.

Preitner N, Damiola F, Lopez-Molina L, Zakany J, Duboule D, Albrecht U, Schibler U. (2002) The orphan nuclear receptor REV-ERBα controls circadian transcription within the positive limb of the mammalian circadian oscillator. *Cell* 110(2): 251–260.

Prendergast BJ, Renstrom RA, Nelson RJ. (2004) Genetic analyses of a seasonal interval timer. *J Biol Rhythms* 19(4): 298–311.

Price JL, Blau J, Rothenfluh A, Abodeely M, Kloss B, Young MW. (1998) *Double-time* is a novel *Drosophila* clock gene that regulates PERIOD protein accumulation. *Cell* 94(1): 83–95.

Prolo LM, Takahashi JS, Herzog ED. (2005) Circadian rhythm generation and entrainment in astrocytes. *J Neurosci* 25(2): 404–408.

Pye EK. (1971) Periodicities in intermediary metabolism. In: *Biochronometry*. Menaker M, ed. Washington, DC: National Academy of Sciences, pp. 623–636.

Ralph MR, Menaker M. (1988) A mutation of the circadian system in golden hamsters. *Science* 241(4870): 1225–1227.

Rao KP. (1967) Biochemical correlates of temperature acclimation. In: *Molecular Mechanisms of Temperature Adaptation*. Prosser CL, ed. Washington, DC: American Association for the Advancement of Science, pp. 227–244.

Reddy P, Zehring WA, Wheeler DA, Pirrotta V, Hadfield C, Hall JC, Rosbash M. (1984) Molecular analysis of the *period* locus in *Drosophila melanogaster* and identification of a transcript involved in biological rhythms. *Cell* 38(3): 701–710.

Rensing L, Mohsenzadeh S, Ruoff P, Meyer U. (1997) Temperature compensation of the circadian period length—a special case among general homeostatic mechanisms of gene expression? *Chronobiol Intl* 14(5): 481–498.

Rensing L, Meyer-Grahle U, Ruoff P. (2001) Biological timing and the clock metaphor: oscillatory and hourglass mechanisms. *Chronobiol Intl* 18(3): 329–369 (Review).

Rinaldi S, Scheffer M. (2000) Geometric analysis of ecological models with slow and fast processes. *Ecosystems* 3: 507–521.

Robertson RN. (1983) *The Lively Membranes*. Cambridge: Cambridge University Press, 206 pp.

Roenneberg T, Daan S, Merrow M. (2003) The art of entrainment. *J Biol Rhythms* 18(3): 183–194 (Review).

Ruoff P, Vinsjevik M, Monnerjahn S, Rensing L. (1999) The Goodwin oscillator: on the importance of degradation reaction in the circadian clock. *J Biol Rhythms* 14(6): 469–479 (Review).

Rutila JE, Suri V, Le M, So WV, Rosbash M, Hall JC. (1998) CYCLE is a second bHLH-PAS clock protein essential for circadian rhythmicity and transcription of Drosophila period and timeless. *Cell* 93(5): 805–814.

Salisbury FB, Ross CW. (1992) *Plant Physiology*, 4th edn. Belmont: Wadsworth, 682 pp.

Salomé PA, McClung CR. (2004) The *Arabidopsis thaliana* clock. *J Biol Rhythms* 19(5): 425–435.

Salomé PA, McClung CR. (2005) *PSEUDO-RESPONSE REGULATOR 7* and *9* are partially redundant genes essential for the temperature responsiveness of the *Arabidopsis* circadian clock. *Plant Cell* 17(3): 791–803.

Sargent ML, Briggs WR, Woodward DO. (1966) Circadian nature of a rhythm expressed by an invertaseless strain of *Neurospora crassa*. *Plant Physiol* 41(8): 1343–1349.

Sargent ML, Briggs WR. (1967) The effects of light on circadian rhythm of conidiation in *Neurospora*. *Plant Physiol* 42: 1504–1510.

Satter RL, Galston AW. (1971) Potassium flux: a common feature of *Albizzia* leaflet movement controlled by phytochrome or endogenous rhythm. *Science* 174: 518–520.

Satter RL, Galston AW. (1981) Mechanisms of control of leaf movements. *Ann Rev Plant Physiol* 32: 83–110.

Schäffer R, Ramsay N, Samach A, Corden S, Putterill J, Carré IA, Coupland G. (1998) The *late elongated hypocotyl* mutation of *Arabidopsis* disrupts circadian rhythms and the photoperiodic control of flowering. *Cell* 93(7): 1219–1229.

Scheffer M. (1989) Alternative stable states in eutrophic shallow freshwater systems: a minimal model. *Hydrobiol Bull* 23: 73–83.

Scheffer M, Rinaldi S, Kuznetsov YA. (2000) Effects of fish on plankton dynamics: a theoretical analysis. *Can J Fish Aquat Sci* 57: 1208–1219.

Scheffer M, Szabo S, Gragnani A, Van Nes EH, Rinaldi S, Kautsky N, Norberg J, Roijackers RM, Franken RJ. (2003) Floating plant dominance as a stable state. *Proc Natl Acad Sci USA* 100(7): 4040–4045.

Schibler U, Ripperger J, Brown SA. (2003) Peripheral circadian oscillators in mammals: time and food. *J Biol Rhythms* 18(3): 250–260.

Schmitz O, Katayama M, Williams SB, Kondo T, Golden SS. (2000) CikA, a bacteriophytochrome that resets the cyanobacterial circadian clock. *Science* 289(5480): 765–768.

Schweiger H-G, Schweiger M. (1977) Circadian rhythms in unicellular organisms: an endeavour to explain the molecular mechanism. *Intl Rev Cytol* 51: 315–342 (Review).

Science Magazine Editors. (1997) Breakthrough of the year. The runners-up (#3: Keeping time). *Science* 278(5346): 2039–2042.

Science Magazine Editors. (1998) Breakthrough of the year. The runners-up (#1: A remarkable year for clocks). *Science* 282(5397): 2157–2161.

Sehgal A, Rothenfluh-Hilfiker A, Hunter-Ensor M, Chen Y, Myers MP, Young MW. (1995) Rhythmic expression of *timeless*: a basis for promoting circadian cycles in *period* gene autoregulation. *Science* 270(5237): 808–810.

Shaw MK, Ingraham JL. (1965) Fatty acid composition of *Escherichia coli* as a possible controlling factor of the minimal growth temperature. *J Bacteriol* 90(1): 141–146.

Shearman LP, Zylka MJ, Weaver DR, Kolakowski LF Jr, Reppert SM. (1997) Two *period* homologs: circadian expression and photic regulation in the suprachiasmatic nuclei. *Neuron* 19(6): 1261–1269.

Sidote D, Majercak J, Parikh V, Edery I. (1998) Differential effects of light and heat on the *Drosophila* circadian clock proteins PER and TIM. *Mol Cell Biol* 18(4): 204–213.

Singer SJ, Nicolson GL. (1972) The fluid mosaic model of the structure of cell membranes. *Science* 175(23): 720–730.

Somero GN. (1995) Proteins and temperature. *Ann Rev Physiol* 57: 43–68 (Review).

Somers DE, Schultz TF, Milnamow M, Kay SA. (2000) ZEITLUPE encodes a novel clock–associated PAS protein from *Arabidopsis*. *Cell* 101(3): 319–329.

Stanewsky R. (2002) Clock mechanisms in Drosophila. *Cell Tissue Res* 309(1): 11–26 (Review).

Steeves TDL, King DP, Zhao Y, Sangoram AM, Du F, Bowcock AM, Moore RY, Takahashi JS. (1999) Molecular cloning and characterization of the human *CLOCK* gene: expression in the suprachiasmatic nuclei. *Genomics* 57(2): 189–200.

Stetson MH, Elliott JA, Menaker M. (1975) Photoperiodic regulation of hamster testis: circadian sensitivity to the effects of light. *Biol Reprod* 13(3): 329–339.

Stewart I. (2000) The Lorenz attractor exists. *Nature* 406(6799): 948–949.

Strubbe JH, Spiteri NJ, Alingh Prins AJ. (1986) Effect of skeleton photoperiod and food availability on the circadian pattern of feeding and drinking in rats. *Physiol Behav* 36(4): 647–651.

Sun ZS, Albrecht U, Zuchenko O, Bailey J, Eichele G, Lee CC. (1997) *RIGUI*, a putative mammalian ortholog of the *Drosophila period* gene. *Cell* 90(6): 1003–1011.

Sweeny BM, Hastings JW. (1960) Effects of temperature upon diurnal rhythms. In: *Biological Clocks. Cold Spring Harbor Symposia on Quantitative Biology*, Vol. 25. New York: The Biological Laboratory, pp. 87–104.

Sweeney BM. (1974a) The temporal regulation of morphogenesis in plants, hourglass and oscillator. *Brookhaven Symp Biol* 25: 95–110.

Sweeney BM. (1974b) A physiological model for circadian rhythms derived from the *Acetabularia* rhythm paradoxes. *Intl J Chronobiol* 2(1): 25–33.

Sweeney B, Borgese MB. (1989) A circadian rhythm in cell division in a prokaryote the cyanobacterium *Synechococcus*. *J Phycol* 25(1): 183–186.

Takahashi JS, Pinto LH, Vitaterna MH. (1994) Forward and reverse genetic approaches to behavior in the mouse. *Science* 264(5166): 1724–1733 (Review).

Taylor BL, Zhulin IB. (1999) PAS domains: interval sensors of oxygen, redox potential, and light. *Microbiol Mol Biol Rev* 63(2): 479–506 (Review).

Tei H, Okamura H, Shigeyoshi Y, Fukuhara C, Ozawa R, Hirose M, Sakaki Y. (1997) Circadian oscillation of a mammalian homologue of the *Drosophila period* gene. *Nature* 389(6650): 512–516.

Tomita J, Nakajima M, Kondo T, Iwasaki H. (2005) No transcription-translation feedback in circadian rhythm of KaiC phosphorylation. *Science* 307(5707): 251–254.

Tóth R, Kevei E, Hall A, Millar AJ, Nagy F, Kozma-Bognár L. (2001) Circadian clock-regulated expression of phytochrome and cryptochrome genes in Arabidopsis. *Plant Physiol* 127(4): 1607–1616.

Tsinkalovsky O, Smaaland R, Rosenlund B, Sothern RB, Hirt A, Eiken HG, Steine S, Badiee A, Foss Abrahamsen J, Laerum OD. (2005) Circadian variations of clock gene expression in CD34+ progenitor cells in the human bone marrow. (*in preparation*).

Tucker W. (1999) The Lorenz attractor exists. *CR Acad Sci* Paris 328(1): 1197–1202.

Tyson JJ, Kauffman S. (1975) Control of mitosis by a continuous biochemical oscillation: synchronization, spatially inhomogeneous oscillations. *J Math Biol* 1: 289–310.

Underwood H, Calaban M. (1987) Pineal melatonin rhythms in the lizard *Anolis carolinensis*: I. Response to light and temperature cycles. *J Biol Rhythms* 2(3): 179–193.

van der Pol B, van der Mark J. (1928) The heartbeat considered as a relaxation oscillation and an electrical model of the heart. *Phil Mag* Ser. 7, 6: 763–775.

Vitaterna MH, King DP, Chang A-M, Kornhauser JM, Lowrey PL, McDonald JD, Dove WF, Pinto LH, Turek FW, Takahashi JS. (1994) Mutagenesis and mapping of a mouse gene, *Clock*, essential for circadian behavior. *Science* 264(5159): 719–725.

Wagner E, Frosch S, Deitzer GF. (1974) Metabolic control of photoperiodic time measurement. *J Interdiscipl Cycle Res* 5(3–4): 240–246.

Wang ZY, Tobin EM. (1998) Constitutive expression of the *CIRCADIAN CLOCK ASSOCIATED 1 (CCA1)* gene disrupts circadian rhythms and suppresses its own expression. *Cell* 93(1): 1207–1217.

Went FW. (1960) Photo- and thermoperiodic effects in plant growth. In: *Biological Clocks. Cold Spring Harbor Symposia on Quantitative Biology*, Vol. 25. New York: Cold Spring Harbor, pp. 221–230.

Went FW. (1974) Reflections and speculations. *Ann Rev Plant Physiol* 25: 1–26.

Whitrow GJ. (1988) *Time in History*. Oxford: Oxford University Press, 217 pp.

Williams JA, Sehgal A. (2001) Molecular components of the *Drosophila* circadian clock. *Ann Rev Physiol* 63: 729–755 (Review).

Winfree AT. (1972) Spiral waves of chemical activity. *Science* 175: 634–636.

Winfree AT. (1987) *The Timing of Biological Clocks*. New York: Scientific American Books, Inc., 199 pp.

Winfree AT. (2001) *The Geometry of Biological Time*, 2nd edn. New York: Springer, 777 pp.

Xu Y, Mori T, Johnson CH. (2000) Circadian clock-protein expression in cyanobacteria: rhythms and phase setting. *EMBO J* 19(13): 3349–3357.

Xu Y, Mori T, Johnson CH. (2003) Cyanobacterial circadian clockwork: roles of KaiA, KaiB and the *kaiBC* promoter in regulating KaiC. *EMBO J* 22(9): 2117–2126.

Yin L, Lazar MA. (2005) The orphan nuclear receptor Rev-erbα recruits the N-CoR/Histone Deacetylase 3 corepressor to regulate the circadian *Bmal1* gene. *Mol Endocrinol* 19(6): 1454–1459.

Young MW. (1998) The molecular control of circadian behavioral rhythms and their entrainment in *Drosophila*. *Ann Rev Biochem* 67: 135–152 (Review).

Yu Q, Jacquier AC, Citri Y, Hamblen M, Hall JC, Rosbash M. (1987) Molecular mapping of point mutations in the period gene that stop or speed up biological clocks in *Drosophila melanogaster*. *Proc Natl Acad Sci USA* 84(3): 784–788.

6
Tidal and Lunar Rhythms

"The illimitable, silent, never-resting thing called Time, rolling, rushing on, swift, silent, like an all-embracing ocean-tide, on which we and all the universe swim like exhalations, like apparitions which are, and then are not: this is forever very literally a miracle;...."
— Thomas Carlyle (1795–1881), Scottish essayist, historian

Introduction

Ancient people acknowledged the two sources of natural light as the greater (the Sun) and the lesser (the Moon): *"And God made two great lights; the greater light to rule the day, and the lesser light to rule the night..."* (Genesis 1: 16. Bible, King James Version).

Throughout this book, and either directly or indirectly, we have emphasized the "greater," the solar-generated 24-h day and its role in the synchronization of biological rhythms.

We now turn our attention to the "lesser," the lunar cycles and the biological variables having periods that are similar to them. Within the vast collection of scientific literature we find numerous articles that describe the association of the Moon with the physical components of the Earth (e.g., tides). We find papers also about possible relationships between the Moon and biological variables, especially the behavior of marine[1] invertebrates that live in or near intertidal zones. Because the Moon revolves around the Earth while the Earth is rotating on its axis and revolving around the Sun, a number of physical rhythms are generated. Included in this list are the cycles of the tide (12.4 h), the lunar day (24.8 h), and the lunar month (29.5 days). Biological variables that have these approximate periods are often referred to as having *circatidal* (12.4 h), *circalunidian* (24.8 h), and *circalunar* (29.5 days) rhythms. Theoretically, these three periods

---

[1] Marine (adjective), as used here, relates to oceans, saltwater seas, and estuarine regions where rivers flow into the sea. Marine organisms live in the ocean, sea, or tidal regions for all or part of their life. Terrestrial organisms, on the other hand, live on land. Amphibians can be found on both land and water.

fall within the ultradian, circadian, and infradian domains, but due to their uniqueness or correlation to lunar-generated frequencies, they warrant special consideration.

Because of the close relationship between lunar-based cycles, such as tides, and biological variables that have similar cycles, the objectives of this chapter are to introduce and discuss: (1) some of the influences and characteristics of the Moon as they relate to light and tides; (2) the characteristics of circatidal, circalunidian, and circalunar rhythms and the role of these rhythms in the life of various marine organisms; (3) biological variables found in terrestrial organisms that have periods similar to those of lunar-generated cycles; and (4) some circatidal oscillator hypotheses.

Moon and Light

The two most profound influences of the Moon on our environment are the light that it reflects from the Sun and its influence on tides. Both moonlight and tides affect the rhythmic behavior and development of many organisms (cf. Neuman, 1981).

The amount of light reflected from the Moon follows a cycle of *lunar phases*, a predictable pattern of changes and times when the Moon rises in the east and sets in the west. A complete cycle from one new Moon phase to the next is completed in about 29.5 days, the duration of which is called the *lunar* or *synodic month*. The *sidereal month*[2] is actually about 27.32 days, but since the Earth is also revolving around the Sun, two additional days are required before the same phase can be viewed from the Earth (see Figure 3.6 in Chapter 3 on Time). The mean solar day is 24 h, but the *lunar day* is 24.8 h, which explains why the Moon appears to rise an average of 50–51 min later each day.

Compared to the Sun, which provides an *illuminance*[3] of about 1,000 lux 10 min after sunrise (cf. Bünning, 1969) and about 108,000 lux of direct sunlight at noon in the northern hemisphere (cf. Salisbury & Ross, 1992), the illuminance of full moonlight is less than 1 lux. For example, in tropical regions near the equator, illuminance values during full Moon are only 0.2–0.3 lux (cf. Erkert, 1974). Also, the amounts of light reaching photoreceptors will, to a certain degree, depend upon how high the Moon is, as well as the positions and locations of receptive structures (i.e., leaf or eye).

Light reflected from the Moon has the same visible spectrum as sunlight, but a little redder. There is also some variation in the color index relative to phase, with the full Moon being bluer and becoming slightly redder during the quarter phases

---

[2] The sidereal month is the duration of time needed for the Moon to return to the same location based upon a distant star.

[3] Units, such as lux and foot-candle (1 ft-c = 10.76 lux) are no longer used extensively in research, since they are based upon the sensitivity of the human eye and not on wavelengths. Light radiation quantities are now more appropriately expressed as irradiance or fluence and in accord to the Système Internationale.

TABLE 6.1. Energy levels of light obtained during nearly full Moon at Logan, Utah from July 26 to July 28 when the Moon was approximately 30° above the southern horizon.[a]

Color	Wavelength (nm)	Light level (mW m^2 nm^1)
Blue	425	0.00114
Green	550	0.00230
Red	660	0.00242
Far-Red	725	0.00219

[a]Estimated from a spectral distribution plot in Figure 11 in Salisbury (1981).

(Gehrels et al., 1964; Kopal, 1969). Overall, the quality of full moonlight appears to be quite close to that experienced just after sunset (Salisbury, 1981), a significant feature since photoreceptors of oscillators, such as the phytochromes, cryptochromes, and other pigments, depend not upon the total energy of the visible spectrum, but rather upon the energy of specific wavelengths (Table 6.1). Furthermore, it is the light present at dawn and dusk, not the changes in irradiance during day and night that organisms utilize for circadian photoentrainment (cf. Salisbury & Ross, 1992; Panda et al., 2003).

A direct association with moonlight on any phase of plant development remains unanswered, although some phytochrome-controlled responses require extremely low irradiance levels (Mandoli & Briggs, 1981). On the one hand, there are a limited number of reports that suggest a slight photoperiodic response of moonlight to floral induction (Von Gaertner & Braunroth, 1935; Kadman-Zahavi & Peiper, 1987), including a possible association with the amplitude of leaf movement rhythms (Bünning & Moser, 1969; Bünning, 1971; Bünning, 1979). However, these effects and relationships have not been fully substantiated in related studies (cf. Salisbury, 1981; Fuhrman, 1992; Salisbury & Ross, 1992; Panda et al., 2003). This is not to say, as will be discussed later, that rhythms with periods falling in the range of the lunar period do not exist for some physiological processes in plants. Unlike the situation for terrestrial plants, the response of various animals to moonlight is more definitive. A number of events drawn from a diverse group of terrestrial animals occur in conjunction with moonlight (Table 6.2). Often the amplitude of a circadian pattern is modulated by increased or decreased activity during increasing or decreasing moonlight.

Visible light of sufficient energy and spectral quality is absorbed by pigments, be it from the Moon or any other source. While the daily light/dark cycle serves as a major synchronizer of circadian rhythms, biological rhythms are possible without the involvement of pigments. Locomotor activity, which is one of the many variables studied in marine organisms, normally oscillates in synchrony with the tides, not LD cycles. Although this rhythm can be synchronized to daily LD cycles in the laboratory, tidal effects, such as wave action, mechanical agitation, and other physical forces serve as dominant synchronizers in a natural setting (cf. Enright, 1965; Morgan, 1991).

TABLE 6.2. Animals displaying modulations in activity relative to the phase of the Moon.

Animal	Behavior and phase	Reference
Night monkey (*Aotus trivirgatus*)	Active at dusk and dawn near new Moon, active through the night near full Moon	Erkert, 1974
Leaf-nosed bat (*Artibeus lituratus*)	Moonlight decreases activity	Erkert, 1974
Egyptian fruit bat (*Rousettus aegyptiacus*)	Maximum activity when Moon waxes (becomes brighter)	Erkert, 1974
Leaf-nosed bat (*Phyllostomus hasatatus*)	Maximum activity when Moon wanes (becomes dimmer)	Erkert, 1974
Guppy (*Lebistes reticulatus*)	Dorsal light reaction: sensitivity greatest to yellow light at full Moon and to violet light at new Moon	Lang, 1977
Glass eels (*Anguilla australis* and *A. dieffenbachii*)	Annual migration influenced by tides and moonlight	Jellyman & Lambert, 2003
Mayfly (*Povilla adusta*)	Emergence near full Moon	Hardtland-Rowe, 1955
Ant lion (*Myrmeleon obscurus*)	Peak in pit-building activity near full Moon	Youthed & Moran, 1969
Nocturnal bee (*Sphecodogastra texana*)	More nocturnal foraging between new Moon and full Moon	Kerfoot, 1967

Moon and Tides

Perhaps it is safe to say that many people know that the Moon has something to do with tides. Oceans comprise about 70% of the surface area of the Earth and people living close to the ocean are especially aware that the tide fluctuates systematically between a high- and low-water level. When these fluctuations are especially large, they can affect boating and the appearance and disappearance of land masses and the flora and fauna that inhabit such intertidal zones (Figure 6.1).

High and Low Tides

The timing and magnitude of tides not only depend upon the position of the Moon and Sun, but also upon the geographic location, type of shoreline, ocean bottom, wind, and barometric pressure. Tables that contain the schedule of dates and times for high and low tides (e.g., tables for various sites are readily available on the internet), show that tides rise and fall twice each day.[4] A global view of the Earth

---

[4] There are some exceptions where coasts experience only one noticeable high and low tide per day, such as the northern shores of the Gulf of Mexico (Barnwell, 1976).

FIGURE 6.1. High and low tides that reoccur every 12.4-h can affect boating and the appearance and disappearance of land masses and the flora and fauna that inhabit intertidal zones. Photographs taken at the Alta Fjord, Norway, June 28, 2004 between 10:50 and 11:47 h (near high tide) and 6.2 h later between 17:07 and 17:54 h (near low tide) (Photos by R. Sothern).

reveals two high tides and two low tides occurring simultaneously (Figure 6.2A). One high tide occurs approximately in line with the Moon overhead and the other on the opposite side of the Earth. Similarly, two low tides are present also at these times, but located 90° from the high tides. However, the actual occurrence of high and low tides, as well as the mixing of two semi-diurnal tides to create a single-lunar diurnal tide, is much more involved than what we have described here (Barnwell, 1976).

A. Tide at a given location on earth (x) at 4 different times (a-e) during one lunar cycle of 24.8h

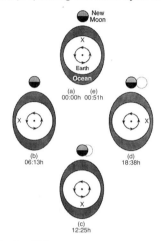

B. Spring Tides

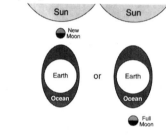

C. Neap Tides

FIGURE 6.2. Diagrams illustrating the influence of the Moon on ocean tides occurring on the Earth (not drawn to scale): (*A*) The relationship of the tide to a given location on Earth (*x*) at four different times (*a–e*) during one lunar cycle of 24.8 h; (*B*) Spring tides; (*C*) Neap tides attributed to the influence of the locations of the Sun and Moon.

To better understand how tides are caused, one should view the relationship between the Earth and Moon as a system that is affected by the gravitational forces of the Sun and the Moon on the Earth. According to Newton's law, the force of gravitation decreases with distance.[5] Therefore, the region of the Earth nearest the Moon is subjected to a greater attraction or "pull," and the waters of the oceans in that area bulge toward the Moon (Figure 6.3).

While the gravitational force accounts for the bulge on the side that faces the Moon, a centrifugal force dominates in producing the bulge on the opposite side.

---

[5] In more precise terms, the force of gravitation between two masses is inversely proportional to the square of the distance between them.

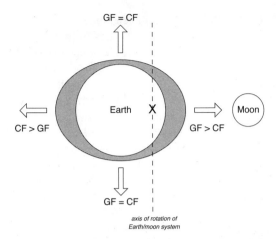

FIGURE 6.3. Diagram illustrating the Earth–Moon system and the influence of the gravitational force (GF) and centrifugal force (CF) on the two tidal bulges. X = center of mass of Earth/Moon gravitational system. Based upon Figure 21.3 from Salisbury & Ross, 1992.

Within the Earth–Moon system, the mass of the Earth is much greater than the mass of the Moon, and thus their common center of mass (gravity) lies well within the surface of the Earth that is closest to the Moon. Both the Earth and Moon revolve around this center of gravity. Therefore, on the side away from the Moon, the resulting centrifugal force exceeds the gravitational force (pull) of the Moon and the water bulges away (Figure 6.3).

As the Earth revolves around the Sun, the axis of rotation is tilted by about 23.5° to a line perpendicular to its orbit[6] (Figure 6.4). Therefore, the path followed

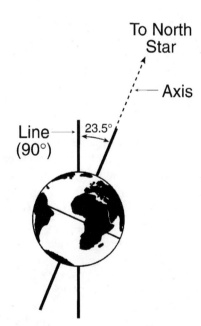

FIGURE 6.4. Diagrams illustrating the 23.5° tilt of the Earth relative to a vertical line drawn perpendicular to the ecliptic plane.

---

[6] Many of the globes that are sold commercially are mounted to show this tilt and could be used, along with a carpenter's square, to visualize the concepts presented in this section.

by the Moon as it completes one full revolution of 27.3 days (= one tropical month[7]) around the Earth varies from 28°30´ north latitude to 28°30´ south latitude. In other words, the angle of declination changes. Except when the Moon is over the equator (0°), the amplitudes of the two tides at a given coast during a lunar day (24.8 h) do not remain equal. To illustrate this change, we can select a location in the Northern Hemisphere at a time when the Moon is overhead and the bulge of the tide is at its crest.[8] After 12.4 h, the Earth has rotated 180° relative to the Moon and the location selected earlier will experience again a high tide. However, because of the 23.5° tilt of the Earth's axis, the location will no longer be situated at the crest of the bulge (amplitude) and consequently, this second high tide will be slightly lower than that which occurred 12.4 h earlier. This effect is greater on some coastlines than others and introduces the diurnal inequality in the heights of successive tides (cf. Barnwell, 1976).

Spring and Neap Tides

Missing from the overall scenario that we have discussed thus far in reference to the physical forces that are associated with the tides, is the Sun. Because the Moon is closer to the Earth than the Sun, the gravitational attraction between the Moon and Earth is greater. During the new Moon phase, which occurs when the Sun, Moon, and Earth are lined up in such a sequence, the gravitational pull of the Sun combines with the gravitational pull of the Moon to cause much higher tides. These higher than normal tides are called *spring tides* (Figure 6.2B). When the gravitational pull of the Sun is at right angle to the gravitational pull of the Moon (near the first and third quarter phases of the Moon), the high tides are lower than normal and the low tides are higher than normal. These conditions are referred to as *neap tides* (Figure 6.2C). The duration between two successive neap tides is about 14.8 days, as is the duration between two spring tides.

Earth Tide

Not only do the oceans respond to these physical forces, but so too does the terrestrial mass of the Earth itself in a phenomenon known as "Earth tides" (see Essay 6.1). Except for material covered in Essay 6.1, the focus of the text that follows, as well as the term tide, will be in reference to ocean and sea.

Essay 6.1 (by WLK): Earth Tides

My fascination of Earth tides commenced when I read a brief note in a book that stated that twice each day the tallest buildings in New York City rose and fell over 30 cm, and

---

[7] The tropical month, 27.321582 days (i.e., 27 days 7 h 43 min 5 s), is 7 s shorter than the sidereal month and is the time between the passage of the Moon through a given celestial longitude.

[8] In reality, there is a lag in the tide that is measured in hours following the Moon's passage since the gravitational pull from the Moon does not have an immediate effect on the sea water due to friction between the sea water and the sea floor.

tilted about 5 cm in response to Earth tides. I was in awe that such an event, which was dependent upon the Moon/Earth relationship, was taking place in a city that I had often visited without my noticing it.[9] In reality, however, it was happening simultaneously over the entire state of New York and its neighboring states and the surrounding landmass. The phenomenon moves as a wave from east to west, the land surface tilting first toward the west and then toward the east. Just as one does not notice the rise and fall of the water level when on a ship that is out at sea, similarly, when on land, one does not notice the rise and fall of the earth's surface. During times of full Moon and new Moon, the amplitude is almost 50 cm (20 inches), but depending upon the latitude the rise and fall may be much less (e.g., Germany, 30 cm (12 inches)) (cf. Lecolazet, 1977).

A major premise for understanding how Earth tides are produced is the fact that the direction of the vertical is not constant and will vary under the attraction of the lunar–solar system.[10] Just as we have illustrated for the ocean tides in Figure 6.3, the effects of gravitational and centrifugal forces, as well as the common center of the lunar–Earth system, change in accord with the movements of the Earth and Moon.

A simple, everyday device for measuring the vertical consists of a small weight suspended by a string, and when allowed to swing, represents a pendulum in motion. In an earlier chapter (see Chapter 3 on Time) we described the pendulum as being one of the first timing devices used for the quantification of oscillations. When modified to be an integral component of an instrument known as the *horizontal pendulum*, it can be used to demonstrate the rhythmic deviations of the vertical.[11] In addition to the horizontal pendulum, other instruments and procedures have been developed to study Earth tides. One of the classical procedures, as well as modern modifications of it, has been to monitor the rise and fall of water or other liquids within wells,[12] or in pipes buried in the Earth.[13] The relationship between underground water levels and Earth tides is associated with the compression and expansion of terrestrial materials, not the lunar gravitational attraction of water as in ocean tides.

Historically, one of the earliest hints or observations of changes recurring within the crust of the Earth that relate to Earth tides was noted by Pliny the Elder[14] in his *Historia Naturalis* (cf. Melchior, 1966). In some wells, the water level would be in synchrony

---

[9] After much searching, I located the source of the brief statement that I had read, and found that it had been based upon an article that had appeared in The New York Times (Sullivan, 1981). In any event, one would have to have very sharp senses to detect such movements.

[10] This important discovery occurred during the mid to late 1800s, but for all practical purposes the deviation is so slight that most engineers and carpenters do not factor it into their calculation of the vertical.

[11] The horizontal pendulum was one of the basic instruments used to study Earth tides (Melchoir, 1966).

[12] Locations where water levels in wells have been monitored include Wisconsin, Iowa, and Utah in the USA, and Belgium.

[13] Michelson & Gale (1919) recorded tidal tilt in Wisconsin with a device that was constructed with pipes.

[14] Pliny the Elder, a Roman senator, natural historian, and scientist, was born in 23 A.D. and died in 79 A.D. during the Vesuvius eruption. He published the world's first encyclopedia, *Historia Naturalis*, in 77 A.D.

with the high and low tides, while in others it would be out of phase. Wells situated near the ocean may be in synchrony with the ocean tides, due to the pressure of the ocean tide on the Earth near it. Wells located further away may indicate a high tide (water level) in the well when the lunar–solar attraction indicates a low tide (cf. Melchoir, 1966). This could be caused by the compression of the water-bearing region of the Earth's crust at low-tide squeezing water into the well and the subsequent rise in the level of the water in the well (Pekeris, 1940). Likewise, the expansion of the crust at high tide causes the water level in the well to fall. The situation is much more complex, however, than stated here.

The positioning of various instruments, which themselves are attached to the crust that tilts and bulges, must be factored into the equations used for monitoring and analyzing Earth tides. Quantification of the phenomenon of Earth tides is steeped in theory, mathematics, and physics, much of which extends beyond the context of this essay.[15] Yet, it is a topic that should not be ignored herein. Life has evolved, and continues, in the presence of the various periodicities of Earth tides. In addition to semidiurnal and diurnal periods, periods of longer duration, such as fortnightly, semiannual, 4.4 years, and 18.6 years, are components of known tidal frequencies (cf. Rinehart, 1976). Biological variables having periods that are similar to the first three or four periodicities in this list are inherent in some of the organisms that are included in this chapter.

In this section we have highlighted the approximate periods for three prominent physical cycles: 24.8 h for the lunar day; 29.5 days for the lunar month; and 14.8 days for the one-half lunar cycle, sometimes referred to as the semi-monthly cyclic, or the bimodal spring tide peaks of the full synodic lunar month. The periodicities are especially relevant to the study of biological rhythms of marine organisms.

Marine Organisms

Circatidal and circalunar rhythms of marine organisms have a number of characteristics (Table 6.3), some of which are similar to those of circadian rhythms, while others reflect unique behavioral and physiological properties of their endogenous timing system (cf. Palmer, 1974).

Circatidal Rhythms

Many of the variables in marine organisms that have been monitored under conditions of either continuous illumination (LL) or darkness (DD) and constant temperature oscillate with periods that are both circadian and circatidal. Circatidal rhythms of activity, like circadian rhythms of activity, free-run under constant conditions (e.g., LL or DD and constant temperature). They differ from the circadian rhythms, however, by the fact that they are synchronized by tides. While tides *per se* represent the exogenous oscillator, a number of physical components

---

[15] Some other areas of study that are relevant to Earth tides include earthquakes, volcanic activity, geyser activity, geology, astronomy, and space travel.

TABLE 6.3. Features associated with characteristics of circatidal (CT) and circalunar (CL) rhythms observed for some variables in marine organisms.

Feature	Comments and examples
Endogenous (free-running)	Activity rhythms observed for isopod under constant nontidal conditions for CT (24 h 55 min) and CL (26–33 days) (Enright, 1972)
Agents of entrainment or synchronization	Examples relative to tides include wave action or mechanical stimulation by water currents may entrain the circatidal rhythm of activity (Enright, 1965); other agents include hydrostatic pressure, temperature, inundation, and salinity (Morgan, 1991)
Temperature compensation	Locomotor tidal rhythms of the crab, *Carcinus maenus* (Williams & Naylor, 1969)
Multiple periodicities and oscillators	Multiple frequencies or periods may be observed for a single variable in a single organism; variable having circadian periods may be present; one variable may display multiple frequencies
Chemicals	Chemicals, such as ethanol and deuterium oxide, may lengthen the period and/or cause a phase shift in certain organisms (e.g., isopod) (Enright, 1971a,b).

comprise a tide, one or more of which could be a primary synchronizer. Included among possible components of synchronization and entrainment within the tidal complex are factors such as wave action or mechanical agitation, changes in various types of pressure (e.g., hydrostatic pressure, gases, etc.), temperature, light levels, and salinity (cf. Morgan, 1991).

Results from laboratory studies with the sand-beach isopod *Excirolana chiltoni* have shown that mechanical stimulation by water currents similar to those encountered on a beach may entrain the circatidal rhythm of locomotor activity (Enright, 1965). Similarly, wave action simulated in an aquarium has entrained a circatidal rhythm in the swimming activity of the blenny (*Lipophrys*), a fish found along rocky shores (Morgan, 1991). Hydrostatic pressure associated with the rise and fall of tides has been shown under laboratory conditions to entrain the circatidal activity rhythms of a number of marine animals (Morgan, 1984, 1991).

Crab Activity

"*Time and tide wait for no one*" is a truism upon which depends the very existence of various organisms living in intertidal zones. The fiddler crab *Uca pugilator* (Bosc) is a prime example. During low tide, the crabs emerge from their underwater burrows, scramble along the sandy beaches, and engage in activities, such as feeding, defending a territory, and attracting a mate (Figure 6.5). But, does this occur just because the water level changes or is there an inherent clocking mechanism present within the crab? The natural setting of a beach and tidal region,

FIGURE 6.5. Fiddler crab (*Uca pugnax*) on beach where it searches for food and a mate (Photograph provided by Professor F.H. Barnwell, University of Minnesota).

however, are not always the ideal place for biologists to address such a question. To best test hypotheses about the activity patterns of these animals and to more accurately monitor their behavior, a laboratory setting where the locomotor activity (movements) of individuals can be monitored under controlled or rearranged environmental conditions, takes precedence (Figure 6.6).

Results from the analyses of activity patterns of fiddler crabs have clearly demonstrated the presence of circatidal motor activity rhythms, which may persist for as long as 5 weeks under LL conditions (Barnwell, 1966). In addition to the circatidal rhythm, a free-running circadian rhythm, which is displayed as a single daily peak, becomes evident and occurs later each day for both *U. pugilator* and *U. minax*. In nature, these two species are primarily more active during the normal dark span.

Circadian vs. Circatidal

Rhythms of crabs may be influenced by a number of factors, including their habitat. Thousands of species of crabs exist in nature and may be found in different habitats. Some, such as the shore crab, *Carcinus maenas* (L.), occur in either tidal or nontidal locations. A comparison of locomotor activity rhythms of *C. maenas* collected from nontidal docks, which typically display a nocturnal peak in activity, with those of shore crabs from an intertidal zone, which typically display tidal-related activity patterns, have revealed some interesting features (Naylor, 1960). When "dock" crabs were maintained in the laboratory under LD, nocturnal peaks were observed, while under LL, they displayed peaks in activity each day that occur about 50 min later each day. This drift was absent, however, if the crabs were maintained under a light/dark regimen. When the "shore" crabs were removed from their tidal environment and maintained in a nontidal aquarium under LD, their activity rhythm was at first in phase with the shore tidal cycle. After a number of weeks, the tidal component became suppressed and their behavior resembled that of the "dock" population while under both LD (nocturnal activity) and LL (phase drifting).

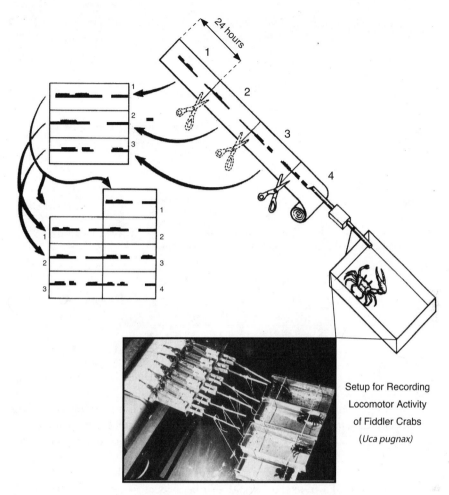

FIGURE 6.6. Diagram illustrating an early method for monitoring the activity of a crab in a box, which will tilt up or down as the crab moves and cause a pen to record the activity in a series vertical lines on a rotating strip of paper. Each 24-h span can be cut from the strip chart and arranged in a column, 1 day per row (inset photograph provided by Professor F.H. Barnwell, University of Minnesota).

Other Organisms

Circatidal rhythms have been observed in other organisms (Table 6.4), ranging from microscopic algae and small worms that accumulate on the surface of water (Bohn, 1903; Martin, 1907; Palmer & Round, 1967) to littoral[16] fish (Gibson, 1965). An example of the latter is the small intertidal fish *Coryphoblennius galerita* (L.) whose swimming activity consists of "hops" and "rests" as it moves

---

[16] Littoral (adjective) as used here refers to fish living near the shore of a sea.

TABLE 6.4. Activity rhythms of marine organisms that have circatidal rhythms.

Organism	Comments	Reference
Fiddler crab (*Uca* sp.)	Rhythms in behavior and physiology have been studied extensively	Palmer, 1973, 1995a
Sand-beach amphipod (*Synchelidium* sp.)	Buried in sand during low tide; rides up and down beaches with tides	Enright, 1963
Sand-beach isopod (*Excirolana chiltoni*)	Buried in sand except during 4–6 h after tide crest on beach; mechanical stimulation of wave action can entrain rhythm	Enright, 1965
Pink shrimp (*Penaeus duorarum*)	Swimming activity of postlarval and juvenile shrimp have phase relation with tidal and daily cycles	Hughes, 1972
Intertidal fish (*Coryphoblennius galerita*)	Circatidal rhythm observed in LD and DD in phase with tidal cycle	Gibson, 1970
Littoral fish (*Blennius pholis*)	Circatidal rhythm observed in LL (12.5 h) and DD (12.56 h)	Gibson, 1965
Diatoms (single-celled algae) (*Hantzschia virgata*)	Complex vertical-migration rhythm (24.8 h) persists in LL	Palmer & Round, 1967

from one place to another. When monitored in the laboratory, these littoral fish display circatidal periods in their activity under both LD and DD (Gibson, 1970). The first of the two circatidal peaks is usually the greatest and has been attributed to a weak circadian component that occurs in the late morning (Gibson, 1970). Because *C. galerita* normally feed on the appendages (cirri) of barnacles when they become available during high tide, their activity rhythms may be related to feeding.

Because feeding can be a rhythmic behavior that is in synchrony with the tides, it follows that a tidal frequency would also be evident in certain physiological variables. An example is found in the blood sugar levels of the shore crab *Carcinus maenas* (Parvathy Rajan et al., 1979) and the shrimp *Crangon vulgaris* (Poolsanguan & Uglow, 1974). Lowest levels of blood sugars in shore crabs, which were maintained in a nontidal seawater aquarium with a natural LD photoperiod and assayed within 3 days of capture, occurred at times corresponding to the times of high tide in their previous tidal habitat. Similarly, blood sugar levels of shrimp have shown a circatidal pattern, with higher levels being associated with times of low tide (Poolsanguan & Uglow, 1974). This rhythm persisted under both DD and LL, although the levels were generally higher in DD (Figure 6.7).

Reproduction

For many marine invertebrates (Table 6.5), especially those found within or near the tropics, rhythms in reproductive events such as mating, egg, and sperm formation, spawning, and hatching (release of larvae) have periods and/or phases that

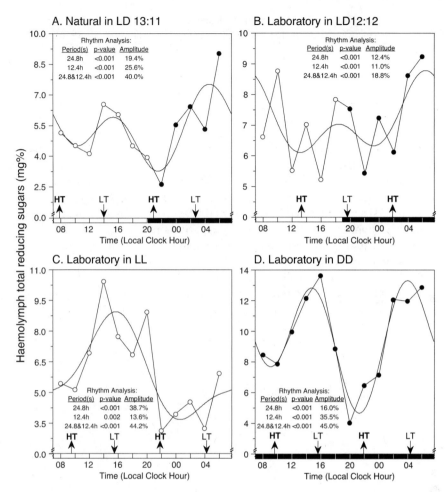

FIGURE 6.7. Tidal rhythms (24.8 and 12.4 h) detected in blood sugar of shrimp (*Crangon vulgaris*). Peak values were associated with high tides (HT) regardless if shrimp were fresh caught during a natural photoperiod of LD 13:11 (*panel A*); or kept in a laboratory under LD 12:12 for 3 weeks (*panel B*), or in constant light (LL, *panel C*) or constant darkness (DD, *panel D*) for 4 weeks. Manipulation of the photoperiod did not abolish the tidal rhythms, but the 12.4-h period was more prominent (showed a higher amplitude) in the natural setting (A) and during constant darkness (D) (redrawn and analyzed from Figures 1, 2, and 3 in Poolsanguan & Uglow, 1974).

appear to be associated with the phase of the Moon. The events may occur only fortnightly,[17] monthly, or just once a year. For example, a prominent and reproducible annual cycle in gonadal weight is found in a number of marine

---

[17] Fortnight, a span of 14 days, is a noun that is no longer used extensively in day to day conversations. Fortnightly, the adjective, describes something that occurs once in a fortnight. Semi-monthly is a more commonly used term.

6. Tidal and Lunar Rhythms

TABLE 6.5. Marine invertebrates that contain some species that display reproductive rhythms associated with a lunar cycle and tides (Pearse, 1990).

Clams	Flies
Corals	Marine worms
Crabs	Snails
Crustaceans	Sponges

invertebrates, including a sea urchin, a mollusk, and a sea star (Figure 6.8), monitored in their natural habitat (Pearse et al., 1985; Halberg et al., 1987). In the laboratory, an annual cycle in ovarian weight for the Jamuna River catfish has also been documented under natural LD conditions (Figure 6.8) and this annual rhythm persisted with a free-running period when the catfish were kept in either continuous light or darkness for several years (Sundararaj et al., 1973, 1982).

Often the fortnightly and monthly cycles may be linked to an annual and/or photoperiodic cycle. The same may apply to the daily cycle that may be biologically linked or mathematically correlated to the other two cycles. A prime example is found in the marine worms *Typosyllis prolifera*, which display reproductive

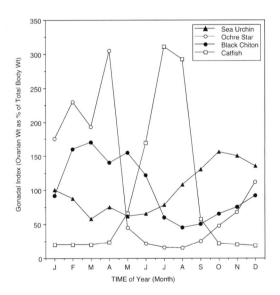

FIGURE 6.8. Circannual rhythms in gonadal weight of some marine invertebrates. Monthly mean gonadal weights shown as percent of body weight (transformed to percent of overall mean for comparisons; SE's not shown for clarity). Data collected near Pacific Grove California between January 1954 and December 1958 for purple sea urchin (*Strongylocentrotus purpuratus*), black chiton (*Katharina tunicata*), and ochre sea star (*Pisaster pchraceus*) and from the Jamuna River near Dehli, India between February 1966 and September 1967 for a river catfish (*Hetropneustes fossilis*) (redrawn from Figures 1 and 2 in Halberg et al., 1987).

variables in three different domains of periodicity (Franke, 1986). The sequence of events that comprise the reproductive history of this worm is different from what is common for most animals. The specific epithet[18] of the species *prolifera* provides a clue that the worm is prolific. During its life history, the worm sometimes appears as "two worms in one." Briefly, this species reproduces by a process known as stolonization, a situation where the rear (posterior) portion of the worm develops into a reproductive individual called the stolon.[19] When the egg- or sperm-laden stolons are detached from their parental stocks they swim to the surface of the water, swarm, spawn, and then die. The released gametes display swarming activity also, which results in the union of the egg and sperm and the beginning of embryonic development. Unlike the stolon-type individuals, the parent stock can continue to live and regenerate new stolons for a number of times before it dies.

Under natural conditions of the northern Adriatic Sea, reproduction of *T. prolifera* occurs primarily from late March to early October. The number of worms in the population that are in the reproductive state displays an annual rhythm that has a peak in May (Franke, 1986). Under these natural conditions, the release of stolons follows both a daily and a monthly rhythm. Based upon when they can be collected at the water surface, the daily release takes place within an hour after sunrise[20] and is dependent also upon the process of photoperiodism (length of the light span). A lunar-monthly cycle of stolonization for individuals under constant conditions free-runs with a mean period of about 31 days with a Q_{10} of 1.04. As for the role of moonlight, it has been suggested that in the case of *T. prolifera*, it may be serving as a possible synchronizer that entrains the individual rhythms (Franke, 1986).

One of the best known examples of reproduction occurring at a specific phase of the Moon is the spawning behavior of a marine worm that inhabits the coral reefs of the Samoan Islands. The marine worm *Eunice viridis* Gray, commonly known as the palolo in the Samoan language, produces posterior segments called the epitoke. These structures are about 20 cm long and contain either eggs or sperm. They are released near or after midnight, depending upon the location of the islands (Caspers, 1984). The much shorter anterior portion of the body remains in the coral habitat and regenerates the rest of the body. The epitoke is a favorite food collected by the islanders, who for centuries have known when to be

---

[18] The species name of an organism consists of two parts, the genus (e.g., *Typosyllis*) and the specific epithet (e.g., *prolifera*). The first name can be used alone to indicate the genus, but the specific epithet should not be used alone, since many species may have the same specific epithet. For example, *vulgaris* is the specific epithet for many common plants and animals, such as bean (*Phaseolus vulgaris*), beet (*Beta vulgaris*), lilac (*Syringa vulgaris*), shrimp (*Crangon vulgaris*), common octopus (*Octopus vulgaris*), European starling (*Sturnus vulgaris*), and Red or European squirrel (*Sciurus vulgaris*).

[19] The term "stolon" is used also to describe the above ground horizontal stems of strawberries (*Fragaria vesca*) from which new plants arise.

[20] Results from laboratory experiments (unpublished), indicate that it is not under circadian control (Franke, 1986).

prepared for the appearance of the epitokous segments.[21] The casting off of these segments takes place during the third quarter phase of the Moon in October or November[22] and is limited to a duration of 3 days (Caspers, 1984).

Hatching represents the process by which larvae are released from eggs. Timing is important not only for the survival of larvae released into the environment, but to the adult female who must venture into an environment where she may be more vulnerable to predation. For the fiddler crab, *Uca pugilator*, the combination of night and high tide may represent the best situation for hatching. For the occurrence of hatching to take place at such a time, it must be synchronized in phase with both solar and tidal cycles. Results from experiments conducted under LL and constant temperature have indicated a free-running period of about 25 h (Bergin, 1981). This endogenous rhythm of timing may have a number of adaptive advantages, including minimizing the predation of the adult females as they walk to the edge of the water to deposit their hatching eggs, reducing the predation of larvae, and the flushing of larvae out with the tide (DeCoursey, 1979; Bergin, 1981; Backwell et al., 1998).

Color Change

Recurring changes in the color of cells, tissues, and organs are found in diverse groups of organisms and often associated with time-dependent physiological processes. We know that changes in color may be associated with the seasons, as in the case of the fur or feathers of animals and the leaves of plants. In addition to these seasonal changes, which are under the control of photoperiodism, there are changes in the level and/or in the intracellular dispersal of specific pigments that recur much faster. In some marine organisms, such as the fiddler crab (*Uca pugnax*), changes in color arise from special pigment carrier cells called chromatophores.[23] When the pigments are concentrated into condensed spots, the animal appears to be lighter in color, in contrast to being darker when the pigments are dispersed.

Rhythmic changes in color have been studied most extensively in crabs, especially the chromatophores that are located in the hypodermis of the merus[24] of the

---

[21] According to Miller & Pen (1959), who analyzed samples of palolo, it has high nutritive value.

[22] The reason that both months are mentioned for the emergence of the "palolo" is due to the difference between the lengths of the solar months and the lunar month, and thus when the calendar date coincides with the third quarter phase of the Moon.

[23] *Note*: It is easy to confuse the word "chromatophore" with the word "chromophore." The light-absorbing portion of the pigment is called a *chromophore* (Gr. *chroma* = color, *phoros* = carrying), while a special cell of an animal that carries the pigment(s) is called a *chromatophore*. There are various types of chromatophores, and in at least some animals, they provide a good mechanism for camouflage, advertisement, and temperature regulation.

[24] Crabs have five pair of appendages (four pairs of walking legs and one pair of legs with pinchers). The merus represents the broadest part of the leg, which is positioned to be more or less parallel to the ground. The hypodermis, which can be observed without dissecting or injuring the organism, is a layer of cells located below the cuticle (exterior covering).

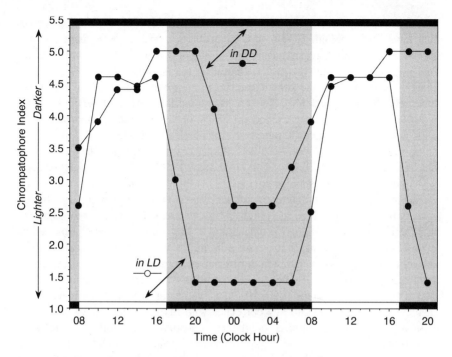

FIGURE 6.9. Generalized diagram illustrating color change based upon the mean dispersion of pigment in black chromatophores on the menus of the walking legs of two groups of juvenile *Carcinus maenus* (L.) crabs maintained for 35 days in either LD or DD prior to being scored at 2-h intervals under similar conditions (LD [open circles] or DD [closed circles]) over a duration of 36 h (values recalculated and illustration redrawn from Powell, 1962). Stage 1 represents maximum concentration of pigment and stage 5 the maximum dispersion of the pigment (cf. Hogben & Slome, 1931).

walking legs. The chromatophores can be observed manually with the aid of a microscope and stage of the dispersion of the pigment scored according to a scale ranging from maximum concentration to maximum dispersion.[25] Some of the more common colors of the chromatophores include black, red, white, and yellow. Because crabs have prominent circatidal rhythms of activity and these colors change during the 24-h day, we can now ask whether the changes are circadian, circatidal, circalunar, or possibly best characterized as displaying multiple frequencies.

For many of the crabs that have been studied, the daily and lunar rhythms observed under natural conditions continue in both DD and LL (cf. review by Palmer, 1973), although there may be changes in the pattern of the rhythm as illustrated for juvenile green crabs (Figure 6.9). While changes in color that have been observed in various species definitely have the characteristics of circadian rhythms (e.g., free-running in constant conditions) (Powell, 1962, 1966;

---

[25] Stage 1 represents the maximum concentration of the pigment, and stage 5 the maximum dispersion (Hogben & Slome, 1931).

Barnwell, 1968), the circatidal feature is not as definitive, at least for the species that have been studied. Herein lies the problem. First, the very nature of the research is labor intensive, requiring the scoring of visual observations at specified intervals (e.g., 2 h intervals) over a duration that should span many weeks or months. Second, often the genera, the species, and the environmental conditions have been different. And third, there could be a masking of one of the tidal components (e.g., high tide) when the second 12.4-h peak should also occur during the dark phase, a time when the pigment remains dispersed. However, results from some extensive studies conducted with fiddler crabs in the mid-1950s (Brown et al., 1953; Brown, 1954; Hines, 1954; Fingerman, 1956) have indicated a tidal component that appears as a supplement to the daily rhythm in the color change pattern.

As has already been noted, pigments serve as photoreceptors in the eyes of animals. Crustaceans, such as the crayfish *Procambarus*, the shore crab *Carcinus*, and the hermit crabs *Pagurus samuelis* and *P. granosimanus*, have been studied extensively in this regard. In the case of hermit crabs, the proximal pigment[26] of the compound eye is concentrated during darkness (dark-adapted), but not during light (light adapted) (Ball, 1968). The migration of retinal pigments in the compound eyes of hermit crabs (Ball, 1968), as well as in crayfish (Aréchiga, 1977), displays a circadian rhythm that correlates with rhythmic dispersal activity patterns of these crabs in response to light. The circadian oscillations in the locomotor activity of *P. granosimanus* persist for a number of days following the removal of eyestalks, although the eyestalks may participate in entrainment (Page & Larimer, 1975a,b). In the case of *Uca pugilator*, eyestalkless crabs appear to exhibit a daily rhythm of color change in LL, although melanin dispersion was less than in those with eyestalks (Fingerman & Yamamoto, 1964).

But what does all of this have to do with the life of these marine organisms? Rhythmic changes in dispersal or movements of pigments could be an adaptive feature relating to antipredation strategies and social signaling (cf. DeCoursey, 1983). Included here would be various behavioral activities, such as safer foraging on beaches, defending territories, and courtship (e.g., a male crab that becomes brighter in color to attract a female).

Terrestrial Organisms

Events or activities of terrestrial organisms that are in phase with or synchronized by lunar cycles abound in folklore[27] and myth. Separating fact from fiction is not easy and studies that address these matters are viewed by many scientists with a

---

[26] These pigments actually serve as opaque screens around the photoreceptors. They are not involved in photoreception, *per se*. During low light/darkness, they withdraw, exposing more of the photoreceptors for dim light detection.

[27] A folk belief by inhabitants of the Shetland Islands (200 km north of Scotland and 6° below the Arctic Circle) was that during the long, dark winters, moonlight should never fall on the face of someone asleep since they noted an "unsettling effect" after an hour of such nocturnal light during a "sensitive" time in the 24-h day (Sinclair, 1987).

certain degree of skepticism. This skepticism may explain, at least in part, why the topic of lunar cycles is either absent or minimized in some of the well-known books on biological rhythms.

The Menstrual Cycle

The female menstrual cycle is the most obvious recurring cycle that has a period that is close to that of a lunar cycle. The Swedish chemist, Svante Arrhenius (1859–1927), reported that menstruation (based upon 12,000 cases), as well as epileptic seizures, exhibited a periodicity between 25.9 and 27.9 days (Arrhenius, 1898; Dunea, 1993). Although a period of 28 days is often cited as being the average length, results from a number of more recent studies indicate that the mean period of the menstrual cycle is about 29.5 days (Presser, 1974; Cutler, 1980).

Having stated this, however, there does not appear to be agreement relative to Moon phase and menses. For example, in two studies totaling 439 women who were not using birth control pills, there was a tendency toward more menstruation onsets between the first and last quarter phases of the Moon (Cutler, 1980).[28] On the other hand, another study of 826 young women found that a large proportion of menstruations occurred around the new Moon (Law, 1986). It is possible that both studies are correct, since one report in the early part of the 20th century stated that "*a significant number of menstrual cycles started at full or new Moon.*" (Guthmann & Ostwald, 1936) and another that menstruation frequency was accentuated at the new *and* full Moon (cf. Kleitman, 1949). Regardless of any discrepancy between Moon phase and menses, a number of other variables in humans have been reported to vary with the phases of the Moon (Table 6.6).

Atmospheric Tides

Starting in the mid-1950s and continuing into the late 1970s, a number of studies found associations between biological rhythms and various lunar-day and solar-day "atmospheric tides," such as geomagnetism, electrostatic field, and background radiation (cf. Brown, 1970). The processes examined ranged from the respiration rates of potato tubers (*Solanum tuberosum*), carrot roots (*Daucus carota*), and chicken embryos (*Gallus gallus*), to the running activity of hamsters (*Cricetus*) and the response of planarians (*Dugesia dorotocephala*) to light. The planarian studies that were conducted for 5–6 days each week over 5 years in a laboratory are of interest here, since the orientation movements of these flatworms in response to light displayed a monthly and annual cycle, with minimum phototaxis (turning away from a light source) occurring near full Moon every month and in June every year (Brown & Park, 1975).

---

[28] A similar phase relationship has been observed in apes and monkeys, which also have menstrual cycles (cited by Smolensky & Lamberg, 2000).

TABLE 6.6. Variables in humans reported to vary with the phases of the Moon.

Variable	Phase and behavior	Reference
Births (5,927,978 cases)	More between last quarter and new Moon	Guillon et al., 1986
Accidental poisonings (2215 patients)	Greater near new Moon for women	Buckley et al., 1993
Meal and alcohol intake (694 adults)	Meal size increase and alcohol decrease near full Moon	de Castro & Pearcey, 1995
Urinary retention (815 Patients)	Incidence greater during new Moon	Payne et al., 1989
Psychopathology in schizophrenia (n = 56)	Deterioration near full Moon	Barr, 2000
Aggressive behaviors (homicides, aggravated assaults, suicide, fatal traffic accidents)	Aggression greater around full Moon; psychiatric admissions greater during first quarter	Lieber, 1978
Myocardial infarction	Maximal near new Moon	Sha et al., 1989
Emergency room visits for animal bites (1621 cases)	Peak near full Moon	Bhattacharjee et al., 2000
Atrial fibrillation (one man, 127 episodes)	Nadir (low) near full Moon	Mikulecky & Valachova, 1996
Gout attacks (55 patients, 126 attacks)	Highest peaks under the new and full Moon	Mikulecky & Rovensky, 1999
Spontaneous pneumothorax (244 patients)	Maximum cases one week before and after new Moon	Sok et al., 2001

Insects

Some variables in insects have also been associated with certain Moon phases. For the mosquito (*Culex (Melanoconion) caudelli*) in the West Indies, the span of the nocturnal peak (22:00 h to 04:00 h) in biting activity was the longest and the number that were collected per 24 h was the greatest during the first quarter (Chadee & Tikasingh, 1989). Predatory activity of female mites (*Typhlodromus pyri Scheuten*) has an about 14-day rhythm, with a pronounced decrease near the full Moon (Mikulecky & Zemek, 1992). Also, analysis of hemolymph lipid concentrations in worker honeybees showed that triacylglycerols, steroids, and body weight were greater near new Moon than at full Moon (Mikulecky & Bounias, 1997).

Plants

Variables in seeds, stems, and leaves have been associated with lunar cycles. Results from studies of the daily imbibition (absorption) of water by bean seeds (Figure 6.10) have shown a circannual rhythm, as well as a circaseptan rhythm

FIGURE 6.10. The process of imbibition (uptake of water) of pole beans (*Phaseolus vulgaris*) is illustrated by placing 50 seeds into a beaker, adding 100 ml of water, and observing the results after 24 h. Photos by R. Sothern.

with a phase relationship to the lunar cycle—peaks preceded each lunar phase and imbibition was greatest from just before full Moon through the last quarter (Brown & Chow, 1973; Spruyt et al., 1987). Lunar-associated rhythmic variations in stem diameter of spruce trees have been found in the field and in a controlled environment, with a double-peaked wave of about 25 h caused by accumulation of water in the cell walls associated and synchronous with gravimetric tides (Zürcher et al., 1998).

Another lunar association has been reported for elongation of stems (cm/day) measured every 3–4 days, where periods of 14.7 days and 29.5 days were found in a number of plant species present in a grassy area of a botanical garden (Abrami, 1972). Palm leaves harvested near the full Moon have higher total cellulose and lower calcium concentrations than at other phases, demonstrating that fluctuations relative to the lunar cycle may maximize durability and minimize foliage herbivory (Vogt et al., 2002). While results from these studies, as well as from others (e.g., Maw, 1967) indicate correlations between lunar phases or cycles with certain physiological and developmental processes of terrestrial plants, their direct relationship, if any, have not been fully established.

Lunar/Tidal Clock Hypotheses

The most common external cycles associated with our environment, and thereby possibly serving as synchronizers for biological rhythms, are the 12.4-h tide, the 24-h solar day, the 24.8-h lunar day, the 29.5-days lunar month, and the year. There are other environmental cycles, but it is upon these five cycles that most clock models or hypotheses are based.

Circadian vs. Circalunidian

Without an extensive elaboration on specific lunar-based models, there is the commonality of circadian involvement, and thus a *circadian oscillator*. Endogenous cycles of about 12.4-h could result from the superposition of two rhythms with

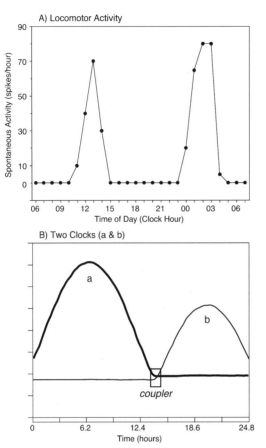

FIGURE 6.11. One circatidal oscillator or two circalunidian clocks? (*A*) Example of fiddler crab locomotor activity showing dual peaks 12.4 h apart; (*B*) Diagram of circalunidian clock hypothesis consisting of two 24.8-h clocks (redrawn from Figures 1 and 6 in Palmer, 1995b).

periods of about 24.8-h (Bünning, 1973), not synchronized normally by LD cycles, but by the physical forces of tides (cf. Bünning, 1979). Something similar, although with an emphasis upon a phase component, has been envisioned in a *circalunidian-clock* hypothesis where two 24.8-h clocks, which are fundamentally one and the same, are coupled together in antiphase to generate the rhythms (Palmer & Williams, 1986; Palmer, 1995a, 2000). This is somewhat easier to understand or picture if one remembers that when locomotor activity rhythms of many of the marine crabs are plotted, there are two peaks, about 12.4-h apart (Figure 11A). In the circalunidian-clock model, each peak is actually a display of one 24.8-h pattern. In other words, each of the two rhythms is driven by its own clock and is normally coupled at the trough between the two peaks in each 24.8-h pattern (Figure 11B).

Interacting Oscillators?

Using the locomotor activity rhythm of the shore crab (*Carcinus maenas*), a "circatidal/circadian" clock hypothesis has been proposed, which involves interacting circadian (about 24 h) and true circatidal (about 12.4 h) oscillators (Naylor, 1996,

1997). While this model serves well for the shore crab and its habitat, there is some question as to the accuracy of a 12.4-h clock that would be synchronized by the tides.

In reality, a 12.4-h period represents a mean value, whereas in nature, deviations of up to 3 h may exist between successive tides. However, if the deviations of the tide from a 24.8-h interval are plotted, the deviations are much less. In other words, a 24.8-h clock would be a more accurate timer (Palmer, 1995a, 1997). Studies of the clam, *Austrovenus stutchburyi*, the crab, *Macrophthalmus hirtipes* (Williams, 1998), and the sand beach isopod, *Cirolana cookii* (Mehta & Lewis, 2000), concluded that these intertidal dwellers lacked a circadian timer and that the endogenous tidal rhythms were controlled by a dual circalundian clock, thus supporting the applicability of the circalunidian clock model hypothesis for tidal rhythms.

Take-Home Message

The Moon is responsible for oscillations in ocean tides and Earth tides, and a number of animals on Earth have adapted to them. Periods similar to lunar-generated periods of 12.4 h (circatidal), 24.8 h (circalunidian), and 29.5 days (circalunar) can be found in a wide variety of biological processes in marine and terrestrial organisms, including fish, amphibians, insects, plants, birds, and humans. A number of hypotheses and models have been proposed for these rhythms.

References

Abrami G. (1972) Correlations between lunar phases and rhythmicities in plant growth under field conditions. *Can J Bot* 50: 2157–2166.

Aréchiga H. (1977) Circadian rhythmicity in the nervous system of crustaceans. *Fed Proc* 36(7): 2036–2041.

Arrhenius S. (1898) [Cosmic influences on physiological phenomena.] [Swedish]. *Skand Archiv Physiol* 8: 367.

Backwell PRY, O'Hara PD, Christy JH. (1998) Prey availability and selective foraging in shorebirds. *Anim Behav* 55(6): 1659–1667.

Ball EE. (1968) Activity patterns and retinal pigment migration in *Pagurus* (Decapoda, Paguridae). *Crustaceana* 14: 302–306.

Barnwell FH. (1966) Daily and tidal patterns of activity in individual fiddler crabs (Genus *Uca*) from the Woods Hole region. *Biol Bull* 130: 1–17.

Barnwell FH. (1968) Comparative aspects of the chromatophoric responses to light and temperature in fiddler crabs of the genus *Uca*. *Biol Bull* 134(2): 221–234.

Barnwell FH. (1976) Variation in the form of the tide and some problems it poses for biological timing systems. In: *Biological Rhythms in the Marine Environment*. DeCoursey PJ, ed. Columbia, SC: University of South Carolina Press, pp. 161–187.

Barr W. (2000) Lunacy revisited. The influence of the moon on mental health and quality of life. *J Psychosoc Nurs Ment Health Serv* 38(5): 28–35.

Bergin ME. (1981) Hatching rhythms in *Uca pugilator* (Decapoda: Brachyura). *Marine Biol* 63: 151–158.

Bhattacharjee C, Bradley P, Smith M, Scally AJ, Wilson BJ. (2000) Do animals bite more during a full moon? Retrospective observational analysis. *BMJ* 321(7276): 1559–1561.

Bohn G. (1903) Sur les movements oscillatoires des *Convoluta roscoffensis*. *CR Acad Sci (Paris)* 137: 576–578.

Brown FA Jr, Fingerman M, Sandeen M, Webb HM. (1953) Persistent diurnal and tidal rhythms of color change in the fiddler crab, *Uca pugnax*. *J Exp Zool* 123: 29–60.

Brown FA Jr. (1954) Biological clocks and the fiddler crab. *Scient Amer* 190(4): 34–37.

Brown FA Jr. (1970) Hypothesis of environmental timing of the clock. In: *The Biological Clock: Two Views*. Brown FA Jr, Hastings JW, Palmer JD, eds. New York: Academic Press, pp. 13–59.

Brown FA Jr, Chow CS. (1973) Lunar-correlated variations in water uptake by bean seeds. *Biol Bull* 145: 265–278.

Brown FA Jr, Park YH. (1975) A persistent monthly variation in responses of planarians to light, and its annual modulation. *Intl J Chronobiol* 3: 57–62.

Buckley NA, Whyte IM, Dawson AH. (1993) There are days ... and moons. Self-poisoning is not lunacy. *Med J Aust* 159(11–12): 786–789.

Bünning E. (1969) [The importance of circadian leaf movements for the precision of day-length measurement] [German]. *Plant (Berl)* 86: 209–217.

Bünning E, Moser I. (1969) Interference of moonlight with the photoperiodic measurement of time by plants, and their adaptive reaction. *Proc Natl Acad Sci USA* 62: 1018–1022.

Bünning E. (1971) The adaptive value of circadian leaf movements. In: *Biochronometry*. Menaker M, ed. Washington, DC: Natl Acad Sci, pp. 203–211.

Bünning E. (1973) *The Physiological Clock*, 3rd edn. Berlin: Springer-Verlag, 258 pp.

Bünning E. (1979) Circadian rhythms, light, and photoperiodism: a re-evaluation. *Bot Mag Tokyo* 92: 89–103.

Caspers H. (1984) Spawning periodicity and habitat of the palolo worm *Eunice viridis* (Polychaeta: Eunicidae) in the Samoan Islands. *Marine Biol* 79: 229–236.

Chadee DD, Tikasingh ES. (1989) Diel biting activity of Culex (*Melanoconion*) caudelli in Trinidad, West Indies. *Med Vet Entomol* 3(3): 231–237.

Cutler WB. (1980) Lunar and menstrual phase locking. *Amer J Obstet Gynecol* 137: 834–839.

de Castro JM, Pearcey SM. (1995) Lunar rhythms of the meal and alcohol intake of humans. *Physiol Behav* 57(3): 439–444.

DeCoursey PJ. (1979) Egg-hatching rhythms in three species of fiddler crabs. In: *Cyclic Phenomena in Marine Plants and Animals*. Naylor E, Hartnoll RG, eds. Oxford: Pergamon, pp. 399–406.

DeCoursey PJ. (1983) Biological timing. In: *The Biology of Crustacea, Vol. 7: Behavior and Ecology*. Vernberg FJ, Vernberg WB, eds. New York: Academic Press, pp. 107–162.

Dunea G. (1993) Moon over Slovakia. *BMJ* 307: 1363.

Enright JT. (1963) The tidal rhythm of activity of a sand-beach amphipod. *Z Vergl Physiol* 46: 276–313.

Enright JT. (1965) Entrainment of a tidal rhythm. *Science* 147: 864–866.

Enright JT. (1971a) Heavy water slows biological timing processes. *Z Vergl Physiol* 72: 1–16.

Enright JT. (1971b) The internal clock of drunken isopods. *Z Vergl Physiol* 75: 332–346.

Enright JT. (1972) A virtuoso isopod. Circa-lunar rhythms and their tidal fine structure. *J Comp Physiol* 77: 141–162.

Erkert H-G. (1974) [The influence of moonlight on the activity patterns of night-active mammals] [German]. *Oecologia (Berl)* 14: 269–287.

Fingerman M. (1956) Phase difference in the tidal rhythms of color change of two species of fiddler crab. *Biol Bull* 110: 274–290.

Fingerman M, Yamamoto Y. (1964) Daily rhythm of color change in eyestalkless fiddler crabs, *Uca pugilator* (abstract). *Amer Zool* 4(3): 334.

Franke H-D. (1986) The role of light and endogenous factors in the timing of the reproductive cycle of *Typosyllis prolifera* and some other polychaetes. *Amer Zool* 26: 433–445.

Fuhrman M. (1992) *Phase Relationships of Circadian Oscillations in the Leaves of Soybean (Glycine Max L. Merr) Cultivars of Differing Maturity Groups Relative to Photoperiodic Sensitivity*. Thesis. University of Minnesota.

Gehrels T, Coffeen T, Owings D. (1964) Wavelength dependence of polarization: III. The lunar surface. *Astronom J* 69: 826–852.

Gibson RN. (1965) Rhythmic activity in littoral fish. *Nature* 207: 544–545.

Gibson RN. (1970) The tidal rhythm of activity of *Coryphoblennius galerita* (L.) (Teleostei, Blennidiidae). *Anim Behav* 18: 539–543.

Guillon P, Guillon D, Lansac J, Soutoul JH, Bertrand P, Hornecker JP. (1986) [Births, fertility, rhythms and lunar cycle. A statistical study of 5,927,978 births.] [French]. *J Gynecol Obstet Biol Reprod (Paris)* 15(3): 265–271.

Guthmann H, Ostwald D. (1936) Menstruation und Mond. *Manschrift für Geburtsch und Gynekologie* 103: 232–235.

Halberg Fcn, Halberg F, Sothern RB, Pearse JS, Pearse VB, Shankaraiah K, Giese AC. (1987) Consistent synchronization and circaseptennian (about 7-yearly) modulation of circannual gonadal index rhythm of two marine invertebrates. In: *Advances in Chronobiology–Part A*. Pauly JE, Scheving LE, eds. New York: Alan R Liss, pp. 225–238.

Hardtland-Rowe R. (1955) Lunar rhythm in the emergence of an ephemeropteran. *Nature* 176: 657.

Hines MN. (1954) A tidal rhythm in behavior of melanophores in autotomized legs of *Uca pugnax*. *Biol Bull* 107: 386–396.

Hogben LT, Slome D. (1931) The pigmentary effector system: VI. The dual character of endocrine coordination in amphibian colour change. *Proc Royal Soc (London), Ser B* 108: 10–53.

Hughes DA. (1972) On the endogenous control of tide-associated displacements of pink shrimp, *Penaeus duorarum* Burkenroad. *Biol Bull* 142: 271–280.

Jellyman DJ, Lambert PW. (2003) Factors affecting recruitment of glass eels into the Grey River, New Zealand. *J Fish Biol* 63: 1067–1079.

Kadman-Zahavi A, Peiper D. (1987) Effects of moonlight on flower induction in *Pharbitis nil*, using a single dark period. *Ann Bot* 6: 621–623.

Kerfoot WB. (1967) The lunar periodicity of *Sphecodogastra texana*, a nocturnal bee (Hymenoptera: Halictidae). *Anim Behav* 15: 479–486.

Kleitman N. (1949) Biological rhythms and cycles. *Physiol Rev* 29(1): 1–29 (see p. 18).

Kopal Z. (1969) *The Moon*. Dordrecht, The Netherlands: D. Reidl Publishing Company.

Lang HJ. (1977) Lunar periodicity of colour sense of fish. *J Interdiscipl Cycle Res* 8(3–4): 317–321.

Law SP. (1986) The regulation of menstrual cycle and its relationship to the moon. *Acta Obstet Gynecol Scand* 65(1): 45–48.

Lecolazet R. (1977) Section physical geodesy, permanent commission on Earth tides. In: *Proceedings of 8th International Symposium on Earth Tides*. Bonn, September 19–24, 1977. Bonn: Bonatz & Melchior, pp. 23–29.

Lieber AL. (1978) Human aggression and the lunar synodic cycle. *J Clin Psychiatry* 39(5): 385–392.

Mandoli DF, Briggs WR. (1981) Phytochrome control of two low-irradiance responses in etiolated oat seedlings. *Plant Physiol* 67: 733–739.
Martin L. (1907) La mémoire chez *Convoluta roscoffensis*. *CR Acad Sci (Paris)* 145: 555–557.
Maw MG. (1967) Periodicities in the influences of air ions on the growth of garden cress *Lepidium sativum* C. *Can J Plant Sci* 47: 499–505.
Melchior P. (1966) *The Earth Tides*. Oxford: Pergamon, 458 pp.
Mehta TS, Lewis RD. (2000) Quantitative tests of a dual circalunidian clock model for tidal rhythmicity in the sand beach isopod *Cirolana cookii*. *Chronobiol Intl* 17(1): 29–41.
Michelson AA, Gale HG. (1919) The rigidity of the earth. *Astrophy J* L: 330–345.
Mikulecky M, Zemek R. (1992) Does the moon influence the predatory activity of mites? *Experientia* 48(5): 530–532.
Mikulecky M, Valachova A. (1996) Lunar influence on atrial fibrillation? *Braz J Med Biol Res* 29(8): 1073–1075.
Mikulecky M, Bounias M. (1997) Worker honeybee hemolymph lipid composition and synodic lunar cycle periodicities. *Braz J Med Biol Res* 30(2): 275–279.
Mikulecky M, Rovensky J. (2000) Gout attacks and lunar cycle. *Med Hypoth* 55(1): 24–25.
Miller CD, Pen F. (1959) Composition and nutritive value of Palolo (*Palolo siciliensi* Grube). *Pacif Sci* 13: 191–194.
Morgan E. (1984) The pressure-responses of marine invertebrates: a psychophysical perspective. *Zool J Linn Soc* 80: 209–230.
Morgan E. (1991) An appraisal of tidal activity rhythms. *Chronobiol Intl* 8(4): 283–306.
Naylor E. (1960) Locomotor rhythms in *Carcinus maenas* (L.) from non-tidal conditions. *J Exp Biol* 37: 481–488.
Naylor E. (1996) Crab clockwork: the case for interactive circatidal and circadian oscillators controlling rhythmic locomotor activity of *Carcinus maenas*. *Chronobiol Intl* 13(3): 153–161.
Naylor E. (1997) Crab clocks rewound. *Chronobiol Intl* 14(4): 427–430.
Neuman D. (1981) Tidal and lunar rhythms. In: *Handbook of Bahvioral Neurobiology, Vol. 4. Biological Rhythms*. Aschoff J, ed. New York: Plenum Press, pp. 351–380.
Page TL, Larimer JL. (1975a) Neural control of circadian rhythmicity in the crayfish: I. The locomotor activity rhythm. *J Comp Physiol* 97: 59–80.
Page TL, Larimer JL. (1975b) Neural control of circadian rhythmicity in the crayfish: II. The ERG amplitude rhythm. *J Comp Physiol* 97: 81–96.
Palmer JD, Round FE. (1967) Persistent, vertical-migration rhythms in benthic microflora: VI. The tidal and diurnal nature of the rhythm in the diatom *Hantzschia virgata*. *Biol Bull* 132: 44–55.
Palmer JD. (1973) Tidal rhythms: the clock control of the rhythmic physiology of marine organisms. *Biol Rev* 48: 377–418.
Palmer JD. (1974) *Biological Clocks in Marine Organisms. The Control of Physiological and Behavioral Tidal Rhythms*. New York: John Wiley & Sons, 173 pp.
Palmer JD, Williams BG. (1986) Comparative studies of tidal rhythms: II. The dual clock control of the locomotor rhythms of two decapod crustaceans. *Mar Behav Physiol* 12: 269–278.
Palmer JD. (1995a) *The Biological Rhythms and Clocks of Intertidal Animals*. New York: Oxford University Press, 217 pp.
Palmer JD. (1995b) Review of the dual-clock control of tidal rhythms and the hypothesis that the same clock governs both circatidal and circadian rhythms. *Chronobiol Intl* 12(5): 299–310.

Palmer JD. (1997) Dueling hypotheses: circatidal versus circalunidian battle basics–second engagement. *Chronobiol Intl* 14(4): 431–433.
Palmer JD. (2000) The clocks controlling the tide-associated rhythms of intertidal animals. *Bioessays* 22(1): 32–37 (Review).
Panda S, Hogenesch JB, Kay SA. (2003) Circadian light input in plants, flies and mammals. *Novatis Found Symp* 253: 73–82; Discussion 82–88, 102–109, 281–284.
Parvathy Rajan K, Kharour HH, Lockwood APM. (1979) Rhythmic cycles of blood sugar concentrations in the crab *Carcinus maenas*. In: *Cyclic Phenomena in Marine Plants and Animals*. Naylor E, Hartnoll RG, eds. New York: Pergamon, pp. 451–458.
Payne SR, Deardon DJ, Abercrombie GF, Carlson GL. (1989) Urinary retention and the lunisolar cycle: is it a lunatic phenomenon? *BMJ* 299(6715): 1560–1562.
Pearse JS, Pearse VB, Giese AC, Sothern RB, Halberg F. (1985) Circannual rhythm with similar timing characterizes gonadal index of a marine invertebrate (ochre star) studied 30 years apart (abstract). *Chronobiologia* 12(3): 264.
Pearse JS. (1990) Lunar reproductive rhythms in marine invertebrates: maximizing fertilization? In: *Advances in Invertebrate Reproduction 5*. Hoshi M, Yamashita O, eds. New York: Elsevier, pp. 311–316.
Pekeris CL. (1940) Notes on tides in wells. In: *American Geophysical Union Transactions*. Part II: 21st Annual Meeting, April, 1940. Washington, DC: National Research Council, pp. 212–213.
Poolsanguan B, Uglow RF. (1974) Quantitative changes in blood sugar levels of *Crangon vulgaris*. *J Comp Physiol* 93: 1–6.
Powell BL. (1962) Types, distribution and rhythmical behaviour of the chromatophores of juvenile *Carcinus maenas* (L.). *J Anim Ecol* 31: 251–161.
Powell BL. (1966) The control of the 24 hour rhythm of colour change in juvenile *Carcinus maenas* (L.). *Proc R Ir Acad [B]* 64(21): 379–399.
Presser HB. (1974) Temporal data relating to the human menstrual cycle. In: *Biorhythms and Human Reproduction*. Ferin M, Halberg F, Richert RM, Vande Wiele R, eds. New York: John Wiley & Sons, Inc., pp. 145–160.
Rinehart JS. (1976) Influence of tidal strain on geophysical phenomena. In: *Proceedings of 7th International Symposium on Earth Tides*. Szadeczky-Kardoss G, ed. Stuttgart: Nägele Obermiller, pp. 181–185.
Salisbury FB. (1981) Twilight effect: initiating dark measurement in photoperiodism of *Xanthium*. *Plant Physiol* 67: 1230–1238.
Salisbury FB, Ross CW. (1992) *Plant Physiology*, 4th edn. Belmont: Wadsworth, 682 pp.
Sha LR, Xu NT, Song XH, Zhang LP, Zhang Y. (1989) Lunar phases, myocardial infarction and hemorrheological character. A Western medical study combined with appraisal of the related traditional Chinese medical theory. *Chin Med J (Engl)* 102(9): 722–725.
Sinclair RM. (1987) Moonlight and circadian rhythms. *Science* 235(4785): 145.
Smolensky M, Lamberg L. (2000) *The Body Clock Guide to Better Health*. New York: Henry Holt & Co., 428 pp.
Sok M, Mikulecky M, Erzen J. (2001) Onset of spontaneous pneumothorax and the synodic lunar cycle. *Med Hypoth* 57(5): 638–641.
Spruyt E, Verbelen J-P, De Greef JA. (1987) Expression of circaseptan and circannual rhythmicity in the imbibition of dry stored bean seeds. *Plant Physiol* 84: 707–710.
Sullivan W. (1981) Land tides may affect earth's core of rotation. The New York Times, 23 August 1981 (Sunday), Late City Final Edition. Section 1, Part 2, page 56.
Sundararaj BI, Vasal S, Halberg F. (1973) Circannual rhythmic ovarian recrudescence in the catfish, *Heteropneustes fossilis*. *Intl J Chronobiol* 1: 362–363.

Sundararaj BI, Vasal S, Halberg F. (1982) Circannual rhythmic ovarian recrudescence in the catfish, *Heteropneustes fossilis* (Bloch). In: *Toward Chronopharmacology*. Takahashi R, Halberg F, Walker C, eds. New York: Pergamon, pp. 319–337.

Vogt KA, Beard KH, Hammann S, Palmiotto JO, Vogt DJ, Scatena FN, Hecht BP. (2002) Indigenous knowledge informing management of tropical forests: the link between rhythms in plant secondary chemistry and lunar cycles. *Ambio* 31(6): 485–490.

Von Gaertner T, Braunroth E. (1935) [About the influence of moonlight on flowering date of long- and short-day plants.] [German]. *Botan Centralblatt Abt* A53: 554–563.

Williams BG, Naylor E. (1969) Synchronization of the locomotor tidal rhythms of *Carcinus*. *J Exp Biol* 51: 715–725.

Williams BG. (1998) The lack of circadian timing in two intertidal invertebrates and its significance in the circatidal/circalunidian debate. *Chronobiol Intl* 15(3): 205–218.

Youthed GJ, Moran VC. (1969) The lunar-day activity rhythm of myrmeleontid larvae. *J Insect Physiol* 15: 1259–1271.

Zürcher E, Cantiani M-G, Sorbetti-Guerri F, Michel D. (1998) Tree stem diameters fluctuate with tide. *Nature* 392: 665–666.

7
Sexuality and Reproduction

"The time of the seasons and the constellations. The time of milking and the time of harvest. The time of the coupling of man and woman. And that of beasts."
— *T.S. Eliot (1888–1965), US-born/British poet, critic*

Introduction

Reproduction, which is more easily defined than sexuality, is a process by which new organisms (offspring) of the same kind are produced. This occurs through the union of mature male and female sex cells (sperm and egg), called *gametes*, or in some cases, by asexual means (e.g., without sexual action).[1] *Sexuality*, on the other hand, refers to the status of sexual activity, perception, or interest. For much of our society, these features tend to obscure the full extent of sexuality, which in addition includes the capacity to bring together in one individual, a new recombination of genes that originate from two different individuals (cf. Purves et al., 1998). Therefore, the full extent of sexuality includes all four features: activity, perception, interest, and the recombination of genes.

What makes sexuality so interesting and diverse is the broad extent of its influence on life and on how it is viewed in society. As will be evident in this chapter, sexuality and reproduction are part of all four integrating disciplines of biology: evolution, genetics, development, and biological rhythms (Ehret, 1980). Depending upon the organism, sexuality can be expressed in a number of ways, including anatomic and morphological features (form), color, levels of specific chemicals, behavioral traits, and responses to environmental signals (Figure 7.1 and Table 7.1). The expression of sexuality is a fundamental characteristic of life and

---

[1] Asexual reproduction does not involve the fusion of gametes and occurs along with sexual reproduction in many groups of organisms (plants, bacteria, algae, fungi, and animals). For example, most strawberry plants (*Fragaria ananassa*) in a garden arise from stems (stolens), not from seeds. Green peach aphids (*Myzus persicae*) have a unique stage where females, in the absence of male aphids and their sperm, give birth to female nymphs by a process known as parthenogenesis (virgin birth).

FIGURE 7.1. Sexuality can be expressed in color (male and female cardinals) and in anatomical and morphological features (male and female moose). Photos by Marcia Koukkari (top) and R. Sothern (bottom).

is present in nearly all major groups of organisms, even in one-celled microorganisms, such as bacteria (see Essay 7.1)

TABLE 7.1. Examples of features by which sexuality can be expressed in certain organisms.

Feature	Sex example
Anatomy	Male deer have antlers and testes, while females lack antlers and have ovaries
Morphology	Male Scottish Blackface sheep have longer and more curved horns than do the females
Color	Male cardinal bird is reddish and female is brownish
Behavior	Female cat protects kittens
Chemicals	Female moth emits pheromones to attract males
Environment	Male to female ratio affected by temperature in some fish, reptiles, and amphibians

Essay 7.1 (by WLK): Parasexuality

Bacteria are a special group of organisms, not only in regard to being prokaryotic, but also in regard to sexuality. They reproduce asexually, a single cell dividing by binary fission into two cells. Under optimal conditions, the generation time can be rapid (e.g., about 20 min in *E. coli,*). However, there also can be sexual activity within a population (Figure 7.2). The complete sexual act requires about 100 min, which is

Introduction 239

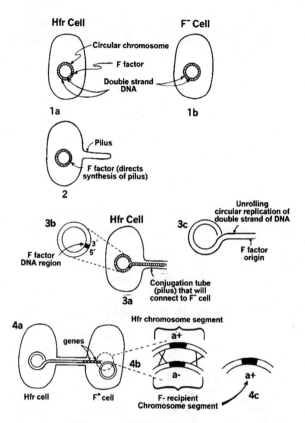

FIGURE 7.2. Genetic exchange in bacteria does not involve meiosis or fertilization and is referred to as "parasexuality." Illustrated here is mating between strains, an Hfr (high-frequency recombination) donor cell (1a) and a recipient F⁻ cell (1b) of *E. coli*. A fertility factor on the circular chromosome of the Hfr cell directs the formation of a pilus (2) or conjugation tube (3a), which connects to the F⁻ cell (4a). DNA synthesis occurs (3b) whereas the unrolling circular replication of the double-stranded DNA with various genes are transferred to the F⁻ cell. During crossing (4b = double crosses), a Hfr chromosome segment (e.g., gene a^+ and a^-) is exchanged with that of the F⁻ recipient chromosome (4c). Based on Snustad & Simmons (2003).

four times longer than the normal life span of the bacterium (cf. Hayes, 1966a, b).[2] Because the genetic exchange in bacteria is unique in that it does not involve meiosis or fertilization, it is more commonly referred to as *parasexuality*.[3]

---

[2] An interesting account of the discoverer of bacteria, Anton van Leeuwenhoek (1632–1723) not being surprised about the existence of sexual activity in bacteria, as well as observations and studies of the topic by others, can be found in a classic paper by Hayes (1966b).

[3] Parasexuality has broad implications in the genetics of bacteria, such as drug resistance, and in the transmission of viruses (for a general overview, see Snustad & Simmons, 2003)

7. Sexuality and Reproduction

Many aspects of sexuality and reproduction, be they the recombination of genes, the courtship of animals, maturation of the ovum, or the tumescence of the penis, have a rhythmic temporal organization. The objectives of this chapter are to: (1) review the basics of sexuality and reproduction, especially the recombination of genes that regulate not only morphological features (e.g., eye color, sex, and body size), but also the characteristics of biological rhythms; (2) examine rhythmicity in courtship, mating, and photoperiodism; and (3) explore the rhythmic nature of human sexuality and reproduction.

Nuclear Division and Genetics

At the very core of sexuality is genetics, because genes contain the information to be expressed in the progeny.[4] This genetic information is called the *genome*. Sexuality involves the recombination of genes from two different parents; genes occur in chromosomes, which are found in nuclei.[5] Essay 7.2 provides an overview of the basics of cell division and genetics.

Essay 7.2 (by WLK): *Mitosis, Meiosis, and the Punnett Square*

Cell Division — Cells are the basic structural and functional units of life. Most cells are small (e.g., 1 to 100 μm).[6] They contain cytoplasm (living matter), a nucleus (if eukaryote), and various organic molecules (e.g., nucleic acids, proteins, etc.). In cells of fungi, protista,[7] plants and animals, the nucleus and organelles, such as mitochondria, are enclosed by membranes. When cells of these organisms divide to produce new cells, the division of the nucleus is called *mitosis*, and the division of the cytoplasm is known as *cytokinesis*. Together, mitosis and cytokinesis represent a phase in the life history or cycle of the cell that is known as the *M-phase* (M = mitotic). A cell enters the M-phase after much of the molecular activity required for the preparation of division has taken place. The preparatory phase is called the *interphase* and includes activities such as the replication of DNA, synthesis of proteins, and the condensing of chromosomes. During mitosis, the chromosomes are first duplicated and then divided so that each of the two daughter cells will have the same identical chromosomes and genes that were present in the original nucleus. All cells having nuclei potentially contain copies of all of the genes present in the organism. However, only some of the genes need be expressed at any given stage of development.

---

[4] The hereditary material is DNA, which is primarily located in the nucleus of the cell. The molecular biology of how the genetic information is transcribed, translated, and expressed is described and illustrated in Essay 5.2 in the Chapter 5 on Models.

[5] DNA and special types of chromosomes are found also in mitochondria and chloroplasts.

[6] One micrometer (μm) equals 10^{-6} meters (m) and one millimeter (mm) equals 10^{-3} meters. Microscopes are required to observe cells, although there are some exceptions, such as the eggs of birds. For example, an egg of a domestic chicken (*Gallus gallus*) is a large single cell.

[7] Protista represents a kingdom of organisms that includes unicellular algae and protozoans, such as *Euglena*.

Meiosis and Gametogenesis — The process of meiosis differs from mitosis in that the diploid nucleus divides to produce 4 haploid daughter cells (Figure 7.3). In eukaryotes, chromosomes occur in pairs called *homologs* (Gr. *homos* = same) that contain the same kinds of genes (e.g., color). During interphase I (the first phase of *meiosis*), these two chromosomes pair along their entire length and replicate so that each chromosome now consists of two chromatids (the Latin *id* = daughter of). Next, the chromatids of one homolog align themselves with the *chromatids* of the other homolog in a manner that fosters *crossing over*, which leads to the exchange of pieces of chromatids between homologs (prophase I). Soon the nuclear membrane disappears and the homologous chromosomes, each of which still appears as two chromatids, become randomly aligned at the equatorial plate (metaphase I). The homologs are pulled toward opposite poles (anaphase I), the results of which are that each daughter nuclei will contain one of each of the original homologs (telophase I). The nuclei are now *haploid* (n) since they contain one-half the number of chromosomes present in the parent ($2n$). However, the total amount of DNA in the nucleus is not reduced during meiosis I. The reduction of DNA to one-half its normal level occurs during meiosis II, where nuclear division again takes place, but without the replication of DNA.

In summary, a number of important things are accomplished during meiosis. Perhaps, the two that most closely relate to sexuality are (1) the *redistribution of genes*, which is accomplished by the crossing over of chromatids (prophase I) and the random distribution of chromosomes during metaphase I, and (2) the production of *four haploid cells* that differ from each other. In addition, the number of chromosomes has been reduced from the diploid to the haploid state, and each haploid cell has one of each type of chromosomes (e.g., one of the homologs) (Figure 7.3). Some of the stages of meiosis in actual plant tissue are shown in Figure 7.4.

A gene may have more than one different form and these are called *alleles*. A *diploid* cell ($2n$) typically has two alleles of the gene. One occurs on each of the homologous chromosomes that constitute a pair. Following meiosis, each haploid gamete (n) will have one allele of each gene. The production of the egg (*oogenesis*) and sperm (*spermatogenesis*), which depend upon meiosis, occurs by a process that is called *gametogenesis*.

Punnett Square — The Punnett square[8] is a grid of small squares that is extremely helpful in delineating possible combinations of alleles and thus predicting the traits of offspring. The Punnett square works well for demonstrating how the recombination and segregation of genes can control a trait, such as the color of some flowers, which was one of the many traits studied by Mendel (1865)[9] in his classic research with peas (*Pisum*). It also represents an excellent teaching tool in helping to understand how the circadian period was first shown to be an inherited trait (Bünning, 1932, 1935).

---

[8] A grid that displays the results of a genetic cross was devised by Reginald C. Punnett (1875–1967), a British geneticist who worked with feather color traits of chickens to separate male and female chicks soon after hatching.

[9] The work of Gregor Mendel (1822–1884) is discussed extensively in most introductory biology books and will not be elaborated upon here. The background knowledge already acquired from this essay about meiosis, chromosomes, and genes was not known when Mendel conducted his experiments. What are now called dominant and recessive "genes" were referred to as dominant and recessive "characters" by Mendel.

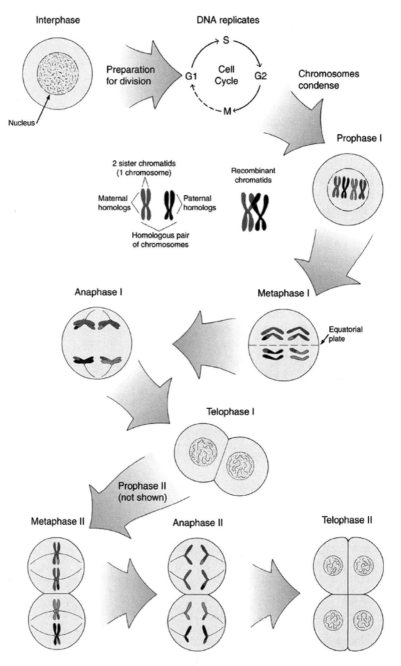

FIGURE 7.3. A generalized scheme for meiosis commencing with interphase and ending with telophase II.

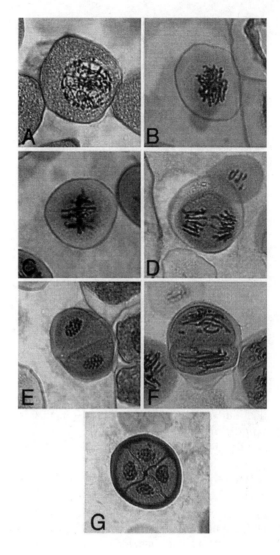

FIGURE 7.4. Various stages of meiosis illustrated by light micrographs (480×) for a Lily (*Lilium*). Individual cells in (*A*) early prophase, (*B*) prometaphase I, (*C*) metaphase, (*D*) anaphase I, (*E*) prophase II, (*F*) metaphase II, and (*G*) tetrad products of meiosis (each of the four cells with a nucleus with a haploid (*n*) number of chromosomes).

Biologists have often used individual letters of the alphabet to denote alleles, an uppercase letter for a dominant trait (e.g., A) and a lowercase letter for a recessive trait (e.g., a).[10] As has already been stated, individuals possess two alleles of a gene. Individuals may be homozygous, in which case the two identical alleles are noted by identical letters of the same case (AA or aa) or, they may be heterozygous (Aa). The genetic composition represented by the two alleles is referred to as the *genotype* (e.g., AA or Aa), while the trait that is expressed is called the *phenotype* (e.g., eye or hair color, smooth skin, etc.) (Figure 7.5). In addition to the use of letters to denote alleles,

---

[10] In a monohybrid cross (single trait), the trait that appears in the F_1 (first filial generation) is called the dominant trait. The recessive trait reappears in the F_2 (see Figure 7.5B).

A) F₁ Generation

	A	A
a	Aa	Aa
a	Aa	Aa

FIGURE 7.5. Punnett Square illustrating the recombination of genes in the F₁ (A) and F₂ (B) generations. Note the differences in genotype and phenotype ratios are found in F₂, but not in F₁.

B) F₂ Generation

	A	a
A	AA	Aa
a	Aa	aa

Genotype Ratio: 1:2:1
Phenotype Ratio: 3:1

geneticists use the letter P to indicate the parental generation and the letter F₁ for the first filial (offspring) generation and F₂ for the second filial generation.

When Mendel examined the trait for flower color of peas (*Pisum*), he found a dominant trait and a recessive trait. The genotypic ratio of the F₂ was 1:2:1 and a phenotypic ratio was 3:1. The 3:1 phenotypic ratio was indicative of the dominance of the purple trait over the recessive white trait.

The nuclei of cells that produce gametes (sperm and egg) undergo a special type of division called *meiosis* which results in two daughter cells having only half as many chromosomes (n) as the mother cell ($2n$). When the nucleus divides (*mitosis*) in other types of cells, the chromosomes are replicated and each of the two daughter cells receives the same number of chromosomes ($2n$) and genes as were present in the mother cell.[11]

One of the first reports of a daily periodicity in cell division was based upon studies with the root tips of onion (*Allium cepa*) bulbs (Kellicott, 1904), wherein nuclear division is relatively easy to observe (Figure 7.6). However, many of the early studies with onion and other species of plants did not include constant environmental conditions and long series of sampling times. In 1973, a study of mitosis in the development of stomata[12] in epidermal tissue of onion (*Allium cepa*)

---

[11] As in prokaryotes (e.g., bacteria), mitochondria and chloroplasts contain DNA, but the transmission of mitochondrial and chloroplast genomes to daughter cells occurs during cytokinesis. During cytokinesis, these organelles divide by being "pinched" into two parts. Each daughter cell must receive at least one new mitochondrion or chloroplast to have the genome to reproduce these organelles.

[12] Stomata is the plural for stoma (Greek, *stoma* = mouth) and refers to small pores, each of which is formed between two specialized cells, called guard cells. These pores are located in the epidermis (outer layer of tissue) of leaves and allow gases to move into (e.g., CO_2) and out of (e.g., water vapor) the plant. The movement of water vapor out of plants, called transpiration, displays a circadian rhythm (Nixon, et al., 1987).

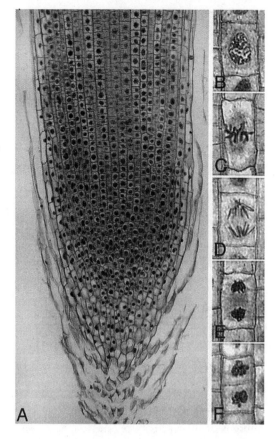

FIGURE 7.6. Various stages of mitosis can be seen in cells of an onion (*Alium cepa*) root tip. (*A*) Longitudinal section of onion root tip (120x). Individual cells (480x) in: prophase (*B*), metaphase (*C*), anaphase (*D*), telophase (*E*), and cell division (cytokinesis) (*F*) (courtesy of Mark Sanders, Program Director, Imaging Center, University of Minnesota).

showed circadian variations in LD for both asymmetrical (involved with lengthening) and symmetrical (involved with widening) divisions (Zeiger & Cardemil, 1973). Our analysis of these data detected both a 12-h and a 24-h period (Figure 7.7) with peaks occurring at 15:00 h and 03:00 h for symmetrical division and a single peak between 21:00 h and 00:00 h (during darkness) for asymmetrical mitoses, which is when the greatest elongation rate is usually observed in stem growth of higher plants (cf. Millet & Bonnet, 1990; Bertram & Karlsen, 1994). Studies of the shoot apices from seedlings of an annual weed (*Chenopodium rubrum*) monitored in continuous light and temperature indicate that circadian mitotic rhythms, at least in some plants, could be endogenous (King, 1975), although some questions about the circadian nature of the clocking mechanism remain to be answered (Hillman, 1976).

The vast amount of literature on the topic, especially in the area of medicine, underscores the importance of cell division and proliferation in matters pertaining especially to the treatment of cancer and bone-marrow procedures (see Chapter 11 on Clinical Medicine). Circadian rhythms of cell division and mitotic indices for various tissues in a number of animal species, including humans, have been well

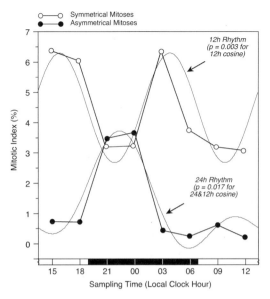

FIGURE 7.7. Circadian patterns in the mitotic index in the epidermis of the cotyledon of *Alium cepa* seedlings during 137–159 h after germination. Asymmetrical mitoses, peaking during darkness, are associated with elongation, while symmetrical mitoses result in widening of the cotyledon. (Our analysis of data in Table 2 in Zeigler & Cardemill, 1973).

documented (cf. Scheving et al., 1984). A few examples of peak times for mitotic rhythms in humans are presented in Table 7.2. A review of a large number of studies shows a peak in DNA synthesis (S-phase) in the epidermal[13] tissue of humans around 14:00 h, while the bulk of the cell division (M-phase) occurs between 22:00 h and 02:00 h (Brown, 1991).[14] Interestingly, rats and mice, which are nocturnal, display the same sequencing in their epidermal mitotic circadian rhythms as is found in diurnally active humans. That is, the peaks occur during the same portions of their activity regime (S-phase peaks in mid-activity, while M-phase peaks at the beginning of the resting span) (Figure 7.8).

TABLE 7.2. Peak times for mitotic phases in various tissues in humans.

Cell-cycle variable	Site	24-h peak (h)^a	Reference
S-phase (3HTdr uptake)	Rectal mucosa	07:04 (03:20–11:28)	Buchi et al., 1991
S-phase (3HTdr uptake)	Bone marrow	11:21 (09:16–13:28)	Mauer, 1965
S-phase (cells)	Bone marrow	13:04 (09:32–16:04)	Smaaland et al., 1991
G1/S-phase (Cyclin E)	Oral mucosa	14:59 (14:08–15:48)	Bjarnason et al., 1999
M-phase (mitotic index)	Bone marrow	21:09 (19:08–23:12)	Mauer, 1965
M-phase (Cyclin B1)	Oral mucosa	21:13 (18:36–23:56)	Bjarnason et al., 1999
Mitoses (mitotic index)	Skin	00:00–03:00	Scheving, 1959
Mitoses (mitotic index)	Skin	00:00–03:00	Fisher, 1968

^{a}95% confidence limits in parentheses (…) if peak determined by cosinor analysis.

---

[13] Epidermis (Greek *epi* on + *derma* skin) is an outer layer of tissue, which in flowering plants, is a single layer that arises from the protoderm (primary meristem). In animals, epidermal tissues form the outer covering of the body, as well as that of the digestive tract and lungs.

[14] A diagram of cell-cycle stages appears at the top of Figure 7.3 (discussed later).

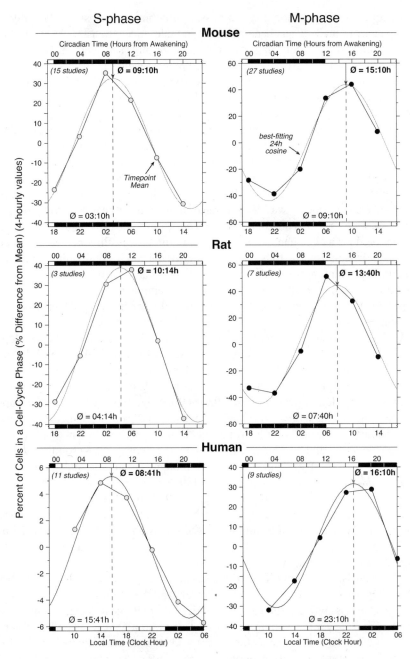

FIGURE 7.8. Epidermal mitotic rhythms in nocturnally active rodents and diurnally active humans are similarly timed in relation to the rest-activity cycle. Dark bar (Lights off) = activity span for rodents, resting span for humans. Rhythm p value ≤ 0.01 for each series, Ø = acrophase (peak of fitted 24-h cosine) computed from local midnight (bottom) or from awakening (*top*, in bold) (Our analysis of data from Tables 1–6 in a literature summary by Brown, 1991).

Fewer studies have been done on possible rhythms in meiosis. A 24-h rhythmic pattern in meiosis has been recorded in male mice (Rienstein et al., 1998). Meiotic indices (% of cells in first meiotic division and % of spermatids) in testes of mice living in an LD14:10 cycle showed highest values in the first half of the light (resting) span and lowest values in early dark (onset of activity). Interestingly, after 15 days in continuous light (LL) a 12-h cycle was detected, but not a 24-h cycle, possibly implicating the importance of a 24-h LD cycle as a gating mechanism that controls the approximate 12-day overall meiotic process in these mice. However, after more than 2 weeks in free-running conditions, individual mice may also have become desynchronized from each other.

A unique situation appears to exist, at least in some organisms, between biological rhythms and cell division. It is known as the *circadian-infradian rule*, or the *G-E-T effect*, since it was demonstrated in three unicellular eukaryotic organisms: *Gonyaulax*, *Euglena*, and *Tetrahymena*. Briefly, this rule states that a cell cannot produce a circadian output if its generation time is ultradian, but it can do so if cell division is close to being infradian[15] or the cell is not dividing. In multicellular organisms, however, the temporal organization is such that cells of a given tissue may be in the ultradian mode, but the communication network may be based upon what is being sent from other tissues that are in the circadian mode (cf. Hillman, 1976).

Within the context of this chapter, mitosis is closely allied with development, while meiosis is a prerequisite to sexuality that makes possible the recombination of genes from two individuals. The recombination of genes resulting in offspring that express circadian periods under free-running conditions was first demonstrated in the 1930s (Bünning, 1932, 1935). Two groups of bean plants (*Phaseolus multiflorus*) were selected from a population to have either short (23-h) or long (26-h) periods in their leaf movements under continuous darkness and 20°C temperature. When each of the two groups of plants was "selfed" (i.e., sperm and egg were from plants having the same period) for four successive generations, the periods of the progeny (offspring) continued to be the same as that of their parents (Figure 7.9A). In other words, the trait for periodicity was homozygous (Gr. *homos* = same; *zygotos* = yoked, joined together). However, when Bünning crossed plants having short periods (23 h) with plants having long periods (26 h), the F_1 generation had intermediate periods (Figure 7.9B).

The expression of a trait in the F_1 generation that is intermediate differs from what Mendel observed for the color of flowers and the condition of the seed coat for peas (see Essay 7.2). Students of genetics interpret this to mean that both the short period (t_{23}) and the long period (t_{26}) were dominant traits. When the F_1 plants were selfed, the F_2 generation displayed close to the predicted 1:2:1 ratio for both the genotype and phenotype (Figure 7.9B). The actual data presented by Bünning (1935) are shown in Figure 7.10. The results from this study with

---

[15] Note: an infradian mode can be induced by variables such as changes in temperature or nutrition (Ehret, 1980).

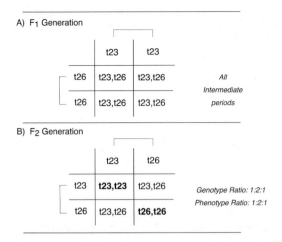

FIGURE 7.9. (A) Punnett square illustrating the F_1 generation following the cross of plants that produce short circadian periods with plants that produce long circadian periods. The homozygous dominant alleles are presented here as $t_{23}\, t_{23}$ and $t_{26}\, t_{26}$. Note: In this instance, we have not chosen letters, rather the two alleles are represented as t_{23} and t_{26}. Based upon the actual results of Bünning experiments (Bünning, 1935), there was a lack of dominance and all members of the F_1 display an intermediate circadian periodicity. (B) Punnett square illustrating the F_2 generation following the selfing of F_1 plants. The heterozygous alleles are presented here as $t_{23}\, t_{26}$. The homozygous alleles are $t_{23}\, t_{23}$ and $t_{26}\, t_{26}$. Both the genotype and phenotype are present in a 1:2:1 ratio. One-fourth of the plants display short circadian period, one-half of the population has an intermediate circadian period and one-fourth has a long circadian period (Bünning, 1935).

beans provided the first scientific demonstration that circadian periodicity is an inherited trait.

By the early 1970s, the location of mutations on chromosomes that control periodicity had been reported for animals, fungi, and plants. In the case of fruit flies, the expression of periodicity has been identified for ultradian, circadian, and infradian rhythms (cf. Hall, 1994). Results from studies with both the *period* (*per*)[16] gene of *Drosophila* and the *frequency* (*frq*) gene of *Neurospora* show that the mRNA oscillates with a circadian periodicity, as does the PER protein that results from the *per* gene expression (Hardin et al., 1992; Aronson et al., 1994; Iwasaki & Thomas, 1997). In humans, the relative RNA expression of h*per1* and h*Bmal1* in oral mucosa and in skin also have a circadian expression, peaking after awakening and late in the evening, respectively (Bjarnason et al., 2001). The peak in h*per1* expression coincides with a peak in cells in the G_1-phase and precedes the peak in cells in S-phase of the mitotic cycle, while the peak in h*Bmal1* coincides with the peak for cells in M-phase of the mitotic cycle in human oral mucosa (Figure 7.11) (Bjarnason et al., 1999).

---

[16] *Per* and *frq* are names of specific clock genes that control circadian rhythms and are written in italics.

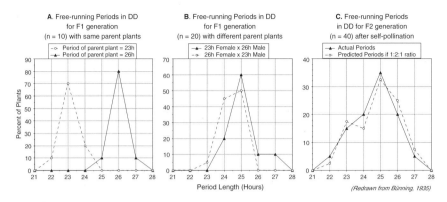

FIGURE 7.10. Bünning's genetic results illustrating the percent of bean plants *(Phaseolus multiforus)* having various free-running periods in constant darkness (DD). (*A*) Two populations ($n = 10$), one having shorter than 24-h periods and the other having longer than 24-h periods; (*B*) The F_1 generation from crossing 23-h period females with 26-h period males (solid line) or crossing 23-h period males with 26-h period females (dashed line) ($n = 20$/group); (*C*) The F_2 generation showing the actual results (solid line) and the predicted theoretical results (dashed line) after self-pollination ($n = 40$/group) (redrawn from Figures 3, 4, and 5 of Bünning, 1935).

Sex and Reproduction: The Difference

From general biology, one learns that *sex* is the combination of genes from two cells, while *reproduction* is the formation or development of an individual, or in some cases, the formation of various structures, such as organelles, cells, tissues and organs. The process of reproduction that includes sex is called *sexual reproduction*, an important feature in genetic diversity and evolution.

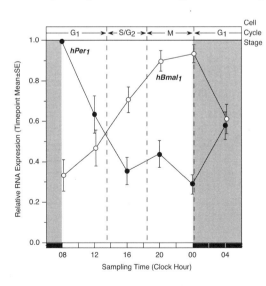

FIGURE 7.11. Circadian expression of clock genes in humans. Relative RNA expression of $hPer_1$ and $hBmal_1$ in oral mucosa coincides with progression of cell cycle events (G_1 and M). Data from 4-h tissue biopsies from eight healthy men normalized from 0 to 1, with highest value = 1. Dark bar = sleep and/or rest (redrawn from Figure 4 in Bjarnason et al., 2001).

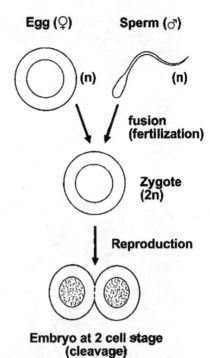

FIGURE 7.12. Fertilization and reproduction in mammals.

Sexual reproduction involves the production of haploid cells[17] called *gametes* (Greek *gamein* = to marry) that are known as *sperm* and *eggs* (or *ova*). In mammals they are produced by germ cells in the *testes* and *ovaries*. Without meiosis, the fusing of sperm and egg would result in the doubling of chromosomes, but with meiosis, the fusing restores the original chromosome number of the parent cells. *Fertilization* is the union of the nucleus of the haploid egg (n) with the nucleus of the haploid sperm (n) to form a diploid *zygote* ($2n$) (Greek *zygōtos* = joined together), a single-celled structure that marks the beginning of an individual and is followed by the process of reproduction and the formation of an embryo (Figure 7.12). This process is typical of most organisms, although in flowering plants, a second sperm is involved. One sperm fuses with an egg and the other fuses with two additional nuclei to produce a food source for the developing embryo (see Essay 7.3).

Essay 7.3 *(by WLK)*: *Artificial Hybridization and How Sex Produces Both "Lunch" and an Embryo*

The male and female parts of flowers, such as the Lily (Figure 7.13) are conspicuous and relatively easy to observe. However, the studies by Mendel with peas (*Pisum*)

---

[17] Haploid ($1n$) cells have single sets of chromosomes in their nuclei, whereas in diploid ($2n$) cells, the chromosomes occur in pairs (see Essay 12.2).

FIGURE 7.13. Flower of lilies (*Lilium sp.*) showing male and female structures (see Essay 7.3). Photo by R. Sothern.

and Bünning with beans (*Phaseolus*) relied upon flowers in which the male (stamens) and female (carpels) parts are enclosed by the petals until after fertilization, thus reducing the chances for cross-pollination by other plants. The flower, reproductive structures (stamens and carpels) and the fruit are shown for a pea in Figure 7.14. Fertilization in flowering plants is much different than that which occurs in humans and other animals. Perhaps, the biggest difference is that two sperms are involved, a process known as *double fertilization*. When a pollen grain from the anther (Figure 7.15A) lands upon the stigma (Figure 7.15B), the generative cell divides to produce two sperms. The pollen tube grows down the style and the two sperms are discharged into the embryo sac (Figure 7.15C).[18] One fuses with the egg, forming the diploid zygote ($2n$) and subsequently the *embryo*. The other sperm fuses with the polar nuclei, a union that leads to a type of cellular "lunch box" known as

---

[18] The history or life cycle of plants may include an alternation of generations. In plants, such as certain moss species, both the gametophyte *(phyte* = Greek word for plant *phyton)*

Sex and Reproduction: The Difference 253

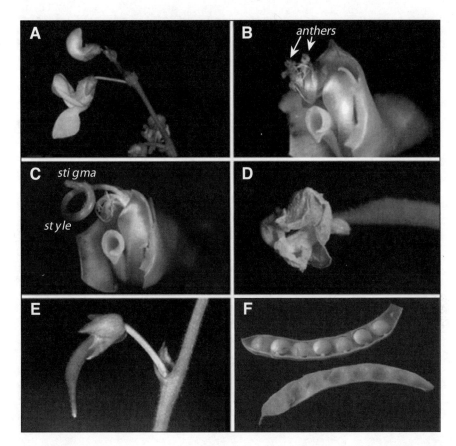

FIGURE 7.14. A pea (*Pisum*) from flower to fruit and seed: (*A*) flower, (*B*) parts of petals removed to show stamens with anthers, (*C*) stigma and style of carpel (anthers removed), (*D*) senescence of petals, (*E*) developing fruit, and (*F*) mature fruit (legume) opened to show seeds.

the *endosperm*. This 3n tissue[19] serves as a food supply for the embryo and is the white-colored material you see in popcorn (Figure 7.16). In the immature fruit of grains, such as oats and corn, it is a milk-like substance. While we are able to easily see the starchy endosperm in the fruits of corn and oats, in plants such as peas and beans, the endosperm has been absorbed and not visible by the time that the fruit and seed[20] are mature.

---

generation (1n tissue) and the sporophyte (2n tissue) generation are visible to the naked eye. However, in flowering plants, the male gametophyte is the pollen grain and female gametophyte is the embryo sac, both of which must be viewed under magnification (microscope).

[19] *Note*: The sperm is 1n and when it fuses with the two polar nuclei (1n + 1n), the result is a triploid nucleus (3n).

[20] A fruit is a mature ovary and a seed is a mature ovule. When one sows corn, it is the entire fruit that is being planted (e.g., the wall of the ovary and seed coat are fused, forming a fruit called a *grain* or *caryopsis*). In the case of peas and beans, the seeds are located in a pod, a type of fruit that botanists call a legume.

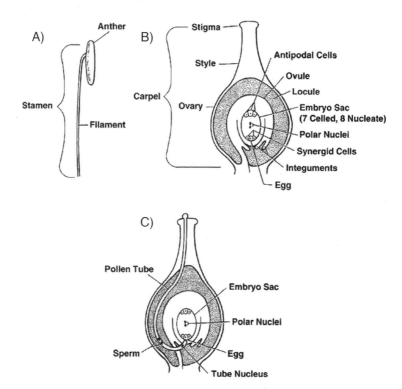

FIGURE 7.15. Reproductive parts of a flower illustrating: (*A*) the stamen, (*B*) the carpel, and (*C*) the pollen tube and embryo sac.

FIGURE 7.16. Popcorn (*Zea*) illustrates the result of double fertilization: one sperm fertilized an egg that developed into an embryo, while a second sperm fused with the polar nuclei and produced the endosperm, which is seen as the white portion in corn that is popped.

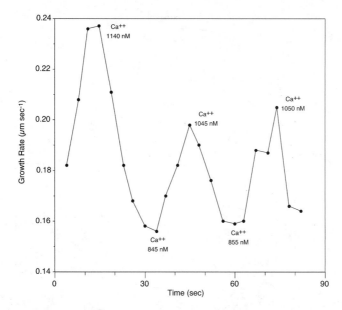

FIGURE 7.17. Pollen tube growth (essential for sexual reproduction) and free calcium (Ca^{++}) (essential for this process) show "in phase" ultradian oscillations at the tip of a pollen tube in *Lilium longiflorum* (redrawn from Figure 1 in Holdaway-Clarke et al., 1997).

The diagram in Figure 7.15B illustrates an ovary that contains one ovule. However, the ovary of the pea and bean plant contains many ovules, each of which has an egg and polar nuclei that are fertilized by sperm. Therefore, when geneticists conduct artificial hybridization experiments, as did Mendel and Bünning, each embryo will have alleles of genes that may or may not be the same as found in other embryos located within the same ovary.

Numerous biological rhythms are associated with pollination and fertilization, from the movements of petals that surround reproductive parts (stamens and carpels) to the emission of volatile substances that attract pollinators (see Chapter 8 on Agriculture and Natural Resources). Even the growing tip of the pollen tube and the calcium content within it oscillate (Figure 7.17).

Asexual Reproduction

It would be easy to assume that the egg and sperm are necessary for reproduction, but asexual reproduction (without sex) is common in plants, fungi, bacteria, and even some animals. A cyclic asexual reproductive process in the fungus *Neurospora crassa* has provided one of the most important systems for studying the mechanisms of circadian clocks (see Chapter 5 on Models). During a stage in the life history of *Neurospora*, haploid cells known as conidia (spores) germinate and produce new haploid organisms (see Essay 7.4).

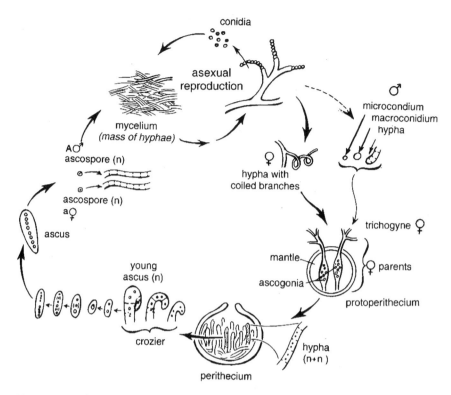

FIGURE 7.18. Life history of *Neurospora crassa*. See Essay 7.4 for a description of events.

Essay 7.4 (by WLK): An Abbreviated Life History of Neurospora crassa

Neurospora crassa is a pink mold that belongs to a group (phylum, from Greek *phylon* = class, race, or stock) of fungi known as the Ascomycota, the sac (ascus) fungi (Greek *askos* = bladder). Sometimes it can be found on bread[21] or as a contaminant in many laboratories. *Neurospora* is credited with initiating the foundations for biochemical genetics and molecular biology (cf. Davis & Perkins, 2002). The life history of *Neurospora* can be viewed as a cycle (cf. Fincham & Day, 1965), starting and ending with spores in an ascus (pl = asci) (see Figure 7.18). These spores are called ascospores and represent one of the three types of spores produced. When an ascospore lands on a suitable substrate, it germinates and gives rise to a filamentous strand called a hypha (pl = hyphae). Soon, a mass of hyphae, which is now called the mycelium, is produced. Both the ascospores and the hyphae that they produce have names or labels that signify sexuality or potential mating strains. Often, one of the strains is referred to as the male, micro, or + type, while the other is referred to as the female, macro, or − type. Morphologically, the two are indistinguishable.

Under natural environmental conditions, hyphal development may proceed along two different paths of reproduction: sexual or asexual. In the case of asexual reproduction,

---

[21] *Neurospora crassa* should not be confused with *Rhizopus stolonifer*, the black bread mold that belongs to the phylum Zygomycota.

the undifferentiated vegetative (white) hyphae can differentiate to form aerial (upright) structures, which are pinkish or yellow-orange in color. The aerial hyphae give rise to spores called conidia, and depending upon the potential mating type, either microconidia or macroconidia. When the conidia are dispersed, they germinate, produce hyphae, and form mycelia. This asexual cycle, which can keep recurring, is the one that has been used in the laboratory for studying circadian rhythms.

The sexual path commences with the transfer of nuclei from the parent, which may be a microconidium, a macroconidium, or a section of hypha. Meanwhile, female hyphae develop coil-like branches and produce an outer mantle (protoperithecium) in which arise multinucleates containing ascogonia with protruding trichogynes. If you remember the story of a women named Rapunzel who let out her hair from a tower window for her lover to climb and reach her chamber, you will remember the trichogynes (Gr. *trich* = hair, *gyne* = women). They are spike-like hairs where the male can come in contact with the female for the exchange of nuclei.

The cell walls of *Neurospora* hyphae have pores (septate hyphae) that allow for the movement of nuclei. Thus, more than one nucleus may be found in a hyphal cell. Following the transfer of nuclei to the archegonium, developing hyphae cells contain nuclei from both parents, but since fusion has not occurred, the status is known as the dikaryotic stage. The protoperithecium develops into a flask-like fruiting body called the perithecium. Here, the tips of some hyphae form a hook called a crozier and seal off the two ends of the subapical cell leaving it to have two nuclei, one from each parent. This cell is the beginning of the ascus, where, following the fusion of nuclei (karyogamy), meiosis, and subsequent mitotic division, an ascus containing eight ascospores is produced.

The asexual formation of these spores continues with a circadian period when maintained in glass tubes for days (see Figure 1.4 in Chapter 1), and later each generation of organisms can be monitored and assayed for chemical components. To study the circadian clock, the medium at one end of the tube is inoculated with a suspension of conidia spores. After a day in LL, the location of the "growing" front of mycelium is marked on the outside of the glass and the tube is transferred to DD. This sets the rhythm (clock) to be at a given circadian time (CT12), and the mycelium that will subsequently be produced develops as an undifferentiated mass of hyphae advancing along the tube. Later, morphological differentiation is induced in the form of "aerial-oriented" hyphae that produce spores. The entire asexual process from the undifferentiated to the differentiated stage of development is rhythmic and recurs with a free-running period of about 21.5 h (cf. Gooch et al., 2004).

Courtship and Mating

Prior to fertilization, most species of animals engage in *mating* or *courtship*[22] behavior, which brings together the potential parents, and subsequently, the two gametes. Usually, courtship precedes mating and is typical for many species of

---

[22] Unlike in the past century, the uses of the terms courtship and courting in conversational English have been replaced by other terms (e.g., dating). Briefly, courtship (noun) refers to the act of courting, while to court (verb) relates to various social activities that precede engagement, marriage, or mating.

animals from insects to mammals. For humans, the romance of courtship has been portrayed in the theater, as well as in "old-time" films, by a male singing to his female in waiting. In reality, this type of courtship behavior is rare in modern society. However, for the male fruit fly, a "song" is part of the courtship ritual that precedes mating and involves a series of behavioral movements. A number of courtship genes has been identified and used as a means to better understand the process of courtship and mating in these insects (*Drosophila*)[23] (Hall, 1994). Both sexual orientation[24] and courtship behavior of male *Drosophila* are regulated by a gene called *fruitless* (*fru*), which functions specifically in the central nervous system (Ryner et al., 1996). Courtship for fruit flies does not occur in flight, but instead the male orients toward the female and will follow her if she is walking. During one of a series of maneuvers that includes tapping her with his forewings, the male fruit fly extends one of his wings (see Figure 2.11 in Chapter 2 on General Features of Rhythms), and by its vibration, produces a species-specific courtship "love song." For *D. melanogaster*, pulses of tone are produced at intervals of about 34 ms, with about one minute between series of pulses (Kyriacou & Hall, 1980). This expression of an ultradian courtship rhythm is controlled by the *per* gene, the same gene that regulates the circadian rhythms of eclosion[25] and locomotor activity (Konopka & Benzer, 1971). The ultradian and circadian expressions of periodicity parallel each other. In other words, the *per* allele that has a shorter circadian period (per^s) also has a shorter ultradian period (about 41.5 s), and the *per* allele that has a longer circadian period (per^l) has a longer ultradian period (about 82.1 s) (Kyriacou & Hall, 1980).

In mayflies (Ephemeroptera [*ephemera* = short-lived; *ptera* = wings]), their adult life revolves solely on sexuality: no food, no drink, only copulation. Prior to their sexual adult stage (imago), they live underwater as larvae. The adults, on the other hand, are terrestrial and live for only a short span of time, often less than a day. Their mouthparts are considered to be nonfunctional and they do not appear to feed. The form (morphology) of their large front wings is fan-like, consisting of a series of corrugations and well-adapted to the nuptial flights,[26] which involve flying upward and then coasting or gently flying downward (cf. Edmunds &

---

[23] *Drosophila melanogaster* is one of the important organisms used in genetic studies because of its small size, short generation time (e.g., 25 generations in one year), ease of culture, and an extensive background of information that is known about this organism.

[24] *Fru* (fruitless) mutant males show abnormal or absent later steps of courtship (e.g., singing, copulation) and are sterile. They court both males and females, and when grouped together, each male courts and is courted. Females are not affected by *fru* mutations (Ryner et al., 1996).

[25] The life history of the fly includes four prominent stages and occurs in the following sequence: egg, larva, pupa, and adult. Eclosion occurs when the adult fly emerges from the pupal case.

[26] Perhaps, you remember how you can fold an ordinary sheet of paper to be strong enough to be a fan or be able to extend a retractable carpenter's ruler out to a great distance. Similarly, the corrugations that give the may fly wing a fan-like form, provide strength.

Traver, 1954). Generally, a female will fly into a swarm of males, whereupon she is seized by a male and copulation commences; this lasts in the range of 10–40 s in a number of species (cf. Morgan, 1913; Cooke, 1940; Brink, 1957). The emergence of the adult is cyclic, occurring at approximately the same time each year. After mating and the laying of eggs (oviposition), the adults die (see Chapter 8 on Natural Resources and Agriculture).

Photoperiodism and Sexuality

The occurrence of courtship and mating for many groups of animals is rhythmic,[27] associated with the seasons of the year, and thus, with photoperiodism. In these animals, a number of physiological and structural changes occur during and prior to courtship and mating. Among the more common examples are the development of larger gonads in birds, formation of antlers in deer, migration of fish to spawn, and estrus in mammals (e.g., sheep and goats).

Because of the dominating influence of the biology of humans in the life sciences, it is not too uncommon to overlook the topic of photoperiodism in the study of biological rhythms, especially in matters pertaining to sexuality. However, the use of the term "sexuality" is not restricted to humans and other animals. It appeared in the title of one of the first papers to emphasize the role of daylength (photoperiod) in reproduction (Tournois, 1914), a paper that focused not on animals, but on the flowering of plants.

Diet

In addition to photoperiodism, chemical signals from the natural food supply may also influence reproduction. For example, it was originally noted that chemical signals in actively growing plant food resources may act as a reproductive triggering stimulus, as was shown for a natural population of montane meadow voles (*Microtus montanus*) that were fed sprouted wheat grass in the winter (Negus & Berger, 1971). A melatonin-related compound, 6-MBOA (6-methoxybenzazolinone), which occurs naturally in the actively growing food supply, was identified and shown to increase ovarian weight when fed to laboratory mice (Sanders et al., 1971). When voles were fed rolled oats coated with 6-MBOA in the winter nonbreeding season, there was a dramatic increase in the pregnancy rate in females and in testicular size in the males (Berger et al., 1971). This cueing mechanism was suggested as more profitable for reproductive effort than photoperiod among herbivorous species that live less than a year or species that inhabit highly unpredictable environments, such as rabbits, kangaroos, desert rodents, and other microtine rodents (Berger et al., 1981).

In addition to tall fescue grass or winter wheat, a variety of other edible plants (e.g., rice, sweet corn, oats, Japanese radish) have also been shown to contain

---

[27] There is evidence that circadian rhythms in sexual behavior are reversed for some diurnal and nocturnal species, such as the rat (Mahoney & Smale, 2005).

Flowers

Events and rituals in our society associated with sexuality and reproduction, such as birth, marriage, and courtship, are often acknowledged with gifts of flowers. While flowers may serve as expressions of love, honor, and remembrance for humans, they are a representation of sexuality and reproduction that occurs in about 235,000 species (phylum Anthophyta). Starting with the process of floral induction and continuing through seed germination, the reproductive phase in the life history of flowering plants illustrates a spectrum of biological rhythms with periods that span a range from seconds to over a century (e.g., the intervals between flowering can span from decades to over 100 years for some species of bamboo).

Often the production of flowers depends upon interactions between circadian rhythmicity and photoperiodism. Leaves are the photoreceptors, but the message must be transferred to growing regions known as apical meristems, which are located at the tips of shoots and axillary buds. This transition from vegetative development to reproductive development is rhythmically-controlled in many species, such as Biloxi soybean (*Glycine max*) (Figure 4.13 in Chapter 4 on Photoperiodism) (Hamner, 1960), Japanese morning glory (*Pharbitis nil*) (Takimoto & Hamner, 1964), and *Chenopodium rubrum* (Cumming, 1967, 1969).

Much of the beauty of a flower is contributed to the conspicuous-colored petals, which are sterile parts, but in terms of sexuality, can serve to attract pollinators. Of special interest is the fact that the petals of some species open and close over the course of the 24-h day. The phase of these circadian rhythms differs among species, a feature that makes possible the floral clocks of nature (see Chapter 3 on Time). Some plants also rely on odors to attract pollinators and the production of these volatiles may be rhythmic. Both the opening and closing of flowers and production of odor in the Night Blooming Jessamine[28] (*Cestrum nocturnum*) have a free-running period of 27 h to 28 h (Overland, 1960). Under natural conditions, the flowers are open at night and the powerful fragrant odor attracts night-flying insects.

Pollination is followed by fertilization (Figure 7.14) and depends upon the growth of a pollen tube, in which the two sperm cells are produced. In lily (Holdaway-Clarke et al., 1997), and perhaps in other species as well, the growth of the pollen tube displays a high-frequency ultradian rhythm (Figure 7.17).

After fertilization and the development of the embryo, two familiar structures emerge: the fruit, which is the mature ovary, and the seed, which is a mature ovule. Both have variables that may display rhythms. For example, changes in the diameter of the orange (*Citrus sinensis* cv. Valencia) display a daily rhythm (Elfving

---

[28] This native plant of the Caribbean, also called Jasmine or Lady of the Night, produces a delicate perfume at night from July to October.

& Kaufmann, 1972), while seeds of a number of species display an annual rhythm in germination (Bünning & Müssle, 1951; Ludwig et al., 1982). For seeds to germinate, however, they require the imbibition (uptake) of water and rhythms having periods of about a week, month, and a year have been reported for this process in *Phaseolus* (Spruyt et al., 1987, 1988).

Finally, the cyclic nature of the many variables that are part of the temporal organization of sexuality and reproduction in flowering plants depends also upon the cyclic nature and synchronization of other organisms and the environment. This includes such things as the activity cycles of insects and other animal pollinators, as well as the seasonal time cues of the solar day. In fact, a cyclic environment is often required for normal development of flower clusters (inflorescence) (cf. Bünning, 1962).

Rhythmic Phases of Sexual Behavior in Humans

Human sexual behavior involves a network of biological rhythms, ranging from the ultradian to the infradian. Some are influenced by society and convenience, while others are associated with recurring biological events.

Activity

Times of sexual activity can be seen in statistical data that list times when sexual intercourse more commonly occurs during the day, week, month, and year (Table 7.3) (Smolensky & Lamberg, 2000). Monthly changes in estrogens may enhance a woman's desire for sexual interactions during midcycle when she is most likely to conceive (Adams et al., 1978). For men, an increased desire for sexual activity during autumn (fall) may be influenced by an increase in the male sex hormone testosterone, which also occurs in the fall (Smals et al., 1976; Dabbs, 1990). It is of interest that an increase in birthrate in late summer and early fall occurs about nine months after the reported late fall and winter peak in human sexual activity in the USA and in countries at a similar latitude (Roennenberg & Aschoff, 1990a,b).

Disease

Several sexually transmitted diseases show a yearly rhythm, with new cases peaking in summer for genital infections with herpes simplex virus (Sumaya et al.,

TABLE 7.3. Most common times for human sexual intercourse.

Time frame	More activity	Reference
Day	17:00–23:00 h, 08:00 h	Reinberg & Lagoguey, 1978; Leonard & Ross, 1997
Week	Weekend (Sat.–Sun.)	Udry & Morris, 1970; Palmer et al., 1982
Month	Near ovulation	Adams et al., 1978
Year	Fall	Reinberg & Lagoguey, 1978

1980), in late summer for gonorrhea and trichomonas, and in the fall for syphilis[29] (cf. Smolensky, 1981). Contact between individuals from different populations has also been reported to synchronize an about 10-year cycle in the annual incidence of primary and secondary syphilis across the USA (Grassly et al., 2005). Although sexual behavior influences the overall number of people infected, the 8–11-year cycle in syphilis infection was shown to reflect an intrinsic property of the disease itself, since similar cycling was not observed in gonorrhea, another sexually transmitted disease that occurs at ten times the rate observed for syphilis.

The Menstrual Cycle

The monthly female menstrual cycle of humans is probably the most familiar of all the cycles that are associated with sexuality.[30] While a relatively small number of women may be aware of their own circadian cycles that are associated with sexuality, nearly all are aware of a periodic bleeding (*menses* or *menstruation*) from the uterus that occurs about every four weeks during their reproductive years. The first bleeding (called *menarche*) occurs at the onset of puberty, usually when the female is 11 to 13 years of age, and unless interrupted by pregnancy, disease, strenuous activity, etc., continues to recur until *menopause* (about 50 years of age).

Duration and Phase

As is the situation for many physiological rhythms, the *menstrual cycle* includes a number of phases (Essay 7.5), which are regulated by hormones produced in the pituitary, hypothalamus, ovaries, follicles, and the corpus luteum (lining of the uterus) in a continuous feedback loop (Figure 7.19) (Lein, 1979; Reilly, 2000).

> ***Essay 7.5 (by RBS): Brief Physiology of Menstrual Cycle Events and Phases***
>
> By the strict definition presented in Chapter 2, a period represents the length of time required to complete a cycle. In matters pertaining to the menstrual cycle, however, the term period is commonly used to denote the phase during which bleeding occurs. Also, it is customary to count the length of a menstrual cycle beginning with day 1 of menstruation, a phase (span) that usually lasts 4–5 days. During this span there is blood loss,[31] along with discharge of the surface of the endometrial lining of the uterus. When

---

[29] Syphilis infection most likely occurs in late summer since the onset of symptoms leading to a diagnosis of syphilis requires several weeks after contact.

[30] In chronobiology, the term "circatrigintan" for cycles of about 30 days is sometimes preferred over "menstrual" because some aspects of the system are found in postmenopausal and premenopausal females (cf. Haus and Touitou, 1992).

[31] Normal blood loss is about 40 ml (range 25–65 ml), but in some instances can be more or less (Reilly, 2000).

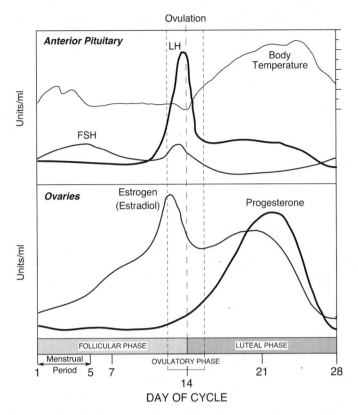

FIGURE 7.19. Generalized scheme for levels of hormones during a typical menstrual cycle (composite figure derived from Dyrenfurth et al., 1974; Lein, 1979; and Reilly, 2000).

menses ceases, estrogen production by the ovaries increases to renew the endometrial lining, while follicle-stimulating hormone (FSH) from the pituitary stimulates the maturation of an ovum (and occasionally more than one). This is called the *follicular phase* of the menstrual cycle. During this time span, body temperature and progesterone levels are decreasing.

When estrogen, in a form called estradiol (E_2), reaches a certain level, usually about mid-cycle (day 14), it causes the hypothalamus to produce a sharp increase in luteinizing hormone (LH), which triggers the release of a mature egg from the follicle.[32] There is also a slight dip in body temperature at mid-cycle that is associated with the release of the egg. The mature egg is received by the fallopian tube and moves down to the uterus. This begins the *luteal phase* or second half of the menstrual cycle, wherein FSH and LH levels drop dramatically, estrogen levels decrease by half, body temperature rises and progesterone also increases, being produced by the lining of the altered follicle, now called the *corpus luteum* (Latin *luteus* = saffron-yellow).

For conception to occur, an egg needs to be fertilized within 24 h of its release from the ovary (Reilly, 2000). If fertilization occurs, the ovum embeds itself in the lining

---

[32] Usually the two ovaries alternate in releasing an egg.

TABLE 7.4. Timing of variables that show both 24-h and monthly rhythms in women.

Variable[a]	24-h peak (h)	Reference(s)	Monthly[b] peak	Reference
FSH	05:12–06:20	Haus et al., 1988	Mid-follicular	Dyrenfurth et al., 1974
Estrogen	04:20–13:24	Haus et al., 1988	Late-follicular	Dyrenfurth et al., 1974
LH	02:40–04:12	Haus et al., 1988	Mid-cycle	Dyrenfurth et al., 1974
Progesterone	07:24–12:08	Haus et al., 1988	Mid-luteal	Dyrenfurth et al., 1974
Temperature	Late afternoon	Baker et al., 2001	Mid-luteal	Dyrenfurth et al., 1974
Pulse	Afternoon-early evening	Hermida, 1999	Mid- to late luteal	Engel & Hildebrandt, 1974
Blood pressure	Afternoon and early evening	Hermida, 1999	Late luteal	Engel & Hildebrandt, 1974

[a]FSH = follicle-stimulating hormone; LH = luteinizing hormone.
[b]Monthly = one menstrual cycle.

(endometrium) of the uterus and starts developing. However, if fertilization does not occur, the corpus luteum will begin to regress by day 21, progesterone and estrogen levels begin to fall, the unfertilized egg passes from the body "unnoticed," and about two-thirds of the lining of the endometrium breaks down and is shed during the start of menstruation on day 28 (Reilly, 2000; Smolensky & Lamberg, 2001). This bleeding marks the beginning of a new cycle, with a renewed increase in FSH to promote follicle development and the next ovarian cycle.[33]

A large number of variables, such as hormones, tissues, and cells that are part of the ovarian and uterine system, oscillate in synchrony with the menstrual cycle. Furthermore, a number of variables also display both 24-h and monthly rhythms (Table 7.4). It is interesting that the occurrence of hot flashes (internal sensations of upper body heat) that are associated with hormone fluctuations during the cessation of the menstrual cycle (menopause) and with surgically-induced menopause show a circadian rhythm, with a major peak in the late afternoon and evening, a time when body temperature itself is normally highest (Albright et al., 1989; Freedman et al., 1995). A minor secondary peak is often evident in the morning near awakening.

While the length of most menstrual cycles ranges from 25–30 days (mean from 21 studies = 29.495 days), it is not uncommon for cycles to be longer or shorter than this range (Presser, 1974). The length of the menstrual cycle can differ among individuals and even for the same individual (Figure 7.20) (cf. Sothern et al., 1993). In one study, nearly half (46%) of women, ages 18–40 years, who recorded four or more cycles and reported that they were "regular," had cycle lengths that differed by 7 days or more, while 20% had a range that differed by 14 days or more (Creinin et al., 2004). Some of these inter- and intra-subject differences in cycle length can be related to the age of an individual (Figure 7.21), the sequence of consecutive cycles, time of year, physical activity, stress, health, changing time zones, etc. The mean length appears to be longer near menarche and menopause and shorter and more consistent in between (Figure 7.21)

---

[33] A woman ovulates about 400 times during her reproductive years (Smolensky & Lamberg, 2001).

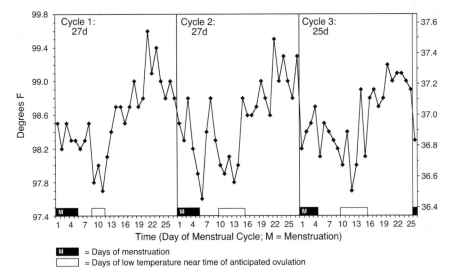

FIGURE 7.20. Variation in menstrual cycle length and accompanying body temperature over three consecutive menstrual periods in a healthy woman (age 45 years) (depiction of subset of data presented in Sothern et al., 1993).

(Presser, 1974). In addition, cycle lengths are 3 days, on average, in January (winter) than in the April (spring) (Reilly & Binkley, 1981). Activities such as working at odd hours, rapidly changing time zones, stress, and illness can also alter the length of the menstrual cycle and/or result in anovulatory cycles (no egg released). Variations in menstrual cycle length could have clinical impact on contraceptive practice and research and pregnancy-related care (Creinin et al., 2004), and possibly other treatments.

The duration of specific phases within the menstrual cycle may also differ. The length of the follicular (preovulatory) phase is more variable than the length of the luteal (postovulatory) phase, which is shorter and consistently close to 14 days in length (Presser, 1974). This can be seen in Figure 7.22, where a woman's body-temperature patterns over three cycles of differing lengths (23 days, 27 days, or 31 days) are aligned either forward (poor alignment during luteal phase) or backward (better alignment during luteal phase) from the onset of menstruation (cf. Sothern et al., 1993).

While 28 days are often viewed as the average length of the cycle, results from studies indicate that the mean or median of the menstrual cycle is about 29.5 days, which are the same as the period for the lunar cycle (Cutler, 1980; McClintock, 1971). Because of the similarity between a physical and biological periodicity, one could question if there is also a correlation between phases. In fact, there are results from some studies that suggest a possible relationship between the two. When a large population of women who were not using birth control pills recorded the onset of menses for 14 weeks, ovulation tended to be more prevalent

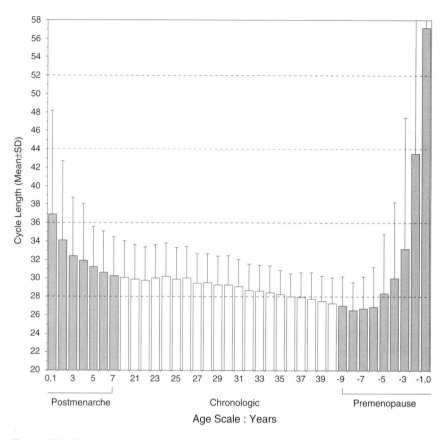

FIGURE 7.21. Variations in the mean length of menstrual cycle by age (data from Table 4 in Presser, 1974). *Note*: cycle lengths are very irregular during both postmenarchal and premenopausal years.

during the dark phase of the lunar cycle (i.e., near new Moon), with menses (flow) commencing during the light half of the lunar cycle, near full Moon in a majority of the women (Cutler, 1980). This phase relationship between the lunar and menstrual cycle stages is especially evident in apes and monkeys, which also have menstrual cycles.[34]

---

[34] Females of various other mammalian species have a sexual cycle where estrus occurs several times a year. Some examples for average cycle length include: chimpanzee, 36 days; red howler monkey, 29.5 days; lowland gorilla, 28–38 days; marmoset, 28.3 days; macaque, 27–29 days; rhesus monkey, 28 days; capuchin monkey, 21 days; black-handed spider monkey, 20–23 days; sow, 21days; cow, 20 days; ewe, 16 days; guinea pig, 15 days; rat and mouse, 4–6 days; dog, 2/year (cf., Reinberg & Ghata, 1962; Wright & Bush, 1977; Shaikh et al., 1978; Jenkin et al., 1980; Dahl et al., 1987; Harter & Erkert, 1993; Herrick et al., 2000; Campbell et al., 2001).

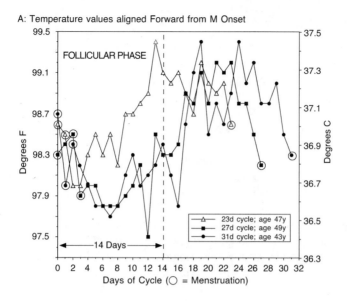

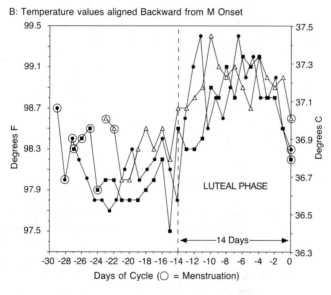

FIGURE 7.22. Morning body temperatures over menstrual cycles of different lengths reveal better alignment with second half of cycle when arranged backward from onset of menstruation (= day 0) (depiction of subset of data presented in Sothern et al., 1993).

Social Synchronization

One of the many interesting features of the menstrual cycle is *social synchronization*, whereby the phases of the rhythms of several individuals become synchronized. This has been documented for female room-mates and close friends attending college (McClintock, 1971) and universities (Graham & McGrew, 1980; Quadagno et al., 1981) and also among females having close associations in other types of institutions (e.g., prison, work groups) and within families. Various statistical tests have shown that factors other than close associations (e.g., photoperiod, food,) are not causing the synchronization. Apparently, the synchronization is dependent upon socialization, and indications are that the causal agent is an odor or volatile substance belonging to a group of chemicals known as *pheromones*.[35]

When axillary secretions were obtained from female donors and subsequently applied just under the nose on the upper lips of female recipients,[36] menstrual cycle synchrony often occurred between recipients and donor (Preti et al., 1986; Quadagno et al., 1981). There may be multiple pheromones with opposing effects, since axillary compounds from donors in their follicular phase produced a shortening of the follicular phase of the menstrual cycle in recipient women, but a lengthening of the follicular phase, and hence a longer cycle, was found in recipient women when the donor axillary compound was applied near ovulation (Stern & McClintock, 1998). In addition, greater sexual arousal and positive mood was reported when women were exposed to male fragrance during the periovulatory phase of their menstrual cycle (Graham et al., 2000). Research is continuing to investigate the possibility that steroids may also act as pheromones to influence or modulate human behavior by eliciting a chemosensory response in a region of the nose that subsequently affects hormone levels (McClintock, 1998; Jacob & McClintock, 2000).

Sexual Activity and Birth Control

Phases within the menstrual cycle have been closely associated with times of sexual activity, or lack thereof, which in turn can be strongly linked to the rhythmic levels of biological molecules, as well as to society, culture, and religion. In some primitive cultures, women were isolated from men and lived in special huts during menses. They were even considered to be dangerous to men during their menses phase, to the extent that in some cultures contact during and for several

---

[35] Pheromones (from Greek *pherein,* to carry, and *hormon,* to stir up or excite) are airborne chemicals that are produced and released into the environment by members of a species to communicate and affect the physiology or behavior of other members of the same species.

[36] Secretions from axillary regions (arm pits) were collected onto cotton pads, extracted with ethanol and stored prior to application. None of the women reported smelling anything other than the alcohol used to carry the compounds, indicating the presence and subconscious detection of human pheromones, most likely via receptors in the nose.

days after menses was forbidden (cf. Lein, 1979). Today up to 70% of men and women report that they abstain from sexual relations during this time (Barnhart et al., 1995). In addition to matters related to hygiene, abstaining from sexual relations during menstruation may help to avoid certain health conditions, including heavier and/or longer flow (Cutler et al., 1996), an increased chance of endometriosis (Filer & Wu, 1989), and an increase in the risk of contracting a sexually-transmitted disease (Tanfer & Aral, 1996).

An increase in female-initiated sexual activity near ovulation has been reported for married women not using oral contraceptives (Adams et al., 1978). Two issues must be addressed, however, in regards to any cyclic implications of a woman's sexual drive (libido) or activity. First, individual differences among females and the complexity of the endocrine system are such that all generalizations in regard to sexual drive must be viewed with caution. Second, a variety of factors, such as the desires of the woman's mate, the cultural, social and economic status, and level of education, may influence sexual practices (cf. Lein, 1979).

For couples desiring to have children, or for those who for religious or other reasons must rely solely on a "natural" contraceptive method, the so-called "rhythm method" provides only a rough guide for sexual activity. Several authors have reported that a fertile "window" occurs during a six-day span ending with the day of ovulation and that intercourse during this time is more likely to result in conception since sperm can survive three to five days in a woman's body (Wilcox et al., 1995; Dunson et al., 1999). Even though the probability of conception may drop to near zero on the day after ovulation, there are reports of sporadic late or early ovulations, such that conception can occur during days 6–21 of the menstrual cycle in a few women, rather than days 10–17, as identified by clinical guidelines (Wilcox et al., 2000).

Mood and behavior can also be influenced by the menstrual cycle. Some women experience discomfort or even pain at certain phases of their cycle. For example, in young women in their 20's with normal menstrual cycles, positive moods were more pronounced in the follicular and ovulatory phases, while negative moods were higher in the luteal phase preceding and during menses (Figure 7.23) (Reilly, 2000). For women who enter hospitals for mental (psychiatric) illness, 47% of them were admitted just before and during the menstrual days (Targum et al., 1991).

Because the body temperature of a woman drops on the day of ovulation and rises the day after, a basal body thermometer and calendar have been used to approximate the time of ovulation. Tracking the cycle appears simple, but can be imprecise, especially in regard to the duration of an ovum (2–3 days after ovulation), viability of sperm (up to 6 days after deposit), time of day of temperature reading, health-related issues[37] and the procedure for monitoring temperature, including site (Figure 7.24) (cf. Sothern et al., 1993), accuracy and type of thermometer.

---

[37] A number of years ago, two female students in one of our biological rhythms classes elected to monitor their basal body temperature relative to predicting their time of ovulation. Ironically, both had their wisdom teeth extracted during the course of their study, causing an elevation in temperature and a failure to observe a significant "monthly" temperature rhythm.

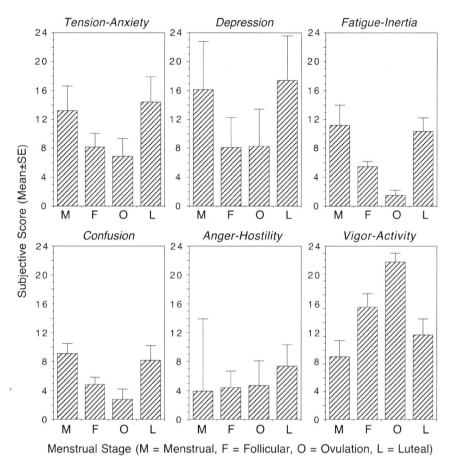

FIGURE 7.23. Mood factors change consistently with menstrual cycle phase in young women in their 20s with normal menstrual cycles (data from Table 1 in Reilly, 2000).

Unless the length of the cycles is absolutely constant, the exact time or phase of ovulation cannot be predicted with 100% accuracy. Aside from abstinence, which is the most certain method for preventing fertilization and pregnancy, the various methods of contraception (e.g., condoms, birth-control pills, and vasectomy) block one or more parts of the cycle, such as fertilization, ovulation, or gestation. Birth-control pills that contain small amounts of estrogen and progestin (or progestin only in a so-called mini-pill) act to prevent ovulation by inducing the body to function as though an egg has already been released or that an embryo has been implanted. For example, some types of pills are typically taken on a 21/7 regimen, with active pills taken for 21 days, followed by a 4th week on an iron pill or no pill at all.[38] Breast-feeding can also play an important role in contraception by delay-

---

[38] Another type of birth-control pill is available that reduces the number of menstrual periods in a year to only once each season (Kalb, 2003). This pill contains the same hormonal

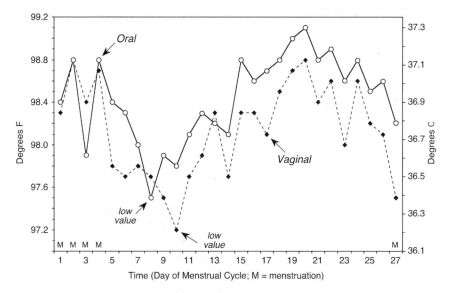

FIGURE 7.24. Morning body temperature from two sites (oral and vagina) in a healthy woman (age 42 years) reveals similar changes during one menstrual cycle (depiction of subset of data presented in Sothern et al., 1993).

ing the onset of ovulation after a birth and thereby prolonging the interval between successive births (Short, 1984).

Birth-control pills alter not only the course of hormones during a natural menstrual cycle, but other cyclic variables as well. For example, the use of birth-control pills has been shown to reduce maximal anaerobic power (maximal cycling power) in both the mid-follicular and mid-luteal menstrual phases, and to reduce overall anaerobic power when compared with noncontraceptive users (Giacomoni & Falgairette, 1999). A circadian (Reinberg et al., 1996) or circannual rhythm (Weydahl et al., 1998) in the hormone melatonin normally present could not be demonstrated in contraceptive users. In addition, the 24-h means for systolic blood pressure (BP), heart rate, skin blood flow, and transepidermal water loss were greater, while cortisol was lower, with the acrophases significantly delayed to later in the evening for systolic BP and heart rate, in contraceptive users when compared with nonusers (Reinberg et al., 1996). Also, women who use oral contraceptives did not show the female-initiated increase in sexual activity at mid-cycle that has been observed in nonusers (Adams et al., 1978).

---

ingredients as the conventional pill, but is taken for 84 days, followed by 7 days of no pill. While many women welcome the reduction in monthly bleeding from 13 to 4 times a year, it remains to be seen what effect this form of menstrual suppression will have on the normal overall rhythmic interaction of cycles with ultradian, circadian, and other infradian periods in a woman's body, since it does not follow the period of a woman's natural menstrual cycle.

Primary and Secondary Sex-Related Rhythms in Men

Although more obvious and well-known in the female, human males also have a "gonadal" rhythmicity ranging from ultradian to circadian to infradian (about–monthly and yearly) in many primary and secondary aspects related to their sexuality. Primary aspects include rhythms in hormone and sperm-production, while secondary aspects include cycles in variables such as body weight, grip strength, emotions, body hair growth, and sexual activity.

Ultradian and Circadian Cycles

Ultradian rhythms having periods close to 95 min, with a range from about 50–130 min, span a diverse realm of biological variables found in bacteria, fungi, plants, humans, and other animals (Koukkari et al., 1997). Of these variables, the REM/non-REM sleep cycles (REM = Rapid Eye-Movements) may be the best known. Occurring concurrently with the REM phase of the cycle in virtually all healthy males is a phenomenon called *penile tumescence*. About 80% to 95% of the REM spans may be accompanied by an erection of various degrees in young men (Fisher et al., 1965; Karacan et al., 1976). These erections during sleep occur with a higher frequency for teens than they do for males in their 60s and 70s (Karacan et al., 1976). Although not mentioned earlier in this chapter, nocturnal genital arousal during REM has also been reported to occur in women (Goldstein et al., 2000). Another example of an ultradian rhythm in men is the release of luteinizing hormone (LH) in repetitive bursts from the pituitary, with an average cycle length of 1.9–2.3 h (Nankin & Troen, 1974).

Many, if not all, physiological and psychological variables in men show a circadian rhythm (Kanabrocki et al., 1973, 1990; Haus et al., 1988), but the extent to which these variables correlate with hormonal ones is still open to debate. A statistically significant 24-h change in concentrations of androgens[39] and other sex-related hormones has been found in blood of mature human males (Table 7.5). Testosterone, the most potent naturally occurring androgen, has been shown to reach highest levels from about 04:00 h until noon, with an average peak time near awakening. This is also the time for highest levels of epitestosterone, androstenedione, FSH (follicle-stimulating hormone), LH, and prolactin. Plasma sex-binding globulin, which binds to testosterone, is maximal around 14:00 h, while PSA (prostate-specific antigen), a measure of prostate epithelial cell activity, is maximal around 17:00 h, near the time of lowest circulating testosterone levels (see Table 7.5). In fact, production of PSA may be affected by testosterone. In a study of 11 men ages 46–72 years, not only were the overall 24-h means in serum PSA and total testosterone inversely correlated with each other at $p<0.001$, but the circadian patterns were also nearly inverted: testosterone reached a peak at 08:38 h and PSA at 17:02 h (Mermall et al., 1995). These data suggested different functional states over a 24-h span for the prostate due to changes in testosterone levels.

---

[39] Androgens are substances that can induce masculine characteristics.

TABLE 7.5. Circadian peak times for reproductive hormones in blood of men.[a]

Variable	No. of men	Time of peak (h)	Reference
Testosterone	7	07:02	Clair et al., 1985
"	124	07:40	Haus et al., 1988
"	11	08:38	Mermall et al., 1995
Epitestosterone	14	10:43	Lagoguey et al., 1972
Androstenedione	3	07:00	Crafts et al., 1968
FSH	23	05:00	Faiman & Ryan, 1967
"	13	07:30	Leyendecker et al., 1970
LH	13	08:00	Leyendecker et al., 1970
"	26	02:32	Haus et al., 1988
Prolactin	7	05:00	Nokin et al., 1972
"	27	04:20	Haus et al., 1988
Sex-binding hormone	7	14:06	Clair et al., 1985
Prostate-specific antigen (PSA)	11	17:02	Mermall et al., 1995

[a]Peaks determined by time of highest mean value or peak of fitted 24-h cosine.

Infradian Cycles

The possibility of monthly rhythms being present in human males was referred to in the 17th century, first by the Italian physician Santorio Santorio based upon his own body weight, and later in a letter about bleeding and pain incurred by an Irish inn-keeper (see Essay 7.6).

Essay 7.6 (by RBS): 17th Century Notes of Monthly Rhythms in Males

Monthly rhythms in body weight in the human male were documented as early as 1614 when the Italian physician Santorio Santorio (1561–1636) pointed out that while monthly changes in women are taken note of by everyone, similar alterations in men were ignored (Fidanza, 1989). He may very well have been the first to provide evidence for a possible monthly cycle in men. He monitored his own body weight throughout 30 years under "moderate" conditions of life, noting that the extent of change was about 1–2 pounds. Santorio, born in 1561, would have been 53 years of age when he published his small book in 1657 reporting his observations on body weight changes and insensible perspiration that he observed on himself for 30 years. Thus, he would have been 22- or 23-years old when he began this project. The original Latin publication (Santorio S. (1614) *De Statica Medicina.* Hagae-Comitis, ex typographia, A. Vlaco) was later translated into Italian, English, French, and German.

Menstrual bleeding as an indicator of a rhythmic variable is by far the most common infradian rhythm, but is specific to females. By contrast, periodic bleeding in the human male has been a curiosity at most, since a short letter in the 17th century referred to a man who bled from one of his fingers about once a month (Ash, 1685). Walter Walsh, a 43-year-old Irish innkeeper, experienced pain in his right arm and bleeding from the end of his forefinger about once a month. This about-monthly phenomenon continued over the course of 12 years until his death at age 55 years. The bleeding certainly cannot be associated with a menstrual cycle. However, as will be discussed later in this chapter, infradian cycles for pain have been reported for both females and males. Even though blood was the variable, it need not be related to reproduction, yet this anecdote has inspired others to ponder the possibility of about-monthly changes in human males. Does the human male also exhibit longer than 24-h cycles that relate to sexuality and reproduction? The complexity of this repeatedly discussed problem is enhanced by the fact that no easy marker, such as menstruation, has been identified to assess any low frequency physiological, hormonal, or psychosocial changes that may characterize the human male.

Compared to what is known about infradian rhythms in females, especially the menstrual cycle, our knowledge of infradian rhythms in the human male is based upon a small number of studies. Due to the lack of a distinct marker, such as monthly bleeding (see Essay 7.5), any about-monthly changes in men are inconspicuous or vague and are commonly assumed to be of little widespread importance. Therefore, results from the relatively few studies presented here, although statistically significant, need to be confirmed whenever results from future studies become available. Nevertheless, a number of variables, including those that are listed in Table 7.6 and illustrated in Figure 7.25 have been found to display an infradian period in human males.

Body Weight

As noted in Essay 7.5, Santorio Santorio noted a monthly rhythm in his body weight (Fidanza, 1989). Statistical analyses of daily body weight that was recorded simultaneously with beard growth (see below) by a 24-year-old male over a span of 100 days indicated the presence of a 31-day cycle for body weight (Table 7.6) (Sothern, 1974). In a later study by the same subject, who was then 30 years old, periods of 22.5 days, 45 days, and 1 year were found when body weight was measured every day in the morning for one entire year (Figure 7.26) (Sothern, unpublished[40]). Of interest, the change over the course of the year, as determined by the amplitude, was much greater than for the about-20-day cycle (1.7 lbs vs. 0.4 lbs, respectively). When 10 young medical students were studied for 75 days, oscillations in their body weight were noted with shorter periods in the range of 11–13 days (Kühl et al., 1974).

Grip Strength

Among the factors involved in building muscle mass during development are male hormones such as testosterone. For this reason, the grip strength of hands has been suggested as a variable that might reflect changes in male hormones. Based upon results from studies that spanned 26 days to more than 100 days (Table 7.6), mean periods in grip strength ranged from about 13 days in a group of 10 male medical students[41] to 19 days in a young man whose beard growth (mentioned above) showed a 16–18-day rhythm during the same 100-day study span (Sothern, 1974).

---

[40] *Note*: this subject has monitored his daily body weight for more than 30 years.

[41] These results were part of a larger study where a circadian rhythm in grip strength was detected with peaks in the late afternoon.

TABLE 7.6. Some variables monitored in men indicate statistically significant infradian rhythms in the range of ~7–30 days.

Variable	No. of men[a]	Period(s)	Comments[b]	Reference
Body weight	10	11–13 days	Data 5/day for 75 days	Kühl et al., 1974
	1	31 days	Daily for 100 days	Sothern, 1974
	1	22.5 days, 45 days	Daily for 365 days	Sothern, unpublished
Grip strength	10	11–15 days	Data 5/day for 75 days	Kühl et al., 1974
	1	13–16 days	Data 5/day for 26 days	Kühl et al., 1974
	1	17–19 days	Data 5–6/day for 100 days	Sothern, 1974
	1	21.5 days, 47 days	Data 5–6/day for 365 days	Sothern, unpublished
Cutaneous pain	?	25 days	Young men for 40–100 days	Procacci et al., 1974
	7	32 days	Older men for 75 days	Procacci et al., 1974
17-Ketosteroids	1	17–21 days, 30 days	24-h urines for 15 years	Halberg et al., 1965
Estrogen	5	8–10 days	Urine for 16–90 days	Exley & Corker, 1965
Testosterone	9	3–5 days, 12–18 days	Urine collected 10–45 days	Harkness, 1974
	1	4.5 days, 8.5 days	Urine collected for 60 days	Sothern, unpublished
	1	20 days	Urine for 42 days; married	Corker & Exley, 1968
	1	9 days	Urine for 42 days; single	Corker & Exley, 1968
	20	20–22 days	Blood every 2 days for 2 months	Doering et al., 1975
Mood	1	38 days	Data for 3 months in manic-depressive, age 24 years	Bryson & Martin, 1954
	1	16 days, 38 days	5/6 data/day for 7 months	Sothern, 1974
	1	21 days	Data for 160 days in 22-year man with periodic catatonia	Simpson et al., 1974
Neutrophil appendages	1	27–28 days	3 days blood samples for 4.5 months	Manson, 1965
Facial sebum	2	10–15 days	Data for 33 days and 47 days	Harkness, 1974
Beard growth	1	33 days	Data for 6 months, age 39 years	Kihlstrom, 1971
	1	16.5 days	Data for 7 months, age 27 years	Sothern, 1974
	1	24 days	Data for 2.25 years, age 24–26 years	Sothern, unpublished
	1	5 days, 14 days, 25 days	Data for 5.5 months, age 55 years	Levine et al., 1974
Sexual activity	2	3.5 days	Data over 45 days	Fox et al., 1972
	1	7 days and 3.5 days	Data for 2.5 years, age 45–47 years	Hamburger et al., 1974
	1	4.75 days	Data for 2.25 years, age 21–23 years	Hamburger et al., 1974

[a]Healthy unless indicated.
[b]Sample type, duration of sampling, additional subject information.

276   7. Sexuality and Reproduction

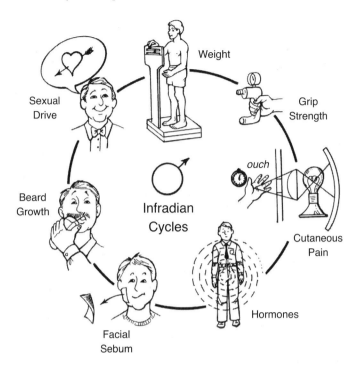

FIGURE 7.25. Examples of variables that display infradian rhythms in men.

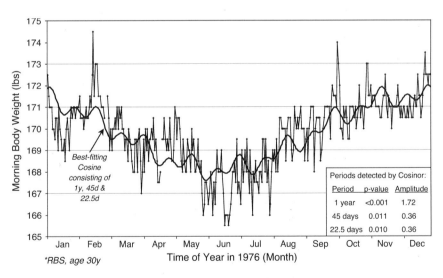

FIGURE 7.26. Infradian oscillations in morning body weight in a healthy man (RBS, age 30 years). The 1-year amplitude is nearly five times that found for periods of 45 days and 22.5 days (Sothern, unpublished).

Cutaneous Pain

Cutaneous pain threshold in women, which has been shown to be higher in the first half of menstrual cycles, continues to cycle in some post-menopausal women with a period of about 30 days (Procacci et al., 1974). When men ages 19–34 years were tested every day for 40–100 days for cutaneous pain threshold caused by heat, a 25-day rhythm was reported (Procacci et al., 1974). Even though the studies were performed two years apart, the authors felt the peaks (acrophases) at 25 days were synchronous. In the same study, an average period of 32 days for pain threshold was found in a group of seven older men, ages 56–82 years, measured every 3 days for 75 days.

Hormones

Statistical analysis of urinary 17-ketosteroids[42] collected and assayed from a single subject[43] over a span of 15 years has revealed a rhythmic component in the 17–21 days region, with other periodicities found around 7 days, 30 days, and 1 year (Halberg et al., 1965). There appears to be support for the possible presence of infradian periods in other variables as well, such as an 8–10-day cycle in urinary estrogen (total amount of estrone, estradiol, and estriol) and 17-oxosteroid excretion (Exley & Corker, 1966). The main male sex hormone, testosterone, shows an annual cycle (Dabbs, 1990), as well as infradian cycles with periods ranging from a few days to about 3 weeks (see Essay 7.7).

> ### Essay 7.7 (by RBS): *More About Infradians in Male Hormones*
>
> Exley and Corker (1965) suggested that there was a similarity in period (9±1 days) for urinary estrogen and steroid excretion and in their phase in relation to each other for all subjects, even though the men studied were of different ages (24–50 years) and differed in both their overall hormone concentrations and amplitudes. After considering the two possible sources of this cyclical phenomenon, the authors ruled out the steroids of adrenal origin and concluded that the most likely source of the estrogen cycles was the testosterone metabolites, thought to be secreted by Leydig cells in the gonads. Thus, these authors envisioned the possibility of a sex cycle in the human male and went on to hypothesize a correlation between cycles of the human male and female. Noting that both gonadotropins LH and FSH are present and assist in the sexual process of both sexes, they reviewed patterns of urinary LH and FSH excretion in published data related to the female menstrual cycle and noted that three peaks of greater and lesser amplitude with periods of 8–11 days were consistently evident, corresponding to ovulation, the luteal maximum and menstruation. These changes in female hormone activity were interpreted by the authors as corresponding to the cycle lengths in their study of men

---

[42] About two-thirds of the total 17-ketosteroids represent metabolites originating from the adrenals and the remaining one-third are of testicular origin (Halberg & Hamburger, 1964).

[43] The renowned Danish gonadotropins researcher, Dr. Christian Hamburger (1904–2002), is one of the few known data sources for a study of low-frequency hormonal rhythms in the male.

and suggested to them the existence of a neural regulation of human male hormones similar to that regulating the female's hormonal cycle.

These authors continued this line of thought and presented results of a longitudinal study of the daily 24-h excretion of urinary testosterone for 42 days in two healthy men in whom they sought similar evidence of cycling (Corker & Exley, 1968). They noted a large variation in the testosterone output from day to day. Subject A (married) had no clear 8–10-days cycle, although macroscopic inspection suggested the occurrence of fluctuations with a period of about 20 days. Subject B (single) had an average interval of about 9 days between peaks, in agreement with earlier results by the same authors, who concluded that the probable cyclic production of testosterone is not easily uncovered due to the complex metabolism of this hormone from origin to urine. One of us (RBS, age 26 years) also collected urine samples for 60 days during beard growth collection (see below) in order to look at 24-h testosterone excretion and multiple oscillations were found, with periods of 4.5 days and 8.5 days best describing them. In nine men studied over spans of 10–45 days, testosterone peaks in urine fell into two groups, with average period lengths between 3-and 5 days and between 12-and 18 days (Harkness, 1974). When testosterone levels in blood were studied every 2nd day for 2 months in 20 healthy men, a cluster of cycles was found for periods around 20–22 days (Doering et al., 1975). These authors hypothesized that cycles in testosterone of 20–30 days might approximate the female menstrual cycle. A more recent study of salivary testosterone levels in 53 young men and women reported a peak near the periovulatory phase of the women's menstrual cycle and an "analogical" cycle in men, concluding that there are infradian ("circalunar") fluctuations in both sexes (Celec et al., 2002).

Others have also suggested similar control mechanisms for both the obvious ovarian cycle and the less obvious changes in the male "sex" cycle. An infradian rhythm in the frequency of androgen-induced nucleus C-appendages (small clubs, hooks, threads, and the like) of neutrophil leukocytes has been reported (Månson, 1965). It is pertinent that these C-appendages increase in frequency with sexual maturity in the male rat (Zsifkovits et al., 1959). Development of their frequency in the adult animal can be accelerated by administering testosterone propionate, while castration of adult male rats reduces the frequency of these appendages. Moreover, progesterone prevents the frequency of C-appendages from increasing, and when estrone is given with progesterone, the frequency of these appendages reportedly decreases (Jobst, 1960). In women, testosterone propionate leads to an increase in C-appendages (Pfeifer, 1962). Our cosinor analysis found a best-fitting period of 27.2 days in neutrophil C-appendage data collected every 3 days over 4.5 months from a 33-yearr-old man, in agreement with the author's interpretation that his data revealed *"a highly probable cyclic variation"* with an average period of 27.2 days in the frequency of neutrophils with one or more appendages (Manson, 1965). [We also analyzed data for the total number of C-appendages and found a best-fitting period of 27.7 days (Figure 7.27).] The author interpreted his results as seeming to indicate *"that there is a male sexual cycle, four weeks in length."*

Emotions

Low-frequency rhythms in emotional states, such as those encountered in alternating mania and depression, may give clues to the normal functions of the neuroendocrines, notably, the hypothalamic-pituitary-adrenal-gonadal interactions. However, apart from circadian and circannual information, results on infradian

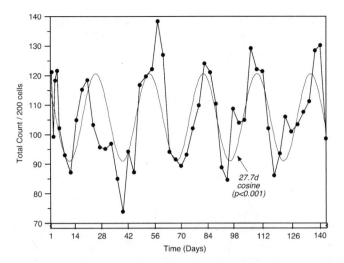

FIGURE 7.27. A "monthly" (~27–28 days) rhythm in androgen-induced (and therefore presumably sex-related) count of nucleus C-appendages (e.g., clubs, hooks, and threads) in blood neutrophils of a healthy man measured every 3 days for 143 days (our analysis of data from Figure 2 in Mänson, 1965).

variations in mood in men is lacking.[44] A 38-day cycle in mental state ratings correlated with eosinophil counts and excretion of 17-ketosteroids in a 24-year-old man with manic-depressive psychosis who was studied for over 3 months, with all variables apparently reaching peaks at the same time (Bryson & Martin, 1954). More extensive data were obtained on a 22-year-old man suffering from periodic catatonia with a recurrent pattern of 21 days (Gjessing, 1966). Cosinor analysis of these data mapped an about 20-day system, with the stupor state associated with peaks of temperature, chloride, body weight, urine volume, pH, and respiratory quotient, whereas urine phosphoric acid, nitrogen output, SO_4, CO_2, hemoglobin, white blood cells, titrable acid, O_2 consumption, systolic and diastolic blood pressure, pulse and motor activity all reached peaks during or before the excitement stage (Simpson et al., 1974). One healthy young man found periods of 16 days and 38 days when mood was rated 5–6 times each day for 7.5 months (Sothern, 1974).

Facial Sebum

When facial fatty oil (sebum) excretion was studied in two men for 33 and 47 days, cycles were found with peaks 10–15 days after the broad peak of testosterone excretion measured at the same time, indicating sebum excretion as a possible target function and its potential usefulness as a bioassay for androgens in

---

[44] Most young men (and women) in our biological rhythms course rate their mood and vigor for about 3 days as part of an exercise in monitoring their own body rhythms. While a circadian rhythm in mood is almost always found, the ratings do not continue long enough to test for infradian periods.

Beard Growth and Body Hair

Beard growth is often associated with male hormones. Beard growth collected daily by a 26-year-old man over a span of 60 days showed rhythms of 20 days and 6 days. Urinary testosterone was also determined for each 24-h span, and periods of 8.5 days and 4.5 days were found. While no direct correlation between cycles in testosterone and beard growth could be seen, the overall levels of these two variables seemed to be inversely related over the 60-day observation span (Figure 7.28) (Sothern, unpublished observations). It is possible that beard growth may be an unheeded parameter of hormonal activity in the pituitary-adrenal-gonadal circuit. That the beard is a secondary sex characteristic of the human male is attested to by the observation that under normal conditions it achieves full development in mature males, being dependent upon gonadal secretions for development and in some instances for maintenance. For example, little beard growth, if any, is seen in males who had been castrated before the age of 16 years, with a progressively heavier beard the longer the interval between sexual maturation and castration, though significantly less heavy than in intact males (Hamilton, 1958). Stimulation of whisker growth has been reported by androgenic treatment in eunuchs, as well as in older men, suggesting that growth of the beard sums up various stimuli impinging upon the organism. Thus, gonadal endocrine stimulation probably contributes to endogenous beard growth stimulation, although it may not be the sole or invariably most critical determinant thereof.

It has been suggested that male sexual hair, which is dependent upon levels of steroid hormones, is a remnant of the once elaborate display of the mature male, serving the purpose of sexual attraction (Rook, 1965). Indeed, cyclic activity of the hair follicle has led some researchers to suggest it as a relic of moulting, originally linked to the reproductive cycle and the seasons and still observable in wild animals, but now freed from such association in domesticated animals and in humans (Ebling, 1965; Rook, 1965). Nevertheless, it has been suggested that beard growth may offer an opportunity to study circadian and infradian rhythms in man (Chaykin, 1986).

An anonymous investigator in 1970 was led to suspect that resumption of sexual activity after isolation on a remote island was the stimulus for his reported increase in beard growth (Anon, 1970). Another author dissociated beard growth from sexual activity determined by the ovulatory phase of the sexual partner, since his wife used an oral contraceptive on an exact 28-day schedule and his collection of 24-h beard weights for 6 months revealed a 33-day cycle (Kihlstrom, 1971).

One of us (RBS) at the age 24 years reported a 16.5-day rhythm in 7 months of daily beard growth (Sothern, 1974). Analysis of the first 60 days of this series detected oscillations at two frequencies: about 6 days and 18 days (Figure 7.29). A close look at the peaks in these data indicates that the cycle lengths fluctuated

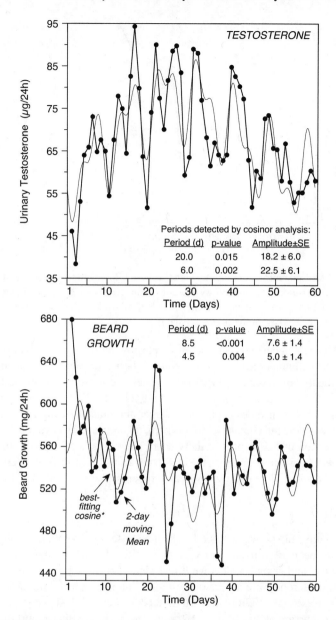

FIGURE 7.28. Multiple periods found in urinary testosterone (μg/24 h) and 24-h beard weights collected for 60 days (January 11–March 11, 1973) by a healthy man (RBS, age 27 years) (Sothern, unpublished).

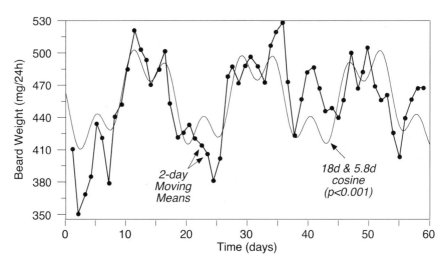

FIGURE 7.29. Multiple infradian rhythms in the region of 20 days and 6 days found in 24-h beard growth measured over 60 days by a healthy man (RBS, age 24 years). Twenty-four-hour rates and 2-day moving means were calculated to adjust for intervals longer or shorter than 24 h between shaves (Sothern, unpublished).

considerably, with underlying periodicities quite obvious though unstable, as can be seen from the still numerous deviations from the best-fitting curve. When this time series eventually covered 2.25 years and was analyzed for rhythms, the best-fitting period was 24 days with an amplitude of 7.2 mg/24 h. In addition, a yearly rhythm in beard growth was also detected, with a larger amplitude (9.1 mg/24 h and with highest values in the fall (Figure 7.30) (Sothern, unpublished observations). Macroscopic inspection of 5.5 months of daily beard weights of a 55-year-old man also reveals several oscillations in the range of 4–6 days and 13–15 days, in addition to a period of 30 days or more that was originally reported (Levine et al., 1974).

Sexual Activity

No clear relationship between blood levels of testosterone and human male sexual activity has been established (Fox et al., 1972). When plasma testosterone concentrations and sexual outlets were recorded over 45 days in two men, peak values in testosterone were detected at irregular intervals which bore no apparent relationship to sexual activity—testosterone peaked every 10–14 days, with sexual outlets occurring at an average of 3.5-day intervals. Another report on a man in his 50s and another man in his 30s reported average periods of 7 days and 3.5 days, respectively, in sexual outlets (Hamburger et al., 1985).

In another study concerning sexual activity and urinary steroids (estrone, 17-oxysteroids and 17-hydroxycorticoids), five males were studied for 90 days (Dewhurst, 1969). Dominant cycles of 9±1 days in all steroid excretions were noted, yet no relationship between these cycles and sexual activity on the same day or on up to 10 days in either direction could be detected. Even though the data showed a periodicity in male hormone production, the author concluded that while sex hormones trigger

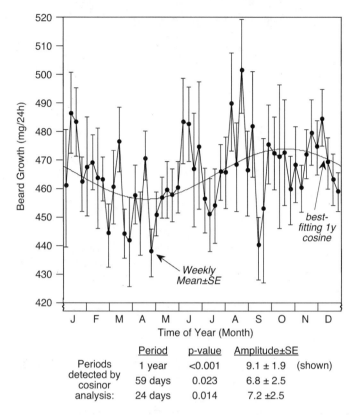

FIGURE 7.30. Low-frequency rhythms in 24-h beard growth of a young man (RBS, age 24 years at start), who collected 723 morning beard weights over 2.25 years (1971-03-02–1973-06-01) (Sothern, unpublished).

sex behavior patterns in lower animals, sexual desire in the human male bears no direct relationship to the quantity of hormone, as any excess is excreted.

Conceptions in humans seem to be more frequent at certain times of the year (Roenneberg & Aschoff, 1990a,b). In temperate climates, an increase in conceptions in the late fall and winter seemingly follows a peak in serum testosterone in men in the fall and winter (Dabbs, 1990). Sperm production has also been reported to vary over the course of the year (Table 7.7) (Levine, 1999). Parameters such as sperm count, concentration, volume, and linear motility all were generally found to be higher in late winter and spring and lowest in the summer. These findings strongly suggest that human testicular function is influenced by time of year. Since the cells in the testis that undergo meiosis take about a week for prophase I and about a month for the entire meiotic cycle (Purves et al., 1998), it is possible that sperm parameters may also undergo a monthly cycle. In addition, a study of 54 men limited to two times of day (7 AM vs. 5 PM) found higher number and concentration of sperm and greater number of spermatozoa with progressive linear motility in the afternoon than the morning, suggesting a: *"diurnal rhythm in sperm quality....that may prove useful for spontaneous and assisted conceptions"* (Cagnacci et al., 1999).

TABLE 7.7. Annual peak times for semen quality of men.

Semen Variable	No. of men (samples)	Time of high values	Time of low values	Reference
Sperm count, sperm concentration	4435 (4435)	February–March	September	Tjoa et al., 1982
Sperm count	1566 (1566)	Late winter, early spring	–	Mortimer et al., 1983
Volume, sperm density, total count	52 (110)	Late winter, spring	–	Spira, 1984
Young spermatids, sperm motility	1 (12)	June–July	–	Abbaticchio et al., 1987
Volume, sperm count, sperm motility	260 (260)	April–May	–	Reinberg et al., 1988
Sperm count	2697 (2697)	Winter–spring	Summer	Politoff et al., 1989
Count	4196 (4196)	Late winter, early spring	Summer	Saint Pol et al., 1989
Sperm concentration, total count, concentration of motile sperm	131 (262)	Winter	Summer	Levine et al., 1990
% Rapid sperm, progressive straight line velocity	2065 (2065)	Spring	–	Centola & Eberly, 1999
Concentration, total count	1927 (1927)	Spring	Summer	Gyllenborg et al., 1999

Take-Home Message

Sexuality and reproduction have a temporal organization, with variables that span the ultradian, circadian, and infradian domains of biological rhythms. These oscillations are present throughout a genetically-based structural hierarchy that extends from organelles to ecosystems. Knowledge of the rhythms of sexuality and reproduction help us to better understand the timing of various processes, activities and events associated with development, behavior, and society.

References

Abbaticchio G, de Fini M, Giagulli VA, Santoro G, Vendola G, Giorgino R. (1987) Circannual rhythms in reproductive functions of human males, correlations among hormones and hormone-dependent parameters. *Andrologia* 19(3): 353–361.

Adams DB, Gold AR, Burt AD. (1978) Rise in female-initiated activity at ovulation and its suppression by oral contraceptives. *New Engl J Med* 299(21): 1145–1150.

Albright DL, Voda AM, Smolensky MH, His B, Decker M. (1989) Circadian rhythms in hot flashes in natural and surgically induced menopause. *Chronobiol Intl* 6: 279–284.

Anon. (1970) Effects of sexual activity on beard growth in one man. *Nature* 226: 869–870.
Aronson BD, Johnson KA, Loros JJ, Dunlap JC. (1994) Negative feedback defining a circadian clock: autoregulation of the clock gene frequency. *Science* 263(5153): 1578–1584.
Ash Mr. (1685) An extract of *J of the Soc at Dublin* giving an account of a periodical evacuation of blood at the end of one of the fingers. *Philosoph Trans* 15: 989–990.
Baker FC, Waner JI, Vieira EF, Taylor SR, Driver HS, Mitchell D. (2001) Sleep and 24 hour body temperatures: a comparison in young men, naturally cycling women and women taking hormonal contraceptives. *J Physiol* 530(3): 565–574.
Barnhart K, Furman I, Devoto L. (1995) Attitudes and practice of couples regarding sexual relations during menses and spotting. *Contraception* 51(2): 93–98.
Berger PJ, Negus NC, Sanders EH, Gardner PD. (1981) Chemical triggering of reproduction in *Microtus montanus*. *Science* 214(4516): 69–70.
Bertram L, Karlsen P. (1994) Patterns in stem elongation rate in chrysanthemum and tomato plants in relation to irradiance and day-night temperature. *Sci Hort* 58: 139–150.
Bjarnason GA, Jordan R, Sothern RB. (1999) Circadian variation in the expression of cell-cycle proteins in human oral epithelium. *Amer J Path* 154(2): 613–622.
Bjarnason GA, Jordan R, Wood PA, Li Q, Lincoln D, Sothern RB, Hrushesky WJM, Ben-David Y. (2001) Circadian expression of clock genes in human oral mucosa and skin: association with specific cell cycle phases. *Amer J Path* 158(5): 1793–1801.
Blomquist CH, Holt Jr JP. (1992) Chronobiology of the hypothalamic-pituitary-adrenal axis in men and women. In: *Biologic Rhythms in Clinical and Laboratory Medicine*. Touitou Y, Haus E, eds. Berlin: Springer-Verlag, pp. 313–329.
Brandt WH. (1953) Zonation in a prolineless strain of *Neurospora*. *Mycologia* 45: 194–208.
Brink P. (1957) Reproductive system and mating in Ephemeroptera. *Opusc Entomol* 22(1): 1–37.
Brown WR. (1991) A review and mathematical analysis of circadian rhythms in cell proliferation in mouse, rat, and human epidermis. *J Invest Dermatol* 97: 273–280.
Bryson RW, Martin DF. (1954) 17-Ketosteroid excretion in a case of manic-depressive psychosis. *Lancet* 267(6834): 365–367.
Buchi KN, Moore JG, Hrushesky WJ, Sothern RB, Rubin NH. (1991) Circadian rhythm of cellular proliferation in the human rectal mucosa. *Gastroenterol* 101(2): 410–415.
Bünning E. (1932) Über die Erblichkeit der tagesperiodizität bei den Phaseolus-Blättern. *Jahrbücher Wiss Bot* 77: 283–320.
Bünning E. (1935) Zur Kenntnis der erblichen Tagesperiodizität bei den Primärblättern von Phaseolus multiflorus. In: *Jahrbücher für Wissenschaftliche Botanik*. Leipzig: Gebrüder Borntraeger, pp. 411–418.
Bünning E, Müssle L (1951) Der Verlauf der endogenen Jahresrhythmik in Samen unter dem Einfluss verschiedenartiger Aussenfaktoren. *Z Naturforsch* 6b: 108–112.
Bünning E. (1962) Mechanism in circadian rhythms: functional and pathological changes resulting from beats and from rhythm abnormalities. In: *Rhythmic Functions in the Living System* (Wolf W, ed). *Ann NY Acad Sci* 98(4): 901–915.
Burton JL, Cunliffe WJ, Shuster S. (1970) Circadian rhythm in sebum excretion. *Brit J Dermatol* 82(5): 497–501.
Cagnacci A, Maxia N, Volpe A. (1999) Diurnal variation of semen quality in human males. *Human Reprod* 14(1): 106–109.
Campbell CJ, Shideler SE, Todd HE, Lasley BL. (2001) Fecal analysis of ovarian cycles in female black-handed spider monkeys (*Ateles geoffroyi*). *Amer J Primatol* 54(2): 79–89.
Celec P, Ostatnikova D, Putz Z, Kudels M. (2002) The circalunar cycle of salivary testosterone and the visual-spatial performance. *Bratisl Lek Listy* 103(2): 59–69.

Centola GM, Eberly S. (1999) Seasonal variations and age-related changes in human sperm count, motility, motion parameters, morphology, and white blood cell concentration. *Fertil Steril* 72(5): 803–808.

Chaykin S. (1986) Beard growth: a window for observing circadian and infradian rhythms of men. *Chronobiologia* 13(2): 163–165.

Clair P, Claustrat B, Jordan D, Dechaud H, Sassolas G. (1985) Daily variations of plasma sex hormone-binding globulin binding capacity, testosterone and luteinizing hormone concentrations in healthy rested adult males. *Horm Res* 21(4): 220–223.

Cooke HG. (1940) Observations on mating flights of the mayfly *Stenonemavicarium* (Ephemerida). *Entmol News* 51: 12–14.

Corker CS, Exley D. (1968) Daily changes in urinary testosterone levels of the human male. *J Endocrinol* 40(2): 255–256.

Crafts R, Llerena LA, Guevara A, Lobotsky J, Lloyd CW. (1968) Plasma androgens and 17-hydroxycorticosteroids throughout the day in submarine personnel. *Steroids* 12(1): 151–163.

Creinin MD, Keverline S, Meyn LA. (2004) How regular is regular? An analysis of menstrual cycle regularity. *Contraception* 70(4): 289–292.

Crews D. (1994) Animal sexuality. *Sci Amer* 270: 108–114.

Cumming BG. (1967) Circadian rhythmic flowering responses in *Chenopodium rubrum*: effects of glucose and sucrose. *Can J Bot* 45: 2173–2193.

Cumming BG. (1968) *Chenopodium rubrum* L. and related species. In: *The Induction of Flowering. Some Case Histories*. Evans LT, ed. Ithaca, NY: Cornell Univ Press, pp. 157–185.

Cutler WB. (1980) Lunar and menstrual phase locking. *Amer J Obstet Gynecol* 137: 834–839.

Cutler WB, Friedman E, McCoy NL. (1996) Coitus and menstruation in perimenopausal women. *J Psychosom Obst Gynecol* 17(3): 149–157.

Dabbs JM Jr. (1990) Age and seasonal variation in serum testosterone concentration among men. *Chronobiol Intl* 7(3): 245–249.

Dahl KD, Czekala NM, Lim P, Hsueh AJ. (1987) Monitoring the menstrual cycle of humans and lowland gorillas based on urinary profiles of bioactive follicle-stimulating hormone and steroid metabolites. *J Clin Endocrinol Metab* 64(3): 486–493.

Davis RH, Perkins DD. (2002) Timeline: Neurospora: a model of model microbes. *Nat Rev Genet* 3(5): 397–403.

Dewan EM. (1967) On the possibility of a perfect rhythm method of birth control by periodic light stimulation. *Amer J Obst Gynecol* 99(7): 1016–1019.

Dewhurst K. (1969) Sexual activity and urinary steroids in man with special reference to male homosexuality. *Brit J Psychiat* 115(529): 1413–1415.

Doering CH, Kraemer HC, Brodie KH, Hamburg DA. (1975) A cycle in plasma testosterone in the human male. *J Clin Endocrinol Met* 40: 492–500.

Dunson DB, Baird DD, Wilcox AJ, Weinberg CR. (1999) Day-specific probabilities of clinical pregnancy based on two studies with imperfect measures of ovulation. *Human Reprod* 14(7): 1835–1839.

Dyrenfurth I, Jewelewicz R, Warren M, Ferin M, Vande Wiele RL. (1974) Temporal relationships of hormonal variables in the menstrual cycle. In: *Biorhythms and Human Reproduction*. Ferin M, Halberg F, Richert RM, Vande Wiele R, eds. New York: John Wiley and Sons, Inc, pp. 171–201.

Ebling FJ. (1965) Factors affecting periodicity of hair follicles. In: *Biology of the Skin and Hair Growth*. Lyne AG, Short BF, eds. New York: Elsevier: 806 pp,

Edmunds GF Jr, Traver JR. (1954) The flight mechanics and evolution of the wings of Ephemeroptera, with notes on the archetype insect wing. *J Wash Acad Sci* 44(12): 390–400.

Ehret CF. (1980) On circadian cybernetics, and the innate and genetic nature of circadian rhythms. In: *Chronobiology: Principles and Applications to Shifts in Schedules*. NATO

Advanced Study Institutes Series D(3). Scheving LE, Halberg F, eds. Alphen aan den Rijn: Sijthoff & Noordhoff: 109–125.
Elfving DC, Kaufmann MR. (1972) Diurnal and seasonal effects of environment on plant water relations and fruit diameter of citrus. *J Amer Soc Hort Sci* 97(5): 566–570.
Engel P, Hildebrandt G. (1974) Rhythmic variations in reaction time, heart rate, and blood pressure at different durations of the menstrual cycle. In: *Biorhythms and Human Reproduction*. Ferin M, Halberg F, Richert RM, Vande Wiele R, eds. New York: John Wiley & Sons, Inc., pp. 325–333.
Exley D, Corker CS. (1966) The human male cycle of urinary oestrone and 17-oxosteroids. *J Endocrinol* 35(1): 83–99.
Faiman C, Ryan RJ. (1967) Diurnal cycle in serum concentrations of follicle-stimulating hormone in men. *Nature* 215(103): 857.
Fidanza F. (1989) The "De Medicina Statica" by Santorio Santorio. *Nutrition* 5(4): 227–228.
Filer RB, Wu CH. (1989) Coitus during menses. Its effect on endometriosis and pelvic inflammatory disease. *J Reprod Med* 34(11): 887–890.
Fincham JRS, Day PR. (1965) *Fungal Genetics*, 2nd edn. Oxford: Blackwell, 326 pp.
Fisher C, Gross J, Zuch J. (1965) Cycle of penile erection synchronous with dreaming (REM) sleep. *Arch Gen Psychiatry* 12: 29–45.
Fisher LB. (1968) The diurnal mitotic cycle in the human epidermis. *Brit J Dermatol* 80: 75–80.
Fox CA, Ismail AA, Love DN, Kirkham KE, Loraine JA. (1972) Studies on the relationship between plasma testosterone levels and human sexual activity. *J Endocrinol* 52(1): 51–58.
Freedman RR, Norton D, Woodward S, Cornélissen G. (1995) Core body temperature and circadian rhythm of hot flashes in menopausal women. *J Clin Endocrinol Metab* 80(8): 2354–2358.
Giacomoni M, Falgairette G. (1999) Influence of menstrual cycle phase and oral contraceptive use on the time-of-day effect on maximal anaerobic power. *Biol Rhythm Res* 30(5): 583–591.
Gjessing R. (1966) Beiträge zur Kenntnis der Pathophysiologie des katatonen Stupors. Mitteilung III: Über periodisch rezdivierende katatone Erregung, mit Kritischem Beginn und Abschluss. *Arch Psychiatr* 104: 355–416.
Goldstein I and Working Group Study Central Mechanisms Erectile Dysfunction. (2000) Male sexual circuitry. *Sci Amer* 283: 70–75.
Gooch VD, Freeman L, Lakin-Thomas PL. (2004) Time-lapse analysis of the circadian rhythms of conidiation and growth rate in *Neurospora*. *J Biol Rhythms* 19(6): 493–503.
Graham CA, McGrew WC. (1980) Menstrual synchrony in female undergraduates living on a coeducational campus. *Psychoneuroimmunol* 5: 245–252.
Graham CA, Janssen E, Sanders SA. (2000) Effects of fragrance on female sexual arousal and mood across the menstrual cycle. *Psychophysiol* 37: 76–84.
Grassly NC, Fraser C, Garnett GP. (2005) Host immunity and synchronized epidemics of syphilis across the United States. *Nature* 433(7024): 417–421.
Gyllenborg J, Skakkebaek NE, Nielsen NC, Keiding N, Giwercman A. (1999) Secular and seasonal changes in semen quality among young Danish men: a statistical analysis of semen samples from 1927 donor candidates during 1977–1995. *Intl J Androl* 22(1): 28–36.
Halberg F, Hamburger C. (1964) 17-Ketosteroid and volume of human urine. *Minn Med* 47: 916–925.
Halberg F, Engeli M, Hamburger C, Hillman D. (1965) Spectral resolution of low-frequency small-amplitude rhythms in excreted 17-ketosteroid; probable androgen induced circaseptan desynchronization. *Acta Endocrinol (Kbh)* 103(Suppl): 5–54.
Hall J. (1994) The mating of a fly. *Science* 264(5166): 1702–1714 (Review).
Hamburger C, Sothern RB, Halberg F. (1985) Circaseptan and circasemiseptan aspects of human male sexual activity (Abstract). *Chronobiologia* 12: 250.

Hamilton JB. (1958) Age, sex and genetic factors in the regulation of hair growth in man: a comparison of Caucasian and Japanese populations. In: *The Biology of Hair Growth*. Montagna W, ed. New York: Academic Press, pp. 399–433.

Hamner KC. (1960) Photoperiodism and circadian rhythms. In: *Biological Clocks. Cold Spring Harbor Symposia on Quantitative Biology*, Vol 25. New York: Long Island Biol Assoc, pp. 269–277.

Hardin PE, Hall JC, Rosbash M. (1992) Circadian oscillations in period gene mRNA levels are transcriptionally regulated. *Proc Natl Acad Sci USA* 89(24): 11711–11715.

Harkness RS. (1974) Variations in testosterone excretion by man. In: *Biorhythms and Human Reproduction*. Ferin M, Halberg F, Richert RM, Vande Wiele R, eds. New York: John Wiley & Sons, Inc., pp. 469–478.

Harter L, Erkert HG. (1993) Alteration of circadian period length does not influence the ovarian cycle length in common marmosets, *Callithrix j jacchus* (primates). *Chronobiol Intl* 10(3): 165–175.

Hattori A, Migitaka H, Iigo M, Itoh M, Yamamoto K, Ohtani-Kaneko R, Hara M, Suzuki T, Reiter RJ. (1995) Identification of melatonin in plants and its effects on plasma melatonin levels and binding to melatonin receptors in vertebrates. *Biochem Mol Biol Intl* 35(3): 627–634.

Haus E, Nicolau GY, Lakatua D, Sackett-Lundeen L. (1988) Reference values for chronopharmacology. In: *Annual Review of Chronopharmacology* Vol 4. Reinberg A, Smolensky M, Labrecque G, eds. Oxford: Pergamon, pp. 333–424.

Haus E, Touitou I. (1992) Principles of clinical chronobiology. In: *Biologic Rhythms in Clinical and Laboratory Medicine*. Touitou Y, Haus E, eds. Berlin: Springer-Verlag, pp. 6–34.

Hayes W. (1966a) Genetical Society Mendel Lecture: Sex factors and viruses. *Proc Royal Soc (Series B)* 164: 230–245.

Hayes W. (1966b) The Leeuwenhoek Lecture, 1965. Some controversial aspects of bacterial sexuality. *Proc Royal Soc* (Series B) 165: 1–19.

Hermida RC. (1999) Time-qualified reference values for 24h ambulatory blood pressure monitoring. *Blood Press Monitoring* 4(3–4): 137–147.

Herrick JR, Agoramoorthy G, Rudran R, Harder JD. (2000) Urinary progesterone in free-ranging red howler monkeys (*Alouatta seniculus*): preliminary observations of the estrous cycle and gestation. *Amer J Primatol* 51(4): 257–263.

Hillman WS. (1976) Biological rhythms and physiological timing. *Ann Rev Plant Physiol* 27: 159–179.

Holdaway-Clarke TL, Feijo JA, Hackett GR, Kunkel JG, Hepler PK. (1997) Pollen tube growth and the intracellular cytosolic calcium gradient oscillate in phase while extracellular calcium influx is delayed. *The Plant Cell* 9: 1999–2010.

Iwasaki K, Thomas JH. (1997) Genetics in rhythm. *Trends in Genetics* 13: 111–115.

Jacob S, McClintock MK. (2000) Psychological state and mood effects of steroidal chemosignals in women and men. *Horm Behav* 37(1): 57–78.

Jenkin G, Mitchell MD, Hopkins P, Matthews CD, Thorburn GD. (1980) Concentrations of melatonin in the plasma of the rhesus monkey (*Macaca mulatta*). *J Endocrinol* 84(3): 489–494.

Jobst K, Mehes K, Zsifkovits S. (1960) [On the hormonal relations of sex specific nuclear characteristics of leukocytes of rats.] [German]. *Endokrinologie* 39: 277–282.

Kalb C. (2003) Farewell to 'Aunt Flo.' *Newsweek* Feb. 3, p. 48.

Kanabrocki EL, Scheving LE, Halberg F, Brewer RL, Bird TJ. (1973) *Circadian Variations in Presumably Healthy Young Soldiers*. Springfield, VA: Natl Tech Inform Service, US Dept. Commerce, 56 pp.

Kanabrocki EL, Sothern RB, Scheving LE, Vesely DL, Tsai TH, Shelstad J, Cournoyer C, Greco J, Mermall H, Nemchausky BM, Bushnell DL, Kaplan E, Kahn S, Augustine G, Holmes E, Rumbyrt J, Sturtevant RP, Sturtevant F, Bremner F, Third JLHC, McCormick JB, Mudd CA, Dawson S, Sackett-Lundeen L, Haus E, Halberg F, Pauly JE, Olwin JH. (1990) Reference values for circadian rhythms of 98 variables in clinically healthy men in fifth decade of life. *Chronobiol Intl* 7(5/6): 445–461.

Karacan I, Salis PJ, Thornby JI, Williams RL. (1976) The ontogeny of nocturnal penile tumescence. *Waking and Sleeping* 1: 27–44.

Kellicott WE. (1904) The daily periodicity of cell-division and of elongation in the root of Allium. *Bull Torrey Bot Club* 31: 529–550.

Kihlstrom JE. (1971) A monthly variation in beard growth in one man. *Life Sciences* 10(6): 321–324.

King RW. (1975) Multiple circadian rhythms regulate photoperiodic flowering responses in *Chenopodium rubrum*. *Can J Bot* 53: 2631–2638.

Konopka RJ, Benzer S. (1971) Clock mutants of *Drosophila melanogaster*. *Proc Natl Acad Sci USA* 68: 2112–2116.

Koukkari WL, Bingham C, Hobb, JD, Duke SH. (1997) In search of a biological hour. *J Plant Physiol* 151: 352–357.

Kühl JFW, Lee JK, Halberg F, Haus E, Günther R, Knapp E. (1974) Circadian and lower frequency rhythms in male grip strength and body weight. In: *Biorhythms and Human Reproduction*. Ferin M, Halberg F, Richert RM, Vande Wiele R, eds. New York: John Wiley & Sons, Inc., pp. 529–548.

Kyriacou CP, Hall JC. (1980) Circadian rhythm mutations in *Drosophila melanogaster* affect short-term fluctuations in the male's courtship song. *Proc Natl Acad Sci USA* 77(11): 6729–6733.

Lagoguey M, Dray F, Chauffournier JM, Reinberg A. (1972) [Circadian and circannual rhythms of testosterone and epitestosterone glucuronides in healthy human adults.] [French]. *Comptes Rendus Hebdomadaires des Seances de l Academie des Sciences - D: Sciences Naturelles* 274(25): 3435–3437.

Lein A. (1979) *The Cycling Female. Her Menstrual Rhythm*. San Francisco: WH Freeman, 135 pp.

Leonard L, Ross MW. (1997) The last sexual encounter: the contextualization of sexual risk behavior. *Intl J STD & AIDS* 8(10): 643–645.

Levine H, Halberg F, Sothern RB, Bartter FC, Meyer WJ, Delea CS. (1974) Circadian phase-shifting with and without geographic displacement. In: *Biorhythms and Human Reproduction*. Ferin M, Halberg F, Richert RM, Vande Wiele R, eds. New York: John Wiley & Sons, Inc., pp. 557–574.

Levine RJ, Mathew RM, Chenault CB, Brown MH, Hurtt ME, Bentley KS, Mohr KL, Working PK. (1990) Differences in the quality of semen in outdoor workers during summer and winter. *New Engl J Med* 323: 12–16.

Levine RJ. (1999) Seasonal variation of semen quality and fertility. *Scand J Work, Environ Health* 25 (Suppl 1): 34–37 (Review); Discussion 76–78.

Leyendecker G, Saxena BB. (1970) [Daily fluctuation of the follicle-stimulating (FSH) and luteinizing (LH) hormone in the human plasma.] [German]. *Klin Woch* 48(4): 236–238.

Lin MC, Kripke DF, Parry BL, Berga SL. (1990) Night light alters menstrual cycles. *Psychiat Res* 33: 135–138.

Ludwig H, Hinze E, Junges W. (1982) Endogene Rhythmen des Keimverhaltens der Samen von Kartoffeln, insbesondere von *Solanum acaule*. *Seed Sci Technol* 10: 77–86.

Mahoney MM, Smale L. (2005) A daily rhythm in mating behavior in a diurnal murid rodent *Arvicanthis niloticus*. *Horm Behav* 47(1): 8–13.

Månson JC. (1965) Cyclic variations of the frequency of neutrophil leucocytes with "androgen induced" nucleus appendages in an adult man. *Life Sci* 4: 329–334.

Mauer AM. (1965) Diurnal variation of proliferative activity in the human bone marrow. *Blood* 26: 1–7.

McClintock MK. (1971) Menstrual synchrony and suppression. *Nature* 229: 244–245.

McClintock MK. (1984) Estrous synchrony: modulation of ovarian cycle length by female pheromones. *Physiol & Behav* 32(5): 701–705.

McClintock MK. (1998) Whither menstrual synchrony? *Ann Rev Sex Res* 9: 77–95.

McClintock MK. (1998) On the nature of mammalian and human pheromones. *Ann NY Acad Sci* 855: 390–392.

Mendel G. (1865) [Experiments in plant hybridization.] [German]. *Verhandlungen Naturforsch Vereines in Brünn*, IV, pp. 3–47.

Mermall H, Sothern RB, Kanabrocki EL, Quadri SF, Bremner FW, Nemchausky BM, Scheving LE. (1995) Temporal (circadian) and functional relationship between prostate-specific antigen and testosterone in healthy men. *Urology* 46(1): 45–53.

Millet B, Bonnet B. (1990) Growth rhythms in higher plants. In: *Prog Clin Biol Res 341B: Chronobiology: Its Role in Clinical Medicine, General Biology, and Agriculture, Part B.* Hayes DK, Pauly JE, Reiter RJ, eds. New York: Wiley-Liss, Inc., pp. 835–851.

Morgan AH. (1913) A contribution to the biology of may-flies. *Ann Entomol Soc Amer* 6: 371–413.

Mortimer D, Templeton AA, Lenton EA, Coleman RA. (1983) Annual patterns of human sperm production and semen quality. *Arch Androl* 10(1): 1–5.

Nankin HR, Troen P. (1974) Oscillatory changes in LH secretion in men. In: *Biorhythms and Human Reproduction.* Ferin M, Halberg F, Richert RM, Vande Wiele R, eds. New York: John Wiley & Sons, Inc., pp. 457–468.

Negus NC, Berger PJ. (1977) Experimental triggering of reproduction in a natural population of *Microtus montanus*. *Science* 196(4295): 1230–1231.

Nixon EH, Markhart III AH, Koukkari WL. (1987) Stomatal aperture oscillations of *Abutilon theophrasti* Medic. and *Hordeum vulgare* L. examined by three techniques. In: *Advances in Chronobiology, Part A.* Pauly JE, Scheving LE, eds. New York: Alan R. Liss, Inc., pp. 67–80.

Nokin J, Vekemans M, L'Hermite M, Robyn C. (1972) Circadian periodicity of serum prolactin concentration in man. *Br Med J* 3(826): 561–562.

Overland L. (1960) Endogenous rhythm in opening and odor of flowers of *Cestrum nocturnum*. *Amer J Bot* 47: 378–382.

Palmer JD, Udry JR, Morris NM. (1982) Diurnal and weekly, but no lunar rhythms in humans copulation. *Hum Biol* 54(1): 111–121.

Pfeifer GW. (1962) Do androgens influence the nuclear sexing or polymorphic sexing? *Acta Cytol* 6: 122–126.

Politoff L, Birkhauser M, Almendral A, Zorn A. (1989) New data confirming a circannual rhythm in spermatogenesis. *Fertil Steril* 52(3): 486–489.

Presser HB. (1974) Temporal data relating to the human menstrual cycle. In: *Biorhythms and Human Reproduction.* Ferin M, Halberg F, Richert RM, Vande Wiele R, eds. New York: John Wiley & Sons, Inc., pp. 145–160.

Preti G, Cutler WB, Garcia CR, Huggins GR, Lawley HJ. (1986) Human axillary secretions influence women's menstrual cycles: the role of donor extract of females. *Horm Behav* 20: 474–482.

Procacci P, Della Corte M, Zoppi M, Maresca M. (1974) Rhythmic changes of the cutaneous pain threshold in man. A general review. *Chronobiologia* 1(1): 77–96.

Purves WK, Orians GH, Heller HC, Sadava D. (1998) *Life: The Science of Biology*, 5th edn. Sunderland, MA: Sinauer, 1243 pp.
Quadagno DM, Shubeita HE, Deck J, Francoeur D. (1981) Influence of male social contacts, exercise and all-female living conditions on the menstrual cycle. *Psychoneuroimmunol* 6(3): 239–244.
Reilly T, Binkley S. (1981) The menstrual rhythm. *Psychoneuroendocrinol* 6(2): 181–184.
Reilly T. (2000) The menstrual cycle and human performance. An overview. *Biol Rhythm Res* 31(1): 29–40.
Reinberg A, Ghata J. (1964) *Biological Rhythms*. New York: Walker, pp. 55.
Reinberg A, Lagoguey M. (1978) Circadian and circannual rhythms in sexual activity and plasma hormones (FSH, LH, testosterone) of five human males. *Arch Sex Behav* 7(1): 13–30.
Reinberg A, Smolensky MH, Hallek M, Smith KD, Steinberger E. (1988) Annual variation in semen characteristics and plasma hormone levels in men undergoing vasectomy. *Fertil Steril* 49(2): 309–315.
Reinberg AE, Touitou Y, Soudant E, Bernard D, Bazin R, Mechkouri M. (1996) Oral contraceptives alter circadian rhythm parameters of cortisol, melatonin, blood pressure, heart rate, skin blood flow, transepidermal water loss, and skin amino acids of healthy young women. *Chronobiol Intl* 13(3): 199–211.
Rienstein S, Dotan A, Avivi L, Ashkenazi I. (1998) Daily rhythms in male mice meiosis. *Chronobiol Intl* 15(1): 13–20.
Roenneberg T, Aschoff J. (1990a) Annual rhythm of human reproduction: I. Biology, sociology, or both. *J Biol Rhythms* 5: 195–216.
Roenneberg T, Aschoff J. (1990b) Annual rhythm of human reproduction: II. Environmental correlations. *J Biol Rhythms* 5: 217–239.
Rook A. (1965) Endocrine influences on hair growth. *Brit Med J* 5435: 609–614.
Ryner LC, Goodwin SF, Castrillon DH, Anand A, Villella A, Baker BS, Hall JC, Taylor BJ, Wasserman SA. (1996) Control of male sexual behavior and sexual orientation in *Drosophila* by the *fruitless* gene. *Cell* 87: 1079–1089.
Saint Pol P, Beuscart R, Leroy-Martin B, Hermand E, Jablonski W. (1989) Circannual rhythms of sperm parameters of fertile men. *Fertil Steril* 51(6): 1030–1033.
Sanders EH, Gardner PD, Berger PJ, Negus NC. (1981) 6-Methoxybenzoxazolinone: a plant derivative that stimulates reproduction in *Microtus montanus*. *Science* 214(4516): 67–69.
Scheving LE. (1959) Mitotic activity in the human epidermis. *Anat Rec* 135: 7–20.
Scheving LE. (1984) Chronobiology of cell proliferation in mammals. Implications for basic research and cancer chemotherapy. In: *Cell Cycle Clocks*. Edmunds Jr. LN, ed. New York: Marcel Dekker, pp. 455–500.
Shaikh AA, Naqvi RH, Shaikh SA. (1978) Concentrations of oestradiol-17beta and progesterone in the peripheral plasma of the cynomolgus monkey (*Macaca fascicularis*) in relation to the length of the menstrual cycle and its component phases. *J Endocrinol* 79(1): 1–7.
Short RV. (1984) Breast feeding. *Sci Amer* 250: 35–41.
Simpson HW, Gjessing L, Fleck A, Kühl J, Halberg F. (1974) Phase analysis of the somatic and mental variables of Gjessing's case 2484 of intermittent catatonia. In: *Chronobiology*. Scheving LE, Halberg F, Pauly JE, eds. Tokyo: Igaku Shoin Ltd., pp. 535–539.
Smaaland R, Laerum OD, Lote K, Sletvold O, Sothern RB, Bjerknes R. (1991) DNA synthesis in human bone marrow is circadian stage dependent. *Blood* 77(12): 2603–2611.
Smals AG, Kloppenborg PW, Benraad TJ. (1976) Circannual cycle in plasma testosterone levels in man. *J Clin Endocrinol Metab* 42(5): 979–982.

Smolensky MH. (1981) Chronobiologic factors related to the epidemiology of human reproduction. In: *Research on Fertility and Sterility.* Cortés-Prieto J, Campos da Paz A, Neves-e-Castro M, eds. Lancaster: MTP Press Ltd., pp. 157–181.

Smolensky M, Lamberg L. (2000) *The Body Clock Guide To Better Health.* New York: Henry Holt & Co., 428 pp.

Snustad DP, Simmons MJ. (2003) *Principles of Genetics,* 3rd edn. New York: John Wiley & Sons, Inc., 840 pp.

Sothern RB. (1974) Low frequency rhythms in the beard growth of a man. In: *Chronobiology.* Scheving LE, Halberg F, Pauly JE, eds. Tokyo: Igaku Shoin Ltd., pp. 241–244.

Sothern RB, Slover GPT, Morris RW. (1993) Circannual and menstrual rhythm characteristics in manic episodes and body temperature. *Biol Psychiat* 33(3): 194–203.

Spira A. (1984) Seasonal variations in sperm characteristics. *Arch Androl* 12(Suppl): 23–28.

Spruyt E, Verbelen J-P, De Greef JA. (1987) Expression of circaseptan and circannual rhythmicity in the imbibition of dry stored bean seeds. *Plant Physiol* 84: 707–710.

Spruyt E, Verbelen J-P, De Greef JA. (1988) Ultradian and circannual rhythmicity in germination of *Phaseolus* seeds. *J Plant Physiol* 132: 234–238.

Stern K, McClintock MK. (1998) Regulation of ovulation by human pheromones. *Nature* 392(6672): 177–179.

Sumaya CV, Marx J, Ullis K. (1980) Genital infections with herpes simplex virus in a university student population. *Sex Transm Dis* 7(1): 16–20.

Tanfer K, Aral SO. (1996) Sexual intercourse during menstruation and self-reported sexually transmitted disease history among women. *Sexually Transm Dis* 23(5): 395–401.

Takimoto A, Hamner KC. (1964) Effect of temperature and preconditioning on photoperiodic response of *Pharbitis nil. Plant Physiol* 39: 1024–1030.

Targum SD, Caputo KP, Ball SK. (1991) Menstrual cycle phase and psychiatric admissions. *J Affect Disord* 22(1–2): 49–53.

Tjoa WS, Smolensky MH, Hsi B, Steinberger E, Smith KD. (1982) Circannual rhythm in human sperm count revealed by serially independent sampling. *Fertil Steril* 38: 454–459.

Tournois J. (1914) Études sur la sexualité du Houblon. *Ann Sci Nat (Bot)* 19: 49–191.

Udry JR, Morris NM. (1970) Frequency of intercourse by day of the week. *J Sex Res* 6: 229–234.

Weydahl A, Sothern RB, Wetterberg L. (1998) Daily and yearly variation in salivary melatonin in the subarctic. In: *Biological Clocks. Mechanisms and Applications.* Touitou Y, ed. Amsterdam: Elsevier Science BV, pp. 333–336.

Wilcox AJ, Weinberg CR, Baird DD. (1995) Timing of sexual intercourse in relation to ovulation. Effects on the probability of conception, survival of the pregnancy, and sex of the baby. *New Engl J Med* 333(23): 1517–1521.

Wilcox AJ, Dunson D, Baird DD. (2000) The timing of the "fertile window" in the menstrual cycle: day specific estimates from a prospective study. *Brit Med J* 321(7271): 1259–1262.

Wright EM Jr, Bush DE. (1977) The reproductive cycle of the capuchin (*Cebus apella*). *Lab Anim Sci* 27(5 Pt 1): 651–654.

Zeiger E, Cardemil L. (1973) Cell kinetics, stomatal differentiation, and diurnal rhythm in *Allium cepa. Dev Biol* 32(1): 179–188.

Zsifkovits S, Mehes K, Jobst K. (1959) Effect of sexual maturation and castration on the sex chromatin pattern in the male rat. *Nature* 184: 1239–1240.

8
Natural Resources and Agriculture

> *"To every thing there is a season, and a time*
> *to every purpose under the heaven:*
> *A time to be born, and a time to die; a time to plant,*
> *and a time to pluck up that which is planted..."*
> —Ecclesiastes 3:1–3 (Bible, King James Version)

Introduction

Long before our current clocks and calendars came into use and the endogenous characteristics of biological rhythms became known, there was an emphasis already upon the timing of events (Ecclesiastes 3). Some of the earliest written records pertaining to the rhythmic nature of life are found within the domains of natural resources and agriculture. Very likely, early humans were aware of cyclic activity, such as seasonal migration, folding and unfolding of leaves and flowers, and activity and rest in animals. Most of these early observations of *cyclic* changes were probably viewed as passive responses to the cyclic environment, for it was not until the twentieth century that the scientific foundations of biological rhythms became established (see Chapter 3 on Time).

Applications and implications of biological rhythms abound in matters pertaining to natural resources and agriculture, two disciplines that share a number of unifying components. Natural resources deal with substances, products and organisms formed by "natural" means. Commonly cited examples include water, minerals, fossil fuels, and forests, living things and all that sustains them.[1] Most free-living undomesticated plants and animals are seen primarily as natural resources.[2] Typically in agriculture (*ager* = field, *cultura* = cultivation), the focus

---

[1] Included here are various species of organisms and the items they require for normal development and reproduction, such as essential mineral elements (e.g., P, K, Mg, Ca, Fe, etc.), water, gases (e.g., O_2, CO_2, N_2), and light. This list also includes items such as the energy sources for *chemosynthetic autotrophs* and the need of a dark span for normal development (see Chapter 2).

[2] Often, it is difficult to distinguish whether a species or an environmental issue falls within the domain of agriculture or natural resources. For example, the Department of Forestry at the University of Minnesota resides in the College of Natural Resources, but at the

is on the care, culture and management of domesticated plants and animals, as well as their related pests.

Regardless of whether the species is domesticated or undomesticated, prey or predator, native or foreign, free-living or in captivity, its physiology has a temporal organization of rhythmic components. Because different species of organisms coexist in the same location, various temporal phase relationships for behaviors and functions exist within specific habitats and ecosystems, helping to reduce conflict and enhance survival. Each of the sections in this chapter will focus on implications and applications of biological rhythms in natural resources and/or agriculture (Koukkari et al., 1990).

Photoperiodism

One of the more obvious methods to have an annual plant produce flowers a month or two earlier than normal would be to sow the seeds a month or two earlier. However, with some species or cultivars, sowing seeds earlier is not the answer, even when they are supplied with sufficient levels of nutrients, water, heat, and light. The missing factor is the timing of light and darkness,[3] a process known as *photoperiodism* (see Chapter 4 on Photoperiodism). Both the term photoperiodism and its practical applications in the production of food, fiber and fuel were introduced by scientists working with plants in an agricultural setting (see Essay 8.1). What was demonstrated for agricultural species (e.g., soybeans, tobacco) raised in cultivated fields and in greenhouses, provided, at least in part, clues to better understanding natural changes in the development of organisms that were correlated with the seasons of the year.

Essay 8.1 (by WLK): Photoperiodism as a Basic Principle of Biology and Its Applications

In the fields and buildings near Washington, DC, two scientists employed by the US Department of Agriculture, WW Garner and HA Allard conducted their classical studies with Maryland Mammoth tobacco and Biloxi soybeans (Garner & Allard, 1920). They had observed that during the summer months, Maryland Mammoth tobacco plants would grow to be 10–15 feet (3–4.6 meters) tall, but would not produce flowers. However, the plants would flower during the winter months when cultured without the use of artificial light in a greenhouse (Figure 8.1).

---

federal level, forestry is administered in the US Department of Agriculture. A state may have a Department of Natural Resources (DNR), which is responsible for the management of wildlife, such as fish, birds, deer, etc., while matters pertaining to domestic plants and animals are administered by a Department of Agriculture.

[3] The emphasis here is on the timing of light and darkness, although there are examples where the timing of water (Halaban & Hillman, 1970), nutrients (Halaban & Hillman, 1971), and temperature will influence flowering.

FIGURE 8.1. Photographs showing (1) early light-tight building (*top*) used to control the length of light and darkness received by plants; and (2) Maryland Mammoth tobacco plants (*bottom*) with which the process of photoperiodism was introduced (Garner & Allard, 1920). Plants maintained in greenhouses bloomed when on short days (*left*) but not on long days (*right*) (Photographs courtesy of USDA).

TABLE 8.1. Examples of events or variables present in the terrestrial, aerial, or aquatic environment that are associated with photoperiodism.

Environment	Animals	Plants
Terrestrial	Estrus cycle (female mammals)	Dormancy (trees)
Aerial	Migration (birds)	Pollen (flowers)
Aquatic	Spawning (fish)	Flowering (*Lemna* sp)

The situation with Biloxi soybeans[4] was equally astounding. The first blossoms appeared in September, regardless of whether the seeds had been sown in April, May, June, or July (Table 4.2 in Chapter 4 on Photoperiodism). Studies, first with Maryland Mammoth and soybeans, and later with various species of plants, ushered in a basic principle of biology known as *photoperiodism*. Although the concept had already been suggested by 1907 in relation to the migration of birds (cf. Rowan, 1926) and by 1910 in relation to flowering of plants (Tournois, 1914), Garner and Allard provided quantitative evidence for the process. Results from studies initiated in this Department of Agriculture laboratory on the duration and timing of light and darkness (*photoperiodism*) and subsequent contributions on the role of light quality on plant development (*photomorphogenesis*) and the pigment *phytochrome*, rank among the great contributions to science. The applications of *photoperiodism* have contributed significantly to our economy, bringing in billions of dollars in benefits (cf. Sage, 1992). Benefits derived from the applications of the knowledge of *photoperiodism* are worldwide in scope, evident in the production and availability of food, gifts of remembrance and thanksgiving, the color of foliage in gardens, parks and forests, the songs of birds, mating calls of animals, and the management and culture of organisms in a variety of settings.

For those who venture into gardens, parks and the great outdoors, the effects of *photoperiodism* are ever present, often seen or sensed by the presence and absence of certain species or by their stage of development, behavior, and/or color (Figure 8.2). Photoperiodic responses are evident on land, in the air, and in the water (Table 8.1). The migration of animals, changes in pelage (fur or hair) or feather colors, the dormancy of plants, and the induction of reproductive processes are among the more common events readily observed in nature (cf. Hendricks, 1956). During certain seasons of the year, our senses may detect odors that arise from various photoperiodically induced flowers or our immune system may detect pollen[5] released into the atmosphere.

Changes in the development and behavior of organisms that parallel the seasons are usually taken for granted. Perhaps it is only through human interference, such as transplanting or moving plants and animals to different latitudes or to buildings

---

[4] Garner & Allard (1920) used the soybean cultivar (Peking) earlier and included a number of other cultivars of soybean in their studies.

[5] Pollen (Latin for fine flour) is a collective term for a mass of mature microspores. Each microspore is commonly called a pollen grain. The pollen grain contains the male *gametophyte*, a structure that produces the two gametes (sperm) that are involved the process of fertilization (see Chapter 7 on Sexuality and Reproduction).

FIGURE 8.2. Example of photoperiodic response to short days for several animals in which their summer darker color has changed to white during winter. Photos (clockwise from upper left) of Arctic quail (*Lagopus mutus*), hare (*Lepus timidus*), and fox (*Alopex lagopus*) courtesy of Andi Weydahl, Alta College, Norway; photo of weasel (*Mustela frenata*) courtesy of James Ford Bell Museum of Natural History, University of Minnesota

or controlled environmental chambers, that the importance of *photoperiodism* is appreciated. *Photoperiodism*, especially its role in *circadian* and *circannual* rhythms, is so important in the temporal organization of life that one entire chapter of this book has been assigned to it.

Thermoperiodism and Temperature Cycles

Equally important as the LD cycle, if not more so in certain organisms, is the 24-h cycle in temperature (Went 1957, 1962, 1974; Liu et al., 1998). The promotion of development by a change between the day (light span) and night (dark span) temperature is called *thermoperiodism* (Went, 1944).[6] Plants such

---

[6] Thermoperiodism as proposed by Went (1944) *"includes all effects of a temperature differential between light and dark periods on responses of the plant."*

as California goldfields (*Baeria chrysostoma*) die if the night temperature is as warm (e.g., 26.5°C) as the day temperature (Lewis & Went, 1945). In fact, for a number of California annual plants, a day temperature of 26.5°C has not been lethal unless accompanied by a 26.5°C night temperature (Lewis & Went, 1945). On the other hand, the vegetative growth (dry weight) of many common species does not appear to be affected adversely by an optimum constant temperature (Friend & Helson, 1976).

The entire matter of what constitutes an optimum temperature is complex, since even the upper and lower surfaces of a single tissue[7] may have different optimal temperatures (Wood, 1953). An important feature in all of this is, however, that the underlying relationship between *photoperiodism* and *thermoperiodism* can be controlled by a common *circadian oscillator*. Such is the case in tomato (*Solanum lycopersicum*), where the combination of continuous illumination and constant temperature is harmful (Arthur & Harvill, 1937; Hillman, 1956) and leads to abnormal development (see Figure 2.12 in Chapter 2 on Characteristics). Yet, only one—either light with dark or high temperature with low temperature—needs to oscillate in the circadian range to maintain normal development. Effects such as these are not so much the need for a dark span to alternate with a light span, as it is for a requirement of the exogenous cycle, whether LD or temperature, to be in the circadian domain (Hillman, 1956, Went, 1974).[8] Because a temperature cycle can sometimes be just as effective as a light/dark cycle, the internal clocking mechanism need not be directly linked with a pigment system (Went, 1960).

Vernalization

In some cases, either a temperature cycle or a single span at a given temperature may be a prerequisite to circadian expression (e.g., flowering) or a photoperiodic response. A prime example of the effects of a single long span of cold temperature on rhythmic timing is illustrated by a process known as *vernalization*. Briefly, vernalization is a low temperature requirement for flowering and is often associated with annual plants, such as winter barley and wheat,[9] as well as

---

[7] The sepals and petals of crocus and tulip flowers are indistinguishable and are therefore collectively referred to as the perianth (or tepals). The opening and closing movements of the flower (perianth) in these species is due to different growth rates of mesophyll cells on the two sides; the outer surface cells have an optimum temperature for growth that is about 10°C lower than the cells of the inner surface (Wood, 1953). Growth movements such as these that are induced by temperature are called *thermonasty*.

[8] The situation is more complex than presented here, since there is an effect relative to the stage of development or age, pretreatment, and other interactions, all of which have been discussed by the authors cited.

[9] Seeds of winter strains are sown in the late summer or early autumn. The seeds germinate and the young seedlings are exposed to low temperature (winter), flowering is induced, and the grain is harvested in early summer. Spring wheat, on the other hand does not require vernalization and is sown in the spring and the grain harvested in the autumn.

with some biennials and perennials. Japanese radish (*Raphanus sativus* L.) displays a circannual rhythm for the number of days from germination to the opening of flowers (Yoo & Uemoto, 1976). However, the circannual rhythm is not observed if the seeds have been vernalized (see Figure 4.15 in Chapter 4 on Photoperiodism). The Japanese radish flowers more rapidly under long days, a photoperiodic response that is often associated with vernalization. In fact, many plants that require long days for floral induction either require low-temperature vernalization (e.g., winter barley '*Hordeum vulgare*' and sugar beet *Beta saccharifera*) or the response to flower is accelerated by vernalization (Salisbury & Ross, 1992).

The applications of vernalization are obvious in agriculture, although they may be limited in scope to some horticultural and floricultural practices (cf. Hillman, 1962). The main reason for its limited use in agronomy can be attributed to the success of breeding programs that have introduced cultivars suited to various geographical regions. The implications in nature of a cold span before floral induction are especially evident for plants that have evolved in geographical regions that have cold winter temperatures. In addition to annual plants (e.g., some cereal grasses), many perennials and biennials require cold for the flowering process. In the case of biennial plants, a short growing season allows the plant to be vegetative the first year, and after a winter of cold temperature, to flower and produce seeds the second year. The "memory" of a cold environment appears to involve a floral inhibitor that becomes "down-regulated," but returns to a high level in the next generation (Michaels & Amasino, 2000). In other words, seeds originating from the flowers that were induced following vernalization will require an exposure to cold for photoperiodic induction and subsequent flowering.[10]

Historically, the implications of vernalization are much broader than its temporal role in the physiology of plant development. Ostentatiously, vernalization entered the realm of political ideology in the Soviet Union during the early and mid decades of the 1900s, when the academician, T.D. Lysenko (1898–1976) advocated the agronomic applications of vernalization and promulgated the misconception that the low-temperature effect becomes an inheritable trait (Lysenko, 1954). The adoption of an official stance by the government that changes attributed to the environment are directly inherited had serious repercussions that led to the expulsion or demotion of geneticists and other scientists who opposed this view (cf. Caspari & Marchak, 1965). From the 1930s and into the 1960s, the research efforts of geneticists in Russia and other former Soviet bloc countries suffered from these false suppositions that were imposed upon their research.[11]

---

[10] In addition to seeds, there are examples where the location for vernalization or the receptors of cold may be in the buds and/or meristems.

[11] Caspari and Marshak (1965) provide an interesting summary of the background events that led to the rise and fall of "Lysenkoism" in the Soviet Union.

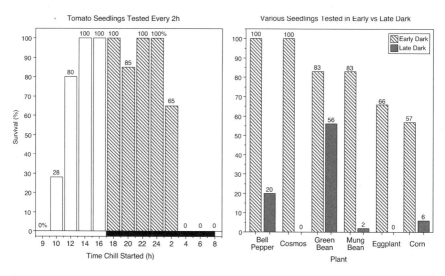

FIGURE 8.3. Chilling sensitivity as gauged by survival of cold-sensitive plants. Seedlings, 2-4 weeks of age, were exposed to 2°C for 48–99 h beginning every 2 h around-the-clock for tomatos (*left*) or beginning in early dark (16:00–22:00 h) or late dark (06:00–07:00 h) for various plants (*right*). Lights on 08:00–17:00 h (adapted from King et al., 1982).

Temperature Compensation

Temperature compensation is one of the major characteristics of circadian rhythms (Chapter 2). While the effects of temperature on the biological clock are minimal ($Q_{10} = 0.8$ to 1.2), the response of certain organisms to both high and low temperatures is rhythmic. For example, the leaves of *Kalanchoe blossfeldiana* plants under LD 12:12 have been shown to be severely injured by a high temperature (e.g., 46°C for 0.5 h) during the first half of the light span and only slightly injured by the same high temperature in the first half of the dark span (Schwemmle & Lang, 1959; Schwemmle, 1960). This rhythm of sensitivity or resistance to heat continues with a free-running circadian period in DD.

A number of species of cold-sensitive plants also show a rhythmic response to low temperatures, with more injury to seedlings when cold (2°C for 48–96 h) is delivered beginning near the end of the daily D-span and affected least or not at all in the second half of the daily L-span and first half of the D-span (Figure 8.3) (King et al., 1982). Soybeans are also least sensitive to cold (-10°C for 4 min) late in the daily L-span (Couderchet & Koukkari, 1987).

Migration

Migration refers to the back and forth recurring annual movements of groups of organisms from one geographic location to another. Some of the great spectacles of nature that are as striking today as they no doubt were in the Stone Age (Rowan, 1926) include annual mass migrations by birds (Figure 8.4) and butterflies in the

FIGURE 8.4. Seasonal migrations of geese (*Branta*) and other animals are linked to photoperiodism and circannual rhythms. Photos by R. Sothern.

air, by whales, lobsters, sea turtles, salmon and other fish in the water, and by many mammals on land, including reindeer in the Arctic regions and the massive herds of animals in Africa, where the act of migration is repeated twice annually.

Birds

In 1926 it was suggested that annual migrations of birds and fish depended upon an internal physiological factor and an external environmental factor provided by the precise change in the duration of daylight (Rowan, 1926). Since then, interactions between biological rhythms and migration have been studied most extensively in birds. The annual cycles of migratory variables in birds, such as restlessness, fattening, and molt, have been shown to be controlled by an endogenous circannual clock (Gwinner, 1977, 2003). The synchronizer appears to be the annual rhythm of the recurring seasonal changes in the length of L and D spans (photoperiod). Under controlled laboratory conditions, annual rhythms in migratory variables for a number of birds have been found to continue, but with free-running circannual periods that deviate from being exactly one year. In the case of garden warblers (*Sylvia borin*) and blackcaps (*S. atricapilla*) that have been maintained under controlled laboratory conditions with constant LD regimens of either 10:14, 12:12, or 16:8 for three years, the average free-running molt rhythm was about 320 days (Berthold et al., 1972).

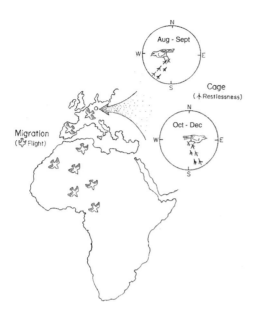

FIGURE 8.5. Drawing illustrating the migration route (wings) of garden warblers and the direction of their perch-hopping activity (feet) when maintained in a circular cage (inserts). Illustrations modified from Figure 14 in Gwinner, 1977 and Figure 3 in Gwinner & Wiltschko, 1980.

In addition to the endogenous rhythmic control of the temporal aspects of migration in birds, there also appears to be a spatial or directional endogenous circannual rhythm (Gwinner, 1977, 2003). Garden warblers that breed in central and western Europe and spend the winter in Africa leave their breeding grounds by first flying southwest and then changing to a south or southeast direction (Figure 8.5). This change in direction has been observed also in the nocturnal perch-hopping activity of caged garden warblers maintained in a laboratory under controlled conditions (see Figure 7.14 in Chapter 7 on Sexuality). In other words, the restlessness exhibited by perch-hopping occurred on average to be southwesterly during August and September and toward the southeast during October to December (Figure 8.5) (Gwinner & Wiltschko, 1978, 1980).

Butterflies

The monarch butterfly (*Danaus plexippus*) (Figure 8.6) of North America migrates annually. In the autumn, millions of monarchs travel as much as 3,600 km (>1600 miles) from north-central and northeastern areas in the United States and southern Canada to reach their overwintering grounds in central Mexico (e.g., State of Michoacan) (Brower, 1995, 1996) (Figure 8.7). Results from studies in Minnesota with adult monarchs reared in the laboratory under controlled conditions, as well as with specimens collected from fields, indicate that the induction of *diapause* during late summer appears to be involved (Goehring & Oberhauser, 2002). Diapause, which is a stage of arrested development or dormancy in insects whereby reproduction is delayed, is often attributed to photoperiodism.

FIGURE 8.6. Monarch butterfly, a species that migrates between Mexico (State of Michoacan) and the northern USA and southern Canada. Photo by R. Sothern.

In the case of monarchs, reproduction and the laying of eggs ceases during summer, and resumes after overwintering in Mexico and migrating north in the spring upon reaching the Gulf Coast states. Two or more short-lived generations are produced over the summer in their northern-most breeding ranges. These migration rhythms of the monarch butterfly are all the more remarkable since there are four generations of butterfly produced each year. Thus, the butterfly that migrates south is not the one that migrated north, but each carries the same genetic information that is passed on through successive generations.

Migration timing mechanisms have been studied less extensively in monarchs than in birds.[12] Recently, however, a properly functioning circadian clock in the

FIGURE 8.7. Map illustrating the late summer or autumn migration of monarch butterflies from Northern America to their overwintering areas (modified from Brower, 1995). The western population migrates between the Rocky Mountains and the California coastline, while the much larger eastern population migrates to central Mexico.

---

[12] A recent surge of interest in the biology of monarchs, including their migration, has occurred nationally in the United States for K-12 grade teachers and students through a program known as "Monarchs in the Classroom." (Oberhauser, 1999). Dr. Karen Oberhauser (Department of Ecology, Evolution and Behavioral Biology, University of Minnesota, St. Paul, MN 55108) directs this program (see http://www.monarchlab.umn.edu).

monarch has been found to be necessary for successful migration. Studies of *eclosion* (emergence of adult from pupal case) and levels of RNA of the clock gene *period* showed prominent circadian rhythms in LD and DD, but not under LL (Froy et al., 2003). Migratory flight orientation was also affected (disoriented) by LL, as well as by UV interference filters, indicating that there are certain light input pathways for appropriate oriented flight behavior and entrainment of the circadian clock. This study was the first to show that molecular gears of the circadian clock are likely a genetic component underlying migratory behavior.

Organisms that migrate, as well as the animals and plants associated with the routes of migration, are often viewed as natural resources. Knowingly or unknowingly, matters pertaining to time of year and the natural rhythms of migration are deeply entrenched in the regulations, management practices, and laws of local, federal and international governing bodies. These agencies focus not only on the management and protection of species, but also on the issuance of fishing and hunting permits (see sections on Gardens and Outdoor Hobbies below).

Pest Management and Agents of Stress

Plants and animals may be protected from harmful pests and diseases by their own defense mechanisms, by symbiotic interactions, or through human intervention, such as various cultural practices and the application of pesticides. In addition to pests and disease-causing agents, animals and plants are exposed to various agents of stress, such as cold, heat and chemicals. Often the responses of plants and animals to pests and to agents of stress oscillate, typically with a circadian periodicity (Table 8.2).

Herbicides

The effectiveness of herbicides on regulating or inhibiting plant growth and development can be related to the time of day that the herbicide is applied (Koukkari & Warde, 1985). The concept that chemicals or other treatments are tolerated better at some times of day than others, is well-known in medicine (see Chapter 11 of Clinical Medicine).[13] At least 20 herbicides have been shown to display significant time-of-day effects upon plants (cf. Table 1 in Martinson et al., 2002; Miller et al., 2003), albeit with different peak times of efficacy depending upon their mode of action relative to the physiology, morphology and the developmental stage of the plant. Circadian rhythmicity in the response to herbicides under LD and free-running conditions has been demonstrated for cotton (*Gossypium hirsutum*) (Rikin et al., 1984) and velvetleaf (*Abutilon theophrasti*) (Koukkari & Johnson, 1979; Koukkari & Johnson, unpublished).

---

[13] Known as "hours of diminished resistance" (Halberg, 1960) and "hours of changing responsiveness or susceptibility" (Reinberg, 1967). This concept of "chronotolerance" has also been demonstrated in agriculture.

TABLE 8.2. Examples of the rhythmic response of plants and animals to agents of stress and to pests, relative to the time of day.

Agent	Organism	Comments	Reference
High temperature	*Kalanchoë blossfeldiana*	Leaf more sensitive to injury during early L and less in mid D	Schwemmle, 1960; Schwemmle & Lange, 1959
Low temperature	Soybean (*Glycine max*)	Least sensitive during late L span	Couderchet & Koukkari, 1987
Volatile chemicals	Tobacco (*Nicotinia tabacum*)	Plants release volatile compounds during D to repel female moths	De Moraes et al., 2001
Mechanical	Bean (*Phaseolus vulgaris*)	Touching by hand or wind during light span reduces elongation	Anderson-Bernadas et al. 1997; Sothern et al., 2001
Herbicides	Many species of plants	Timing depends upon herbicide and method of application	Riken. et al. 1984; Koukkari & Warde, 1985; Martinson et al., 2002;
Insecticides	Many species	Timing depends upon the insect and chemical	Cole & Adkisson, 1964; Eesa et al., 1987; Onyeocha & Fuzeau-Braesch, 1991
Sound (noise)	Mouse	More convulsions and deaths in early activity span	Halberg et al., 1955

Peak times of efficacy are not always similar even for a given herbicide, since it may depend upon the type of herbicide used, method of application, and the prevailing environmental conditions. When velvetleaf plants were sprayed with bentazon[14] from a nozzle located above the foliage, more of the chemical reached the foliage and better control of the weed occurred during the day, a time span when the leaves were in a more horizontal position (Doran & Anderson, 1976). Less chemical reached the leaves and poorer control of the weed occurred during darkness, a time when the leaves were in a more vertical position. However, when the herbicide was applied with a pipette to specific locations of the leaf, less damage occurred to plants treated during the light span (Koukkari & Johnson, 1976).

Oscillations in the response of plants to herbicides have been attributed to many factors, including the phases of other rhythms (gating), such as leaf movements (mentioned above), phloem transport (Hendrix & Huber, 1986), translocation of photosynthate (Weaver & Nylund, 1963), and stomatal opening (Schuster, 1970). Metabolic oscillation in processes of photosynthesis, respiration, and transpiration (cf. Koukkari & Warde, 1985) could also be involved.

While environmental conditions such as light, temperature and relative humidity can affect the efficacy of a herbicide, large within-day differences in weed control were still present after adjusting for these and other variables when four

---

[14] 3-isopropyl-1H-2,1,3-benzothiadiazin-4(3H)-one 2,2-dioxide.

FIGURE 8.8. A drawing illustrating a horsedrawn hopperdozer in action and comments supplied by EB Forbes that evening is better than earlier in the day to catch (control) grasshoppers present in wheat fields (Forbes, 1901) (illustrated in *Minn Sci* 1975: 31(1): 81.)

herbicides widely used today[15] were tested in a natural setting (Miller et al., 2003). Because of the relationship that exists between herbicide application and the rhythms of efficacy, it should be possible to time chemical applications to agronomically-important crops so as to reduce the amount of herbicide necessary to achieve the maximum effect,[16] and also provide for a safer environment.

Pest Control

The practical application of selecting the best time of day to control pests relative to some cyclic phenomenon was used before the principles of photoperiodism and chronobiology were introduced. One example is found in the archives of the University of Minnesota Agricultural Experiment Station, where a recommendation was offered in 1901 to wheat farmers in the Red River Valley in northwestern Minnesota (USA) who were trying to control grasshoppers (Figure 8.8). The entomologist E.B. Forbes suggested that evening was the best time to control grasshoppers, since it was the time of day when grasshoppers would crawl up the

---

[15] The four herbicides [tradename] were: (glyphosate [Roundup Ultra™], an amino acid synthesis inhibitor; glufosinate [Liberty™], a glutamine synthetase inhibitor; fomesafen [Flexstar™], a protoporphyrinogen oxidase inhibitor; and chlorimuron ethyl [Classic™], an acetolactate synthase inhibitor). Each showed maximum weed control near midday.

[16] The timing of herbicide treatment may also apply to the homeowner who occasionally wants to control weeds in the lawn and elsewhere around the house.

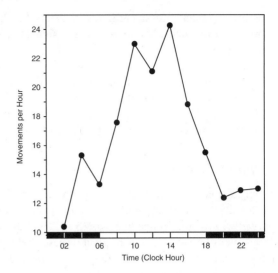

FIGURE 8.9. Grasshoppers (*Romalea microptera*) show a circadian rhythm in locomotor activity, with decreasing activity in late afternoon (composite graph redrawn from Figures 1 and 2 in Fingerman et al., 1958).

stalks of wheat and thus could be captured in a 16-foot (~5 meters) horse-drawn trough containing oil (Figure 8.8) (Forbes, 1901). This recommended time of day coincides with the less active phase of the locomotor activity rhythm of certain grasshoppers (*Romalea microptera*) (Figure 8.9).

Today, there are a large number of chemical insecticides that are more effective or efficient than oil in a trough for controlling insect pests. However, the time of application and the type of insecticide and its mode of action remain important factors. For example, the time of maximum susceptibility of insects, such as the housefly (*Musca domestica*) to the insecticide pyrethrum has also been reported to be in the late afternoon and early evening (cf. Halberg J et al., 1974). Conversely, larvae of the migratory locust (*Locusta migratoria migratorioides*) were twice as susceptible to the toxic effects of the insecticide dieldrin at midnight than during the afternoon (Onyeocha & Fuzeau-Braesch, 1991). Results from a study on the effects of the insecticide methyl parathion have shown that under certain photoperiods (e.g., LD 10:14), 10% of adult boll weevils (*Anthonomus grandis*) treated at dawn are killed compared to 90% of those treated three hours later (Cole & Adkisson, 1964).

Plant Responses to Injury

It appears that long before humans began to realize the importance of rhythms in pest management, plants had evolved their own chemical temporal system of pest management. Results from more recent studies indicate that some plants, without the assistance of humans, may rely upon both the chemicals that they produce and their own internal timing system to control pests that will harm them. Tobacco (*Nicotiana attenuata*) plants that have been damaged by insect herbivores have been shown to release certain volatile chemicals that attract insects that eat the herbivore insects (Kessler & Baldwin, 2001). In other words, by the use of volatile organic compounds, *N. attenuata* plants signal the insect enemies of their pests for assistance.

Another species of tobacco, *Nicotiana tabacum,* has been shown to release herbivore-induced volatiles during both the light and dark span, but some that are emitted exclusively at night are highly repellent to egg-laying moths[17] (*Heliothis virescens*) (De Moraes et al., 2001). The temporal effects are two-fold, since both the phase of moth activity and the phase of volatile emission occur at night.

Plant Diseases

Only a small number of studies have been published on the interactions between plant diseases and rhythms, an area of research that has been called *chronophytopathology*[18] (cf. Kennedy & Koukkari, 1987; Kennedy et al., 1990). Most of the pathogens or causal agents of plant diseases belong to one of three groups: fungi, bacteria, or viruses. Both the fungus *Helminthosporium oryzae* and the disease it causes on rice have variables that are rhythmic (Rathinavel & Sundararajan, 2003). This includes the severity of infection (as quantified by leaf spots and visual ratings) and the germination of conidial spores. All are highest during the early morning between 02:00 h and 06:00 h.

The bacterium *Xanthomonas campestris* pv. *Phaseoli*, which causes common bacterial blight on beans (*Phaseolus vulgaris*), alters the circadian rhythmic movements of leaves. A change in the pattern of movement is evident during the light span of the cycle and continues for a number of cycles prior to the death of the plant (Guillaume et al., 1986). The bacterial pathogens *X. campestris* and *Corynebacterium flaccumfaciens* have been shown to affect the ultradian rhythmic circumnutation movements of bean shoots (Kennedy et al., 1986). When compared to healthy plants, both the amplitude and speed of movement of the infected plants are reduced.

In the case of plant diseases caused by viruses, there are variations in host-virus interactions relative to the time of day (Helms & McIntyre, 1967; Matthews, 1953a). Inoculation of bean (*Phaseolus vulgaris*) leaves with tobacco necrosis virus in the afternoon (during mid-L) resulted in the most lesions, while inoculation near the end of D produced the least (Matthews, 1953a). Similar results have been shown for three different viruses (tobacco mosaic virus, lucerne mosaic virus, and turnip mosaic virus) when they were applied to tobacco plants (*Nicotiana glutinosa*) (Mathews, 1953a). Especially pertinent to our discussion here is the fact that the period of the circadian rhythm in the response may free-run during

---

[17] After hatching, the larvae of these moths are destructive herbivores and would need to compete with larvae hatched from eggs of other moths.

[18] The focus of *chronopathology* is on changes in the biological time structure of humans relative to functional disorders and disease (cf., Touitou & Haus, 1992). The focus of *chronophytopathology* is similar, but the emphasis is upon plants (phyto), rather than upon humans (Kennedy & Koukkari, 1987).

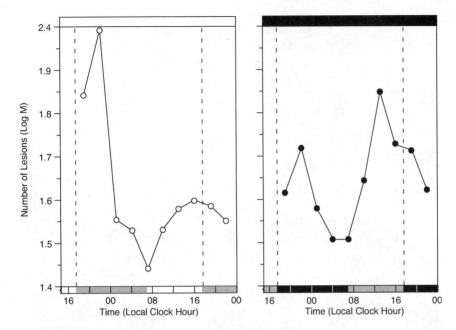

FIGURE 8.10. Circadian rhythm in response of beans (*Phaseolus vulgaris* var. Sydney Wonder) to inoculation with the tobacco necrosis virus at different times persists under constant conditions of continuous light (*left*) or darkness (*right*). Response gauged by the number of local lesions counted 72 h after inoculation. Redrawn from Figure 3 in Matthews, 1953.

continuous light or dark, thereby indicating the endogenous nature of this susceptibility rhythm (Figure 8.10) (Matthews, 1953b; cf. Matthews, 1981).

Production of Produce

It is practically impossible to identify an agricultural product, crop, or animal that has not involved variables that were rhythmic. Fruits, vegetables, cereals, meats and beverages are all linked to the temporal rhythmic nature of life (Figure 8.11). Farm products representing samples from some of the major categories of agriculture, along with their rhythmic component or variable, are listed in Table 8.3 and Table 8.4. Similar relationships also exist in nature for plants, animals, fungi, various single-celled organisms, and even some bacteria. In other words, if the subject is an organism, or if the product in question is of biological origin, then either directly or indirectly, biological rhythms were part of the temporal organization that led to its development.

Some types of produce, especially those derived from fishing and hunting, come from the oceans, lakes, rivers, and lands that are within the bounds of natural resources. The rhythmic variables referred to in the previous paragraph for agricultural products, also exist in the organisms that comprise the "catch." In recent years, some of the commercial aspects of fishing and hunting have shifted

FIGURE 8.11. Examples of agricultural products and organisms that display biological rhythms: (*A*) barley, (*B*) onions, (*C*) beans, (*D*) potatoes, (*E*) apples, (*F*) chickens, (*G*) honeybees, (*H*) sheep, (*I*) cows, (*J*) horses. *Note*: specific variables are presented in Tables 8.1–8.4 and Tables in Chapter 9 on Veterinary Medicine. Photos courtesy of David Hansen, University of Minnesota Agricultural Experiment Station.

to be more "farm-like," with fish being reared in enclosed or artificial ponds, and land-dwelling animals, such as bison, elk and game birds, confined behind fences.

Fisheries and Aquaculture

Understanding photoperiodic and circadian systems in fish may be helpful in the management of fisheries and aquaculture, since the physiology and behavior of fish have ultradian, circadian, tidal, monthly, and annual rhythmic components

TABLE 8.3. Some major categories of agricultural crops (Type) and an example of a species within the category that has a variable that displays a biological rhythm.

Type	Species	Rhythmic Variable	Reference
Cereals	Barley (*Hordeum vulgare*)	Loss of water	Nixon et al., 1987
	Wheat (*Triticum aestivum*)	Nitrate reductase activity	Upcroft & Done, 1972
	Rice (*Oryza sativa*)	Rubisco activase mRNA level	To et al., 1999
Forage	White clover (*Trifolium repens*)	Leaf movements	Scott & Gullline, 1972
	Alfalfa (*Medicago sativa*)	Invertase activity in nodules	Henson et al., 1986
Sugar	Sugar cane (*Saccharum officinarum*)	Acid invertase activity in storage internode cells	Slack, 1965
Oils	Soybean (*Glycine max*)	Induction of flowering	Hamner, 1960
Fuel	Corn (*Zea mays*)	Enzymes of nitrate reduction and ammonia assimilation	Duke et al., 1978
Fruits	Orange (*Citrus sinensis*)	Fruit diameter	Elfving & Kaufmann, 1972
Potatoes	Potatoes (*Solanum tuberosum*)	Tuber circumference	Stark & Halderson, 1987
Vegetables	Tomato (*Solanum lypopersicum*)	Growth change in diameter	Ehret & Ho, 1986
Fiber	Cotton (*Gossypium hirsutum*)	Sensitivity to herbicide Abscisic acid levels Leaf movements	Rikin et al., 1984 McMichael & Hanny, 1977 Miller, 1975
Timber	Pine (*Pinus sylvestris*) Willow (*Salix fragilis*) Silk tree (*Albizzia julibrissin*)	Indole-3-acetic acid content of seedlings Xylem water flow Leaf movements	Sandberg et al., 1982 Cermak et al., 1984 Koukkari, 1974
Rubber	Rubber (*Hevea brasiliensis*)	Rubber content in latex and activity of latex enzyme	Wititsuwannakul, 1986
Coffee, tea, and cocoa	Tea (*Camellia sinensis*)	Photosynthesis rate, stomatal conductance and intercellular CO_2 concentration	Mohotti & Lawlor, 2002
Tobacco	Tobacco (*Nicotiana tabacum*)	Emission of volatiles	De Moraes et al., 2001

(cf. Ali, 1992). In many fish species, some individuals are nocturnal and others are diurnal. However, circadian rhythms in some individuals have been shown to display plasticity by switching from nocturnal to diurnal activity patterns and vice versa during migration,[19] larval drift, spawning and parenting, and in response to several environmental conditions, including light, temperature, predation, shoal size, or food availability (cf. Reebs, 2002). Fish with a more rigid, less plastic circadian system tend to be marine species found in the deep sea where the environment is more stable, while fish in freshwater habitats need to respond with more

---

[19] Some species of birds that are usually diurnally-active also change their activity patterns to fly at night during their spring and fall migration.

TABLE 8.4. Some major categories (type) of farm animals and an example of a biological variable for the species that displays a rhythm.

Type (species)	Variable	Comments	Reference
Holstein cattle (*Bos taurus*)	Udder and body temperature	Ultradian (90 min) and daily (24 h)	Bitman et al., 1984
Hogs	Melatonin	Circadian peak during usual dark span	Lewczuk & Przybylska-Gornowicz, 2000
Poultry	Body temperature and feeding activity	Circadian (25.2 h or 25.3 h in LL)	Kadono & Usami, 1981
Sheep	Testosterone (in rams)	Annual peak May-July	Gomes & Joyce, 1975
Fish (many species)	Activity	Many circadian studies in LD	Ali, 1992; Spieler, 1990
Horses (*Equus caballus*)	Cortisol secretion	Ultradian (15 min)	Drake & Evans, 1978

plasticity and less rigidly to changes in daily events (e.g., food availability, predation risk) in an unstable freshwater environment.

A better understanding of the interaction of photoperiod, melatonin, and other factors such as temperature and nutrition, should enable the production of farmed species of fish of consistent size and quality all-year-round (cf. Bromage et al., 2001). For example, several reproductive hormones, including LH and testosterone, showed a circadian pattern in caged sea bass (*Dicentrarchus labrax*) under a natural photoperiod in December, but a long photoperiod (18L:8D) suppressed circulating melatonin, delayed the daily peak in LH storage and release and suppressed *gonadal* development (Bayarri et al., 2004).

Atlantic salmon (*Salmo salar*) maintained under natural and reversed photoperiods show circadian patterns in circulating melatonin levels that are inversely related to the light levels (Randall et al., 1995). The seasonally-changing circadian pattern in melatonin reflects the prevailing daylength, thereby encoding accurate information on daily and yearly time to the salmon. Studies have shown that the timing of *smoltification* (the adaptation of a fish from freshwater to the osmotic conditions of seawater) can be affected by the manipulation of melatonin levels (Porter et al., 1998). When salmon were *pinealectomized* at the winter solstice, the onset of smoltification was delayed by three weeks and resulted in lower survival rates by March. However, smoltification in pinealectomized salmon that also received a melatonin implant was advanced by three weeks. In addition, when young salmon received melatonin implants in June, they were significantly larger than controls one month later and 92% of them were able to enter smoltification at one year of age, rather than the normal two years. Thus, implantation produced a growth advantage within a month and an advance in seawater adaptability.

While photoperiodism is often studied in fish, other aspects of their circadian lifestyle are not. These include 24-h rhythms in behaviors (e.g., locomotor activity, feeding, shoaling, agonistic behavior, sexuality-related activities) and physiology (e.g., circulating hormones, enzymes, electrolytes, other substances). In addition, the reaction of the organism to external stimuli, such as pharmaceuticals, hormones, handling, temperature and feeding, can depend upon the time of day

(Spieler, 1990). Some routine tasks could be done at specific times within the day or seasonally in order to disturb the fish less and improve yield. For example, when channel catfish (*Ictalurus punctatus*) were fed either on 07:30 h or 16:00 h (while on an LD12:12 photoperiod with L-on at 06:00 h), those fed at 07:30 h had higher weights, lengths and food conversion efficiencies, while those fed at 16:00 h had 36% more abdominal fat (Noeske-Hallin et al., 1985). Interestingly, another group of catfish fed half the daily food ration at each time (07:30 h & 16:00 h) had weights similar to fish fed the single early meal and also high abdominal fat similar to fish fed the single late meal. Thus, by changing all or part of daily feeding to later in the normal work day, the same amount of food resulted in a significant gain in growth and weight that could translate into profit for the fish farmer (Spieler, 1990).

Weather Patterns and Agriculture

There is evidence that cyclic weather phenomena affects plant life. A teleconnection weather pattern (a phenomenon in which weather patterns in one region of the world influence the weather patterns in a distant location) can occur on many time scales, lasting from a few days to several centuries. The El Niño-Southern Oscillation (ENSO) has a major influence on rainfall in the southern hemisphere, and in recent years considerable progress has been made in forecasting crop yields from information on the ENSO in southern hemisphere countries such as wheat in Australia (Rimmington & Nicholls, 1993) and maize in Zimbabwe (Cane et al., 1994).

The annual mean quality of wheat grain in the United Kingdom, as determined by the Hagberg Falling Number (HFN), which is a measure of alpha-amylase activity,[20] followed a cyclical pattern with a period of ~8 years. This cycle was close to that of the January and February mean of an index of the North Atlantic Oscillation (NAO), with lower values of the NAO index associated with lower HFN (Kettlewell et al., 1999). These ~8-year cycles had a significant economic impact, since they were accompanied by inverse cycles in the percent of imported wheat and the premium paid for a loaf of bread (Figure 8.12). A high HFN (low alpha-amylase activity) is desirable for bread baking and when HFN is low, wheat must be imported from abroad and mixed with the local product to achieve desired baking results.

Gardens

In many of the books that are published on biological rhythms, even if their emphasis is upon animal and human physiology (e.g., Moore-Ede et al., 1982) or clinical and laboratory medicine (Reinberg & Smolensky, 1983; Touitou & Haus,

---

[20] Alpha-amylase is an enzyme that catalyzes the degradation of starch. Twenty-four hour rhythms in amylase activity have been reported in several photosynthetic tissues and the peaks of these rhythms are inversely correlated with rhythms in starch content (cf. Henson & Duke, 1990).

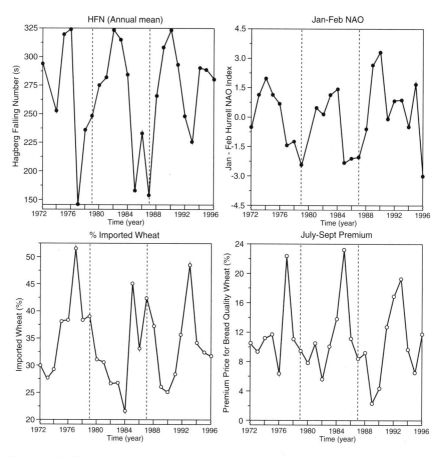

FIGURE 8.12. There may be an association between cycles in weather patterns and cycles in crop quality. Here, an about 8-year teleconnection association is demonstrated between the North Atlantic Oscillation (NAO) in January and February and wheat quality and economics in Great Britain over a 25-year span. Vertical dashed lines at end of years of low NAO index. Hagberg Falling Number (HFN) is an inverse measure of alpha amylase activity in wheat grain, with a high HFN desirable for bread baking (Kettlewell et al., 1999).

1992), reference is made to the floral clock of Linnaeus (see Figure 3.11 in Chapter 3 on Time).[21] The relative time of day (Figure 8.13) and/or season (Figure 8.14) of flowering can be approximated, though broadly, by the position and the presence of certain flowers[22] and leaves.

---

[21] Supposedly, one could tell the time by observing when a flower or inflorescence of a given species would be open or closed. However, attempts that were made in Botanical Gardens to construct such a clock never met with success because all of the plants listed by Linnaeus do not flower during the same season (Kerner von Marilaun, 1895).

[22] Goldsmith and Hafenrichter (1932) included a number of species in their book that show changes in floral structures relative to days and clock hours.

FIGURE 8.13. Position, presence and/or orientation of certain flowers can approximate the time of day. For example, a dandelion *(Taraxacum)* inflorescence (a flower cluster) is open throughout the day, but is already [mostly] closed at least a half an hour before sunset (*middle panel*). Photos taken by R. Sothern on Sept 25–26. Sunrise was at 07:03 AM and sunset at 07:06 PM local time.

2 PM

6:30 PM

9 AM

Gardens are often designed to include an assortment of plants that have different photoperiodic requirements for floral induction. However, some may have been transplanted from containers where they were originally raised commercially and induced to flower by photoperiod manipulations that occurred elsewhere (e.g., in a greenhouse). The commercial potted plant industry relies on photoperiod manipulation, such as cloth covers, supplemental light, or light breaks during darkness to induce flowering on a year-round basis. For example, during the year when the span of daylight is long, the induction of flowering in chrysanthemum (*Dendranthemum grandiflora* Ramat), a short-day plant, is achieved by pulling a cloth over the plants to extend the dark span and thereby provide short-day conditions.

An increase in the utilization of controlled photoperiods has been driven primarily by both the demands of sellers and buyers to have plants be in bloom when

FIGURE 8.14. Different flowers and fruits are present in different seasons. While there is overlap between some seasons, flowers found in the temperate climate of Minnesota include tulips (*Tulipa*) and apple (*Malus*) in spring, roses (*Rosa*) in summer and chrysanthemums (*Chrysanthemum*) in fall. Photos by R. Sothern.

they are marketed for home gardeners in the spring and late summer, and a desire among growers to hasten flowering in order to reduce their costs of production (cf. Erwin & Warner, 2002). Likewise, overhead lights are commonly used to supplement and increase the duration of the light span in autumn and winter for the induction of flowers in long-day plants or to maintain short-day plants in the vegetative state.

FIGURE 8.15. A hobby such as birding (watching birds at a feeder or in a natural setting) can reveal the presence of rhythmic behavior, such as times of feeding, courtship, migration, etc. Photos by R. Sothern.

Outdoor Hobbies

A vast number of hobbies are based upon the rhythmic variables present in nature. These include, among others, birding, fishing, photography, travel, cooking, camping, searching for mushrooms and collecting maple sap.

Birding

A common hobby, which has a strong rhythmic basis and is pursued by individuals in both urban and rural areas, is *birding*.[23] Interest in this fascinating hobby has been spurred by the availability of easy-to-use field guides, binoculars, and inexpensive feeding stations (Figure 8.15). Birders observe time-dependent events

---

[23] *Birding* is the preferred term and replaces the phrase "bird watching." The word *birder* is a name often applied to those who watch and take an interest in birds. *Ornithology* is a discipline of biology that focuses on birds and an ornithologist is a specialist or professional biologist who studies birds.

or activities, such as recurring changes in behavior, feeding, mating, brooding, raising of young, and migration. All of these activities are variables of a rhythmic nature. Some phases of activity may also be related to the phases of circadian and circannual rhythms of other organisms, such as predator/prey interactions, activity cycles of insects, opening of floral structures and production of flowers and fruits.

Fishing

Recreational fishing[24] and hunting are hobbies greatly entwined in the rhythmic nature of life. For example, fishing for the rainbow smelt (*Osmeus mordax*) in the Great Lakes regions of N. America occurs only during spawning runs into tributary streams over a few weeks in late winter and early spring, with best smelt-dipping time generally within a few hours of midnight. Salmon fishing in many areas of the world is restricted to the annual spawning runs, such as the mid-summer spawning runs in Norway. A prime example of how deeply biological rhythms can be entrenched in hobbies is illustrated in the sport of fly-fishing,[25] especially for trout (Essay 8.2 and Figure 8.16).

Essay 8.2 (by WLK): Fly-fishing for Trout

Biological rhythms are a common and natural phenomenon upon which the sport of fly-fishing for trout is based. Aquatic insects provide one of the chief food sources for trout, which the fly-fisher[26] tries to imitate with an artificial pattern known as the "fly" (Figure 8.16). Many who are skilled in the sport seek knowledge of the life history and the stages in the development of aquatic insects (Figure 8.17), crustaceans (e.g., *Gammarus* sp. "scuds") and other organisms that are preyed upon by trout and salmon.

Two of the stages in the life history of insects, eclosion and adult activity, which were emphasized earlier in this book as variables that contributed greatly to our knowledge of circadian rhythms and biological clocks (see chapter on Time and Clocks), are also important in the sport of fly-fishing. Species of aquatic insects are present in a stream throughout the year as larvae, nymphs or pupae, but the duration for the adult stage is relatively short. For example, entomologists have given the group (order) name, *Ephemeroptera* (Ephemeral = remaining or living for a short time), to mayflies because the adult stage is relatively short. Depending upon the

---

[24] In a number of states in the United States, the economic impact of sport fishing for all species exceeds $1 billion (Maharaj & Carpenter, 1996).

[25] About 20% of stream trout anglers use flies when fishing (personal communication, M.A. Ebbers, Minnesota Dept of Natural Resources).

[26] Fly-fisher is the modern term of preference that has replaced "fly fisherman." Historically, two women made a tremendous impact upon fly-fishing that is still appreciated. Dame Juliana Berners is credited with writing the first book in English on angling (cf. Berners, 1496) and Mary Orvis Marbury (1988) for writing an excellent book first published in 1892 entitled "Favorite Flies and Their Histories."

FIGURE 8.16. Science and Art are combined in the sport of fly-fishing. From left, beginning at the top, the insects include an adult caddisfly (imago), mayfly (subimago), mayfly (imago), caddisfly (larva), and mayfly (nymph). The artificial imitations for these insects appear at the far right, along with a series of five steps for tying the imitation of a subimago. Included with the tools of science are calendar (season), clock (time of day), collection net, vial (collected specimen), book (dichotomous key for identifying insects), field book (for personal records) and thermometer (for water and air temperature). Tools illustrated include hook, scissors, hair, thread, yarn, dubbing material, feathers, and polarized sun glasses (to see in the water). A fly rod is shown in the center and two casting techniques (overhead and roll) are shown at the top. Reading the stream and its characteristics complete the illustration. (©WL Koukkari)

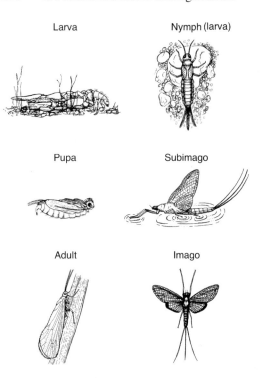

FIGURE 8.17. Life history of a caddisfly (*left*) and mayfly (*right*). Most time is spent under water as a larva or nymph, followed by a transformation stage as a pupa (caddisfly) or subimago (mayfly) and a brief adult stage. (©WL Koukkari)

species, the duration of the adult stage may be less than a day, a few days or longer, but usually less than a month. Tables similar to those compiled by chronobiologists in the field of medicine to indicate peak phases of circadian and circannual rhythms are compiled also by fly-fishers to indicate the approximate times when adults of various genera (Table 8.5) and species (Table 8.6) are likely to be present (cf. Humphrey & Shogren, 1995). Just as medical doctors and chronobiologists should know when phases and amplitudes of circadian variables occur in their practice or research, so too should fly-fishers know when stages and peaks occur for various activities that are indicative of the insects that they try to simulate. For example, my records show that during late summer, the spinner fall[27] for one group of mayflies (*Tricorythodes* spp.) on an Eastern Wisconsin stream occurs usually between mid-morning and noon.

Caddisflies (*Trichoptera*), unlike mayflies (*Ephemeroptera*) and stoneflies (*Plecoptera*), have a pupa stage and therefore undergo the process of eclosion. While a chronobiologist can easily monitor the daily phase of eclosion in *Drosophila* that occurs in a laboratory glass dish, it is much more difficult for anglers to monitor the eclosion of

---

[27] The term *spinner* is used by fly-fishers when referring to the mature adult mayfly, which entomologists and biologists call the *imago*. The behavior of adult mayflies to fall, land or settle on the water surface after mating is called the *spinner fall* by anglers. This is another phase in the life history of mayflies when anglers try to imitate these spent insects, which are readily fed upon by trout.

TABLE 8.5. Time of year when seven select adult aquatic insects and grasshoppers are likely to be present in Minnesota. Each ◊ represents a week. Names of artificial flies that are used to represent the insect are also listed. *Note*: Only the genus is listed, but there are many species that trout selectively feed upon. Species, as well as genera, are not necessarily found in the same stream or in the same region. For example, the genus *Rhyacophila* appears to be confined to the northeastern part of the state.

Insect	Mar.	Apr.	May	June	July	Aug.	Sept.	Artificial fly name
Mayfly (*Baetis*)	◊◊◊◊	◊◊◊◊	◊◊◊◊	◊◊◊◊	◊◊◊◊	◊◊◊◊	◊◊◊◊	Blue-winged Olive
Midges (many species)[a]	◊◊◊◊	◊◊◊◊	◊◊◊◊	◊◊◊◊	◊◊◊◊	◊◊◊◊	◊◊◊◊	Griffith's Gnat
Mayfly (*Ephemerella*)		◊◊◊◊	◊◊◊◊	◊◊◊◊	◊◊			Dark Hendrickson, Sulphur, Adams
Caddisfly (*Brachycentrus*)		◊◊	◊◊◊◊					Henryville Special, Hairwing Caddis
Caddisfly (*Rhyacophila*)				◊	◊◊◊◊	◊◊◊◊	◊	Henryville Special, Hairwing Caddis
Caddisfly (*Pycnopsyche*)						◊	◊◊◊◊	Henryville Special, Hairwing Caddis
Mayfly (*Tricorythodes*)				◊◊	◊◊◊◊	◊◊◊◊	◊◊◊◊	Trico
Grasshoppers (Numerous species)				◊◊◊◊	◊◊◊◊	◊◊◊◊	◊◊◊◊	Hoppers

[a]Midges belong to the Order *Dipter*, as do houseflies and mosquitoes. Many of the midges are in the family, Chironomidae.

caddisflies in a stream.[28] By examining the feeding behavior of trout (e.g., the manner in which they move toward the surface of the water and the nature of the splash they produce when capturing an insect), as well as the use of artificial patterns that depict eclosion, anglers are able to apply their knowledge of eclosion to their fishing tactics. For example, an artificial fly pattern that is tied with yarn that sparkles simulates the air bubbles that are encased within the pupal sheath during eclosion (LaFontaine, 1981).

TABLE 8.6. Dates when adult species of *Brachycentrus* caddisflies have been collected in Minnesota.[a]

Species	Time span of collected specimens
B. occidentalis	From April 25 to May 6
B. amercanus	From May 30 to July 27
B. numerosus	May 8

[a]The time spans presented are based upon specimens maintained in the Insect Museum of the University of Minnesota and data complied by David C. Houghton.

---

[28] Colin Pittendrigh (1919–1996) is well known for his contributions to the study of circadian rhythms, including the eclosion rhythms of *Drosophila*. He was also a fly-fisher, who, as Ernest Schwiebert (1973) described, became "sold" on using a "little green-bodied wet fly" pupal caddisfly imitation of *Rhyacophila basalis* after watching Schwiebert land several trout while using this fly on an eastern US stream. This particular fly was the right one for that time of day and year.

Many fly-fishers prefer to fish when adult insects are active, since they can use dry flies that float on the surface of the water. However, most large trout feed more heavily on the immature stages of insects and other organisms that are found at greater depths. Thus, anglers interested in catching larger fish, try to simulate the stages and species of various bottom-dwelling organisms. Their preferred times to fish occur during the peaks of a *diel* (i.e., 24-h) rhythm known as *drift*. The rhythmic down-stream transport or drift of mayfly nymphs, caddisfly larvae, stonefly nymphs, and other invertebrates is a common feature of streams in various parts of the world (Waters, 1972; Sagar & Glova, 1992). The reasons for drift are not entirely known, although factors such as nocturnal feeding and crowding may play a key role in drift behavior (Elliott, 1967). Mayflies, especially the genus, *Baetis* (see Figure 8.16), as well as the crustacean, *Gammarus*, are among the more common organisms that have been observed in drift samples. Highest numbers of these organisms and most other aquatic insects occur during the night (Waters, 1972). There are some exceptions, including certain species of caddisflies, although most members of this order (*Trichoptera*) of insects drift at night. The distance traveled varies with the species, stream, substrate, etc. and is often in the range of 10 to 100 meters (~30 to 300 feet) per night, although the distance may be as short as one to five meters (cf. Waters, 1972). Results from an extensive study of the rhythmic drift of *Baetis* and *Gammarus,* where exclosures were used to shield from light or external artificial lights were used at night, indicate that the rhythm of drift of these two organisms may be exogenously-controlled (Holt & Waters, 1967). Usually there are two peaks, a major one occurring at the beginning of darkness and a minor one before dawn (Waters, 1972). These two phases also represent the best time to catch large fish, especially by using artificial fly patterns that simulate nymphs, larvae, and scuds.

Rural and Urban Development

In the past, it was common for cities and villages to be confined to small tracts of land and separated from each other by wilderness, forests or farms. Today, natural or unsettled areas are often confined to small tracts of land that are dispersed within sprawling cities and urban areas or between large farms. Wooded lots, marshes, wetlands and natural areas that were part of the rural community have been eliminated or reduced in size as the tillable acreage has increased and urban developments have ravaged the natural landscape. Such changes, as well as cultural practices and other factors,[29] have greatly disrupted the natural food web, as well as the composition and diversity of species native to the area. While it has been relatively easy to see what has been happening to the structural organization of the rural ecosystem, changes to the temporal organization in native species, especially animals, have not been as evident.

Telemetry Tracking Systems

Much of our knowledge of circadian, ultradian, and even infradian rhythms of animals has been obtained from laboratory studies where organisms have been

---

[29] Major contributing factors have been the use of various pesticides and the introduction of non-native species.

FIGURE 8.18. Animals that have been fitted with radio collars used for the study of activity patterns under natural field conditions. Collar clearly visible in the photo of the gray squirrel. Photos provided by Dr. J. Tester from the University of Minnesota Cedar Creek Natural History Area in Bethel, Minnesota.

confined to cages, running wheels, pans, or enclosed areas. Under such conditions, there is great regularity and uniformity in rhythm characteristics (e.g., phase and pattern of activity). Does this uniformity carry over to organisms that are relatively unrestricted? The breakthrough that enabled scientists to address this question was the development of automated radio-tracking systems.[30] Results from studies where radio-telemetry systems have been used to monitor free-ranging animals under natural conditions (Figure 8.18) indicate great plasticity in response to changes in the environment (Tester, 1987). Five of the many species monitored will serve as examples: muskrats (*Ondatra zibethicus*), red foxes (V*ulpes vulpes*), gray squirrels (*Sciurus carolinsis*), snowshoe hares (*Lepus americanus*), and ruffed grouse (*Bonasa umbellus*) (Figala et al., 1984; Figala & Tester, 1986; Tester, 1987)

---

[30] During the early Sputnik era (ca. 1957), more information was being monitored and collected for a dog that was orbiting the earth, than for any wild animal that was living free in the forests and prairies of Minnesota. This inspired scientists like John Tester at the University of Minnesota, one of the early investigators of the rhythms of animals in their natural environment, to construct and develop radio-tracking systems for wild animals for use at the Cedar Creek Natural History Area in east-central Minnesota.

Muskrats

Muskrats are common animals that inhabit marshy areas. The number of spans of activity per day changes with the seasons of the year (Stolen, 1974). During August, free-ranging muskrats display four spans of activity evenly distributed over the 24-h cycle. Later in the year, less activity occurs during daylight, thereby reducing the number of spans to three and finally in winter to two. The two daytime spans are eliminated in the winter and the activity of the muskrat became almost exclusively nocturnal.

Squirrels and Foxes

The length of the activity span of gray squirrels closely follows the duration of daylight. Foxes, on the other hand, hunt primarily at night, digging under plant litter, snow, etc., to feed upon mice and other rodents. However, if there is ice cover over the snow, foxes have been found to change their activity by searching for food during daylight hours. During daylight, they are able to hunt and feed upon gray squirrels, which are day-active. In both captive and wild squirrels there is a marked increase in the length of activity in September that is likely to be related to food gathering and does not appear to follow the exact duration of the photoperiod (Figala & Tester, 1986).

Hare

The snowshoe hare (*Lepus americanus*) is another example of a truly interesting inhabitant of the ecosystem of the northern United States and Canada.[31] During the summer their color is brown, while in the winter it is white, an adaptive photoperiodic response that is illustrated in Figure 4.6 in Chapter 4 on Photoperiodism. The circadian pattern of activity of this night-active species has two features that bear mentioning. First, the onset of activity is associated more closely with sunset, than is the cessation of activity associated with sunrise. Second, the length of the activity span closely follows the annual cycle of daily length of darkness, with a trough occurring near midsummer when nights are short and a peak near midwinter when nights are long (Figure 8.19). An increase in activity from June through August deviates from the expected association with the length of darkness and may be due to reproductive activities or other unknown factors (Figala et al., 1984).

The inclusion of oscillating temporal components in the study of animal behavior under natural conditions of the outdoor environment was superbly illustrated in a telemetry study of snowshoe hares conducted by Tester and colleagues. Prior

---

[31] The snowshoe hare is one of the primary food sources of the North American lynx (*Lynx canadensis*), whose population appears to fluctuate with the hare population (see Figure 5.8 Chapter 5 on Models).

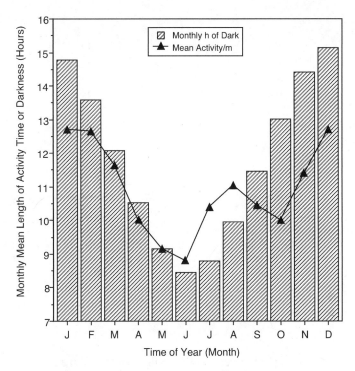

FIGURE 8.19. The length of the nocturnal activity span of the snowshoe hare (*Lepus americanus*) follows an annual cycle closely associated with the length of darkness, with a trough occurring near midsummer and a peak near midwinter. Activity monitored via telemetry for more than two years. Overall monthly means recomputed from Figure 7 in Figala et al., 1984.

to the use of telemetry, the researchers had not been successful in locating young hares, only the mature ones. After the females were captured and tagged with collar-type radio transmitters,[32] they not only located the young, but also learned about the maternal relations between the young and the activities of female in caring for them. A unique relationship became known from the rhythmic plots of the data (Figure 8.20), which showed that each day the female was with her young for only 5 to 10 min (Rongstad & Tester, 1971). This rhythmic event occurred during twilight. As the young developed, they spent the day in separate hiding places, but gathered together for only about 5 to 10 min before the female returned to nurse them. After nursing, they again dispersed and remained solitary until the following evening. In summary, the continuous monitoring of snowshoe hares by radio transmitters provided the means by which the characteristics of a

---

[32] Over the duration of two years during which the study was conducted, hares had to be captured every 3–4 months using mesh nets in order to replace the batteries in their collars.

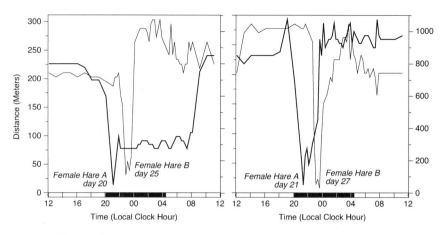

FIGURE 8.20. The distance (location) of two female snowshoe hares from their nursing sites varies rhythmically over the span of a day. Note that each female returned to nurse at about the same time (phase) during the evening twilight, but only for a brief span of about 5–10 min. Shown here are the activity patterns for two females on two separate days for each. Female A returned to nurse each night around 21:00 h, while female B returned slightly later each night at around 23:00 h (redrawn from Rongstad & Tester, 1971).

rhythm in activity were observed and the quantified. One of these characteristics, namely the phase, served as a pivotal reference span for studying behavioral and maternal relations of snowshoe hares under a natural outdoor environment.

Ruffed Grouse

Ruffed grouse are normally diurnal, but exhibit changes in rhythms of activity. These changes appear to be in response to their environment and to behavioral activities that relate to reproduction. For example, in the winter, the males are active for brief spans near sunrise and sunset, feeding on aspen (*Populus*) buds, etc., and resting in the snow between these times. In the spring and extending into summer, the activity span is much longer, and includes drumming.[33] Much of the activity for the female in the summer revolves around incubation and care of the young (Maxson, 1977). Prior to incubation, the activities of the female are associated closely with sunrise and sunset. However, during incubation of the eggs, she leaves the nest only two to three times per day to feed (total of 57 to 70 min/day). Brooding hens move with their chicks during the warmer times of the day, reaching peak activity after sunrise and slowing down before sunset, a pattern that protects the chicks from the dampness and cold (cf. Maxson, 1977). Hens without

---

[33] Drumming is a ritual performed by the male ruffed grouse during the mating season and serves to attract a female and warn away other males. It often takes place on a fallen tree or on a log. The male stands on the log and cups his wings and beats them in the air. The "thump thump" sound starts slow and then speeds up.

chicks start their activity earlier, becoming highly active close to sunrise and not stopping until near sunset and just before darkness.

The Outdoor Laboratory

The use of radio-telemetry in monitoring the rhythmic activity of free-ranging animals provides a means by which the effects of a prevailing environment (temperature, water, food, other species, various agents of stress, etc.) can be incorporated into studies on the temporal organization of individuals, as well as upon an ecosystem. It would have been difficult, if not too costly, to have duplicated the habitat of the muskrat and snowshoe hare in controlled chambers. Yet, results from these two studies and the others have greatly increased our knowledge of biological rhythms, providing a perspective that not only compliments results from laboratory experiments, but also provides a means by which to test some well-known hypotheses for biological rhythms.

As mentioned earlier, in natural settings some animals change from light-activity to dark-activity depending on the season or stage of development, with systematic changes in the phase angle between activity onset and sunrise or sunset (Aschoff, 1964). One of Aschoff's circadian rules states that an increase in the length of the photoperiod can produce an advance in the phase-angle difference in day-active animals in the summer (when the time of light is longer) and a delay in winter (when the time of light is shorter), with the opposite being the case for dark-active animals (see Figure 2.10 in Chapter on Characteristics). The activity of the night-active snowshoe hare appeared to follow this rule (Figala et al., 1984). However, the rule was not found to be true for the day-active gray squirrel in Minnesota (Figala & Tester, 1986).

Temporal Agroecosystems

Currently, the topic of agroecosystems and rural development presents a somewhat bleak picture. Rhythmic patterns of native plants and animals are being modified and this stems from the overpopulation of humans. Farmed areas have become larger, while native habitats, including forests, meadows and wetlands, have been removed, resulting in interruptions in food chains and food webs and a destabilization and decrease in native plant and animal diversity (Figala & Tester, 1990). For example, the mass emergence every 17 years of periodical cicadas is a wonderful phenomenon of nature, but already one brood in the state of Connecticut (USA) no longer exists. Brood XI, documented as far back as 1869, was last seen in 1954, and its demise was hastened by the cutting of forests for land use by humans (Maier, 1996).

Although neither large farms nor the development of rural areas are likely to disappear, natural areas can be reconstructed based on knowledge about the natural cycles in food chains and webs. It is often stated that *"you cannot stop development, you can only direct it."* Directing the reconstruction efforts, therefore, should involve the application of the knowledge that has been acquired from the study of biological rhythms, especially the patterns of feeding, locomotion, reproduction, and migration that are associated with daily and annual cycles of organisms. Ultimately, allowing species to become established in their own time niches may shorten the time needed to produce and maintain a more diverse and stable ecosystem (cf. Tester & Figala, 1990). More detailed information on the ecological significance of biological clocks with regards to the expression of behavior in a variety of nondomesticated species in their natural environment can be found in a special issue on rhythms and ecology in *Biological Rhythm Research* (Marques & Waterhouse, 2004).

Light Pollution

Throughout this text we have emphasized the importance of the LD cycle as the primary synchronizer of circadian rhythms and the timing of these spans as it applies to the process of photoperiodism. As has already been discussed in other chapters, extending the length of the light span or providing a brief exposure to light during darkness can induce biological changes in reproduction, development, and behavior. Of interest, however, is that even at low levels of intensity, light can be harmful to the temporal organization of life if presented at the wrong time. The unnatural luminescence created by outdoor lamps and light coming from buildings that leads to "urban sky glow" at night has been labeled *light pollution*[34] by astronomers and others (cf. Lockwood et al., 1990) (see Figure 10.10 in Chapter 10 on Society). Light pollution can exist in rural areas as well.

Organisms, be they on land, in the water, or in the air, may be subjected to the harmful effects that light pollution has on behavior and development. Animals[35] for which a response to light during night has been shown include zooplankton (*Daphnia*), invertebrates (insects, scuds), amphibians (frogs), reptiles (turtles), fish (trout) and birds (for light pollution effects on humans, see Chapter 10 on Society). As will be discussed in the material that follows, the combination of glass and light can also be deadly for birds, insects, and other animals (Price & Mesure, 1996).

Aquatic Animals

The effects of light pollution on a single species, can and often will have repercussions that will affect the entire food web of an ecosystem. One such system is the fresh-water lake habitat, wherein may be found various types of algae, zooplankton, plants, and animals.

---

[34] Light pollution is sometimes referred to as photopollution.

[35] In addition to animals, the flowering process in a number of species may be either inhibited or induced by the lights that are used in the home environment.

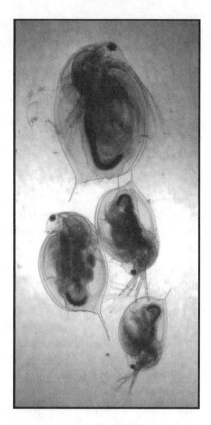

FIGURE 8.21. The small freshwater organism, *Daphnia retrocurva*, exhibits circadian patterns in vertical migration that can be disrupted by light pollution. Photo by R. Sothern.

Vertical Migration

Many fresh water lakes contain zooplankton, such as the water flea *Daphnia* (Figure 8.21). These small aquatic organisms, which are actually more common and numerous than ants are on land, display a daily rhythm in *vertical migration*, relying upon the change in light levels at twilight as an initiating cue for vertical migration (cf. Ringelberg, 1987; Haney, 1993). Although migration is influenced by a number of variables, including temperature, chemical cues, food supply (Haney, 1993), and water clarity (Dodson, 1990), the level of nighttime light influences the height to which *Daphnia* ascend.

For example, the amplitude of migration displays a linear relationship to the percent of the moon illuminated: *Daphnia* rise closer to the surface of the water during the new moon phase (when it appears dark at night) than during a full moon (when it appears bright at night), thus reducing their vulnerability to predation[36] (cf. Dodson, 1990). Especially pertinent to such behavior is that urban lighting may supply an intensity in urban lakes that is similar to that provided during a full moon phase.

---

[36] A brief, but interesting discussion related to research on light pollution can be found in Petersen (2001).

Actually, the average horizontal illuminance in certain areas of an incorporated city can easily reach levels of intensity that are five times that which is seen during full moon (cf. Lockwood et al., 1990). Results from experiments conducted in a lake near Boston, MA (USA) have shown that both the amplitude (e.g., 2 meters lower) and magnitude (e.g., fewer individuals) for the diel vertical migration of *Daphnia* were significantly reduced by brighter lake levels due to urban light pollution (Moore et al., 2001). Because of the effects of light on their vertical migration, there is a potential for less surface algae to be consumed by *Daphnia*, which in turn can lead to blooms of algae and a poorer quality of water.

Drift

In many creeks and rivers there is a downstream transport of insects and other invertebrates known as *drift* (described in Essay 8.2). Aquatic insects, such as the larvae of mayflies (genus: *Baetis*) and amphipods, such as scuds (genus: *Gammarus*), are among the more common organisms found in the drift. Typically, the drift of these organisms follows a 24-h cycle, often referred to as a diel period, with peaks occurring usually during the night.[37]

Depending upon the species, location, and quantification procedures, the distances traveled may vary from less than a meter to over 130 meters (cf. Waters, 1972). Especially relevant is the fact that artificial light at very low levels of intensity (e.g., 0.1 ft candles or 1 lux) can cause cessation of drift for both mayflies and scuds (Holt & Waters, 1966). Conceivably, this interruption of the phase of the rhythm when prey (e.g., insect) should be the highest and the organisms more available to predators (e.g., fish) could affect the dynamics of the aquatic food web.

Trout

The effects of light can also affect the emergence and feeding behavior of fish. Results from a study designed to examine the relationship between light intensity and the density of juvenile rainbow trout in a river[38] have shown that the number of fish emerging at night from their winter concealment in the river substratum is affected by the presence of light during the night span (Contor & Griffith, 1995). The number of fish emerging at night from their concealment decreases during the presence of moonlight, as well as when artificial light from a commercial billboard is present on the riverbank. Not only is the rhythmic presence, occurrence and feeding behavior of invertebrates and fish affected by artificial light at night, but so too are nocturnal frogs, which show a reduction in their ability to detect and consume prey (Buchanan, 1993).

---

[37] One of the exceptions is the day-active drift of the larvae of a caddisfly species (*Oligophlebodes sigma* Milne) in a Utah (USA) stream (Waters, 1968).

[38] The study was conducted on the Henrys Fork of the Snake River, Idaho, USA. The trout were counted by individuals using snorkeling methods (Contor & Griffith, 1995).

Turtles

A dramatic and harmful effect of light pollution on migration or orientation occurs with newly hatched loggerhead turtles (*Caretta caretta*), which must travel on land to reach the sea. On the Mediterranean coast (Peters & Verhoeven, 1994), as in Florida (Salmon et al., 1995), loggerhead turtles come ashore and build their nests on beaches. Normally, the hatchlings emerge at night (Witherington et al., 1990) and the relatively bright glow of the horizon over the ocean serves as a cue for them to immediately crawl toward the sea (Mrosovsky & Kingsmill, 1985; Witherington, 1991). However, city lights disrupt the normal sea-finding behavior of the hatchlings, directing them to travel inland, rather than to the sea (Salmon & Witherington, 1995). Disorientation can occur throughout a nesting beach, subjecting the baby turtles to the perils of urban traffic, predators, and dehydration during the extended duration of time it takes to reach the sea. In some studies, up to 63% of the hatchlings did not show a correct orientation, indicating that the effects of light pollution caused by city lighting can be potentially disastrous on the long term survival of the species (Peters & Verhoeven, 1994), especially as human activities and their settlements result in a loss of the natural darkness that is needed at the beach for normal hatchling sea-finding behavior (Salmon et al., 1995).

Insects

In our global community most people have not seen the vertical migration of water fleas, the drift of invertebrates, or the travels of loggerhead turtles, but they have seen the attraction of moths and other flying insects to light at night.[39] However, light traps have actually not eradicated moth populations, but the light has had an impact on moth circadian rhythms and photoperiodism (Frank, 1988). Light at night can disturb flight, navigation, vision, migration, oviposition, mating and feeding, and interferes with their defensive behavior (Svensson & Rydell, 1998).

Light pollution thus exposes moths to increased predation by birds, bats and spiders. The overall result may be evolutionary modification of behavior and/or disruption or elimination of entire moth populations (Frank, 1988). As reported in a newspaper article (Boudette, 2003), lights shining all night long had a devastating effect on butterfly populations in Innsbrück, Austria. After lights were added to a bridge crossing the river Inn for the 1964 Winter Olympics, butterflies swarmed to the lights during the summer months, but after three years they had all but disappeared. According to Prof. Gerhard Tarmann, who manages the butterfly collection at Innsbruck's Ferdinadeum Museum, the butterflies couldn't lay eggs or find food since they were injured by the hot white lights and/or were too exhausted from all the nighttime flying.[40]

---

[39] Excess "white" light also allows for easier orientation on land by a nocturnal carpenter ant (*Camponotus pennslyvanicus*) (Klotz & Reid, 1993).

[40] Tarmann has noted that there are now maybe eight different species of butterfly in the area when years ago there were perhaps 600. One of his suggestions is to use yellow sodium street lamps in place of the standard mercury lamps, since they do not produce

Birds

Outdoor lighting has the same potential to disrupt circadian rhythms and photoperiodism in birds as it does in insects. This disruption is especially evident during the migrating phase of the circannual rhythm of birds. Unfortunately for songbirds, which mostly migrate at night and fly at lower altitudes, the human influence is far too often deadly.

Migrating birds use a variety of cues for orientation, such as the setting sun, stars, and patterns of polarized light, but they can become confused by brightly-lit skyscrapers and towers. This can result in collisions or exhaustion, especially when birds fly low due to a low cloud ceiling (Cochran & Graber, 1958) and lead to injury or death, not only from collisions, but falling to the ground and becoming victims to predators. Thus, structures lit at night (often only for esthetic reasons) and the presence of lighted windows result in an extremely high level of *anthropogenic*[41] mortality, which for birds is recorded in the hundreds of millions each year in the USA alone (Ogden, 1996).[42]

As is the case for many biological processes, the effects of light and structures are also strongly influenced by environmental factors. This helps to explain why the confusion of migrating birds, which can be attributed to lights on towers, may be more likely to occur when the cloud ceiling is low at night, since both the maximum numbers of birds in flight and the maximum altitudes attained are generally found during the first few hours after sunset (cf. Cochran & Graber, 1958).

Take-Home Message

Biological rhythms are evident in the outdoor environment and impact upon most organisms in nature, as well as the various ecosystems found in rural, urban, and natural areas. Once again, timing can have a significant impact on outcome, be it: (*a*) the duration of light and dark or high and low temperatures that an organism receives; (*b*) light during the night that affects many organisms adversely; (*c*) the annual migration cycles of animals; (*d*) the application of herbicides or pesticides; or (*e*) the enjoyment of hobbies, including gardening, birding, and fishing.

---

the ultraviolet light that attracts the butterflies and other insects. Yellow-coated light bulbs are often marketed as "bug lights" for home use, since they attract fewer bugs when used on the back porch on a summer evening. In addition to having a more positive impact on bug populations, yellow streetlights also save energy and their use therefore indirectly reduces power plant emissions (cf. Boudette, 2003).

[41] Anthropogenic is a key word when discussing light pollution, for it defines the influence of humans on nature.

[42] More birds are killed each year by structural hazards than by high profile catastrophes, such as oil spills.

References

Alheit J, Hagen E. (1997) Long-term climate forcing of European starling and sardine populations. *Fisheries Oceanography* 6: 130–139.

Ali MA. (1992) *Rhythms in Fishes. NATO ASI Series A: Life Sciences*, Vol. 236, Plenum Press, New York, 348 pp.

Anderson-Bernadas C, Cornelissen G, Turner CM, Koukkari WL. (1997) Rhythmic nature of thigmomorphogenesis and thermal stress of *Phaseolus vulgaris* L. shoots. *J Plant Physiol* 151: 575–580.

Arthur JM, Harvill EK. (1937) Plant growth under continuous illumination from sodium vapor lamps supplemented by mercury arc lamps. *Contrib Boyce Thompson Inst* 8: 433–434.

Aschoff J. (1964) Die Tagesperiodik licht- und dunkelaktiver Tiere. *Rev Suisse Zool* 71: 528–558.

Berthold P, Gwinner E, Klein H. (1972) Circannuale Periodik bei Grasmücken I. Periodik des Körpergewichtes, der Mauser und der Nachtunruhe bei *Sylvia atricapilla* und *S. borin* unter verschiedenen konstanten Bedingunger. *J Orniothol* 113: 170–190.

Berners DJ. (1885) *Bibliotheca Curiosa. A Treatyse of Fysshynge Wyth An Angle*. Originally printed by Wynkyn de Worde in 1496. Edited by "Piscator." Privately printed, Edinburgh, 36 pp.

Bitman J, Lefourt A, Wood DL, Stroud B. (1984) Circadian and ultradian temperature rhythms of lactating cows. *J Dairy Sci* 67: 1014–1023.

Boudette NE. (2003) Effort to save butterflies casts Alpine villages in a whole new light — Amateur lepidopterist leads drive to switch to yellow street lamps. *Wall Street J (Europe)*. Brussels: January 29, 2003, pg. A.1.

Brower LP. (1995) Understanding and misunderstanding the migration of the monarch butterfly (Nymphalidae) in North America: 1857–1995. *J Lepid Soc* 49: 304–385.

Brower L. (1996) Monarch butterfly orientation: missing pieces of a magnificent puzzle *J Exp Biol* 199(Pt 1): 93–103.

Buchanan BW. (1993) Effects of enhanced lighting on the behavior of nocturnal frogs. *Animal Behav* 45(5): 893–899.

Cane MA, Eshel G, Buckland RW. (1994) Forecasting Zimbabwean maize yield using eastern equatorial Pacific sea surface temperature. *Nature* 370: 204–205.

Caspari EW, Marshak RE. (1965) The rise and fall of Lysenko. *Science* 149: 275–278.

Catchpole A, Auliciems A. (1999) Southern oscillation and the northern Australian prawn catch. *Intl J Biometeorol* 43: 110–112.

Cermak J, Jenik J, Kucera J, Zidek V. (1984) Xylem water flow in a crack willow tree (*Salix fragilis* L.) in relation to diurnal changes of environment. *Oecologia* 64 (2): 145–151.

Cochran WW, Graber RR. (1958) Attraction of nocturnal migrants by lights on a television tower. *Wilson Bull* 70(4): 378–380.

Cole CL, Adkisson PL. (1964) Daily rhythm in the susceptibility of an insect to a toxic agent. *Science* 144: 1148–1149.

Contor CR, Griffith JS. (1995) Nocturnal emergence of juvenile rainbow trout from winter concealment relative to light intensity. *Hydrobiologia* 299(3): 179–183.

Couderchet M, Koukkari WL. (1987) Daily variations in the sensitivity of soybean seedlings to low temperature. *Chronobiol Intl* 4: 537–541.

De Moraes CM, Mescher MC, Tumlinson JH. (2001) Caterpillar-induced nocturnal plant volatiles repel conspecific females. *Nature* 410: 577–580.

Dodson S. (1990) Predicting diel vertical migration of zooplankton. *Limnol Oceanogr* 35: 1195–1200.

Doran DL, Anderson RN. (1976) Effectiveness of bentazon applied at various times of the day. *Weed Sci* 24: 567–570.

Drake DJ, Evans JW. (1978) Cortisol secretion pattern during prolonged ACTH infusion in dexamethasone treated mares. *J Interdiscipl Cycle Res* 9: 88–96.

Duke SH, Friedrich JW, Schrader LE, Koukkari WL. (1978) Oscillations in the activities of enzymes of nitrate reduction and ammonia assimilation in *Glycine max* and *Zea mays*. *Physiol Plant* 42: 269–276.

Eesa N, Cutkomp LK, Cornelissen G, Halberg F. (1987) Circadian change in Dichloros lethality (LD50) in the cockroach in LD 14:10 and continuous red light. In: *Advances in Chronobiology, Part A*. Pauly JE, Scheving LE, eds. New York: Alan Liss, Inc., pp. 265–279.

Ehret DL, Ho LC. (1986) Effects of osmotic potential in nutrient solution on diurnal growth of tomato fruit. *J Exp Bot* 37(182): 1294–1302.

Elfving DC, Kaufmann MR. (1972) Diurnal and seasonal effects of environment on plant water relations and fruit diameter of citrus. *J Amer Soc Hort Sci* 97: 566–570.

Elliott JM. (1967) The life histories and drifting of the Plecoptera and Ephemeroptera in a Dartmoor stream. *J Anim Ecol* 36: 343–362.

Erwin J, Warner R. (2002) Determination of photoperiodic response group and effect of supplemental irradiance on flowering of several bedding plant species. *ACTA Hort* 580: 95–99.

Figala J, Tester JR, Seim G. (1984) Analysis of the circadian rhythm of a snowshoe hare (*Lepus americanus*, Lagomorpha) from telemetry data. *Vest Cs Spol Zool* 48: 14–23.

Figala J, Tester JR. (1986) Comparison of seasonal rhythms of activity of grey squirrels (*Sciurus carolinensis*, Rodentia) in captivity and in the wild. *Vest Cs Spol Zool* 50: 33–48.

Figala J, Tester JR. (1990) Chronobiology and agroecosystems. In: *Chronobiology: Its Role in Clinical Medicine, General Biology and Agriculture, Part B, Prog Clin Biol Res*, Vol. 341B. Hayes DK, Pauly JE, Reiter RJ, eds. New York: Wiley-Liss, pp. 793–807.

Fingerman M, Lago AD, Lowe ME. (1958) Rhythms of locomotor activity and O_2-Consumption of the grasshopper *Romalea microptera*. *Amer Mid Nat* 59: 58–66.

Forbes EB. (1901) Fight Grasshoppers with oil in your hopperdozer. *Minn Science* 31(1): 81 (Spring 1975).

Forchhammer MC, Post E, Stenseth NC. (1998) Breeding phenology and climate. *Nature* 391: 29–30.

Frank KD. (1988) Impact of outdoor lighting on moths: an assessment. *J Lepidop Soc* 42(2): 63–93.

Friend DJC, Helson VA. (1976) Thermoperiodic effects on the growth and photosynthesis of wheat and other crop plants. *Bot Gaz* 137: 75–84.

Fromentin JM, Planque B. (1996) *Calanus* and environment in the eastern North Atlantic: II. Influence of the North Atlantic Oscillation on *C. finmarchicus* and *C. helgolandicus*. *Mar Ecol Prog Ser* 134: 111–118.

Froy O, Gotter AL, Casselman AL, Reppert SM. (2003) Illuminating the circadian clock in monarch butterfly migration. *Science* 300(5623): 1303–1305.

Garner WW, Allard HA. (1920) Effect of the relative length of day and night and other factors of the environment of growth and reproduction in plants. *J Agric Res* 18: 553–606.

Goehring L, Oberhauser KS. (2002) Effects of photoperiod, temperature and host plant age on induction of reproductive diapause and development time in *Danaus plexippus*. *Ecol Entom* 27: 674–685.

Goldsmith GW, Hafenrichter AL. (1932) *Anthokinetics, The Physiology and Ecology of Floral Movements*. Carnegie Institution of Washington. Publication No. 420, 198 pp.
Gomes WR, Joyce MC. (1975) Seasonal changes in serum testosterone in adult rams. *J Anim Sci* 41(5): 1373–1375.
Guillaume FM, Kennedy BW, Carlson L, Koukkari WL. (1986) Leaf movement alterations on bean plants with common bacterial blight. *Phytopath* 76: 270–272.
Gwinner E. (1977) Circannual migrations in bird migration. *Ann Rev Ecol Syst* 8: 381–405.
Gwinner E, Wiltschko W. (1978) Endogenously controlled changes in migratory direction of the garden warbler, *Sylvia borin*. *J Comp Physiol* 125: 267–273.
Gwinner E, Wiltschko W. (1980) Circannual changes in migratory orientation of the garden warbler, *Sylvia borin*. *Behav Ecol Sociobiol* 7: 73–78.
Gwinner E. (2003) Circannual rhythms in birds. *Curr Opin Neurobiol* 13(6): 770–778.
Halaban R, Hillman WS. (1970) Response of *Lemna perpusilla* to periodic transfer to distilled water. *Plant Physiol* 46: 641–644.
Halaban R, Hillman WS. (1971) Factors affecting the water-sensitive phase of flowering in the short day plant *Lemna perpusilla*. *Plant Physiol* 48: 760–764.
Halberg F, Bittner JJ, Gully RJ, Albrecht PG, Brackney EL. (1955) 24-hour periodicity and audiogenic convulsions in I mice of various ages. *Proc Soc Exp Biol Med* 88(2): 169–173.
Halberg F. (1960) Temporal coordination of physiologic function. In: *Biological Clocks. Cold Spring Harbor Symposia on Quantitative Biology, Cold Spring Harbor, LI*. The Biological Laboratory: New York, Vol. 25, 289–310.
Halberg J, Halberg F, Lee JK, Cutkomp L, Sullivan WN, Hayes DK, Cawley BM, Rosenthal J. (1974) Similar timing of circadian rhythms in sensitivity to pyrethrum of several insects. *Intl J Chronobiol* 2: 291–296.
Hamner KC. (1960) Photoperiodism and circadian rhythms. In: *Biological Clocks. Cold Spring Harbor Symposia on Quantitative Biology*, Vol. 25. New York: Cold Spring Harbor, pp. 269–277.
Haney JF. (1993) Environmental control of diel vertical migration behaviour. *Arch Hydrobiol Beih Ergebn Limnol* 39: 1–17.
Helms K, McIntyre GA. (1967) Light-induced susceptibility of *Phaseolus vulgaris* L. to tobacco mosaic virus infection: II. Daily variation in susceptibility. *Virology* 32(3): 482–488.
Hendricks SB. (1956) Control of growth and reproduction by light and darkness. *Amer Sci* 44(3): 229–247.
Hendrix DL, Huber SC. (1986) diurnal fluctuations in cotton leaf carbon exchange rate, sucrose synthesizing enzymes, leaf carbohydrate content and carbon export. In: *Plant Biology*, Vol. 1. *Phloem Transport*. Cronshaw J, Lucas WJ, Giaquinta RT, eds. New York: Alan R. Liss, pp. 369–373.
Henson CA, Duke SH, Koukkari WL. (1986) Rhythmic oscillations in starch concentration and activities of amylolytic enzymes and invertase in *Medicago sativa* nodules. *Plant Cell Physiol* 27(2): 233–242.
Henson CA, Duke SH. (1990) Oscillations in plant metabolism. In: *Chronobiology: Its Role in Clinical Medicine, General Biology and Agriculture, Part B, Prog Clin Biol Res*, Vol. 341B. Hayes DK, Pauly JE, Reiter RJ, eds. New York: Wiley-Liss, pp. 821–834.
Hillman WS. (1956) Injury of tomato plants by continuous light and unfavorable photoperiodic cycles. *Planta* 114: 119–129.
Hillman WS. (1962) *The Physiology of Flowering*. New York: Holt, Rinehart & Winston, 164 pp.

Holt CS, Waters TF. (1967) Effect of light intensity on the drift of stream invertebrates. *Ecology* 48(2): 225–234.

Humphrey J, Shogren B. (1995) *Wisconsin & Minnesota Trout Streams. A Fly-Angler's Guide*. Woodstock, VT: Backcountry Publication, 263 pp.

Kadono H, Usami E. (1983) Ultradian rhythm of chicken body temperature under continuous light. *Jpn J Vet Sci* 45(3): 401–405.

Kennedy BW, Koukkari WL. (1987) Chronophytopathology. In *Advances in Chronobiology. Part A*. Pauly JE, Scheving LE, eds. New York, Alan R. Liss, Inc., pp. 95–103.

Kennedy BW, Denny R, Carlson L, Koukkari WL. (1986) Effect of bacterial infection on speed and horizontal trajectory of circumnutation in bean shoots. *Phytopath* 76: 712–715.

Kennedy BW, Denny R, Fetzer JL, Hills R. (1990) Modified behavior oscillations in diseased plants and its implication to epidemiology and crop loss assessment. In: *Chronobiology: Its Role in Clinical Medicine, General Biology and Agriculture, Part B, Prog Clin Biol Res,* Vol. 341B. Hayes DK, Pauly JE, Reiter RJ, eds. New York: Wiley-Liss, pp. 867–881.

Kerner von Marilaun A. (1895) *The Natural History of Plants, their Forms, Growth, Reproduction, and Distribution*. New York: H. Holt, p. 215.

Kessler A, Baldwin IT. (2001) Defensive function of herbivore-induced plant volatile emissions in nature. *Science* 291: 2141–2144.

Kettlewell PS, Sothern RB, Koukkari WL. (1999) U.K. wheat quality and economic value are dependent on the North Atlantic Oscillation. *J Cereal Science* 29: 205–209.

King AI, Reid MS, Patterson BD. (1982) Diurnal changes in the chilling sensitivity of seedlings. *Plant Physiol* 70: 211–214.

Klotz JH. Reid BL. (1993) Nocturnal orientation in the black carpenter ant *Componotus pennsylvanicus* Degeer (Hymenoptera: *Formicidae*). *Insectes Sociaux* 40(1): 95–106.

Koukkari WL. (1974) Rhythmic movements of *Albizzia julibrissin* pinnules. In: *Chronobiology*. Scheving LE, Halberg F, Pauly JE, eds. Igaku Shoin Ltd., Tokyo, pp. 676–678.

Koukkari WL, Johnson MA. (1979) Oscillations of leaves of *Abutilon theophrasti* (velvetleaf) and their sensitivity to bentazon in relation to low and high humidity. *Physiol Plant* 47: 158–162.

Koukkari WL, Warde SB. (1985) Rhythms and their relations to hormones. In: *Encyclopedia of Plant Physiology, New Series*, Vol. 11, *Hormonal Regulation of Development III. Role of environmental Factors*. Pharis RP, Reid DM, eds. Berlin: Springer-Verlag, pp. 37–77.

Koukkari WL, Duke SH, Hayes DK. (1990) Biological oscillations and agriculture: A brief introduction. In: *Chronobiology: Its Role in Clinical Medicine, General Biology and Agriculture, Part B, Prog Clin Biol Res*, Vol. 341B. Hayes DK, Pauly JE, Reiter RJ, eds. New York: Wiley-Liss, pp. 785–792.

LaFontaine G. (1981) *Caddisflies*. New York: Nick Lyons Books, 336 pp.

Lewczuk B, Przybylska-Gornowicz B. (2000) The effect of continuous darkness and illumination on the function and the morphology of the pineal gland in the domestic pig: Part I. The effect on plasma melatonin level. *Neuroendocrinol Lett* 21(4): 283–291.

Lewis H, Went FW. (1954) Plant growth under controlled conditions: IV. Response of California annuals to photoperiod and temperature. *Amer J Bot* 32: 1–12.

Liu Y, Merrow M, Loros JJ, Dunlap JC. (1998) How temperature changes reset a circadian oscillator. *Science* 281: 825–829.

Lockwood GW, Floyd RD, Thompson DT. (1990) Sky glow and outdoor lighting trends since 1976 at the Lowell Observatory. *Publ Astron Soc Pac* 162: 481–491.

Lysenko TD. (1954) *Agrobiology, Essays on Problems of Genetics, Plant Breeding and Seed Growing*. Moscow: Foreign Languages Publication, 636 pp.

Maharaj V, Carpenter JE. (1996) *The 1996 Economic Impact of Sport Fishing in Minnesota*. Alexandra, VA: Amer Sportfishing Assoc., 10 pp.

Maier CT. (1996) Connecticut is awaiting return of the periodical cicada. *Frontiers Plant Sci* 48(2): 4–6.

Marbury MO. (1988) *Favorite Flies and Their Histories*. Secaucus, NJ: The Wellfleet Press, 552 pp. [Originally published in 1892.]

Marques M, Waterhouse J, eds. (2004) Rhythms and Ecology - Do chronobiologists still remember Nature? (special issue). *J Biol Rhythms* 35(1/2): 1–170.

Martinson KB, Sothern RB, Koukkari WL, Durgan BR, Gunsolus JL. (2002) Circadian response of annual weeds to Glyphosate and Glufosinate. *Chronobiol Intl* 19(2): 405–422.

Matthews REF. (1953) Factors affecting the production of local lesions by plant viruses: I. Effects of time of day of inoculation. *Ann Appl Biol* 40: 377–383.

Matthews REF. (1953) Factors affecting the production of local lesions by plant viruses: II. Some effects of light, darkness and temperature. *Ann Appl Biol* 40: 556–565.

Matthews REF. (1991) *Plant Virology*, 3rd edn. New York: Academic Press, 835 pp.

Maxson SJ. (1977) Activity patterns of female ruffed grouse during the breeding season. *The Wilson Bulletin* 89: 439–455.

McMichael BL, Hanny BW. (1977) Endogenous levels of abscisic acid in water-stressed cotton leaves. *Agron J* 69: 979–982.

Michaels SD, Amasino RM. (2000) Memories of winter: vernalization and the competence to flower. *Plant Cell Environ* 23: 1145–1153.

Miller CS. (1975) Short interval leaf movements of cotton. *Plant Physiol* 55: 562–566.

Miller R, Martinson KB, Sothern RB, Durgan BR, Gunsolus JL. (2003) Circadian response of annual weeds in a natural setting to high and low application doses of four herbicides with different modes of action. *Chronobiol Intl* 20(2): 299–324.

Mohotti AJ, Lawlor DW. (2002) Diurnal variation of photosynthesis and photoinhibition in tea: effects of irradiance and nitrogen supply during growth in the field. *J Exp Bot* 53(367): 313–322.

Moore MV, Pierce SM, Walsh HM, Kvolvik SK, Lim JD. (2001) Urban light pollution alters the diel migration of *Daphnia*. *Verh Intl Verein Limnol* 27(2): 779–782.

Moore-Ede MC, Sulzman FM, Fuller CA (1982) *The Clocks That Time Us. Physiology of the Circadian Timing System*. Cambridge, MA: Harvard University Press, 448 pp.

Mrosovsky N, Kingsmill SF. (1985) How turtles find the sea. *Z Tierpsychol* 67: 237–256.

Nixon EH, Markhart AH III, Koukkari WL. (1987) Stomatal aperture oscillations of *Abutilon theophrasti* Medic. and *Hordeum vulgare* L. examined by three techniques. In: *Advances in Chronobiology, Part A*. Pauly JE Scheving LE, eds. New York: Alan R. Liss, pp. 67–79.

Noeske-Hallin TA, Spieler RE, Parker NC, Suttle MA. (1985) Feeding time differentially affects fattening and growth of channel catfish. *J Nutr* 115(9): 1228–1232.

Oberhauser KS. (1999) *Monarchs in the Classroom: an inquiry-based curriculum for grades 6–8. 3rd edn*. Univ. Minn.: Dept. Ecol. Evol. Behav., 216 pp.

Ogden LJE. (1996) *Collision Course: The hazards of lighted structures and windows to migrating birds. Special Report*. Toronto: World Wildlife Fund Canada and the Fatal Light Awareness Program, 45 pp.

Onyeocha FA, Fuzeau-Braesch S. (1991) Circadian rhythm changes in toxicity of the insecticide dieldrin on larvae of the migratory locust *Locusta migratoria migratorioides*. *Chronobiol Intl* 8(2): 103–109.

Peters A, Verhoeven KJF. (1994) Impact of artificial lighting on the seaward orientation of hatchling Loggerhhead turtles. *J Herpetol* 28(1): 112–114.

Petersen A. (2001) Night lights. *Amer Sci* 89: 24–25.

Price S, Mesure M. (1996) Preface. In: *Collision Course: The hazards of lighted structures and windows to migrating birds. Special Report*. Ogden LJE, ed. Toronto: World Wildlife Fund Canada and the Fatal Light Awareness Program, 45 pp.

Rathinavel S, Sundararajan KS. (2003) Chronopathological aspects of disease incidence in rice (*Oryza sativa* L). *Chronobiol Intl* 20(1): 81–96.

Reinberg, A. (1967) The hours of changing responsiveness or susceptibility. *Perspect Biol Med* 11: 111–128.

Reinberg A, Smolensky MH. (1983) Introduction to chronobiology. In: *Biological Rhythms and Medicine. Cellular, Metabolic, Physiopathologic, and Pharmacologic Aspects*. Reinberg A, Smolensky MH, eds. New York: Springer-Verlag, pp. 1–21.

Rikin A, St John JB, Wergin WPP, Anderson JD. (1984) Rhythmical changes in the sensitivity of cotton seedlings to herbicides. *Plant Physiol* 76: 297–300.

Rimmington GM, Nicholls N. (1993) Forecasting wheat yields in Australia with the Southern Oscillation Index. *Aust J Agric Res* 44(4): 625–632.

Ringelberg J. (1987) Light induced behaviour in Daphnia. In: *Daphnia*. Peters RH, De Bernardii R, eds. Verbania Palanza: Mem 1st Ital Idrobiol, pp. 285–323.

Rongstad OJ, Tester JR. (1971) Behavior and maternal relations of young snowshoe hares. *J Wildlife Manage* 35(2): 338–346.

Rowan W. (1926) On photoperiodism, reproductive periodicity, and the annual migrations of birds and certain fishes. *Proc Boston Soc Nat Hist* 38: 147–189.

Sagar PM, Glova GJ. (1992) Diel changes in the abundance and size composition of invertebrate drift in five rivers in South Island, New Zealand. *New Zeal J Mar Freshwater Res* 26: 103–114.

Sage LC. (1992) *Pigment of the Imagination, A History of Phytochrome Research*. San Diego: Academic Press, 562 pp.

Salisbury FB, Ross CW. (1992) *Plant Physiology*, 4th edn. Belmont: Wadsworth, 682 pp.

Salmon M, Tolbert MG, Painter DP, Goft M, Reiners R. (1995) Behavior of loggerhead sea turtles on an urban beach. II. Hatchling orientation. *J Herpetol* 29(4): 568–576.

Salmon M, Witherington BE. (1995) Artificial lighting and seafinding by loggerhead hatchlings: Evidence for lunar modulation *Copeia* 4: 931–938.

Sandberg G, Odén P-C, Dunberg A. (1982) Population variation and diurnal changes in the content of indole-3-acetic acid of pine seedlings (*Pinus sylvestris* L.) grown in a controlled environment. *Physiol Plant* 54: 375–380.

Schuster JL. (1990) Plains pricklypear control by night applications of phenoxy herbicides. *Proc South Weed Sci* 23: 245–249.

Schwiebert E. (1973) *Nypmhs. A Complete Guide to Naturals and Their Imitations*. New York: Winchester Press, 339 pp.

Schwemmle B, Lange OL. (1959) Endogen-tagesperiodische schwankungen der hitzresistenz bei *Kalanchoë blossfeldiana*. *Planta* 53: 134–144.

Schwemmle B. (1960) Thermoperiodic effects and circadian rhythms in flowering plants. In: *Biological Clocks. Cold Spring Harbor Symposia on Quantitative Biology*, Vol. 25. New York: Cold Spring Harbor, pp. 239–243.

Scott BIH, Gulline HF. (1972) Natural and forced circadian oscillations in the leaf of *Trifolium repens*. *Aust J Biol Sci* 25: 61–76.

Slack CR. (1965) The physiology of sugar-cane: VIII. Diurnal fluctuations in the activity of soluble invertase in elongating internodes. *Aust J Biol Sci* 18: 781–788.

Sothern RB, Okusami AE, Koukkari WL. (2001) Circadian aspects of wind-induced thigmomorphogenesis on shoot elongation of pole beans (abstract). *Chronobiol Intl* 18(6): 1192–1193.

Spieler RE. (1990) Chronobiology and aquaculture: Neglected opportunities. In: *Chronobiology: Its Role in Clinical Medicine, General Biology and Agriculture, Part B, Prog Clin Biol Res*, Vol. 341B. Hayes DK, Pauly JE, Reiter RJ, eds. New York: Wiley-Liss, pp. 905–920.

Stark JC, Halderson JL. (1987) Measurement of diurnal changes in potato tuber growth. *Amer Potato J* 64: 245–248.

Stolen PD. (1974) *Fall and winter movements and activity of muskrats in East-Central Minnesota*. Master of Science Thesis, Univ. Minn., 74 pp.

Svensson AM, Rydell J. (1998) Mercury vapour lamps interfere with the bat defense of tympanate moth (*Operophtera* spp.; Geometridae). *Anim Behav* 55: 223–226.

Tester JR. (1987) Changes in daily activity rhythms of some free-ranging animals in Minnesota. *Can Field-Naturalist* 101: 13–21.

Tester JR, Figala J. (1990) Effects of biological and environmental factors on activity rhythms of wild animals. In: *Chronobiology: Its Role in Clinical Medicine, General Biology and Agriculture, Part B, Prog Clin Biol Res*, Vol. 341B. Hayes DK, Pauly JE, Reiter RJ, eds. New York: Wiley-Liss, pp. 809–819.

To K-Y, Suen D-F, Chen S-CG. (1999) Molecular characterization of ribulose-1,5-bisphospate carboxylase/oxygenase activase in rice leaves. *Planta* 209: 66–76.

Touitou Y, Haus E. (1992) *Biologic Rhythms in Clinical and Laboratory Medicine*. Berlin: Springer-Verlag, 730 pp.

Tournois J. (1914) Études sur la sexualité du Houblon. *Ann Sci Nat (Bot)* 19: 49–191.

Upcroft JA, Done J. (1972) Evidence for a complex control system for nitrate reductase in wheat leaves. *FEBS Lett* 21(2): 142–144.

Waters TF. (1968) Diurnal periodicity in the drift of a day-active stream invertebrate. *Ecology* 49(1): 152–153.

Waters TF. (1972) The drift of stream insects. *Ann Rev Entomol* 17: 253–272.

Weaver ML, Nylund RE. (1963) Factors influencing the tolerance of peas to MCPA. *Weeds* 11: 142–148.

Went FW. (1944) Plant growth under controlled conditions. II. Thermoperiodicity in growth and fruiting of the tomato. *Amer J Bot* 31: 135–150.

Went FW. (1957) *The Experimental Control of Plant Growth*. New York: Ronald Press, 343 pp.

Went FW. (1960) Photo- and thermoperiodic effects in plant growth. In: *Biological Clocks. Cold Spring Harbor Symposia on Quantitative Biology*, Vol. 25. New York: Cold Spring Harbor, pp. 221–230.

Went FW. (1962) Ecological implications of the autonomous 24-hour rhythm in plants. *Ann NY Acad Sci* 98: 886–875.

Went FW. (1974) Reflections and speculations. *Ann Rev Plant Physiol* 25: 1–26.

Witherington BE, Bjorndal KA, McCabe CM. (1990) Temporal pattern of nocturnal emergence of loggerhead turtle hatchlings from natural nests. *Copeia* 4: 1165–1168.

Witherington BE. (1991) Orientation of hatchling loggerhead turtles at sea off artificially lighted and dark beaches. *J Exp Mar Biol Ecol* 149(1): 1–11.

Wititsuwannakul R. (1986) Diurnal variation of 3-hydroxy-3-methylglutaryl coenzyme A reductase activity in latex of *Hevea brasiliensis* and its relation to rubber content. *Experientia* 42: 44–45.

Wood WML. (1953) Thermonasty in tulip and crocus flowers. *J Exp Bot* 4: 65–77.

Yoo KC, Uemoto S. (1976) Studies on the physiology of bolting and flowering in *Raphanus sativus* L.: II Annual rhythm in readiness to flower in Japanese radish, cultivar 'Wase-shijunichi.' *Plant Cell Physiol* 17: 863–865.

9
Veterinary Medicine

> *"Personally, I have always felt the best doctor in the world is the Veterinarian.*
> *He can't ask his patients what's the matter.*
> *He's got to know."*
> —*Will Rogers (1879–1935), American humorist*

Introduction

The focus of veterinary[1] medicine is on the science and art of treatment and prevention of disease and injury of nonhuman animals, especially domestic ones. Health care services provided by veterinarians are much needed and appreciated in both urban and rural areas. Households in the USA that own pets (dogs, cats, birds, etc.)[2] (Figure 9.1) visit a veterinarian an average of three times per year (AVMA, 2002). Health care is devoted also to livestock (cattle, sheep, swine), horses and poultry, both in the clinic and during visits to farms and ranches.

Many of the applications of biological rhythms, although perhaps not always realized as to cause, date back centuries to the earliest practices of animal husbandry. Knowing and being able to predict when events change and recur is highly advantageous to a veterinarian. Deviations from the predictable rhythmic pattern of physiological variables in animals provides a valuable tool that has diagnostic possibilities and therapeutic advantages in veterinary medicine (cf. Piccione & Caola, 2002). This is especially pertinent when procedures, observations, and treatments need to be performed on animals at various times of day or year.

Rhythms found in pets and livestock span a range that extends from high frequency *ultradian* oscillations (Table 9.1) to the *circadian* (Table 9.2) and *infradian* domains (Table 9.3). Very likely, circadian rhythms, as well as the interactions between circadian and *circannual* rhythms associated with *photoperiodism*, are foremost among the periodicities encountered by a veterinarian. In the sections that follow, biological rhythms in veterinary medicine are addressed as they relate to

---

[1] from (L.) *veterinarius* = pertaining to beasts of burden.
[2] Horses are sometimes included in demographic sources as pets.

342   9. Veterinary Medicine

FIGURE 9.1. Animals that are kept as pets may need veterinary care at one time or another. Photos by R. Sothern (top 3) and Marcia Koukkari (bottom 4).

body temperature and activity, hematology and urology, primary circadian clocks, disease, pests and stress, and reproduction and photoperiodism.

Body Temperature and Activity

Veterinarians use body temperature as a clinical parameter to ascertain health status, to monitor progression of infectious diseases, to detect systemic side-effects after vaccinations, and to define humane endpoints in animal experiments (Hartinger et al., 2003).

TABLE 9.1. Some variables showing ultradian oscillations in pets and livestock.

Animal	Variable	Period	Reference
Cat	Catecholamine release	~10 min and 1 h	Lanzinger et al., 1989
Dog	Cortisol in blood	3 and 90 min[b]	Benton et al., 1990
Dog	Gallbladder bile composition (8 compounds)	~90 min	Camello et al., 1991
Cattle	Udder temperature	80–120 min	Bitman et al., 1984
Chicken	Core Body Temperature	~5 h	Kadono & Usami, 1983
Cockerel	LH[a]	66.4 ± 4.0 min	Chou & Johnson, 1987
Horse	Plasma glucose	~18 min	Evans & Winget, 1974
Monkey	LH (during follicular phase of the menstrual cycle)	~50 min	Norman et al., 1984
Sheep	Core Temperature	90 and 140 min	Mohr & Krzywanek, 1990

[a]Luteinizing hormone.
[b]Sampling was every 15–20 s.

Livestock and most domestic pets are *homeotherms*[3] and exhibit a daily rhythmic variation in body temperature that can range, on average, from 0.5°C to 1.2°C (0.9°F to 2.2°F) (cf. Hahn, 1989). Within this range, some relatively large fluctuations have been attributed to a variety of factors, including activity, metabolism, physiological changes, availability of food, and agents of environmental stress. However, perhaps the most common and often overlooked single factor that underlies recurrent changes in body temperature is the endogenous biological rhythm. With this in mind, knowledge of normal characteristics of circadian and estrus body temperature rhythms in farm animals is important for increasing the accuracy of diagnostic procedures and for optimizing yield in dairy and meat industries (Piccione & Refinetti, 2003a).

Diurnal vs. Nocturnal

Body temperature rhythms have been studied extensively in many of the species commonly seen by a veterinarian (Table 9.4). Generally, body temperatures are associated with activity and are usually highest during the middle or second half of the daily activity span (Figure 9.2). Thus, for animals that are diurnally active, such as monkeys and chickens, their maximum body temperature generally is found during the light span, while the maximum body temperature of nocturnally active animals, such as most rats and mice, occurs during the dark span (cf. Sothern, 1995).

---

[3] Homeotherms (also warm blooded or endotherms) regulate their body temperature independent of the environmental temperature. All other species of animals are classified as ectotherms (also cold blooded or poikilotherms) and depend upon external sources of heat (solar heat) to regulate their body temperature. Nevertheless, rhythms in heat-seeking behavior and locomotor activity have been documented in ectotherms such as the lizards *Iguana iguana* (Tosini & Menaker, 1995) and *Gekko gecko* (Refinetti & Susalka, 1997). A third category, heterotherms, includes some common hibernators which regulate their temperature only at certain times.

TABLE 9.2. Some variables showing circadian oscillations in pets and livestock (during monitoring: LD = Light/Dark cycles present; DD = continuous darkness; LL = continuous light).

Lighting	Animal	Variable	Time(s) of maxima	Reference
LD, LL	Cat	Melatonin in blood	Midday	Reppert et al., 1982
LD	Cattle	Urea in blood	Morning	Lefcourt et al., 1999
LD, DD	Chicken	Pineal N-acetyltransferase enzyme activity in blood	Midday	Binkley, 1976
LD	Dog	Bone formation markers in blood and urine	Early morning	Ladlow et al., 2002
LD	Dog	Gallbladder bile composition (6 compounds[a])	Morning	Camello et al., 1991
LD	Dog	Gallbladder bile composition (chloride)	Evening	Camello et al., 1991
LD, DL	Goldfish	Brain histamine	During L	Burns et al., 2003
LD	Horse	AspAT and AlAT enzyme activity in blood[b]	Near midnight	Komosa et al., 1990
LL, DD	Pig	Melatonin	Usual dark span	Lewczuk & Przybylska-Gornowicz, 2000
LD	Turkey	Gastric motility	During L	Duke & Evanson, 1976

[a]Bilirubin, bile salts, phospholipids, cholesterol, sodium, potassium.
[b]AspAT = aspartate aminotransferase, AlAT = alanine aminotransferase.

Timing of Food

Timing of food has been associated with temperature rhythms in pigs fed once or twice a day, suggesting a possible conditioning to the presentation of food (cf. Ingram & Dauncey, 1985). The circadian rhythm of body temperature continues in young pigs maintained under conditions of constant ambient temperature and illumination and deprived of food for three days (Ingram & Mount, 1973).

TABLE 9.3. Some variables showing infradian oscillations in pets and livestock.

Domain	Animal	Variable	Time(s) of maxima	Reference
Estrous	Cow	Body temperature	Every 18–23 days (mean = 21 days)	Piccione et al., 2003b
28-day menstrual	Monkey	Bone formation and resorption biomarkers in blood	Days 2 to 7	Hotchkiss & Brommage, 2000
Annual	Chicken	Embryo O$_2$ consumption	Spring	Johnson, 1966
"	Dog	Uric acid in blood	Fall	Sothern et al., 1993
"	Goat	Serum cortisol	Winter	Alila-Johanasson et al., 2003
"	Horse	RBC, WBC and Hemoglobin in blood	March	Piccione et al., 2001
"	Sheep (ewe)	Estrous	Late summer–early fall	Karsch et al., 1984
"	Sheep (ram)	Testosterone	May–July	Gomes & Joyce, 1975

TABLE 9.4. Circadian rhythm in body temperature in animals during synchronized LD conditions.

Animal	Site	Time(s) of maxima	Reference
Cat	Brain	Midday with 2–4 h ultradians	Kuwabara et al., 1986
Cow (lactating)	Udder	21:00–02:00 h	Lefcourt et al., 1999
Cow (lactating)	Vagina	Afternoon	Araki et al., 1985
Cow (calves)	Ear	12:00–17:00 h	Macaulay et al., 1995
Chicken	i.p.[a]	During L-span	Hawking et al., 1971; Kadono & Usami, 1983
Dog	Rectum	Late in L-span	Refinetti & Piccione, 2003
Duck	i.p.	During L-span	Hawking et al., 1971
Goat	Rectum	18:00 h; Early-dark in LD12:12	Ayo et al., 1999; Piccione et al., 2003c
Gerbil	i.p.	Midday	Refinetti, 1999
Hamster	i.p.	Midday	Refinetti, 1999
Horse	Rectal	20:00 h; 23:00 h	Stull & Rodiek, 2000; Piccione et al., 2004a
Monkey	Colon	During L-span	Fuller et al., 1979
Mouse	Rectum	Midday (activity) span	Haus & Halberg, 1969
Pig	Rectum	During activity (easily altered by feeding)	Ingram & Dauncey, 1985
Rat	i.p.	Midday (activity) span	Sothern & Halberg, 1979
Sheep (lambs)	i.p.	Early L (07:00–10:00 h)	Davidson & Fewell, 1993
Sheep (adults)	Carotid artery	16:00 h	Mendel & Raghavan, 1964
Sheep (adults)	Rectum	16:00–18:00 h	Mendel & Raghavan, 1964
Squirrel (flying)	i.p.	Midday	Refinetti, 1999
Squirrel (ground)	i.p.	2nd half of L-span	Refinetti, 1999
Tree Shrew	i.p.	Mid-L	Refinetti, 1999

[a]i.p. = Intraperitoneal.

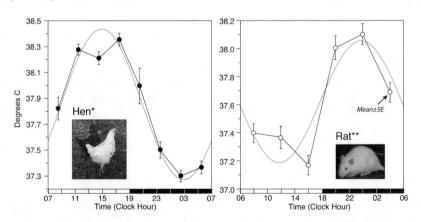

FIGURE 9.2. Examples of circadian rhythms in body temperature of animals. Hen: three-hour means computed from intraperitoneal temperature monitored every 15 min for 2 days in a full-grown hen (L-on 07:00 h–19:00 h) (redrawn from Fig 11 in Hawking et al., 1971). Rat: rectal temperature sampled from 24 female LOU rats (4 rats/timepoint; age 3 months) every week for eight weeks (total = 192 samples) (L-on 06:00 h–18:00 h) (Sothern unpublished). Dark bar = lights off.

However, feeding *ad libitum* (i.e., upon urge) obliterated the rhythm that had been present, suggesting that pigs have an endogenous body temperature rhythm that is easily masked by food intake.

This observation seems to be unique in the domesticated pig, since most animals maintained under constant illumination with continuous access to food maintain a circadian rhythm in body temperature. However, because so many external variables can affect body temperature, especially activity and access to food, the exact characteristics of endogenous rhythms, such as phase, amplitude, and period, can be easily perturbed or *masked* in animals.

Masking

Masking of the endogenous characteristics of circadian rhythms is quite common and can affect ultradian, as well as infradian rhythms. Furthermore, masking is not confined only to temperature (see discussion on masking in mammals and other organisms in Chapter 2 on Rhythm Characteristics) and can involve hormones and other molecules. For example, when mice were fed once a day during the first 4 h of light (L) or dark (D) of an LD 12:12 regimen, the 24 h means and amplitudes for body temperature, serum corticosterone and liver glycogen were all significantly increased when compared with mice that had food continuously available (Nelson et al., 1975). In addition, the peaks (acrophases) in the meal-fed mice were determined by the time of food presentation regardless of its relation to the LD regimen (Figure 9.3). Similar phase adaptations were observed in peak times for liver glycogen and serum glucose when rats were maintained on four different 4-hourly feeding times along the LD regimen (Philippens et al., 1977). Phase adaptations were also observed for body temperature, liver glycogen, plasma insulin, serum glucose, and serum thyroxine when mice were meal-fed at six different 4 h intervals over 24 h (Lakatua et al., 1988).

In the three studies just mentioned, however, the peaks for some variables (corneal mitoses, white blood cells, enzymes) remained synchronized more closely with the LD schedule than the restricted feeding schedules or showed an interaction between the two competing synchronizers. Thus, one should not generalize about the synchronizing effect of meal-timing since all body functions do not react equally to different synchronizers and this can result in alterations in internal phase relations that are normally present during *ad libitum* feeding.

Environment

Depending upon the species and the prevailing environment where the animal is maintained, the 24 h cycle for body temperature can be quite complex and difficult to quantify. Some of the changes in body temperature can be attributed to the influence of instrumentation and measurement site (e.g., intraperitoneal, tympanic, rectal, udder, etc.), the environment (housing, pastures, kennels) and animal activity.

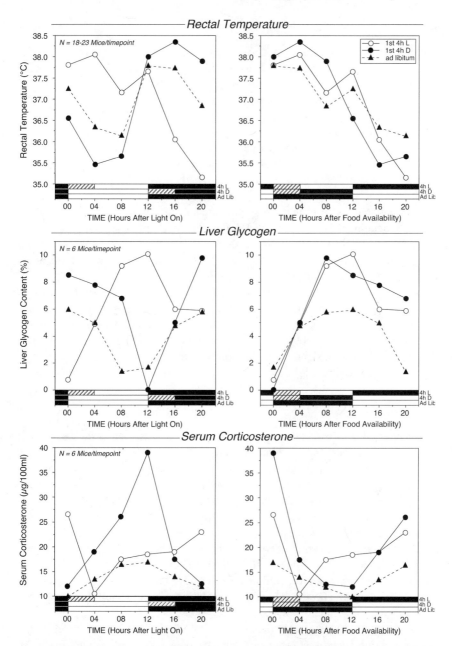

FIGURE 9.3. Timing of a single daily meal can alter (mask) circadian rhythm characteristics of body temperature, liver glycogen and serum corticosterone in mice. Mice allowed access to food during the first 4 h of the L-span or D-span or continuously (ad lib) for 4–6 weeks. Timepoint means (SEs not shown for clarity) shown aligned to onset of light (L) [*left*] and onset of food access [*right*] (mice fed *ad libitum* began eating at the beginning of the D-span). Time of food presentation (hatched bars) determined peak (acrophase) regardless of the relation of food to the LD regimen (light–dark bars). Redrawn from Nelson et al., 1975.

Cattle

For dairy cows, activities such as repeated positional transitions from lying to grazing (Lefcourt, 1990), as well as the physiological and metabolic process associated with lactation, daily feeding(s) and milking (e.g., at 08:00 h and 20:00 h) quite likely factor into the rhythmic profile for body temperature. The normal body temperature rhythm of lactating dairy cows, which has a peak between 18:00 h and midnight and a trough from late morning to early afternoon, displays a range of change over 24 h on the order of 0.5°C to 1.0°C (0.8°F to 1.8°F) (cf. Hahn, 1989). Ultradian oscillations with periods averaging about 100 min between peaks may also appear in the circadian pattern (Lefcourt, 1990).

Encapsulated radio transmitters implanted in the peritoneal cavity near an udder of lactating dairy cows showed changes in hourly means in temperature that ranged from 38.82°C (100.8°F) to 39.09°C (102.4°F), with peaks occurring mostly late at night[4] between 21:00 and 02:00 h and troughs located in the afternoon between 13:00 h and 18:00 h (Lefcourt et al., 1999).

On the other hand, the 24-h tympanic (ear) temperature rhythm of Holstein calves that were in the nursing stage of development, displayed maximum values during the day between 12:00 and 17:00 h and minimum values in the morning between 06:00 h and 09:00 h (Macaulay et al., 1995). Generally, for animals not experiencing stress and feeding *ad libitum,* the 24 h rhythm in body temperature for cattle has a peak in the afternoon or evening (Hahn et al., 1992).

Dogs and Cats

The task of taking the rectal temperature of a pet, such as a cat or dog, is in itself no small feat to accomplish and often arouses the activity of the animal. In many laboratory studies of animals, activity and body temperature are monitored continuously via an implanted sensor (Gegout-Pottie et al., 1999; Hartinger et al., 2003). After recovery from the implant surgery, the sensor usually does not interfere with the animal's normal activities and thus the effect of handling or restraint is minimized, if not removed.

Under such experimental conditions, the activity[5] of dogs appears to be only slightly greater during the light portion of the LD cycle. Over a typical 24 h span, changes in rectal temperature seldom exceed 1°C (e.g., 37.9°C to 39.0°C; 100.2°F to 102.2°F), but do show a tendency to rise during L and to fall during early D (Hawking et al., 1971).

As for cats, circadian variations in activity and body temperature levels are complex and include ultradian cycles that have been reported to range from 1–2 h

---

[4] One of the six cows that were monitored in the study displayed a rhythm with a peak occurring near midday at 12:30 h., which was 180° out of phase with the other animals.

[5] Dogs in a laboratory setting spend little time walking around and tend to sit most of the time and only move their head or a leg.

(Hawking et al., 1971)[6] or 2–4 h (Kuwabara et al., 1986)[7]. Patterns in nocturnal wakefulness and high temperatures in cats display bimodal peaks, which occur near dawn and dusk (Kuwabara et al., 1986).

Poultry

Unlike cats and dogs, the daily rest/activity cycle of free-range chickens is easily observed visually. Chickens feed during L and are at roost during D, with some anticipation of the change from L and D (Figure 9.4). Likewise, the body temperature of poultry has a pronounced 24-h rhythm. For example, higher temperature levels for a hen, as well as for a duck, occur during the active L span[8] (Hawking et al., 1971).

Hematology and Urology

The emphasis of *hematology* is upon blood, while the focus of *urology* is upon urine. Both blood and urine are used extensively for diagnostic purposes in clinical veterinary medicine.

Sampling Blood

Blood is a tissue that is quite distinct from other tissues in that its intercellular matrix is a liquid and the tissue circulates within the body via veins and arteries. In both veterinary clinics and in research studies, samples are collected by venipuncture and sent to commercial or clinical laboratories for hematological determinations of cell types, organic molecules, elements, etc.

Multiple Rhythms

The number of variables found in blood that are known to cycle with a significant ultradian, circadian or circannual periodicity is large. In dogs alone, at least 30 variables have been documented as showing circannual variations in morning baseline values (cf. Sothern et al., 1993). Some examples of rhythms in hematological variables in mammals are presented in Table 9.5.

The rhythm characteristics of the various biochemical molecules that circulate within the vascular system of an animal can be different. One could expect to find differences in amplitude, which depend upon recurring processes of synthesis,

---

[6] A single cat was monitored via an implanted radio-telemetry capsule. About 1.4 times more activity occurred during the day than night and slightly higher temperatures occurred during the night.

[7] The brain temperature of 16 cats maintained in LD 12:12 was monitored via implanted electrodes.

[8] Hawking et al. (1971) studied one hen and one duck using a "radio capsule" inserted into the body.

A) Feeding during light-span

FIGURE 9.4. A daily activity/rest cycle of free-range chickens: (*a*) feeding during the light span, (*b*) returning to housing at end of the light span, and (*c*) roosting during the dark span (photos courtesy of Krishona Bjork-Martinson).

B) Returning to housing at end of light-span

C) Roosting during dark-span

utilization and degradation, but both the period and the phase, as has been discussed relative to changes in temperature and activity, also differ.

In addition, one biological variable may have only one prominent period, while others may have statistically significant periods in two or more domains. For example, in the same lactating cow the plasma concentration of urea was shown to be circadian, cortisol levels were only ultradian, while prolactin levels showed both circadian and ultradian oscillations (Lefcourt, 1990). Sampling intervals, as

TABLE 9.5. Examples of some variables in blood that have been reported to oscillate (cycle) in animals.

Domain	Animal	Variable	Time(s) of maxima	Reference
Ultradian	Cow	Urea	40–80 min between peaks	Lefcourt et al., 1999
"	Horse	Packed cell volume and platelets	Near noon and midnight	Piccione et al., 2001
Circadian	Cow	Insulin	Late afternoon	Lefcourt et al., 1990
"	Dog	Melatonin	Mid-dark (02:00 h)	Sääf et al., 1980
"	Goat	Cholesterol	End of dark in LD12:12	Piccione et al., 2003c
"	Goat	Urea	During dark in LD12:12	Piccione et al., 2003c
"	Horse	Cortisol	Morning	Komosa et al., 1990; Black et al., 1999
"	Horse	Leptin	Mid-dark	Piccione et al., 2004 a
"	Horse	Soluble vitamins (A, D, E, K)	Mid to late afternoon	Piccione et al., 2004 b
"	Mouse	RBC, WBC, hemoglobin, hematocrit	1st half of L (resting) span	Swoyer et al., 1987
"	Pig	Cortisol	Morning	Griffith & Minton, 1991
"	Monkey	Melatonin	During darkness	Jenkin et al., 1980
Infradian (annual)	Dog	Creatinine	May–July	Sothern et al., 1993
"	Horse	Blood cell sedimentation rate	Late spring, early summer	Gill & Kompanowska-Jezierska, 1986
"	Horse	Testosterone	Summer	Byers et al., 1983
"	Mouse	Corticosterone	Winter	Haus & Halberg, 1970

well as the environment and the phase of the synchronizer (L or D span), can also contribute to which periods are detectable.

Peak Times

The phase of a hematological variable, especially the peak or trough, serves as an important parameter, potentially applicable in diagnosis and therapy. Twenty-four hour acrophase charts, which identify the time of peaks for rhythmic variables (Figure 9.5) are available in human clinical medicine (Dawes, 1974; Haus et al., 1988; Kanabrocki et al., 1990; Smaaland & Sothern, 1994; Abrahamsen et al., 1999; Bjarnason et al., 2001), and for several animals. These include *mice* (Halberg & Nelson, 1978; Scheving, 1984; Feuers et al., 1986; Smaaland & Sothern, 1994), *rats* (Ehret et al., 1978; Haus & Halberg, 1980), *monkeys* (Halberg & Nelson, 1978; Moore-Ede & Sulzman, 1981), and *horses* (cf. Komosa et al., 1990). A one-year acrophase chart has also been prepared for *dogs* (Sothern et al., 1993).

Collecting Urine

Urine is a readily available, noninvasive source of rhythmic physical and biochemical information. In veterinary medicine, urine is usually collected through

9. Veterinary Medicine

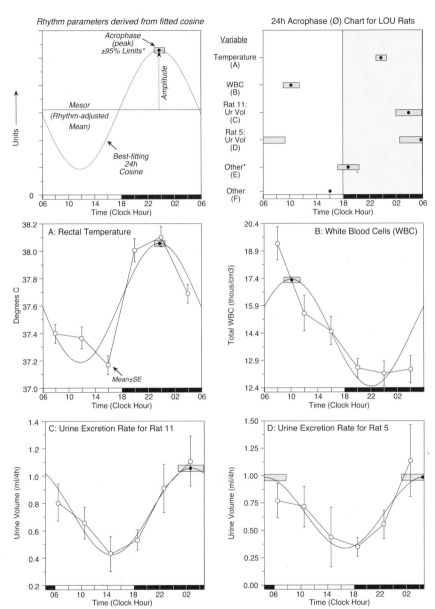

FIGURE 9.5. Using the acrophase (upper left) one can construct an acrophase chart (upper right) in order to compare timing of peak for various variables (shown here for body temperature, white blood cells and urine excretion rate of rats). 95% Limits = Acrophase ($\emptyset$) ± 1.96 SE and shown if rhythm significant at $p \leq 0.05$ from zero-amplitude test. Note: In the acrophase chart, 95% limits (bar) not shown for variable F because of the lack of significance. Dark bar = Lights off. (Sothern, unpublished).

FIGURE 9.6. Metabolic cages for collection of animal urine samples in the laboratory. Whenever an animal urinates, in this case a rat, the liquid flows via a funnel below the grid base of the cage to a collection cup or tube. A new cup is rotated into place at specified intervals (e.g., every 2, 3, or 4 h). Photo by R. Sothern.

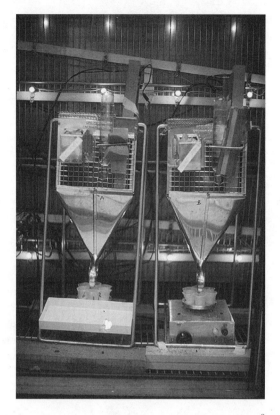

external excretion, often by using a midstream catch upon voiding the bladder.[9] However, needle puncture (cytocentesis) of the bladder of cats, dogs, and other small animals is sometimes used to obtain uncontaminated urine, often needed for the identification of bacteria causing urinary tract infections.

In some laboratory studies with animals, urine can be collected via an indwelling catheter or the animal is housed in a metabolic cage that can collect the entire urine volume excreted over a specific span of time (e.g., 3 h, 4 h, etc.) (Figure 9.6). When using a metabolic cage, successive samples can be collected from animals housed in a cage (e.g., for mice, rats, dogs, etc.) or confined to a chair (e.g., for monkeys), thereby avoiding the trauma associated with catheters and physical restraints that could impede other normal functions.

Excretion Rates

Unlike blood, where its total volume does not change appreciably over 24 h and the concentration of a substance will thus reflect the absolute amount available at the time of sampling, the level of a substance determined in a urine sample will

---

[9] Often, race horses that are routinely tested for drugs and other compounds have been trained to urinate at the sound of a whistle.

TABLE 9.6. Examples of some variables in urine that have been reported to oscillate (cycle) in animals.

Domain	Animal	Variable	Time(s) of maxima	Reference
Ultradian	Dog	Volume, Na, K, osmolality	Peaks every 150–200 min	Gordon & Lavie, 1982
"	Monkey	Potassium	12 h major and minor peaks	Kass et al., 1980
Circadian	Dog	Volume, Na, K, Ca, Mg	10:00–14:00 h	Hartenbower et al., 1974
"	Hamster	Melatonin	During dark span	Stieglitz et al., 1995
"	Horse	Bone resorption markers	02:00–08:00 h	Black et al., 1999
"	Monkey	Potassium	1st half of activity (L) span	Kass et al., 1980
"	Rat	Volume, protein, glomular filtration rate	midday (activity) span	Pons et al., 1996
Infradian (menstrual)	Monkey	3 bone formation and resorption biomarkers	Days 3, 7, & 28 of 28-day cycle	Hotchkiss & Brommage, 2000

depend upon the total volume of the voided sample. While adjustment for total volume may not be necessary for the detection of bacteria or presence of certain drugs or other trace elements, it is necessary for rhythm studies that seek to quantify excretion rates of metabolic substances at different times of the day or year (see Essay 11.1 on adjusting urinary analyte concentrations for volume and collection interval in Chapter 11 on Clinical Medicine).

Urinary Rhythms

Rhythms in urinary variables (Table 9.6) have provided a tool by which the health status and other metabolic information of animals can be quantified, such as for *dogs* (Liesegang et al., 1999, *rabbits* (Jilge & Stahle, 1984; Jilge, 1991), *fetal sheep* (Brace & Moore, 1991), *pregnant sows* (Pol et al., 2002), *rats* (Rabinowitz, et al., 1987; Pons et al., 1994; Schmidt et al., 2001), *monkeys* (Sulzman et al., 1978), *chimpanzees* (Muller & Lipson, 2003), and *gorillas* (Stoinski et al., 2002).

Urine has also been collected around-the-clock from rats[10] to assess excretion rates of amino acids associated with the presence of a tumor (Halberg et al., 1976, 1978) and to evaluate renal toxicity associated with anti-cancer drugs (Levi et al., 1982). Around-the-clock urine collection from rats has also documented a decrease in circadian mean and amplitude in excretion rates of glucose, sodium, and potassium associated with aging (Langevin et al., 1979).

Interpreting a Sample

Usually, only a single blood or urine sample is obtained in a clinic and this could take place at any given time of day (e.g., morning or late evening, etc.) or season of the year. Since the baseline for many hematological and urinary variables is

---

[10] While rats are used extensively in medical research, they are sometimes kept as pets at home.

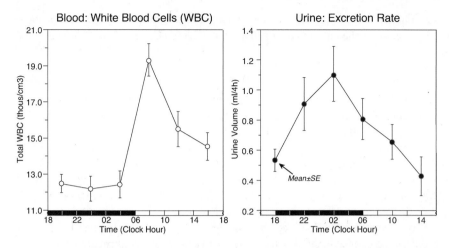

FIGURE 9.7. Examples of circadian patterns found in rat blood and urine. Blood sampled from tail of 24 female LOU rats (4 rats/timepoint; age 3 months) every week for 8 weeks (total = 192 samples). Urine collected every 4 h for 14 days from a female LOU rat (age 3 months) in a metabolic cage (total = 83 samples). Means assigned to midpoint of each 4 h sampling interval. Dark bar = lights off (Sothern, unpublished).

more like a curve than a straight line (Figure 9.7), the correct interpretation of a value will actually depend upon where it falls on the normal curve throughout the day (i.e., *when* the sample was obtained).

Even if the clock hour of sampling is known, however, the correct interpretation or extrapolation of a value present in the sample may depend not only upon the external clock time, but in some cases it may depend on the internal biological time of the animal (cf. Sothern, 1995). For example, the animal in question could have experienced a disruptive effect in its environment, such as an unusual LD exposure (e.g., a very long or short light span, continuous light, or a reversed LD schedule often used for nocturnal animals on display in zoos), extreme environmental temperatures, food availability, and/or social disruptions.

Masking is another factor, as was discussed earlier, when *ad libitum* feeding during continuous light (LL) masked the circadian temperature rhythm in young pigs. However, circadian rhythms continued in plasma cortisol and melatonin in pigs exposed to LL or DD for 14 days, thereby confirming the endogenous basis for the circadian cycles found in these variables during entrainment to a LD schedule (Griffith & Minton, 1991).

A Primary Circadian Oscillator

Individual cells of mammalian tissues can generate circadian rhythms, but a primary oscillator resides in the hypothalamus of the brain (Richter, 1965). This is a basic structure in mammals where many functions are controlled, including internal body temperature, blood pressure, reproduction, sexuality, and the production of neurohormones. The location of the *hypothalamus*, *suprachiasmatic nucleus*, and *pineal gland* in the rat brain is shown in Figure 2.13 in Chapter 2 on Characteristics.

The Suprachiasmatic Nucleus

Within the hypothalamus, there is a site where two clumps of nuclei containing thousands of neurons are located above the optic chiasm and where the two optic nerves cross. This area is called the suprachiasmatic nucleus (SCN). Circadian rhythms are synchronized by light signals received by retinal photoreceptors (rods, cones, and ganglion cells containing *melanopsin*) and projected via a *retinohypothalamic tract* (RHT) to the bilaterally paired SCN.[11] The light information received by the SCN is transduced into neural and hormonal output signals that affect various rhythms of the animal (cf. Lowrey & Takahashi, 2004).

The SCN has been identified as a primary circadian clock of mammals. Results from a large number of studies that led to the discovery and confirmation of the SCN as a primary oscillator have been introduced elsewhere in this text (see Chapter 2 on Rhythm Characteristics) (Richter, 1965; Hendrickson et al., 1972; Moore & Lenn, 1972; Stephan & Zucker, 1972; Moore & Eichler, 1972; Ralph et al., 1990; Sujino, et al., 2003). An extensive and rapidly expanding body of literature on various aspects of the SCN clock continues (cf. special issues: *Chronobiol Intl* 1998:15[5]; *J Biol Rhythms* 2003:18[3].), as investigations into the functions of this major pacemaker are explored at the molecular level, including the description of clock genes (for a discussion of clock genes, see Chapter 5 on Models and Mechanisms).

As is true in many areas of biology, there are reports that raise the possibility of other primary circadian oscillators in mammals outside the brain and retina (Zylka et al., 1998). In addition, it must be emphasized that while peripheral oscillators occur throughout the body of an animal, the SCN of mammals does not drive these other oscillators, but serves as a synchronizer for peripheral clock genes (cf. Hastings et al., 2003; Lowrey & Takahashi, 2004; Panda & Hogenesch, 2004; Yoo et al., 2004).

The Pineal Gland

In birds, the pineal gland functions as a primary circadian pacemaker. Rhythms of both activity and body temperature are abolished in pinealectomized sparrows maintained in DD (Menaker, 1974). Interestingly, the rhythm can again be initiated when the pineal from another sparrow is transplanted into the recipient, although the phase is that of the donor bird (Zimmerman & Menaker, 1979).

In chickens, a circadian rhythmicity in the activity of an enzyme that is involved in the serotonin to melatonin pathway (*N*-acetyltransferase) continues when pineal tissue is maintained *in vitro* in constant darkness for at least 4 days (Binkley et al., 1978; Kasal et al., 1979). Furthermore, a rhythm in melatonin released by

---

[11] See Chapter 2 for discussion of primary circadian clocks and the roles of melanopsin., the RHT and the SCN.

individually isolated chick pineal glands maintained *in vitro* during 2 days in LD also persists over 4 days in DD with a free-running circadian period slightly shorter than 24.0 h and a dampening of the amplitude (cf. Takahashi et al., 1980).

The pineal gland in fish is a direct photosensory organ that is a part of a central neural system (cf. Zachmann et al., 1992; Ekström & Meissl, 1997). Under natural conditions, the pineal releases melatonin rhythmically to convey daily and seasonal photoperiodic information that affects thermoregulation, activity patterns, growth, metabolism, and reproduction. While the photic regulation of melatonin production may be universal in fishes, there is evidence for a lack of universal control of endogenous circadian melatonin rhythms in the pineal and retina, since in some fish, production of melatonin is consistently high when maintained under DD and low when maintained under LL (Iigo et al., 1997). This suggests the presence of two regulatory systems in fish, one depending upon photoperiodic and the other on an internal circadian system. Nevertheless, circadian melatonin rhythms in the pineal gland and retina of all vertebrate classes have been reported, except for a few species, such as the rainbow trout (*Oncorhynchus mykiss*), desert iguana, *Dipososaurus dorsalis*, the gecko, *Christinus marmoratus*, and the frogs, *Rana tigrinia regulosa*, and *Rana pipiens* (cf. Iigo et al., 1997).

Diseases, Pests, and Stress

The rhythmic responses of animals to causal agents of disease, stress, or injury have been studied extensively in the laboratory. Included in the list of causal agents are microorganisms, parasites, chemicals, drugs, and physical agents, such as noise and light (see Table 1.1 in Chapter 1 and Figure 11.14 in Chapter 11 on Clinical Medicine for examples). Besides factors or variables, such as age, sex, health, and dose, which can affect the response of an animal to an external agent, there are at least three or more that pertain to rhythms:

1) The response of the organism to the agent may be rhythmic, be it a drug (e.g., alcohol) or any other external agent (cf. Figure 11.13).
2) The agent may affect the rhythm parameters of a variable, which leads to a shift in phase, or a change in amplitude or period.
3) The agent itself may be cycling and be time-dependent (e.g., a rhythm in the activity levels of a parasite or oscillations in the concentration of a chemical).

Parasites

Host-vector-parasite interactions are a prime example of the adaptive value of biological rhythms (Barrozo et al., 2004). The host, which could be a domesticated pet or free-ranging animal, generally has physiological and behavioral rhythms synchronized to the solar day. The vector, often an insect or arachnid, has an adaptive feeding rhythm synchronized with the rhythms of the host that is most beneficial to it. The parasite (microfilaria) responds to both the rhythms of its host and the potential vector that will feed upon the host.

Microfilariae are the motile embryos of certain nematodes (worms), such as *Wuchereria bancrofti*, that accumulate in the lungs for part of the day and migrate to the bloodstream for the other part in order to be available to continue their life history (cycle) when taken up by a blood-sucking insect (e.g., a mosquito) that will eventually infect another host. Some parasites show 24-h periodicity in cell division and migration from one location to another in the bodies of their hosts. Depending upon the parasite species, the number of microfilariae may be maximal in the blood at night or during the day in different hosts, such as cat, dog, monkey, rabbit, raccoon, mongoose, and several birds (partridge, crow, canary, duck), as well as humans (cf. Table 1 in Hawking, 1962).

In the case of cats and dogs, the integration of biological rhythms with the disciplines of parasitology and hematology underscores the importance of the phase of a rhythm in determining what is present, where it can be found, and when to best obtain a diagnostic sample (Fontes et al., 2000). The number of microfilaria circulating in the blood of a cat oscillates with a circadian periodicity that usually peaks near midnight, with virtually no nematodes being found during the day (Hayasaki et al., 2003). In dogs, the count is often maximal between 18:00 h and 20:00 h (Hawking, 1962). These microfilaria accumulate near the lungs during the day for better access to oxygen (Hawking, 1971), while at night, they swarm in the peripheral blood, a time and a place where they can be sucked up more readily by night-feeding mosquitoes (Pandian, 1994; Hassan et al., 2001).

A seasonal periodicity of microfilarial (counts) that showed a peak in late summer (August–September) has also been observed in dogs (Hawking, 1971). This circannual cycle continued to manifest itself annually over a 4-year observation span of the same dog, even though no new infections occurred.

Bacterial Infections

One of the more common symptoms of disease is an elevated temperature, and since body temperature is also rhythmic, the pattern of the normal body temperature rhythm may also be altered by disease. Such is the case for mastitis, a well-known disease found in lactating dairy cows, which include an elevated temperature as one of its symptoms. The effects of one causal bacterial agent (*Streptococcus agalactiae*) have been observed on the normal rhythmic temperature pattern about 26 h after infection (actually the injection of the organism into a teat) by a prominent elevation of temperature that lasts for 6 h and is higher than the usual 24-h pattern (Lefcourt, 1990).

Seasonal Diseases

When owners of pets and livestock do not provide for the prevention of disease through annual vaccinations or immunization treatments, seasonal cycles of disease occur sporadically within populations. This includes equine encephalitis and West

A) Horse Fly	B) Deer Fly	C) Cat Flea
Tabanus similus	*Hybomitra lasiopthalma*	*Ctenocephalus felis*

FIGURE 9.8. Three common insect pests noted for their activity rhythms in livestock and pets: (a) horse fly (*Tabanus similus*), (b) deer fly (*Hybomitra lasiopthalma*), and (c) cat flea (*Ctenocephalides felis*). *Note*: Insects not shown to scale.

Nile virus in horses and heartworm in dogs.[12] It is possible that the efficacy of some vaccinations could be seasonal or circannual[13] (cf. Shifrine et al., 1980a,b), although much more needs to be learned about seasonal and circannual rhythms in efficacy.

Seasonal changes in immune function may also play a role in impact of the parasite on the host. For example, several temperate bird species show an annual change in immune function (increased spleen mass and T-cell-mediated immunity) timed to their reproductive season as a response to a similarly timed annual peak in increased virulence of ecto- and endoparasites (Moller et al., 2003).

The incidence of some noninfectious diseases are also circannual or seasonal. A circannual rhythm for spontaneous canine diabetes mellitus in pet dogs, which peaks in January and February, is well documented in a 3-year study conducted in Wisconsin, USA (Atkins & MacDonald, 1987). Of added interest is the fact that these results were similar to what has been reported for the incidence of insulin-dependent diabetes mellitus in humans (Fleegler et al., 1979). In some ways, the origins and causes (etiologies) of the two types of diabetes are dissimilar. However, the seasonal incidence of occurrence for both species is quite similar in that the onset was 2.5- to 4-fold greater in the winter (January–February) when compared with the summer (June–August).

Flies

Many of the biological agents that are classified as pests can produce stress in animals. The effects of blood-feeding flies (Tabanidae) serve as an example (Figure 9.8). Many attack horses, cattle, deer, and other mammals, including

---

[12] All three diseases are transmitted by mosquitoes. The first two are viral diseases. Heartworm is caused by a parasite (e.g., *Dirofilaria immitis*), which as microfilariae, remains active in the blood stream, but requires mosquitoes as an intermediate host to complete its life history of development.

[13] An immune response to a standard lymphocyte stimulation test in clinically normal beagle dogs was up to 110% greater in the summer (June–October) than in the winter (January) (Shifrine et al., 1980a,b).

humans. In cattle they adversely affect milk and meat production, the quality of leather, and can transmit diseases (cf. Drummond et al., 1988).

Blood-feeding flies disrupt the grazing of livestock and cause animals to gather in bunches, which in turn contributes to heat stress. The flight activity and occurrence of two species, *Hybomitra lasiopthalma* (deer fly) and *Tabanus similis* (horse fly), which occur in Manitoba (Canada), North Dakota (USA), and elsewhere, peak near noon, and as can be expected, their populations are highest in summer (Hayes & Meyer, 1990).

Fleas

Daily rhythms in activity also occur for some of the pests found on pets. The common cat flea, *Ctenocephalides felis*, is a prime example, although it is so small (Figure 9.8, panel C) and quick, practically no one notices it. Nevertheless, their locomotor movements peak near the end of the light span (Koehler et al., 1990). Egg production of cat fleas is also rhythmic, being highest from midnight to 03:00 h (Kern et al., 1992).

What is also intriguing is that the response of a number of insect pests to chemical control is rhythmic (Cole & Adkisson, 1964; Halberg et al., 1974; Eesa et al., 1987; Onyeocha & Fuzeau-Braesch, 1991). This suggests that the timing of a treatment can play a role in the outcome. The concepts of timing of sampling (chronodiagnosis) and the timing of treatment (chronotherapy) are discussed in more detail in Chapter 11 on Clinical Medicine.

Reproduction and Photoperiodism[14]

In veterinary medicine the applications of biological rhythms in hematology, urology, and pathology are just beginning to be introduced. On the other hand, and whether realized or not, the applications of biological rhythms to matters pertaining to reproduction and estrus have been used for decades, if not centuries.

Interactions that exist among the photoinductive features of circadian rhythms with the endogenous features of the circannual rhythms of reproduction and migration, as well as the interactions between rhythms *per se* and photoperiodism, provide a complex panorama seen in nature[15] and quite often manipulated by the intervention of humans. The list of possible interventions is diverse and includes factors such as artificial insemination, drug treatments, the use of electric lamps, the

---

[14] Photoperiodism is defined as a response of an organism to the timing and duration of light and dark. An entire chapter is devoted to photoperiodism, and this process is also discussed in Chapters 7 (Sexuality and Reproduction) and 8 (Natural Resources).

[15] Aspects of daily and seasonal endocrine rhythms and associations with photoperiod and reproduction have been reported for many species of birds, mammals, and other animals (cf. Assenmacher & Farner, 1978).

construction of light-tight buildings, and management practices that are driven more by market conditions than by the need to provide a better quality of life for animals.

Photoperiod

Reproduction in mammals is seldom controlled by a single environmental factor since many environmental factors (e.g., photoperiod, temperature, food availability, stress, social cues) can interact in complex ways to alter a mammal's reproductive potential (Bronson, 1989). However, in nature, and unlike temperature and other factors, the recurring seasonal changes in the lengths of daylight and darkness are dependable and precise. Many farm animals, pets, and domestic birds that are seasonal breeders rely upon this environmental cue, which is known as the *photoperiod*. Actually, it is the timing and duration of both the light (photo) span and the dark (scoto) span that serve as the cue, a feature that is more easily demonstrated under controlled environmental conditions provided by light-tight buildings and chambers.

Circannual reproductive cycles are synchronized by light/dark (LD) spans, regardless of whether estrus is induced by long photoperiods (long L span with a short D span) or by short photoperiods (short L span with a long D span). In nature, a noninductive photoperiod[16] serves as a natural contraceptive.[17]

Seasonal reproduction is also an important strategy that has evolved to ensure survival of a species. The birth of young in the spring is common for many groups of animals, regardless of the length of their gestation span or when the females are fertile. Typical examples of gestation lengths are about 5 months for sheep and about 11 months for the horse.

Melatonin

The hormone melatonin is a chemical messenger that is synthesized during darkness by the pineal gland. The pineal gland either directly or indirectly responds to changes in the environmental light–dark conditions. In several species (bony fishes, amphibians, turtles, lizards, birds), the pineal gland is a direct photosensory organ. However, in mammals, pineal cells (pinealocytes) gradually evolved from being direct photoreceptor cells to a nonsensory neuroendocrine organ under photoperiod control via opsins and other phototransduction-related proteins (cf. Menaker et al., 1997; Ekstöm & Meissl, 2003). Melatonin levels function to allow seasonally reproductive animals to perceive changes in length of the day and thereby synchronize reproductive activity to the optimal time of the year (cf. Matthews et al., 1993). It is actually the duration of melatonin secretion that is the critical daylength signal for

---

[16] A noninductive photoperiod for an animal that is classified as a short-day breeder would be a 24-h cycle that includes a long span of daylight followed by a short span of darkness.

[17] Natural selection, through maximizing the efficiency of reproduction and limiting breeding to certain times of year has been called "Nature's Contraceptive" (cf. Lincoln & Short, 1980).

seasonal rhythm organization (Arendt, 1998) (for a discussion of positive and adverse effects of melatonin, see Chapter 11 on Clinical Medicine).

Subcutaneous implants of this hormone that mimic the winter photoperiod are being used in veterinary clinical medicine to induce an advance of the natural cyclic reproductive activities of several domestic animals. In fact, a commercial melatonin implant is marketed in many countries (cf. Chemineau & Malpaux, 1998). In combination with artificial photoperiods, appropriately timed melatonin implants have been used to manipulate out-of-season reproductive activity in several species, including *sheep* (Shelton, 1990; Martin, 1995; Zuniga et al., 2002), *pigs* (Shelton, 1990; Paterson & Foldes, 1994), *milk goats* (du Preez et al., 2001), *pony mares* (Peltier, et al., 1998), and captive *white-tailed deer* (Osborn et al., 2000).

Domestic Fowl

The physiology of reproduction and development in domestic fowl is regulated greatly through the manipulation of the light and dark spans of the 24-h day. Here again, as with plants and insects, skeleton photoperiods (see Chapter 4 on Photoperiodism) can be substituted for the normal light span.

Briefly, for a LD14:10 regimen (i.e., 14 h L followed by 10 h D), the skeleton counter part could be 2L:4D:8L:10D. Under such a skeleton schedule, the hen would respond physiologically as if there was a continuous 14 h span of light (cf. Ernst et al., 1987; Ernst, 1989); a manipulation that has helped in understanding the mechanism of photoperiodism (cf. Morris, 1979).

In domestic birds, light-dependent mechanisms interact with the circadian timing system and affect the rate of egg-laying and the size of the egg (Morris, 1979). In order to obtain greater yields in commercial production, the lengths of the photoperiod can often be changed as hens mature. For example, from the time they are hatched to 18 weeks, pullets (hens less than one year of age) can be maintained under short light span conditions (e.g., LD6:18), after which through successive increments a 17-h L span (LD17:7) is achieved (cf. Morris, 1979).

For breeding turkey hens, a typical photoperiod schedule is: 0–18 weeks in LD14:10, 18–30 weeks in LD6:18, and after 30 weeks in LD14:10 (Noll & Halawani, 1995). For toms (male turkeys) used for artificial insemination, the photoperiod schedule is: 0–16 weeks in LD10:14 and 16 weeks to end of production in LD12:12 or a gradual increase to a maximum of LD16:8 by end of production (Sharp, 1995). While the laying of fertilized eggs follows a daily cycle, it is not necessary for insemination to occur daily since the female is able to store sperm in storage tubules present at the junction of the uterus and vagina (Noll & Otis, 1995).

The use of supplemental light and various combinations of L and D spans to affect physiological processes of reproduction and development are more typically used for domestic fowl than for livestock. Yet there are reports where the use of supplemental lighting has resulted in greater yields of milk and less fat in commercial dairy herds[18] (Stanisiewski et al., 1985).

---

[18] Typically, dairy cows are bred about yearly to maintain lactation, although there are reports where hormones have been used to maintain a cow in lactation for a number of years.

Sheep

Seasonal strategies involving photoperiodism and three general kinds of timing mechanisms (entrainable circadian and circannual clocks and an interval timer) are also evident in mammalian reproductive processes (cf. Bronson, 1989). Seasonal reproduction has been studied extensively in sheep, a highly photoperiodic short-day breeder (Figure 9.9). The seasonal reproductive process in ewes is highly complex, involving neural and endocrine response systems, the presence of photoreceptors, pathways or tracts, and negative feedback. Included within such a complex process are the SCN and the pineal, as well as melatonin and other hormones. The circannual rhythm of reproduction in ewes is timed to a circadian rhythm of melatonin, which regulates a 16-d estrus cycle (Karsch et al., 1984).

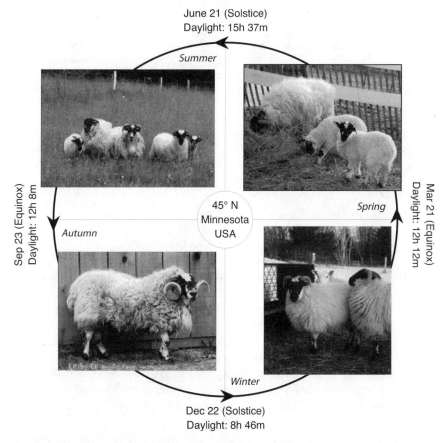

FIGURE 9.9. Seasonal reproduction cycle in sheep, a short-day breeder. Mating occurs in autumn, when the ram becomes sexually active, the ewes are pregnant over winter, give birth in the spring, and lambs grow in the summer. Photos by Marcia Koukkari.

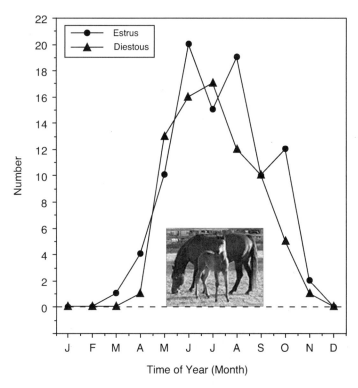

FIGURE 9.10. Seasonal changes in the characteristics of two aspects of fertility—estrus and diestrus—in 14 mares monitored in Wisconsin (derived from Table 3 in Ginther, 1974) (photo of mare and foal courtesy of Krishona Bjork-Martinson, University of Minnesota).

Horses

In Temperate Zones, the process of ovulation in horses is seasonal (Ginther, 1974, 1975). The number of ovulatory estrus periods, their length, as well as the number of diestrous periods and interovulatory intervals reach maximum values in the summer and decrease during autumn and winter (Figure 9.10).

Long photoperiods (e.g., 16 h), either natural or supplied by lamps, not only hasten the onset of the breeding season, but result in the length of hair and the depth of coat to be less (Kooistra & Ginther, 1975). Initiating a 16-h photoperiod in December (Michigan, USA) for anestrous brood mares results in the estrous cycles beginning within about two months (Oxender, 1977).[19] If followed by a successful breeding, this could result in mares foaling months earlier than for mares maintained under "normal" natural ambient conditions.

---

[19] In order to obtain 16 h of light, a skeleton photoperiod is widely used in France, wherein lights are turned on again for two hours during the night (personal communication, Scott Madill, DVM, University of Minnesota).

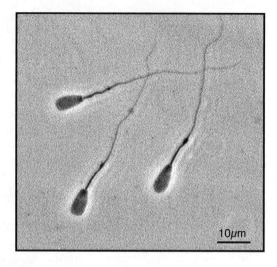

FIGURE 9.11. Annual cycles have been found in male testicular functions, such as bull sperm (shown) used for artificial insemination. Light micrograph scale bar is 10 μm (courtesy of University of Minnesota College of Biological Sciences Imaging Center).

Pigs and Goats

In the case of domestic pigs, it has been suggested that 16-h photoperiods could be used to alter circadian melatonin secretion patterns and thereby reduce the effect of season on reproduction (Tast et al., 2001). Similarly, exposure of male Creole goats to 2.5 months of long days (16 h L) from November 1 to January 15 (followed by melatonin implants) has been shown to induce intense sexual activity during the nonbreeding season (February to April vs. July to November) in northern Mexico (Delgadillo et al., 2001).

Artificial Insemination

Artificial insemination is one of the more common interventions into the natural reproductive cycle of livestock (cows, horse, milk goats) and turkeys, and more recently for pets, such as dogs. The practice of each dairy farm having its own bull for breeding is a thing of the past, since most breeding is now done by artificial insemination (cf. Kiddy, 1979; Lefcourt, 1990).

Two features, recognizing the estrus cycle and the availability of viable sperm (Figure 9.11) for insemination, are key components. On a national basis, failing to recognize or forgetting to watch for signs of estrus is costly, easily exceeding hundreds of millions of dollars per year in the USA.[20]

Semen Quality and Season

Semen for artificial insemination is collected by aseptic techniques from cattle, horses, dogs, poultry, and less commonly from other animals. Depending upon

---

[20] For dairy cows alone, the costs are far-reaching and measured in millions of dollars per year (cf. Gerrits et al., 1979).

the species, the semen can be either frozen and saved for years, shipped and used when needed (bulls), or the semen must be used within a given number of days.

Already in the early 1970s, using the DNA content of spermatozoa as a marker, an annual cycle was indicated in bulls (Krzanowski, 1974), with the mean DNA content reaching levels that were 12–13% higher in autumn then they were in the spring. Sperm fertility rates in dairy bulls (Finnish Holstein-Friesian breed) is also best in autumn (June to October) and lowest in winter (January to March) (Alm et al., 2001).

Significant annual rhythms for semen volume, motility and concentrations of sperm have been reported for stallions, with peaks located in spring or summer for ejaculate volume and in autumn for sperm motility and concentration (Byers et al., 1983; Araujo et al., 1996). A significant seasonal effect is also found in turkey toms, with higher rates of fertility occurring in the spring and summer (Sexton, 1986).

While such seasonal rhythms in the quality, quantity, and other characteristics of semen are documented, however, being able to preserve semen by freezing or other methods and assessing its quality so it can be used at the best time remains an important area of study (cf. Sexton, 1979; Bakst & Wishart, 1995; Sutovsky et al., 2003).

Implications

Because of the rhythmic, and thus predictable, temporal organization of physiological and pathological processes and systems, the implications of biological rhythms in clinical veterinary medicine are extensive. Present and future applications of biological rhythms in clinical veterinary medicine will most likely parallel the advances already underway in human medicine (discussed in Chapter 11 on Clinical Medicine). For example, the study of endogenous rhythms in the physiology of domestic animals is leading to an increased understanding of basic processes that regulate reproduction, including the use of exogenous melatonin and manipulation of the photoperiod on development *in utero* and postnatally and on pelage (wool, fur, hair) production (Paterson & Foldes, 1994).

Take-Home Message

Animals seen in veterinary medicine have a time-dependent rhythmic organization in their biological functions and sensitivity to external agents. Such rhythmic, and thus predictable, variations can serve as potential tools or *markers* that provide insight into physiological and pathological processes, which have diagnostic, therapeutic, and economic value.

References

Abrahamsen JF, Sothern RB, Sandberg S, Aakvaag A, Laerum OD, Smaaland R. (1999) Circadian variations in human peripheral blood on days with and without bone marrow sampling and relation to bone marrow cell proliferation. *Biol Rhythm Res* 30(1): 29–53.

Alila-Johanasson A, Erikson L, Soveri T, Laakso ML. (2003) Serum cortisol levels in goats exhibit seasonal but not daily rhythmicity. *Chronobiol Intl* 20(1): 65–79.

Alm, K, Taponen J, Dahlbom M, Tuunainen E, Koskinen E, Andersson M. (2001) A novel automated fluorometric assay to evaluate sperm viability and fertility in dairy bulls. *Theriogenol* 56(4): 677–684.

American Veterinary Medical Association. (2002) *U.S. Pet Ownership and Demographics Sourcebook.*

Araki CT, Nakamura RM, Kam LWG, Clark NL. (1985) Diurnal temperature patterns of early lactating cows with milking parlor cooling. *J Dairy Sci* 68: 1496–1501.

Araujo JF, Righini AS, Fleury JJ, Caldas MC, Costa-Neto JB, Marques N. (1996) Seasonal rhythm of semen characteristics of a Brazilian breed ("Mangalarga") stallion. *Chronobiol Intl* 13(6): 477–485.

Arendt J. (1998) Melatonin and the pineal gland: influence on mammalian seasonal and circadian physiology. *Rev Reprod* 3(1): 13–22.

Assenmacher I, Farner DS, eds. (1978) *Environmental Endocrinology.* Berlin: Springer-Verlag, 334 pp.

Atkins CE, Macdonald MJ. (1987) Canine diabetes mellitus has a seasonal incidence: implications relevant to human diabetes. *Diabetes Res* 5(2): 83–87.

Ayo JO, Oladele SB, Ngam S, Fayomi A, Afolayan SB. (1999) Diurnal fluctuations in rectal temperature of the Red Sokoto goat during the harmattan season. *Res Vet Sci* 66(1): 7–9.

Bakst MR, Wishart GJ. (1995) *Proc 1st Intl Symp Artificial Insemination of Poultry.* Bakst MR, Wishart GJ, eds. Savoy, IL: The Poultry Science Assoc., Inc., 297 pp.

Barrozo RB, Schilman PE, Minoli SA, Lazzari CR. (2004) Daily rhythms in disease-vector insects. *Biol Rhythm Res* 35(1/2): 79–92.

Benton LA, Yates FE. (1990) Ultradian adrenocortical and circulatory oscillations in conscious dogs. *Amer J Physiol* 258(3 Pt 2): R578–590.

Binkley S. (1976) Pineal gland biorhythms: N-acetyltransferase in chickens and rats. *Fed Proc* 35: 2347–2352.

Binkley SA, Riebman JB, Reilly KB. (1978) The pineal gland: a biological clock *in vitro*. *Science* 202(4373): 1198–1201.

Bitman J, Lefcourt A, Wood DL, Stroud B. (1984) Circadian and ultradian temperature rhythms of lactating dairy cows. *J Dairy Sci* 67: 1014–1023.

Bjarnason GA, Jordan R, Wood PA, Li Q, Lincoln D, Sothern RB, Hrushesky WJM, Ben-David Y. (2001) Circadian expression of clock genes in human oral mucosa and skin: association with specific cell cycle phases. *Amer J Path* 158(5): 1793–1801.

Black A, Schoknecht PA, Ralston SL, Shapses SA. (1999) Diurnal variation and age differences in the biochemical markers of bone turnover in horses. *J Anim Sci* 77: 75–83.

Brace RA, Moore TR. (1991) Diurnal rhythms in fetal urine flow, vascular pressures, and heart rate in sheep. *Amer J Physiol* 261(4 Pt 2): R1015–R1021.

Bronson FH. (1989) *Mammalian Reproductive Biology.* Chicago: University of Chicago Press, 325 pp.

Burns TA, Huston JP, Spieler RE. (2003) Circadian variation of brain histamine in goldfish. *Brain Res Bull* 59(4): 299–301.

Byers SW, Dowsett KF, Glover TD. (1983) Seasonal and circadian changes of testosterone levels in the peripheral blood plasma of stallions and their relation to semen quality. *J Endocrinol* 99(1): 141–150.

Camello PJ, Pozo MJ, Salido GM, Madrid JA. (1991) Ultradian rhythms in canine gallbladder bile composition. *J Interdiscipl Cycle Res* 22(3): 281–291.

Chemineau P, Malpaux B. (1998) [Melatonin and reproduction in domestic farm animals.] [French]. *Therapie* 53(5): 445–452 (Review).

Chou H-F, Johnson AL. (1987) Luteinizing hormone secretion from anterior pituitary cells of the cockerel: evidence for an ultradian rhythm. *Poultry Sci* 66(4): 732–740.

Cole CL, Adkisson PL. (1964) Daily rhythm in the susceptibility of an insect to a toxic agent. *Science* 144: 1148–1149.

Davidson TL, Fewell JE. (1993) Ontogeny of a circadian rhythm in body temperature in newborn lambs reared independently of maternal time cues. *J Dev Physiol* 19(2): 51–56.

Dawes C. (1974) Circadian and circannual maps for human saliva. In: *Chronobiology*. Scheving LE, Halberg F, Pauly JE, eds. Tokyo: Igaku-Shoin, Ltd., pp. 224–227.

Delgadillo JA, Carrillo E, Moran J, Duarte G, Chemineau P, Maplaux B. (2001) Induction of sexual activity of male creole goats in subtropical northern Mexico using long days and melatonin. *J Anim Sci* 79(9): 2245–2252.

Drummond RO, George JE, Kunz SE. (1988) *Control of Arthropod Pests of Livestock: A Review of Technology*. Boca Raton: CRC Press, 245 pp.

Duke GE, Evanson OA. (1976) Diurnal cycles of gastric motility in normal and fasted turkeys. *Poultry Sci* 55: 1802–1807.

du Preez ER, Donkin EF, Boyazoglu PA, Rautenbach GH, Barry DM, Schoeman HS. (2001) Out-of-season breeding of milk goats–the effect of light treatment, melatonin and breed. *J S Afr Vet Assoc* 72(4): 228–231.

Eesa N, Cutkomp LK, Cornelissen G, Halberg F. (1987) Circadian change in Dichloros lethality (LD50) in the cockroach in LD 14:10 and continuous red light. In: *Advances in Chronobiology, Part A*. Pauly JE, Scheving LE, eds. New York: Alan Liss, Inc., pp. 265–279.

Ehret CE, Groh KR, Meinert JC. (1978) Circadian dyschronism and chronotypic ecophilia as factors in aging and longevity. In: *Aging and Biological Rhythms*. Samis HV, Capobianco S, eds. London: Plenum Press, pp. 185–213.

Ekström P, Meissl H. (1997) The pineal organ of teleost fishes. *Rev Fish Biol Fisheries* 7(2): 199–284.

Ekström P, Meissl H. (2003) Evolution of photosensory pineal organs in new light: the fate of neuroendocrine photoreceptors. *Philos Trans R Soc Lond B* 358(1438): 1679–1700 (Review).

Ernst RA, Millam JR, Mather FB. (1987) Review of life-history lighting programs for commercial laying fowls. *World's Poultry Sci J* 43: 45–55.

Ernst R. (1989) Fine-tuning flock management: lighting. *Egg Industry* 95: 8–9.

Evans JW, Winget CM. (1974) Equine glucose circadian and ultradian rhythms (Abstract). *J Anim Sci* 39(1): 207.

Feuers RJ, Delongchamp RR, Scheving LE, Casciano DA, Tsai TH, Pauly JE. (1986) The effects of various lighting schedules upon the circadian rhythms of 23 liver or brain enzymes of C57BL/6J mice. *Chronobiol Intl* 3(4): 221–235.

Fleegler FM, Rogers KD, Drash A, Rosenbloom AL, Travis LB, Court JM. (1979) Age, sex, and season of onset of juvenile diabetes in different geographic areas. *Pediatrics* 63(3): 374–379.

Fontes G, Rocha EM, Brito AC, Fireman FA, Antunes CM. (2000) The microfilarial periodicity of *Wuchereria bancrofti* in north-eastern Brazil. *Ann Trop Med Parasitol* 94(4): 373–379.

Fuller CA, Lydic R, Sulzman FM, Albers HE, Tepper B, Moore-Ede MC. (1981) Circadian rhythm of body temperature persists after suprachiasmatic lesions in the squirrel monkey. *Amer J Physiol* 241(5): R385–R391.

Gegout-Pottie P, Philippe L, Simonon MA, Guingamp C, Gillet P, Netter P, Terlain B. (1999) Biotelemtery: an original approach to experimental models of inflammation. *Inflamm Res* 48(8): 417–424.

Gerrits RJ, Blosser TH, Purchase HG, Terrill CE, Warwick EJ. (1979) Economics of improving reproductive efficiency in farm animals. In: *Animal Reproduction* (Beltsville Symposia in Agricultural Research 3), Hawk E, ed. New York: Allanheld, Osmun & Co., pp. 413–421.

Gill J, Kompanowska-Jezierska E. (1986) Seasonal changes in the red blood cell indices in Arabian brood mares and their foals. *Comp Biochem Physiol* 83A(4): 643–651.

Ginther OJ. (1974) Occurrence of anestrus, estrus, diestrus and ovulation over a 12-month period in mares. *Amer J Vet Res* 35(9): 1173–1179

Ginther OJ. (1979) Reproductive seasonality and regulation of LH and FSH in pony mares. In: *Animal Reproduction* (Beltsville Symposia in Agricultural Research 3). Hawk E, ed. New York: Allanheld, Osmun & Co., pp. 291–304.

Gomes WR, Joyce MC. (1975) Seasonal changes in serum testosterone in adult rams. *J Anim Sci* 41(5): 1373–1375.

Gordon C, Lavie P. (1982) Ultradian rhythms in renal excretion in dogs. *Life Sci* 31(24): 2727–2734.

Griffith MK, Minton JE. (1991) Free-running rhythms of adrenocorticotropic hormone (ACTH), cortisol and melatonin in pigs. *Domest Anim Endocrinol* 8(2): 201–208.

Hahn GL (1989) Body temperature rhythms in farm animals - a review and reassessment relative to environmental influences. In: *Proc 11th Intl Soc Biometeorol Congress*. Driscoll D, Box EO, eds. The Hague: SPB Academic Publication, pp. 325–337.

Hahn GL, Chen YR, Nienaber JA, Eigenberg RA, Parkhurst AM. (1992) Characterizing animal stress through fractal analysis of thermoregulatory responses. *J Therm Biol* 17: 115–120.

Halberg J, Halberg F, Lee JK, Cutkomp L, Sullivan WN, Hayes DK, Cawley BM, Rosenthal J. (1974) Similar timing of circadian rhythms in sensitivity to pyrethrum of several insects. *Intl J Chronobiol* 2: 291–296.

Halberg F, Gehrke CW, Zinneman HH, Kuo K, Nelson WL, Dubey DP, Cadotte LM, Haus E, Scheving LE, Good RA, Rosenberg A, Soong L, Bazin H, Abdel-Monem MM. (1976) Circadian rhythms in polyamine excretion by rats bearing an immunocytoma. *Chronobiologia* 3(4): 309–322.

Halberg F, Gehrke CW, Kuo K, Nelson WL, Sothern RB, Cadotte LM, Haus E, Scheving LE. (1978) Immunocytoma effect upon circadian variation in murine urinary excretion of beta-aminoisobutyric acid, beta-alanine, phenylalanine and tyrosine. *Chronobiologia* 5(3): 263–276.

Halberg F, Nelson W. (1978) Chronobiologic optimization of aging. In: *Aging and Biological Rhythms*. Samis HV, Capobianco S, eds. London: Plenum Press, pp. 5–55.

Hartenbower DL, Friedler RM, Coburn JW, Massry SG, Sellers A. (1974) Spontaneous variations in electrolyte excretion in the awake dog. *Proc Soc Exp Biol Med* 45(2): 648–653.

Hartinger J, Kulbs D, Volkers P, Cussler K. (2003) Suitability of temperature-sensitive transponders to measure body temperature during animal experiments required for regulatory tests. *ALTEX* 20(2): 65–70.

Hassan AA, Rahman WA, Rashid MZ, Shahrem MR, Adanan CR. (2001) Composition and biting activity of *Anopheles* (*Diptera: Culicidae*) attracted to human bait in a malaria endemic village in peninsular Malaysia near the Thailand border. *J Vector Ecol* 26(1): 70–75.

Haus E, Halberg F. (1969) Phase-shifting of circadian rhythms in rectal temperature, serum corticosterone and liver glycogen of the male C-mouse. *Rass Neur Veg* 23(3): 83–112.

Haus E, Halberg F. (1970) Circannual rhythm in level and timing of serum corticosterone in standardized inbred mature C-mice. *Environ Res* 3(2): 81–106.

Haus E, Halberg F. (1980) The circadian time structure. In: *Chronobiology: Principles and Applications to Shifts in Schedules*. Scheving LE, Halberg F, eds. Alphen aan den Rijn: Sijthoff & Noordhoff, pp. 47–94.

Haus E, Nicolau GY, Lakatua D, Sackett-Lundeen L. (1988) Reference values for chronopharmacology. In: *Annual Review of Chronopharmacology*, Vol 4. Reinberg A, Smolensky M, Labrecque G, eds. Oxford: Pergamon Press, pp. 333–424.

Hawking F. (1962) Microfilaria infestation as an instance of periodic phenomena seen in host-parasite relationships. *Ann NY Acad Sci* 98(4): 940–953.

Hawking F. (1971) Circadian rhythms of parasites. *J Interdiscipl Cycle Res* 2(2): 157–160.

Hawking F, Lobban MC, Gamage K, Worms MJ. (1971) Circadian rhythms (activity, temperature, urine and microfilariae) in dog, cat, hen, duck, *Thamnomys* and *Gerbillus*. *J Interdiscipl Cycle Res* 2(4): 455–473.

Hayasaki M, Okajima J, Song KH, Shiramizu K. (2003) Diurnal variation in microfilaremia in a cat experimentally infected with the larvae of *Dirofilaria immitis*. *Vet Parasitol* 111(2–3): 267–271.

Hayes RM, Meyer HJ. (1990) Daily and seasonal flight activity of Tabanidae (Diptera) in the North Dakota sandhills. In: *Chronobiology: Its Role in Clinical Medicine, General Biology, and Agriculture, Part B*. Hayes DK, Pauly JE, Reiter RJ, eds. New York: Wiley-Liss, Inc., pp. 683–690.

Hendrickson AE, Wagoner N, Cowan WM. (1972) An autoradiographic and electron microscopic study of retino-hypothalamic connections. *Z Zellforsch Mikrosk Anat* 135(1): 1–26.

Hotchkiss CE, Brommage R. (2000) Changes in bone turnover during the menstrual cycle in cynomolgus monkeys. *Calcif Tissue Intl* 66(3): 224–228.

Iigo M, Hara M, Ohtani-Kaneko R, Hirata K, Tabata M, Aida K. (1997) Photic and circadian regulations of melatonin rhythms in fishes. *Biol Signals* 6(4–6): 225–232.

Ingram DL, Mount LE. (1973) The effects of food intake and fasting on 24-hourly variations in body temperature in the young pig. *Pflügers Arch* 339(4): 299–304.

Ingram DL, Dauncey MJ. (1985) Circadian rhythms in the pig. *Comp Biochem Physiol* 82A(1): 1–5.

Jenkin G, Mitchell MD, Hopkins P, Matthews CD, Thorburn GD. (1980) Concentrations of melatonin in the plasma of the rhesus monkey (*Macaca mulatta*). *J Endocrinol* 84(3): 489–494.

Jilge B, Stahle H. (1984) The internal synchronization of five circadian functions of the rabbit. *Chronobiol Intl* 1(3): 195–204.

Jilge B. (1991) The rabbit: a diurnal or a nocturnal animal? *J Exp Anim Sci* 34(5–6): 170–183.

Johnson L. (1966) Diurnal patterns of metabolic variations in chick embryos. *Biol Bull* 131: 308–322.

Kadono H, Usami E. (1983) Ultradian rhythm of chicken body temperature under continuous light. *Jpn J Vet Sci* 45(3): 401–405.

Kanabrocki EL, Sothern RB, Scheving LE, Vesely DL, Tsai TH, Shelstad J, Cournoyer C, Greco J, Mermall H, Nemchausky BM, Bushnell DL, Kaplan E, Kahn S, Augustine G,

Holmes E, Rumbyrt J, Sturtevant RP, Sturtevant F, Bremner F, Third JLHC, McCormick JB, Mudd CA, Dawson S, Sackett-Lundeen L, Haus E, Halberg F, Pauly JE, Olwin JH. (1990) Reference values for circadian rhythms of 98 variables in clinically healthy men in fifth decade of life. *Chronobiol Intl* 7(5/6): 445–461.

Karsch FJ, Bittman EL, Foster DL, Goodman RL, Legan SJ, Robinson JE. (1984) Neuroendocrine basis of seasonal reproduction. *Recent Prog Horm Res* 40: 185–232 (Review).

Kasal CA, Menaker M, Perez-Polo JR. (1979) Circadian clock in culture: N-acetyltransferase activity of chick pineal glands oscillates *in vitro*. *Science* 203(4381): 656–658.

Kass DA, Sulzman FM, Fuller CA, Moore-Ede MC. (1980) Are ultradian and circadian rhythms in renal potassium excretion related? *Chronobiologia* 7: 343–355.

Kern WH Jr, Koehler PG, Patterson RS. (1992) Diel patterns of cat flea (Siphonaptera: Pulicidae) egg and fecal deposition. *J Med Entomol* 29(2): 203–206.

Kiddy CA. (1979) Estrus detection in dairy cattle. In: *Animal Reproduction*. Hawk H, ed. New York: John Wiley, pp. 77–89.

Koehler PG, Leppla NC, Patterson RS. (1990) Circadian rhythm in the cat flea, *Ctenocephalides felis* (Siphonaptera: Pulicidae). In: *Chronobiology: Its Role in Clinical Medicine, General Biology, and Agriculture, part B*. Hayes DK, Pauly JE, Reiter RJ, eds. New York: Wiley-Liss, Inc., pp. 661–665.

Komosa M, Flisinska-Bojanowska A, Gill J. (1990) Diurnal changes in the haemoglobin level, red blood cell number and mean corpuscular haemoglobin in foals during the first 13 weeks of life and in their lactating mothers. *Comp Biochem Physiol A* 96(1): 151–155.

Kooistra LH, Ginther OJ. (1975) Effect of photoperiod on reproductive activity and hair in mares. *Amer J Vet Res* 36(10): 1413–1419.

Krzanowski M. (1974) short- and long-term rhythms in testicular function in the bull. In: *Biorhythms and Human Reproduction*. Ferin M, Halberg F, Richart RM, Vande Wiele RL, eds. New York: John Wiley & Sons, Inc., pp. 447–457.

Kuwabara N, Seki K, Aoki K. (1986) Circadian, sleep and brain temperature rhythms in cats under sustained daily light–dark cycles and constant darkness. *Physiol Behav* 38: 283–289.

Ladlow JF, Hoffmann WE, Breur GJ, Richardson DC, Allen MJ. (2002) Biological variability in serum and urinary indices of bone formation and resorption in dogs. *Calcif Tissue Intl* 70: 186–193.

Lakatua DJ, Haus M, Berge C, Sackett-Lundeen L, Haus E. (1988) Diet and mealtiming as circadian synchronizers. In: *Annual Review of Chronopharmacology*, Vol 5. Reinberg A, Smolensky M, Labrecque G, eds. Oxford: Pergamon Press, pp. 303–306.

Langevin T, Halberg F, Fishbein SJ, Sothern RB, Scheving LE, Goetz F, Anderson GE, Bazin H. (1979) Circadian amplitude and mesor decrease from young adulthood to maturity in hourly urinary excretion rates of glucose, sodium, potassium and volume by male LOU rats (Abstract). *Chronobiologia* 6: 125–126.

Lanzinger I, Kobilanski C, Philippu A. (1989) Pattern of catecholamine release in the nucleus tractus solitarii of the cat. *Naunyn Schmiedebergs Arch Pharmacol* 339(3): 298–301.

Lefcourt AM. (1990) Circadian and ultradian rhythms in ruminants: Relevance to farming and science. In: *Chronobiology: Its Role in Clinical Medicine, General Biology, and Agriculture, part B*. Hayes DK, Pauly JE, Reiter RJ, eds. New York: Wiley-Liss, Inc., pp. 735–753.

Lefcourt AM, Huntington JB, Akers RM, Wood DL, Bitman J. (1999) Circadian and ultradian rhythms of body temperature and peripheral concentrations of insulin and nitrogen in lactating dairy cows. *Domestic Anim Endocrinol* 16(1): 41–55.

Levi F, Hrushesky WJ, Borch RF, Pleasants ME, Kennedy BJ, Halberg F. (1982) Cisplatin urinary pharmacokinetics and nephrotoxicity: a common circadian mechanism. *Cancer Treat Rep* 66(11): 1933–1938.

Lewczuk B, Przybylska-Gornowicz B. (2000) The effect of continuous darkness and illumination on the function and the morphology of the pineal gland in the domestic pig: Part I. The effect on plasma melatonin level. *Neuroendocrinol Lett* 21(4): 283–291.

Liesegang A, Reutter R, Sassi ML, Risteli J, Kraenzlin M, Riond JL, Wanner M. (1999) Diurnal variation in concentrations of various markers of bone metabolism in dogs. *Amer J Vet Res* 60(8): 949–953.

Lincoln GA, Short RV. (1980) Seasonal breeding: nature's contraceptive. In: *Recent Progress in Hormone Research*, Vol. 36. Greep RO, ed. New York: Academic Press, pp. 1–52.

Lowrey PL, Takahashi JS. (2004) Mammalian circadian biology: elucidating genome-wide levels of temporal organization. *Ann Rev Genomics Hum Genet* 5: 407–441 (Review).

Macaulay AS, Hahn GL, Clark DH, Sisson DV. (1995) Comparison of calf housing types and tympanic temperature rhythms in Holstein calves. *J Dairy Sci* 78: 856–862.

Martin GB. (1995) Reproductive research on farm animals for Australia – some long-distance goals. *Reprod Fertil Dev* 7(5): 967–982 (Review).

Matthews CD, Guerin MV, Deed JR. (1993) Melatonin and photoperiodic time measurement: seasonal breeding in the sheep. *J Pineal Res* 14(3): 105–116 (Review).

Menaker M. (1974) Aspects of the physiology of circadian rhythmicity in the vertebrate nervous system. In: *The Neurosciences: Third Study Program*. Schmitt FO, Worden FG, eds. Cambridge, Mass: MIT Press, pp. 479–489.

Menaker M, Moreira LF, Tosini G. (1997) Evolution of circadian organization in vertebrates. *Braz J Med Biol Res* 30(3): 305–313.

Mendel VE, Raghavan GV. (1964) A study of diurnal temperature patterns in sheep. *J Physiol* 174: 206–216.

Mohr E, Krzywanek H. (1990) Variations of core-temperature rhythms in unrestrained sheep. *Physiol Behav* 48(3): 467–473.

Moller AP, Erritzoe J, Saino N. (2003) Seasonal changes in immune response and parasite impact on hosts. *Amer Nat* 161(4): 657–671 (e-pub 2003 March 28).

Moore RY, Eichler VB. (1972) Loss of a circadian adrenal corticosterone rhythm following suprachiasmatic lesions in the rat. *Brain Res* 42(1): 201–206.

Moore RY, Lenn NJ. (1972) A retinohypothalamic projection in the rat. *J Comp Neurol* 146(1): 1–14.

Moore-Ede MC, Sulzman FM. (1981) Internal temporal order. In: *Biological Rhythms, Handbook of Behavioral Neurobiology,* Vol. 4. Aschoff J, ed. New York: Plenum Press, pp. 215–241.

Morris TR. (1979) The influence of light on ovulation in domestic birds. In: *Animal Reproduction* (Beltsville Symposia in Agricultural Research 3). Hawk E, ed. New York: Allanheld, Osmun & Company, pp. 307–322.

Muller MN, Lipson SF. (2003) Diurnal patterns of urinary steroid excretion in wild chimpanzees. *Amer J Primatol* 60(4): 161–166.

Nelson W, Scheving L, Halberg F. (1975) Circadian rhythms in mice fed a single daily meal at different stages of lighting regimen. *J Nutr* 105(2): 171–184.

Noll SL, Halawani ME. (1995) Lighting control for turkey breeders. *Minn Ext Serv*, St. Paul, MN: Univ. of Minn., 2 pp.

Noll SL, Otis JS. (1995) Artificial insemination for turkey breeders. *Minn Ext Serv*, St. Paul, MN: Univ. of Minn., 2 pp.

Norman RL, Lindstrom SA, Bangsberg D, Ellinwood WE, Gliessman P, Spies HG. (1984) Pulsatile secretion of luteinizing hormone during the menstrual cycle of rhesus macaques. *Endocrinology* 115(1): 261–266.

Onyeocha FA, Fuzeau-Braesch S. (1991) Circadian rhythm changes in toxicity of the insecticide dieldrin on larvae of the migratory locust *Locusta migratoria migratorioides*. *Chronobiol Intl* 8(2): 103–109.

Osborn DA, Gassett JW, Miller KV, Lance WR. (2000) Out-of-season breeding of captive white-tailed deer. *Theriogenology* 54(4): 611–619.

Oxender WD, Noden PA, Hafs HD. (1977) Estrus, ovulation, and serum progesterone, estradiol, and LH concentrations in mares after an increased photoperiod during winter. *Amer J Vet Res* 38(2): 203–207.

Panda S, Hogenesch JB. (2004) It's all in the timing: many clocks, many outputs. *J Biol Rhythms* 19(5): 374–387.

Pandian RS. (1994) Circadian rhythm in the biting behaviour of a mosquito *Armigeres subalbatus (Coquillett)*. *Indian J Exp Biol* 32(4): 256–260.

Paterson AM, Foldes A. (1994) Melatonin and farm animals: endogenous rhythms and exogenous applications. *J Pineal Res* 16(4): 167–177 (Review).

Peltier MR, Robinson G, Sharp DC. (1998) Effects of melatonin implants in pony mares: 2. Long-term effects. *Theriogenology* 49(6): 1125–1142.

Philippens KM, von Mayersbach H, Scheving LE. (1977) Effects of the scheduling of meal-feeding at different phases of the circadian system in rats. *J Nutr* 107(2): 176–193.

Piccione G, Assenza A, Fazio F, Giudice E, Caola G. (2001) Different periodicities of some haematological parameters in exercise-leaded athletic horses and sedentary horses. *J Equine Sci* 12(1): 17–23.

Piccione G, Caola G. (2002) Biological rhythm in livestock. *J Vet Sci* 3(3): 145–157 (Review).

Piccione G, Refinetti R. (2003a) Thermal chronobiology of domestic animals. *Front Biosci* 8: s258–s264 (Review).

Piccione G, Caola G, Refinetti R. (2003b) Daily and estrous rhythmicity of body temperature in domestic cattle. *BMC Physiol* 3: 7 (*www.biomedcentral.com/1472–6793/3/7*).

Piccione G, Caola G, Refinetti R. (2003c) Circadian rhythms of body temperature and liver function in fed and food-deprived goats. *Comp Biochem Physiol. Part A, Molec Integ Physiol* 134(3): 563–72.

Piccione G, Bertolucci C, Foa A, Caola G. (2004a) Influence of fasting and exercise on the daily rhythm of serum leptin in the horse. *Chronobiol Intl* 21(3): 405–417.

Piccione G, Assenza A, Grasso F, Caola G. (2004b) Daily rhythm of circulating fat soluble vitamin concentration (A, D, E and K) in the horse. *J Circadian Rhythms* 2(1): 3 (*www.circadianrhythms.com/content/2/1/3*).

Pol F, Courboulay V, Cotte JP, Martrenchar A, Hay M, Mormede P. (2002) Urinary cortisol as an additional tool to assess the welfare of pregnant sows kept in two types of housing. *Vet Res* 33(1): 13–22.

Pons M, Tranchot J, L'Azou B, Cambar J. (1994) Circadian rhythms of renal hemodynamics in unanesthetized, unrestrained rats. *Chronobiol Intl* 11(5): 301–308.

Pons M, Forpomés O, Espagnet S, Cambar J. (1996) Relationship between circadian changes in renal hemodynamics and circadian changes in urinary glycosaminoglycan excretion in normal rats. *Chronobiol Intl* 13(5): 349–358.

Rabinowitz L, Berlin R, Yamauchi H. (1987) Plasma potassium and diurnal cyclic potassium excretion in the rat. *Amer J Physiol* 253(6 Pt 2): F1178–F1181.

Ralph MR, Foster RG, Davis FC, Menaker M. (1990) Transplanted suprachiasmatic nucleus determines circadian period. *Science* 247(4945): 975–978.

Refinetti R, Susalka SJ. (1997) Circadian rhythm of temperature selection in a nocturnal lizard. *Physiol Behav* 62(2): 331–336.

Refinetti R. (1999) Relationship between the daily rhythms of locomotor activity and body temperature in eight mammalian species. *Amer J Physiol* 277: R1493–R1500.

Refinetti R, Piccione G. (2003) Daily rhythmicity of body temperature in the dog. *J Vet Med Sci* 65(8): 935–937.

Reppert SM, Coleman RJ, Heath HW, Keutmann HT. (1982) Circadian properties of vasopressin and melatonin rhythms in cat cerebrospinal fluid. *Amer J Physiol* 243: E489–E498.

Richter CP. (1965) *Biological Clocks in Medicine and Psychiatry*. Springfield, IL: CC Thomas.

Richter CP. (1967) Sleep and activity: their relation to the 24-hour clock. In: *Sleep and Altered States of Consciousness. Res Publ Assoc Nerv Ment Dis*, Vol 45. Baltimore: Williams and Wilkins, pp. 8–29.

Rietveld WJ. (1992) The suprachiasmatic nucleus and other pacemakers. In: *Biological Clocks. Mechanisms and Applications*. Touitou Y, ed. Amsterdam: Elsevier, pp. 55–64.

Sääf J, Wetterberg L, Bäckström M, Sundwall A. (1980) Melatonin administration to dogs. *J Neural Trans* 49: 281–285.

Scheving LE. (1984) Chronobiology of cell proliferation in mammals. Implications for basic research and cancer chemotherapy. In: *Cell Cycle Clocks*. Edmunds Jr. LN, ed. New York: Marcel Dekker, pp. 455–500.

Schmidt F, Yoshimura Y, Ni R-X, Kneesel S, Constantinou CE. (2001) Influence of gender on the diurnal variation of urine production and micturition characteristics of the rat. *Neurol Urodynam* 20(3): 287–295.

Sexton TJ (1979) Preservation of poultry semen - a review. In: *Animal Reproduction* (Beltsville Symposia in Agricultural Research 3). Hawk E, ed. New York: Allanheld, Osmun & Company, pp. 159–170.

Sexton TJ. (1986) Relationship of the number of spermatozoa inseminated to fertility of turkey semen stored 6 h at 5 degrees C. *Brit Poult Sci* 27(2): 237–245.

Sharp J. (1995) Managing turkeys for semen production. In: *Proc 1st Intl Symp Artificial Insemination of Poultry*. Bakst MR, Wishart GJ, eds. Savoy, IL: The Poultry Science Association, Inc., pp. 39–50.

Shelton JN. (1990) Reproductive technology in animal production. *Sci Tech* 9(3): 825-845 (Review).

Shifrine M, Rosenblatt LS, Taylor N, Hetherington NW, Matthews VJ, Wilson FD. (1980a) Seasonal variations in lectin-induced lymphocyte transformation in beagle dogs. *J Interdiscipl Cycle Res* 11: 219–231.

Shifrine M, Taylor NJ, Rosenblatt LS, Wilson FD. (1980b) Seasonal variation in cell-mediated immunity of clinically normal dogs. *Exp Hematol* 8: 318–326.

Smaaland R, Sothern RB. (1994) Cytokinetic basis for circadian pharmacodynamics: Circadian cytokinetics of murine and human bone marrow and human cancer. In: *Circadian Cancer Therapy*. Hrushesky WJM, ed. Boca Raton: CRC Press, pp. 119–163.

Sothern RB, Halberg F. (1979) Timing of circadian core temperature rhythm in rats on 5 lighting schedules with different photofractions (Abstract). *Chronobiologia* 6: 158–159.

Sothern RB, Farber MS, Gruber SA. (1993) Circannual variations in baseline blood values of dogs. *Chronobiol Intl* 10(5): 364–382.

Sothern RB. (1995) Time of day versus internal circadian timing references. *J Infus Chemother* 5(1): 24–30.

Stanisiewski EP, Mellenberger RW, Anderson CR, Tucker HA. (1985) Effect of photoperiod on milk yield and milk fat in commercial dairy herds. *J Dairy Sci* 68: 1134–1140.

Stephan FK, Zucker I. (1972) Circadian rhythms in drinking behavior and locomotor activity of rats are eliminated by hypothalamic lesions. *Proc Natl Acad Sci USA* 69(6): 1583–1586.

Stieglitz A, Spiegelhalter F, Klante G, Heldmaier G. (1995) Urinary 6-sulphatoxymelatonin excretion reflects pineal melatonin secretion in the *Djungarian hamster* (*Phodopus sungorus*). *J Pineal Res* 18(2): 69–76.

Stoinski TS, Czekala N, Lukas KE, Maple TL. (2002) Urinary androgen and corticoid levels in captive, male Western Lowland gorillas (*Gorilla g. gorilla*): age- and social group-related differences. *Amer J Primatol* 56: 73–87.

Stull CL, Rodiek AV. (2000) Physiological responses of horses to 24 hours of transportation using a commercial van during summer conditions. *J Anim Sci* 78: 1458–1466.

Sulzman FM, Fuller CA, Hiles LG, Moore-Ede MC. (1978) Circadian rhythm dissociation in an environment with conflicting temporal information. *Amer J Physiol* 235(3): R175–R180.

Sujino M, Masumoto KH, Yamaguchi S, van der Horst GT, Okamura H, Inouye ST. (2003) Suprachiasmatic nucleus grafts restore circadian behavioral rhythms of genetically arrhythmic mice. *Curr Biol* 13(8): 664–668.

Sutovsky P, Turner RM, Hameed S, Sutovsky M. (2003) Differential ubiquitination of stallion sperm proteins: possible implications for infertility and reproductive seasonality. *Biol Reprod* 68(2): 688–698.

Swoyer J, Haus E, Sackett-Lundeen L. (1987) Circadian reference values for hematologic parameters in several strains of mice. In: *Prog Clin Biol Res*, Vol. 227A. New York: Alan R. Liss, Inc., pp. 281–296.

Takahashi JS, Hamm H, Menaker M. (1980) Circadian rhythms of melatonin release from individual superfused chicken pineal glands *in vitro*. *Proc Natl Acad Sci USA* 77(4): 2319–2322.

Tast A, Love RJ, Evans G, Telsfer S, Giles R, Nicholls P, Voultsios A, Kennaway DJ. (2001) The pattern of melatonin secretion is rhythmic in the domestic pig and responds rapidly to changes in daylength. *J Pineal Res* 31(4): 294–300.

Tosini G, Menaker M. (1995) Circadian rhythm of body temperature in an ectotherm (*Iguana iguana*). *J Biol Rhythms* 10(3): 248–255.

Yoo SH, Yamazaki S, Lowrey PL, Shimomura K, Ko CH, Buhr ED, Siepka SM, Hong HK, Oh WJ, Yoo OJ, Menaker M, Takahashi JS. (2004) PERIOD2::LUCIFERASE real-time reporting of circadian dynamics reveals persistent circadian oscillations in mouse peripheral tissues. *Proc Natl Acad Sci USA* 101(15): 5339–5346.

Zachmann A, Ali MA, Falcón J. (1992) Melatonin and its effects in fishes: an overview. In: *Rhythms in Fishes*. Ali MA, ed. New York: Plenum, pp. 149–165.

Zimmerman NH, Menaker M. (1979) The pineal gland: a pacemaker within the circadian system of the house sparrow. *Proc Natl Acad Sci USA* 76(2): 999–1003.

Zuniga O, Forcada F, Abecia JA. (2002) The effect of melatonin implants on the response to the male effect and on the subsequent cyclicity of *Rasa Aragonesa* ewes implanted in April. *Anim Reprod Sci* 72(3–4): 165–174.

Zylka MJ, Shearman LP, Weaver DR, Reppert SM. (1998) Three period homologs in mammals: differential light responses in the suprachiasmatic circadian clock and oscillating transcripts outside of brain. *Neuron* 20(6): 1103–1110.

10
Society

*"The clock, not the steam engine, is the key machine of the modern industrial age.
...: even today no other machine is so ubiquitous."*
— *Lewis Mumford (1895–1990),*
American social philosopher

Introduction

Where you find people you will find watches, clocks, calendars, and schedules. They may be displayed for all to see indoors or outdoors or carried on one's person. In our modern technological society, external clocks and schedules have become the control or phase setters of *when* something should or should not be done. These inanimate devices may be either in or out of phase with the biological clocks and rhythms of our body, which over millions of years, have adapted to the motions of our planet with its environmental cues as to when an activity should best occur. Circadian rhythms have been said to dominate our existence (Martin & Banks-Schlegel, 1998) and they may well be one of the most common, but overlooked phenomena in society (cf. Luce, 1970). Unfortunately, biological rhythms in the context of human society are seldom acknowledged, realized or understood, even though we often observe synchronized social behaviors in communities of other organisms, including certain insects, birds, fish, and mammals.

The objective of this chapter is to introduce some of the implications and applications of biological rhythms in human society. We start with past and present, a comparison of how the biological rhythms of individuals are more adapted for the activities of life in past centuries than in our current society. This leads to topics that must be accounted for in today's society, such as social synchronization and photoperiodicity as it may relate to aggression and violence, night and shiftwork, the global workplace, sports and performance, travel, meal-timing, and the effect of unnatural luminescence from society (light pollution) on the temporal organization of humans and other animals. Because all that has been learned about biological rhythms and their characteristics has not been accurately conveyed to the general population, a discussion of birth-date-based biorhythms as deceptive- or pseudo-science concludes the chapter.

Past and Present

Often, we have a tendency to view the past as being better for human existence than the present, such as: times were simpler then, the good old days, a slower pace of life, less stress, etc. Certainly, many of the advances of our society in regard to health, education, comfort, communication and transportation that have occurred since the mid-1940s and 1950s strongly refute any such notion about the past being better.

In many ways our society has been and continues to be restructured in accord with the advances in technology. Accompanying these advances are the demands, as well as the options that confront us. Overlooked in this scenario of modern technology is the stark realization that in many ways the human body is better adapted to function in the society of the past, a society, which for most of us no longer exists (cf. Moore-Ede, 1986). On a geological or evolutionary time scale of millions of years, the past 20 centuries of human history is only "yesterday."

The Natural Day vs. 24/7

Prior to the early 1900s, the selective advantage to humans to synchronize their activities with the daily and yearly motion of our planet was very evident. Night was partitioned for rest and sleep, while daylight was the span for activity and movement.[1] People had little else to do at night but sleep. It was unusual for people to ever stay awake all night, unless they were a night guard or an astronomer. The 24-h society of today, however, presents a challenge to our internal 24-h clocks that had been anchored by sleep to the natural cycles of light and dark (Rajaratnam & Arendt, 2001). The availability of affordable watches, the use of electric lights,[2] and the subsequent changes that permitted commerce to operate 24 h a day[3] have allowed masses of individuals and society as a whole to program their own daily and yearly schedules (Figure 10.1).

---

[1] In the words of the English poet, William Blake (1757–1827): *"Think in the morning. Act in the noon. Eat in the Evening. Sleep in the Night."* (Blake, 1794).

[2] Paul Jablochkoff, a Russian electrical engineer, illuminated boulevards in Paris with carbon arc lights (called the electric Jablochkoff candles) in 1876. The electric incandescent lamp was developed by Thomas A. Edison in 1879 and quickly became the lamp that was available for use in industrial buildings and homes, making it much easier for more people to remain active after sunset. Prior to electric lighting, candles, oil lamps, and gas fixtures allowed only a small number of people to remain active at night.

[3] Shiftwork became the trend during World War II, especially in the steel industry where blast furnaces could not be effectively shut off and on each day, and in the underground iron and coal mines where light was always required.

FIGURE 10.1. Electric lamps and clocks enabled activities within a society to continue throughout the 24-h solar day.

Today, the student of biological rhythms may concur with Lewis Mumford (1895–1990), an American social philosopher, who noted that: *"one ate, not upon feeling hungry, but when prompted by the clock; one slept, not when one was tired, but when the clock sanctioned it"* (Mumford, 1934). A time-consciousness has become second nature in Western societies, such that "24/7" describes a society where someone or something is available not only at any time of the day, night, or week, but also at any time of year (e.g., a "24/7/365" society).

Time Schedules

Natural periodicities abound on Earth, many of which are very precise and cannot be altered by humans, since these rhythms are literally part of the universe (e.g., solar and lunar cycles). Regardless of whatever schedules society imposes or we impose, we are born with a temporal system of rhythms as part of our genetic structure. However, to function in modern society, we are often forced to rely upon external schedules and clocks, rather than upon our internal rhythms as to when to eat, sleep, and be active, much as an infant does before adults impose a more rigid scheduling.

FIGURE 10.2. Communications across time zones encounter different times and different circadian phases.

Depending upon the circumstances there may be harmony or discord between our internal rhythms and the external clocks of society.[4] Often the activities that we pursue are either consciously or unknowingly out of phase with one or both. While a shift in a daily time schedule may be immediate, a number of days may be required before physiological functions and optimum performance are shifted completely to a new time schedule. Even then, there may be lingering problems that relate to establishing internal synchronization because of intermittent exposures to external synchronizers, such as light and social cues. For example, switching to work the nightshift while living in a home that operates on a diurnal light–dark schedule can be a major problem not only for the shift-worker trying to sleep at odd times and interact with household members who are on a "normal" schedule, but for the household members as well.

Today technology allows us to achieve or change many things that seemed impossible in decades past. To fly faster than the speed of sound and to communicate over great distances are but two examples of advancements in technology that underscore the need for society to be aware of biological rhythms (Figure 10.2).

---

[4] For example, the circadian rhythms or clock of a person who lives in Minnesota and is active during the day and sleeps at night are in synchrony with the local time, but are 7 h out of phase upon arriving in Paris after a rapid airplane flight.

Failure to do so can have negative consequences, such as awakening someone in a different time zone via phone or agreeing to something when being too tired to fully understand what is being asked.[5] Alternatively, modern technology can result in positive consequences when considering a global workplace (see below).

Social Synchronization

Social cues as synchronizers have been observed for various ultradian, circadian, and infradian rhythms. Social synchronization refers to a behavioral rhythm (e.g., sleeping, eating, grooming, drinking, physical activity) being regulated by an external source generated by another individual or some other social condition. It occurs not only in humans, but in many other species, as well, including fish (see Essay 10.1), birds, monkeys, deer, coyotes, hamsters, mice, rats, crabs, and bees (cf. Yellin & Hauty, 1971; Regal & Connolly, 1980; Goel & Lee, 1997).

Essay 10.1 (by RBS): Social Synchrony in Animals

In fish, social organization affects the circadian activity characteristics. When studied under LD in a laboratory, groups of 25 white suckers (*Catastomus commersoni*) were more active during the day, while isolated individuals were more active during the night (Kavaliers, 1980). Similarly, seabass (*Dicentrarchus labrax*) tracked in a natural environment (an outdoor earthen pond) showed the same social facilitation of swimming patterns: groups of 6 to 60 fish showed highest activity during the light span, while the greatest activity for solitary seabass was at dusk and during darkness. (Bégout-Anras et al., 1997). In addition, groups of normally day-active rainbow trout (*Oncorhynchus mykiss*) and normally night-active European catfish (*Silurus glanis*) could synchronize more quickly to changes in photoperiod and feeding schedules than single fish (Bolliet et al., 2001).

An endogenous preening ultradian rhythm with periods of 20–100 min initiated by the dominant alpha animal was reported for mynah birds (*Acridotheres cristatellus*) (Nguyen-Clausen, 1975). Similarly, rhesus monkeys *(Macaca mulatta)* in separate cages and with visual and auditory contact with each other have displayed ultradian activity rhythms that tended to be synchronized by the dominant animal (Yellin & Hauty, 1971).

A dramatic synchrony of a high-frequency ultradian rhythm is displayed by some species of fireflies (*Pteroptyx malaccae* and *Pt. cribellata*) in Malaysia and New Guinea that flash in unison just after twilight with a "circa-second" rhythmic frequency (Hanson, 1978). This phenomenon has also been observed in males of two North American firefly species (*Photuris frontalis* and *Photinus carolinus*) (cf. Moiseff & Copeland, 2000). The synchronized flashing by male fireflies is thought to be involved in courtship by aiding male orientation toward females and vice versa (Buck,

---

[5] One of the grievances voiced by the American colonists was the call for assemblies at odd hours and at distant places by agents of the British king, which they perceived as an attempt to disrupt their sleeping schedules and render them less alert and psychologically vulnerable (Luce, 1970, p. 133).

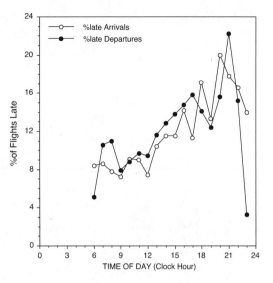

FIGURE 10.3. Circadian variation in late takeoffs and arrivals of aircraft for MSP Intl Airport, Minnesota in September, 1999 (US Dept Transportation, Air Travel Consumer Report, November, 1999). Scheduling a flight in the morning can result in more on-time departures and arrivals by avoiding the congestion found later in the day due to social crowding.

1988). High-frequency ultradian rhythms have also been observed in the aggregation and relay of signals by groups of cells of the slime mold *Dictyostelium discoideum* following cyclic AMP signaling from individual cells (cf. Gross, 1994) (also discussed in Chapter 5 on Models).

Circadian Events

In humans, social conditioning can synchronize circadian rhythms. Meals are often scheduled or prepared only at very specific times. Children may be scolded or punished if they do not go to bed at a time established by their parents, while many adults rely upon alarms to awaken them in the morning (Regal & Connolly, 1980). For reasons such as this, individuals who volunteer to be subjects for isolation studies under constant environmental conditions are stripped of watches, radios, internet access, daily newspapers, and other time-related items, in order to allow their internal rhythms to free-run in the absence of external time cues (see Chapter 2).

Social desynchronization, like social synchronization, can also affect circadian rhythms. Social crowding due to time territoriality may force some individuals to shift their activity to times when there may be less conflict, thereby partitioning individuals within a population.[6] Thus, humans may schedule their air flights to occur earlier in the day when there are fewer delays (Figure 10.3), drive at specific times to avoid the morning and evening rush hours or avoid driving, bicycling or walking near roadways during times when motor vehicle crashes are more likely

---

[6] This has been shown for both rats and lizards (Regal & Connolly, 1980), wherein activity patterns of subordinate animals were shifted or even shown to be free-running with periods different from 24.0 h during crowded conditions.

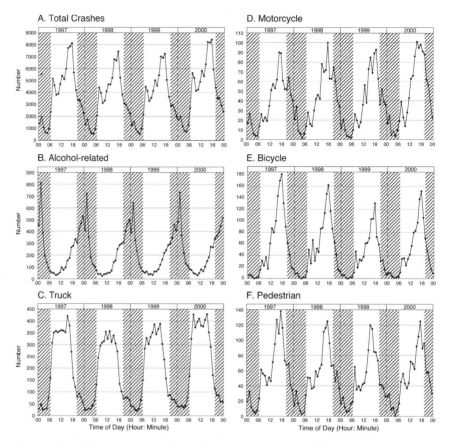

FIGURE 10.4. Circadian patterns for crashes involving a motor vehicle regardless of cause (e.g., distractions, weather, congestion, unsafe speeds, fatigue, etc.). Hourly incidence shown by year (1997–2000) in Minnesota for a motor vehicle crash involving: (*A*) total crashes; (*B*) alcohol; (*C*) trucks (excluding pick-ups and vans); (*D*) motorcycles; (*E*) bicycles; (*F*) pedestrians. Shaded areas = evening or night. See Table 10.1 for acrophases from our cosinor analysis of these data (Min Motor Vehicle Crash Facts, 2000).

to occur (Figure 10.4), shop or dine out when there are fewer crowds, or even work on the internet in the evening or during the night when communication lines are less crowded. The extent to which altered timing of activity affects individual fitness is unknown, but one could speculate that crowding might actually produce jet-lag-like symptoms in some individuals.

Knowledge of population circadian patterns in events can be applied to avoiding certain activities or scheduling of extra workers, such as in hospital emergency rooms (Manfredini et al., 2002). For example, while most types of accidents in adults are more frequent at night or during specific times of the day (Table 10.1), acute poisonings peak in the late afternoon (Manfredini et al., 2002)

TABLE 10.1. Examples of areas where time-of-day effects have been implicated in accidents (traumas) and/or performance errors.

Subjects	Variable	Time(s) of maximum	Reference
Bus drivers	Accidents	07:00–11:00 h, 14:00–18:00 h	Pokorny et al., 1981, 1985
Air force pilots	Pilot error	Early morning	Ribak et al., 1983
Glass and steel workers	Accidents	11:00 h and 03:00 h	Wojtczak-Jarosowa & Jarosz, 1987
Automobile drivers	Single car accidents	Night for <45 years, afternoon for older drivers	Langlois et al., 1985
Truck drivers	Accidents	Night (00:00 –08:00 h)	Harris, 1977
All vehicles	Crashes	14:40–15:28 h	Our 24-h analysis of data in Fig 10.4
Alcohol-related	Crashes	22:28–23:20 h	"
Trucks	Crashes	15:36 –16:28 h	"
Motorcycle	Crashes	16:44 –17:24 h	"
Bicycle	Crashes	15:52 h –16:44 h	"
Pedestrian	Crashes	12:28 –13:00 h	"
Locomotive engineers	Emergency-braking incidents	03:00 h –15:00 h	Hildebrandt et al., 1974
Switchboard operators	Delays in answering calls	Night shift	Browne, 1949
Factory gas meter readers	Logging and calculation errors	Night shift	Bjerner et al.,1955
Medical students and residents	Exposure to blood-borne pathogen by needle stick, laceration or splash	14:00 –15:00 h for actual number of exposures; 00:00 – 01:00 h for hourly exposure rate	Parks et al., 2000
Nurses	Motor vehicle accidents and near-accidents	High rate at end of night shift	Novak & Auvil-Novak, 1996
Adults	Acute poisoning from analgesics, opiates, etc.	Late afternoon	Raymond et al., 1992; Manfredini et al., 2002
Children	Traumas/h from all activities	15:00 –17:00 h	Reinberg et al., 2002

and childhood injury and trauma hospital admissions from all causes peak around 16:00 h, whether on a school day or not (Reinberg et al., 2002). This implies that more pediatric surgeons should be available to work in the emergency room between 4 PM and 8 PM than in the morning and a need for an awareness of the circadian pattern of trauma in children[7] when designing or implementing injury prevention programs.

Conversely, the morning peak in heart attacks (myocardial infarctions or MIs) and pain (ischemia) between 06:00 h and 12:00 h that results in more requests for helicopter transport has led to a call that a "morning-loaded" staffing pattern be implemented for air medical services (Fromm et al., 1992). Similarly, a dramatic increase in the need for urgent cesarean section between 08:00 h and 14:00 h suggests a similar staffing pattern be implemented (Goldstick et al., 2003).

Ultradian and Infradian Events

The social synchronization of most ultradian rhythms is less common than synchronization of circadian rhythms, but has been demonstrated for several animal species (see Essay 10.1).

The most well-known social synchronization of an infradian rhythm is that of the fertility cycle. With regards to humans, menstrual synchrony has been reported among groups of females living together at school, working together, or within a family (McClintock, 1971; Graham & McGrew, 1980; Quadagno et al., 1981). Social regulation of ovulation may occur throughout much of a woman's lifespan (McClintock, 1998), since *pheromones*, which are airborne chemical signals of a female or male individual, can affect the menstrual cycle (Stern & McClintock, 1998).

The ovarian cycles of rats also become synchronized when they live together or even if they live apart, but share a common air supply (McClintock, 1984; Schank & McClintock, 1992). Interestingly, females of some species are known to synchronize males. For example, when Syrian male hamsters (*Mesocricetus auratus*) are housed with females, they show a 4-day activity rhythm which corresponds to the female's estrous cycle, while isolated male hamsters do not show a 4-day activity cycle (Davis et al., 1987). This topic is discussed in greater detail in Chapter 7 on Sexuality.

Aggression and Violence

In a number of animal species, including mammals, aggressive behavior is associated with specific seasons and photoperiod. For example, one study analyzed data on individual violent crimes, such as sexual offenses and aggravated assaults,

---

[7] Since there is a temporal aspect to childhood injury, the term "trauma" is sometimes preferred since "accident" indicates a random occurrence (Reinberg et al., 2002). Indeed, it has been stated that *"injuries are no more likely to occur by chance than diseases."* (Guyer & Gallagher, 1985).

and nonviolent crimes, such as burglary, from three northern hemisphere countries (Denmark, Israel, and the USA) and one southern hemisphere country (Australia) (Schreiber et al., 1997).

While nonviolent crimes appear to be equally distributed throughout the year, individual violent crimes showed an annual rhythm with a peak during summer (July–August in the northern hemisphere and December–February in the southern hemisphere) and a nadir during the two winter months in each hemisphere. Regardless of whether the significant correlation between aggression and photoperiod is dependent upon an endogenous or exogenous mechanism, the implications for crime prevention and law enforcement are a practical matter that confronts society.

Night and Shiftwork

Individuals who work the nightshift are exposed to shifts in the phase of environmental synchronizers, most notably, light and darkness. Because external synchronizers can establish the phase of a circadian rhythm, a number of questions need to be addressed in regard to those who work the nightshift. While it has been stated that the synchronizing effects of environmental factors or cues are identical whether or not geographical area has changed (cf. Reinberg & Smolensky, 1992), there are some obvious differences in the prevalence of mixed groups of synchronizers that a shiftworker is likely to encounter. Thus, a person on the nightshift may work under artificial light sources at night and sleep in a darkened bedroom during much of the day, while most of the activities in the rest of the world are synchronized to the normal light–dark schedule and routines of the geographical area.

Time for Sleep

Sleep is an important span in our daily schedule and the typical time for its onset is usually after dark and up until midnight. In addition, waking up too early can have adverse effects that are associated with sleep deprivation. While sleep is often viewed as a time for rest and restoration, it may have had the adaptive advantage of a behavior control function by protecting early humans from the perils of what could be encountered by those who dared to venture into the outside darkness at night and be subjected to cold temperatures, unseen physical obstacles, or animals of prey (cf. Webb, 1971). Most likely, early humans slept or rested through most of the dark span each day, using dawn and dusk as time cues.

Today, with the availability of electric lights and a faster pace of life, most people probably do not receive enough sleep and the problem seems to be becoming more acute. In the year 1900, Americans slept close to 9 h each night, on average, while in the year 2000 the average was only 6 h 54 min during the work week and 7 h 34 min on the weekend—shorter than the recommended 8 h/night (National

Sleep Foundation, 2001).[8] This suggests that many people today may be chronically sleep-deprived and subject to some of the same health and performance problems as shiftworkers. Even early school start times for adolescents (07:20 h in 10th grade vs. 08:25 h in 9th grade) are associated with significant sleep deprivation and daytime sleepiness (Carskadon et al., 1998).[9]

Problems with Shiftwork

In developed countries, 10–27% of the workforce is involved with nocturnal jobs, either continuously or on rotating schedules (Czeisler et al., 1982; Moore-Ede, 1985; Reinberg & Smolensky, 1992; Rajaratnam & Arendt, 2001). Under such conditions, a worker may need to respond at a time when he or she would normally be asleep, creating a conflict between the inherent protective mechanism of not responding at night, and the responsiveness demanded by the job.

Theoretically, bright lights may help such workers function at night, but often the job occurs in a dimly lit area. Up to 80% of night workers report problems with fatigue, disturbed sleep, insomnia, and malaise (discomfort) (Moore-Ede, 1982), with increasing age, especially around age 45 years, and *diurnal type* (i.e., morningness) associated with more difficulties in adjusting to shiftwork (Åkerstedt & Torsvall, 1981). Fatigue is a major concern, and with it, the safety of those who depend upon the operation and all that could be lost by human error.[10]

Data obtained from nurses who worked for at least one year on either day work or shiftwork alternating among day, evening and night shifts, revealed that 80.2% of shiftworkers, but only 23.5% of day workers, reported sleep disorders (Barak et al., 1995). Because nursing demands attention, concentration, and coordinated manual skills, chronic sleep disorders might have major implications in the planning of work schedules.

The concerns of chronic sleep deprivation extend to many areas of society. Sometimes a change from an undesirable schedule (e.g., a 12-h night shift) back to a more usual schedule (e.g., an 8-h night shift) alleviates problems, such as decreased alertness and poor sleep habits, as has been reported for petrochemical

---

[8] Total sleep time/24 h typically decreases with age in humans as follows: age (hours of sleep): 1–3 days (11–23 h), 4–26 weeks (12–15 h), 2–4 years (10–15.5 h), 9–20 years (7.5–12 h), 30–65 years (6.5–10 h), 66–90 (4–12 h). For comparison purposes, the length of sleep per 24 h for adults of other selected species is: opossum, 19.4 h; bat, 18.0 h; cat, 14.4 h; rat, 13.5 h; pig, 13.2 h; chimpanzee; 11.0 h; mole, 8.4 h (cf. Webb, 1971). In addition to total sleep time, other sleep parameters that decrease with age in humans include sleep efficiency, percentage of slow-wave and REM sleep, and REM latency, while sleep latency, percentage of stage 1 and stage 2 sleep and wake after sleep onset all significantly increase with age (cf. Ohayon et al., 2004).

[9] While students woke earlier on school days in the 10th grade, they did not go to bed earlier and thus slept less than in the 9th grade.

[10] Nuclear accidents at 3-Mile Island and Chernobyl, the Exxon Valdez oil spill and the cyanide leak in Bhopal, India all occurred in the middle of the night and were attributed to human error (see footnote 5 in Chapter 1).

workers in Brazil (Fischer et al., 2000). Even when the issue is based upon leisure, rather than work, such as a late-night party or outdoor activity, humans and their lights are apt to discover that they have placed themselves into the active phase of certain pests, such as certain mosquitoes, which feed more heavily beginning at dusk.

The implications of biological rhythms are evident in the workplace, and so too are the applications. To assume that an individual can perform any task at the same level of competence or efficiency at any time of the day or night is erroneous (Reinberg & Smolensky, 1992). Based upon what has been learned about the circadian rhythms of performance, night work can be viewed as an accident risk factor. Results from a study of 4,645 injury accidents in a large engineering company over the span of a year in which workers rotated on a three-shift system showed that the risk of injury on the night shift was 1.23 times higher than on the morning shift and 1.82 times higher if the work was self-paced (Smith et al., 1994).

Adjusting to Shiftwork

Bridging the gap between basic research on the physiology of biological rhythms and the day-to-day operations of the workplace does occur,[11] and has been successfully demonstrated when the properties and characteristics of the human circadian system, such as the direction of phase-shifting (e.g., when work hours start and end), duration of the shift (8 h vs. 12 h), the range of entrainment (e.g., how long workers remain on a set schedule), health and social tolerance to shiftwork, and the free-running period of the circadian rhythm (e.g., 25 h) have been taken into account (Czeisler et al., 1982). For example, work schedules that rotate by phase-delay (e.g., the next shift begins later) and have an extended duration between rotations in schedule appear to be the most compatible with the circadian system, resulting in greater satisfaction, improved health and safety among workers, and greater productivity and profit for management.

Other approaches to deal with tolerance and performance of shiftwork include: (*a*) *rapid rotation* (one or two days); (*b*) *dedicated shifts* (maintain a worker on a permanent schedule); and (*c*) *slow rotation* (1–5 weeks) that is supplemented by techniques for synchronization, such as times of feeding and types of food (e.g., protein, carbohydrate) (Ehret, 1981). However, a second job or social obligations during the day could still contribute to a decrease in performance during the night shift.

Similarly, social and family life affects morning shiftworkers who might stay up too late even though they need to get up early for work. Starting the morning shift at 06:00 h rather than 05:00 h at an oil refinery resulted in workers that were less sleep-deprived since they no longer needed to wake up at 03:50 h in order to be at work on time (Reinberg, 1979). Several European navies have found that

---

[11] A discussion with numerous examples from various areas of society that focus on the application of the knowledge of circadian rhythms to technology and the workplace can be found in "The Twenty Four Hour Society" by Martin Moore-Ede (1993).

splitting the night shift from 20:00 h to 08:00 h into three sections of 4 h each and having a sailor rotate among shifts from day to day (begin at 20:00 h the first night, at 00:00 h the second and 04:00 h the third) alleviates sleep deprivation and circadian disruption. This is probably because a few good hours of sleep each night "anchors" the biologic rhythms (Minors & Waterhouse, 1981, 1983). In fact, the phasing of circadian rhythms can be maintained at a chosen time by using 4 h spans of anchor sleep at the same time of day (Minors & Waterhouse, 1992).

Regarding underlying physiology, investigators found that individuals highly tolerant to shiftwork exhibited larger amplitude circadian rhythms in several body functions (oral temperature, grip strength, heart rate) and slower shifts of the circadian acrophases than individuals who were poorly tolerant of shiftwork (Reinberg et al., 1978; Reinberg & Smolensky, 1992), resulting in less internal desynchronization of rhythms during the adjustment to the new schedule. Additional reports on night and shiftwork concerns and observations can be found in a special issue on equity and working time in the journal *Chronobiology International* (Fischer et al., 2004). This issue summarizes an International Symposium of Night and Shiftwork held in November 2003 and contains articles discussing topics on working time and health, such as flexible working hours, sleep parameters, sickness and depression, nutritional status, alcohol, napping, and indexes to assess work schedules.

The Global Workplace

The speed, as well as the available technology, by which individuals throughout the world can communicate with each other is astonishing. Telephones, computers, and electronic mail have made voice, and sometimes video, communications almost instantaneous. With the exception of certain astronomical or physical properties of the universe, the electronic and physical components of global communications function with unimpeded efficiency through the day and night. The same constant efficiency does not apply to humans.

Communication

For the most part, we are synchronized to be in phase with our own geographical location, the place where we send and receive messages. If we are aroused to communicate near dawn, the chances are that we will be less vigilant or awake than if it occurred at 11:00 h. The duration of time that is required for direct communications may be measured in fractions of a second, but the duration of time that could exist between the phases of the biological rhythms for individuals who are in communication with each other may be in the order of hours. In other words, one individual may be more alert and attentive than the other who is in a different time zone (Figure 10.2). This difference in phase is one of the problems facing international corporations when they attempt telecommunication conferences among persons located in different parts of the world. For example, you may have a brilliant new idea to discuss with a colleague in Beijing, but if it's 15:00 h where you are (e.g., central USA), it's 03:00 h in Asia. Global time differences may also impact decision making by politicians or physicians via phone from opposite sides of the globe. For example, using telemedicine, an x-ray taken in the middle of the night could be sent to and read by an expert radiologist on the other side of the world, where he/she could be at peak performance, rather than rely on a sleepy local radiologist.

Work Schedules and Outsourcing

As an alternative or modification to shiftwork, the 24 h solar day work/rest schedule could be divided into either two daily 12 h shifts or into three daily 8 h shifts, but having the 12 h or 8 h spans assigned according to longitude time zones. This would allow the 12 h or 8 h shift to be in synchrony with the activity phases of the circadian rhythms for all the workers. This type of synchrony and division of work is already occurring among scientists located in different countries or regions when they cooperate on joint research projects. After a "normal" day of work for a scientist, numerical data, typed messages, or drafts of papers can be sent easily by e-mail or the internet between 17:00h–20:00h from one location (latitude) to a second location where the local time is early morning (e.g., Europe to the N. America, USA to India, etc.). Later in the evening at location 2, revisions, analyses, or new results are returned to the first location to be available at the beginning of the next day in the first location. The work on the project can continue on practically a 24-h schedule, but the circadian rhythms of all the participants are not disrupted. This same type of temporal outsourcing, which directly takes advantage of the circadian rhythms of activity/rest for greater productivity, applies to a vast array of professions and businesses, including medicine,[12] marketing,[13] finance, etc.

Sports and Performance

Since a circadian rhythm has been demonstrated in virtually every physiological and psychological variable, it is reasonable to expect that biological rhythms should play a role in physical exercise, sports, and performance (Table 10.2). In fact, there is now enough information available to suggest that sports performance is influenced by time of day and underlying endogenous mechanisms, with optimal performance dependent upon type of activity (e.g., sports requiring physical efforts is best later in the day) (cf. Drust et al., 2005).

---

[12] The notes dictated by medical doctors and nurses at the end of their workday can be sent electronically and transcribed in another global time zone, returned, and be ready to be reviewed and filed the next morning.

[13] Sometimes when we purchase a product during "off hours" or late at night, the order may be processed by an individual who is working their normal dayshift located in a different time zone.

TABLE 10.2. Examples of sporting-related events and activities where time-of-day effects have been implicated in influencing performance (cf. Winget, 1985).

Sporting or related event	Parameter	Time of best performance	Reference
Anaerobic cycle leg exercise	Maximal, peak, and mean power	~18:00 h	Souissi et al., 2004
Biking	Tolerance to high intensity exercise	22:00 h	Reilly & Baxter, 1983
Biking	Knee extensor torque	19:30 h	Callard et al., 2000
Fencing	Best scores related to speed	12:00 ±1 h	Reinberg et al., 1985
Golf	Optimal psychophysiological conditions	08:00–16:00 h	Rossi et al., 1983
Rowing	Performance	Evening	Conroy & O'Brien, 1974
Rowing	O_2 consumption and heart rate at blood lactate threshold	Evening	Forsyth & Reilly, 2004
Running	Performance	Evening	Conroy & O'Brien, 1974
Shotput	Performance	Evening	Conroy & O'Brien, 1974
Soccer	Endurance	17:00–21:00 h	Reilly & Walsh, 1981
Sprint (stair run)	Anaerobic power	Late afternoon	Reilly & Down, 1992
Standing broad jump	Anaerobic power	Late afternoon	Reilly & Down, 1992
Swimming	Speed	17:00–18:00 h	Rodahl et al, 1976
Swimming	100-min and 400-min time	22:00 h	Baxter & Reilly, 1983
Swimming	Trunk flexibility	13:30 h	Baxter & Reilly, 1983
Swimming	Peak power output	Late afternoon	Reilly & Marshall, 1991
Tennis	Velocity, accuracy	13:00–18:00 h	Bloom, 1989; Atkinson & Spears, 1998
Water polo	Optimal psychophysiological conditions	20:00–23:00 h	Rossi et al., 1983

Many variables relevant to athletic performance peak in the mid to late afternoon (Winget et al., 1985). These include oxygen consumption, respiration, heart rate and recovery rate, muscle blood flow, forearm blood flow, minimum sweating threshold, grip strength, minimal anxiety, minimal fatigue, and body temperature. Circadian rhythms have been shown in human muscular efficiency, with maximal efficiency found at 18:00 h for elbow flexor torque in physical education students (Gauthier et al., 1996) and at 19:00 h for knee extensor torque in competitive ultradistance cyclists (Callard et al., 2000). Anaerobic power at 21:00 h and anaerobic capacity at 15:00 h and 21:00 h were significantly higher than at 03:00 h or 09:00 h in nine college-aged male athletes (Hill & Smith, 1991). Even in the case of men not in training for sports, treadmill endurance time was 2% longer in the evening between 19:30—20:30 h than in the morning between 07:30—08:30 h (Burgoon et al., 1992). Thus, some of the early pioneers in sleep and performance research were correct when they wrote: *"Contrary to public opinion, one is not at his best, or most wide-awake, in the morning. Immediately upon getting up, one's performance is, as a rule, about as poor as it was the night before, prior to going to bed."* (Kleitman & Jackson, 1950).

Body Temperature and Performance Variables

Circadian patterns for body temperature and simple tasks involving motor activity (e.g., card sorting, copying, code substitution) or performance requiring memory, muscular coordination, and prolonged vigilance, all with maxima in the afternoon and minima at night, suggests a causal relationship between performance and body temperature (Kleitman & Jackson, 1950; Kleitman, 1963).

However, circadian rhythms in performance may not always exactly parallel the body temperature rhythm under normal sleep–wake conditions due to interactions between the circadian system and time since waking, but they invariably do when the sleep/wake cycle is suspended during a constant routine (Monk & Carrier, 1998). In addition, differences in the "best time" to perform a task do exist and appear to depend upon a number of factors,[14] such as the nature of the task, age, level of practice, and if the person is a morning or evening type individual[15] (Folkard, 1983; Carrier & Monk, 2000).

A number of other variables that contribute to athletic performance and display circadian rhythmic peaks that occur during the afternoon and evening between 14:00 h and 20:00 h include neuromuscular coordination, manual dexterity, reaction time, vigilance or monitoring, cognitive ability, short-term memory, and various other psychomotor tasks, such as letter cancellation and card sorting (Winget et al., 1992). Because these variables reach their peak around the same time of day as does body temperature, it has been postulated that one should be able to determine peak performance time ±3 h by self-measuring oral temperature over several days and locating the time of peak values in graphs of the collected data (Winget et al., 1992).

A Time to Train or Win

Results from a number of studies support the view that there is a circadian advantage or enhancement to athletic performance if workouts and events are held between 12:00 h and 21:00 h (Winget et al., 1985). Anecdotally, it has been said that sports performance improves throughout the day and that few world records are ever set in the morning (Reilly, 1992; Eichner, 1994). Indeed, most reports on athletic competitions in individual or team sports have concluded that triumphs or best performance is most likely to occur later in the day or in the evening. While

---

[14] One study tested 63 boys and 56 girls 9–11 years of age who performed tasks such as sprinting (running speed) and a skill game (ball and cup) at fixed clock hours (08:30, 10:30, 13:30, and 15:30 h) while at school (Huguet et al., 1995). It was pointed out that while the end of the afternoon was most favorable for strength performance overall, tests for agility and speed in children showed a second peak in the late morning and that both the end of the morning and the end of the afternoon would be good times for physical and sports activities in children of this age.

[15] The concept of *chronotype* (morningness vs. eveningness) is discussed in Chapter 12 on Autorhythmometry.

many additional factors will influence the outcome of athletic competition, such as the extent of training, lack of sleep, pain, fatigue, mental state, general health, and chronotype (morning vs. evening person), there thus seems to be a circadian advantage when competing at certain times of the day, at least for well-trained athletes.[16]

Jet Lag and Professional Sports

Another factor that has been shown to affect sports performance is jet lag, whether caused by shiftwork or crossing time zones rapidly via jet travel, and its accompanying circadian desynchrony (Atkinson & Reilly, 1996). After as few as three time zone changes, the body's circadian rhythms will be in a state of internal desynchronization during adjustment to a new time schedule, resulting in symptoms associated with jet lag.

When the complete season records for 1991–1993 of the 19 North American major league baseball teams based in cities on the West Coast (WC) and East Coast (EC) were compared, it was found that the EC teams had a significantly better winning record, not only due to home field advantage, but to previous transcontinental travel within the past two days by WC teams (Recht et al., 1995).[17] This effect was only found when the travel was eastward, since EC teams traveling to the west coast were not affected by jet lag. Presumably, WC teams often played a game in the East that was too early for their peak performance (e.g., a game beginning at 12:00 h on Eastern (EC) Time would be 3 h early (09:00 h) on Pacific (WC) time).

In a 25-year retrospective analysis of all Monday Night Football games by EC vs. WC teams in the USA between 1970 and 1994, jet lag was also implicated as influencing the outcome of the game (Smith et al., 1997), but in favor of the WC teams. WC teams won more often and by more points than EC teams when playing on the East Coast, since the EC teams started the game at 21:00 h (late in the day for the EC players), while the WC teams essentially started playing when their internal clock was set 3 h earlier (18:00 h Pacific Time) and were thus playing closer to the time of peak athletic performance (e.g., late afternoon).

Allowing for Jet Lag

It has been suggested that competitive sports should consider the necessity of allowing several days for resynchronization following time zone changes. For example, when crossing 6 or more time zones, the following schedule was suggested for Japanese athletes: (*a*) no competitions or hard training within the first 3 days of arrival at a new destination; (*b*) only light training over the next 3 days; and (*c*) only beginning on day 7 of arrival should competitions be

---

[16] Some coaches view chronobiological principles as the ultimate weapon that can give their athletes a winning edge (Winget, 1992).

[17] The data showed that the home team scored ~1.24 more runs than usual when a visiting team has crossed three time zones eastward within the previous two days.

scheduled, with best performance expected beginning 10 days after arrival (Sasaki, 1980).

Such a schedule may be of crucial importance for ultra-athletes having to cross several time zones in order to train and compete in the Olympics or World Games. This conclusion was confirmed for male members of the German Olympic gymnastic team who were studied before and after flights from Frankfurt to Atlanta (a 6 h delay in time zones from the east to the west) and from Munich to Osaka, Japan (an 8 h advance in time zones) (Lemmer et al., 2002). When monitoring jet lag symptoms, training performance, and several physiologic variables (oral temperature, grip strength, blood pressure, heart rate, salivary cortisol, and melatonin), all functions were disturbed on the first day of arrival and remained so for at least 5 days after westward travel and 7 days after eastward travel. The authors concluded that jet lag symptoms and rhythm disturbances in physiological functions due to time-zone transitions (in spite of twice daily exercise regimens) need to be considered in highly competitive athletes. They thus recommended that for competition in a different time zone, athletes arrive at least two weeks in advance of competitions to overcome the effects of jet lag.

In light of jet lag induced changes in the cardiovascular system, early arrival at the new time zones is also suggested for anyone scheduling meetings abroad (Lemmer et al., 2002). This has implications for diplomats and business persons who must schedule meetings abroad, sometimes on the same day that they arrive, since they will probably not be at peak physical and psychological performance due to effects of jet lag. In addition to athletes (Reilly, 1992), an appropriate flight itinerary chosen to reduce the negative effects of jet lag has thus been suggested for coaches, as well as for academics, politicians,[18] and others attending meetings in a different time zone (Waterhouse et al., 2002).

Travel on the Earth's Surface

Today we can travel rapidly at great speeds on land, water, and in the air, often across many time zones in a relatively short duration of time. When this occurs, the phases of our biological rhythms must be either advanced or delayed and our rhythms resynchronized to the local time upon arrival at the new destination.

---

[18] Political leaders and diplomats often face the problems associated with jet lag due to time zone changes associated with travel. Former Secretary of State Henry Kissinger tried to retire an hour earlier each night and rise earlier before an important meeting overseas. Politicians can also arrive early and allow several days to adjust, as was the case of President Eisenhower's visit to Geneva in 1955 to meet with the Soviet leader Nikita Khrushchev, or keep a home schedule and arrange meetings accordingly while in a foreign country, as was usually the case for President Lyndon Johnson (cf. Ehret & Scanlon, 1983). *Note*: Traveling by ship from your own time zone is the most advantageous way to avoid jet lag, since the slow travel by ship allows for a natural resynchronizing to a new local time zone.

However, the impact of travel upon our biological rhythms is not only limited to the crossings of time zones, a topic that is commonly discussed relative to jet lag, but also upon the time of day when travel occurs.

Driver Fatigue and Vehicle Accidents

Long, undemanding and monotonous driving easily facilitates fatigue and sleepiness. Those who have driven an automobile on an overnight trip may be aware that the toughest time to stay awake is near dawn. Based upon the personal experience of many individuals, this is the span when one may resort to an array of activities, such as rolling down the windows to allow cold air to blow on your face, drinking a caffeinated beverage, singing, etc., all done to "fight" the urge to sleep. While such gimmicks may be helpful, one cannot escape the physiological need for sleep and the consequences of fatigue, which is an inherent part of our circadian system.

The endogenous rhythms of vigilance, performance, and the sleep/wake cycle, especially as they relate to driver fatigue, have strong implications to society. The stark realization for drivers of motor and other vehicles is that the risk for accidents due specifically to fatigue has been shown to display a rhythm, with a nocturnal peak near 03:00 h and a secondary diurnal peak around noon (Reinberg & Smolensky, 1992; also cf. Moore-Ede, 1993). Peaks in accidents at specific times of day or night have also been shown to be the case for truck drivers (Harris, 1977), locomotive engineers (Hildebrandt et al., 1974), bus drivers (Pokorny et al., 1985), and air force pilots (Ribak et al., 1983), among others (Table 10.1).

In the USA, it has been estimated that every truck will be involved in at least one sleep-related accident during its lifetime (cf. Horne & Reyner, 1999). Based upon results from the analysis of interstate truck accident data, a circadian effect has been shown, with approximately twice as many accidents occurring for dozing drivers between midnight and 08:00 h than in the other 16 h of the day (Harris, 1977). Both accident ratios and accident frequencies over a 4-year span in Texas showed similar rhythms for single car and truck accidents (Langlois et al, 1985). Again, this was attributed to driver fatigue compounded by the difficulty of seeing at night. Not only was a very prominent 24 h pattern detected with an acrophase at 03:00 h, but a high-amplitude 7-day rhythm was found, with a weekend peak. These patterns were virtually identical for urban and rural settings.

The phase of a circadian rhythm, monotony and accumulating lack of sleep all play a role in automobile accidents. In a study conducted in the Netherlands (Riemersma et al., 1977), eleven 20-year-old university students performed vigilance tasks while driving an automobile along a four-lane triangular course of about 56 km ten times between 22:00 h and 06:00 h. When compared to a similar drive at 20:00 h (prior to sleep) or at 06:00 h (after a night's sleep), there were significantly more incorrect responses with regard to vigilance markers, such as lane position, reaction speed, steering wheel deflection, maintenance of speed, missed signals, and reporting kilometrage. In addition, heart rate variability increased toward the end of the night. Apparently, driving all night and the combined effect of lack of sleep on one's circadian rhythms increases risk factors that could lead to an accident. Age can also be a factor, with drivers younger than age 30 years most at risk for a sleep-related crash (cf. Horne & Baulk, 2004), while an early afternoon peak in traffic accidents has been reported for older drivers (cf. Horne & Reyner, 1999).

Studies have also been conducted to examine the incidence of traffic crashes that might be attributed to the 1 h time shifts that occur when there is a change to and from daylight saving time (Lambe & Cummings, 2000). Based upon accident data covering a 16-year span in Sweden, it was concluded that the time shift did not have a measurable impact on incidence rate ratio on the Monday following the time changes compared with the preceding or following Monday. However, the direction of change was suggestive of an effect due to less sleep in the spring (ratio = 1.11) as compared to an extra hour of sleep in the fall (ratio = 0.98).

Alcohol, Driving, and Fatigue

Although not generally realized, sleepiness and its effect on mental alertness has surpassed alcohol[19] and drugs as the greatest cause of accidents in all modes of transportation (Åkerstedt, 2000; Rajaratnam & Arendt, 2001). In addition, alcohol consumption or sleepiness produce similar decrements on driving performance and can impose a greater risk for a driving accident when combined (Arnedt et al., 2000). In a study monitoring simulated driving performance, decrements in driving were similar for subjects following a prolonged 20-h wakefulness span or ingestion of alcohol to reach a blood concentration of 0.08%, with the combination of wakefulness and alcohol slightly worse than either alone on driving performance. Interestingly, 18.5 h, 21 h, or 24 h of sustained wakefulness can produce decrements in driving performance comparable to blood alcohol concentrations of 0.05%, 0.08%, and 0.10%, respectively, suggesting that a sleepiness metric scale could be created (Arnedt et al., 2000, 2001). However, even moderate levels of alcohol intake can interact with sleepiness to exacerbate driving impairment (Arnedt et al., 2001), especially during the natural afternoon "dip" in alertness, placing the driver at a greater risk of an accident after alcohol at lunchtime (Horne et al., 2004).

In the United States alone, driver fatigue-related costs to society from accidents associated with work schedules, night activity and inadequate sleep are calculated in billions of dollars (Moore-Ede, 1993). For example, in one case where a woman was killed after her car was hit from behind by a tractor-trailer, her family was awarded US $24 million from the driver's employer for violating hours of work regulations that led to driver fatigue (cf. Rajaratnam & Arendt, 2001). In another case in England, a man fell asleep at the wheel of his car at 06:00 h, the car went off the road and caused a train collision that killed 10 people and injured

---

[19] In 1997, alcohol accounted for 7% of motor vehicle crashes in the USA, while estimates of crashes related to sleepiness range from 1–3% to 35–42% (cf. Arnedt et al., 2000) or 15–20% (Åkerstedt, 2000). This estimate is around 10% in the UK (Horne et al., 2003).

more than 70 passengers. He was subsequently convicted of causing death by dangerous driving, received a 5-year jail sentence and the case resulted in the largest motor insurance claim in U.K. history (Rajaratnam & Jones, 2004). Litigation for such accidents is likely to increase since it was concluded that knowledge of sleepiness indicates responsibility on the part of the defendant and that continuing to drive is voluntary conduct. Indeed, studies comparing subjective sleepiness and EEG changes during simulated driving after sleep deprivation have documented that drivers had knowledge of their actual neurophysiological level of sleepiness while making driving errors (Horne & Baulk, 2004).

When one includes the large numbers of deaths and injuries with these monetary losses, the application of what is known about the characteristics and rules of circadian rhythmicity to reduce accidents becomes a practical problem that can often be resolved through similar educational and prescribed approaches that have been applied to alcohol while driving or during work hours. (e.g., if sleepy (or drunk), especially if "fighting" sleep, don't drive!). In order to reduce liability with regard to public safety (and increase health and productivity) employers could take steps to reduce sleepiness and fatigue in the workplace (Rajaratnam, 2001).

The Post-Lunch Dip

An interesting finding that supplements what is known about driver fatigue comes from a relatively large study of locomotive drivers of the Federal German Railway (Hildebrandt et al., 1974). Here again there was the expected nocturnal maximum poor performance peak that occurred at about 03:00 h. In addition, a second peak occurred about 12 h later, the so-called "post lunch dip." Although there is a general tendency to relate a post lunch dip feeling of being tired or sleepy to a recently consumed lunch, the dip has been found to occur even if a meal is not consumed (Monk et al., 1996). In the case of the locomotive drivers and perhaps for other drivers as well, the earlier the work schedule starts for the driver, the more pronounced is the post lunch dip.

Animal Activity and Vehicle Accidents

In addition to the human fatigue factor in vehicle-related accidents, an increase in on-road encounters with animals can result from the circadian patterns in the animals' activity. For example, the highest crash rate between vehicles and more than 20,000 moose and white-tailed deer observed in Finland occurred in the first hour after sunset, suggesting a need for increased driver vigilance during this time (Haikonen & Summala, 2001).

Travel Above the Earth's Surface

Transmeridian flights (east to west or west to east) that rapidly cross time zones create a set of symptoms (poor quality and altered patterns of sleep, persisting fatigue, changes in mood and behavior, and digestive problems) known as *jet lag* (see Chapter 2).

Jet Lag

Recovery from jet lag involves shifting the phases of circadian rhythms from the original time zone to be in phase with the prevailing synchronizers of the new local environment. The duration of time required to completely shift the biological rhythms of humans depends upon a number of factors, including the direction of travel, the variable to be shifted, time of travel, number of time zones crossed, the rigidity of the sleep-wake cycle upon arrival,[20] and differences among individuals and within the same individual from trip to trip (Figure 10.5). For example, after crossing six time zones, the sleep–wake rhythm synchronizes within about 2 to 5 days and the body temperature in about 4 to 7 days (Reinberg & Smolensky, 1992).

About 80–85% of the people who have been monitored after changing to a new time zone synchronized faster if their activity phase was delayed (east to west travel, which results in a long first day of shift) than if it was advanced (west to east travel, which results in a short first day of shift). The remaining 15–20% adjusted faster after a phase advance than a phase delay (Reinberg & Smolensky, 1992). Many approaches have been suggested to help alleviate the symptoms of jet lag following transmeridian flights or shiftwork, including diet, bright light, exercise, varying shift schedules, and use of substances, such as melatonin (cf. Waterhouse & Redfern, 1997).[21]

Results from animal studies have shown that *social cues* can increase reentrainment rates following shifts of the lighting schedule under laboratory conditions. This was true for male golden hamsters exposed to female olfactory and auditory cues after a phase-advance of the light–dark schedule (Honrado & Mrosovsky, 1989). Female, but not male, South American rodents (*Octodon degus*) resynchronized their circadian temperature and activity rhythms 20–35% faster when housed with male or female *O. degus* that were already entrained to

---

[20] The living schedule may be tightly constrained when going in one direction and not the other. For example, a trip to Europe from the USA may necessitate getting up every day at a precise time (e.g., 07:00 h) in order to attend a meeting at 08:00 h or beginning a sight-seeing trip and thus forcing the 7 h advance phase shift to occur as quickly as possible. Upon returning to the USA, one may not be forced to stay up until the usual time of retiring (e.g., midnight), but can retire and get up earlier than normal for the first few days home, even if work or school do not start until several hours later in the morning. Since the 7 h delayed phase shift is not constrained by a strict sleep-wake schedule, it will occur more gradually.

[21] When traveling west to east, jet lag is often lessened if one can sleep on the plane after the meal, avoid caffeine, alcohol, and movies and use of substances, such as melatonin (cf. Arendt & Deacon, 1997), arrive at the final destination in the afternoon rather than early morning, stay awake by being exposed to sunlight, walking, drinking caffeine, or whatever it takes to be active until finally retiring in the evening. Jet lag is not as severe when traveling north or south if time zones do not change, but a traveler may notice some physiological changes due to longer or shorter days.

398   10. Society

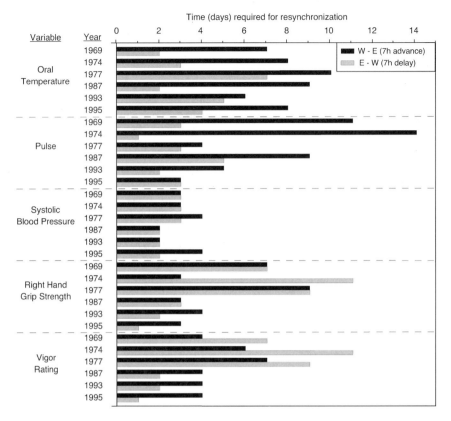

FIGURE 10.5. Days required for circadian resynchronization after 6 transmeridian flights from Minnesota to Central Europe (7-h phase advance) and from Central Europe to Minnesota (7-h phase delay). Healthy man (RBS, age 22 years in 1969) performed 5–6 self-measurements each day. Single cosinor procedure was used to determine 24-h acrophase (peak). Slower adjustment was always seen in oral temperature and pulse after an eastbound flight, while adjustment sometimes occurred faster after an eastbound flight in the other variables. The sleep-wake schedule following each advance was always more tightly structured, while after each delay it was less structured, making strict comparisons of adjustment times tenuous. Exogenous melatonin was used during the eastbound flight in 1995. Destinations: Italy in 1969, 1974, 1977, and 1995; the Netherlands in 1987; Norway in 1993; all return flights were from Germany (Sothern, unpublished).

the new lighting schedule (Goel & Lee, 1995a,b). When housed alone, males actually shifted faster than females.

Life in Space

Spaceflights are also a concern in matters such as circadian sleep/wake synchronizer schedules, light levels and timing of departure and arrival. Already, on multiple-shift missions, astronauts are preconditioned to different sleep/wake schedules prior to a spaceflight so that sleep onset is staggered (e.g., by 8 h or 12 h) among groups of individuals who will then be well-rested and at their peak performance at different times over each 24 h span.

Results from studies in space on astronauts and cosmonauts have shown a remarkable resiliency of the circadian body temperature and melatonin rhythms, with no change in amplitude, phasing or free-running when strict circadian scheduling was maintained (Monk et al., 1998). However, certain aspects of sleep quality and quantity (e.g., shorter, more disturbed, REM latency shorter) were affected (Gundel et al., 1993; Monk et al., 1998).

Future space missions, including time spent on space stations, lunar bases and interplanetary journeys (e.g., to Mars), where people and the plants and animals they bring along will spend months or even years away from time cues on Earth, may have to consider not only circadian, but circannual and perhaps other rhythmic schedules, as well.[22] In addition, one will have to consider how (or whether) to best maintain a 24-h and/or yearly schedule when on a distant planet, which might have very different "days" or "years." For example, in Earth units, Mars rotates once every 24.66 h and its year is 687 days long, while Jupiter rotates on its axis once every 9.93 h and rotates around the sun once every 11.86 years (Astronomical Almanac, 2001).[23]

Travel Beneath the Seas

Adherence to a 24-h rest-activity schedule is or at least should be a concern for those who travel beneath the seas. At one time the work schedule for submarine crewmen in the American or British Navy was 12 h in length and consisted of 4 h on, 8 h off work, since there was a need to rest after the 4-h shift due to the strenuous physical labor involved. Because each sailor was on a rotating watch, with six different watches every 72 h, this meant that a sailor never slept more than 7 h at a time.

The pioneer in sleep research, Nathaniel Kleitman (1895–1999), spent 2 weeks onboard the submarine USS Dogfish, where he studied 74 crew members either on a fixed three-watch 12 h schedule (4 h on/8 h off) or on a shore-type 24-h schedule (cooks, messmen, some officers). The quality of sleep on board was rated as poor by 69% of the men on shifts and 78% needed up to 2 h to become fully alert after awakening. In addition, body temperature measured by nine of the men showed that each subject on a 12-h watch schedule showed two 12 h, but unequally bimodal, curves every 24 h. However, the temperature peaks were differently timed between sailors depending upon the watch schedule, implying that there would be differences in alertness and performance in an emergency situation (Figure 10.6) (Kleitman, 1949). The author later concluded that these 12 h watch

---

[22] The schedules for life in outer space would not be complete without a plan for rhythmic scheduling, not only for humans, but for all organisms on board.

[23] For practical purposes, NASA has adopted the term "sol" to designate a Martian solar day and there are 24 "maurs" in one sol. A planetary body outside of our solar system will undoubtedly have lengths of its day and year that will differ from lengths on Earth.

400   10. Society

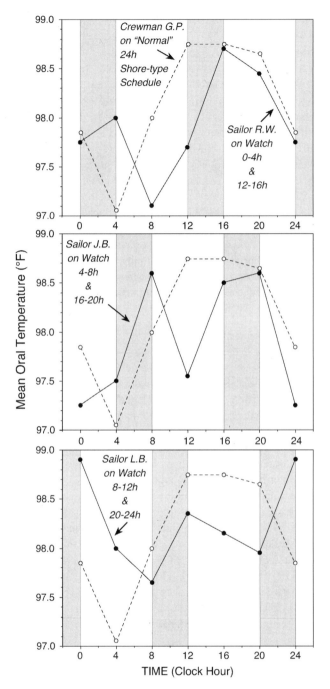

FIGURE 10.6. Body temperature of submarine crewmen on 12-h watch schedules shows 12-h cycles artificially superimposed upon the underlying 24-h cycle in body temperature. Redrawn from Fig 3 in Kleitman, 1949.

schedules were "*unphysiological*" and the 12-h cycle was artificially superimposed upon the underlying 24-h cycle in body temperature (Kleitman, 1963). In another study of twelve submarine sailors on rapidly rotating 4 h watches, body temperature rhythms departed from their original 24 h pattern, with a noticeable dampening of the circadian amplitude (Colquhoun et al., 1978). These results could well be indicative of free-running circadian rhythms in the body temperature.

Today, the rest/activity schedule for most personnel aboard American submarines is usually 18 h, consisting of 6 h duty alternating with 12 h off duty, thereby allowing more time to sleep and/or rest between watches.[24] Nevertheless, studies of nuclear submariners on 18 h watch schedules for up to 10 weeks have shown disruptions of 24 h cycles (especially reductions in amplitude) for body temperature, pulse, blood pressure, and mood parameters (Schaefer et al., 1979; Naitoh et al., 1983). The sleep–wakefulness cycle, however, seemed to entrain well to the 18-h schedule. Since peak times for body temperature became widely dispersed, just as on an 8-h schedule, a desynchronization between an individual's activity cycle and internal physiologic cycles can still occur on an 18-h routine, implying possible chronic health and performance consequences (Naitoh et al., 1983).

Adding to findings of desynchronized rhythms during submerged patrols, the circadian rhythm in melatonin in submariners living on an 18-h schedule for up to 6 weeks was shown to free-run with an average period of 24.35 h (Kelly et al., 1999). As a result, the peak in melatonin had shifted by more than 13 h, suggesting that the human circadian system may be unable to synchronize to such a short "day" of 18 h. All of this occurs in spite of the availability of social synchronization cues, such as meals being served at the same times each day, presence of clocks, and contact with some personnel living on 24 h schedules. In addition, one submariner who lived on a 24-h work schedule, but had the night shift, also free-ran with a period longer than 24.0 h ($\tau = 24.39$ h), indicating a kind of social isolation from those on the 18-h workdays. The phenomenon of individual differences in free-running circadian rhythms during social isolation was also noticed in subjects that were studied while living together under temporal isolation in an underground facility in Germany (Wever, 1979).

Mealtimes and Health

Daily exogenous cycles of feeding may synchronize circadian rhythms (see Chapter 2 on General Features of Rhythms). In addition, the timing of meals and their dietary implications for losing weight strongly display a rhythmic component. Diets are a pertinent topic, not only in the United States where obesity is

---

[24] Compared to hammocks in a large mess area in the past or sleeping bunks too narrow and too close together, both of which made daytime sleep nearly impossible, separate sleeping areas currently available provide more privacy for sleeping.

rampant and dietary products and plans represent a multimillion dollar industry, but in developing countries and the third-world, such as in Africa, Asia and elsewhere, where large numbers of people are faced with hunger and starvation.

In rats and mice, timing of food availability (single meals at different times over the 24 hours) affected whether they lived or died, altered their rhythm patterns and either prolonged or shortened their lifespan (see Essay 10.2).

Essay 10.2 (by RBS): Preclinical Meal-Timing Studies

Studies with large numbers of mice and rats, where lighting schedules and access to food could be strictly controlled, provided the framework for later meal-timing studies with humans.

Single Meals and Survival–It has been shown with mice on restricted diets, that the timing of food intake may actually play a role in survival or early mortality. In two separate studies, groups of 20–28 singly-housed male or female mice were allowed to eat only once a day for four hours at either the beginning of the daily light (L; resting) or dark (D; activity) span. By the 3rd or 4th day on this restricted feeding regimen, some mice began to die. After 10 days, most (90%) of those mice allowed access to food during early-L were dead, but only half (40–50%) of the early-D mice had died (none of the ad libitum fed control mice had died) (Nelson et al., 1973). Interestingly, housing four mice together or raising the room temperature prevented any deaths in the meal-fed mice, indicating that body heat loss was a critical factor, since the mice were often seen huddled together in a group, thereby reducing heat loss and caloric requirement.

Single Meals Alter Rhythms–In addition to affecting mortality, meal-timing also altered circadian rhythms in mice and rats. Circadian patterns in body temperature, serum corticosterone, and liver glycogen, but not corneal mitoses, showed higher amplitudes in meal-fed mice than in mice feeding *ad libitum* and the relationships between the peak times was different depending upon the time of food presentation (see Figure 9.3 in Chapter 9 on Veterinary Medicine). The authors concluded that: "...*optimal nutrition may depend not only on what is eaten, but on when it is eaten in relation to other schedules and demands.*" (Nelson et al., 1975).

In rats meal-fed for 4 h at four different times over 24 h, peaks in liver glycogen and serum glucose, but not corneal mitoses, were also synchronized to the times of restricted feeding, while peaks in serum corticosterone and several enzymes reflected an interaction between feeding times and the LD schedule (Philippens et al., 1977).

When mice were meal-fed at one of six meal times 4 h apart, as well as fed different diet compositions (increased carbohydrate, protein or lipids), it was determined that meal-timing and not diet regimen determined the location of the circadian acrophase for variables such as body temperature, plasma insulin, serum glucose, serum thyroxine and liver glycogen, with only slight acrophase shifts in circulating white blood cells and serum corticosterone (Lakatua et al., 1988).

Thus, the peak times for some body functions are readily altered by meal-timing, while others are not, thereby leading to altered internal phase relationships. From a practical point of view, studies in mice have shown that time-limited feeding may lead to changes in phase relationships of circadian rhythms in cell proliferation vs. other body functions and it was suggested that such alterations in phase relationships might be used to advantage in the treatment of certain cancers (Pauly et al., 1976; Lakatua et al., 1983).

Single Meals Prolong Lifespan–In addition, restricted meal-feeding schedules and a reduction in calories available[25] significantly prolonged the lifespan of mice (Nelson & Halberg, 1986; Duffy et al., 1999) and rats (Duffy et al., 2001). For rats on lifespan studies, a caloric-restricted diet presented during the first half of the daily activity span (5 h after D-onset for nocturnally-active rodents) was also better at maintaining circadian rhythms more closely synchronized with those of rats on *ad lib* diets (Duffy et al., 1997). It was pointed out, however, that the effects of food restriction on altered relations among rhythmic variables in mice and/or rats are not necessarily involved with the effect on survival (Nelson, 1988) and that the beneficial effects of chronic caloric restriction might be controlled via mechanisms involving the hypothalamic-pituitary-adrenal axis (e.g., metabolic pathways, lower body temperature, hormonal regulation, higher DNA repair) (Duffy et al., 1990, 1997).

It was also shown that tumor-bearing rats receiving anticancer chemotherapy who were fed early in their resting span (during L) lived significantly longer than *ad libitum* fed rats bearing the same tumor, suggesting that caloric restriction had an effect on tumor growth and susceptibility (Nelson et al., 1977). More recently, overall survival of untreated tumor-bearing mice was significantly prolonged (and tumor progression slowed) when food was presented during early L, when compared with mice fed during D or *ad libitum* (Wu et al., 2004). The authors hypothesized that meal-timing during L modified circadian clock function and signaling pathways in tumor cells and host peripheral tissues such that the tumor became more susceptible to host-mediated control.

The fact that in mice and rats, a calorie appears to be utilized differently depending upon *when* it is consumed, is relevant to humans. For most of us, our eating habits depend not only upon hunger, but also food supply, social habits and convenience. A high percentage of the American population consumes 50–75% of their daily calories during their evening meal,[26] a time of day when energy expenditure is lowest (Cohn, 1964).

This may be especially characteristic of obese individuals, many of whom exhibit a night-eating syndrome (NES), a unique stress-related eating, sleeping and mood disorder associated with altered circadian patterns in several neuroendocrine functions and that is characterized by overeating into the night[27] (cf. Stunkard & Allison, 2002). NES is also present in about 1.5% of nonobese subjects and the onset of night-eating has been shown to precede the diagnosis of obesity, indicating that NES may be a risk factor for obesity (Marshall et al., 2004). Studies of meal sizes and intervals between meals in humans during social isolation have shown that most subjects continued to eat three meals per waking span during free-running conditions, suggesting that a mechanism similar to that which

---

[25] Calories were reduced by 25% for mice and 10, 25, or 40% for rats from standard calories consumed by *ad lib* fed animals.

[26] Large meals in the evening are also typical in some other countries, such as Spain and Italy, while other countries, such as Germany, tend to have the largest meal at midday.

[27] Binge eating disorder (BED) consists of eating a large amount of food during a discrete time period, while NES is characterized by a greater frequency of nighttime awakenings and a smaller portion of food ingested at these times (Stunkard & Allison, 2002).

regulates the timing of sleep and wake also regulates the timing of meals (Aschoff et al., 1986; Green et al., 1987).

While some authors have advocated the intake of more protein in the morning and more carbohydrate in the evening in order to better control weight (cf. Baker & Baar, 2000), only a few reports concerning the applications or implications of meal-timing as they relate to diets, hormone levels[28] and changes in human rhythms or body weight have appeared. For example, after eating three meals a day on a regimen delayed by 6.5 h from a control schedule, the circadian peak in cholesterol synthesis was delayed more than 8.5 h (from 22:00 h to 04:36 h) (Cella et al., 1995).

The rise in serum leptin, an adipose tissue hormone, was greater following an evening meal than the same meal taken in the morning (Elimam & Marcus, 2002), while leptin was shifted by 5–7 h from its usual peak at 00:00 h following a 6.5-h meal shift regimen (Schoeller et al., 1997).

A report on obese subjects eating a single daily meal at either 10:00 h or 18:00 h for 3 days or 18 days found a higher lipid oxidation and lower cholesterol oxidation following the evening meal, but no effect on weight loss after the short study spans (Sensi & Capani, 1987). Two reports on humans studied for longer spans are worth noting, however, because the results from these studies are fascinating and could have broad implications from diet therapy to international nutrition. At the very least, they indicate the need for further research in this area.

Changes in Body Weight and Rhythms

A series of studies on human meal-timing were conducted at the University of Minnesota. Initially, three students were recruited to eat a single 2,000-calorie meal in the morning or evening each day for one week, after which a similar meal was consumed at the opposite time for one week.[29] The subjects, one male and two females, attended classes during the day and could pursue their other daily activities, but otherwise lived and slept in a clinical research center in the hospital, where they were weighed three times each day. Each subject lost weight during the week of eating breakfast-only and gained weight during the week of eating dinner-only (Figure 10.7) (Halberg, 1983).

Ultimately, this study was extended to include a total of six students. All six subjects lost weight when eating breakfast-only and four gained weight when

---

[28] Eating only once or twice a day has been shown to significantly elevate serum cholesterol levels and the risk of atherosclerosis in birds, rabbits, rats, dogs, and monkeys, indicating that meal frequency can affect a disease to which humans are prone (cf. Cohn, 1964). This led the author to suggest that six or more meals a day may be beneficial in reducing some metabolic diseases.

[29] The 2,000 calories consisted of 50% carbohydrate, 15% protein, and 35% fat. The content of each meal was varied, but the calories always totaled 2,000 calories, even if it meant extra pats of butter on a muffin at breakfast or on a piece of bread at dinner. One subject started by eating breakfast-only, two started on dinner-only.

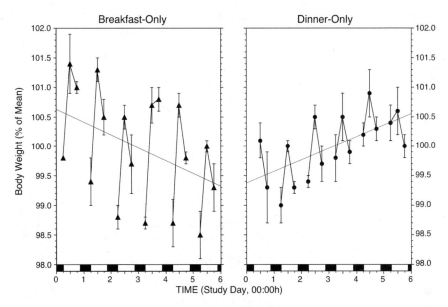

FIGURE 10.7. Average body weight change eating 2,000 kcal meal for 1 week as breakfast and 1 week as dinner-only. Combined body weight data from 3 students eating single daily meals, even though calories consumed within each 24-h period were the same, all subjects lost weight while eating breakfast-only and gained while eating dinner-only. Redrawn from Fig 9 in Halberg, 1974.

eating dinner-only. Two subjects lost weight on both schedules, but lost more weight eating breakfast only (Halberg, 1983).[30] Rhythms in blood and urine sampled around-the-clock from these subjects on day 7 of each diet schedule showed changes in peak times (as was seen earlier in the meal-fed mice—see Figure 9.3 in Chapter 9 on Veterinary Medicine) for some variables (serum insulin, iron, and plasma glucagon) (Figure 10.8) and not others (urinary cyclic AMP), indicating meal-induced changes in the relationship between peak times for many internal body rhythms (Goetz et al., 1976), which may in turn have contributed to the differences in weight gain or loss between the two single meal regimens (Hirsch et al., 1975).

Equally astonishing results were found in a second study, where 19 subjects (5 males and 7 females in Minnesota and 7 males in Connecticut, each within acceptable weight for their age) were allowed a single, free-choice meal daily for 6 weeks. They could eat as much or as little as they liked, but only once a day as either breakfast or dinner for 3 weeks, after which they switched to the opposite meal for the second 3 weeks (Graeber et al., 1978).[31] Breakfast meant that within one hour of

---

[30] The relative weight loss eating breakfast only as compared to dinner only was 1.2 kg (2.6 lbs) per week.

[31] One of this books authors (RBS) participated in this study as a subject. He lost 2.755 lb/week while eating breakfast only and 0.067 lb/week while eating dinner only.

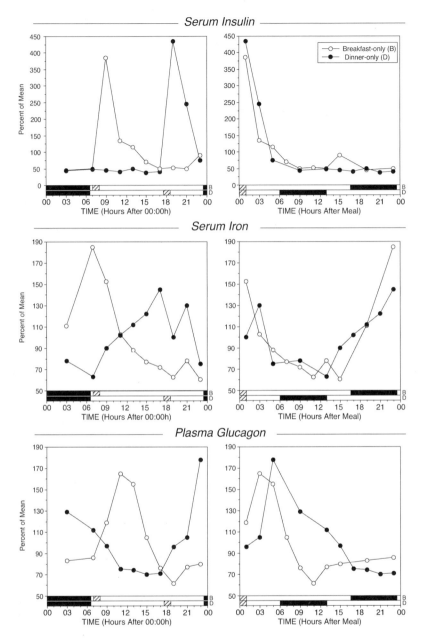

FIGURE 10.8. Timing of a single daily meal can influence circadian rhythm characteristics in humans. Rhythms in blood and urine sampled around-the-clock from three young men on day 7 of eating breakfast or dinner-only showed changes in peak times for some variables (serum insulin, iron, and glucagon) and not others (urinary cyclic AMP, not shown), indicating meal-induced changes in the relationship between peak times for many internal body rhythms (redrawn from Goetz et al., 1976).

awakening, the subject had one hour to eat as much as he/she desired, while dinner meant the subject fasted for at least 12 h after awakening before eating for up to an hour. Time and food quantities consumed were always recorded. Subjects chose from frozen and canned foods of known calorie, protein, fat, and carbohydrate content (each subject ate these foods *ad libitum* for 4 weeks prior to the single meal study so that they would be able to select foods that they really enjoyed during the single meal studies). There were five subjects who ate *ad libitum* as controls throughout the entire project. Throughout this study, each subject performed 5–6 self-measurements of several physiologic variables daily during waking and went about their usual activities. On the last day of each 3-week diet schedule, the subjects in Minnesota provided 4-hourly blood and urine samples around-the-clock.

As in the fixed 2,000 calorie study, all subjects lost more weight when eating breakfast-only than dinner-only, while a few subjects actually gained weight eating dinner-only (the difference in weekly body weight change was about 2.2 lbs (1 kg) between breakfast and dinner, while controls eating *ad libitum* more or less maintained a steady weight) (Graeber et al., 1978; Halberg, 1983). On average, subjects ate less on breakfast-only, but it was shown that the differential loss of weight was more closely related to the timing of meals than to the decrease in the number of calories consumed. In other words, timing was more important than the size of the meal or which meal they started eating during the study, since most weight loss occurred when individuals were in the breakfast-only group (Figure 10.9).[32]

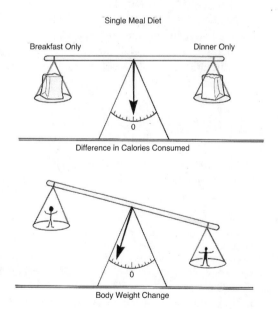

FIGURE 10.9. Dieting by time of day—the same amount (calories) of food consumed as breakfast-only or dinner-only (top) may result in a weight loss or gain (bottom).

---

[32] Maximal weight gain was also shown when dark-active catfish were fed a single daily meal late in their activity span (Sundararaj et al., 1982).

Just as in the fixed calorie single meal study, meal-timing substantially altered many of the internal body rhythms, such as in blood pressure, heart rate, white blood cells, serum iron, and insulin (but no shift was found in performance rhythms), indicating that hormonal and other physiological rhythms were manipulated by meal-timing. Since the metabolic fate of a meal is thus determined by meal-timing, weight loss or gain can be one of the side-effects.

Food: What, How Much, and When

It has been noted that the challenge of maintaining a healthy body weight in the modern world is to provide the knowledge, cognitive skills, and incentives to better manage body weight in an environment where physical activity is minimal and food is abundant (Peters et al., 2002). In light of the human experiments just summarized, perhaps this educational process should also add "when" to "what" and "how much" food is consumed.

Indeed, recent articles, while not specifically studying meal-timing, have reported on the effect on body weight if calories are consumed in the morning or the evening. In a study of nearly 3,000 obese individuals, the vast majority that maintained a significant weight loss for at least a year reported regularly eating breakfast (Wyatt et al., 2002). Another study of overweight individuals that consumed a liquid diet for at least a month concluded that increased food intake later in the day (night-eating) contributed to less weight loss (Gluck et al., 2001). Depending upon what the results from future research holds, the implication of manipulating meal-timing, as well as the number of calories a person eats, may someday find their way into international nutrition programs[33] and the diet fads of the general public.[34] In addition, meal-timing is of interest with regards to particular difficulties of night workers (e.g., appetite, digestion, and metabolism) (Waterhouse et al., 1997).

Light Pollution

The prevalent trend in human society to use artificial light has reached ruinous levels throughout much of the world such that nocturnal sky brightness has increased many-fold from natural conditions (Figure 10.10). This unnatural luminescence

---

[33] A gift of food to a third world country may not only include instructions as to proper preparation, but also *the best time of day* to eat the contents, such as "late in the day" or "near sundown" if the goal is to gain or maintain body weight. From a practical standpoint, however, the food may need to be distributed late in the day in order to force a starving person to adhere to this schedule.

[34] For example, a weight-control diet might suggest drinking a can of liquid diet for breakfast, another for lunch and then to eat a sensible dinner. But based upon the meal-timing results mentioned above, this reduced calorie scheduling might result in better weight control (or loss) on an opposite schedule consisting of a sensible breakfast followed by cans of liquid diet at lunch and dinner.

FIGURE 10.10. Lights of earth as seen from space today, while appearing as a beautiful sight, may be evidence of large-scale light pollution that was not present during the millions of years of human evolution. (Credit for this composite image of Earth at night: NASA - Goddard Space Flight Center Scientific Visualization Studio).

has been labeled *light pollution* and exists in both rural and urban areas. Illuminating the night and thereby extending the "day" long past sunset is an ever-increasing dilemma. It is perpetuated by schedules for work and leisure, by recreational activities, personal concerns, and the lingering fear of darkness and its analogy to evil and malice.[35] While some activities, such as those that relate to work, safety (e.g., traffic lights, guide lights for boats and airplanes), criminal behavior, health, and survival, certainly justify the use of unshielded artificial light, there is often an excessive use of it.

Effects on Melatonin Production

Perhaps of mounting concern is our ambivalent ignorance or disrespect for the natural light–dark cycle of the 24-h solar day and its effect on biological rhythms. The effects that light pollution has on humans are often mediated by the hormone *melatonin*, a chemical messenger that is produced during darkness by the pineal gland, a pea-sized gland located in the center of the brain (see Figure 11.8 in Chapter 11 on Clinical Medicine). Melatonin (MLT) serves as a chemical indicator of darkness in mammals, birds, fish, reptiles, and insects, and is an important component of the human timing system (Cassone & Natesan, 1997) for circadian, and possibly infradian (e.g., circannual), rhythms. In addition to synchronizing circadian clocks, MLT has strong antioxidant properties (Reiter et al., 2003). MLT has also been found in plants,[36] although its physiological role in plants has not been fully established (Caniato et al., 2003).

---

[35] Examples can be found in the great books of religion (e.g., the Bible, John 9:4–5) and literature (e.g., works by Shakespeare, including Macbeth and Hamlet).

[36] Plants do not have a pineal gland, although melatonin has been found in various structures and organs of a number of species (Balzer & Hardeland, 1996; Hardeland et al., 1996; Kolár et al., 1997).

While originally it was thought that very bright light of approximately 1500–2500 lux[37] was necessary to maximally suppress MLT, studies have demonstrated that ordinary room light of 250 lux can significantly suppress MLT secretion when applied in the evening and that the duration of inhibition is intensity-dependent (Trinder et al, 1996). Exposure to artificial light of moderate intensity (180 lux) during the middle of the night has been shown to phase-shift human circadian rhythms of MLT and cortisol levels (Boivin & Czeisler, 1998), which demonstrates that the circadian entrainment mechanism of humans, like those of other mammals, is sensitive to low levels of light (Boivin et al., 1996).

In addition to the timing, intensity, and duration of exposure, the MLT response to light at night is also dependent upon wavelength, with very dim light (0.1 lux) in the range of 446–477 nm causing the most suppression of MLT production (Brainard et al., 2001). Peak sensitivity of the human circadian pacemaker, as denoted by maximum MLT suppression, was later found after exposure to blue light at 555 nm, confirming that MLT suppression is primarily blue wavelength dependent (Lockley et al., 2003; Figueiro et al., 2004).

Effects on Clinical Health

As societies have industrialized, there is something about modern life that is associated with an increase in certain health problems, such as breast cancer, yet known risk factors are associated with less than half of breast cancer cases (cf. Stevens, 2005). Ill-timed artificial lighting and the lack of sunlight may cause circadian disruption that can lead to health problems (Erren et al., 2003). At the hormonal level, the production of MLT is a circadian event driven by light and dark. It is likely that the pineal gland responds to too much or too little light at certain times of the day by alterations in the production of MLT. This may have a profound, yet mostly overlooked, effect on human health. For example, light increased resting heart rate in 17 healthy men and women depending on the intensity of the light and time of day, with the strongest effects in the middle of the night and in the early morning (Scheer et al., 1999). Bright light can also reset the human circadian pacemaker that controls circadian rhythms in physiologic, behavioral, and cognitive functions (Czeisler et al., 1986; Khalsa et al., 2003). Bright light in the evening delays the onset of MLT production (Kennaway et al., 1987; Deacon & Arendt, 1994; Gronfier et al., 2004), while bright light in the morning advances the onset of nighttime MLT production (Lewy et al., 1987; Danilenko et al., 2000).

Light thus has the ability to "act like a drug" and, as such, has become a public health issue in the industrialized world (Pauley, 2004). Areas possibly affected by changes in melatonin production include endocrine functions associated with

---

[37] Units such as lux and foot-candle (10.76 lux = 1.0 ft-c) are presented or defined relative to the sensitivity of the human eye. Because organisms, such as plants, respond in ways that are different than the human eye, a description of the source (e.g., lamp) and wavelengths should be given.

puberty, psychiatric illness, stress-related disorders, immune responses, and carcinogenesis (Piccoli et al., 1991). Prolonged reduction of daylight exposure may also impact occupational health. For example, an increase in the rate of skin cancer (malignant melanoma) was found in airline pilots flying over five time zones or more on a routine basis (Rafnasson et al., 2000). Exposure to excess light and the resulting disturbance in the circadian rhythm in melatonin was discussed as a likely contributing factor. The relationship between melatonin production and human health is discussed further in Chapter 11 on Clinical Medicine.

Better Lighting Practices

Humans are subjected to both indoor and outdoor light pollution. Regulations in the United States and in a few other places in the world have started to shield individuals from the harmful effects of cigarette smoke and various gasses, but not from light.[38] Seldom, if ever, is there an attempt to shield the light so it goes downward, not up or sideways, and thereby prevent light from escaping into the environment.[39] The effects of light pollution on birds, insects, and other animals are discussed in Chapter 8 on Natural Resources and Agriculture. The artificial light we use indoors during the day, as well as at night, usually differs from natural light in both intensity (usually lower) and the spectrum of wavelengths (shorter). While prior human exposure to artificial light at night came from sources such as flames of an orange-red fire or the yellow light of candles, gas lamps or incandescent bulbs, today's lights emit more blue light from outdoor signs and security lamps, as well as indoor television screens, computer monitors, and reading lights (Pauley, 2004).

Based upon mounting evidence, better lighting practices have been suggested as preventive measures in order to reduce the potential health risk to humans of inappropriate lighting (Pauley, 2004). These include using full spectrum lighting during the daytime (e.g., inside homes, offices, hospitals, nursing homes, during shiftwork) and using non-blue indoor and outdoor lighting at night (e.g., incandescent rather than fluorescent lights, low sodium vapor street lights). In addition, it was suggested that night workers wear orange lens glasses (known as Blue Blocker sunglasses) on their way home from work to prevent the suppression of melatonin that would ordinarily occur with bright light exposure in the morning (Sasseville et al., 2003). Rather than wearing very dark goggles that block the full light spectrum evenly and possibly diminish vision, orange lens glasses filter out blue–green wavelengths of the spectrum and thereby reduce the ability of morning light to resynchronize a night worker's biological clock from a nighttime to a daytime schedule.

---

[38] Reference here is to visible light, and not to harmful short wavelength ultraviolet light or longer infrared.

[39] It has been estimated that more than a billion dollars are wasted each year in the USA on nocturnal lighting that serves no purpose for safety, security, or utility (The Prairie Astronomy Club, 2001).

Pseudoscience: Birthdate-Based Biorhythms

Over the years when university students have selected a topic for an oral report that concerns the rhythms of humans, they have occasionally found literature that describes the importance of three biorhythms. These particular biorhythms are described as having periods of 23 days, 28 days, and 33 days and their characteristics (e.g., peak or phase) are based upon one's date of birth (Thommen, 1973). To some extent, the biorhythm concept has confused and alienated scientists and the public alike, who have subsequently negated and ignored the study of biological rhythms and their relevance to all life. It is thus important to provide an overview of the development of this theory (Essay 10.3) and its subsequent debunking in the remaining text of this section.

Essay 10.3. (by RBS): Development of the Biorhythm "Theory"

What has become known as the "theory of biorhythm" was developed in the late 19th century and early 20th century by several individuals in Europe. Hermann Swoboda (1873–1963), a psychologist at the University of Vienna, noticed periodicities in data he collected with regards to the recurrence of fevers, illness outbreaks, asthma attacks, heart attacks, development in babies, and pain and swelling of tissues after insect bites (cf. Gittelson, 1977). He noted that the periodic intervals were often 23 days or 28 days and published his findings in a book entitled "Die Perioden des Menschlichen Lebens" (The Periodicity in Man's Life) in 1904. After concluding that these cycles might have a fixed origin at birth, he later developed a slide rule to find the critical days in the life of any person whose birthdate was known. In 1909 he published an instruction booklet entitled "Die Kritischen Tage des Menschen" (The Critical Days of Man) designed to encourage other scientists and medical doctors to record the mathematical biorhythms. He was also one of the first to study rhythmical development in cancer.

Wilhelm Fliess (1859–1928), a nose and throat specialist in Berlin, also noticed periodic outbreaks of fevers and deaths and in his book entitled "Der Ablauf Des Lebens" (The Course of Life) in 1906 he claimed that 23- and 28-day rhythms were fundamental to life (cf. Gittelson, 1977). Fliess believed that each individual inherits both male and female sexual characteristics and thus has elements of bisexuality in their makeup.[40] He deduced that the 23-day cycle originates in muscle fibers or cells and is the masculine rhythm. This cycle affects physical strength, endurance, energy, resistance, and physical confidence, thereby affecting the overall physical condition of humans. The 28-day rhythm was ascribed to the rhythmical changes of the feminine inheritance and originates in the nervous system. This cycle influences creativity, feelings, reaction time, love, cooperation, emotions and sensitivity. While periodicities in behavior and disease were often recognized, the medical profession showed little interest in the 23- and 28-day biorhythm theories due to the vast amount of mathematics involved, a problem that is still common in modern biological rhythm research.

In the 1920s in Innsbruck, an engineer, Alfred Teltscher, described a 33-day cycle in the intellectual performance of students (Thommen, 1973). Others thought this could result from periodic secretions of glands, such as the thyroid, that affect brain cells. This

---

[40] For example, breast nipples on the male are one of the physical marks left as a reminder of man's dual sex evolution.

33-day cycle is said to influence clear thinking, memory, and mental responses, such as the ability to absorb new material, studying, and creative thinking. In the early 1930s, R. Hersey and M. Bennett at the University of Pennsylvania also reported finding 33–36 day rhythms in the emotions of workers in railroad shops.

The 23-, 28-, and 33-day cycles in the "biorhythm theory" are said to be initiated by the intense stimulation of all senses and organs at birth. Despite the ups and downs of an individual's life, the theory claims that these three cycles persist with mathematical precision to the minute throughout life. The cycles repeat throughout one's lifetime and are only all back at the same starting point every 58 years plus 67 or 68 days (23 days × 28 days × 33 days = 21,252 days). During the repeated ups and downs of each cycle, the switchover days when a cycle is going from low to high or high to low are considered critical days, because a person is believed to show a greater degree of instability on those days. During a full biorhythm span of 58 year 68 days, there are 4,327 days on which a switch or changeover takes place, resulting in 79.6% mixed rhythm days and 20.4% critical days. However, the 23- and 28-day rhythms are considered more important than the 33-day intellectual cycle, and they start a new biorhythmic year with a simultaneous upswing every 644 days (1 year, 9 months). During this time frame, 85% of days are mixed rhythm and 15% are critical.

While the "theory" states that critical days in themselves are not dangerous, there are days when one's reaction to the environment may bring about a critical situation. Accidents are not predicted, but on 20% of the days and under trying circumstances, a person is potentially in a more dangerous situation than at other times. Thus, "biorhythm theory" claims to predict potential human disposition on a particular day and give a forewarning by referring to charts showing the ups and downs of the 3 cycles for a particular individual with reference to his or her birthdate. Several devices were created for this purpose, including a slide rule developed by Swoboda between 1904 and 1909, calculation tables developed by Alfred Judt in Bremen in the late 1920s, hand-operated calculators with built-in tables by Hans Frueh in Switzerland in the 1940s and tables of key numbers for date of birth and year of birth (Thommen, 1973). Computer printouts of these curves were popular in the 1970s and can still be found on the Internet.

Lack of Scientific Support

The existence of a 23-day physical cycle, a 28-day emotional cycle, and a 33-day mental (intellectual) cycle seems plausible when given the example of magazine articles of successful sporting events (Gittelson, 1977) or disastrous things that have happened in accordance to what could have been predicted by these cycles according to biorhythm "theory" (Thommen, 1973). Specific evidence originally supporting these three periods was reviewed, however, and found to be very limited (Ahlgren, 1974). Biorhythm theory was compared to palmistry and I. Ching, wherein there is so much flexibility in interpreting ups and downs and critical days that practitioners of biorhythm theory are given *"almost complete inventive freedom"* to interpret results and thereby prove this *"crackpot theory"* (Ahlgren, 1974).

Several subsequent scientific studies of the biorhythm "theory" all reached the conclusion that issues of safety, performance, health, and death do not appear to support the claims of critical days that these cycles are supposed to predict

(Rodgers et al., 1974; Khalil & Kurucz, 1977; Kurucz & Khalil, 1977; Salisbury & Ross, 1992). Of the scientific studies designed to test the validity of the 3-cycle/critical days biorhythm "theory" after its resurgent interest in the 1970s, a significant effect could not be supported in any area where it was claimed to exert an effect. These included (1) accidents: by aircraft pilots (Wolcott et al., 1977a,b), highway crashes (Shaffer et al., 1978), mining accidents (Persinger et al., 1978), and forest worker injuries (Slama, 1981); (2) performance: in archery (Haywood, 1979), reaction time (Wolcott et al., 1979), quizzes by students throughout a semester (Englund & Naitoh, 1980) or track and field world records between 1913 and 1977 (Quigley, 1982); and (3) medicine: day of death (Feinleib & Fabsitz, 1978), asthma symptoms & medication use (Strumpf et al., 1978), 5-year survival after irradiation for cervical cancer (Kucera et al., 1980) or psychiatric hospitalizations (Winstead et al., 1981). The latter authors concluded that: *"..., belief in the theory is likely to be an expression of superstitious behavior"* and was *"much too simplistic to account for the complexities of everyday life."* (Winstead et al., 1981)

Rigidity vs. Elasticity of Infradian Periods

With regard to physiologic systems, the biorhythm hypothesis does not allow for phase-shifting or period alteration of the three precise periods, which repeatedly happens throughout one's lifetime due to disease, changes in living schedules, transmeridian flights and/or development. For example, most infradian (e.g., monthly) rhythms gradually emerge in early childhood, making it unlikely that a period and phase could be invariably linked to the precise date of one's birth.

Regarding a 28-day cycle, variations associated with the menstrual cycle do not appear until a female reaches puberty in her teens, while later in life this cycle becomes more irregular near menopause and wanes thereafter. It is, of course, well-known that the length of a woman's menstrual cycle is not rigidly fixed throughout her fertile years, but reoccurs within a range of several days. In addition, the length of the menstrual cycle can be reset by trauma or transmeridian flights and is interrupted during pregnancy or certain unfavorable states of health (e.g., anorexia). In any event, the average length of a woman's menstrual cycle is 29.1 days (Presser, 1974) and not 28 days as claimed by biorhythm proponents.

It is intriguing, however, that the masculine cycle was claimed to be 23 days in length, since cycles near 20 days have been found in men for a few variables, including urinary steroid excretion over 15 years (17-21 days) (Halberg et al., 1965) and beard growth over 7 months (16.5 days) (Sothern, 1974). However, the beard growth in another male was reported to show a 33-day rhythm over 6 months (Kihlstrom, 1971). Clearly these infradian rhythms are not of a fixed length and need more study as to their biological significance.

An Oversimplification of Rhythms

The use of biorhythms was first popularized in the area of sports, which attracted the attention of the public (Gittelson, 1977), but unfortunately the "biorhythm theory" may have ultimately fostered some adverse consequences regarding the acceptance of a serious study of biological rhythms. Perhaps the birthdate-based biorhythm notion is best compared to astrology and its horoscopes, which unlike astronomy is not a science, but relies heavily on subjective interpretation (as mentioned above).

While the biorhythm theory promotes the idea that there are significant biological rhythms at various levels of physiology and over a wide frequency, it oversimplifies the complex interactions we now know to exist among ultradian, circadian, and infradian periodicities in all living things. Perhaps the biggest contribution of the biorhythm theory is that its proponents called attention to little-studied infradian periodicities and experimented with numbers as a tool in deciphering the rhythmic relationship between biology and mathematics, much as modern-day chronobiologists continue to do today. There is as yet no alternative to collecting and analyzing data to document a biological cycle's true length and characteristics.

Take-Home Message

In addition to synchronization by natural LD cycles, a number of biological rhythms can be synchronized by schedules and routines imposed by society, such as times of activity (work), meals, recreation, and sleep. Likewise, social cues can serve as synchronizers for various rhythms, such as the menstrual cycle. External clocks and calendars imposed by society may be either in or out of phase with the internal biological clocks and rhythms of the human body. Characteristics of biological rhythms (phase and amplitude) can be disturbed or altered by events associated with our modern 24/7/365 society, such as shiftwork, time zone changes (jet lag), and lack of sleep. External and internal phase differences can have negative or positive effects in areas of society, such as work, health, sports, and communications. In addition, the unnatural exposure to light at night (light pollution) can alter normal physiological rhythms of humans and other organisms.

References

Ahlgren A. (1974) Biorhythms – Letter to the editor. *Intl J Chronbiol* 2(1): 107–109.
Åkerstedt T-L. (1981) Shift work: shift-dependent well-being and individual differences. *Ergonomics* 24(4): 265–273.
Åkerstedt T. (2000) Consensus statement: fatigue and accidents in transport operations. *J Sleep Res* 9(4): 395.
Arendt J, Deacon S. (1997) Treatment of circadian rhythm disorders–melatonin. *Chronobiol Intl* 14(2): 185–204. (Review).
Arnedt JT, Wilde GJ, Munt PW, MacLean AW. (2000) Simulated driving performance following prolonged wakefulness and alcohol consumption: separate and combined contributions to impairment. *J Sleep Res* 9(3): 233–241.
Arnedt JT, Wilde GJ, Munt PW, MacLean AW. (2001) How do prolonged wakefulness and alcohol compare in the decrements they produce on a simulated driving task? *Accid Anal Prev* 33(3): 337–344.

Aschoff J, von Goetz C, Wildgruber C, Wever RA. (1986) Meal timing in humans during isolation without time cues. *J Biol Rhythms* 1(2): 151–162.

Astronomical Almanac. (2001) Washington, Dept of Defense: Navy Dept, Naval Observatory, Nautical Almanac Office (facts also available at: http://nssdc.gsfc.nasa.gov/planetary/factsheet).

Atkinson G, Reilly T. (1996) Circadian variation in sports performance. *Sports Med* 21: 292–312.

Atkinson G, Speirs L. (1998) Diurnal variation in tennis service. *Perc Motor Skills* 86(3 Pt 2): 1335–1338.

Baker SM, Baar K. (2000) *The Circadian Prescription*. New York: Perigree, 227 pp.

Balzer I, Hardeland R. (1996) Melatonin in algae and higher plants – possible new roles as a phytohormone and antioxidant. *Bot Acta* 109: 180–183.

Barak Y, Achiron A, Lampl Y, Gilad R, Ring A, Elizar A, Sarova-Pinhas I. (1995) Sleep disturbances among female nurses comparing shift to day work. *Chronobiol Intl* 12(5): 345–350.

Baxter C, Reilly T. (1983) Influence of time of day on all-out swimming. *Br J Sports Med* 17(2): 122–127.

Bégout Anras M-L, Lagardére J-P, Lafaye J-Y. (1997) Diel activity rhythm of seabass tracked in a natural environment: group effects on swimming patterns and amplitudes. *Can J Aquat Sci* 54: 162–168.

Bjerner B, Holm A, Swensson A. (1955) Diurnal variation in mental performance. *Br J Ind Med* 12: 103–110.

Blake W. (1794) *The Marriage of Heaven and Hell*. London: J.M. Dent & Sons.

Bloom M. (1989) Tennis in good time. *World Tennis* (Feb): 70–71.

Boivin DB, Duffy JF, Kronauer RE, Czeisler CA. (1996) Dose-response relationships for resetting of human circadian clock by light. *Nature* 379: 540–542.

Boivin DB, Czeisler CA. (1998) Resetting of circadian melatonin and cortisol rhythms by ordinary room light. *Neuroreport* 9(5): 779–782.

Bolliet V, Aranda A, Boujard T. (2001) Demand-feeding rhythm in rainbow trout and European catfish. Synchronisation by photoperiod and food availability. *Physiol Behav* 73(4): 625–633.

Brainard GC, Hanifin JP, Greeson JM, Byrne B, Glickman G, Gerner E, Rollag MD. (2001) Action spectrum for melatonin regulation in humans: evidence for a novel circadian photoreceptor. *J Neurosci* 21(16): 6405–6412.

Browne RC. (1949) The day and night performance of teleprinter switchboard operators. *Occup Psychol* 23: 121–126.

Buck J. (1988) Synchronous rhythmic flashing of fireflies. II. *Q Rev Biol* 63(3): 265–289 (Review).

Burgoon PW, Holland GJ, Loy SF, Vincent WJ. (1992) A comparison of morning and evening 'types' during maximum exercise. *J Appl Sport Sci Res* 6(2): 115–119.

Callard D, Davenne D, Gauthier A, Lagarde D, Van Hoecke J. (2000) Circadian rhythms in human muscular efficiency: continuous physical exercise versus continuous rest. A crossover study. *Chronobiol Intl* 17(5): 693–704.

Caniato R, Filippini R, Piovan A, Puricelli L, Borsarini A, Cappelletti EM.(2003) Melatonin in plants. *Adv Exp Med Biol* 527: 593–597 (Review).

Carrier J, Monk TH. (2000) Circadian rhythms of performance: new trends. *Chronobiol Intl* 17(6): 719–732.

Carskadon MA, Wolfson AR, Acebo C, Tzischinsky O, Seifer R. (1998) Adolescent sleep patterns, circadian timing, and sleepiness at a transition to early school days. *Sleep* 21(8): 871–881.

Cassone VM, Natesan AK. (1997) Time and time again: the phylogeny of melatonin as a transducer of biological time. *J Biol Rhythms* 12(6): 489–497.
Cella LK, Van Cauter E, Schoeller DA. (1995) Effect of meal timing on diurnal rhythm of human cholesterol synthesis. *Amer J Physiol* 269(5 Pt 1): E878–E883.
Cohn C. (1964) Feeding patterns and some aspects of cholesterol metabolism. *Fed Proc* 23: 76–81.
Colquhoun WP, Paine MW, Fort A. (1978) Circadian rhythm of body temperature during prolonged undersea voyages. *Aviat Space Environ Med* 49(5): 671–678.
Conroy RT, O'Brien M. (1974) Diurnal variation in athletic performance. *J Physiol* 236(1): 51P.
Czeisler, CA, Moore-Ede MC, Coleman RM. (1982) Rotating shift work schedules that disrupt sleep are improved by applying circadian principles. *Science* 217: 460–463.
Czeisler CA, Allan JS, Storigatz SH, Ronda JM, Sánchez R, Ríos CD, Freitag WO, Richardson GS, Kronauer RE. (1986) Bright light resets the human circadian pacemaker independent of the timing of the sleep-wake cycle. *Science* 233: 667–671.
Danilenko KV, Wirz-Justice A, Krauchi K, Cajochen C, Weber JM, Fairhurst S, Terman M. (2000) Phase advance after one or three simulated dawns in humans. *Chronobiol Intl* 17(5): 659–668.
Davis FC, Stice S, Menaker M. (1987) Activity and reproductive state in the hamster: independent control by social stimuli and a circadian pacemaker. *Physiol Behav* 40(5): 583–590.
Deacon SJ, Arendt J. (1994) Phase-shifts in melatonin, 6-sulphatoxymelatonin and alertness rhythms after treatment with moderately bright light at night. *Clin Endocrinol (Oxf)* 40(3): 413–420.
Drust B, Waterhouse J, Atkinson G, Edwards B, Reilly T. (2005) Circadian rhythms in sports performance—an update. *Chronobiol Intl* 22(1): 21–44 (Review).
Duffy PH, Feuers RJ, Hart RW. (1990) Effect of chronic caloric restriction on the circadian regulation of physiological and behavioral variables in old male B6C3F1 mice. *Chronobiol Intl* 7(4): 291–303.
Duffy PH, Leakey JE, Pipkin JL, Turturro A, Hart RW. (1997) The physiologic, neurologic, and behavioral effects of caloric restriction related to aging, disease, and environmental factors. *Environ Res* 73(1–2): 242–248.
Duffy PH, Seng JE, Lewis SM, Mayhugh MA, Aidoo A, Hattan DG, Casciano DA, Feuers RJ. (2001) The effects of different levels of dietary restriction on aging and survival in the Sprague-Dawley rat: implications for chronic studies. *Aging (Milano)* 13(4): 263–272.
Ehret CF. (1981) New approaches to chronohygiene for the shift worker in the nuclear power industry. In: *Night and Shiftwork. Biological and Social*. Reinberg A, Vieux N, Andlauer P, eds. Oxford: Pergamon, pp. 263–270.
Ehret CF, Waller-Scanlon L. (1983) *Overcoming Jet Lag*. New York: Berkley, 160 pp.
Eichner ER. (1994) Circadian Rhythms. The latest word on health and performance. *The Physician Sports Med* 22(10): 82–93.
Elimam A, Marcus C. (2002) Meal timing, fasting and glucocorticoids interplay in serum leptin concentrations and diurnal profile. *Eur J Endocrinol* 147(2): 181–188.
Englund CE, Naitoh P. (1980) An attempted validation study of the birthdate-based biorhythm (BBB) hypothesis. *Aviat Space Environ Med* 51(6): 583–590.
Erren TC, Reiter RJ, Piekarski C. (2003) Light, timing of biological rhythms, and chronodisruption in man. *Naturwiss* 90: 485–494.

Feinleib M, Fabsitz R. (1978) Do biorhythms influence day of death? *New Engl J Med* 298: 1153.

Figueiro MG, Bullough JD, Parsons RH, Rea MS. (2004) Preliminary evidence for spectral opponency in the suppression of melatonin by light in humans. *Neuroreport* 15(2): 313–316.

Fischer FM, Moreno CRC, Borges FNS, Louzada FM. (2000) Implementation of 12-hour shifts in a Brazilian petrochemical plant: impact on sleep and alertness. *Chronobiol Intl* 17(4): 521–537.

Fischer FM, Moreno CRC, Rotenberg L, eds. (2004) Equity and working time: a challenge to achieve. *Chronobiol Intl* 21(6): 813–1077 (Special Issue on Night and Shift Work).

Folkard S. (1983) Diurnal variation in human performance. In: *Stress and Fatigue in Human Performance*. Hockey GRJ, ed. Chichester: Wiley, pp. 245–272.

Forsyth JJ, Reilly T. (2004) Circadian rhythms in blood lactate concentration during incremental ergometer rowing. *Eur J Appl Physiol* 92(1–2): 69–74.

Fromm RE Jr, Levine RL, Pepe PE. (1992) Circadian variation in the time of request for helicopter transport of cardiac patients. *Ann Emerg Med* 21(10): 1196–1199.

Gauthier A, Davenne D, Martin A, Cometti G, Van Hoecke J. (1996) Diurnal rhythm of the muscular performance of elbow flexors during isometric contractions. *Chronobiol Intl* 13(2): 135–146.

Gittelson B. (1977) *Biorhythm Sports Forecasting*. New York: Arco, 238 pp.

Gluck ME, Geliebter A, Satav T. (2001) Night eating syndrome is associated with depression, low self-esteem, reduced daytime hunger, and less weight loss in obese outpatients. *Obes Res* 9(4): 264–267.

Goel N, Lee TM. (1995a) Sex differences and effects of social cues on rate of resynchronization in *Octodon degus*. *Physiol Behav* 58(2): 205–213.

Goel N, Lee TM. (1995b) Social cues accelerate reentrainment of circadian rhythms in diurnal female *Octodon degus* (Rodentia-Octodontidae). *Chronobiol Intl* 12(5): 311–323.

Goel N, Lee TM. (1997) Social cues modulate free-running circadian activity rhythms in the diurnal rodent, *Octodon degus*. *Amer J Physiol* 273(2 Pt 2): R797–R804.

Goetz FC, Bishop J, Halberg F, Sothern RB, Brunning R, Senske B, Greenberg B, Minors D, Stoney P, Smith ID, Rosen GD, Cressey D, Haus E, Apfelbaum M. (1976) Timing of single daily meal influences relations among human circadian rhythms in urinary cyclic AMP and hemic glucagon, insulin and iron. *Experientia* 32(8): 1081–1084.

Goldstick O, Weissman A, Drugan A. (2003) The circadian rhythm of "urgent" operative deliveries. *Isr Med Assoc J* 5(8): 564–566.

Graeber RC, Gatty R, Halberg F, Levine H. (1978) Human Eating Behavior: Preferences, Consumption Patterns and Biorhythms. *NATICK/TR-78/022 Technical Reports*, US Army, 287 pp.

Graham CA, McGrew WC. (1980) Menstrual synchrony in female undergraduates living on a coeducational campus. *Psychoneuroendocrinol* 5(3): 245–52.

Green J, Pollak CP, Smith GP. (1987) Meal size and intermeal interval in human subjects in time isolation. *Physiol Behav* 41(2): 141–147.

Gronfier C, Wright KP Jr, Kronauer RE, Jewett ME, Czeisler CA. (2004) Efficacy of a single sequence of intermittent bright light pulses for delaying circadian phase in humans. *Amer J Physiol Endocrinol Metab* 287(1): E174–E181.

Gross JD. (1994) Developmental decisions in *Dictyostelium discoideum*. *Microbiol Rev* 58(3): 330–351 (Review).

Gundel A, Nalishiti V, Reucher E, Vejvoda M, Zulley J. (1993) Sleep and circadian rhythm during a short space mission. *Clin Invest* 71(9): 718–724.
Guyer B, Gallagher SS. (1985) An approach to the epidemiology of childhood injuries. *Pediatr Clin North Amer* 32(1): 5–15.
Haikonen H, Summala H. (2001) Deer-vehicle crashes: extensive peak at 1 hour after sunset. *Amer J Prev Med* 21(3): 209–213.
Halberg F, Engeli M, Hamburger C, Hillman D. (1965) Spectral resolution of low-frequency, small-amplitude rhythms in excreted 17-ketosteroid; probable androgen-induced circaseptan desynchronization. *Acta Endocrinol (Kbh.)* Suppl 103: 5–54.
Halberg F. (1974) Protection by timing treatment according to bodily rhythms – an analogy to protection by scrubbing before surgery. *Chronobiologia* 1 (Suppl 1): 27–68.
Halberg F. (1983) Chronobiology and nutrition. *Contemporary Nutr* 8(9): 2 pp.
Hanson FE. (1978) Comparative studies of firefly pacemakers. *Fed Proc* 37(8): 2158–2164.
Hardeland R, Balzer I, Fuhrberg B, Behrmann G. (1996) Melatonin in unicellular organisms and plants. In: *Melatonin: A Universal Photoperiodic Signal with Diverse Actions*. Tang PL, Pang SF, Reiter RJ, eds. Basel: Karger, pp. 1–6.
Harris W. (1977) Fatigue, circadian rhythm, and truck accidents. In: *Vigilance. Theory. Operational Performance, and Physiological Correlates*. Mackie RR, ed. New York: Plenum Press, pp. 133–146.
Haywood KM. (1979) Skill performance on biorhythm theory's physically critical day. *Percep Motor Skills* 48(2): 373–374.
Hildebrandt G, Rohmert W, Rutenfranz J. (1974) 12 and 24 h rhythms in error frequency of locomotive drivers and the influence of tiredness. *Intl J Chronobiol* 4: 175–180.
Hill DW, Smith JC. (1991) Circadian rhythm in anaerobic power and capacity. *Can J Sport Sci* 16(1): 30–32.
Hirsch EE, Halberg E, Halberg F, Goetz F, Cressey D, Wendt H, Sothern R, Haus E, Stoney P, Minors D, Rosen G, Hill B, Hilleren M, Barett K. (1975) Body weight change during 1 week on a single daily 2000-calorie meal consumed as breakfast (B) or dinner (D) (Abstract). *Chronobiologia* 2 (Suppl 1): 31–32.
Honrado G, Mrosovsky N. (1989) Arousal by sexual stimulii accelerates the re-entrainment of hamsters to phase advanced light–dark cycles. *Behav Ecol Sociobiol* 25: 57–63.
Horne J, Reyner L. (1999) Vehicle accidents related to sleep: a review. *Occup Environ Med* 56(5): 289–294 (Review).
Horne JA, Reyner LA, Barrett PR. (2003) Driving impairment due to sleepiness is exacerbated by low alcohol intake. *Occup Environ Med* 60(9): 689–692.
Horne JA, Baulk SD. (2004) Awareness of sleepiness when driving. *Psychophysiol* 41(1): 161–165.
Huguet G, Touitou Y, Reinberg A. (1995) Diurnal changes in sport performance of 9- to 11-year-old school children. *Chronobiol Intl* 12(5): 351–362.
Kavaliers M. (1980) Circadian activity of the white sucker, *Catostomus commersoni*: comparison of individual and shoaling fish. *Can J Zool* 58(8): 1399–1403.
Kelly TL, Neri DF, Grill JT, Ryman D, Hunt PD, Dijk DJ, Shanahan TL, Czeisler CA. (1999) Nonentrained circadian rhythms of melatonin in submariners scheduled to an 18-hour day. *J Biol Rhythms* 14(3): 190–196.
Kennaway DJ, Earl CR, Shaw PF, Royles P, Carbone F, Webb H. (1987) Phase delay of the rhythm of 6-sulphatoxy melatonin excretion by artificial light. *J Pineal Res* 4(3): 315–320.

Khalil TM, Kurucz CN. (1977) The influence of 'biorhythm' on accident occurrence and performance. *Ergonomics* 20(4): 389–398.

Khalsa SB, Jewett ME, Cajochen C, Czeisler CA. (2003) A phase response curve to single bright light pulses in human subjects. *J Physiol* 549(Pt 3): 945–952.

Kihlström JE. (1971) A monthly variation in beard growth in one man. *Life Sciences* 10: 321–324.

Kleitman N, Jackson DP. (1950) Body temperature and performance under different routines. *J Appl Physiol* 3: 309–328.

Kleitman, N. (1949) The sleep–wakefulness cycle of submarine personnel. In: *A Survey Report on Human Factors in Undersea Warfare*. Washington, DC: Committee on Undersea Warfare, National Research Council, pp. 329–341.

Kleitman N. (1963) *Sleep and Wakefulness*. Chicago: University of Chicago Press, 552 pp.

Kolár J, Macháchová I, Eder J, Prinsen E, van Dongen W, van Onckelen H, Illnernova H. (1997) Melatonin: occurrence and daily rhythm in *Chenopodium rubrum*. *Phytochem* 44(8): 1407–1413.

Kucera H, Riss P, Weghaupt K. (1980) [Biorhythm theory and primary irradiation of inoperable cancer of the cervix.] [German]. *Strahlentherapie* 156(7): 453–456.

Kurucz CN, Khalil TM. (1977) Probability models for analyzing the effects of biorhythms on accident occurrence. *J Safety Res* 9: 150–158.

Lakatua DJ, White M, Sackett-Lundeen LL, Haus E. (1983) Change in phase relations of circadian rhythms in cell proliferation induced by time-limited feeding in BALB/c X DBA/2F1 mice bearing a transplantable Harding-Passey tumor. *Cancer Res* 43(9): 4068–4072.

Lakatua DJ, Haus M, Berge C, Sackett-Lundeen L, Haus E. (1988) Diet and mealtiming as circadian synchronizers. In: *Ann Rev Chronopharmacol*, Vol. 5. Reinberg A, Smolensky M, Labrecque G, eds. Oxford: Pergamon Press, pp. 307–309.

Lambe M, Cummings P. (2000) The shift to and from daylight savings time and motor vehicle crashes. *Accid Anal Prev* 32(4): 609–611.

Langlois PH, Smolensky MH, His BP, Weir FW. (1985) Temporal patterns of reported single-vehicle car and truck accidents in Texas, U.S.A. during 1980–1983. *Chronobiol Intl* 2(2): 131–140.

Lemmer B, Kern R-I, Nold G, Lohrer H. (2002) Jet lag in athletes after eastward and westward time-zone transition. *Chronobiol Intl* 19(4): 743–764.

Lewy AJ, Sack RL, Miller S, Hoban TM. (1987) Antidepressant and circadian phase-shifting effects of light. *Science* 235: 352–354.

Lockley SW, Brainard GC, Czeisler CA. (2003) High sensitivity of the human circadian melatonin rhythm to resetting by short wavelength light. *J Clin Endocrinol Metab* 88(9): 4502–4505.

Luce GG. (1970) Biological Rhythms in Psychiatry and Medicine. US Public Health Services Publication No. 2088, 183 pp.

Manfredini R, La Cecilia O, Boari B, Steliu J, Michelinidagger V, Carlidagger P, Zanotti C, Bigoni M, Gallerani M. (2002) Circadian pattern of emergency calls: implications for ED organization. *Amer J Emerg Med* 20(4): 282–286.

Marshall HM, Allison KC, O'Reardon JP, Birketvedt G, Stunkard AJ. (2004) Night eating syndrome among nonobese persons. *Intl J Eat Disord* 35(2): 217–222.

Martin RJ, Banks-Schlegel S. (1998) Chronobiology of asthma. *Amer J Respir Crit Care Med* 158(3): 1002–1007.

McClintock MK. (1971) Menstrual synchrony and suppression. *Nature* 229: 244–245.

McClintock MK. (1984) Estrous synchrony: modulation of ovarian cycle length by female pheromones. *Physiol Behav* 32(5): 701–705.

McClintock MK. (1998) Whither menstrual synchrony? *Ann Rev Sex Res* 9: 77–95.
Minnesota Motor Vehicle Crash Facts. (1997–2000) Office of Traffic Safety, St. Paul, MN. (http://www.dps.state.mn.us/trasafe).
Minors DS, Waterhouse JM. (1981) Anchor sleep as a synchronizer of rhythms on abnormal routines. *Intl J Chronobiol* 7(3):165–88.
Minors DS, Waterhouse JM. (1983) Does "anchor sleep" entrain circadian rhythms? Evidence from constant routine studies. *J Physiol* 345: 451–467.
Minors DS, Waterhouse JM. (1983) The impact of irregular sleep-wake schedules on circadian rhythms and the role of "anchor" sleep. In: *Why We Nap. Evolution, Chronobiology, and Functions of Polyphasic and Ultrashort Sleep.* Stampi C, ed. Boston: Birkhauser, pp. 82–101.
Moiseff A, Copeland J. (2000) A new type of synchronized flashing in a North American firefly. *J Insect Behav* 13(4): 597–612.
Monk TH, Buysse DJ, Reynolds III CF, Kupfer DJ. (1996) Circadian determinants of the postlunch dip in performance. *Chronobiol Intl* 13(2): 123–133.
Monk TH, Carrier J. (1998) A parallelism between human body temperature and performance independent of the endogenous circadian pacemaker. *J Biol Rhythms* 13: 113–122.
Monk TH, Buysse DJ, Billy BD, Kennedy KS, Willrich LM. (1998) Sleep and circadian rhythms in four orbiting astronauts. *J Biol Rhythms* 13(3): 188–201.
Moore-Ede, MC. (1982) What hath night to do with sleep? *Nat History* 91(9): 22 and 24.
Moore-Ede, MC. (1985) Medical implications of shift-work. *Ann Rev Med* 36: 607–617.
Moore-Ede, MC. (1986) Jet lag, shift work, and maladaption. *NIPS* 1: 156–160.
Moore-Ede, MC. (1993) *The Twenty Four Hour Society*. Reading, PA: Addison-Wesley Publication, 230 pp.
Mumford L. (1934) *Technics and Civilization*. London: Routledge and Kegan Paul, 495 pp (see pp. 14–17).
Naitoh P, Beare AN, Biersner RJ, Englund CE. (1983) Altered circadian periodicities in oral temperature and mood in men on an 18-hour work/rest cycle during a nuclear submarine patrol. *Intl J Chronobiol* 8(3): 149–173.
National Sleep Foundation. (2001) 2000 Omnibus Sleep in America Poll. Washington, DC: National Sleep Foundation.
Nelson W, Cadotte L, Halberg F. (1973) Circadian timing of single daily "meal" affects survival of mice. *Proc Soc Exp Biol Med* 144(3): 766–769.
Nelson W, Scheving L, Halberg F. (1975) Circadian rhythms in mice fed a single daily meal at different stages of lighting regimen. *J Nutr* 105(2): 171–184.
Nelson WL, Halberg F, Zinneman H, Haus E, Scheving L, Bazin H. (1977) Survival of immunocytoma-bearing rats affected by meal-feeding and timing of adriamycin administration. In: *Proc XII Intl Conf Intl Soc Chronobiol*, Washington, DC. Rome: Il Ponte, pp. 443–447.
Nelson W, Halberg F. (1986) Meal-timing, circadian rhythms and life span of mice. *J Nutr* 116(11): 2244–2253.
Nelson W. (1988) Food restriction, circadian disorder and longevity of rats and mice. *J Nutr* 118(3): 286–289 (Review).
Nguyen-Clausen A. (1975) Zum Aktivitätsverlauf des verhaltensmusters putzen bei einer Gruppe von Haubenmainahs, *Acridotheres cristatellus* (GM). *Behaviour* 53: 91–108.
Novak RD, Auvil-Novak SE. (1996) Focus group evaluation of night nurse shiftwork difficulties and coping strategies. *Chronobiol Intl* 13(6): 457–463.
Ohayon MM, Carskadon MA, Guilleminault C, Vitiello MV. (2004) Meta-analysis of quantitative sleep parameters from childhood to old age in healthy individuals: developing normative sleep values across the human lifespan. *Sleep* 27(7): 1255–1273.

Parks DK, Yetman RJ, McNeese MC, Burau K, Smolensky MH. (2000) Day-night pattern in accidental exposures to blood-borne pathogens among medical students and residents. *Chronobiol Intl* 17(1): 61–70.

Pauly JE, Scheving LE, Burns ER, Tsai TH. (1976) Circadian rhythm in DNA synthesis in mouse thymus: effect of altered lighting regimens, restricted feeding and presence of Ehrlich ascites tumor. *Anat Rec* 184(3): 275–284.

Pauley SM. (2004) Lighting for the human circadian clock: recent research indicates that lighting has become a public health issue. *Med Hypoth* 63(4): 588–596.

Persinger MA, Cooke WJ, Janes JT. (1978) No evidence for relationship between biorhythms and industrial accidents. *Percep Motor Skills* 46(2): 423–426.

Peters JC, Wyatt HR, Donahoo WT, Hill JO. (2002) From instinct to intellect: the challenge of maintaining healthy weight in the modern world. *Obes Rev* 3(2): 69–74 (Review).

Philippens KM, von Mayersbach H, Scheving LE. (1977) Effects of the scheduling of meal-feeding at different phases of the circadian system in rats. *J Nutr* 107(2): 176–193.

Piccoli B, Parazzoli S, Zaniboni A, Demartini G, Frashini F. (1991) [Non-visual effects of light mediated via the optical route: review of the literature and implications for occupational medicine.] [Italian]. *Medicina del Lavoro* 82(3): 213–232.

Pokorny MLI, Blom DHJ, van Leeuwen P. (1985) Diurnal variations of bus drivers' accident risk: 'Time of day or time of task?' In: *Accidents of Bus Drivers. An epidemiological approach.* Blom DHJ, Pokorny MLI, eds. Leiden: Instituut voor Praeventieve Gezondheidszorg TNO, pp. 124–135.

Pokorny MLI, Blom DHJ, van Leeuwen P. (1981) Analysis of traffic accident data from bus drivers – An alternative approach (I). In: *Night and Shiftwork. Biological and Social.* Reinberg A, Vieux N, Andlauer P, eds. Oxford: Pergamon Press, pp. 271–278.

Presser HB. (1974) Temporal data relating to the human menstrual cycle. In: *Biorhythms and Human Reproduction.* Ferin M, Halberg F, Richart RM, Vande Wiele RL. eds. New York: John Wiley & Sons, Inc., pp. 145–160.

Quadagno DM, Shubeita HE, Deck J, Francoeur D. (1981) Influence of male social contacts, exercise and all-female living conditions on the menstrual cycle. *Psychoneuroimmunol* 6(3): 239–244.

Quigley BM. (1982) "Biorhythms" and men's track and field world records. *Med Sci Sports Exer* 14(4): 303–307.

Rafnasson V, Hrafnkelsson J, Tulinius H. (2000) Incidence of cancer among commercial airline pilots. *Occup Environ Med* 57: 175–179.

Rajaratnam SM. (2001) Legal issues in accidents caused by sleepiness. *J Hum Ergol (Tokyo)* 30(1–2): 107–111.

Rajaratnam SM, Arendt J. (2001) Health in a 24-h society. *Lancet* 358(9286): 999–1005 (Review).

Rajaratnam SM, Jones CB. (2004) Lessons about sleepiness and driving from the Selby rail disaster case: *R v Gary Neil Hart. Chronobiol Intl* 21(6): 1073–1077.

Raymond RC, Warren M, Morris RW, Leikin JB. (1992) Periodicity of presentations of drugs of abuse and overdose in an emergency department. *J Toxicol Clin Toxicol* 30(3): 467–478.

Recht LD, Lew RA, Schwartz WJ. (1995) Baseball teams beaten by jet lag. *Nature* 377: 583.

Regal PJ, Connolly MS. (1980) Social influences on biological rhythms. *Behaviour* 72: 171–198.

Reilly T, Walsh TJ. (1981) Physiological, psychological and performance measures during an endurance record for 5-a-side soccer play. *Br J Sports Med* 15(2): 122–128.

Reilly T, Baxter C. (1983) Influence of time of day on reactions to cycling at a fixed high intensity. *Br J Sports Med* 17(2): 128–130.

Reilly T, Down A. (1986) Circadian variation in the standing broad jump. *Percept Motor Skills* 62: 830.

Reilly T, Marshall TJ. (1991) Circadian rhythms in power output on a swim bench. *J Swim Res* 7(2): 11–13.

Reilly T. (1992) Circadian rhythms in muscular activities. In: *Muscle Fatigue Mechanisms in Exercise and Training*. Marconnet P, Komi PV, Saltin B, Sejersted OM, eds. Basel: Karger, pp. 218–238.

Reinberg A, Vieux N, Ghata J, Charrmont AJ, Laporte A. (1978) Is the rhythm amplitude related to the ability to phase-shift circadian rhythms of shift-workers? *J Physiol (Paris)* 74(4): 405–409.

Reinberg A, ed. (1979) Chronobiological field studies of oil refinery shift workers. *Chronobiologia* 6(Suppl 1): 1–119.

Reinberg A, Proux S, Bartal JP, Lévi F, Bicakova-Rocher A. (1985) Circadian rhythms in competitive sabre fencers: internal desynchronization and performance. *Chronobiol Intl* 2(3): 195–201.

Reinberg AE, Smolensky MH. (1992) Night and shift work and transmeridian and space flights. In: *Biologic Rhythms in Clinical and Laboratory Medicine*. Touitou Y, Haus E, eds. Berlin: Springer-Verlag, pp. 243–255.

Reinberg O, Reinberg A, Téhard B, Mechkouri M. (2002) Accidents in children do not happen at random: predictable time-of-day incidence of childhood trauma. *Chronobiol Intl* 19(3): 615–631.

Reiter RJ, Tan DX, Mayo JC, Sainz RM, Leon J, Czarnocki Z. (2003) Melatonin as an antioxidant: biochemical mechanisms and pathophysiological implications in humans. *Acta Biochim* Pol 50(4): 1129–1146 (Review).

Ribak J, Ashkenazi IE, Klepfish A, Augar D, Tall J, Kallner B, Noyman Y. (1983) Diurnal rhythmicity and airforce flight accidents due to pilot error. *Aviat Space Environ Med* 54: 1096–1099.

Riemersma JBJ, Sanders AF, Wildervanck C, Gaillard AW. (1977) Performance decrement during prolonged night driving. In: *Vigilance Theory. Operational Performance, and Physiological Correlates*. Mackie RR, ed. New York: Plenum Press, pp. 41–58.

Rodahl A, O'Brien M, Firth RG. (1976) Diurnal variation in performance of competitive swimmers. *J Sports Med* 16(1): 72–76.

Rodgers CW, Sprinkle RL, Lindberg FH. (1974) Biorhythms: three tests of the predictive validity of the 'critical days' hypothesis. *Intl J Chronobiol* 2(3): 247–252.

Rossi B, Zani A, Mecacci L. (1983) Diurnal individual differences and performance levels in some sports activities. *Percep Motor Skills* 57: 27–30.

Salisbury FB, Ross CW. (1992) *Plant Physiology*. 4th edn. Belmont, CA: Wadsworth, 682 pp.

Sasaki T. (1980) Effect of jet lag on sports performance. In: *Chronobiology: Principles and Application to Shifts in Schedules*. Scheving LE, Halberg F, eds. Alphen aan den Rijn: Sijthoff & Noordhoff, pp. 417–431.

Sasseville A, Paquet N, Sévigny J, Hébert M. (2003) The use of orange lens glasses to block the suppression of melatonin by bright light. (abstract) *Chronobiol Intl* 20(6): 1184–1185.

Schaefer KE, Kerr CM, Buss D, Haus E. (1979) Effect of 18-h watch schedules on circadian cycles of physiological functions during submarine patrols. *Undersea Biomed Res* 6 Suppl: S81–S90.

Schank JC, McClintock MK. (1992) A coupled-oscillator of ovarian-cycle synchrony among female rats. *J Theor Biol* 157(3): 317–362.

Scheer FAJL, van Doornen LJP, Buijs RM. (1999) Light and diurnal cycle affect human heart rate: possible role for the circadian pacemaker. *J Biol Rhythms* 14(3): 202–212.

Schoeller DA, Cella LK, Sinha MK, Caro JF. (1997) Entrainment of the diurnal rhythm of plasma leptin to meal timing. *J Clin Invest* 100(7): 1882–1887.

Schreiber G, Avissar S, Tzahor Z, Barak-Glantz I, Grisaru N. (1997) Photoperiodicity and annual rhythms of wars and violent crimes. *Med Hypoths* 48: 89–96.

Sensi S, Capani F. (1987) Chronobiological aspects of weight loss in obesity: effects of different meal timing regimens. *Chronobiol Intl* 4(2): 251–261.

Shaffer JW, Schmidt CW Jr, Slotowitz HI, Fisher RS. (1978) Biorhythms and highway crashes. Are they related? *Arch Gen Psychiatry* 35(1): 41–46.

Slama O. (1981) The influence of "biorhythm" on the incidence of injuries among forest workers. *Eur J Appl Physiol Occup Physiol* 47(4): 331–335.

Smith L, Folkard S, Poole CJ. (1994) Increased injuries on night shift. *Lancet* 344(8930): 1137–1139.

Smith RS, Guilleminault C, Efron B. (1997) Circadian rhythms and enhanced athletic performance in the National Football League. *Sleep* 20(5): 362–365.

Sothern RB. (1974) Low-frequency rhythms in the beard growth of a man. In: *Chronobiology*. Scheving LE, Halberg F, Pauly JE, eds. Tokyo: Igaku Shoin, pp. 241–244.

Souissi N, Gauthier A, Sesboue B, Larue J, Davenne D. (2004) Circadian rhythms in two types of anaerobic cycle leg exercise: force-velocity and 30-s Wingate tests. *Intl J Sports Med* 25(1): 14–19.

Stern K, McClintock MK. (1998) Regulation of ovulation by human pheromones. *Nature* 392(6672): 177–179.

Stevens RG. (2005) Circadian disruption and breast cancer. *Epidemiology* 16(2): 254–258.

Strumpf IJ, Simmons MS, Sayre JW, Tashkin DP. (1978) Biorhythm theory and asthma. *Ann Allergy* 41(6): 330–332.

Stunkard AJ, Allison KC. (2003) Two forms of disordered eating in obesity: binge eating and night eating. *Intl J Obes Relat Metab Disord* 27(1): 1–12 (Review).

Sundararaj BI, Nath P, Halberg F. (1982) Circadian meal timing in relation to lighting schedule optimizes catfish body weight gain. *J Nutr* 112(6): 1085–1097.

The Prairie Astronomy Club. (2001) Light Pollution Fact Sheet. (*http://www.prairieastronomyclub.org/light.htm*).

Thommen G. (1973) *Is This Your Day? How Biorhythm Helps You Determine Your Life Cycles*. New York: Crown Publishers, 160 pp.

Trinder J, Armstrong SM, O'Brien C, Luke D, Martin MJ. (1996) Inhibition of melatonin secretion onset by low levels of illumination. *J Sleep Res* 5: 77–82.

US Dept Transportation (1999) Air Travel Consumer Report, November 1999. (*http://airconsumer.ost.dot.gov/reports/index.htm*).

Waterhouse J, Minors D, Atkinson G, Benton D. (1997) Chronobiology and meal times: internal and external factors. *Br J Nutr* (Suppl): S29–S38.

Waterhouse J, Redfern P. (1997) Jet lag and shiftwork: Current perspectives on intervention and treatment. *Chronobiol Intl* 14(2): 89–229.

Waterhouse J, Edwards B, Nevill A, Carvalho S, Atkinson G, Buckley P, Reilly T, Godfrey R, Ramsay R. (2002) Identifying some determinants of "jet lag" and its symptoms: a study of athletes and other travelers. *Br J Sports Med* 36(1): 54–60.

Webb WB. (1971) Sleep as a biorhythm. In: *Biological Rhythms and Human Performance*. Colquhoun WP, ed. London: Academic Press, pp. 149–177.

Weitzman ED, Czeisler CA, Moore-Ede M. (1979) Sleep-wake, neuroendocrine and body temperature circadian rhythm under entrained and non-entrained (free-running) conditions in man. In: *Biological Rhythms and Their Central Mechanism.* Suda M, Hagaishi P, Nakagawa H, eds. Amsterdam: Elsevier, pp. 199–227.

Wever RA. (1979) *The Circadian System of Man: Results of Experiments under Temporal Isolation.* New York: Springer-Verlag, 276 pp.

Winget CM, DeRoshia CW, Holley DC. (1985) Circadian rhythms and athletic performance. *Med Sci Sports Exer* 17: 498–516.

Winget CM, Soliman MRI, Holley DC, Meylor JS. (1992) Chronobiology of physical performance and sports medicine. In: *Biologic Rhythms in Clinical and Laboratory Medicine.* Touitou Y, Haus E, eds. Heidelberg: Springer-Verlag, pp. 230–242.

Winstead DK, Schwartz BD, Bertrand WE. (1981) Biorhythms: fact or superstition? *Amer J Psychiat* 138(9): 1188–1192.

Wojtczak-Jarosozwa J, Jarosz D. (1987) Chronohygienic and chronosocial aspects of industrial accidents. In: *Advances in Chronobiology. Part B.* Pauly JE, Scheving LE, eds. New York: Alan R. Liss, pp. 415–426.

Wolcott JH, Hanson CA, Foster WC, Kay T. (1979) Correlation of choice reaction time performance with biorhythmic criticality and cycle phase. *Aviat Space Environ Med* 50(1): 34–39.

Wolcott JH, McMeekin RR. (1977a) Correlation of general aviation accidents with the biorhythm theory. *Human Factors* 19: 283–293.

Wolcott JH, McMeekin RR, Burgin RE, Yanowitch RE. (1977b) Correlation of occurrence of aircraft accidents with biorhythmic criticality and cycle phase in U.S. Air Force, U.S. Army, and civil aviation pilots. *Aviation Space Environ Med* 48(10): 976–983.

Wu MW, Li XM, Xian LJ, Levi F. (2004) Effects of meal timing on tumor progression in mice. *Life Sci* 75(10): 1181–1193.

Wyatt HR, Grunwald GK, Mosca CL, Klem ML, Wing RR, Hill JO. (2002) Long-term weight loss and breakfast in subjects in the National Weight Control Registry. *Obes Res* 10(2): 78–82.

Yellin AM, Hauty GT. (1971) Activity cycles of the rhesus monkey (*Macaca mulatta*) under several experimental conditions, both in isolation and in a group situation. *J Interdiscipl Cycle Res* 2: 475–490.

11
Clinical Medicine

"One was wrong to consider only the influence of the night on our body; this is to look for only half of an important revolution in the course of our economy: one must, therefore, study the effect of the [day-night alternation] as a whole."
— *Julien-Joseph Virey (1775–1846), French Pharmacist*

Introduction

The importance of biological rhythms to medicine was pointed out in a thesis by Julien-Joseph Virey in 1814: *"This successive rotation of our functions each day, . . . is it not aimed to establish a regular periodicity, likely to be innate . . . ?"* Virey is credited as being the first person to (1) suggest that the effects of some drugs, such as narcotics, vary according to the time of day of dosing (*"Any medication is not equally indicated at all clock hours, and here the diurnal periodicity needs to be taken into account."*) and (2) document daily and annual rhythms in human mortality (Reinberg et al., 2001).

It is a common practice to group various body functions and related structures (cells, tissues and organs) into systems (Figure 11.1). Within each of these systems, changes in amounts, levels or activity display a prominent biological time structure as part of a spectrum of oscillations whose rhythm characteristics (periods, amplitudes, phases) have a significant role in the practice of *clinical medicine*. Alterations in these three rhythm characteristics can contribute to subjective and objective evaluations, an assessment of the situation, and the plan of action (e.g., kind, amount, and timing of treatment).

The primary objective of this chapter is to introduce some of the implications and applications of biological rhythms in clinical medicine, as was done in a separate chapter for veterinary medicine. While it is impossible to adequately cover every medical area in this chapter, we try to cover the main concepts related to *chronomedicine* and discuss several medical areas where the concept of timing can be or is being applied.

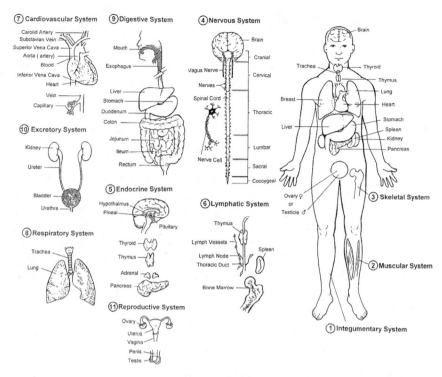

FIGURE 11.1. Normal functioning of every system of the human body has been shown to be characterized by circadian and/or other (e.g., ultradian, infradian) rhythms. See Tables 11.1 to 11.12 for an overview of rhythms for variables by system.

Circadian Rhythms in Health

Most humans entrain to a selected light–dark schedule that is circadian, in spite of artificial lights which enable modern humans to behave independently from the effects of natural darkness (Honma et al., 2003). This is because in the presence of environmental cues, such as a light–dark cycle and social cues, free-running circadian rhythms are rare in nature and this allows the biological clock to fine-tune body functions to specific times of the day and night (Roenneberg et al., 2003). This entrainment allows for generalizations as to characteristics of circadian rhythms across groups of humans.

As discussed in Chapter 12 on *autorhythmometry* and mentioned elsewhere throughout this book, virtually every body function in humans (Figure 11.1) has been shown to display a circadian and/or other rhythm in healthy individuals (cf. Kanabrocki et al., 1983, 1990a; Touitou & Haus, 1994), which persists into old age, often with a reduction in amplitude (Casale & de Nicola, 1984; Haus et al., 1989; Haus & Touitou, 1997). Many aspects of human performance and efficiency also show a circadian pattern that usually closely follows that of body temperature (cf. Colquhoun, 1972; Rutenfranz & Colquhoun, 1979). These rhythms generally have known times of highest and lowest values in relation to an individual's sleep–wake (light–dark) schedule, which is usually the dominant synchronizer for the body clock.

In diurnally active persons, some hormones are highest upon awakening in the morning (e.g., cortisol), others are highest in the afternoon or early evening (e.g., gastrin, testosterone), and others, such as TSH (thyroid-stimulating hormone), melatonin and prolactin, reach their peak levels just before or during sleep (Figure 11.2). In the skin, protective functions are highest during the day, while renewal and diverse metabolic processes are highest at night (cf. Henry et al., 2002). At the physiologic level, body temperature, heart rate and blood pressure are generally highest in the afternoon, while hearing and pain sensitivity are most acute in the evening. Peaks and troughs are also quite consistent for a variable whether it is sampled in blood, saliva or urine,[1] as demonstrated for cortisol and melatonin sampled in healthy men as part of the Medical Chronobiology Aging Project (MCAP) 1993 study[2] (Figure 11.3) (Sothern et al., 1994). It must be noted, however, that while substances in blood and saliva are reported as a concentration per volume (e.g., per ml, per cu cm, etc.), analytes determined in urine need to be re-expressed as a rate over time (e.g., per hour, per 3 h, etc.) and assigned to the midpoint of the collection interval for correct interpretation of the actual amount and rhythmic pattern present in the variable being studied (see Essay 11.1).

Essay 11.1. (by RBS): Adjusting Urinary Concentrations for Volume and Time

While it is a noninvasive procedure, many people are squeamish about collecting a urine sample, even though it is a readily available source of information on biochemical rhythms that avoids any trauma associated with a needle and does not interfere with other normal functions. Urine samples can be collected around-the-clock at specific intervals (e.g., every 3 h or 4 h) or "upon urge." Urine collected over several hours will usually smooth any high-frequency fluctuations, allowing for description

---

[1] However, concentrations can differ dramatically between these body fluids. For example, for the hormones shown in Figure 11.3, the cortisol values in saliva were only 4% of those found in blood (0.40 μg/dl vs. 10.1 μg/dl) and for melatonin they were only 88% (12.4 ng/dl vs. 14.1 ng/dl). Others have confirmed the use of salivary melatonin as a circadian phase marker (Voultsios et al., 1997) and that lower levels in saliva reflect the same circadian pattern as the protein-free fraction of melatonin in plasma (Kennaway & Voultsios, 1998). Concentrations in urine are expressed as rate/hr (e.g., 2.8 μg/h for cortisol and 338 ng/h for melatonin) and assigned to the middle of the collection interval. Nevertheless, the circadian patterns reveal similar timing for a hormone in each fluid.

[2] The Medical Chronobiology Aging Project (MCAP) has studied groups of men for circadian rhythms vs. aging in more than 100 variables in blood, urine, saliva and whole body physiology for 24 h in May 1969, 1979, 1988, 1993, 1998 and 2003. These studies were directed by Dr. EL Kanabrocki and Prof. LE Scheving† beginning in San Antonio, TX (in 1969) and subsequently at the Hines VA Hospital near Chicago, IL. One of this book's co-authors (RBS) participated in the 1988, 1993, and 2003 studies.

Circadian Rhythms in Health 429

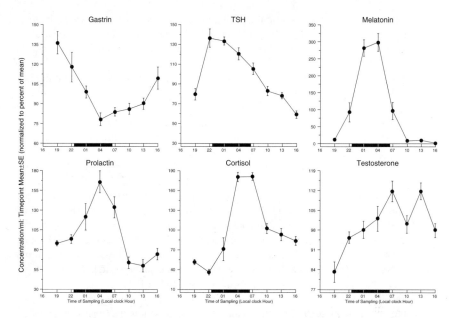

FIGURE 11.2. Circadian patterns in several hormones in the blood of healthy men reveal different times of high and low values for each hormone. Nine healthy men, ages 46–65 years, provided a blood sample every 3 h for 24 h on May 14–15, 1993 as part of the Medical Chronobiology Aging Project (MCAP). TSH = thyroid-stimulating hormone. Dark bar = sleep span. (Kanabrocki et al., unpublished, used by permission).

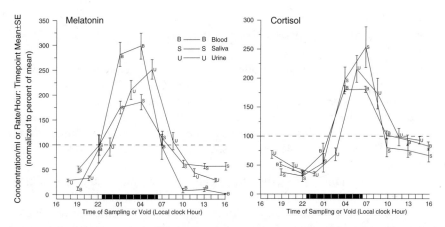

FIGURE 11.3. Consistency of circadian peaks and troughs is demonstrated for cortisol and melatonin sampled simultaneously in different body fluids (blood, saliva or urine) (data from Sothern et al., 1994). All data normalized to percent of individual mean for comparison of circadian patterns. Nine healthy men, ages 46–65 years, collected urine and provided a blood and saliva sample every 3 h for 24 h as part of the Medical Chronobiology Aging Project (MCAP) study in May 1993. Urine value assigned to midpoint of collection interval. Dark bar = sleep span. (cf. Sothern et al., 1994).

of circadian, but usually not ultradian, rhythms. As opposed to blood, where the total volume does not change appreciably throughout the day and night, the concentration of an analyte in urine is only a first step in determining the actual amount and rate of excretion.

Detection of a substance in urine indicates its presence, but the correct quantification of how much of the substance was excreted depends upon two things: the total volume produced by voiding and the time interval over which the urine was collected. For example, a potassium concentration of 3.0 mEq/L in each of two 4-h urine samples collected 12 h apart may lead to the conclusion of no time-related difference between the samples. However, if the total volume of one sample was 100 ml and the other was 200 ml, the actual rate of potassium excretion in the two examples would differ by 100% (e.g., 3.0 mEq/L × 0.1 L = 0.30 mEq/4h vs. 3.0 mEq/L × 0.2 L = 0.60 mEq/4 h). In addition, the total amount of an analyte is often further adjusted to a common time interval, usually one hour, if collection intervals varied (e.g., 1.5 h, 3 h, 6 h, etc.). This is accomplished by dividing the amount of analyte first adjusted for volume by the hours of the collection interval and expressing the analyte amount as rate/hour. Finally, each rate/hr needs to be assigned to the midpoint of the interval over which the urine was produced for the correct time assignment (e.g., an excretion rate in a urine sample collected between 08:00 and 12:00 h would be assigned to 10:00 h).

Rather than excretion rate/hour, urinary analytes are sometimes expressed in relation to the amount of creatinine (i.e., /mg creat) also present in the same sample. However, urinary creatinine itself has been reported to show a rhythmic time-structure in humans that is both circadian (peak late in the day) (Sothern et al., 1974; Kanabrocki et al., 1990a) and circannual (peak in late summer and early fall) (Sothern et al., 1981a; Haus et al., 1988), thereby possibly confounding the interpretation of results from this technique.

Overview of Rhythms in Body Systems

An overview of circadian and/or other rhythms documented in the various body systems displayed in Figure 11.1 is presented in Tables 11.1–12. These tables include a list of variables in a particular body system, the period investigated, and the time(s) or time span of estimated maximum values for the following systems: integumentary (skin), muscular, skeletal, nervous, endocrine, immune/lymphatic, cardiovascular, respiratory, digestive, excretory (urine), reproductive, and whole body.

In addition, numerous studies have demonstrated that many coagulation parameters in humans exhibit circadian patterns, thereby making the times of peaks and troughs in the daily cycles of these variables somewhat predictable in diurnally active individuals (Table 11.13). The peaks that occur later in the day for platelets and prothrombin time (PT), into the late evening hours for activated partial thrombin time (APTT), and overnight for thrombin time (TT) play an integral role in maintaining the continuous, though 24-h rhythmic, equilibrium of blood rheology (flow). It would appear that the early morning changes in blood levels of coagulation-related variables, including a drop in fibrinolytic activity and thrombin time generation, the splitting off of fibrinopeptide-A from fibrinogen

TABLE 11.1. Examples of biological rhythms related to the human integumentary system.

Variable	Period	Time(s) or range of maxima	Author	Year
Skin mitoses	24 h	Early dark	Scheving	1959
Skin reaction to house dust	24 h	16:02–01:08 h	Reinberg et al.	1969
Skin reaction to penicillin	24 h	18:48–04:00 h	Reinberg et al.	1969
Skin reaction to histamine	24 h	21:44–00:48 h	Reinberg et al.	1969
Muscular blood flow (calf)[a]	24 h	14:00–15:00 h	Bartoli et al.	1975
Brilliance of complexion of facial skin in women	24 h	11:00 h	Reinberg et al.	1990a
Texture of facial skin in women	24 h	19:30 h	Reinberg et al.	1990a
Appearance of facial skin in women	24 h	19:30 h	Reinberg et al.	1990a
Changes in size of facial skin corneocytes in women	24 h	23:00 h	Reinberg et al.	1990a
N of actively secreting sebaceous follicles (face)	24 h	14:00 h	Verschoore et al.	1993
Sebum excretion (face)	24 h	09:48–16:68 h	Le Fur et al.	2001
Transepidermal water loss (face)	24 h	09:08–14:08 h	Le Fur et al.	2001
Transepidermal water loss (forearm)	24 h	01:30–10:30 h	Le Fur et al.	2001
Basal skin blood flow (arm)	24 h	12:20–22:36 h	Yosipovitch et al.	2004
Red intensity (face)	24 h	12:00–16:00 h	Latreille et al.	2004
Surface pH (face)	24 h	12:00–16:00 h	Latreille et al.	2004
Capacitance (forearm)	24 h	20:00–00:00 h	Latreille et al.	2004

[a] In patients with intermittent claudication due to arteriosclerosis obliterans.

and the sluggishness of the RBCs due to adhesiveness and aggregation of platelets, culminate in an increase in blood viscosity. Morning increases or decreases in coagulation-related variables may be more than coincidental, since the timing of these changes coincides with the greatest incidence of cardiovascular morbidity and mortality (see below).

Reports of rhythms in many additional variables are available in the literature. For example, circadian rhythms in more than 100 variables related to the human immune system have been summarized in a similar tabular format (Sothern & Roitman-Johnson, 2001). Also, circadian rhythms in humans have been documented for clock genes in skin and oral mucosa (Bjarnason et al., 2001), peripheral blood mononuclear and polymorphonuclear cells (Boivin et al.,

TABLE 11.2. Examples of biological rhythms related to the human muscular system.

Variable	Period	Time(s) or range of maxima	Author	Year
Muscular endurance: finger	24 h	00:00–03:00 h	Bochnik	1958
Grip strength (men)	24 h	15:00–17:40 h	Kühl et al.	1974
Grip strength (men)	Infradian	~13 days	Kühl et al.	1974
Muscle (calf) blood flow	24 h	00:00 h	Rieck & Damm	1975
Neuromuscular efficiency: elbow flexor torque	24 h	18:00 h	Davenne & Gauthier	1998
Neuromuscular efficiency: isometric knee extensor torque	24 h	19:10 h	Callard et al.	2000

TABLE 11.3. Examples of biological rhythms related to the human skeletal system.

Variable	Period	Time(s) or range of maxima	Author	Year
Calcium	24 h	08:04–10:52 h	Haus et al.	1988
Calcium	7 days	Sun. 15:00–Weds. 05:00	Haus et al.	1988
Calcium	1 year	October 3–December 8	Haus et al.	1988
Tooth eruption rate	24 h	Overnight	Lee & Proffit	1995
Serum bone-resorbing activity	24 h	03:00 h	Lakatos et al.	1995
Lumbar intervertebral distance	24 h	Morning	Ledsome et al.	1996
Lumbar intervertebral disc volume	24 h	Morning	Roberts et al.	1998
Femoral articular knee cartilage thickness	24 h	Morning	Waterton et al.	2000
Bone marrow (BM) mitotic index	24 h	00:00 h	Mauer	1965
BM: total cells in DNA S-phase	24 h	09:32 h–16:04 h	Smaaland et al.	1991
BM: myeloid progenitor cells (CFU-GM[a])	24 h	11:00–16:08 h	Smaaland et al.	1992
BM: glutathione content	24 h	08:35 h	Smaaland & Laerum	1992
BM: total cells in DNA S-phase	1 year	July 10–September 15	Sothern et al.	1995
BM: mononuclear cells in S-phase	24 h	08:52–14:28 h	Abrahamsen et al.	1997
BM: erythroid cells in S-phase	24 h	10:32–15:32 h	Abrahamsen et al.	1997
BM: CD34+ progenitor cells	24 h	10:24–14:48 h	Abrahamsen et al.	1998

[a] Colony-forming unit-granulocyte-macrophage.

2003; Kusanagi et al., 2004) and bone marrow (Tsinkalovsky et al., in preparation) (Table 11.14). Twenty-four-hour acrophase charts have been prepared for identifying the location of peaks in a number of variables monitored in clinical medicine (cf. Dawes, 1974a; Haus et al., 1988, 1990; Lévi et al., 1988a; Kanabrocki et al., 1990b; Rivera-Coll et al., 1993; Bremner et al., 2000).

TABLE 11.4. Examples of biological rhythms related to the human nervous system.

Variable	Period	Time(s) or range of maxima [or minima]	Author	Year
Choice reaction time: auditory	24 h	[17:00–20:00 h]	Pöppel et al.	1970
Choice reaction time: visual	24 h	[15:00–20:00 h]	Pöppel et al.	1970
Cutaneous pain threshold: thermal	24 h	06:34 h	Procacci et al.	1974
Cutaneous pain threshold: needle prick	24 h	03:00 h	Strempl	1977
Vigilance (% signals detected)	24 h	16:28–18:26 h	Froberg	1977
Motor nerve conduction velocity	24 h	17:42–23:30 h	Ferrario et al.	1980
Sensory nerve conduction velocity	24 h	20:48–02:30 h	Ferrario et al.	1980
Subjective alertness	24 h	15:56 h	Monk et al.	1983

TABLE 11.5. Examples of biological rhythms in the human endocrine system.

Variable[a]	Period	Time(s) or range of maxima	Author	Year
ACTH	24 h	09:15 h	Veldhuis et al.	1994
ACTH secretory spikes	~15 spikes/24 h	Median = 60 min duration	Van Cauter & Honinckx	1985
Beta-endorphin	24 h	06:26 h	Veldhuis et al.	1994
Cortisol	24 h	07:28–09:24 h	Haus et al.	1988
Cortisol	7 days	Sat 03:00–Mon 06:00 h	Haus et al.	1988
Fibrinogen	7 days	Sat 12:00–Sun 21:08 h	Kanabrocki et al.	1995
Fibrinogen	1 year	April 29–July 9	Kanabrocki et al.	1995
Fibrinogen	24 h	08:16–11:20 h	Kanabrocki et al.	1999
FSH	24 h	22:53 h	Veldhuis et al.	1994
Gastrin	24 h	17:44–22:36 h	Kanabrocki et al.	1987
GH	24 h	00:32 h	Veldhuis et al.	1994
GH secretory spikes	~4 spikes/24 h	Median = 75 min duration	Van Cauter & Honinckx	1985
Insulin	24 h	14:03–16:00 h	Kanabrocki et al.	1987
Leptin	24 h	00:00–06:00 h	Sinha et al.	1996
LH	24 h	00:00 h	Veldhuis et al.	1994
Melatonin	Menstrual	Near menstruation	Wetterberg et al.	1976
Melatonin	24 h	00:24–04:02 h	Kanabrocki et al.	1987
Melatonin	1 year	Winter (January)	Stokkan & Reiter	1994
Prolactin	24 h	01:51 h	Veldhuis et al.	1994
Prolactin secretory spikes	~12 spikes/24 h	Median = 75 min duration	Van Cauter & Honinckx	1985
TSH	24 h	00:20–02:36 h	Kanabrocki et al.	1987
TSH secretory spikes	~18 spikes/24 h	Median = 45 min duration	Van Cauter & Honinckx	1985

[a] LH = luteinizing hormone, FSH = follicle-stimulating hormone, TSH = thyroid stimulating hormone, GH = growth hormone, ACTH = adrenocorticotropic hormone.

What and When is Normal?

The many naturally occurring daily and other body rhythms that have evolved as adjustments to environmental changes due to the solar day and year and lunar month have a regulating influence on the "normal" functioning of the body's many processes. Humans and other animals have genetically based biological clocks (cf. Rensing et al., 2001) that are present in individual cells,[3] resulting in semi-autonomous oscillators in many peripheral tissues that can be more or less coordinated by the suprachiasmatic nucleus (SCN) in the anterior hypothalamus (cf. Shirakawa et al., 2001). A time-dependent control of homeostasis for different endocrine, physiological, and behavioral functions is thus controlled by this master clock (the SCN) that sends a message of time throughout the body (Perreau-Lenz et al., 2004) (see discussion later in this chapter and in Chapter 2

---

[3] It is estimated that 1,000 or more genes are clock-controlled in any given peripheral cell type (Schibler et al., 2003).

TABLE 11.6. Examples of biological rhythms related to the human immune/lymphatic system.

Variable[a]	Period	Time(s) or range of maxima	Author	Year
Activated T-cells	24 h	20:30–03:06 h	Born et al.	1997
Alpha-2-globulins	24 h	10:28–15:32 h	Casale & de Nicola	1984
Basophils	24 h	18:40–01:56 h	Kanabrocki et al.	1990
Beta-globulins	24 h	10:44–14:40 h	Casale & de Nicola	1984
B-lymphocytes	24 h	20:00 h	Lévi et al.	1988a
DPD activity of mononuclear cells	24 h	21:48–02:24 h	Tuchman et al.	1988
Eosinophils	24 h	00:56–03:32 h	Haus et al.	1992
Gamma-globulins	24 h	07:52–15:44 h	Casale & de Nicola	1984
Immunoglobulin-A (IgA)	24 h	09:04–17:24 h	Casale & de Nicola	1984
Immunoglobulin-G (IgG)	24 h	08:24–15:04 h	Casale & de Nicola	1984
Immunoglobulin-M (IgM)	24 h	10:16–15:44 h	Casale & de Nicola	1984
Interferon-gamma	24 h	20:30–04:00 h	Petrovsky & Harrison	1997
Interleukin-1	24 h	21:00 h	Petrovsky et al.	1998
Interleukin-2	24 h	12:48 h	Young et al.	1995
Interleukin-6	24 h	01:24–03:12 h	Sothern et al.	1995
Interleukin-10	24 h	20:00–02:00 h	Petrovsky & Harrison	1997
Interleukin-12	24 h	23:00 h	Petrovsky et al.	1998
Lymphocytes	24 h	00:04–01:08 h	Haus et al.	1992
MIF	24 h	Late morning	Petrovsky et al.	2003
Monocytes	24 h	18:16–22:56 h	Haus et al.	1992
NK Cells	24 h	17:01–19:15 h	Born et al.	1997
Neutrophils	24 h	17:44–21:20 h	Haus et al.	1992
Platelets	24 h	15:36–18:16 h	Kanabrocki et al.	1999
Polymorphonuclear phagocytic activity	24 h	22:18–03:32 h	Mechart et al.	1992
T-lymphocytes	24 h	00:00–04:00 h	Lévi et al.	1988a
Total white blood cells	24 h	21:56–23:08 h	Haus et al.	1992
TNFα	24 h	21:30 h	Petrovsky et al.	1998

[a]DPD = Dihydropyrimidine Dehydrogenase; MIF = Macrophage Migration Inhibitory Factor; NK = Natural Killer; TNF = Tumor Necrosis Factor.

on Rhythm Characteristics). Rhythms in these functions, however, have different times of their highest and lowest values. Rhythms are a normal characteristic of health, and while they continue during illness, their alteration may be involved with or reveal health problems (e.g., values may be too high or too low at unusual times or the shape of the circadian pattern might change).

In the scientific literature there is a wide diversity of normal values for physiological parameters in humans and other species. However, usually only a single value for each variable is listed as "reasonably normal" for the majority of the population (Davies & Morris, 1993). Such "ballpark" figures, however, do not provide error estimates to compute a normal range (e.g., standard deviations or standard errors to compute 95% confidence limits), nor do they take into account differences for sex or time of day or year (Sothern & Gruber, 1994). In addition to normal or clinical reference values published in the literature, hospitals often

TABLE 11.7. Examples of biological rhythms related to the human cardiovascular system.

Variable	Period	Time(s) or range of maxima	Author	Year
WHOLE BODY				
Heart rate (HR)	24 h	15:16–19:48 h	Kanabrocki et al.	1990
Systolic blood pressure (SBP)	24 h	14:20–17:44 h	Kanabrocki et al.	1990
SBP (females)	~28 days	Late follicular phase	Freedman et al.	1974
SBP	1 year	December 19–February 17	Sothern	1994
Diastolic blood pressure (DBP)	24 h	14:40–21:00 h	Kanabrocki et al.	1990
DBP	~28 days	Late follicular phase	Freedman et al.	1974
DBP	1 year	November 4–February 5	Sothern	1994
Double product (SBP × HR)	24 h	Afternoon	Hermida et al.	2001
CIRCULATION				
Albumin	7 days	Sun. 10:00–Weds. 10:00 h	Haus et al.	1988
Atrial natriuretic factor	24 h	05:28–07:00 h	Vesely et al.	1990
Hematocrit	24 h	10:20–11:48 h	Haus et al.	1988
Hematocrit	7 days	Sun. 04:00–Tues. 15:00 h	Haus et al.	1988
Hemoglobin	24 h	11:00–12:24 h	Haus et al.	1988
Hemoglobin	7 days	Sun. 14:00–Mon. 00:00 h	Haus et al.	1988
Iron	24 h	09:28–13:28 h	Casale & de Nicola	1984
Platelets	24 h	17:16–20:40 h	Haus et al.	1988
Red blood cells	24 h	10:20–11:48 h	Haus et al.	1988
Red blood cells	7 days	Sat. 22:00–Mon. 13:00 h	Haus et al.	1988
Sodium	24 h	12:36 h	Haus et al.	1988
Transferrin	24 h	11:48–14:52 h	Casale & de Nicola	1984
Uric acid	24 h	08:56–14:20 h	Haus et al.	1988

create their own tables of normal values generated from their clinical laboratories and these are used by medical personnel at that institution in making diagnoses.

When a sample of blood, urine, saliva, or tissue is taken from the body or a physiological measurement such as temperature or blood pressure is obtained, the values measured are used to diagnose the presence of normality, abnormality or disease. This is done by comparing them with a range of normal values

TABLE 11.8. Examples of biological rhythms in the human respiratory system.

Variable	Period	Time(s) or range of maxima	Author	Year
Airway resistance	24 h	00:00–03:52 h	Smolensky & Halberg	1977
Dynamic lung compliance	24 h	09:03–16:39 h	Gaultier et al.	1977
Forced expiratory volume	24 h	14:00 h	Guberan et al.	1969
Functional residual capacity	24 h	08:00 h	Kerr	1973
Lung mucociliary clearance	24 h	12:00–18:00 h	Bateman et al.	1978
Lung resistance	24 h	00:25–10:54 h	Gaultier et al.	1977
Peak flow	24 h	13:00–15:00 h	Smolensky & Halberg	1977
Specific airway conductance	24 h	12:00 h	Kerr	1973
Vital capacity	24 h	08:12–15:04 h	Smolensky & Halberg	1977

TABLE 11.9. Examples of biological rhythms related to the human digestive system.

Variable	Period	Time(s) or range of maxima	Author	Year
Gastric acidity in duodenal ulcer patients	24 h	16:30–02:30 h	Moore & Halberg	1986
Gastric acidity in health	24 h	20:00–23:00 h	Moore & Halberg	1986
High density lipoprotein CHOL[a]	24 h	08:52–16:52 h	Casale & de Nicola	1984
High density lipoprotein CHOL	1 year	Winter	Blüher et al.	2001
Lipoprotein(a)	24 h	07:24–10:04 h	Bremner et al.	2000
Low density lipoprotein CHOL	1 year	Winter	Gordon et al.	1988
Low density lipoprotein CHOL	24 h	05:40–10:00 h	Bremner et al.	2000
Plasma gastrin levels in health	24 h	20:00 h	Moore & Wolfe	1974
Propagation velocity of the migrating motor complex in proximal small bowel	24 h	Midday	Kumar et al.	1986
Rectal mucosa DNA synthesis	24 h	03:20–11:28 h	Buchi et al.	1991
Rectal mucosa DNA synthesis	1 year	October 12–November 29	Sothern et al.	1995
Salivary flow rate	24 h	15:22–17:30 h	Dawes	1974a,b
Salivary flow rate	1 year	Winter	cf. Dawes	1974a,b
Small intestine epithelium regeneration	5 days	Approx. every 5 days	Vander et al.	1998
Total bilirubin	24 h	11:40–15:04 h	Casale & de Nicola	1984
Total CHOL (F)	1 year	Winter	Garde et al.	2000
Total CHOL	24 h	04:52–15:24 h	Casale & de Nicola	1984
Triglycerides	24 h	18:04–23:00 h	Bremner et al.	1990
Very low density lipoprotein CHOL	24 h	18:00–23:08 h	Bremner et al.	1990

[a] CHOL = cholesterol.

(e.g., ±2SD from the mean) obtained in healthy people. If a newly measured value is within this range, it is considered "normal," and if it is outside the normal range (i.e., too high or too low), it is considered abnormal, possibly indicating illness or disease and signaling that follow-up and/or treatment may be necessary. This is almost always done without regard for the time of day a sample is obtained, which is usually not precisely recorded, since it is assumed that values are more or less constant around a homeostatic setpoint and time-of-day doesn't matter. Thus, if the value obtained is located within the upper and lower range of similar values obtained for a population of healthy people, it is determined to be normal, regardless of the time of day it was obtained.

But integrating information about the right amount, the right substance and the right place also depends upon information on timing. It was stated at a major gathering of chronobiologists in 1960 that in physiology, biochemistry, pharmacology and other sciences: "... *equally pertinent [to the physiologist] is the recognition of rules governing provisions for the 'right time' as a function of physiologic state and ecologic condition. Information on circadian time need not take second place to the concern for compounds, doses, sites of action, age effects, etc.*" (Halberg, 1960). In relation to clinical medicine, in making diagnoses or decisions, it is now known that "*when*" something is done can be as important as "*what*" and "*how much*."

TABLE 11.10. Examples of biological rhythms in the human excretory system (urine).

Variable	Period	Time(s) or range of maxima	Author	Year
17-Ketosteroids	7 days	Tues 04:00–Weds 17:00 h	Haus et al.	1988
Calcium	24 h	22:48–07:32 h	Haus et al.	1988
Calcium	1 year	September 8–November 21	Haus et al.	1988
Chloride	24 h	15:16–17:12 h	Kanabrocki et al.	1988
Cortisol	24 h	07:00–10:24 h	Kanabrocki et al.	1990a
Creatinine	24 h	19:36–22:28 h	Haus et al.	1988
Creatinine	7 days	Thur. 12:00–Sat. 03:00 h	Haus et al.	1988
Creatinine	1 year	July 16–November 23	Haus et al.	1988
Cyclic-AMP	7 days	Sat. 08:00–Sun. 16:00 h	Haus et al.	1988
Dopamine	24 h	13:12–22:48 h	Kanabrocki et al.	1990a
Glucose	24 h	14:52–17:08 h	Kanabrocki et al.	1988
Melatonin	24 h	00:40–03:44 h	Kanabrocki et al.	1990a
Osmolality	24 h	17:16–02:56 h	Kanabrocki et al.	1990a
Phosphorus	24 h	19:14–22:52 h	Kanabrocki et al.	1988
Potassium	24 h	13:44–17:40 h	Haus et al.	1988
Sodium	24 h	17:40–21:44 h	Haus et al.	1988
Sodium	7 days	Fri. 23:00–Sun. 13:00 h	Haus et al.	1988
Specific gravity	24 h	13:08–18:08 h	Kanabrocki et al.	1990a
Total protein	24 h	14:14–17:00 h	Kanabrocki et al.	1990c
Total solids	24 h	15:32–22:56 h	Kanabrocki et al.	1990a
Urea nitrogen	24 h	13:56–16:24 h	Kanabrocki et al.	1988
Volume	24 h	13:12–22:08 h	Haus et al.	1988
Volume	7 days	Fri. 21:00–Sat. 00:00 h	Haus et al.	1988

Most equate 98.6°F (37.0°C) as the normal body temperature, which is the value often used by physicians to determine if a fever is present. As discussed further in Chapter 12 on Autorhythmometry, 37.0°C was actually the mean value obtained from over one million axillary temperatures when more than 25,000 patients were measured repeatedly throughout the day and night in Germany in the 19th century (cf. Mackowiak et al., 1992). Even though this study performed more than a century ago found a *predictable* range that extended from a low of 36.2°C (97.2°F) between 02:00 h and 08:00 h to a peak of 37.5°C (99.5°F) between 16:00 and 21:00 h, the concept of 37.0°C (98.6°F) as the most normal temperature *at any time of day* persists. However, due to the underlying circadian rhythm in body temperature, 98.6°F in the middle of the afternoon would be a perfectly normal temperature *for that time of day*, but the same value at 04:00 h could indicate a fever[4] (Figure 11.4).

In addition, modern medical sources vary in what should be considered the upper limit of normal for oral temperature, ranging from 98.6°F (37.0°C) to 100.4°F (38.0°C) (see Figure 11.4). Thus, if 98.6°F was considered the upper limit of normal, many of the values shown in Figure 11.4 would erroneously be considered abnormal, since the subject was healthy throughout the study span. Additional graphs displaying typical circadian changes in human body temperature, heart rate and blood

---

[4] Fever is commonly defined as any temperature that is above "normal."

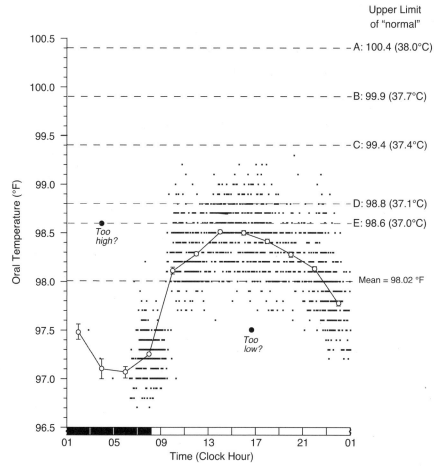

FIGURE 11.4. What and when is a clinical value normal? Comparison of oral temperatures (small dots, n = 1827; open circles = 2-h means) self-measured about 5 times/day over one whole year (2002) by a healthy man (RBS, age 55 years) vs. suggested upper limits of normal (Sothern, unpublished). Dark bar = average sleep span. Time of sampling may result in the same value being too low at one time, too high at another (black dots), yet still within a "normal" range that is commonly used in medicine: upper limits of normal from: (*A*) Textbook of Physiology, 1989; (*B*) Malowiak et al., JAMA, 1992; (*C*) Harrisons Principles of Internal Medicine, 1991; (*D*) Textbook of Medicine and Physiology, 1991; (*E*) Fever: Basic Mechanisms & Management, 1991.

pressure, as well as a discussion of the importance of body site in measuring temperature and blood pressure can be found in Chapter 12 on Autorhythmometry.

Time-Specified Normal Limits

Just as for body temperature, the concept of a circadian range for *normal* values can be applied to nearly every other body function as well. Normal limits for

Overview of Rhythms in Body Systems 439

TABLE 11.11. Examples of biological rhythms in the human reproductive system.

Variable	Period	Time(s) or range of maxima	Author	Year
FEMALES				
Estradiol	24 h	07:20–16:04 h	Haus et al.	1988
Estriol	24 h	13:36–19:16 h	Haus et al.	1988
Estrone	24 h	04:20–13:24 h	Haus et al.	1988
FSH	24 h	02:04–04:08 h	Haus et al.	1988
LH	24 h	22:56–05:00 h	Haus et al.	1988
Ovulation	~28 days	Ovulation at mid-cycle	Reinberg & Ghata	1964
Progesterone	24 h	04:32–07:40 h	Haus et al.	1988
Prolactin	24 h	02:28–03:44 h	Haus et al.	1988
Prolactin	1 year	March–May	Garde et al.	2000
Testosterone	1 year	July–August	Garde et al.	2000
MALES				
Dehydroepiandrosterone	24 h	15:00 h	Verschoore et al.	1993
Delta-4-Androstenedione	24 h	10:00 h	Verschoore et al.	1993
FSH	24 h	04:12–06:20 h	Haus et al.	1988
LH	24 h	00:48–06:24 h	Haus et al.	1988
Prolactin	24 h	02:03–04:36 h	Kanabrocki et al.	1987
Prostate-specific antigen	24 h	02:40–07:24 h	Mermall et al.	1995
Semen quality[a]	1 year	Winter	Levine et al.	1990
Testosterone	Infradian	3–5 days and 13–18 days	Harkness	1974
Testosterone	Infradian	20–22 days	Doering et al.	1974
Testosterone	1 year	November 26–January 21	Dabbs	1990
Testosterone	24 h	07:12–10:04 h	Mermall et al.	1995

[a] Sperm concentration, total sperm, concentration of motile sperm.

TABLE 11.12. Examples of biological rhythms in the human whole body.

Variable	Period	Time(s) or range of maxima	Author	Year
Body composition[a]	24 h	22:52–00:32 h	Cugini et al.	1996
Core-periphery heat exchange	24 h	18:00–21:00 h	Smith	1969
Heat production and loss[b]	24 h	~17:00–21:00 h	Aschoff et al.	1974
Metabolic rate (exhaled O_2 and CO_2)	24 h	17:25–00:05 h	Little & Rummel	1971
Oxygen saturation (SpO_2)	24 h	13:32–16:36 h	Chu et al.	2001
Sweat rate (thigh)	24 h	Daytime	Tokura et al.	1979
Temperature: ear	24 h	14:23–15:52 h	Sothern	Unpubl.
Temperature: oral	24 h	16:25–17:17 h	Sothern	1974a
Temperature: rectal	24 h	15:48–17:44 h	Mills et al.	1978
Temperature: urine	24 h	16:19–17:26 h	Sothern	1974a

[a] Variables measured = body weight, lean and fat body mass and ratio, body cell mass, extracellular mass, total body water, intra- and extra-cellular body water, Na and K exchangeable pool & ratio.
[b] Heat loss lags behind heat production by about 1.2 h.

TABLE 11.13. Sequence of circadian timing for some blood coagulation-related variables in humans.

Variable	Time(s) of maxima	Author	Year
Activated partial thrombin time (APTT)	18:44– 01:40 h	Haus et al.	1990
Alpha1 antitrypsin	10:52 h	Fornasari et al.	1981
Alpha-1-antitrypsin	08:20– 11:36 h	Haus et al.	1990
Alpha2-macroglobulin	08:53 h	Fornasari et al.	1981
Antithrombin III	10:36– 17:56 h	Casale et al.	1983
Beta-thromboglobulin	02:00– 06:00 h	Ogawa et al.	1989
C1 inhibitor	09:26 h	Fornasari et al.	1981
Euglobulin clot lysis time	16:00– 22:00 h	Grimaudo et al.	1988
Euglobulin precipitate	12:50 h	Fornasari et al.	1981
Factor VIII	03:56– 13:00 h	Haus et al.	1990
Fibrinogen	06:08– 11:40 h	Haus et al.	1990
Fibrinopeptide A	02:00 –06:00 h	Ogawa et al.	1989
Howell's time	21:00 h	Petralito et al.	1982
Plasminogen activator inhibitor type 1 (PAI-1)	03:00 h	Andreotti & Kluft	1991
Platelet adhesiveness	08:16– 13:20 h	Haus et al.	1990
Platelet aggregation	11:00 h	Brezinski et al.	1988
Platelet factor 4	22:00– 02:00 h	Ogawa et al.	1989
Platelets	17:16– 20:40 h	Haus et al.	1988
Prothrombin time (PT)	13:52– 19:56 h	Haus et al.	1990
Thrombin time (TT)	01:00– 07:36 h	Haus et al.	1990
Tissue-type plasminogen activator (tPA)	15:00– 18:00 h	Andreotti & Kluft	1991
Urokinase	12:51 h	Fornasari et al	1981
Viscosity	10:00– 13:00 h	Ehrly & Jung	1973

TABLE 11.14. Examples of 24-h rhythms related to the human cellular clock genes.

Clock gene	Site	Time(s) or range of maxima	Author	Year
hPer1	Oral mucosa	07:16– 09:40 h	Bjarnason et al.	2001
hPer1	Skin	11:44– 17:44 h	Bjarnason et al.	2001
hPer1	Blood [mononuclear cells]	09:06– 12:23 h	Boivin et al.	2003
hPer1	Blood [mononuclear cells]	05:55– 09:11 h	Kusanagi et al.	2004
hPer1	Blood [polymorphonuclear cells]	06:23– 09:03 h	Kusanagi et al.	2004
hPer1	BM [CD34+ cells]	06:52– 10:24 h	Tsinkalovsky et al.	In prep.
hPer2	Blood [mononuclear cells]	06:43– 12:35 h	Boivin et al.	2003
hPer2	BM [CD34+ cells]	07:32 – 13:52 h	Tsinkalovsky et al.	In prep.
hPer3	Blood [mononuclear cells]	06:23– 12:28 h	Boivin et al.	2003
hBmal1	Oral mucosa	21:00– 22:20 h	Bjarnason et al.	2001
hBmal1	Skin	20:48– 23:40 h	Bjarnason et al.	2001
hCry1	Oral mucosa	15:36– 18:36 h	Bjarnason et al.	2001
hCry2	BM [CD34+ cells]	08:00 –15:00 h	Tsinkalovsky et al.	In prep.
hDec1	Blood [mononuclear cells]	06:19– 12:19 h	Boivin et al.	2003

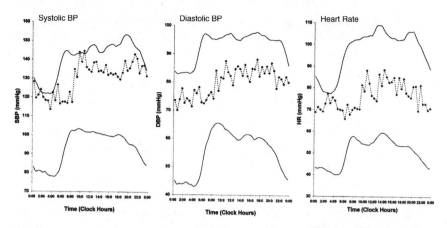

FIGURE 11.5. Example of hourly means over 24 h for blood pressure (BP) and heart rate (HR) in a man compared with time-specified normal limits (Chronodesms). The subject (RBS, age 56 years) wore an ambulatory monitor that measured BP and HR every 15-min around-the-clock for 6 days as part of the Medical Chronobiology Aging Project (MCAP) in May 2003 (Sothern, unpublished). Hourly means (solid squares) computed from 288 values shown against BP and HR chronodesms (upper and lower normal limits for age-matched reference group provided by the Halberg Chronobiology Center, University of Minnesota).

different portions of the 24 hours have been derived for several clinical variables in *blood* (cortisol [Nelson et al., 1983], growth hormone [Hermida et al., 1997a], white blood cells [Swoyer et al., 1984], plasma renin activity, aldosterone [Cugini et al., 1985]), in *urine* (electrolytes, trace elements) (Kanabrocki et al., 1983), and for systolic and diastolic *blood pressure* (Otsuka & Halberg, 1994; Hermida et al., 1999). Upper and lower limits of normal for time spans throughout the day and night are called *chronodesms* (from *chronos* = time, *desmos* = band) (Halberg et al., 1978) and procedures for computing so-called "time-specified tolerance intervals" over 24 h have been published using blood pressure as an example (Hermida & Fernandez, 1996; Hermida, 1999; Fernandez & Hermida, 2000). Some chronodesms are shown in Figure 11.5 along with hourly means over 24 h for blood pressure and heart rate of a clinically healthy man. Except for a couple of hourly means barely above the normal upper limit for systolic pressure, all means are within the upper and lower normal limits for individuals in his age group.

Circadian Rhythms in Symptoms and Disease

Twenty-four-hour variations are also present in the expression of symptoms of illness, often since many diseases get worse at night. For example, acute pulmonary edema (APE) occurs more frequently at night (Manfredini et al., 2000). Classical signs of APE, a possible triggering mechanism for cardiovascular morbidity and mortality, include unceasing cough, perspiration and the feeling of suffocation. The literature contains countless references concerning rhythmic aspects of diseases.

Cold and influenza symptoms (cough, sneezing, runny or blocked nose) are more prominent during the first hours after awakening (Smolensky et al., 1995). An increase upon awakening in intensity of nasal symptoms and other manifestations associated with asthma was first described in 1865 by A. Trousseau (cf. Gelfand, 2004). In general, allergic conditions such as asthma, allergic rhinitis, hay fever, hives and other allergies, display a circadian rhythm in symptoms, with itching, sneezing, wheezing, nasal congestion, runny nose, breaking out in rashes and bronchial constriction more severe at night, during sleep and/or near awakening, than during the day. Similarly, symptoms of rheumatoid- or osteo-arthritis (pain, stiffness and dexterity) are worse at night and improve during the day (Bellamy et al., 1990, 1991, 2002; Cutolo et al., 2003). Even the symptoms (pain, stiffness) of fibromyalgia, a widespread musculoskeletal pain and fatigue disorder for which the cause is still unknown, are worse at night for individuals when overall symptoms scores are above a certain threshold (Bellamy et al., 2004).[5]

An increase in migraine attacks occurs in the morning with a peak near awakening and a dramatic decrease in attacks at night (Solomon, 1992; Fox & Davis, 1998). Peptic (digestive) ulcers, however, are more likely to perforate and require hospitalization around noon if the site is in the intestines (duodenal) or in the stomach (gastric), with a secondary peak around midnight if the site is gastric (Svanes et al., 1998). Table 11.15 contains a listing of 24-h rhythmic patterns in perception of spontaneous or induced pain by humans.[6]

Birth and Death

Epidemiological studies have shown that the likelihood of being born or dying fluctuates over the course of the day. Labor onset occurs more frequently around midnight and a peak in births is found in the early morning hours (see Figure 1.7 in Chapter 1), while more complications (induced births, stillbirths) occur near midday (cf. Kaiser & Halberg, 1962; Smolensky et al., 1972).

Morbidity (symptoms) and mortality from cardiovascular disease have repeatedly been shown to be more prevalent in the morning hours for the incidence of angina and ischemic episodes (heart pain) (cf. Bogaty & Waters, 1989), myocardial infarction (heart attack), sudden cardiac death and stroke (Table 11.16). Sudden death from pulmonary embolism (blood clots in the lung) was also reported to be circadian rhythmic with a peak in the late morning (Gallerini et al., 1992).

---

[5] A weekly variation was also observed as a group phenomenon for all 21 women with fibromyalgia, with higher values for pain, stiffness and fatigue on Sunday & Monday and lower on Friday (Bellamy et al., 2004)

[6] A circannual rhythm in pain threshold of a healthy front tooth of a man was also observed when the minimum cold sensitivity test was performed 8–14 times/month over three years: pain threshold was minimum in May and maximum in October–November (Pöllman & Harris, 1978).

TABLE 11.15. 24-hour rhythmic patterns in the perception of pain by humans.

Type of pain	Comments	Author(s)	Year
SPONTANEOUS			
Angina pectoris	More painful episodes 02:00–06:00 h	von Armin et al.	1985
Arthritis–osteoarthritis	Pain in knee greatest in evening	Bellamy et al.	1990
Arthritis–rheumatoid	Pain in hand lowest in afternoon	Bellamy et al.	1991
Dental–toothache	Onset during early morning 03:00–08:00 h	Pöllman & Harris	1978
Headache–cluster	Highest onset 23:00–06:00 h	Symonds; Ekbom	1956; 1970
Headache–migraine	Highest onset 20:00–08:00 h	Ostfeld; Waters & O'Connor	1963; 1971
Headache–non-migrainous	Highest onset 08:00 h	Waters & O'Connor	1971
Intractable[a]	Intensity increased in evening	Folkard et al.	1976
Limb ischemia in peripheral artery disease	Night	Bartoli et al.	1975
Post-surgical cholecystectomy	Greatest need for narcotics to relieve pain 20:00–00:00 h	O'Donoghue & Rubin	1977
Post-surgical gynecological cancer	Greater need for morphine 04:00–08:00 h	Auvil-Novak et al.	1988
Sickle cell vaso-occlusive	Afternoon to early evening	Auvil-Novak et al.	1996
Sore arm 6 h after influenza vaccination	Greatest following vaccination at 12:00 h	Langlois et al.	1986
Sore arm from Hepatitis B vaccination	More severe pain after injection 13:00–15:00 h	Pöllman & Pöllman	1988
INDUCED			
Cutaneous–cold water	Threshold highest at 12:00 h	Strempel	1977
Cutaneous–needle	Threshold highest at 03:00 h	Strempel	1977
Cutaneous–radiant heat	Threshold highest 04:12–08:11 h	Procacci et al.	1974
Dental–cold	Threshold lowest 00:00–03:00 h	Pöllman	1984
Dull frontal headache	Pain sensitivity from pressure cuff on pericranial musculature greatest at 02:00 h	Gobel & Cordes	1990
Electrical stimulation	Muscle reflex threshold lowest at 01:00 h	Bourdallé-Badie et al.	1990b
Electrical stimulation	Pain sensation greatest at 05:00 h	Bourdallé-Badie et al.	1990b
Electrical stimulus	Tooth sensitivity lowest 00:00–03:00 h	Pöllman	1984
Tourniquet (on forearm)	Highest 1min pain score at 00:00 h	Koch & Raschka	2004

[a] Cannot be controlled by medication.

Cardiovascular Disease

Studies on thousands of patients with coronary artery disease have documented a circadian pattern in the incidence of irregularities in heart function and ischemic symptom onset. Symptoms are generally greatest after midnight and into the morning hours until early afternoon (Araki et al., 1983; Waters et al., 1984; Thompson et al., 1985, 1991; Rocco et al., 1987; Lucente et al., 1988; Ogawa et al.,

TABLE 11.16. Circadian (24-h) timing for human mortality from cardiovascular disease.

Diagnosis[a]	Study Site	N	Time(s) of maxima	Author	Year
CVD	Sweden	400	09:08–12:48 h	Johansson et al.	1990
IHD	USA (NY)	1251	08:00 h	Mitler et al.	1987
MI	USA 1940	9877	05:00–15:36 h	Smolensky et al.	1972
SCD	USA (MA)	2203	07:00–11:00 h	Muller et al.	1987
SCD	USA (MA)	429	07:00–09:00 h	Willich et al.[b]	1987
SCD	Italy	269	05:00 h	Pasqualetti et al.	1990
SCD	Romania	86	08:00–10:00 h	Nicolau et al.	1991
SCD	USA (MN)	94	07:00–13:00 h	Maron et al.	1994
SCD	Greece	223	09:00–12:00 h	Goudevenous et al.	1995
SCD	Japan	264	00:00–06:00 h	Kawamura et al.	1999
SCD	Germany	24061	06:00–12:00 h	Arntz et al.	2000

[a] CVD = Cerebrovascular Disease (stroke); IHD = Ischemic Heart Disease; MI = Myocardial Infarction; SCD = Sudden Cardiac Death.
[b] It was later determined that there is a relative risk of 2.6 for onset of SCD during the initial 3-h interval after awakening (Willich et al., 1992).

1989; Lanza et al., 1999). Often the peak is near the time of awakening, as has been documented for subarachnoid hemorrhage (brain aneurysm) (Feigin et al., 2002). Acute aortic aneurysm rupture or dissection, a life-threatening emergency associated with high mortality, shows a higher frequency from 06:00 to 12:00 h, with a smaller secondary peak in the evening (cf. Manfredini et al., 2004).

Similarly, death from ischemic heart disease (IHD), shows a higher frequency during the same morning hours (Table 11.16). However, while a single peak in the morning after awakening for the onset of infarction symptoms or sudden cardiac death (cf. Cohen et al., 1997) or stroke (cf. Elliott, 1998) is generally observed when summarizing results from large populations, the circadian pattern can change in individuals with different and/or additional risk factors. Upon closer examination of the study populations, a smaller secondary peak in the evening is often observed in the circadian patterns of cardiovascular events (cf. Manfredini et al., 2000). For example, while patients with no history of coronary heart failure (CHF) experienced the onset of a myocardial infarction (MI) between 06:00 h and 12:00 h, patients with a history of CHF have been found to be at greater risk for the onset of an MI between 18:00 h and 00:00 h (Hjalmarson et al., 1989). The secondary peak often found in the evening may thus reflect subsets of patients with differing risk factors.

With regard to mechanisms underlying cardiovascular symptomatology, an increase in basal vascular tone in the morning was documented in 12 healthy subjects (Panza et al., 1991). The authors hypothesized that a morning increase in alpha-sympathetic vasoconstrictor activity could be a possible contributing factor in the increased incidence of cardiovascular events in the hours after awakening. A decrease in endothelial function, as measured by brachial artery flow-mediated endothelium-dependent vasodilation, is also associated with an increased risk for cardiovascular events, and was lowest in the early morning after awakening, when compared with midday or evening (Otto et al., 2004). To some extent, the medical profession has accepted the concept of circadian variation in cardiovascular risk factors with their peaks in the morning hours (Muller, 1999). However, many questions remain unanswered, not only as to its degree of functional significance, but to its influence on and interaction with a number of other factors contributing to acute cardiovascular disorders (Muller & Tofler, 1991; Johnstone et al., 1996).

Recognition of chronobiological markers for such events would be helpful for interpreting and guiding effective treatments. As mentioned above, it is now thought that circadian rhythms in several variables play a pathophysiological role in acute cardiovascular morbidity and mortality, with increases in blood clotting variables, such as platelet aggregation, fibrinogen and blood viscosity accompanying the rapid morning rise in blood pressure (cf. Table 11.13; Kanabrocki et al., 1999; Figure 3, White, 2003; Manfredini et al., 2004).

Circannual Rhythms in Health

It has long been observed that animals breed and reproduce at certain times of the year favorable to their survival (usually in the spring or fall), that some migrate on a seasonal schedule and/or change the color of their hair or hibernate during the cold months (see Chapters 4 on Photoperiodism, 7 on Sexuality, and 9 on Veterinary Medicine). Through epidemiological and longitudinal data collection, it has become apparent that time of year influences not only quantitative characteristics of animals, but of humans, as well (Reinberg & Lagoguey, 1978; Halberg et al., 1981; Haus et al., 1988; Sothern, 1994).

For example, growth of human head and body hair was reported to increase through the summer months, reaching a maximum rate in September (0.545 mm/day) and then decreasing to a minimum in January (0.305 mm/day) (cf. Reinberg & Ghata, 1964). Some functions of human testes wax and wane over the year, such as semen quality that has lowest sperm concentration in the summer from July to September and greatest concentration in the fall and winter (cf. Levine, 1999). Testosterone levels in healthy men are also highest in the late fall and winter (Dabbs, 1990). A peak in human conception at higher latitudes coincides with the vernal (spring) equinox, accounting for more births in the fall (Roenneberg & Aschoff, 1990a,b). Additional aspects of seasonality and human sexuality are discussed in Chapters 4 (Photoperiodism) and 7 (Sexuality).

Large seasonal variations in levels of several immune defense mechanisms, including all white blood cell subtypes, have been reported in humans (cf. Touitou & Haus, 1994). In addition, DNA synthesis in the bone marrow, which manufactures new white blood cells, showed a circannual rhythm in healthy male medical students, with a peak in late summer and early fall (Sothern et al., 1995b). Blood pressure also shows a seasonal variation, with higher values in winter than summer (Brennan et al., 1982; Sothern, 1994). One-year acrophase charts are available for identifying the location of peaks in some variables commonly monitored in blood, urine and saliva in clinical medicine (cf. Reinberg & Lagoguey, 1978; Halberg et al., 1983; Nicolau et al., 1984; Haus et al., 1988).

Circannual Rhythms in Symptoms and Disease

Different infectious diseases are remarkably consistent from year to year in the timing and duration of outbreaks occurring simultaneously across widely dispersed geographic locations that have different weather patterns and diverse populations. Annual patterns for some illnesses can display different peak times depending on whether the disease is of bacterial or viral origin (Hejl, 1977; Pöllman, 1982).

Peaks in clinical illness are usually located in the *winter* for influenza, pneumonia, and rotavirus infection (e.g., gastroenteritis and diarrhea), in *spring* for measles and respiratory syncytial viruses (e.g., bronchiolitis or pneumonia, moderate-to-severe cold-like symptoms), in *summer* for poliomyelitis and other enterovirus infections (e.g., coxsackie and echo viruses, responsible for "summer colds," a flu-like illness with fever and muscle aches, or an illness with rash), and in *fall* for parainfluenza virus type 1 infections associated with or responsible for some respiratory infections (e.g., acute expiratory wheezing in children) (cf. Dowell, 2001). Listeriosis, a serious infection caused by eating food contaminated with the bacterium *Listeria monocytogenes*, is more common in the summer and fall in humans, but in the spring in farm animals (Busch, 1971; Moore & Zehmer, 1973).

Some diseases of the oral mucosa show two peaks a year, one in the spring and another in the fall, if they are of viral origin (recurrent aphthous ulcers, herpes simplex labialis, erythema exsudativum multiforme) or a single peak in the fall or winter if they are of bacterial origin (cervico-facial actinomycosis, acute necrotizing ulcerative gingivitis) (Pöllman, 1982). Other childhood infectious diseases, such as chicken pox, mumps, and rubella (measles), are more frequently diagnosed in the late winter and spring than at other times of the year (Smolensky, 1983). In a 7-year study, a circannual rhythm in rubella antibody concentrations, with highest values found in late spring, was found in a woman, 23 years of age at start, who had been naturally infected at an earlier age (Rosenblatt et al., 1982). This finding suggests a circannual cycle in immunity and the possibility of similar annual rhythms in antibody titers against other infections. Of note, multi-year oscillations in annual rates of primary and secondary syphilis infections across the USA have been documented over a 60-year span, with average peaks between 8 years and 11 years (Grassly et al., 2005). This about-10-year rhythm is thought to reflect an interaction between a partially protective endogenous human immune response and the natural dynamics of syphilis infection. Similar oscillations were not observed in gonorrhoea infections, where a protective immunity is absent, and therefore is more likely to reflect increases and decreases in sexual behavior.

Seasonality has also been described for the incidence, detection or treatment of many other clinical conditions. Dermatological office visits for several skin diseases influenced by ultraviolet light, including seborrheic dermatitis, acne and psoriasis, are higher in the spring (Hancox et al., 2004). Sexually transmitted diseases show a circannual rhythm in incidence, with more new cases diagnosed in late summer for genital infections with herpes simplex virus (Sumaya et al.,

1980), in late summer for gonorrhea and trichomonas, and in the fall for syphilis[7] (cf. Smolensky, 1981).

In addition, a seasonal incidence of cervical pathology (infections and cancer) has been reported, with greater detection rates in the summer months (Rietveld et al., 1988, 1997). A higher incidence of fungal/yeast infections was found in the spring and summer, while more viral infections were found in the spring for herpes simplex and in the summer for herpes zoster (shingles) (Termorshuizen et al., 2003), the latter confirming earlier studies (Glynn et al., 1990; Gallerani & Manfredini, 2000). However, recurrent attacks of herpetic ocular infections exhibited a peak in the winter in 541 patients followed over 15 years (Gamus et al., 1995).

Humans can harbor a number of latent/persistent viruses[8] from childhood that may show seasonal variations in their levels of reactivation and shedding. This has been shown for a herpesvirus (Epstein-Barr, EBV) and a polyomavirus (JCV), where peaks in viral reactivation and shedding in 30 healthy adults were noted in November to January for EBV and in March-May and October for EBV (Ling et al., 2003). Cytomegalovirus may also exhibit a seasonal peak of activity and has been linked with early mortality (peak in winter) from liver transplants (Singh et al., 2001).

An increase in host susceptibility to a particular pathogen may underlie annual variations in infectious disease outbreaks, with changes in host physiology (e.g., characteristics of mucosal surfaces, epithelial receptors, leukocyte numbers) contributing to increased susceptibility to a particular pathogen (cf. Dowell, 2001). This hypothesis was supported in a study where volunteers were treated intranasally with live influenza vaccine in the summer and winter (Shadrin et al., 1977). Fevers developed in 7% of the subjects treated in the winter, but <1% in the summer and antibodies increased 1.5 times more frequently in the winter than summer. Since people are relatively resistant to a disease if exposed in the off-season, pathogens may be present in humans year-round and an increase in morbidity (e.g., an epidemic) appears only when susceptibility, possibly photoperiod-driven, is high in certain seasons, rather than due to pathogens physically being transmitted from person-to-person and/or across geographic locations (cf. Dowell, 2001).

Circannual cycling in immunity may also play a role in the seasonality of certain tumors, including bladder and breast cancers (Cohen et al., 1983; Hostmark et al., 1984), melanoma (Swerdlow, 1985), and Hodgkin's Disease (Newell et al., 1985). Thyroid tumors were more frequently diagnosed in the late autumn and winter, with size and proliferation indicators (tumor cells in S-phase [DNA synthesis] or G_2M-phase [nuclear division] of the cell cycle) also showing highest values in autumn and winter, indicating that these tumors were growing faster at one time of the year than another (Akslen & Sothern, 1998).

A circannual rhythm in the incidence of bleeding ulcer has been documented, with a peak in November–February, indicating a greater risk from peptic ulcer

---

[7] The onset of symptoms leading to a diagnosis of syphilis requires several weeks after contact, so a peak in syphilis infection most likely occurs in late summer.

[8] The herpesvirus Epstein-Barr is estimated to infect >90%, cytomegalovirus 50–90% and polyomaviruses ≥70% of adults (Ling et al., 2003).

disease in the winter (Marbella et al., 1988). However, a six-month rhythm with two peaks in the year, one in the late spring and one in late fall, was reported for the incidence of perforated gastric ulcers over a 55-year span in Norway (Svanes et al., 1998). Similarly, six-month rhythms have been documented in humans for the incidence of oral pathology (canker sores, herpes, various lesions), epilepsy incidence, levels of circulating melatonin, and cell density of vasopressin-containing neurons in the SCN, each with peaks in the spring and fall (cf. Cornélissen et al., 2003).

Cardiovascular Disease

Cardiovascular variables, events and symptoms have been shown to be not only predictably more frequent at certain times of the day, but also at certain times of the year. For example, more shock episodes by an implantable cardioverter defibrillator due to malignant ventricular arrhythmias in 233 patients occurred during January and fewest occurred in June (Muller et al., 2003). With regard to liver disease, a peak in variceal hemorrhage in patients with portal hypertension caused by liver cirrhosis was also found in the winter (Sato et al., 1998). The greatest number of spontaneous intracerebral hemorrhages consistently occurred in January and February over a 6-year observation span in Minnesota (Ramirez-Lassepas et al., 1980). Deaths from cerebrovascular disease in England–Wales exhibited highest mortality in the winter, which was comparable to peaks in deaths from related diseases, such as ischemic heart disease, hypertensive disease, pneumonia, and bronchitis (Haberman et al., 1981).

There is wide agreement in the literature that the distribution of cardiovascular mortality over the year shows a seasonal (circannual) pattern with peaks generally found in the winter. While extreme environmental temperatures may override the endogenous mechanisms involved in cardiovascular disease, mortality from acute MI, stroke and sudden cardiac death has almost exclusively been reported to be maximal in the winter and early spring months in the more than 240,000 cases from all over the world (summarized in (Table 11.17–top). In addition, while the exact mechanism is still not known, the rate of sudden infant death syndrome (SIDS) is also highest in the winter months in both the northern and southern hemispheres (Douglas et al., 1996).

One often-overlooked time component in cardiovascular events is the day of the week. For example, while the risk of subarachnoid hemorrhage (brain aneurysm) is increased in winter and spring, there is also an increased risk on Sunday (Feigin et al., 2002). While exogenous and endogenous factors have not been fully elucidated, reports on cardiovascular mortality indicate in over 31,000 cases that these events were more frequent between Saturday and Monday (Table 11.17–bottom). While little information on weekly variations in blood variables exists, fibrinolytic activity, which is lower in individuals who develop ischemic heart disease, has been reported to be reduced between Friday and Monday (Imeson et al., 1987), while cholesterol, hematocrit and RBC levels were reported to reach peak levels between Sunday and Tuesday (Haus et al., 1988). Other medical conditions can also display weekly and yearly rhythms in their occurrence, such as migraine

TABLE 11.17. Infradian timing for human morbidity or mortality from cardiovascular disease.

Diagnosis[a]	Study site	N	Time(s) of maxima	Author	Year
Circannual (365 days)					
CAD	USA (Hawaii)	10110	Winter (March)	Seto et al.	1998
CVD	Canada	15300	December–January	Gordon	1966
CVD	France	5387	February–March	Reinberg et al.	1973
CVD	England	30679	Winter months	Gill et al.	1988
CVD	Japan	308	November–March	Shinkawa et al.	1990
CVD	England	18262	December–March	Tsementzis et al.	1991
IHD	Scotland	65491	November–April	Dunnigan & Harland	1970
MI	USA (TX)	3971	December–March	Schnur	1956
MI	France	3346	January–March	Reinberg et al.	1973
MI	N. Ireland	68683	January–March	Crawford et al.	2003
SCD	USA (MN)	1054	December–February	Beard et al.	1982
SCD	Italy	269	October–January	Pasqualetti et al.	1990
SCD	Japan	264	April	Kawamura et al.	1999
SCD	Germany	24061	December–February	Arntz et al.	2000
Circaseptan (7 days)					
CVD	Sweden	496	Tues.	Johansson et al.	1990
IHD	Scotland	2088	Sat.–Mon.	Imeson et al.	1987
SCD	Romania	86	Mon.	Nicolau et al.	1991
SCD	Canada (MAN)	152	Mon.	Rabkin et al.	1980
SCD	USA (MN)	1054	Sat.	Beard et al.	1982
SCD	Italy	269	Mon.	Pasqualetti et al.	1990
SCD	Germany	2636	Mon.	Willich et al.	1994
SCD	Japan	264	Sun., Sat.	Kawamura et al.	1999
SCD	Germany	24061	Mon.	Arntz et al.	2000
SCD	USA (CA)	400[b]	Mon.	Karnik et al.	2001
OHCA	Austria	1498	Mon.	Gruska et al.	2005

[a] Includes fatal and/or nonfatal hospital admissions for CAD = Coronary Artery Disease; CVD = Cerebrovascular Disease (stroke); IHD = Ischemic Heart Disease; MI = Myocardial Infarction; OHCA = Out of Hospital Cardiac Arrest; SCD = Sudden Cardiac Death.
[b] Patients with end-stage renal disease who suffered cardiac arrest during kidney hemodialysis.

headache attack, which has been shown to occur for frequently on Sunday and in January (Cugini et al., 1990).

The Coagulation System

Parameters of the coagulation system of the blood, such as platelets and fibrinogen, have been reported to differ predictably throughout the year (Kanabrocki et al., 1995, Maes et al., 1995; Hermida et al., 2003a). Fibrinogen showed highest values in late winter and early spring months in hospitalized male military veterans (Kanabrocki et al., 1995) and in more than 1,000 male and female hypertensive patients (Hermida et al., 2003a). Fibrinogen and C-reactive protein, both acute-phase proteins linked to infections associated with cardiovascular disease, also showed highest levels in the winter (February) in elderly subjects (Crawford et al., 2000). Similarly, fibrinogen and factor VII clotting activity were greater in the winter and correlated with self-reported respiratory infections that activate acute-phase protein responses (Woodhouse et al., 1994).

Seasonal variation in cardiovascular risk-related variables may play a role in circannual rhythms reported in symptoms and/or deaths from coronary artery disease, including the incidence of heart attacks (myocardial infarction), sudden cardiac death, and strokes. For example, a higher incidence of deep vein thrombosis is found in winter, a time associated with cold and relative hypercoagulability (Gallerani et al., 2004). Peaks in cardiovascular mortality are generally located in winter or spring in both hemispheres (see Figure 1.9 and Table 11.17–top).

Cholesterol

Total cholesterol levels in serum are usually reported to be higher in the winter months (Paloheimo, 1961; Thomas et al., 1961; Sasaki et al., 1983; Garde et al., 2000). Winter peaks have also been reported for high density and low density lipoprotein cholesterol (HDL, LDL) and triglycerides (Harlap et al., 1982; Gordon et al., 1988) and for apolipoprotein A-II and B (Fager et al., 1982; Mustad et al., 1996), However, there are inconsistencies concerning the location of seasonal peaks for apoproteins that suggest an influence of diet, lifestyle and geographic location (Buxtorf et al., 1988; Kristal-Boneh et al., 1993; Mustad et al., 1996).

Two landmark studies carried out in the late 1950s were among the first to document seasonal changes in cholesterol levels in healthy individuals. One study carried out in Finland measured total cholesterol levels once a month in the morning in 45 policemen and 35 male prison inmates (mean age was 33 years for each group) (Paloheimo, 1961). Individual cholesterol levels varied across the seasons an average of 58 mg/dl (range 19–128 mg/dl) in the policemen and 71 mg/dl (range 39–124 mg/dl) in the prisoners. Higher cholesterol levels were found for both groups in the winter and lower levels in late spring, with overall lower levels found in the prisoners, where diet may have been more stable throughout the year (Figure 11.6–left). After normalizing individual data to percent of mean, both groups displayed a similar seasonal pattern (Figure 11.6–right) that was significant by our cosinor analysis using a 1-year period (not shown).

Another study carried out in the USA monitored cholesterol levels once a month in the morning for a year in 24 healthy male prisoners, ages 22–28 years (Thomas et al., 1961). When subjects were separated into groups with low and high overall cholesterol levels (less than or greater than 225 mg/dl), a highly significant seasonal variation was found in each group by our 1-year cosinor analysis (Figure 11.7–left). Highest cholesterol levels were found in winter and lowest levels in late spring, summer and early autumn. Of interest, the overall seasonal pattern found in cholesterol was strikingly similar to the seasonal pattern in monthly deaths from coronary heart disease during the year of sampling (1958) (Figure 11.7–right). More recently, a cyclic seasonal variation in total cholesterol has also been demonstrated for women in Denmark, where significantly higher values were also found in the winter (Garde et al., 2000).

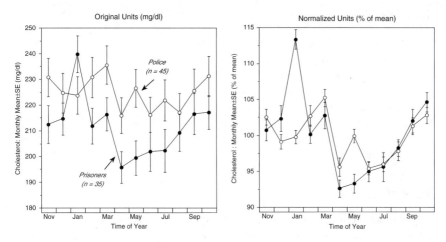

FIGURE 11.6. Circannual variation found in total cholesterol levels sampled once a month in the morning in 45 policemen and 35 male prison inmates in Finland (redrawn from data presented in Paloheimo, 1961). Higher cholesterol levels were found for both groups in the winter months and lower levels in late spring; lower levels were found in the prisoners, where diet may have been more stable throughout the year (*left*). After normalizing individual data to percent of mean, both groups displayed a similar seasonal pattern (*right*) that was significant by our cosinor analysis using a 1-year cosine (not shown).

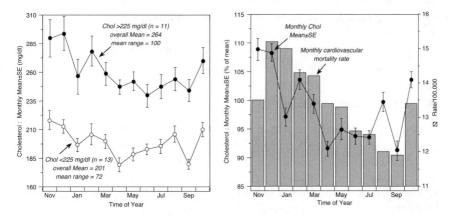

FIGURE 11.7. Circannual variation in total cholesterol levels sampled once a month in the morning in 24 healthy male prisoners in Maryland, USA; men with low or high overall cholesterol levels (less than or greater than 225 mg/dl) showed a highly significant yearly rhythm by our 1-year cosinor analysis (*left*) (redrawn from data presented in Thomas et al., 1961). Cholesterol levels were higher in the winter months and lower levels in late spring, summer and early autumn for each group separately (*left*) and combined (*right*). Of interest, the overall seasonal pattern found in cholesterol was strikingly similar to the seasonal pattern in monthly deaths from coronary heart disease during the same time span (the year 1958) of sampling (*right*).

Respiratory Illness

Respiratory morbidity (e.g., pneumonia, influenza, common cold, tonsillitis, acute bronchitis, and emphysema) and mortality have been reported to peak in winter in both hemispheres (cf. Smolensky et al., 1972; Haberman et al., 1981). There are more visits to the emergency department, especially for children <13 years, in late summer and fall for acute asthma episodes (Silverman et al., 2003). However, death rates were highest in the summer for individuals under age 44 years and in the winter for older subjects, suggesting differences in immune circannual functioning related to age (Weiss, 1990; Fleming et al., 2000).

Mental Disorders

The annual pattern of depression and suicidal behavior generally shows a major peak in the spring, with sometimes a secondary peak in the fall (Lester, 1971; Eastwood & Peacocke, 1976; Meares et al., 1981). This annual pattern may be related to biologically determined factors, such as circannual changes in hormone levels and light quality (e.g., duration and intensity), rather than sociologically determined factors.

A peak in the daily psychiatric inpatient population at a large medical center in Wisconsin was found in February when nearly 12,000 inpatient days over a 2-year span were analyzed (Temte, 1989). At a large hospital in Germany, the total number of anti-depressant drug prescriptions peaked in the spring and fall, in keeping with several reports on the epidemiology of depressive disorders (Koch et al., 2003). Average monthly suicides, hospital admissions for depression, as well as incidence of electroshock therapy, also showed a bimodal distribution, with peaks in the spring and fall in Ontario, Canada (Eastwood & Peacocke, 1976). See Table 4.9 for additional examples of human behaviors or disorders that have been found to correlate with the time of year.

Seasonal Affective Disorder

Season and depression have been mentioned in writings for more than 2,000 years and there has been a recent resurgence of interest in annual depression (cf. Wehr & Rosenthal, 1989). *SAD* (seasonal affective disorder) has been identified as a syndrome characterized by a major depression that recurs at the same time each year (Rosenthal et al, 1984), usually just before and during the winter season (Jacobsen et al., 1987), with a full spontaneous remission during spring and summer. In addition to depression, winter SAD symptoms include an increased duration of sleep (hypersomnia) and latency, fatigue, decreased energy, overeating, carbohydrate craving, weight gain, daytime drowsiness, difficulty concentrating, agitation, anxiety, and headaches (Kasper et al., 1988). Although most SAD individuals do not require hospitalization or absences from the work, their symptoms can affect work performance and public safety. SAD occurs most commonly in women, with onset usually in early adulthood (Attar-Levy, 1997). A seasonal pattern of feeling worse in the winter was found in more than 90% of normal respondents to a survey, indicating that SAD may reflect the extreme effects of a seasonality that commonly affects the general population (Kasper et al., 1989). This agrees with the finding that infradian cycles in mood were found in groups of patients with affective disorders and controls, but the amplitudes in the monthly to seasonal range were greater in the patient group (Eastwood et al., 1985). SAD can also occur in the summer (Wehr et al., 1987) and is more likely to be associated with a decreased appetite and lack of sleep (insomnia) than in winter SAD (Wehr et al., 1991).

SAD is thought to be the result of inadequate light reception and/or processing through the eyes related to decreased duration of sunlight during winter months (Oren et al., 1991). Rates of winter SAD and subsyndromal (less severe) S-SAD were found to be significantly higher at more northern latitudes (Rosen et al., 1990) and the mean prevalence of SAD is two times higher in N. America than Europe (cf. Mersch et al., 1999a).[9] Thus, vulnerability by some individuals to short photoperiods associated with the time of later onset to morning light exposure may be related to depression and winter SAD (Oren et al., 1994). Even though different environmental factors, including photoperiod, influence these incidence rates of SAD, a biological disposition to SAD has been shown in twins in Australia, indicating a significant genetic influence on seasonal mood and behavioral changes (Madden et al., 1996).

Bright artificial light phototherapy, usually in the morning and at a higher intensity than usually present in the home or workplace, has an anti-depressant effect and can often lessen or reverse many of the symptoms of SAD, with the rate of improvement associated with increasing duration and light intensity (cf. Rosenthal et al., 1988). Patients often report sustained benefit from phototherapy over months and years, with more than 80% of individuals typically using phototherapy in January-February-March (Oren et al., 1991). Light exposure in the winter can also be increased informally by spending more time outdoors, increasing house lighting and being near an office window (SAD is also discussed in Chapter 4 on Photoperiodism).

The Menstrual Cycle[10]

Variations associated with the menstrual cycle have been called *circatrigintan* (about 30 days) or *circamensual*. In 1897, Giles described the first changes in basal (early morning) body temperature (BBT) over the course of a woman's menstrual cycle and by 1905, Van de Velde presented evidence that the fluctuation of BBT was linked with the function of the ovaries (cf. Vollman, 1974).

---

[9] Surveys conducted around the world have reported the incidence of winter SAD & S-SAD to be between 1.7% and 5.5% in 9–19 year old children in Washington, DC (Swedo et al., 1996), 4–10% in Maryland (Kasper et al., 1989), 13–20% in college students in northern New England (Low & Feissner, 1998), 3–9% in the Netherlands (Mersch et al., 1999b), 10–18% in northern Finland (Saarijarvi et al., 1999), and <2% of adults in Japan (Ozaki, 1995). Subsyndromal SAD rates were reported to increase from 10.5% in summer to 28.4% in the winter in Antarctica (Palinkus et al., 1996), while cases of SAD have also been reported from India, a tropical country at a latitude of 26°45'N (Srivastava & Sharma, 1998).

[10] Additional discussion and information on sexuality-related female and male rhythms can be found in the chapter on Sexuality and Reproduction.

Many hypothalamic, pituitary and gonadal hormones that regulate the various phases[11] of the menstrual cycle have been shown to vary predictably during the menstrual cycle (Dyrenfurth et al., 1974). The circamensual cycling of these gonadotropins is well known to be important in the maintenance of optimal ovarian function (Yen et al., 1974). It is now known that estrogens reach a peak near mid-cycle, just prior to ovulation, luteinizing hormone and follicle-stimulating hormone peak at ovulation, and progesterone, with its accompanying increase in body temperature, peaks during the luteal phase following ovulation (cf. Figure 7.19 and Table 7.4 in Chapter 7 on Sexuality and Reproduction). Melatonin also shows menstrual cycle variations, with highest values near menstruation and lowest values near ovulation (Wetterberg et al., 1976).

In addition, circamensual rhythms have been described for a large number of neurological, cardiovascular (e.g., blood pressure, heart rate), respiratory, hematologic, enzymatic and hormonal biological activities, and for several aspects of cervical and endometrial physiology (Dyrenfurth et al., 1974; Reinberg & Smolensky, 1974; Bisdee et al., 1989). Acrophase charts presented in four figures in a review by Reinberg & Smolensky (1974) are useful in locating estimated times during a menstrual cycle for peaks in dozens of the aforementioned variables.

Disorders

Several disorders, both physiological and psychological, have been linked to the menstrual cycle (cf. Figure 7.23 in Chapter 7 on Sexuality and Reproduction). Most functional disorders are premenstrual or perimenstrual,[12] including headache, migraine, tiredness, depression, irritability, breast discomfort, and increased symptoms of asthma, epilepsy, rheumatoid arthritis, irritable bowel syndrome, and diabetes (cf. Hejl, 1977; Dalton, 1984; Brush & Goudsmit, 1988; Case & Reid, 1998, 2001; Fox & Davis, 1998). There are also some disorders near midcycle, such as ovulation pain (*Mittelschmerz*) and midcycle migraine (cf. Chapter 4 in Brush & Goudsmit, 1988).

In addition, the days preceding and during menstruation have been associated with an increase in psychopathologic behavior, including psychiatric hospitalizations, suicide attempts, sickness in industry and accidents (O'Conner et al., 1974; Sothern et al., 1993). These disorders may arise from mood-hormonal-metabolic changes due to alterations or an imbalance among rhythms in the various sex and stress hormones over the course of the menstrual cycle. The menstrual cycle itself can also be disturbed (e.g., anovulatory cycles, inadequate luteal phase, impaired follicular development) by hormonal imbalances due to behaviors and disease, such as dieting, weight loss, excess exercise, anorexia and bulimia nervosa (cf. Pirke et al., 1989)

---

[11] The phases of a typical menstrual cycle are: menstrual (days 1–5), follicular (days 6–12), ovulatory (days 13–15), luteal (days 16–23), and premenstrual (days 24–28).

[12] Symptoms of the so-called "premenstrual syndrome" or PMS occur almost exclusively in the second half of the menstrual cycle, increase in severity towards menstruation and are usually relieved by the menstrual flow onset.

Medical Procedures

The outcome of medical procedures may also be influenced by the time during the menstrual cycle when they are performed. For example, the fertility cycle coordination of cellular immunity was shown to play a role in the outcome of breast cancer surgery in both mice and humans. When surgery was performed near the time of estrus (fertile stage) in mice, there were more than *twice* as many cures (i.e., no recurrence of tumors or metastases to the lungs) compared with the infertile stages (Ratajczak et al., 1988).

Extending this finding to pre-menopausal women with breast cancer (BrCa), the 5–12 years disease-free survival was more than *fourfold* better when the surgery was performed near midcycle than near menstruation (Hrushesky et al., 1989a). Overall and recurrence-free survival at 10 years was also significantly higher in 249 women with operable BrCa whose operation was *3–12 days* after their last menstrual period vs. other days of their cycle (46% vs. 16%) (Badwe et al., 1991). Better survival following *luteal* phase surgery was confirmed whether estrogen receptor or progesterone receptor status of the tumor was positive or negative in 112 pre-menopausal women (Cooper et al., 1999). There is also a *decreased* risk of vascular invasion by the tumor following BrCa *luteal* phase surgery (Fentiman, 2002). An overview of published studies confirmed that timing BrCa surgery during the menstrual cycle favorably affects outcome, with 10-year survival for node positive cases undergoing *luteal* phase surgery at 78% vs. 33% following follicular phase surgery (Fentiman et al., 1994; Fentiman, 2002). On the basis of this review, the authors suggested that *"surgeons should consider altering their practice now."* (Fentiman, 2002).

However, it was pointed out that more has to be learned about irregular cycle lengths, the endocrine balance and other tumor dynamics before a stage within the menstrual cycle can be identified with adequate certainty that will more broadly benefit the outcome of timed BrCa biopsy or surgery for both pre- and postmenopausal women (Hrushesky, 1996; Hagen & Hrushesky, 1998). Along these lines, a raised level of progesterone (normally high during the luteal menstrual phase) at the time of surgery, but not estradiol (normally high during the follicular menstrual phase), has been shown to favorably affect the prognosis of BrCa surgery in nearly 500 pre-menopausal women (Badwe et al., 1994; Mohr et al., 1996). Thus, using an independent hormonal measurement of progesterone may be helpful in the classification of correct menstrual cycle status, especially if a woman misreports her last menstrual period or if she is undergoing anovular cycles. Artificially changing the hormonal milieu (e.g., raising progesterone levels) prior to surgery has been suggested as a possibility to improve the prognosis of BrCa surgery outcome, even for postmenopausal women (Fentiman, 2002).

Male Cycles

While less is known about monthly cycles in men, there have been a few reports of infradian variations between three and twenty-five days in human males, including androgen-induced nucleus *C-appendages of neutrophil leukocytes* (Månson, 1965), *17-ketosteroid excretion* (Halberg & Hamburger, 1964; Halberg et al., 1965), *testosterone excretion* (Harkness, 1974), and *grip strength, body weight* and *beard growth* (Kühl et al., 1974; Levine et al., 1974; Sothern, 1974b; Chaykin, 1986). Such low-frequency rhythms in men constitute a potentially important, yet overlooked source of variability that needs further study to determine their role or application in the practice of clinical medicine. Additional discussion about male cycles, including seasonal variations, can be found in Chapter 7 on Sexuality and Reproduction.

Melatonin and Human Health

The hormone melatonin serves as a chemical messenger of darkness in all species studied to date, including birds, fish, reptiles, insects, and mammals, and is an important component of the timing systems for circadian, and possibly infradian, rhythms (Cassone & Natesan, 1997; Arendt, 1997). As such, the pattern of secretion of this chemical signal of darkness helps to induce nighttime behaviors. As mentioned in Chapter 10 on Society, changes in normal melatonin production can be caused by light at night (light pollution) and this can have a profound, yet mostly overlooked, effect on human health. Medical areas that can be affected by changes in melatonin production include endocrine functions associated with puberty, psychiatric illness, stress-related disorders, immune responses, and carcinogenesis (Piccoli et al., 1991).

Darkness and Melatonin

The daily external environmental cycle of a light span followed by a dark span synchronizes circadian rhythms. The cycle provides a signal sent to the pineal gland,[13] which is located near the hypothalamus in the brain[14] of humans (Figure 11.8) and other animals, to produce melatonin[15] during darkness and to inhibit its production during light (cf. Reiter, 1986, 1991). Light exerts an immediate suppression of the secretion of pineal melatonin in mammals, which is normally produced at night when humans are sleeping (Lewy et al., 1980). Humans, however, need a higher intensity of light (1500 lux or more) for melatonin suppression than do other mammals (e.g., <10 lux for the rat).

---

[13] This melatonin-producing gland is sometimes called the "third eye" in lizards and birds since it is located just below the scalp and senses light directly from penetration of light through the scalp and its covering of skin or feathers.

[14] While most, if not all, melatonin reaching peripheral sites is from the pineal, melatonin is also synthesized in the retina, the gastrointestinal tract and the Harderian gland in the orbit of the eye (e.g., of a cat) (cf., Arendt, 1997).

[15] Serotonin in the pineal is converted to melatonin by processes of acetylation and methylation.

FIGURE 11.8. Location of suprachiasmatic nucleii (SCN) and pineal gland, structures in the human brain involved with the body clock. Tryptophan, an amino acid, moves into the pineal from the blood stream, where it is converted to serotonin and melatonin and released back into the bloodstream. The SCN receives light signals from the eye and the pineal gland subsequently receives signals to produce melatonin during darkness and inhibit its production during light.

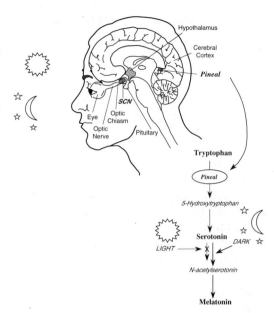

Underscoring the importance of darkness to melatonin production, late bedtimes can significantly phase-delay the melatonin rhythm when compared with early bedtimes, even if the wakeup time is the same (Burgess & Eastman, 2004). Totally blind individuals, who lack light perception, can have free-running rhythms (e.g, 24.9 h rather than precisely 24.0 h) for temperature, cortisol and other body functions and may suffer from severe sleep–wake problems (Miles et al., 1977; Sack & Lewy, 2001). This occurs in spite of the fact that they may be living active lives in a society with an abundance of time cues for entraining circadian rhythms (Sack et al., 1992). A correct low dose of melatonin, however, when administered to blind people near bedtime, has been shown to entrain the free-running rhythms to a circadian period of 24.0 h (Sack et al., 1991; Lockley et al, 2000; Lewy et al., 2002, 2004; Hack et al., 2003). Melatonin has also been shown to be helpful in the treatment of sleep disorders and alleviating the effects of jet lag, especially following time zone advances (east-bound flights) (Skene et al., 1996, 1999; Parry, 2002; Herxheimer & Waterhouse, 2003; Beaumont et al., 2004). These results help to substantiate the relationship between an external light–dark cycle and the role of melatonin in helping to regulate the circadian timing system of humans.

There are indications that under natural conditions, more melatonin is produced in winter (when there is more darkness) than in summer (when there is more light). Support for this view comes from results of a study where individuals were monitored in each of the four seasons in an area of the world (at 70°N in Norway) that receives 24 h of no sunlight in mid-winter and 24 h of total sunlight in the summer. In this study, seasonal differences in melatonin were found, with highest levels in January and lowest levels in June, with a phase-delayed, broader circadian overnight peak in June (Stokkan & Reiter, 1994).

Artificial lighting may counteract the effect of seasonal changes in photoperiod, since no winter-summer differences were found in the durations of nocturnal melatonin secretion, rising morning levels of cortisol or low body temperature during sleep in normal men studied in an urban environment in the eastern USA at 39°N (Wehr et al., 1995). When forehead illumination was monitored by 10 subjects in San Diego, total exposure to bright light (>2,000 lux) did not differ over 24 h between summer and winter and never exceeded more than 90 min, far less than the 3 h–8 h deemed necessary to maximally synchronize human circadian rhythms (Savides et al., 1986). Similarly, illumination measured by a wrist monitor worn by 16 subjects in Montreal found that while bright light exposure was longer in summer than in winter (2.6 h vs. 0.4 h), subjects spent more than 50% of their time in illumination dimmer than 100 lux in both seasons (Hébert et al., 1998). Thus, artificial light and the lack of natural light exposure may interfere with the normal response of organisms (including humans) to seasonal changes in the photoperiod that otherwise would result in changes in the physiology of hormone secretions and other biological functions (Reiter, 1986).

Sexuality

In view of the profound photoperiodic-related effects of melatonin on the reproductive physiology of nonhuman mammals, such as to suppress or influence pubertal development and seasonal reproduction, some effects by melatonin on sexuality are likely to exist in humans (Reiter, 1991). There are indications that the mediation of melatonin by light of certain intensities can shorten and regularize menstrual cycles among women with long or irregular cycles (Dewan, 1967; Lin et al., 1990) or reduce luteal phase dysphoric disorder (depression) (Parry et al., 1993, 1997). For example, menstrual cycles were shortened when a 100-watt bulb was left on for the entire night following days 13–17 of the menstrual cycle, but dim red placebo light produced no difference in menstrual cycle length (Lin et al., 1990).

While there have been suggestions that timed illumination might contribute to rhythm-based contraceptive methods if ovulation could be regularized, it follows that inappropriately timed night lights could also inadvertently alter menstrual cycles and contribute to fertility problems or irregularities. For example, some menstrual irregularity and ovulatory variability could be related to the reduction in daily variations in illumination that no longer resemble the natural photoperiod that most humans experience from day to day as a result of living in modern society (Kripke & Gregg, 1990) and from season to season (Hébert et al., 1988). Much more needs to be learned about the effect of light on menstrual cycles, menarche and menopause.

Males are also affected by light, and again some of these effects may involve melatonin. Men with a low sperm count (oligozoospermia, where sperm count is <10–15 million/ml), show a rise in melatonin *before* darkness and *elevated* levels of this hormone at night compared with normally fertile men, suggesting that light and dark may play a role in certain aspects of male sexuality (Karasek et al., 1990).

There is some evidence for an interaction between melatonin, excess light, and sexual development. Continuous light accelerated mammary gland development in rats such that at 141 days of age, 58% of females in LL, but *none* of the animals in LD, had gross evidence of lactation and well-developed mammary tissue, both of which are not normally observed in virgin animals (Anderson et al., 2000). Polycystic ovaries also occurred in 8 out of 9 rats after 75 days in LL, but *none* of the 6 rats kept in LD (Baldissera et al., 1991). Twenty-four-hour lighting was proposed as a simple experimental model for the study of ovarian physiopathology in rats in that the time to polycystic development could be shortened. It is tempting to speculate that there may be some corollary here between rats developing sexually at an early age when kept in continuous light and human females reaching menarche (and possibly males reaching puberty) at an earlier and earlier age when living in the light-contaminated society of today.

Immune Function

Short days associated with seasonal changes in photoperiod have been shown to bolster immune function in both laboratory and field studies of animals (cf. Nelson et al., 1995). There is substantial evidence that seasonal changes in immune function of the host, rather than seasonal changes in a parasite or pathogen, are responsible for seasonal fluctuations in disease and death rates in mammals. It is therefore possible that changes in photoperiods associated with the seasons could affect human immune function and clinical disease (Nelson & Blom, 1994).

Since melatonin mediates many of the immunological effects of photoperiod, it is possible that unnatural changes in lighting that alter normal timing and levels of melatonin could have profound effects on the etiology (cause) and progression of diseases in humans and other animals. For example, there is a parallel between the marked increase in cancer incidence and the increase in light exposure that took place over the 100 years between 1890 and 1990, especially in the developed countries (Kerenyi et al., 1990). While other risk factors have also increased, including air pollution, smoking, diet, alcohol, occupational exposures and stress, increased light exposure can decrease melatonin production by the pineal gland, thereby compromising immune function and diminishing its anti-cancer effect (Tamarkin et al., 1985; Kerenyi et al., 1990; Nelson et al., 1995; Pang et al., 1998; Blask et al., 1999).

Melatonin has been shown to be a direct free-radical scavenger and have indirect antioxidant properties via the stimulation of antioxidative enzymes (cf. Tan et al., 2003). Melatonin is also an immunomodulatory agent that can augment immune responses (cf. Vijayalaxmi et al., 2003). In addition, its ubiquitous presence in the animal kingdom, melatonin is a molecule that is widely produced throughout the plant kingdom, from bacteria to edible plants (Hattori et al., 1995; Reiter et al., 1995). This means that melatonin consumed in foodstuffs or via medicinal herbs, such as feverfew (*Tanacetum parthenium*), St. John's Wort (*Hypericum perforatum*), and Huang-qin (*Scutellaria baicalensis*) (Murch et al., 1997), can result in measurable levels in the circulation. In addition to endogenously produced melatonin, dietary melatonin present in comparatively high amounts in certain vegetables, fruits, seeds, rice, wheat and medicinal herbs may therefore increase the actions of melatonin mentioned above (Reiter, 1998; Reiter & Tan, 2002). The use of exogenous (dietary or pharmacological) melatonin has clinical implications for the areas of sleep, jet lag, immune function and certain diseases or other processes where excessive free radical generation occurs (Reiter, 2003). For example, melatonin has been shown to play a role in forestalling tissue damage in aging and in a wide variety of disease states, including cancer (Vijayalaxmi, et al., 2003; Reiter, et al., 2003; Reiter, 2004) and dementias (e.g., Alzheimer's Disease, Parkinson's Disease) (Reiter et al., 2004) and stroke (Reiter et al., 2005). Melatonin added to therapy regimens also reduces toxicity from a broad spectrum of drugs (Reiter et al., 2002), while synergistically enhancing their beneficial effects (Reiter et al., 2002).

Light, Melatonin and Cancer

Too much light at the wrong time can suppress the production of melatonin and lead to a physiologic imbalance of biological rhythms, a so-called chronodisruption, that may have adverse health effects (Erren et al., 2003). Pre-clinical evidence is mounting that there is a connection with this phenomenon and cancer development. There have been several reports on the enhancement of tumor growth when rats and mice were kept in continuous light (LL) or long days, in which a longer exposure to light suppresses melatonin production. In one study, the size of a meth-A sarcoma in mice standardized to a LD12:12 regimen was significantly *reduced* when melatonin was administered 10 h into the 12-h light span, thereby artificially extending exposure to melatonin by several hours beyond that normally produced during the dark. However, when melatonin was given in the middle of the 12-h dark span, the tumors actually *increased* in size. Compared with controls two weeks after treatment, tumors of mice treated with melatonin during mid-dark were ~1.4 cm^2 larger, while tumors of those mice treated during late light were ~2.6 cm^2 smaller (Sanchez, 1993).

When mice housed under either short (LD8:16) or long (LD16:8) days were injected with the chemical carcinogen DMBA, 89% of the long-day mice, but not a single short-day mouse, developed tumors (Nelson & Blom, 1994). When rats were housed under LL, liver cancers were induced by the chemical diethylnitrosamine in 95% of the animals, compared with 72% of the rats housed under LD12:12 and macroscopic nodules on the liver surface were larger and more numerous in LL than in LD (van den Heiligenberg et al., 1999). In this study, melatonin levels in the LL group were fourfold *lower* than in the LD group, leading to a light-induced promoting effect on liver carcinogenesis. The growth of a mouse colon adenocarcinoma was also more effectively inhibited by chemical treatment (with difluoromethylornithine) when mice were kept on a short-day schedule of LD6:18 (more darkness than light) than in LD12:12 (equal portions of light and dark) (Waldrop et al., 1989a). It was later found that the colons of the mice housed in more darkness (LD8:16) displayed a significant circadian rhythm in DNA synthesis, while the mice housed under LD12:12 did not (Waldrop et al, 1989b).

Light and dark can also have a modifying effect upon cancer rates in second generation rats depending on the LD regimen given their mothers during gestation (Beniashvili et al., 2001). When pregnant females were treated with a cancer-causing chemical (N-nitrosoethylurea), offspring of mothers housed under LL (which suppressed melatonin) showed a significant *increase* in total tumors of the nervous system and kidney and a *shorter* lifespan than offspring of mothers housed under LD12:12. On the other hand, offspring of mothers housed in DD (where melatonin was not suppressed) showed a 2.4-fold *decrease* in total tumors and lived *longer* than the LD12:12 offspring.

Light Leaks at Night

Small amounts of light during a "normal" dark span can alter melatonin production in rats. This was shown when growth of a liver tumor was compared in rats housed under LD12:12 (Group 1), LD24:0 (Group 2), or in a light-contaminated dark phase LD12:12 (0.2 lux of light during D) (Group 3) (Dauchy et al., 1997, 1999). Prior to tumor implantation, rats in Group 1 showed a normal surge of melatonin in mid-D, rats in Group 2 showed no melatonin rhythm due to no darkness, while rats in Group 3 showed a suppressed melatonin production due to the low light present at night. Linoleic acid, an important tumor growth stimulant, was also increased in Groups 2 and 3. When compared with rats in Group 1, liver tumor growth rates were not only 105% *greater* in Group 2, but also 81% *greater* in Group 3. Levels of two tumor growth-promoting compounds (linoleic acid and 13-hydroxyoctadecadienoic acid) that are normally suppressed by the nocturnal surge of melatonin were also greatly increased in Groups 2 and 3. This study showed that minimal light contamination during the normal dark phase inhibited melatonin secretion and increased the rate of tumor growth, lipid uptake and metabolism. The authors cautioned about "light leaks" in animal rooms during the dark phase possibly contaminating experimental outcomes.

By extrapolation, "light leaks" during the night may have as yet unrecognized consequences to human health and development due to the suppressive effects of light at unnatural times on melatonin production. Several papers have discussed the influence of nighttime light (i.e., exposure during graveyard shift work or sleeping in a bright bedroom, but not if a light was briefly turned on during interrupted sleep) on melatonin production as a possible risk factor for human breast cancer (Stevens et al., 1992; Stevens & Davis, 1996; Davis et al., 2001; Schernhammer et al., 2001). Excess light has also been cited as a strong risk factor for the development of leukemia and other cancers in children (Erren, 2005).

Additional biological evidence has linked constant bright light with melatonin suppression and breast cancer. In a pre-clinical study, an increased growth rate of tissue-isolated human breast cancer xenografts in nude rats was noted: tumor sizes increased seven times *faster* in rats after transfer to LL than in rats remaining in LD12:12 (Blask et al., 2003). In totally blind humans, most of which do not show a bright light-induced reduction in melatonin, but not for severely visually impaired individuals, cancer rates are *reduced*. In one study, totally blind women who had no light perception at all via the optic pathway were *half* as likely to develop breast cancer as sighted women (Hahn, 1991, 1998). In another study of totally blind men and women, the rate for all cancers was *reduced* by about 30% from sighted individuals (Feychting et al., 1998).

A symposium held on 2002 reported many results and research perspectives of light at night, the endocrine system (including the pineal and melatonin) and cancer incidence in humans (Cologne Light Symposium, 2002). The literature now convincingly supports the hypothesis that melatonin suppression at night may be linked to the increased rates of certain cancers in the developed world and that light at night has become a public health issue (cf. Pauley, 2004) (see additional discussion of light pollution in Chapter 10 on Society). There are several potential mechanisms associated with light at night that might interact to increase the risk of cancer and other diseases. These are that light at night reduces melatonin production, disrupts estrogen signaling, disrupts immune functions, and may disrupt circadian gene function in the SCN, thereby altering mammary tissue cell-cycle regulation (cf. Stevens, 2005). Circadian disruption by ill-timed lighting during early developmental stages, such as *in utero* and adolescence, may also affect the risk of cancer later in life. While more studies need to include men, there is growing and consistent preclinical, as well as observational evidence in humans, for a positive association between melatonin suppression (using shift work as a surrogate for light exposure at night) and breast, and possibly colorectal, cancer risk (cf. Schernhammer & Schulmeister, 2004).

When to Sample?

Knowledge of rhythms has a diagnostic and therapeutic role in medicine. This was recognized in the 1960s by one of the modern pioneers in biological rhythms and clocks research, the psychobiologist, Curt Paul Richter (1894–1988): "–*human beings and most animals harbor not only one but many timing devices that are no less impressive in their way than those [clocks] made by man . . . . These timing devices merit great interest . . . because of the light they may throw on the functioning of various organs of the body . . . .*" (Richter, 1965).

Because nearly every variable measured in humans has been found to be circadian rhythmic, it follows that the body is in a quasi-predictable state of flux throughout the day and night. Knowing when to sample a rhythmic variable can be cost-effective, both in time and in procuring greater quantities[16] and/or meaningful results. This concept is called *chronodiagnosis*.

---

[16] Several years ago an immunologist told me (RBS) that a procedure was being developed to energize T-cell lymphocytes using ultraviolet light and return them to the same individual, but they were having trouble harvesting enough cells to make the procedure worthwhile. When asked what time the blood samples were obtained, he said usually in the morning when the clinical lab was fully staffed. I suggested that they try to shift this sampling time to midnight or later, when a peak in circulating T-cells has been reported (Lévi et al., 1988b), but was told that the clinical lab would not be able to process the cells at that time. I suggested that someone on the night staff could obtain the blood sample at night and store it in a refrigerator to be available when the personnel in the clinical lab arrived in the morning. I don't know if this was ever tried.

Diagnosing Normal Levels

Just as outdoor aquatic streams can yield different quantities of fish at different times of day, so too does the blood stream yield different quantities of substances at different times of day (Figure 11.9). For example, "fishing" for white blood cells (WBCs) in the bloodstream during the day may "harvest" only a few of the "prey," while many, many more WBCs and their subtypes (e.g., neutrophils, T-cell lymphocytes, eosinophils, basophils, etc.) can be obtained when the sampling is at night, just as many, many more of certain kinds of fish (e.g., the smelt running at night mentioned in Chapter 8 on the Outdoor Environment) can also be netted in a stream of water at night.

This analogy can also be applied to the correct timing of sampling for other substances (or fish!) that may be greater in number at other times during the day. Because of rhythms, therefore, knowing when to sample is likely to influence the

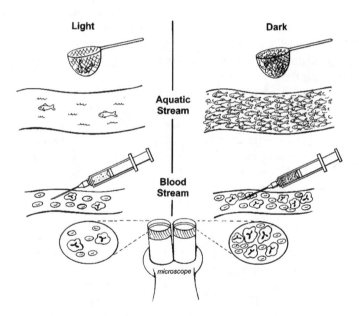

FIGURE 11.9. "Fishing" in two streams: one for blood cells and the other for certain fish can yield more of each at one time of day than another. Fishing in water or sampling in a blood stream during the day "harvests" only a few of the "prey," while many, many more WBCs and their subtypes can be obtained when the sampling is at night, just as many, many more of certain kinds of fish (e.g., the Great Lakes Rainbow smelt running at night mentioned in Chapter 8 on the Outdoor Environment) can also be netted at night. Of course, this analogy can be applied to sampling for other substances that may be highest during the day, as well as fish with diurnal activity patterns. Similarly, time of year can influence the harvest–more WBCs are usually obtained in late summer and early fall, while the smelt "run" occurs in the spring.

outcome of medical and other procedures.[17] In the circadian domain, certain values and functions may be *predictably higher* at one time of the day and they can be *predictably lower* 12 h later and vice versa. However, as mentioned at the beginning of this chapter, these "silent" variations over 24 h are often ignored by the medical community since they are usually within what is considered "normal" upper and lower limits (Figure 11.2).

In diurnally active humans, lower values in the morning and higher values in the evening reflect the usual circadian rhythm in circulating WBCs. Thus, WBCs in blood sampled at 10:00 h may be less than 6,000 cell/cm^3, while 12 h later at 22:00 h total WBCs in a new blood sample may be 50% or more greater in number (e.g., 9,000 cells/cm^3) in the same individual (see Figure 11.10–upper left panel). This range of change is a completely normal daily event in health, since human bone marrow (BM) furiously produces new blood cells[18] in a circadian-rhythmic pattern every day (Smaaland & Sothern, 1994). As a result, there are more BM cells in DNA S-phase at midday than at other times, with an increase in cells in mitosis in the evening and the accompanying release of new daughter cells into the bloodstream at night. Since most of these cells live for only 6 h to 14 h after entering the bloodstream (Sothern & Roitman-Johnson, 2001), there are usually more WBCs in the circulation during the night (during sleep) than during the day in diurnally active humans, while the reverse is true for nocturnally active rats and mice.

A large, predictable change over the course of a day in WBCs has been shown to occur with comparable timing in groups of diurnally active, clinically healthy individuals studied in several locations around the world (Halberg et al., 1977a). On an individual basis, the circadian patterns in total WBC counts and each subset of cells (neutrophils, lymphocytes, monocytes, eosinophils, basophils) appear to be reproducible for an adult man studied 10 years apart (Figure 11.10). WBC rhythms for a group of ten men can be seen in Figure 11.11, where, after normalizing individual blood profiles to percent of mean, distinct, high amplitude circadian patterns are clearly present (cf. Kanabrocki et al., 1990a).

Diagnosing Infectious Agents

In addition to rhythms in the production of WBCs and other hormones by the body, an agent to be diagnosed may itself be cycling and be time-dependent (e.g., activity levels of a parasite or virus). For example, studies of humans in

---

[17] One report even suggested that results from a genetic screening test for Tay-Sachs Disease could be influenced by the time of day of blood sampling (Peleg et al., 1993). When comparing 62 pairs of morning vs. afternoon WBC samples from the same individuals, sampling between 15:00–18:00 h yielded a more reliable genotype assignment (5.2% were inconclusive in the 08:30–11:00 h sample), an important finding for diagnosing parents as carriers of this fatal disease.

[18] The BM produces, on average, two million red cells, two million platelets, and 700,000 granulocytes (WBCs, etc) *every second*, resulting in an estimated more than 170 *billion* each of platelets and red cells and more than 60 *billion* white cells every 24 h (cf., Figure 2 in Sothern & Roitman-Johnson, 2001).

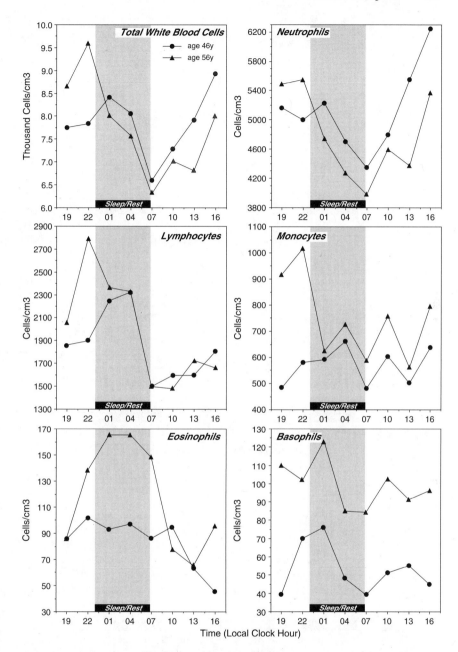

FIGURE 11.10. Reproducibility of circadian patterns in white blood cells in blood sampled every 3 h for 24 h in a healthy man on two occasions ten years apart (RBS, age 46 and 56 years, participated in the MCAP 24-h studies in 1993 and 2003). Most values are lower during the day and higher in the evening and overnight during rest/sleep (Sothern, unpublished).

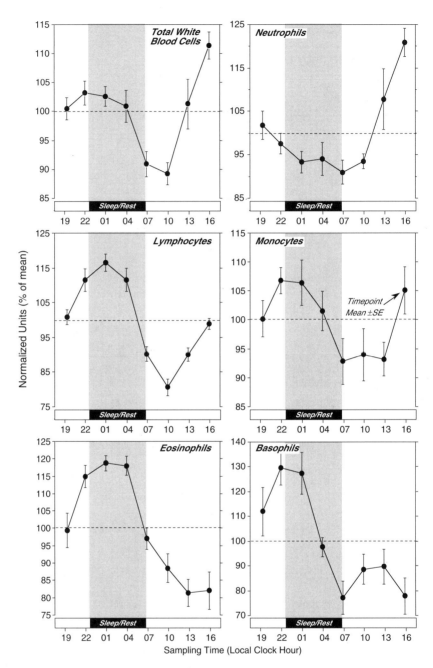

FIGURE 11.11. Circadian patterns for total white blood cells (WBC) and levels of subsets (neutrophils, lymphocytes, eosinophils, monocytes, basophils) in a group of ten men participating in the MCAP study in May 1993 (blood sampled every 3 h for 24 h) (modified from Figure 5 in Sothern & Roitman-Johnson, 2001). Data normalized to percent of mean for each subject before computing timepoint means. Shaded areas = sleep and/or rest.

Kenya (Gatika et al., 1994), Tanzania (Simonsen et al., 1997) and Brazil (Fontes et al., 2000) have shown a nocturnal circadian periodicity in the number of microfilaria (parasitic worms) in the blood. Highest counts in the circulation of the microfilaria, *Wuchereria bancrofti*, were always found near midnight and virtually no nematodes were found during the day. However, the count for the microfilaria, *Loa Loa*, is maximal near midday and minimal at midnight (Hawking, 1962). These studies indicate the importance of taking sampling time into consideration in sampling for certain infections (e.g., blood should be sampled between 22:00 h and 03:00 h for *W. bancrofti* detection) (Fontes et al., 2000).

Diagnosing Abnormal Levels

Knowing the phases of rhythms can also be used to diagnose abnormal levels of a substance if the normal pattern has been previously documented, in which case a single sample that is properly timed (best phase) may be adequate to indicate if a substance is too high or low. In a clinical setting, over- or under-production of cortisol by the adrenal (a small gland that sits on top of the kidneys and secretes steroids) may be diagnosed with a single sample of blood at the right time, since adrenal corticosteroids in healthy humans have been shown to be at very high levels in the morning hours and at very low levels at night (Bartter et al., 1962). A very low cortisol level in the morning can be used to diagnose adrenal *hypo*function (Addison's Disease), since such a patient has *low* cortisol levels over the 24 h, while a very high level of cortisol at night may indicate adrenal *hyper*function (Cushing's Syndrome), since these patients have *high* levels of cortisol throughout the 24 h (Bartter, 1974).

It has also been suggested that only two blood sampling times within the usual hospital working day, one at 08:00 h (near the usual daily peak) and the other at 16:00 h, are sufficient to determine that a cancer patient in good general condition is displaying the usual circadian pattern in cortisol (Mormont et al., 1998). If cortisol is higher at 08:00 h than at 16:00 h, as is the finding for the normal circadian pattern in cortisol, the patient can thus be treated on a pre-designed timed-treatment protocol. Otherwise, a full circadian sampling profile is warranted to test for an altered circadian pattern in order to adjust the timing of the treatment schedule to best suit that particular individual.

Using Rhythm Characteristics in Diagnosis

The alteration of characteristics (waveform, amplitude, acrophase) of a circadian rhythm can also indicate an abnormality. For example, a high body temperature early in the day or late at night may indicate an illness of bacterial or viral origin, respectively (Hejl, 1977), while changes in the sleep–wake cycle may accompany clinical depression or seasonal affective disorder (SAD).[19]

---

[19] A network of interacting causal loops, including disturbances in behavioral state, neurotransmitters, predisposing factors and circadian rhythms, may play a role in rhythm disturbances often associated with depression (cf., Rosenwasser & Wirz-Justice, 1997).

Changes in the circadian temperature patterns of a breast, not only by elevated levels but more *ultradian* oscillations, may indicate the presence of a small tumor (Simpson, 1987).

Symptoms of asthma, perhaps the most rhythmic of all diseases (Smolensky et al., 1986b), are usually worse at night (cf. Martin & Banks-Schlegel, 1998). While lung capacity in general is reduced during sleep, the normal 24-h amplitude for peak expiratory flow (PEF) in healthy subjects is about 5% of the 24-h mean, but is much greater in asthmatics (up to 25% of the 24 h mean for mild cases and $\geq$50% in severe cases), since PEF levels in asthmatics drop precipitously overnight (Smolensky & Halberg, 1977; Smolensky et al., 1986a). When values for PEF fall more than 10% during sleep from their daytime levels, persons are identified as "night dippers" and often require additional medical treatment.

The 24-h pattern in human blood pressure (BP) is usually characterized by low values during sleep, an early morning increase and a plateau while awake and active. In addition to the overall elevation of BP in hypertension, alterations of the 24-h rhythm characteristics can also be useful in assessing cardiovascular risk. For example, hypertensives that display the typical nocturnal BP decrease are termed "dippers," while those displaying a blunted or absent nocturnal BP decrease are termed "non-dippers" (White, 2000).[20] Using a mean difference of 5–10 mmHg between daytime and nighttime BP to categorize a "dipper," hypertensives who were "non-dippers" were eight times more likely to experience a stroke than dippers (O'Brien et al., 1988). It is thought that a flattened 24-h BP profile due to nondipping can result in a longer duration of exposure to high blood pressure, resulting in increased cardiovascular complications over the long term (Pickering, 1990; Verdecchia et al., 1991).

In patients with suspected high blood pressure or with uncomplicated essential hypertension, the circadian amplitude may also increase prior to the clinical diagnosis of hypertension, with blood pressure being above acceptable limits only at certain times of the day (so-called "amplitude-hypertension") (Halberg et al., 1984; Shinagawa et al., 2001). An excessive circadian amplitude in blood pressure in hypertensives with elevated blood pressure (MESOR-hypertension) has been identified as a high risk factor for stroke and kidney damage (nephropathy), irrespective of the presence of other known risk factors, such as age and additional disease conditions (Otsuka et al., 1997, 1999). This condition of circadian blood pressure overswinging, called CHAT (Circadian Hyper-Amplitude-Tension) (Halberg et al., 2002), is defined as a consistent increase in 24-h amplitude above a mapped threshold for a peer group, can indicate a greater risk for stroke than the 24-h average blood pressure.

In addition, blood pressure in secondary hypertension (e.g., due to renal disease or diabetes) may show the lack of the usual nocturnal decline or even an increase during sleep (when blood pressure normally decreases) (Portaluppi et al., 1991).

---

[20] In addition to the non-dipping phenomenon in BP often found in individuals with essential or secondary hypertension, non-dipping has also been observed in normotensive subjects (Cugini et al., 1998).

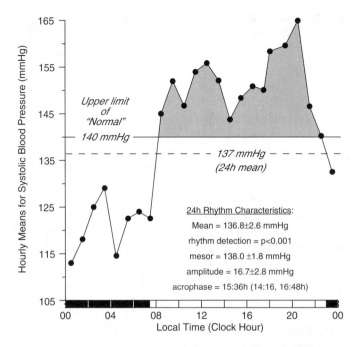

FIGURE 11.12. Example of circadian variation in systolic blood pressure (SBP) of a woman (age 45 years) who was "amplitude-hypertensive," but not "24-h mean" hypertensive. Subject (under treatment for psychophysiological problems, but not hypertension) wore ambulatory BP monitor March 21–22, 2000. A 24-h mean of 136.8 mmHg indicated normal overall blood pressure, while an amplitude (A) of 16.8 mmHg above a rhythm-adjusted mean (mesor, M) of 138 mmHg indicated a diagnosis of amplitude-hypertension (M + A = 155 mmHg). A Hyperbaric Index (HBI) can also be computed by multiplying "mmHg × hours" to represent an integrated measure of total excess pressure (shaded area) based on all readings above a set upper limit of normal (e.g., 140 mmHg for SBP). Dark bars = sleep (Sothern, unpublished).

Conversely, atrial natriuretic peptide (ANP), a hormone involved with BP control, failed to show the usual nocturnal increase in patients with hypertension due to chronic renal failure, suggesting a mechanism of autonomic dysfunction between rhythms in ANP and BP (Portaluppi et al., 1992).

Another feature that can be derived from a 24-h pattern in blood pressure is the calculation of the total excess pressure above a certain limit. Thus, to better define hypertension, total excess blood pressure over 24 h can be computed as a *hyperbaric index* (HBI), which is the area of excess above the upper limit of normal, whether that is defined by a single number (e.g., 140 mmHg for SBP) (Figure 11.12) or as time-varying upper limits of normal (from chronodesms mentioned above) (Figure 11.3) (Hermida et al., 2000, 2002a). The HBI was shown to be a more accurate indicator of gestational hypertension (sensitivity >96%) than the 24-h mean (sensitivity <70%), yet the 24-h mean is still the most common approach to diagnosing hypertension based on ambulatory monitoring (Hermida et al., 2000, 2003d).

Hours of Changing Resistance

Does it matter what time of day a medical procedure (e.g., an allergy test, an injection, etc.) is performed? In a view known as *reactive homeostasis*, a stimulus is thought to provoke similar effects at any time of the day (Figure 11.13–left). The chronobiologic view, however, holds that a periodic system responds differently to the same stimulus at different times of the day (Figure 11.13–right). This has been called *predictive homeostasis* (Moore-Ede, 1986). Many experiments have been performed that overwhelmingly disprove the former and confirm the latter (Haus et al., 1974c), where drugs or compounds can be tolerated at one time without any obvious side-effect, while 12 h earlier or later the same dose can cause much harm or even death due to the body's lowered resistance to the toxic effects of the drug. This concept of *chronotoxicity* is also known as the *"hours of diminished resistance"* (Halberg, 1960) or *"hours of changing responsiveness or susceptibility"* (Reinberg, 1967). The differences in drug effectiveness or toxicity can be brought about by circadian rhythms in rates of absorption, metabolism, excretion and/or interactions with hormones (Moore Ede, 1973).

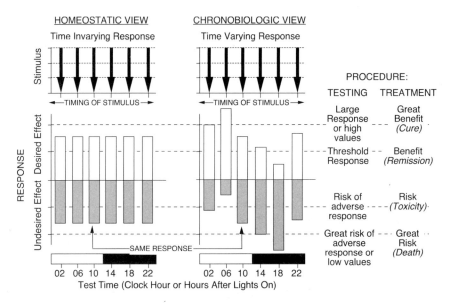

FIGURE 11.13. Scheme of homeostatic vs. chronobiologic response to the same procedure at different times of the day. In the homeostatic view [known as "reactive" homeostasis] (*left*), it is thought that a reaction or result will be the same at all times of the day. In some cases, this is a balance between a desired effect and an undesired effect, such as that from a drug. In reality, however, the result varies throughout the day [known as "predictive" homeostasis] (*right*) and reflects an underlying rhythm (chronobiologic view). Thus, the same procedure (e.g., dose of a drug or sampling time) can result in a greater result at one time or a lesser result at another (Modified from Figure 7 in Sothern & Roitman-Johnson, 2001).

Early Pre-Clinical Findings

Chronotoxicity has been recognized since the 1950s, when the number of convulsions and deaths in mice differed dramatically after a one-minute exposure to identical auditory stimulation at different times of the day (Halberg et al., 1955). For mice confined to a metal laundry tub, 49% convulsed and died following exposure early in the daily dark (active) span to loud ringing doorbells attached to the sides of a tub, while only 6% convulsed and 2% died if exposed to the same sound early in the daily light (resting) span. Large differences in survival also were observed following exposure of mice to *E. coli* endotoxin at different times of day (Halberg & Stephens, 1958; Halberg et al., 1960). Fewer than 10% of the mice died following exposure in the middle of the daily dark span, while more than 70% mortality occurred following the same exposure 8–12 h earlier in the daily light span.[21] Also in the 1950s, a susceptibility rhythm of mice to a toxic dose of ethanol was observed, with more mice succumbing following exposure near the end of the daily light span and the beginning of the daily dark span (Haus & Halberg, 1959).

In a dramatic display of host chronotoxicity, an all or none phenomenon in deaths from whole body x-irradiation was found when different groups of male mice were exposed to the same dose of radiation (550 r) at one of six circadian stages 4 h apart (Haus et al., 1974a). Eight days after exposure, all mice irradiated in the middle of the daily dark (active) span were *dead*, while all mice exposed late in the light (resting) span were *alive*. Mean survival time was longest following irradiation during late light or early dark, while shortest times were observed when mice were treated in the middle or late dark span.

A similarly timed rhythm was also demonstrated in the susceptibility of nucleated bone marrow cells to whole body irradiation in mice, leading the authors to suggest that survival of hematopoietic stem cells (bone marrow cells giving rise to new red and white blood cells) may be the underlying cause of differences in the survival or mortality of mice irradiated at different circadian times (Haus et al., 1974b). Decades later, circadian rhythms in DNA S-phase and mitoses in the bone marrow of rats and mice were indeed shown, confirming the hypothesis (cf. Smaaland & Sothern, 1994). Chronotoxicity has been reported for a wide range of agents and species, including mortality of mice from insulin, ethanol, amphetamine, strychnine, nicotine, phenobarbital, and anti-cancer drugs (cf. Scheving et al., 1986).

Time-Related Responses to Anti-Cancer Drugs

Examples of circadian rhythms in survival of healthy mice that received the same potentially lethal dose of nine different anti-cancer drugs at different times around-the-clock can be seen in Figure 11.14. A significant circadian variation

---

[21] When healthy men were exposed to *Salmonella abortus equi* endotoxin at either 09:00 h or 19:00 h and followed for 11 h, the evening treatment resulted in *double* the increase in body temperature and plasma ACTH and cortisol levels compared with responses to morning treatment (Pollmacher et al., 1996).

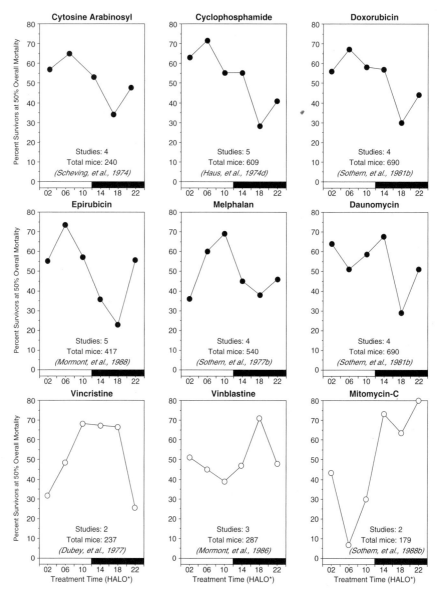

FIGURE 11.14. Examples of chronotoxicity: survival of mice from a single dose of an anticancer drug given at different circadian stages (HALO = hours after lights on). Separate subgroups of healthy young male or female mice were treated at 5 or 6 different times in relation to a 12-h L –12-h D schedule with one of nine different drugs. Percent survival rate computed when 50% of mice in each study had died. Dark bar = daily dark span. Solid dots = more survivors during rest (light) span, open circles = more survivors during activity (dark) span (data summarized from the literature cited in each panel)

was found for each drug, with peak survival rates depending upon the mode of action of each drug. For example, drugs that target the DNA S-phase of the cell cycle were better tolerated during the resting (light) portion of the 24 h, which corresponds to a time of relatively low bone marrow proliferative activity. Chronotoxicity has now been studied for more than 30 different anti-cancer drugs (see below). One report even showed that the toxicity rhythm persisted during continuous light in mice treated with doxorubicin (Sothern et al., 1988a), making it likely that this circadian phenomenon is endogenous and perhaps characteristic of other anti-cancer drugs.

Stage of Rhythm vs. Time of Day

Pre-clinical observations of experimental results are often used to guide human studies. However, in extrapolating findings from mice to humans, one must remember that humans and rodents live on opposite light/dark-activity schedules, with diurnally active humans awake by day when nocturnally active rats and mice sleep. Thus, a peak at 20:00 h in mouse serum corticosterone and a peak at 08:00 h in human plasma cortisol, while seemingly 12 h apart with regards to external local time, both occur 2 h after awakening/activity onset and therefore are not differently timed in relation to the internal body clock set by the rest/activity schedule of each organism. This was also the case for bone marrow DNA S-phase, where peak activity was observed at 23:08 h for mice and 13:04 h for humans, which was about 5–6 h after the onset of activity for both species (cf. Sothern, 1995).

However, it is quite likely that the majority of treatment doses and schedules suggested for humans that are derived from results on rats or mice are obtained from testing performed during the resting span of the rodents' 24-h cycle (e.g. during the first half of the investigator's daily activity (i.e., 06:00–12:00 h), rather than in the first half of the animal's activity span (18:00–24:00 h). Thus, only one stage of the circadian system, which may or may not be an appropriate test time in terms of "best" or "worst" effects, is usually tested in rodents. If results are used to guide treatment of humans at the same clock hour (i.e., in the morning), but without adjustment for the sleep/wake differences between the two species, humans will be treated at a completely opposite circadian stage (early activity rather than early rest), possibly with suboptimal or unexpected results. Thus, it is the *"stage of rhythm"* and not *"time of day"* which would be expected to ultimately result in a successful chronotherapy when extrapolating from preclinical findings to humans (Sothern, 1995).

Varying Positive or Negative Effects

It is now clear that administration of certain treatments without regard to physiological rhythms will result in varying positive or negative effects that can themselves be rhythmic (Figure 11.13–left). *So what to do?* With regard to treatments, a reduction in doses at all times in order to reduce toxicity may also reduce

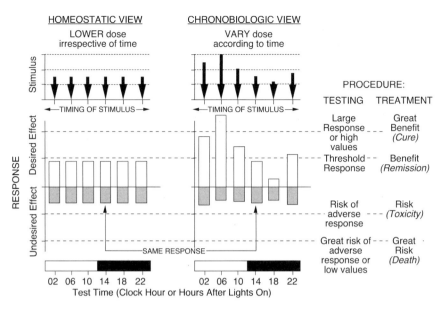

FIGURE 11.15. Schematic showing doses or procedures can be adjusted in order to reduce toxicity. In the homeostatic view (*left*), doses are reduced at all times in order to avoid undue toxicity, but this can reduce the efficacy of the drug. In the chronobiologic view (*right*), doses may be reduced at some times or increased at others to take advantage of the host resistance rhythm. Proper timing can result in greater efficacy of a drug, along with a reduction in undesired side-effects (Modified from Figure 7 in Sothern & Roitman-Johnson, 2001).

efficacy (Figure 11.15–left), whereas adjusting the doses according to the stage of the susceptibility rhythm may reach the desired effect while minimizing the risk from toxicity (Figure 11.15–right). This has far-reaching implications, not only for selecting the best timing of anti-cancer procedures and medications, but for screening procedures, immunotherapy, allergy shots, and many other treatments for a wide variety of illnesses and conditions, as well.

For example, the size of the erythema (redness) resulting from skin testing with a standard dose of histamine can vary more than 100% in the same individual depending upon the time of day (Figure 11.16). Similar circadian patterns have been described for reactions to house dust and grass pollen (Lee et al., 1977). Likewise, responsiveness of bronchial airways following aerosol challenge with various allergens is greatest following evening or overnight tests and less when the same tests are conducted near midday or afternoon (cf. Jarjour, 1999), yet allergy testing is generally not scheduled with this in mind. In addition, the response (e.g., increase in plasma insulin levels) to an oral glucose tolerance test is greater in the morning (Jarrett et al., 1972; Zimmet et al., 1974), and the hypoglycemic action (glucose disappearing rate) in response to an intravenous insulin tolerance test is also greater in the morning than in the afternoon (Gibson & Jarrett, 1972; Sensi & Capani, 1976; Schulz et al., 1983; Morgan et al., 1999).

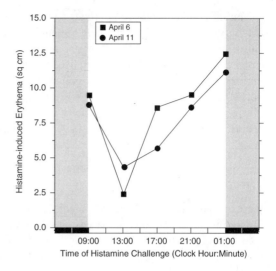

FIGURE 11.16. Circadian pattern in cutaneous response (area of reddening) after histamine challenge at five different times of the day on two occasions in a nonallergic man (RBS, age 28 years). Histamine solution (0.5 cc of 1/500,000 parts) injected subcutaneously into forearm every 4 h during waking-only, with erythema traced after 15 min and size of reddened area computed. Shaded areas = sleep span (Modified from Figure 8 in Sothern & Roitman-Johnson, 2001).

Each of these responses has implications when screening individuals for diabetes (e.g., afternoon or evening testing may uncover defects in glucose metabolism; Jarrett, 1972).

Recovery from a standardized exercise, as gauged by blood glucose levels returning to baseline, has been shown to occur faster in the morning than in the late afternoon (and in winter vs. summer), which has implications not only for conditioned athletes, but also for patients with diabetes mellitus (Weydahl & Sothern, 1995, 1997). In addition, it has been reported that persons should be tested for glaucoma as early in the morning as possible, since intraocular pressure is likely to be highest in the morning shortly after awakening, even in healthy subjects (Drance, 1960; Shue et al., 1994; Pointer, 1997; Qureshi et al., 1999). Intraocular pressure has also been reported to be higher in the winter for ocular hypertensive patients (Qureshi et al., 1999).

Timing Treatment: Chronotherapy

To summarize the information just presented: *an individual's response to medications or other stimuli can be time-dependent and shows a circadian rhythm.* Therefore, information about rhythms can be used to maximize positive and cost-effective outcomes of interventions. By analogy, just as a dentist uses a lead blanket to shield parts of the body from unnecessary radiation during a dental x-ray, so too can *time* be used to shield parts of the body from undesired effects of medical treatments. *Chronotherapy* attempts to time treatment according to biological rhythms in order to achieve the goal of *maximizing* the desired effects and *minimizing* the undesired effects. Chronotherapy includes the best timing of drug treatments, medical and surgical procedures, as well as performance and exercise scheduling.

Timing is all the more important, since many therapies, especially powerful anti-cancer agents, are accompanied by undesired and even life-threatening toxicities, since cytostatic (cell killing) and other effects are exerted in all tissues, diseased or not. Thus, healthy tissue undergoing rapid proliferation, such as the bone marrow, hair and gut, is often damaged by chemotherapy and leads to dose reductions and/or delays in treatment. Therefore, doses of chemotherapy represent an inherent compromise between toxicity in the cancer (the desired target) and the host (the undesired target) while an attempt is made to deliver the maximal dose possible.

Three Times a Day?

For centuries, traditional Chinese medicine has incorporated the concept of timing into treatment for a wide variety of ailments (Wu, 1982). The site of a treatment by acupuncture or moxibustion (heat) or a dose of herbs and other medications will differ depending on the natural cycles of the patient, which may involve the time of day, day of the week, day within the menstrual cycle, phase of the Moon and/or season.

In the current fast-paced world of Western medicine, therapies are, for the most part, administered without regard to biological rhythms and usually at the convenience of the health caregivers or on a schedule that assures patient compliance (e.g., three times a day with meals). This was once called the *"Stumpfsinn des dreimal täglich [stupidity of the three times a day]"* drug administration approach by Arthur Jores (1901–1982), a German internist and endocrinologist who was an early pioneer in exploring the usefulness of timing of treatments according to 24-h rhythms (Jores, 1935). Nearly 70 years ago, he noted that some *"therapies"* work better in the evening than the morning and vice versa and due to this, *"great significance"* should be prescribed accordingly by *"today's Doctor."*

Constant Dosing

Continuous fixed-rate dosing, either by the same dose being administered at equal intervals or by continuous infusion over the 24 h, is thought to eliminate peaks and valleys in blood levels of drugs being administered. In addition, continuous intravenous (i.v.) infusion of some anti-cancer drugs lasting 8 h or more can result in less toxicity and increased antitumor efficacy than a single bolus administration (cf. Thiberville et al., 1994). However, continuous fixed-rate infusion can also present potential dangers. In post-surgical gynecologic oncology patients receiving a continuous fixed-rate infusion of morphine sulfate around-the-clock to control pain, the dose had to be reduced in 42% due to oversedation, 54% developed problems with lung expansion (atelecstasis) and most patients were up at night from inadequate control of their pain (cf. Auvil-Novak, 1997).

Due to the rhythmic time structure of drug pharmacokinetics in the host, it has been documented that constant dosing over 24 h does *not* result in constant blood levels and the times of highs and lows will differ depending upon the compound being infused. For example, constant i.v. dosing of the anti-cancer drugs *5- fluorouracil* (5FU) (Lévi et al., 1988b; Petit et al., 1988; Thiberville et al., 1994; Bressolle et al., 1999), *doxorubicin* (Lévi et al., 1986, Lévi, 1994) or *vindesine* (Focan et al., 1989) resulted in substantial differences in blood levels of these drugs between day and night. The continuous infusion of 5FU at a flat rate over 5 days in seven cancer patients resulted in a nearly 100% difference in plasma levels of this anti-cancer drug between a peak level at 01:00 h and a trough level at 13:00 h (Petit et al., 1988). Flat-rate infusion of doxorubicin in nine women with breast cancer also showed plasma concentrations two-fold higher during the day between 12:00 and 22:00 h, when greatest hematological toxicity is also found, than at night (Sqalli et al., 1988).

Similarly, a circadian rhythm in the anti-coagulant effect of *heparin*, a blood thinner naturally produced by the liver, infused at a constant rate over 24 h was demonstrated in six patients with blood clots (venous thromboembolism) (Decousus et al., 1985). Variations between night and day ranged from 40% to 60% in some standard laboratory anti-coagulation tests, with maxima occurring at night. In patients receiving a constant peridural infusion of *bupivacaine* over 36 h for control of pain following hip prosthesis surgery, plasma levels showed a clear circadian variation, with higher levels in the afternoon (16:40 h) and maximal clearance rates in the morning (06:23 h) (Bruguerolle et al., 1988). Single oral doses of a nonsteroidal anti-inflammatory drug (*ketoprofen*) at 07:00 h resulted in a plasma concentration twice as high than the same dose at 01:00 h (Ollagnier et al., 1987). Substantial within-day variations in plasma *zidovudine* (AZT) levels following steady-state oral dosing every 6 h in human immunodeficiency virus (HIV) infection were observed in three patients, with higher trough levels at night (Sothern et al., 1990).

A constant rate infusion of glucose in healthy volunteers resulted in circulating *glucose* levels during sleep that were 15% above those found during the day (Van Cauter et al., 1989). Continuous infusion of glucose resulted in a higher glucose turnover in the morning between 06:00 h and 09:00 h in both healthy and diabetic subjects (cf. Sensi & Capani, 1976). In another 24-h study of a fixed *glucose & insulin* infusion rate administered to subjects with normal and impaired glucose tolerance, insulin sensitivity (gauged by an insulin/glucose ratio) was higher during the day between 12:00 h and 18:00 h (Schulz et al., 1983). The antibiotic *erythromycin* produced the highest peak, the largest area under the curve and greatest anti-microbial activity when administered around 12:00 h in 24 men treated with 250 mg doses every 6 h for 3 days (DiSanto et al., 1975). This implies that the same dose delivered around midnight would result in less bioavailability and hence less anti-microbial activity. In a pre-clinical study, a circadian rhythm was also shown in theophylline disposition in dogs receiving a constant-rate intravenous infusion of *aminophylline*. Maximum serum concentrations found during the day between 12:00–18:00 h were 25% higher, on average, than the troughs found at night between 00:00–06:00 h (Rackley et al., 1988). This result has important ramifications for the timing of treatment of asthmatic patients.

With regards to ulcer treatment, a constant infusion of an H_2-receptor antagonist drug (*ranitidine*) on 12 ulcer patients was less effective in controlling gastric acidity at night, suggesting a day-night dose-response in gastric acid suppression in spite of a reported steady-state attainment (White et al., 1991). When ranitidine was delivered to six ulcer patients in a sinusoidal infusion, with a higher rate around 19:30 h, gastric acidity was controlled during the times of normal peak acid output (21:00 h to 00:00 h) (Sanders et al., 1991).

Constant dosing at different times of the year may also result in varying outcomes. For example, following the same dose of *heparin* in different months of the year, activated clotting times showed a circannual rhythm with a wintertime peak in patients after coronary bypass surgery (Hodoglugil et al., 2001). Except for a large difference in clotting time due to the use of heparin, the yearly pattern was similar to the pre-surgery pattern in the same patients before heparin administration.

Rhythm-Dependent Effects of Some Drugs

Many drugs have been shown to produce less toxicity, better disease control and more cures at some times of day than others (cf. Reinberg & Smolensky, 1983; Lemmer, 1989; Labrecque & Bélanger, 1991; Redfern & Lemmer, 1997). The concept of *chronopharmacology* encompasses the time of drug administration and the body's response according to the temporal structure of the organism receiving it. The concept of *chronopharmacokinetics* adds time of day as a variable that influences the pharmacokinetics of a drug. This includes rhythmic changes in drug disposition (absorption, distribution, metabolism, elimination) that result from an interaction in processes at the molecular and membrane levels (pharmacokinetics) and rhythms in the desired and undesired effects (susceptibility) (cf. Reinberg & Smolensky, 1982; Bruguerolle, 1998). Drugs with rhythm-dependent effects include analgesics, anti-coagulants, corticosteroids, melatonin, psychobiotics, and anti-hypertensive, anti-ulcer and anti-cancer medications (cf. Tables 1–4 in Reinberg & Smolensky, 1982; Tables 1–2 in Bruguerolle, 1998; Table 3 in Smolensky & Haus, 2001).

Administering Chronotherapy

So how is chronotherapy administered? To achieve proper timing, drugs can be designed to be delivered either by mouth where they are delivered at the best time immediately or a time-delayed coating can allow the drug to be absorbed later, such as during sleep when a drug is taken at bedtime. Implantable or portable drug delivery pumps can be programmed to intravenously deliver increasing and decreasing doses of drugs around-the-clock.

Several dosing schemes shown in Figure 11.17 demonstrate how the same total 24-h dose can be delivered all at once at a certain time or split into equal or unequal doses delivered at different times. For example, a daily low-dose aspirin tablet has been recommended to reduce the risk of cardiovascular morbidity, but the best time of day to take the tablet is usually overlooked.

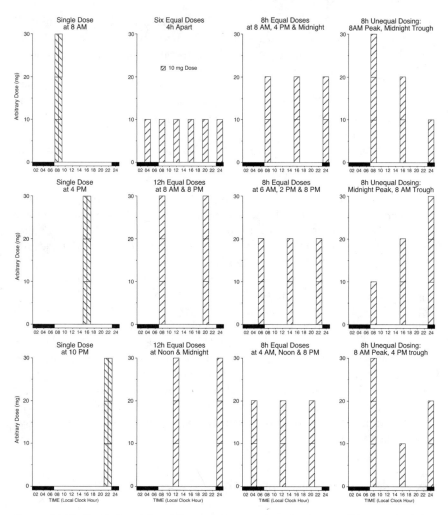

FIGURE 11.17. Examples of chronotherapy schemes, where the same total dose can be delivered all at once or split into two or more treatment times using the same or varying amounts of the drug. Each treatment schedule delivers the same total 24-h dose (60 mg/24 h), yet correct timing may improve efficacy and decrease side-effects. Other schemes can also be devised (e.g., 2/3 of total dose of cortisone delivered at 08:00 h, 1/3 of total dose at 15:00 h and none at night), while continuous infusions can increase and decrease doses centered around times shown in the right column.

However, an aspirin tablet taken once a day has been shown to be more effective in reducing the risk of hypertension when taken before bedtime than at awakening or midday. This was noted by an aspirin-induced reduction in blood pressure following daily bedtime administration in healthy normotensive subjects who took a 500 mg or 100 mg tablet (Hermida et al, 1997b), in untreated hypertensive patients who took a 100 mg tablet (Hermida et al., 2003b) and in pregnant women at high risk of developing preeclampsia (pregnancy-induced hypertension) who took a 100 mg tablet (Hermida et al., 1997c, 2003c). In these studies, there was no significant change in blood pressure when aspirin was taken upon awakening.

The time of an injection and dose can also be specified to achieve maximum benefit. For example, the duration of action of the same dose of an epidurally administered local anesthetic for pain (ropivacaine) during the first stage of labor was shown to be 28% longer when delivered in the afternoon compared with evening administration (Debon et al., 2002). The temporal variation in duration of action of an opioid analgesic (sufentanil) for control of labor pain showed a 12-h rhythm, with acrophases (peaks) near noon and midnight that were 30% higher than minimum values (Debon et al., 2004). One of the 12-h acrophases (12:47 h) coincides with the maximal analgesia duration when the authors reanalyzed the data from their earlier study with ropivacaine (acrophase at 11:58 h). Time-dependencies of several other anesthetics have been recently reviewed, indicating that the timing of drug therapy and time-dependent variations in pain are highly relevant to the practice of pain management (Chassard & Bruguerolle, 2004).

In addition, the best timing of blood pressure medications can depend upon their mode of action. For example, an ACE (angiotension converting enzyme) inhibitor anti-hypertensive medication resulted in better 24 h blood pressure control when taken at 09:00 h than at 21:00 h (Morgan et al., 1997). However, a calcium channel blocker ingested once a day at bedtime in a drug-delay formulation resulted in better overall 24-h BP control (reduced the rate of BP rise upon awakening, normalized daytime BP and modulated the reduction of BP during sleep), than the same dose taken in the morning (White et al., 1995; Smith et al., 2001). The best time for administration of newer long-acting anti-hypertensive drugs awaits study (White, 2003).

Examples of Applied Chronotherapy

Some areas in which proper timing (chronotherapy) of medications or other procedures has been implicated in an improved diagnosis, reduced toxicity and/or an increase in efficacy are listed in Table 11.18 and more conditions, such as Parkinson's Disease (Bruguerolle & Simon, 2002) are under investigation.

Several comprehensive reviews are available on chronotherapeutics applicable to a variety of medical areas (cf. Lemmer, 1989; Reinberg & Lévi, 1990b; Ritschel & Forusz, 1994; Smolensky & Haus, 2001), including *asthma* (Smolensky, 1989; Smolensky & D'Alonzo, 1997) and *allergic rhinitis* (Storms, 2004), *anesthesia* (Bourdallé-Badie et al., 1990a; Chassard & Bruguerolle, 2004), *cancer* (Bjarnason & Hrushesky, 1994; Bjarnason, 1995; Mormont & Levi, 2003), *cardiovascular disease* (Lemmer, 1986; Lemmer & Portaluppi, 1997; Anwar & White, 1998; Glasser, 1999; Poirier et al., 1999; Bridges & Woods, 2001; Smolensky & Haus, 2001; White & LaRocca, 2002), *mental disorders* (Duncan & Wehr, 1988; Nagayama, 1999), *pain and inflammation* (Bruguerolle, 1989; Labrecque &

TABLE 11.18. Some areas in clinical medicine where rhythms in symptoms and/or treatments have been studied.

Adrenal response	Diet, Meal-Timing, Nutrition	Nocturia (bed wetting)
Alcohol	Epilepsy	Pain
Allergies	Gallbladder	Psychiatric disorders
Anesthesiology	Glaucoma	SAD, light–dark effects
Arthritis	Gout	Sexual dysfunction
Asthma, pulmonary disease	HIV infection	Sleep disorders
Blood disorders	Hypertension	Transplants
Cancer	Jet lag, shift work	Ulcer, gastric disorders
Cardiovascular disease	Menstrual cycle	

Reinberg, 1989; Auvil-Novak, 1997, 1999), and *ulcer disease* (Moore & Smolensky, 1990).

In addition, chronotherapy has been applied to *testosterone replacement therapy* (Place, 1988; Place & Nichols, 1991),[22] *malaria* (Cambie et al., 1991; Landau et al., 1991, 1992, 1993), *circadian rhythm sleep disorders* (Campbell et al., 1999; Wyatt, 2004), and *haemodialysis* after renal transplantation (McCormick et al., 2004). Another review concluded that evening administration of vitamin D analogues, such as calcitrol, for the treatment of *secondary hyperthyroidism* in both children and adults results in better control, fewer complications, and allows for higher doses to be used (cf. Sanchez, 2004).

Chronotherapy in cardiovascular disease continues to be an area of intensive study, especially with regards to nighttime dosing of delayed-release drug formulations (Sica et al., 2004; Weber et al., 2004). This treatment protocol has resulted in a decrease in nocturnal myocardial ischemia (Frishman et al., 1999), the prevention of nocturnal hypertension and better overall blood pressure control (cf. Hermida & Smolensky, 2004), as well as better control of the associated progression of diabetic renal disease (Moorthi et al., 2004). Thrombolytic therapy timed late in the day has also been reported to be more efficacious in patients with acute myocardial infarction (Reisin et al., 2004).

For illustrative purposes, the following sections on asthma and cancer will be used to go into more detail about the application of chronotherapy.

Asthma

Circadian differences have been found in the effectiveness of inhaled or injected bronchodilaters, including terbutaline, glucocorticoids and theophylline and other medications (cf. Smolensky et al., 1986b, 1990; Smolensky & D'Alonzo, 1997). It was concluded that proper evaluation of any anti-asthmatic drug should include at least three different drug administration times throughout the 24 h, since a drug

---

[22] With daily usage for over a year, a transdermal patch applied in the morning to the scrotum for hormone replacement therapy in hypogonadal males provided the normal 24-h pattern in testosterone levels (peak level in the morning) and sustained physiologic effects (Place & Nichols, 1991).

that might exhibit only minimal potency when given in the morning may prove more efficacious when administered in the evening.

Studies to date, even of sustained-release preparations, have usually shown that evening, as opposed to morning, treatments are more effective in controlling the symptoms of asthma (Smolensky & D'Alonzo, 1988; D'Alonzo et al., 1990; Milgrom et al., 1990; Steinijans et al., 1990). It was also shown that while on a conventional treatment schedule of theophylline every 12 h and beta-agonist aerosols every 4–6 h, 70% of the use of a steroid inhaler for immediate relief from acute asthma occurred between 20:00 h and 08:00 h (peak use was between 04:00–08:00 h and least was between 16:00–20:00 h), indicating that an equal-interval, equal-dose medication schedule did not effectively control the nocturnal symptoms of asthma (Brown et al., 1990a).

Long-term (3–11 years) dosing with corticoids only in the morning has been shown to be effective in the control of nocturnal asthma and results in a normal cortisol circadian rhythm with no adrenal suppression (Reinberg et al., 1990c). In an attempt to prolong the action of administered steroids, studies have shown that corticosteroid administered in the morning and mid-afternoon (Pincus et al., 1996) or as a single administration between 15:00–17:30 h (Pincus et al., 1997; Vianna & Martin, 1998) also effectively controls airway inflammation without the undue side-effects when steroids are taken at night.

Cancer–Animal Studies

As a prerequisite before chronotherapy of cancer patients, preclinical testing using murine models (rats and mice) with and without tumors are essential. There is now an overwhelming body of experimental evidence in murine models that unequivocally demonstrates that lethal and organ-specific nonlethal chemotherapeutic toxicity, as well as cancer response and/or cure, depends to some extent upon the circadian timing of drug administration (cf. Lévi et al., 1988c; Mormont et al., 1989; Granda & Lévi, 2002).

The earliest demonstration of chronotoxicity from an anti-cancer drug was shown following the administration of cytosine arabinoside (ara-C) at different times of the day (Cardoso et al., 1970). The mortality of healthy male mice from a single dose of ara-C given at one of six different circadian stages of an LD 12:12 schedule followed a circadian pattern. Lowest mortality was found when mice were injected early in the light/resting span, while highest mortality was found late in the light span and early in the dark span, a time we now know of maximal bone marrow DNA S-phase synthesis in mice (cf. Smaaland & Sothern, 1994). This finding was later confirmed in experiments on mice that received repeated injections of ara-C over five to six days, always at the same time of day (Scheving et al., 1974). In further studies, when tumor-bearing animals were treated around-the-clock with ara-C in eight equal doses or with eight doses varying in a sinusoidal manner, such that the peak dose was in the middle of the daily light/resting span, an increased tolerance to ara-C was demonstrated for the sinusoidal treatment (Haus et al., 1972). The survival rate was more than *double* that of the conventionally treated animals even though the *same* total 24-h dose was delivered on each treatment schedule.

In an extension of this work, the location of the peak dose was moved around-the-clock to occur at each of the eight injection times (Haus et al., 1974c). A circadian rhythm gauged by both survival and cures was found, with maximal responses in the second half of the daily light/resting span. Thus, the timing of the highest and lowest doses, and not the pattern of dosing, was the critical factor in survival. When compared with the reference treatment schedule of equal doses, survival of non-leukemic mice was nearly *tripled* (from 25% to 70%) when the peak sinusoidal dose was given during the daily light span and was at least equal or slightly increased when placed in the daily dark/activity span. Cures of leukemic mice were either *doubled* (from 10% to 20%) when the peak dose was placed in the light/resting span (the "right time"), or *halved* (to less than 5%) when peak doses were placed in the second half of the daily dark/activity span (the "wrong time"). Thus, improperly timed chronotherapy was actually less effective than the homeostatic reference schedule in the attempt to cure leukemic mice. These experiments in mice showed that the same number of injections (eight) and total 24-h dose increased tolerance and efficacy with properly timed unequal dose "sinusoidal" chronotherapy when compared with eight equal-dose "homeostatic" injections.

Single (bolus) treatments with an anti-cancer drug also can result in a more favorable outcome at one circadian stage than another. This was shown when tumor regression rates occurred fastest when tumor-bearing rats were treated in the middle of their daily resting (light) span with a single anti-cancer drug (doxorubicin [ADR]) (Figure 11.18) (Good et al., 1977). The anti-tumor effect of this drug was greatest at the same time that it was shown to be least toxic to healthy tissue, as gauged by mortality in healthy mice that was *least* during mid-L and

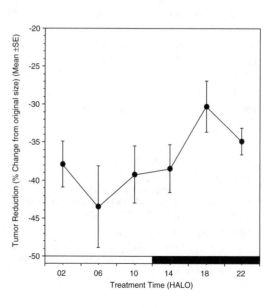

FIGURE 11.18. Differences in tumor regression rate observed 4d after anti-cancer treatment at six different circadian stages. Percent change in tumor size after a single injection of doxorubicin (6 mg/kg i.p.) in male LOU rats bearing a 26d-old immunocytoma (N = 7–11/ timepoint, 57 total). Light/Dark bar = lights on/off for 12 h (see Good et al., 1977; Halberg et al., 1977a).

greatest during mid-D (Sothern et al., 1977a) (see Figure 11.14). Interestingly, tumors in mid-L that regressed fastest also regrew the fastest and vice-versa for tumors treated in mid-D (cf. Figures 6c,d in Halberg et al., 1977b). While the lives of these tumor-bearing rats were significantly prolonged by this single drug treatment, there were ultimately no long-term cures.

Extending the timed treatment of a tumor from a single drug to two drugs (ADR and cisplatin [DDP]), more complete remissions and longer survival were observed for rats bearing an immunocytoma when treated with each drug at its best time: ADR late in the daily light (resting span) and DDP during the daily activity (dark) span. Thus, when ADR was always given at 10HALO (Hours After Lights On) and DDP at one of six circadian stages, a circadian rhythm in survivors was found such that a difference of nearly 100% was found when half of the animals overall had eventually died: 61% were alive when treated with DDP during mid-D (18HALO) vs. only 31% when DDP was delivered in early L (02HALO) (Sothern et al., 1989).[23] The results from this and a second study, where timed-treatments resulted in long-term cures of rats bearing a solid tumor, were subsequently used as a guide in timing human cancer chronotherapy trials at the University of Minnesota (see below).

Cancer–Human Trials

After adjusting to circadian stage (from nocturnal rodents to diurnal humans), the preclinical results just mentioned served as the basis for the first chronotherapeutic schedules with these drugs in clinical trials in humans at the University of Minnesota (Hrushesky, 1985). In a cross-over study, cumulative bone marrow toxicity was reported to be less, the clearance of DDP was greater and there were fewer episodes of nausea when the drugs were delivered at circadian stages comparable to the best times (circadian stages) found in the rats (Schedule A: early morning for ADR, late afternoon for DDP) than when drugs were delivered with the opposite schedule (Schedule B: early morning for DDP, late afternoon for ADR) (Hrushesky et al., 1989b,c). Part of the reduction in toxicity was hypothesized to be due, at least in part, to avoiding times of high bone marrow (BM) and rectal mucosa (RM) toxicity due to their morning or midday peak in DNA S-phase activity (Figure 11.19), which is targeted by doxorubicin.[24]

---

[23] When testing ADR and a different drug (docetaxal) around the clock, the treatment was shown to be more effective (less toxicity, more complete remissions) when both drugs were given simultaneously near the middle of the daily resting (light) span to mice bearing a mammary tumor (Granda et al., 2001). With regard to developing a chronotherapy of human breast cancer, this timing would correspond to the middle of the night in humans.

[24] Of interest, DNA synthesis in cancer cells in 24 patients with non-Hodgkin's lymphoma showed an opposite circadian pattern, with a peak at night (Smaaland et al., 1993), suggesting that timing S-phase specific medications at night might "shield in time" the healthy RM and BM cells from undesired harmful effects from the treatment, while possibly better targeting tumor cells (see Figure 11.19).

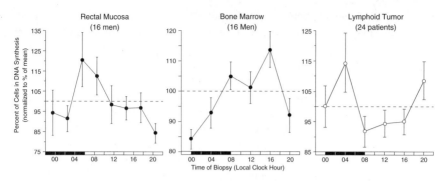

FIGURE 11.19. The continuous, extremely high proliferative capacity of the bone marrow is rivaled by the intestinal mucosa, each of which has been shown to exhibit circadian rhythms in humans. Cells in DNA S-phase show a peak in the morning for intestinal (rectal) mucosa in 16 healthy men and in the early afternoon for bone marrow in 14 healthy men. Of interest, DNA synthesis in cancer cells in 24 patients with non-Hodgkin's lymphoma showed an opposite circadian pattern, with a peak at night (redrawn from Buchi et al., 1991; Smaaland et al., 1991; Smaaland et al., 1993). S-phase specific anticancer drugs damage any cells in S-phase, tumor or healthy, suggesting that timing these medications at night might "shield in time" many of these healthy cells from undesired harmful effects from the treatment, while possibly better targeting tumor cells.

In further studies where each patient remained on a fixed schedule of ADR and DDP (Schedule A or B), there were half as many reductions in dose, one fourth as many delays in treatment, half as many complications (transfusions, infections, bleeding) and greater survival after 5 years for patients receiving Schedule A (50%) than for patients treated on Schedule B (11%). Schedule B-treated patients, however, still performed better than treatment time-unspecified controls (0% survival at 2.5 years). The preclinical and clinical findings on chronotherapy with ADR and DDP have been summarized (cf. Hrushesky et al, 1989a,b).

Several other clinical trials have reported that proper timing of chemotherapy can improve clinical outcome in cancer patients (Hrushesky & Bjarnason, 1993a,b; Bjarnason, 1995; Lévi et al., 1994; Borner, 1999; Giacchetti et al., 1999, 2000; Curé et al., 2002). For example, when patients with various solid tumors received the same combination of drugs (methotrexate or 5-FU, vinblastine, cyclophosphamide) at equal doses, but on two schedules initiated 12 h apart, significant differences in response rates were found (53% vs. 19% major responses) (cf. Focan, 2002a). Patients with ovarian cancer best tolerated another anti-cancer drug, THP-adriamycin, like ADR (doxorubicin), when it was given at 06:00 h (Lévi et al., 1990). These authors also studied 124 patients with lung cancer who received DDP at 18:00 h and etoposide daily either at 06:00 h or 18:00 h and found less hematological toxicity with etoposide at 06:00 h (Lévi & Reinberg, 1990).

For 118 children with acute lymphoblastic leukemia who had achieved complete remission, the 5-year disease-free survival rate was better depending upon the timing of their maintenance chemotherapy with two anti-cancer drugs (6-mercaptopurine and methotrexate): 80% with evening treatment versus 40% morning treatment (Rivard et al., 1985). Eight years later, a re-analysis of the survival data revealed that the risk of relapsing was still 2.6 times greater for the morning treatment schedule (Rivard et al., 1993).

Using optimal chrono-modulated dosing regimens, a lessening of side effects can allow for full, effective doses of anti-cancer agents to be delivered without reductions and/or delays in the next scheduled treatment. For example, drug doses could be significantly increased (dose intensity) and toxicity was reduced when 5-FU and leucovorin were delivered with a peak rate around 21:00–22:00 h on a sinusoidal schedule via a programmable drug pump in patients with metastatic adenocarcinoma (Bjarnason et al., 1993) or rectal cancer (Parulekar et al., 2004). Similarly, when treating patients with colorectal cancer, less toxicity and higher dose intensities were achieved on a chronomodulated schedule of oxaliplatin, 5-FU and leucovorin vs. a fixed (flat) infusion rate (Lévi et al., 1994) or 5-FU and l-folinic acid (Curé et al., 2002). Chronotherapy with 5-FU, folinic acid and carboplatin also showed a high efficacy rate and minor side effects in patients with metastatic cancer confined to the liver (Shimonov et al., 2005).

Chronotherapy has been tested with success in Phase II chemotherapy trials of non-small cell lung and metastatic colorectal cancers (Focan et al., 1995, 1997). A combination of radiation therapy and chrono-modulated 5-FU infusion therapy was also effective in downsizing locally advanced, unresectable rectal carcinoma, thereby making surgery possible (Marsh et al., 1996), and also increased survival in patients with pancreatic adenocarcinoma (Keene et al., 2005). Due to these studies, many ongoing clinical cancer trials beginning in 1998 in Europe have been designed to include a chronotherapy arm (treatment group) (Dogliotti et al., 1998; Coudert et al., 2003).

For more information in the area of cancer and chronotherapy, see a special issue entitled "Cancer Chronotherapeutics" that was published in the journal *Chronobiology International* (Lévi, ed., 2002), and contains articles dealing with experimental and clinical prerequisites, chronotherapy, and economic aspects and quality of life.

Cellular Clocks and Chronotherapy

Several reviews have tied together concepts of cellular clocks, rest-activity cycles, rhythms in healthy and tumor tissues, chronopharmacology of anti-cancer agents and chronobiologic optimization in several clinical cancer trials (Lévi, 1999, 2001; Canaple et al., 2003; Mormont & Lévi, 2003; Lis et al., 2003). Many cancers may be characterized by a lack of circadian control and the development of drugs and other avenues of intervention to modify circadian timekeeping in these cells may improve the treatment of malignancies and other clock-associated disorders (Canaple et al., 2003; Hastings et al., 2003; Lowrey & Takahashi, 2004). For example, disturbances in expression of three *period* genes have been found in breast cancerous cells in women when compared with nearby noncancerous cells, suggesting that while disruption of the normal circadian clock may promote carcinogenesis, it may also provide a molecular basis to guide chronotherapy (Chen et al., 2005). With advances in molecular chronobiology and a better understanding of the role of clock genes[25] and the SCN in cellular coordination and the molecular mechanisms of both normal and malignant tissues,[26] it has been predicted that chronotherapy will become an established therapeutic method that will exceed the limitations and outcomes of current treatments for cancer and many additional diseases (Eriguchi et al., 2003; Lis et al., 2003), including sleep and psychiatric disorders, shift work, jetlag, and aging (cf. Cermakian & Boivin, 2004).

Time-Indicating Genes

It has been suggested that most of the genome is expressed rhythmically across all body tissues, with different phase relationships among circadian output protein in different tissues (Roenneberg & Merrow, 2003). Such a circadian program in cycling would allow for a precise estimation of gene expression in order to better understand circadian biology and its influences on body functions, such as sleep, digestion, and the susceptibility of developing cancer, as well as individual responses to shift work, jet lag, and medical interventions (cf. Roenneberg & Merrow, 2003). In an effort to detect individual body time from a single timepoint assay, a "molecular timetable" has been created based upon peak times for gene RNA expression levels of more than 100 "time-indicating genes" in the mouse (Ueda et al., 2004). The authors speculated that this molecular timetable could be applied to other organisms, including humans, in an effort to use single timepoint sampling to determine body time in an effort to diagnose rhythm disorders and/or personalize medicine and chronotherapy.

Molecular Machinery Underlies Physiology

Perhaps "fixing the clock" via manipulation or treatment of specific malfunctioning parts of the molecular timing mechanism itself by gene therapy or chronotherapy may one day reestablish normal rhythmicity and lessen or reverse the symptoms of specific circadian-related diseases. For example, the sensitivity of mice to the chemotherapeutic agent cyclophosphamide was increased in *Bmal1* and

---

[25] Those genes important for the generation and regulation of circadian rhythms. See Chapter 5 on Models for additional discussion.

[26] For example, an alteration in daily rhythms in motor activity and adrenocortical secretion that are regulated by the SCN is associated with poor survival of patients with breast or metastatic colorectal cancer. When the SCN of tumor-bearing mice was destroyed, the rest-activity cycle was ablated and serum corticosterone rhythm markedly altered. Tumors in these mice grew two to three times faster and survival was shorter than sham-operated mice, demonstrating a functional role of the SCN in tumor growth (Filipski et al., 2003).

Clock/Clock mutants, but decreased in *Cry1/Cry2* double mutants, indicating that the CLOCK/BMAL1 transactivation complex modulates survival of the target cells (Gorbacheva et al., 2005; Green, 2005). Similarly, a mutation of the circadian gene *Period2* in the mouse resulted in increased sensitivity to irradiation (via greater deficiencies in DNA damage responses) and tumor development, suggesting that, in addition to helping to organize the circadian clock, mPER2 is also a tumor suppressor (Fu et al., 2002). Mutations of the human homolog of this gene, h*Per2*, have been identified and associated with familial advanced sleep phase syndrome, wherein variant sleep behavior (e.g., a 4-h advance in sleep, temperature and melatonin rhythms) results from an altered hPER2 expression (Toh et al., 2001).

When mice were subjected to an experimental jet lag by a 6-h phase advance or delay in the lighting schedule, the circadian cycles of m*Per* and m*Cry1* expression in the SCN responded differently (Reddy et al., 2002). Following the advance in lighting, m*Per* reset to the new time very quickly, while there was a gradual resetting for m*Cry1* (in parallel with the gradual resetting of the activity cycle). Following the 6-h delay in lighting, both genes reset their circadian pattern by the second day, as did the activity cycle. This suggests a different molecular response of the clock during advance and delay resetting, with m*Cry1* more involved as a rate-limiting factor in behavioral resetting. Therefore, the *Cry* gene could be a therapeutic target for sleep disturbance, jet lag and shift work (Reddy et al., 2002).

With regard to metabolic syndrome, mutations in *Bmal1* and *Clock* in mice suppressed the circadian variations in glucose and triglycerides and the recovery from insulin-induced hypoglycemia, suggesting that the molecular clock, along with dietary cues, may directly influence glucose homeostasis (Rudic et al., 2004). The authors suggested that not only *what* is eaten, but *when* it is eaten may influence the functional consequences of food (see additional discussion of meal-timing in Chapter 10 on Society). Along these lines, meal-timing has been shown to prolong survival of tumor-bearing mice and it has been suggested that timed feeding schedules could be used to further improve the benefit derived from chronotherapy (Wu et al., 2004).

Marker Rhythms

Whenever possible with all chronotherapy protocols, a physiological variable known to display a prominent circadian rhythm, such as body temperature, total WBCs, serum cortisol or melatonin, or urinary potassium, should be measured as a *marker rhythm*.

For example, the circadian pattern for body temperature under usual entrained conditions is characterized by a peak during mid-activity, a decrease beginning about 3 h before sleep onset, a sharp fall after falling asleep and a rise near the end of sleep (Weitzman et al., 1979). Melatonin in blood, saliva or urine is often thought to be the purest marker rhythm since it best represents the endogenous biological clock and is relatively unaffected by masking factors (other than bright light). Often, melatonin is only monitored during the first several hours of usual darkness, using light sufficiently dim (<50 lux) so as not to mask the secretion, in order to define the onset of the evening rise as a marker (so-called "Dim Light Melatonin Onset" or DLMO). A 24-h profile of melatonin can also provide other useful marker endpoints, such as the acrophase, the time of offset of secretion and the duration of secretion (cf. Arendt, 1997).

A marker rhythm can thus be used to confirm an individual's synchrony to a light–dark schedule and insure and/or guide the correct timing of treatment, as has been demonstrated for cancer patients (Deka et al., 1976; Hermida-Dominguez et al, 1986; Smaaland et al, 1995; Mormont et al, 1998), and even to individualize chronotherapy (Reinberg & Ashkenazi, 1993; Focan, 1995). As mentioned above, a molecular timetable consisting of peak times of RNA expression for more than 100 time-indicating genes may one day allow a single timepoint diagnosis of individual body time for use in guiding chronotherapy (Ueda et al., 2004).

Ambulatory activity monitored via a noninvasive device worn on the wrist has received a lot of attention as a circadian marker rhythm (Brown et al., 1990b). For example, wrist actigraphy has been used in the rhythmometric study of *blind individuals* (Lockley et al., 1999), *preterm neonates* (Korte et al., 2001), *cancer patients* (Mormont et al., 2000, 2002), *liver glycogen storage disease* (Yetman et al., 2002), *hypertension* (Hermida et al., 2002b), *sleep disturbance in liver cirrhosis patients* (Córdoba et al., 1998), *sleep disorders* (Sadeh et al., 1995), and *nocturnal scratching in atopic dermatitis* (Ebata et al., 2001), among others (Minors et al., 1996a,b,c). An example of circadian patterns in whole body activity spanning 5 days is shown for a healthy man in Figure 12.17 in Chapter 12 on Autorhythmometry.

For economic and technical reasons, a reasonable first-step alternative to repeated sampling of blood or urine or use of machines to determine marker rhythms may be to record the times an individual went to bed and got up for the previous week and try to maintain this schedule throughout treatment. A treatment time based on an appropriate quadrant of the 24-h day which has previously proved successful in the treatment of humans or rodents can then be selected according to circadian "stage" (hours after awakening or from midsleep), rather than arbitrary clock hours ("time of day") (Sothern, 1995).

The Medical Community and the Concept of Timing

A strong positive attitude toward the concept of chronotherapy was documented in American physicians and adults in two Gallup surveys conducted in 1996 (Smolensky, 1998). Yet, knowledge of how biological rhythms affect the diagnosis of medical conditions (*When* are symptoms worse? *When* should diagnostics tests be performed?) and the incorporation of chronotherapy (*When* should treatment be given?) into clinical medicine and its correct utilization by physicians, patients, and insurers have been slow to materialize.

Just as a flat-earth society still exists, its logic cannot be deemed based in what we now know to be fact. Similarly, the landscape in human clinical medicine can no longer be considered flat, but consisting of a variety of predictable hills and valleys (oscillations) in biological functions and processes. While extreme outliers at any time are and will continue to be an obvious "red flag," the "flat-earth" concept of single values as boundaries for normal at *all times* needs to be abandoned and replaced with a "round-earth" concept when navigating the area of clinical medicine.[27] In addition, clinical and research papers need to include the time of day and year of sample collection or other measurements in order to properly interpret observations and results.

Chronobiological findings to date strongly suggest that *"when"* is as critical as *"what"* and *"how much"* for interpretation of clinical findings and the treatment of many conditions and illnesses. A common question often raised among medical personnel and the general public is, *"Is chronotherapy and chronodiagnosis practical and/or cost-effective?"* It seems easy enough to take a single aspirin tablet, timed-delay medication or blood pressure medication in the evening at no more cost or effort than in the morning, but physical drug delivery systems might add significant cost. However, when factoring in the cost of equipment and toxicity management in the treatment of cancer, the direct costs for a single chronotherapy cycle are comparable to traditional infusion and/or bolus treatment, since the increased cost of the chronochemotherapy device was offset by a decrease in toxicity management costs (Tampellini et al., 2004). When using drug pumps to deliver either a flat or chronomodulated combination chemotherapy infusion in patients with colorectal cancer, the costs were equivalent, with the added bonus not only of less treatment-associated toxicity, but also an improved quality of life for patients receiving the chronomodulated schedule (Focan, 2002b). There is still a great need for a concerted educational effort in order for the concept of timing of treatments to be fully appreciated and chronotherapy to be correctly applied.

One of the goals of chronotherapy is to have the label on every medication state when it is the best time to take it, if such a time has been documented in clinical studies.[28] It may be necessary for individuals to find out more about their own biological rhythms and chronotherapy and discuss these with their physicians, before the medical community embraces the concept of timing in screening and treatment procedures and incorporates it into routine medical practice.[29] To this end, an extensive overview of rhythms in health, disease and timing of treatments appears in *"The Body Clock Guide to Better Heath"* by Smolensky & Lamberg (2000).

In a book entitled *"Biological Rhythms in Clinical and Laboratory Medicine"* (1994), Touitou & Haus wrote: *"Together with the recent advances in*

---

[27] Not to do so, especially in the area of chronotherapy, may constitute *"ignorance, indolence, or perhaps criminal negligence."* (cf., Halberg et al., 2003).

[28] Such as the directions on bottles of melatonin that state *"To be taken before bedtime."* Certain anti-inflammatory medications for arthritis or osteoarthritis pain and some verapamil formulations (Ca^+ channel blockers) for blood pressure control are also labeled to be taken at night.

[29] Insurers also need to become aware of the proven clinical benefits and cost-effectiveness of chronotherapy (Block, 2002).

chronopharmacology, [the] application of [chronobiologic concepts and methods to clinical medicine] appears now timely and in some areas urgent. . . . The human time structure is a basic fact of our existence, no matter if one wants to study it or not. . . . [Thus,] Chronobiology and its subspecialties, like chronopharmacology, will certainly play an important role in the clinical medicine of the future."

With what we now know about biological rhythms in the early part of the 21st century, the future for a chronomedicine has indeed arrived. Circadian maps have been established for most biologic processes and functions in humans, clock genes have substantiated a built-in, genetic basis for these oscillations and chronotherapy has been shown to be therapeutically and cost-effective.

Take-Home Message

The "normal" functioning of the body's many processes are rhythmic. These rhythms play a role in when we are born, how we develop and live, how well we can perform physically or mentally, the diagnosis of health or disease, our susceptibility to illness or disease, how we respond to medications and when we might die. "Normal" values can change throughout the day, month and year, such that the same value may be too high at one time, too low at another or considered acceptable at other times. For diagnostic and treatment purposes, time of day, or more correctly, stage of rhythm, needs to be taken into account, thereby adding another dimension (time) to our understanding of medicine.

References

Abrahamsen JF, Smaaland R, Sothern RB, Laerum OD. (1997) Circadian cell cycle variations of erythro- and myelopoiesis in humans. *Eur J Hematol* 58: 333–345.

Abrahamsen JF, Smaaland R, Sothern RB, Laerum OD. (1998) Variation in cell yield and proliferative activity of positive selected human CD34+ bone marrow cells along the circadian time scale. *Eur J Hematol* 60: 7–15.

Akslen L, Sothern RB. (1998) Seasonal variations in the presentation and growth of thyroid cancer. *Br J Cancer* 77: 1174–1179.

Anderson LE, Morris JE, Sasser LB, Stevens RG. (2000) Effect of constant light on DMBA mammary tumorigenesis in rats. *Cancer Lett* 148(2): 121–126.

Andreotti FA, Kluft C. (1991) Circadian variation of fibrinolytic activity in blood. *Chronobiol Intl* 8(5): 336–351.

Anwar YA, White WB. (1998) Chronotherapeutics for cardiovascular disease. *Drugs* 55(5): 631–643 (Review).

Araki H, Koiwaya Y, Nakagaki U, Nakamura M. (1983) Diurnal distribution of ST-segment elevation and related arrhythmias in patients with variant angina: a study by ambulatory ECG monitoring. *Circulation* 67(5): 995–1000.

Arendt J. (1997) The pineal gland, circadian rhythms and photoperiodism. In: *Physiology and Pharmacology of Biological Rhythms*. Redfern PH, Lemmer B, eds. Berlin: Springer-Verlag, pp. 375–414.

Arntz HR, Willich SN, Schreiber C, Bruggemann T, Stern R, Schultheiss HP. (2000) Diurnal, weekly and seasonal variation of sudden death. Population-based analysis of 24,061 consecutive cases. *Eur Heart J* 21(4): 315–320.

Aschoff J, Biebach H, Heise A, Schmidt T. (1974) Day–night variation in heat balance. In: *Heat Loss from Animals and Man. Assessment and Control.* Monteith JL, Mount LE, eds. London: Butterworth, pp. 147–172.

Attar-Levy D. (1997) [Seasonal depression.] [French]. *Revue du Praticien* 47(17): 1899–1903.

Auvil-Novak S-E, Novak RD, Smolensky MH, Kavanagh JJ, Kwan JW, Wharton JT. (1988) Twenty-four hour variation in self-administration of morphine sulfate (MS) and hydromorphone (H) by post-surgical gynecologic cancer patients. In: *Ann Rev Chronopharm*, Vol 5. Reinberg A, Smolensky M, Labrecque G, eds. Oxford: Pergamon Press, pp. 343–346.

Auvil-Novak SE, Novak RD, el Sanadi N. (1996) Twenty-four-hour pattern in emergency department presentation for sickle cell vaso-occlusive pain crisis. *Chronobiol Intl* 13(6): 449–456.

Auvil-Novak SE. (1997) A middle-range theory of chronotherapeutic intervention for post-surgical pain. *Nurs Res* 46(2): 66–71.

Auvil-Novak SE. (1999) The chronobiology, chronopharmacology, and chronotherapeutics of pain. *Ann Rev Nurs Res* 17: 133–153 (Review).

Badwe RA, Gregory WM, Chaudary MA, Richards MA, Bentley AE, Rubens RD, Fentiman IS. (1991) Timing of surgery during menstrual cycle and survival of premenopausal women with operable breast cancer. *Lancet* 337(8752): 1261–1264.

Badwe RA, Wang DY, Gregory WM, Fentiman IS, Chaudary MA, Smith P, Richards MA, Rubens RD. (1994) Serum progesterone at the time of surgery and survival in women with premenopausal operable breast cancer. *Eur J Cancer* 30A(4): 445–448.

Baldissera SF, Motta LD, Ameida MC, Antunes-Rodriguez J. (1991) Proposal of an experimental model for the study of polycystic ovaries. *Braz J Med Biol Res* 24(7): 747–751.

Bartoli V, Dorigo B, Tedeschi E, Biti G, Voegelin MR. (1975) Behavior of calf blood flow in normal subjects and in patients with intermittent claudication during a 24-h time span. *Chronobiologia* 2(1): 13–19.

Bartter FC, Delea CS, Halberg F. (1962) A map of blood and urinary changes related to circadian variations in adrenal cortical function in normal subjects. *Annals NY Acad Sci* 98(4): 969–983.

Bartter FC. (1974) Periodicity and medicine. In: *Chronobiology*. Scheving LE, Halberg F, Pauly JE, eds. Tokyo: Igaku Shoin Ltd., pp. 6–13.

Bateman JRM, Pavia D, Clarke SW. (1978) The retention of lung secretions during the night in normal subjects. *Clin Sci* 55(6): 523–527.

Beard CM, Fuster V, Elveback LR. (1982) Daily and seasonal variation in sudden cardiac death, Rochester, Minnesota, 1950–1975. *Mayo Clinic Proc* 57(11): 704–706.

Beaumont M, Batejat D, Pierard C, Van Beers P, Denis JB, Coste O, Doireau P, Chauffard F, French J, Lagarde D. (2004) Caffeine or melatonin effects on sleep and sleepiness after rapid eastward transmeridian travel. *J Appl Physiol* 96(1): 50–58.

Bélanger PM. (1993) Chronopharmacology in drug research and therapy. *Adv Drug Res* 24: 1–80.

Bellamy N, Sothern RB Campbell J. (1990) Rhythmic variations in pain perception in osteoarthritis of the knee. *J Rheumatol* 17(3): 364–372.

Bellamy N, Sothern RB Campbell J, Buchanan WW. (1991) Circadian rhythm in pain, stiffness, and manual dexterity in rheumatoid arthritis: relation between discomfort and disability. *Annals Rheum Dis* 50(4): 243–248.

Bellamy N, Sothern R, Campbell J, Buchanan WW. (2002) Rhythmic variations in pain, stiffness and manual dexterity in hand osteoarthritis. *Annals Rheum Dis* 61(12): 1075–1080.

Bellamy N, Sothern RB, Campbell J. (2004) Aspects of diurnal rhythmicity in pain, stiffness, and fatigue in patients with fibromyalgia. *J Rheumatol* 31(2): 379–389.

Beniashvili DS, Benjamin S, Baturin DA, Anisimov VN. (2001) Effect of light/dark regimen on N-nitrosoethylurea-induced transplacental carcinogenesis in rats. *Cancer Lett* 163(1): 51–57.

Bisdee JT, Garlick PJ, James WPT. (1989) Metabolic changes during the menstrual cycle. *Br J Nutr* 61(3): 641–650.

Bjarnason GA, Kerr IG, Doyle N, Macdonald M, Sone M. (1993) Phase I study of 5-Fluorouracil and Leucovorin by a 14 day circadian infusion in patients with metastatic adenocarcinoma. *Cancer Chemother Pharmacol* 33(3): 221–228.

Bjarnason GA, Hrushesky WJM. (1994) Cancer Chronotherapy. In: *Circadian Cancer Therapy*. Hrushesky WJM, ed. Boca Raton, CRC Press, pp. 241–263.

Bjarnason GA. (1995) Clinical cancer chronotherapy trials: a review. *J Infus Chemother* 5(1): 31–37.

Bjarnason GA, Jordan R, Wood PA, Li Q, Lincoln D, Sothern RB, Hrushesky WJM, Ben-David Y. (2001) Circadian expression of clock genes in human oral mucosa and skin: association with specific cell cycle phases. *Amer J Path* 158(5): 1793–1801.

Blask DE, Sauer LA, Dauchy R, Holowachuk EW, Ruhoff MS. (1999) New actions of melatonin on tumor metabolism and growth. *Biol Signals & Receptors* 8(1–2): 49–55.

Blask DE, Dauchy RT, Sauer L, Krause JA, Brainard GC. (2003) Growth and fatty acid metabolism of human breast cancer (MCF-7) xenografts in nude rats: impact of constant light-induced nocturnal melatonin suppression. *Br Ca Res Treat* 79(3): 313–320.

Block KI. (2002) Chronomodulated chemotherapy: clinical value and possibilities for dissemination in the United States. *Chronobiol Intl* 19(1): 275–287.

Blüher M, Hentschel B, Rassoul F, Richter V. (2001) Influence of dietary intake and physical activity on annual rhythm of human blood cholesterol concentrations. *Chronobiol Intl* 18(3): 541–557.

Bochnik HJ. (1958) Tagesschwangungen der muskulären Leistungsfähigkeit. *Dtsch Zeit Nervenheilkd* 178: 270–275.

Bogaty P, Waters DD. (1989) Possible mechanisms accounting for the temporal distribution of anginal attacks. In: *Chronopharmacology. Cellular and Biochemical Interactions*. Lemmer B, ed. New York: Marcel-Dekker, Inc, pp. 509–524.

Boivin DB, James FO, Wu A, Cho-Park, PF, Xiong H, Sun ZS. (2003) Circadian clock genes oscillate in human peripheral blood mononuclear cells. *Blood* 102(12): 4143–4145.

Born J, Lange T, Hansen K, Molle M, Fehm HL. (1997) Effects of sleep and circadian rhythm on human circulating immune cells. *J Immunol* 158(9): 4454–4464.

Borner MM. (1999) Neoadjuvant chemotherapy for unresectable liver metastases of colorectal cancer-too good to be true? *Ann Oncol* 10: 623–626.

Bourdallé-Badie C, Bruguerolle B, Labrecque G, Robert S, Erny P. (1990a) Biological rhythms in pain and anesthesia. In: *Ann Rev Chronopharmacol*, Vol 6. Reinberg A, Smolensky M, Labrecque G, eds. Oxford: Pergamon Press, pp. 155–181.

Bourdallé-Badie C, Andre M, Pourquier P, Robert S, Cambar J, Erny P. (1990b) Circadian rhythm of pain in man: study by measure of nociceptive flexion reflex. In: *Annual Rev Chronopharm*, Vol 7. Reinberg A, Smolensky M, Labrecque G, eds. Oxford: Pergamon Press, pp. 249–252.

Bremner WF, Sothern RB, Kanabrocki EL, Vahed S, Third JLHC, Scheving LE. (1990) Chronobiological evaluation of 24-hour lipid and lipoprotein changes in middle-aged males. In: *Chronobiology: Its Role in Clinical Medicine, General Biology and Agriculture, Part A*, Hayes DK, Pauly JE, Reiter RJ, eds. New York: Wiley-Liss, Inc., pp. 185–192.

Bremner WF, Sothern RB, Kanabrocki EL, Ryan M, McCormick JB, Dawson S, Connors ES, Rothschild R, Third JL, Vahed S, Nemchausky BM, Shirazi P, Olwin JH. (2000) Relation between circadian patterns in levels of circulating lipoprotein(a), fibrinogen, platelets, and related lipid variables in men. *Amer Heart J* 139(1 Pt 1): 164–173.

Brennan PJ, Greenberg G, Miall WE, Thompson SG. (1982) Seasonal variation in arterial blood pressure. *BMJ* 285: 919–923.

Bressolle F, Joulia JM, Pinguet F, Ychou M, Astre C, Duffour J, Gomeni R. (1999) Circadian rhythm of 5-fluorouracil population pharmacokinetics in patients with metastatic colorectal cancer. *Cancer Chemother Pharmacol* 44(4): 295–302.

Brezinski DA, Tofler GH, Muller JE, Pohjola-Sintonen S, Willich SN, Shafer AI, Czeisler CA, Williams GH. (1988) Morning increase in platelet aggregability. Association with assumption of the upright posture. *Circul* 78(1): 35–40.

Bridges EJ, Woods SL. (2001) Cardiovascular chronobiology: do you know what time it is? *Prog Cardiovasc Nurs* 16(2): 65–79 (Review).

Brown AC, Smolensky MH, D'Alonzo GE. (1990a) Day-night pattern of isoproterenol (ISO) use for relief of acute asthma symptoms. In: *Ann Rev Chronopharmacol*, Vol. 7, Reinberg A, Smolensky M, Labrecque G, eds. Oxford: Pergamon Press, p. 317.

Brown AC, Smolensky MH, D'Alonzo GE, Redman DP. (1990b) Actigraphy: a means of assessing circadian patterns in human activity. *Chronobiol Intl* 7(2): 125–133.

Bruguerolle B, Dupont M, Lebre P, Legre G. (1988) Bupivacaine chronokinetics in man after a peridural constant rate infusion. In: *Ann Rev Chronopharm*, Vol 5. Reinberg A, Smolensky M, Labrecque G, eds. New York: Pergamon Press, pp. 223–226.

Bruguerolle B. (1989) Time dependence of general and local anesthetic drugs. In: *Chronopharmacology: Cellular and Biochemical Interactions*. Lemmer B, ed. New York: Marcel Dekker, pp. 581–596.

Bruguerolle B. (1998) Chronopharmacokinetics. Current status. *Clin Pharmacokinet* 35(2): 83–94.

Bruguerolle B, Simon N. (2002) Biologic rhythms and Parkinson's disease: a chronopharmacologic approach to considering fluctuations in function. *Clin Neuropharmacol* 25(4): 194–201. Review.

Brush MG, Goudsmit EM, eds. (1988) *Functional Disorders of the Menstrual Cycle*. Chichester: John Wiley & Sons, Inc., 303 pp.

Buchi KN, Moore JG, Hrushesky WJM, Sothern RB, Rubin NH. (1991) Circadian rhythm of cellular proliferation in the human rectal mucosa. *Gastroenterol* 101(2): 410–415.

Burgess HJ, Eastman CI. (2004) Early versus late bedtimes phase shift the human dim light melatonin rhythm despite a fixed morning lights on time. *Neurosci Lett* 356(2): 115–118.

Busch LA. (1971) Human listeriosis in the United States, 1967–1969. *J Infect Dis* 123(3): 328–332.

Buxtorf JC, Baudet MF, Martin C, Richard JL, Jacotot B. (1988) Seasonal variations of serum lipids and apoproteins. *Ann Nutr Metab* 32(2): 68–74.

Callard D, Davenne D, Gauthier A, Lagarde D, Van Hoecke J. (2000) Circadian rhythms in human muscular efficiency: Continuous physical exercise versus continuous rest. A crossover study. *Chronobiol Intl* 17(5): 693–704.

Cambie G, Caillard V, Beaute-Lafitte A, Ginsburg H, Chabaud A, Landau I. (1991) Chronotherapy of malaria: identification of drug-sensitive stage of parasite and timing of drug delivery for improved therapy. *Ann Parasitol Hum Comp* 66(1): 14–21.

Campbell SS, Murphy PJ, Van Den Heuvel CJ, Roberts ML, Stauble TN. (1999) Etiology and treatment of intrinsic circadian rhythm sleep disorders. *Sleep Med Rev* 3(3): 179–200.
Canaple L, Kakizawa T, Laudet V. (2003) The days and nights of cancer cells. *Cancer Res* 63(22): 7545–7552.
Cardoso SS, Scheving LE, Halberg F. (1970) Mortality of mice as influenced by the hour of the day of drug (ara-C) administration. *Pharmacologist* 12: 302.
Casale G, Butte M, Pasotti C, Ravecca D, de Nicola P. (1983) Antithrombin III and circadian rhythms in the aged and in myocardial infarction. *Haematologica* 68(5): 615–619.
Casale G, de Nicola P. (1984) Circadian rhythms in the aged: a review. *Arch Gerontol Geriatr* 3: 267–284.
Case AM, Reid RL. (1998) Effects of the menstrual cycle on medical disorders. *Arch Intern Med* 158: 1405–1412.
Case AM, Reid RL. (2001) Menstrual cycle effects on common medical conditions. *Compr Ther* 27(1): 65–71.
Cassone VM, Natesan AK. (1997) Time and time again: the phylogeny of melatonin as a transducer of biological time. *J Biol Rhythms* 12(6): 489–497 (Review).
Cermakian N, Boivin DB. (2003) A molecular perspective of human circadian rhythm disorders. *Brain Res Brain Res Rev* 42(3): 204–220 (Review).
Chassard D, Bruguerolle B. (2004) Chronobiology and anesthesia. *Anesthesiol* 100(2): 413–427 (Review).
Chaykin S. (1986) Beard growth: a window for observing circadian and infradian rhythms of men. *Chronobiologia* 13(2): 163–165.
Chen ST, Choo KB, Hou MF, Yeh KT, Kuo SJ, Chang JG. (2005) Deregulated expression of the *PER1*, *PER2* and *PER3* genes in breast cancers. *Carcinogenesis* 26(7): 1241–1246.
Chu C, Sothern RB, Chu P, Johansson AM. (2001) Circadian characteristics of oxygen saturation monitored via pulse oximetry (Abstract). *Chronobiol Intl* 18(6): 1081–1082.
Cohen P, Wax Y, Modan B. (1983) Seasonality in the occurrence of breast cancer. *Cancer Res* 43(2): 892–896.
Cohen MC, Rohtla KM, Lavery CE, Muller JE, Mittleman MA. (1997) Meta-analysis of the morning excess of acute myocardial infarction and sudden cardiac death. *Amer J Cardiol* 79(11): 1512–1516.
Cologne Light Symposium (2002) Light, endocrine systems and cancer–facts and research perspectives. Proc and abstracts of the Cologne Light Symposium 2002. May 2–3, 2002. Cologne, Germany. *Neuroendocrinol Lett* 23 (Suppl 2): 1–104.
Colquhoun WP, ed. (1972) *Aspects of Human Efficiency. Diurnal Rhythm and the Loss of Sleep.* London: English University Press Ltd., 344 pp.
Cooper LS, Gillett CE, Patel NK, Barnes DM, Fentiman IS. (1999) Survival of premenopausal breast carcinoma patients in relation to menstrual cycle timing of surgery and estrogen receptor/progesterone receptor status of the primary tumor. *Cancer* 86(10): 2053–2058.
Córdoba J, Cabera J, Lataif L, Penev P, Zee P, Blei AT. (1998) High prevalence of sleep disturbance in cirrhosis. *Hepatol* 27(2): 339–345.
Cornélissen G, Halberg F, Pöllmann L, Pöllmann B, Katinas GS, Minne H, Breus T, Sothern RB, Watanabe Y, Tarquini R, Perfetto F, Maggioni C, Wilson D, Gubin D, Otsuka K, Bakken EE. (2003) Circasemiannual chronomics: half-yearly biospheric changes in their own right and as a circannual waveform. *Biomed Pharmacother* 57 (Suppl 1): 45s–54s.

Coudert B, Bjarnason G, Focan C, di Paola ED, Lévi F. (2003) It is time for chronotherapy! *Pathol Biol (Paris)* 51(4): 197–200.

Crawford VL, Sweeney O, Coyle PV, Halliday IM, Stout RW. (2000) The relationship between elevated fibrinogen and markers of infection: a comparison of seasonal cycles. *QJM* 93(11): 745–750.

Crawford VL, McCann M, Stout RW. (2003) Changes in seasonal deaths from myocardial infarction. *QJM* 96(1): 45–52.

Cugini P, Murano G, Lucia P, Letizia C, Scavo D, Halberg F, Cornelissen G, Sothern RB. (1985) Circadian rhythms of plasma renin activity and aldosterone: changes related to age, sex, recumbency and sodium restriction. Chronobiologic specification for reference values. *Chronobiol Intl* 2(4): 267–276.

Cugini P, Romit A, Di Palma L, Giacovazzo M. (1990) Common migraine as a weekly and seasonal headache. *Chronobiol Intl* 7 (5/6): 467–469.

Cugini P, Salandri A, Petrangeli CM, Capodaglio PF, Giovannini C. (1996) Circadian rhythms in human body composition. *Chronobiol Intl* 13(5): 359–371.

Cugini P, Kawasaki T, Coen G, Pellegrino AM, Fontana S, Di Marzo A, Ceccotti P, Lucia P, Petrangeli CM, Leone G. (1998) Who are the non-dippers? A better definition via the blood pressure circadian rhythm. *Clin Ter* 149(5): 343–349.

Curé H, Chevalier V, Adenis A, Tubiana-Mathieu N, Niezgodzki G, Kwiatkowski F, Pezet D, Perpoint B, Coudert B, Focan C, Lévi F, Chipponi J, Chollet P. (2002) Phase II trial of chronomodulated infusion of high-dose fluorouracil and L-folinic acid in previously untreated patients with metastatic colorectal cancer. *J Clin Oncol* 20(5): 1175–1181.

Cutolo M, Seriolo B, Craviotto C, Pizzorni C, Sulli A. (2003) Circadian rhythms in RA. *Ann Rheum Dis* 62(7): 593–596.

Dabbs JM Jr. (1990) Age and seasonal variation in serum testosterone concentration among men. *Chronobiol Intl* 7(3): 245–249.

D'Alonzo GE, Smolensky MH, Gianotti L, Emerson M, Staudinger H, Steinijans V. (1990) Twenty-four hour lung function in adult patients with asthma. Chronoptimized theophylline therapy once-daily dosing in the evening versus conventional twice-daily dosing. *Amer Rev Respir Dis* 142: 84–90.

Dalton KD. (1984) *The Premenstrual Syndrome and Progesterone Therapy, 2nd edn.* London: Wm Heinemann Med Books Ltd., 291 pp.

Dauchy RT, Sauer LA, Blask DE, Vaughan GM. (1997) Light contamination during the dark phase in "photoperiodically-controlled" animal rooms: effect on tumor growth and metabolism in rats. *Lab Anim Sci* 47(5): 511–518.

Dauchy RT. Blask DE. Sauer LA. Brainard GC. Krause JA. (1999) Dim light during darkness stimulates tumor progression by enhancing tumor fatty acid uptake and metabolism. *Cancer Lett* 144(2): 131–136.

Davenne D, Gauthier A. (1998) Location of the mechanisms involved in the circadian rhythm of muscle strength. In: *Biological Clocks. Mechanisms and Applications.* Touitou Y, ed. Amsterdam: Elsevier, pp. 553–556.

Davies B, Morris T. (1993) Physiological parameters in laboratory animals and humans. *Pharmaceut Res* 10(7): 1093–1095.

Davis S, Mirick DK, Stevens RG. (2001) Night shift work, light at night, and risk of breast cancer. *J Natl Cancer Inst* 93(20): 1557–1562.

Dawes C. (1974a) Circadian and circannual maps for human saliva. In: *Chronobiology*, Scheving LE, Halberg F, Pauly J, eds. Tokyo: Igaku-Shoin, pp. 224–227.

Dawes C. (1974b) Rhythms in salivary flow rate and composition. *Intl J Chronobiol* 2: 253–279.

Decousus HA, Croze M, Lévi FA, Jaubert JC, Perpoint BM, Bonadona JFDe, Reinberg A, Queneau PM. (1985) Circadian changes in anticoagulant effect of heparin infused at a constant rate. *Br Med J* 290: 341–344.

Debon R, Chassard D, Duflo F, Boselli E, Bryssine B, Allaouchiche B. (2002) Chronobiology of epidural ropivacaine: variations in the duration of action related to the hour of administration. *Anesthesiol* 96(3): 542–545.

Debon R, Boselli E, Guyot R, Allaouchiche B, Lemmer B, Chassard D. (2004) Chronopharmacology of intrathecal sufentanil for labor analgesia: daily variations in duration of action. *Anesthesiol* 101(4): 978–982.

Deka AC, Chatterjee B, Gupta BD, Balakrishnan C, Dutta TK. (1976) Temperature rhythm – an index of tumour regression and mucositis during the radiation treatment of oral cancers. *Indian J Cancer* 13(1): 44–50.

Dewan EM. (1967) On the possibility of a perfect rhythm method of birth control by periodic light stimulation. *Amer J Obst Gyn* 99(7): 1016–1019.

DiSanto A, Chodos D, Halberg F. (1975) Chronobioavailability of three erythromycin test preparations assessed by each of four indices: time to peak, peak, nadir and area (abstract). *Chronobiologia* 2 (Suppl 1): 17.

Doering CH, Kraemer HC, Brodie KH, Hamburg DA. (1974) A cycle of plasma testosterone in the human male. *JCE & M* 40: 492–500.

Dogliotti L, Tampellini M, Lévi F. (1998) Chronochemotherapy of colorectal cancer. From Villejuif to Europe. In: *Biological Clocks. Mechanisms and Applications.* Touitou Y, ed. Amsterdam: Elsevier, pp. 475–481.

Douglas AS, Allan TM, Helms PJ. (1996) Seasonality and the sudden infant death syndrome during 1987–1999 and 1991–1993 in Australia and Britain. *BMJ* 312(7043): 1381–1383.

Dowell SF. (2001) Seasonal variation in host susceptibility and cycles of certain infectious diseases. *Emerg Infect Dis* 7(3): 369–374.

Drance SM. (1960) The significance of the diurnal tension variations in normal and glaucomatous eyes. *Arch Ophthalmol* 64: 494–501.

Dubbels R, Reiter RJ, Klenke E, Goebel A, Schnakenberg E, Ehlers C, Schiwara HW, Schloot W. (1995) Melatonin in edible plants identified by radioimmunoassay and by high performance liquid chromatography-mass spectrometry. *J Pineal Res* 18(1): 28–31.

Dubey DP, Halberg F, Huq S, Rahman S, Fink H, Sothern RB, Wallach LA, Haus E, Nesbit ME, Theologides A, Scheving LE. (1977) Circadian rhythm with large amplitude in murine tolerance of vincristine. In: *Prevention and Detection of Cancer, Part I, Prevention, Vol. 1, Etiology.* Nieburgs HE, ed. New York/Basel: Marcel Dekker, Inc., pp. 1115–1124.

Duncan WC Jr, Wehr TA. (1988) Pharmacological and non-pharmacological chronotherapies of depression. In: *Ann Rev Chronopharmacol*, Vol 4. Reinberg A, Smolensky M, Labrecque G, eds. Oxford: Pergamon Press, pp. 137–170.

Dunnigan MG, Harland WA, Fyfe T. (1970) Seasonal incidence and mortality of ischemic heart disease. *Lancet* 2(7677): 793–797.

Dyrenfurth I, Jewelewicz R, Warren M, Ferrin M, Vande Wiele RL. (1974) Temporal relationships of hormonal variables in the menstrual cycle. In: *Biorhythms and Human Reproduction.* Ferin M, Halberg F, Richart RM, Vande Wiele RL, eds. New York: John Wiley & Sons, Inc., pp. 171–201.

Eastwood MR, Peacocke J. (1976) Seasonal patterns of suicide, depression and electroconvulsive therapy. *Brit J Psychiatry* 129: 472–475.

Eastwood MR, Whitton JL, Kramer PM, Peter AM. (1985) Infradian rhythms. A comparison of affective disorders and normal persons. *Arch Gen Psychiatry* 42(3): 295–299.

Ebata T, Iwasaki S, Kamide R, Niimura M. (2001) Use of a wrist activity monitor for the measurement of nocturnal scratching in patients with atopic dermatitis. *Br J Dermatol* 144(2): 305–309.

Ehrly AM, Jung G. (1973) Circadian rhythm of human blood viscosity. *Biorheology* 10(4): 577–583.

Ekbom KA. (1970) Patterns of cluster headache with a note on the relations to angina pectoris and peptic ulcer. *Acta Neurol Scand* 46(2): 225–237.

Elliot WJ. (1998) Circadian variation in the timing of stroke onset. A meta-analysis. *Stroke* 29(5): 992–996.

Eriguchi M, Levi F, Hisa T, Yanagie H, Nonaka Y, Takeda Y. (2003) Chronotherapy for cancer. *Biomed Pharmacother* 57(Suppl 1): 92s–95s.

Erren TC, Reiter RJ, Piekarski C. (2003) Light, timing of biological rhythms, and chronodisruption in man. *Naturwiss* 90(11): 485–494 (Review).

Erren TC. (2005) Could visible light contribute to the development of leukaemia and other cancers in children? *Med Hypoth* 64(4): 864–871.

Fager G, Wiklund O, Olofsson SO, Bondjers G. (1982) Seasonal variations in serum lipid and apolipoprotein levels evaluated by periodic regression analyses. *J Chronic Dis* 35(8): 643–648.

Feigin VL, Anderson CS, Rodgers A, Bennett DA. (2002) Subarachnoid haemorrhage occurrence exhibits a temporal pattern – evidence from meta-analysis. *Eur J Neurol* 9(5): 511–516.

Fentiman I, Gregory W, Richards M. (1994) Effects of menstrual phase on surgical treatment of breast cancer. *Lancet* 344(8919): 402.

Fentiman IS. (2002) Timing of surgery for breast cancer. *Intl J Clin Pract* 56(3): 188–190 (Review).

Fernandez JR, Hermida RC. (2000) Computation of model-dependent tolerance bands for ambulatorily monitored blood pressure. *Chronobiol Intl* 17(4): 567–582.

Ferrario VF, Tredici G, Crespi V. (1980) Circadian rhythm in human nerve conduction velocity. *Chronobiologia* 7(2): 205–209.

Feychting M, Osterlund B, Ahlbom A. (1998) Reduced cancer incidence among the blind. *Epidemiol* 9(5): 490–494.

Filipski E, King VM, Li X, Granda TG, Mormont M-C, Liu X, Claustrat B, Hastings MH, Lévi F. (2002) Host circadian clock as a control point in tumor progression. *JNCI* 94(9): 690–697.

Fleming DM, Cross KW, Sunderland R, Ross AM. (2000) Comparison of the seasonal patterns of asthma identified in general practitioner episodes, hospital admissions, and deaths. *Thorax* 55(8): 662–665.

Focan C, Doalto L, Mazy V, Lévi F, Bruguerolle B, Cano JP, Rahmani R, Hecquet B. (1989) [48-hour continuous infusion of vindesine (followed by cisplatin) in advanced lung cancer. Chronopharmacokinetic data and clinical efficacy.] [French]. *Bull Cancer* 76(8): 909–912.

Focan C. (1995) Marker rhythms for cancer chronotherapy. From laboratory animals to human beings. *In Vivo* 9(4): 283–298 (Review).

Focan C, Denis B, Kreutz F, Focan-Henrard D, Lévi F. (1995) Ambulatory chronotherapy with 5-fluorouracil, folinic acid and carboplatin for advanced non-small cell lung cancer. A phase II feasibility trial. *J Infus Chemother* 5(3 Suppl 1): 148–152.

Focan C, Kreutz F, Focan-Henrard D, Moeneclaey N. (2000) Chronotherapy with 5-fluorouracil, folinic acid and carboplatin for metastatic colorectal cancer; an interesting therapeutic index in a phase II trial. *Eur J Cancer* 36(3): 341–347.

Focan C. (2002a) Chronobiological concepts underlying the chronotherapy of human lung cancer. *Chronobiol Intl* 19(1): 253–273 (Review).

Focan C. (2002b) Pharmaco-economic comparative evaluation of combination chronotherapy vs. standard chemotherapy for colorectal cancer. *Chronobiol Intl* 19(1): 289–297.

Folkard S, Glynn CJ, Lloyd JW. (1976) Diurnal variation and individual differences in the perception of intractable pain. *J Psychosom Res* 20(4): 289–301.

Fontes G, Rocha EM, Brito AC, Fireman FA, Antunes CM. (2000) The microfilarial periodicity of *Wuchereria bancrofti* in north-eastern Brazil. *Ann Trop Med Parasitol* 94(4): 373–379.

Fornasari PM, Gratton L, Dolci D, Gamba G, Ascari E, Montalbetti N, Halberg F. (1981) Circadian rhythms of clotting, fibrinolytic activators and inhibitors. In: *Chronobiology, Proc XIII Intl Conf Intl Soc Chronbiol*, Pavia, September 4–7, 1977. Halberg F, Scheving LE, Powell EW, Hayes DK, eds. Milan: Il Ponte, pp. 155–158.

Fox AW, Davis RL. (1998) Migraine chronobiology. *Headache* 38(6): 436–441.

Freedman SH, Ramcharan S, Hoag E. (1974) Some physiological and biochemical measurements over the menstrual cycle. In: *Biorhythms and Human Reproduction*. Ferin M, Halberg F, Richart RM, Vande Wiele RL, eds. New York: John Wiley & Sons, Inc., pp. 259–275.

Frishman WH, Glasser S, Stone P, Deedwania PC, Johnson M, Fakouhi TD. (1999) Comparison of controlled-onset, extended-release *Verapamil* with *Amlodipine* and *Amlodipine* plus *Atenolol* on exercise performance and ambulatory ischemia in patients with chronic stable angina pectoris. *Amer J Cardiol* 83: 507–514.

Froberg JE. (1977) Twenty-four-hour patterns in human performance, subjective and physiological variables, and differences between morning and evening active subjects. *Biol Psychol* 5(2): 119–134.

Fu L, Pelicano H, Liu J, Huang P, Lee C. (2002) The circadian gene *Period2* plays an important role in tumor suppression and DNA damage response *in vivo*. *Cell* 111(1): 41–50.

Gallerani M, Manfredini R, Ricci L, Grandi E, Cappato R, Calo G, Pareschi PL, Fersini C. (1992) Sudden death from pulmonary thromboembolism: chronobiological aspects. *Eur Heart J* 13(5): 661–665. Review.

Gallerani M, Manfredini R. (2000) Seasonal variation in herpes zoster infection. *Br J Dermatol* 142(3): 588–589.

Gallerani M, Boari B, de Toma D, Salmi R, Manfredini R. (2004) Seasonal variation in the occurrence of deep vein thrombosis. *Med Sci Moni* 10(5): CR191–CR196.

Gamus D, Romano A, Sucher E, Ashkenazi IE. (1995) Herpetic eye attacks: variability of circannual rhythms. *Brit J Ophthalmol* 79(1): 50–53.

Garde AH, Hansen AM, Skovgaard LT, Christensen JM. (2000) Seasonal and biological variation of blood concentrations of total cholesterol, dehydroepiandrosterone sulfate, hemoglobin A(1c), IgA, prolactin, and free testosterone in healthy women. *Clin Chem* 46(4): 551–559.

Gatika SM, Fujimaki Y, Njuguna MN, Gachihi GS, Mbugua JM. (1994) The microfilarial pattern of *Wuchereria bancrofti* in Kenya. *J Trop Med Hyg* 97(1): 60–64.

Gaultier C, Reinberg A, Girard F. (1977) Circadian rhythms in lung resistance and dynamic lung compliance of healthy children. *Respir Physiol* 31: 169–182.

Gelfand EW. (2004) Inflammatory mediators in allergic rhinitis. *J Allergy Clin Immunol* 114(5 Suppl): S135–S138 (Review).

Giacchetti S, Itzhaki M, Gruia G, Adam R, Zidani R, Kunstlinger F, Brienza S, Alafaci E, Bertheault-Cvitkovic F, Jasmin C, Reynes M, Bismuth H, Misset JL, Lévi F. (1999)

Long-term survival of patients with unresectable colorectal cancer liver metastases following infusional chemotherapy with 5-fluorouracil, leucovorin, oxaliplatin and surgery. *Ann Oncol* 10: 663–669.

Giacchetti S, Perpoint B, Zidani R, Le Bail N, Faggiuolo R, Focan C, Chollet P, Llory JF, Letourneau Y, Coudert B, Bertheaut-Cvitkovic F, Larregain-Fournier D, Le Rol A, Walter S, Adam R, Misset JL, Lévi F. (2000) Phase III multicenter randomized trial of oxaliplatin added to chronomodulated fluorouracil-leucovorin as first-line treatment of metastatic colorectal cancer. *J Clin Oncol* 18(1): 136–147.

Gibson T, Jarrett RJ. (1972) Diurnal variation in insulin sensitivity. *Lancet* 2(7784): 947–948.

Gill JS, Davies P, Gill SK, Beevers DG. (1988) Wind-chill and the seasonal variation of cerebrovascular disease. *J Clin Epidemiol* 41: 225–230.

Glasser SP. (1999) Circadian variations and chronotherapeutic implications for cardiovascular management: a focus on COER verapamil. *Heart Dis* 1(4): 226–232 (Review).

Glynn C, Crockford G, Gavaghan D, Cardno P, Price D, Miller J. (1990) Epidemiology of shingles. *J R Soc Med* 83(10): 617–619.

Gobel H, Cordes P. (1990) Circadian variation of pain sensitivity in pericranial musculature. *Headache* 30(7): 418–422.

Good RA, Sothern RB, Stoney PJ, Simpson HW, Halberg E, Halberg F. (1977) Circadian state dependence of adriamycin-induced tumor regression and recurrence rates in immunocytoma-bearing LOU rats (Abstract). *Chronobiologia* 4: 174.

Gorbacheva VY, Kondratov RV, Zhang R, Cherukuri S, Gudkov AV, Takahashi JS, Antoch MP. (2005) Circadian sensitivity to the chemotherapeutic agent cyclophosphamide depends on the functional status of the CLOCK/BMAL1 transactivation complex. *PNAS* 102(9): 3407–3412.

Gordon PC. (1966) The epidemiology of cerebrovascular disease in Canada: an analysis of mortality data. *Can Med Assoc J* 95(20): 1004–1011.

Gordon DJ, Hyde J, Trost DC, Whaley FS, Hannan PJ, Jacobs DR, Ekelund LG. (1988) Cyclic seasonal variation in plasma lipid and lipoprotein levels: the Lipid Research Clinics Coronary Primary Prevention Trial Placebo Group. *J Clin Epidemiol* 41(7): 679–689.

Goudevenos JA, Papadimitriou ED, Papathanasiou A, Makis AC, Pappas K, Sideris DA. (1995) Incidence and other epidemiological characteristics of sudden cardiac death in northwest Greece. *Intl J Cardiol* 49(1): 67–75.

Green CB. (2005) Time for chronotherapy? Clock genes dictate sensitivity to cyclophosphamide. *PNAS* 102(10): 3529–3530.

Grimaudo V, Hauert J, Bachmann F, Kruithof EKO. (1988) Diurnal variation of the fibrinolytic system. *Throm Haemost* 59: 495–499.

Granda TG, Filipski E, D'Attino RM, Vrignaud P, Anjo A, Bissery MC, Levi F. (2001) Experimental chronotherapy of mouse mammary adenocarcinoma MA13/C with docetaxel and doxorubicin as single agents and in combination. *Cancer Res* 61(5): 1996–2001.

Granda TG, Levi F. (2002) Tumor-based rhythms of anticancer efficacy in experimental models. *Chronobiol Intl* 19(1): 21–41 (Review).

Grassly NC, Fraser C, Garnett GP. (2005) Host immunity and synchronized epidemics of syphilis across the United States. *Nature* 433(7024): 417–421.

Gruska M, Gaul GB, Winkler M, Levnaic S, Reiter C, Voracek M, Kaff A. (2005) Increased occurrence of out-of-hospital cardiac arrest on Mondays in a community-based study. *Chronobiol Intl* 22(1): 107–120.

Guberan E, Williams MK, Walford J, Smith MM. (1969) Circadian variation in F.E.V. in shift workers. *Brit J Industr Med* 26(2): 121–125.

Haberman S, Capildeo R, Rose FC. (1981) The seasonal variation in mortality from cerebrovascular disease. *J Neurol Sci* 52: 25–36.

Hack LM, Lockley SW, Arendt J, Skene DJ. (2003) The effects of low-dose 0.5-mg melatonin on the free-running circadian rhythms of blind subjects. *J Biol Rhythms* 18(5): 420–429.

Hahn RA. (1991) Profound bilateral blindness and the incidence of breast cancer. *Epidemiology* 2(3): 208–210.

Hahn RA. (1998) Does blindness protect against cancers? *Epidemiology* 9(5): 481–483 (Review).

Halberg F, Bittner JJ, Gully RJ, Albrecht PG, Brackney EL. (1955) 24-hour periodicity and audiogenic convulsions in I mice of various ages. *Proc Soc Exp Biol Med* 88: 169–173.

Halberg F, Stephens AN. (1958) 24-hour periodicity in mortality of C mice from *E. coli* lipopolysaccharide (Abstract 1725). *Fed Proc* 17: 439.

Halberg F. (1960) Temporal coordination of physiologic function. In: *Biological Clocks. Cold Spring Harbor Symposia on Quantitative Biology.* Vol 25. New York: the Biological Laboratory, pp. 289–310.

Halberg F, Johnson E, Brown BW, Bittner JJ. (1960) Susceptibility rhythm to *E. coli* endotoxin and bioassay. *Proc Soc Exp Biol* 103: 142–144.

Halberg F, Hamburger C. (1964) 17-ketosteroid and volume of human urine. Weekly and other changes with low frequency. *Minn Med* 47: 916–925.

Halberg F, Engeli M, Hamburger C, Hillman D. (1965) Spectral resolution of low-frequency, small amplitude rhythms in excreted 17-ketosteroid; probable androgen-induced circaseptan desynchronization. *Acta Endocrinol (Kbh)* Suppl 103: 5–54.

Halberg F, Sothern RB, Roitman B, Halberg E, Halberg Fcn, Mayersbach Hv, Haus E, Scheving LE, Kanabrocki EL, Bartter FC, Delea C, Simpson HW, Tavadia HB, Fleming KA, Hume P, Wilson C. (1977a) Agreement of circadian characteristics for total leucocyte counts in different geographic locations. In: *Proc XII Intl Conf Intl Soc Chronobiology.* Milano: Il Ponte, pp. 3–17.

Halberg F, Gupta BD, Haus E, Halberg E, Deka AC, Nelson W, Sothern RB, Cornélissen G, Lee JK, Lakatua DJ, Scheving LE, Burns ER. (1977b) Steps towards a cancer chronopolytherapy. In: *Proc XIVth Intl Cong of Therapeutics*, Montpellier, France. Paris: L'Expansion Scientifique Francaise, pp. 151–196.

Halberg F, Lee JK, Nelson WL. (1978) Time-qualified reference intervals–chronodesms. *Experientia* 34(6): 713–716.

Halberg F, Cornélissen G, Sothern RB, Wallach LA, Halberg E, Ahlgren A, Kuzel M, Radke A, Barbosa J, Goetz F, Buckley J, Mandel J, Schuman L, Haus E, Lakatua D, Sackett L, Berg H, Wendt HW, Kawasaki T, Ueno K, Uezono K, Matsuoka M, Omae T, Tarquini B, Cagnoni M, Garcia Sainz M, Perez Vega E, Wilson D, Griffiths K, Donati L, Tatti B, Vasta M, Locatelli I, Camagna A, Lauro R, Tritsch G, Wetterberg L. (1981) International geographic studies of oncological interest on chronobiological variables. In: *Neoplasms – Comparative Pathology of Growth in Animals, Plants and Man.* Kaiser H, ed. Baltimore: Williams and Wilkins, pp. 553–596.

Halberg F, Lagoguey M, Reinberg A. (1983) Human circannual rhythms over a broad spectrum of physiological processes. *Intl J Chronobiol* 8(4): 225–268.

Halberg F, Drayer JIM, Cornélissen G, Weber MA. (1984) Cardiovascular reference data for recognizing circadian mesor- and amplitude-hypertension in apparently healthy men. *Chronobiologia* 11: 275–298.

Halberg F, Cornelissen G, Wall D, Otsuka K, Halberg J, Katinas G, Watanabe Y, Halhuber M, Bohn TM, Delmore P, Siegelova J, Homolka P, Fiser B, Dusek J, Sanchez de la Peña S, Maggioni C, Delyukov A, Gorgo Y, Gubin D, Carandente F, Schaffer E, Rhodus N,

Borer K, Sonkowsky RP, Schwartzkopff O. (2002) Engineering and governmental challenge: 7-day/24-hour chronobiologic blood pressure and heart rate screening: Part I. *Biomed Instrum Technol* 36(2): 89–122 (Review).

Halberg F, Cornélissen G, Katinas G, Syutkina EV, Sothern RB, Zaslavskaya R, Halberg J, Halberg Fcn, Watanabe Y, Schwartzkopff O. (2003) Transdisciplinary unifying implications of circadian findings in the 1950s. *J Circadian Rhythms* 1: 2. (*http://www.jcircadianrhythms.com/content/1/1/2*).

Hancox JG, Sheridan SC, Feldman SR, Fleischer AB Jr. (2004) Seasonal variation of dermatologic disease in the USA: a study of office visits from 1990 to 1998. *Intl J Dermatol* 43(1): 6–11.

Harkness RA. (1974) Variations in testosterone excretion in man. In: *Biorhythms and Human Reproduction*. Ferin M, Halberg F, Richart RM, Vande Wiele RL, eds. New York: John Wiley & Sons, Inc., pp. 469–478.

Harlap S, Kark JD, Baras M, Eisenberg S, Stein Y. (1982) Seasonal changes in plasma lipid and lipoprotein levels in Jerusalem. *Isr J Med Sci* 18(11): 1158–1165.

Hastings MH, Reddy AB, Maywood ES. (2003) A clockwork web: circadian timing in brain and periphery, in health and disease. *Nat Rev Neurosci* 4(8): 649–661 (Review).

Hattori A, Migitaka H, Iigo M, Itoh M, Yamamoto K, Ohtani-Kaneko R, Hara M, Suzuki T, Reiter RJ. (1995) Identification of melatonin in plants and its effects on plasma melatonin levels and binding to melatonin receptors in vertebrates. *Biochem Mol Biol Intl* 35(3): 627–634.

Haus E, Halberg F. (1959) 24-hour rhythm in susceptibility of C-mice to a toxic dose of ethanol. *J Appl Physiol* 14(6): 878–880.

Haus E, Halberg F, Scheving LE, Pauly JE, Cardoso SS, Kühl JFW, Sothern RB, Shiotsuka RN, Hwang DS. (1972) Increased tolerance of leukemic mice to arabinosyl cytosine with schedule adjusted to circadian system. *Science* 177(43): 80–82.

Haus E, Halberg F, Loken MK, Kim YS. (1974a) Circadian rhythmometry of mammalian radiosensitivity. In: *Space Radiation Biology*. Tobias CA, Todd P, eds. New York: Academic Press, pp. 435–474.

Haus E, Halberg F, Loken MK. (1974b) Circadian susceptibility-resistance cycle of bone marrow cells to whole body x-irradiation in Balb/c mice. In: *Chronobiology*. Scheving LE, Halberg F, Pauly JE, eds. Tokyo: Igaku-Shoin Ltd., pp. 115–122.

Haus E, Halberg F, Kühl JFW, Lakatua DJ. (1974c) Chronopharmacology in animals. In: *Chronobiological Aspects of Endocrinology*. Aschoff J, Ceresa F, Halberg F, eds. Stuttgart: FK Schattauer-Verlag, pp. 122–156.

Haus E, Fernandes G, Kuhl JF, Yunis EJ, Lee JK, Halberg F. (1974d) Murine circadian susceptibility rhythm to cyclophosphamide. *Chronobiologia* 1(3): 270–277.

Haus E, Nicolau GY, Lakatua D, Sackett-Lundeen L. (1988) Reference values for chronopharmacology. In: *Ann Rev Chronopharm*, Vol. 4. Reinberg A, Smolensky M, Labrecque G, eds. Oxford: Pergamon Press, pp. 333–424.

Haus E, Nicolau G, Lakatua DJ, Sackett-Lundeen L, Petrescu E. (1989) Circadian rhythm parameters of endocrine functions in elderly subjects during the seventh to the ninth decade of life. *Chronobiologia* 16(4): 331–352.

Haus E, Cusulos M, Sackett-Lundeen L, Swoyer J. (1990) Circadian variations in blood coagulation parameters, alpha-antitrypsin antigen and platelet aggregation and retention in clinically healthy subjects. *Chronobiol Intl* 7(3): 203–216.

Haus E, Touitou Y. (1997) Chronobiology of development and aging. In: *Physiology and Pharmacology of Biological Rhythms*. Redfern PH, Lemmer B, eds. Berlin: Springer-Verlag, pp. 95–134.

Hawking F. (1962) Microfilaria infestation as an instance of periodic phenomena seen in host-parasite relationships. *Annals NY Acad Sci* 98(4): 940–953.
Hébert M, Dumont M, Paquet J. (1998) Seasonal and diurnal patterns of human illumination under natural conditions. *Chronobiol Intl* 15(1): 59–70.
Hejl Z. (1977) Daily, lunar, yearly and menstrual cycles and bacterial or viral infections in man. *J Interdiscipl Cycle Res* 8(3–4): 250–253.
Henry F, Arrese JE, Claessens N, Piérard-Franchimont C, Piérard GE. (2002) [Skin and its daily chronobiological clock.] [French]. *Rev Med Liege* 57(10): 661–665 (Review).
Hermida Dominguez RC, Halberg F, Langevin TR. (1986) Serial white blood cell counts and chronochemotherapy according to highest values (macrophases) or by model characteristics (acrophases). In: *Proc 2nd Intl Conf Medico-Social Aspects of Chronobiology*, Florence, October 2, 1984. Halberg F, Reale L, Tarquini B, eds. Rome: Instituto Italiano di Medicina Sociale, pp. 327–343.
Hermida RC, Fernandez JR. (1996) Computation of time-specified tolerance intervals for ambulatorily monitored blood pressure. *Biomed Instrum Technol* 30(3): 257–266.
Hermida RC, Fernández JR, Alonso I, Ayala DE, Garcia L. (1997a) Computation of time-specified tolerance intervals for hybrid time series with nonequidistant sampling, illustrated for plasma growth hormone. *Chronobiol Intl* 14: 409–425.
Hermida RC, Fernández JR, Ayala DE, Mojón A, Iglesias M. (1997b) Influence of aspirin usage on blood pressure: dose and administration-time dependencies. *Chronobiol Intl* 14(6): 619–637.
Hermida RC, Ayala DE, Iglesias M, Mojón A, Silva I, Ucieda R, Fernández JR. (1997c) Time-dependent effects of low-dose aspirin administration on blood pressure in pregnant women. *Hypertension* 39(3): 589–595.
Hermida RC. (1999) Time-qualified reference values for 24 h ambulatory blood pressure monitoring. *Blood Press Monitoring* 4: 137–147. Review.
Hermida RC, Fernandez JR, Mojón A, Ayala DE. (2000) Reproducibility of the hyperbaric index as a measure of blood pressure excess. *Hypertension* 35(1 Pt 1): 118–125.
Hermida RC, Fernandez JR, Ayala DE, Mojón A, Alonso I, Smolensky M. (2001) Circadian rhythm of double (rate-pressure) product in healthy normotensive young subjects. *Chronobiol Intl* 18(3): 475–489.
Hermida RC, Mojón A, Fernandez JR, Alonso I, Ayala DE. (2002a) The tolerance-hyperbaric test: a chronobiologic approach for improved diagnosis of hypertension. *Chronobiol Intl* 19(6): 1183–1211.
Hermida RC, Calvo C, Ayala DE, Mojón A, Lopez JE. (2002b) Relationship between physical activity and blood pressure in dipper and non-dipper hypertensive patients. *J Hypertens* 20(6): 1097–1104.
Hermida RC, Calvo C, Ayala DE, Lopez JE, Fernandez JR, Mojón A, Dominguez MJ, Covelo M. (2003a) Seasonal variation of fibrinogen in dipper and nondipper hypertensive patients. *Circulation* 108(9): 1101–1106.
Hermida RC, Ayala DE, Calvo C, Lopez JE, Fernandez JR, Mojón A, Dominguez MJ, Covelo M. (2003b) Administration time-dependent effects of aspirin on blood pressure in untreated hypertensive patients. *Hypertension* 41(6): 1259–1267.
Hermida RC, Ayala DE, Iglesias M. (2003c) Administration time-dependent influence of aspirin on blood pressure in pregnant women. *Hypertension* 41(3 Pt 2): 651–656.
Hermida RC, Ayala DE, Iglesias M. (2003d) Circadian rhythm of blood pressure challenges office values as the "gold standard" in the diagnosis of gestational hypertension. *Chronobiol Intl* 20(1): 135–156.

Hermida RC, Smolensky MH. (2004) Chronotherapy of hypertension. *Curr Opin Nephrol Hypertens* 13(5): 501–505.

Herxheimer A, Waterhouse J. (2003) The prevention and treatment of jet lag. *BMJ* 326(7384): 296–297.

Hjalmarson A, Gilpin EA, Nicod P, Dittrich H, Henning H, Engler R, Blacky AR, Smith SC Jr, Ricou F, Ross J Jr. (1989) Differing circadian patterns of symptom onset in subgroups of patients with acute myocardial infarction. *Circulation* 80(2): 267–275.

Hodoglugil U, Gunaydin B, Yardim S, Zengil H, Smolensky MH. (2001) Seasonal variation in the effect of a fixed dose of heparin on activated clotting time in patients prepared for open-heart surgery. *Chronobiol Intl* 18(5): 865–873.

Honma K, Hashimoto S, Nakao M, Honma S. (2003) Period and phase adjustments of human circadian rhythms in the real world. *J Biol Rhythms* 18(3): 261–270.

Hostmark J, Laerum OD, Farsund T. (1984) Seasonal variations of symptoms and occurrence of human bladder carcinomas. *Scand J Urol Nephrol* 18(2): 107–111.

Hrushesky WJM. (1985) Circadian timing of cancer chemotherapy. *Science* 228: 73–75.

Hrushesky WJM, Bluming AZ, Gruber SA, Sothern RB. (1989a) Menstrual influence on surgical cure of breast cancer. *Lancet* 8669: 949–952.

Hrushesky WJM, Roemeling Rv, Sothern RB. (1989b) Circadian chronotherapy: from animal experiments to human cancer chemotherapy. In: *Chronopharmacology: Cellular and Biochemical Interactions.* Lemmer B, ed. New York: Marcel Dekker, Inc., pp. 439–473.

Hrushesky WJM, Roemeling Rv, Sothern RB. (1989c) Preclinical and clinical cancer chemotherapy. In: *Biological Rhythms in Clinical Practice.* Arendt J, Minors DS, Waterhouse JM, eds. London: Butterworth & Co. Ltd, pp. 225–252.

Hrushesky WJM, Bjarnason GA. (1993a) Circadian cancer therapy. *J Clin Oncol* 11(7): 1403–1417.

Hrushesky WJM, Bjarnason GA. (1993b) Newer approaches to cancer treatment. The application of circadian chronobiology to cancer chemotherapy. In: *Cancer: Principles and Practice of Oncology,* 4th edn. DeVita VT, Hellman S, Rosenberg SA, eds. Philadelphia: JB Lippincott, pp. 2666–2686.

Hrushesky WJM. (1996) Breast cancer, timing of surgery, and the menstrual cycle: call for prospective trial. *J Women's Health* 5: 555–566.

Imeson JD, Meade TW, Steward GM. (1987) Day by day variation in fibrinolytic activity and in mortality from ischemic heart disease. *Intl J Epidemiol* 16(4): 626–627.

Jacobsen FM, Wehr TA, Sack DA, James SP, Rosenthal NE. (1987) Seasonal affective disorder: a review of the syndrome and its public health implications. *Amer J Pub Health* 77(1): 57–60.

Jarjour NN. (1999) Circadian variation in allergen and nonspecific bronchial responsiveness in asthma. *Chronobiol Intl* 16(5): 631–639.

Jarrett RJ. (1972) Circadian variation in blood glucose levels, in glucose tolerance and in plasma immunoreactive insulin levels. [Multilingual]. *Acta Diabetol Lat* 9(2): 263–275 (Review).

Jarrett RJ, Baker IA, Keen H, Oakley NW. (1972) Diurnal variation in oral glucose tolerance: blood sugar and plasma insulin levels morning, afternoon, and evening. *Brit Med J* 1(794): 199–201.

Johansson BB, Norrving B, Widner H, Wu JY, Halberg F. (1990) Stroke incidence: circadian and circaseptan (about weekly) variations in onset. *Prog Clin Biol Res 341A: Chronobiology: Its Role in Clinical Medicine, General Biology, and Agriculture, Part A.* Hayes DK, Pauly JE, Reiter RJ, eds. New York: Wiley-Liss, Inc., pp. 427–436.

Johnstone MT, Mittleman M, Tofler G, Muller JE. (1996) The pathophysiology of the onset of morning cardiovascular events. *Amer J Hypertens* 9(4 Pt 3): 22S–28S (Review).

Jores A. (1935) Physiologie und Pathologie der 24-Stunden-Rhythmik des Menschen. *Ergeb Inner Med u Kinderheilk* 48: 574–629.

Kaiser IH, Halberg F. (1962) Circadian periodic aspects of birth. *Annals NY Acad Sci* 98(4): 1056–1068.

Kanabrocki EL, Scheving LE, Olwin JH, Marks JS, McCormick JB, Halberg F, Pauly JE, Greco J, DeBartolo M, Nemchausky BA, Kaplan E, Sothern R. (1983) Circadian variations in the urinary excretion of electrolytes and trace elements in man. *Amer J Anat* 166: 121–148.

Kanabrocki EL, Graham L, Veatch R, Greco J, Kaplan E, Nemchausky BA, Halberg F, Sothern R, Scheving LE, Pauly JE, Wetterberg L, Olwin J, Marks GE. (1987) Circadian variations in eleven radioimmunoassay variables in the serum of clinically healthy men. In: *Advances in Chronobiology – Part A,* Proc XVIIth Intl Conf, Intl Soc Chronobiol, Little Rock, AR, Nov. 3–7, 1985. Pauly JE, Scheving LE, eds. New York: Alan R Liss, pp. 317–327.

Kanabrocki EL, Snedeker PW, Zieher SJ, Raymond R, Gordy J, Bird T, Sothern RB, Hrushęsky WJM, Marks G, Olwin JH, Kaplan E. (1988) Circadian characteristics of dialyzable and non-dialyzable human urinary electrolytes, trace elements and total solids. *Chronobiol Intl* 5: 175–184.

Kanabrocki EL, Sothern RB, Scheving LE, Vesely DL, Tsai TH, Shelstad J, Cournoyer C, Greco J, Mermall H, Nemchausky BM, Bushnell DL, Kaplan E, Kahn S, Augustine G, Holmes E, Rumbyrt J, Sturtevant RP, Sturtevant F, Bremner F, Third JLHC, McCormick JB, Mudd CA, Dawson S, Sackett-Lundeen L, Haus E, Halberg F, Pauly JE, Olwin JH. (1990a) Reference values for circadian rhythms of 98 variables in clinically healthy men in fifth decade of life. *Chronobiol Intl* 7(5/6): 445–461.

Kanabrocki EL, Sothern RB, Scheving LE, Vesely DL, Tsai TH, Shelstad J, Cournoyer C, Greco J, Mermall H, Nemchausky BM, Bushnell DL, Kaplan E, Kahn S, Augustine G, Holmes E, Rumbyrt J, Sturtevant RP, Sturtevant F, Bremner F, Third JLHC, McCormick JB, Mudd CA, Dawson S, Olwin JH, Sackett-Lundeen L, Haus E, Halberg F, Pauly JE, Hrushesky WJM. (1990b) Circadian reference data for men in fifth decade of life. In: *Chronobiology: Its Role in Clinical Medicine, General Biology and Agriculture, Part A,* Proc XIXth Intl Conf Intl Soc Chronobiol, Bethesda, MD, June 20–24, 1989. Hayes DK, Pauly JE, Reiter RJ, eds. New York: Wiley-Liss, Inc., pp. 771–781.

Kanabrocki EL, Kanabrocki JA, Sothern RB, Futscher B, Lampo S, Cournoyer C, Rubnitz ME, Zieher SJ, Greco J, Bushnell DL, Tsai TH, Scheving LE, Olwin JH. (1990c) Circadian distribution of proteins in urine from healthy young men. *Chronobiol Intl* 7(5/6): 433–443.

Kanabrocki EL, Sothern RB, Bremner WF, Demakis JG, Bean JT, Ringelstein JG, Riley C, Fabbrini N, Crosby TJ, Mermall H, Third JLHC, Shirazi P, Olwin JH. (1995) Weekly and yearly rhythms in plasma fibrinogen in hospitalized male military veterans. *Amer J Cardiol* 76: 628–631.

Kanabrocki EL, Sothern RB, Messmore HL, Roitman-Johnson B, McCormick JB, Dawson S, Bremner WF, Third JLHC, Nemchausky BA, Shirazi P, Scheving LE. (1999) Circadian inter-relationships among levels of plasma fibrinogen, blood platelets, and serum interleukin-6. *Clin Appl Thromb Hemostasis* 5(1): 37–42.

Karasek M, Pawlikowski M, Nowakowska-Jankiewicz B, Kolodziej-Maciejewka H, Zieleniewski J, Cieslak D, Leidenberger F. (1990) Circadian variations in plasma melatonin, FSH, LH, and prolactin and testosterone levels in infertile men. *J Pineal Res* 9(2): 149–157.

Karnik JA, Young BS, Lew NL, Herget M, Dubinsky C, Lazarus JM, Chertow GM. (2001) Cardiac arrest and sudden death in dialysis units. *Kidney Intl* 60(1): 350–357.

Kasper S, Wehr TA, Rosenthal NE. (1988) [Season-related forms of depression: I. Principles and clinical description of the syndrome.] [German]. *Nervenarzt* 59(4): 191–199 (Review).

Kasper S, Wehr TA, Bartko JJ, Gaist PA, Rosenthal NE. (1989) Epidemiological findings of seasonal changes in mood and behavior. A telephone survey of Montgomery County, Maryland. *Arch Gen Psychiatry* 46(9): 823–833.

Kawamura T, Kondo H, Hirai M, Wakai K, Tamakoshi A, Terazawa T, Osugi S, Ohno M, Okamoto N, Tsuchida T, Ohno Y, Toyama J. (1999) Sudden death in the working population: a collaborative study in central Japan. *Eur Heart J* 20(5): 338–343.

Keene KS, Rich TA, Penberthy DR, Shepard RC, Adams R, Jones RS. (2005) Clinical experience with chronomodulated infusional 5-fluorouracil chemoradiotherapy for pancreatic adenocarcinoma. *Intl J Radiat Oncol Biol Phys* 62(1): 97–103.

Kennaway DJ, Voultsios A. (1998) Circadian rhythm of free melatonin in human plasma. *J Clin Endocrinol Metab* 83(3): 1013–1015.

Kerenyi NA, Pandula E, Feuer G. (1990) Why the incidence of cancer is increasing: the role of "light-pollution." *Med Hypoth* 33(2): 75–78.

Kerr HD. (1973) Diurnal variation of respiratory function independent of air quality: experience with an environmentally controlled exposure chamber for human subjects. *Arch Environ Health* 26(3): 144–152.

Koch HJ, Szecsey A, Jost D, Fischer-Barnicol D, Ibach B. (2003) [Circannual periodicity of prescriptions in a psychiatric hospital.] [German]. *Psychiatr Prax* 30(Suppl 2): 226–228.

Koch HJ, Raschka C. (2004) Diurnal variation of pain perception in young volunteers using the tourniquet pain model. *Chronobiol Intl* 21(1): 171–173.

Korte J, Wulff K, Oppe C, Siegmund R. (2001) Ultradian and circadian activity-rest rhythms of preterm neonates compared to full-term neonates using actigraphic monitoring. *Chronobiol Intl* 18(4): 697–708.

Kripke DF, Gregg L. (1990) Circadian effects of varying environmental lights. In: *Medical Monitoring in the Home and Work Environment*. Miles LE, Broughon RJ, eds. New York: Raven Press, pp. 187–195.

Kristal-Boneh E, Harari G, Green MS. (1993) Circannual variations in blood cholesterol levels. *Chronobiol Intl* 10(1): 37–42.

Kühl JFW, Lee JK, Halberg F, Haus E, Günther R, Knapp E. (1974) Circadian and lower frequency rhythms in male grip strength and body weight. In: *Biorhythms and Human Reproduction*. Ferin M, Halberg F, Richart RM, Vande Wiele RL, eds. New York: John Wiley & Sons, Inc., pp. 529–548.

Kumar D, Wingate D, Ruckebusch Y. (1986) Circadian variation in the propagation velocity of the migrating motor complex. *Gastroenterol* 91: 926–930.

Kusanagi H, Mishima K, Satoh K, Echizenya M, Katoh T, Shimizu T. (2004) Similar profiles in human period1 gene expression in peripheral mononuclear and polymorphonuclear cells. *Neurosci Lett* 365(2): 124–127.

Labrecque G, Reinberg AE. (1989) Chronopharmacology of nonsteroid anti-inflammatory drugs. In: *Chronopharmacology: Cellular and Biochemical Interactions*. Lemmer B, ed. New York: Marcel Dekker, pp. 545–579.

Labrecque G, Bélanger PM. (1991) Biological rhythms in the absorption, distribution, metabolism and excretion of drugs. *Pharmacol Ther* 52(1): 95–107 (Review).

Lakatos P, Blumsohn A, Eastell R, Tarjan G, Shinoda H, Stern PH. (1995) Circadian rhythm of in vitro bone-resorbing activity in human serum. *J Clin Endocrin Metab* 80(11): 3185–3190.

Landau I, Chabaud A, Cambie G, Ginsburg H. (1991) Chronotherapy of malaria: an approach to malaria chemotherapy. *Parasitol Today* 7(12): 350–352.

Landau I, Lepers JP, Ringwald P, Rabarison P, Ginsburg H, Chabaud A. (1992) Chronotherapy of malaria: improved efficacy of timed chloroquine treatment of patients with *Plasmodium falciparum* infections. *Trans R Soc Trop Med Hyg* 86(4): 374–375.

Landau I, Caillard V, Beaute-Lafitte A, Chabaud A. (1993) Chronobiology and chronotherapy of malaria: investigations with murine malaria models. *Parassitologia* 35 Suppl: 55–57 (Review).

Langlois PH, White RF, Glezen WP. (1986) Diurnal variation in human response to influenza vaccination? A pilot study of 125 volunteers. In: *Ann Rev Chronopharmacol*, Vol 3. Reinberg A, Smolensky M, Labrecque G, eds. Oxford: Pergamon Press, p. 123.

Lanza GA, Patti G, Pasceri V, Manolfi M, Sestito A, Lucente M, Crea F, Maseri A. (1999) Circadian distribution of ischemic attacks and ischemia-related ventricular arrhythmias in patients with variant angina. *Cardiologia* 44(10): 913–919 (Review).

Latreille J, Guinot C, Robert-Granie C, Le Fur I, Tenenhaus M, Foulley JL. (2004) Daily variations in skin surface properties using mixed model methodology. *Skin Pharmacol Physiol* 17(3): 133–140.

Ledsome JR, Lessoway V, Susak LE, Gagnon FA, Gagnon R, Wing PC. (1996) Diurnal changes in lumbar intervertebral distance, measured using ultrasound. *Spine* 21(14): 1671–1675.

Lee CF, Proffit WR. (1995) The daily rhythm of tooth eruption. *Amer J Ortho Dentofacial Orthoped* 107(1): 38–47.

Lee RE, Smolensky MH, Leach CS, McGovern JP. (1977) Circadian rhythms in the cutaneous reactivity to histamine and selected antigens, including phase relationship to urinary cortisol excretion. *Ann Allergy* 38(4): 231–236.

Le Fur I, Reinberg A, Lopez S, Morizot F, Mechkouri M, Tschachler E. (2001) Analysis of circadian and ultradian rhythms of skin surface properties of face and forearm of healthy women. *J Invest Dermatol* 117(3): 718–724.

Lemmer B. (1986) The chronopharmacology of cardiovascular medications. In: *Ann Rev Chronopharmacol*, Vol 2. Reinberg A, Smolensky M, Labrecque G, eds. Oxford: Pergamon Press, pp. 199–228.

Lemmer B, ed. (1989) *Chronopharmacology: Cellular and Biochemical Interactions.* New York: Marcel Dekker, 720 pp.

Lemmer B, Portaluppi F. (1997) Chronopharmacology of cardiovascular diseases. In: *Physiology and Pharmacology of Biological Rhythms*. Redfern PH, Lemmer B, eds. Berlin: Springer-Verlag, pp. 251–297.

Lester D. (1971) Seasonal variation in suicidal deaths. *Brit J Psychiat* 118(547): 627–628.

Lévi F, Metzger G, Bailleul, Reinberg A, Mathé G. (1986) Circadian-varying plasma pharmacokinetics of doxorubicin (DOX) despite continuous infusion at constant rate (Abstract 693). *Proc Amer Assoc Cancer Res* 27: 175.

Lévi FA, Canon C, Touitou Y, Sulon J, Mechkouri M, Demey-Ponsard R, Touboul JP, Vannetzel JM, Mowrowicz I, Reinberg A, Mathé G. (1988a) Circadian rhythms in circulating T lymphocyte subtypes, and plasma testosterone, total and free cortisol and in five healthy men. *Clin Exp Immunol* 71: 329–335.

Lévi F, Adam R, Soussan A, Caussanel JP, Benavides M, Misset JL, Burki F, Reinberg A, Smolensky M, Bismuth H, Mathé G. (1988b) Ambulatory 5-day chronotherapy of colorectal cancer with continuous venous infusion of 5-fluorouracil (5-FU) at circadian-modulated rate. Preliminary results. In: *Ann Rev Chronopharm*, Vol 5. Reinberg A, Smolensky M, Labrecque G, eds. New York: Pergamon Press, pp. 419–422.

Lévi F, Boughattas NA, Blazsek I. (1988c) Comparative murine chronotoxicity of anticancer agents and related mechanisms. In: *Ann Rev Chronopharmacol*, Vol 4. Reinberg A, Smolensky M, Labrecque G, eds. New York: Pergamon Press, pp. 283–331.

Lévi F, Benavides M, Chevelle C, Le Saunier F, Bailleul F, Misset J-L, Regensberg C, Vannetzel, J-M, Reinberg A, Mathé G. (1990) Chemotherapy of advanced ovarian cancer with 4(-O-tertahyropyranyl doxorubicin and cisplatin: a randomized phase II trial with an evaluation of circadian timing and dose-intensity. *J Clin Oncol* 8(4): 705–714.

Lévi F, Reinberg A. (1990) Meeting Report. Chronobiology and Chronotherapy of Cancer. *Chronobiol Intl* 7(5/6): 471–474.

Lévi F. (1994) Chronotherapy of cancer: biological basis and clinical application. *Pathol Biol* (Paris) 42(4): 338–341 (Review).

Lévi FA, Zidani R, Vannetzel JM, Perpoint B, Focan C, Faggiuolo R, Chollet P, Garufi C, Itzhaki M, Dogliotti L, Iacobelli S, Adam R, Kumstlinger F, Gastiaburu J, Bismuth H, Jasmin C, Misset J-L. (1994) Chronomodulated versus fixed-infusion-rate delivery of ambulatory chemotherapy with oxaliplatin, fluorouracil, and folinic acid (Leucovorin) in patients with colorectal cancer metastases: a randomized multi-institutional trial. *JNCI* 86: 1608–1617.

Lévi F. (1999) Cancer chronotherapy. *J Pharm Pharmacol* 51: 891–898.

Lévi F. (2001) Circadian chronotherapy for human cancers. Oncol 2(5): 307–315 (Review).

Lévi F. (ed). (2002) Cancer chronotherapeutics. *Chronobiol Intl* 19(1) (special issue): 1–323.

Levine H, Halberg F, Sothern RB, Bartter FC, Meyer WJ, Delea C. (1974) Circadian phase-shifting with and without geographic displacement. In: *Biorhythms and Human Reproduction*. Ferin M, Halberg F, Richart RM, Vande Wiele RL, eds. New York: John Wiley & Sons, Inc., pp. 557–574.

Levine RJ, Methew RM, Chenault B, Brown MH, Hurtt ME, Bentley KS, Mohr KL, Working PK. (1990) Differences in the quality of semen in outdoor workers during summer and winter. *New Engl J Med* 323(1): 12–16.

Levine RJ. (1999) Seasonal variation of semen quality and fertility. *Scand J Work Environ Health* 25(suppl 1): 34–37.

Lewy AJ, Wehr TA, Goodwin FK, Newsome DA, Markey SP. (1980) Light suppresses melatonin secretion in humans. *Science* 210: 1267–1269.

Lewy AJ, Emens JS, Sack RL, Hasler BP, Bernert RA. (2003) Low, but not high, doses of melatonin entrained a free-running blind person with a long circadian period. *Chronobiol Intl* 19(3): 649–658.

Lewy AJ, Emens JS, Bernert RA, Lefler BJ. (2004) Eventual entrainment of the human circadian pacemaker by melatonin is independent of the circadian phase of treatment initiation: clinical implications. *J Biol Rhythms* 19(1): 68–75.

Lin MC. Kripke DF, Parry BL, Berga SL. (1990) Night light alters menstrual cycles. *Psychiat Res* 33(2): 135–138.

Ling PD, Lednicky JA, Keitel WA, Poston DG, White ZS, Peng R, Liu Z, Mehta SK, Pierson DL, Rooney CM, Vilchez RA, Smith EO, Butel JS. (2003) The dynamics of herpesvirus and polyomavirus reactivation and shedding in healthy adults: a 14-month longitudinal study. *J Infect Dis* 187(10): 1571–1580.

Lis CG, Grutsch JF, Wood P, You M, Rich I, Hrushesky WJ. (2003) Circadian timing in cancer treatment: the biological foundation for an integrative approach. *Integr Cancer Ther* 2(2): 105–111 (Review).

Little MA, Rummel JA. (1971) Circadian variations in thermal and metabolic responses to heat exposure. *J Appl Physiol* 31(4): 556–561.

Lockley SW, Skene DJ, Butler LJ, Arendt J. (1999) Sleep and activity rhythms are related to circadian phase in the blind. *Sleep* 22(5): 616–623.

Lockley SW, Skene DJ, James K, Thapan K, Wright J, Arendt J. (2000) Melatonin administration can entrain the free-running circadian system of blind subjects. *J Endocrinol* 164(1): R1–R6.

Low KG, Feissner JM. (1998) Seasonal affective disorder in college students: prevalence and latitude. *J Amer College Health* 47(3): 135–137.

Lowrey PL, Takahashi JS. (2004) Mammalian circadian biology: elucidating genome-wide levels of temporal organization. *Ann Rev Genomics Hum Genet* 5: 407–441 (Review).

Lucente M, Rebuzzi AG, Lanza GA, Tamburi S, Cortellessa MC, Coppola E, Iannarelli M, Manzoli U. (1988) Circadian variation of ventricular tachycardia in acute myocardial infarction. *Amer J Cardiol* 62(10): 670–674.

Mackowiak PA, Wasserman SS, Levine MM. (1992) A critical appraisal of 98.6°C, the upper limit of the normal body temperature, and other legacies of Carl Reinhold August Wunderlich. *JAMA* 268(12): 1578–1580.

Madden PA, Heath AC, Rosenthal NE, Martin NG. (1996) Seasonal changes in mood and behavior. The role of genetic factors. *Arch Gen Psychiatry* 53(1): 47–55.

Maes M, Scharpe S, Cooreman W, Wauters A, Neels H, Verkerk R, De Meyer F, D'Hondt P, Peeters D, Cosyns P. (1995) Components of biological, including seasonal, variation in hematological measurements and plasma fibrinogen concentrations in normal humans. *Experientia* 51(2): 141–149.

Manfredini R, Portaluppi F, Boari B, Salmi R, Fersini C, Gallerani M. (2000) Circadian variation in onset of acute cardiogenic pulmonary edema is independent of patients' features and underlying pathophysiological causes. *Chronobiol Intl* 17(5): 705–715.

Manfredini R, Boari B, Gallerani M, Salmi R, Bossone E, Distante A, Eagle KA, Mehta RH. (2004) Chronobiology of rupture and dissection of aortic aneurysms. *J Vasc Surg* 40(2): 382–388 (Review).

Månson JC. (1965) Cyclic variations of the frequency of neutrophil leucocytes with "androgen induced" nucleus appendages in an adult man. *Life Sci* 4: 329–334.

Marbell A, Graham D, Smolensky M. (1988) Seasonal variation in the incidence of bleeding ulcer. In: *Ann Rev Chronopharm*, Vol 5. Reinberg A, Smolensky M, Labrecque G, eds. Oxford: Pergamon Press, pp. 307–309.

Maron BJ, Kogan J, Proschan MA, Hecht GM, Roberts WC. (1994) Circadian variability in the occurrence of sudden cardiac death in patients with hypertrophic cardiomyopathy. *J Amer Coll Cardiol* 23(6): 1405–1409.

Marsh RW, Chu N-M, Vauthey J-N, Mendenhall WM, Lauwers GY, Bewsher C, Copeland EM. (1996) Preoperative treatment of patients with locally advanced unresectable rectal adenocarcinoma utilizing continuous chronobiologically shaped 5-fluorouracil infusion and radiation therapy. *Cancer* 78: 217–225.

Martin RJ, Banks-Schlegel S. (1998) Chronobiology of asthma. *Amer J Respir Crit Care Med* 158: 1002–1007.

Mauer AM. (1965) Diurnal variation of proliferative activity in the human bone marrow. *Blood* 26(1): 1–7.

McCormick BB, Pierratos A, Fenton S, Jain V, Zaltzman J, Chan CT. (2004) Review of clinical outcomes in nocturnal haemodialysis patients after renal transplantation. *Nephrol Dial Transplant* 19(3): 714–719.

Meares R, Mendelsohn FAO, Milgrom-Friedman J. (1981) A sex difference in the seasonal variation of suicide rate: a single cycle for men, two cycles for women. *Brit J Psychiatry* 138: 321–325.

Melchart D, Martin P, Hallek M, Holzmann M, Jurcic X, Wagner H. (1992) Circadian variation of the phagocytic activity of polymorphonuclear leukocytes and of various other parameters in 13 healthy male adults. *Chronobiol Intl* 9(1): 35–45.

Mermall H, Sothern RB, Kanabrocki EL, Quadri SF, Bremner FW, Nemchausky BM, Scheving LE. (1995) Temporal (circadian) and functional relationship between prostate-specific antigen and testosterone in healthy men. *Urology* 46(1): 45–53.

Mersch PP, Middendorp HM, Bouhuys AL, Beersma DG, van den Hoofdakker RH. (1999a) Seasonal affective disorder and latitude: a review of the literature. *J Affective Disorders* 53(1): 35–48 (Review).

Mersch PP, Middendorp HM, Bouhuys AL, Beersma DG, van den Hoofdakker RH. (1999b) The prevalence of seasonal affective disorder in The Netherlands: a prospective and retrospective study of seasonal mood variation in the general population. *Biol Psychiat* 45(8): 1013–1022.

Miles LE, Raynal DM, Wilson MA. (1977) Blind man living in normal society has circadian rhythms of 24.9 hours. *Science* 198(4315): 421–423.

Milgrom H, Barnhart A, Gaddy J, Bush R, Busse W. (1990) The effect of chronotherapy with a 24h sustained release theophylline preparation (Uniphyl) on A.M. and P.M. differences in airway patency and responsiveness. In: *Ann Rev Chronopharmacol*, Vol 7. Reinberg A, Smolensky M, Labrecque G, eds. Oxford: Pergamon Press, pp. 301–304.

Mills JN, Minors DS, Waterhouse JM. (1978) The effect of sleep upon human circadian rhythms. *Chronobiologia* 5(1): 14–27.

Minors D, Åkerstedt T, Atkinson G, Dahlitz M, Folkard S, Lévi F, Mormont C, Parkes D, Waterhouse J. (1996a) The difference between activity when in bed and out of bed: I. Healthy subjects and selected patients. *Chronobiol Intl* 13(1): 27–34.

Minors D, Folkard S, MacDonald I, Owens D, Sytnik N, Tucker P, Waterhouse J. (1996b) The difference between activity when in bed and out of bed: II. Subjects on 27-hour "days." *Chronobiol Intl* 13(3): 179–190.

Minors D, Waterhouse J, Folkard S, Atkinson G. (1996c) The difference between activity when in bed and out of bed: III. Nurses on night work. *Chronobiol Intl* 13(4): 273–282.

Mitler MM, Hajdukovic RM, Shafor R, Hahn PM, Kripke DF. (1987) When people die. Cause of death vs time of death. *Amer J Med* 82(2): 266–274.

Mohr PE, Wang DY, Gregory WM, Richards MA, Fentiman IS. (1996) Serum progesterone and prognosis in operable breast cancer. *Brit J Cancer* 73(12): 1552–1555.

Monk TH, Folkard S, Leng VC, Weitzman ED. (1983) Circadian rhythms in subjective alertness and core body temperature. *Chronobiologia* 10(1): 49–55.

Moore JG, Wolfe M. (1974) Circadian plasma gastrin patterns in feeding and fasting man. *Digestion* 11: 226–231.

Moore JG, Halberg F. (1986) Circadian rhythm of gastric acid secretion in men with active duodenal ulcer. *Dig Dis Sci* 31(11): 1185–1191.

Moore JG, Smolensky MH. (1990) The chronobiology of peptic ulcer disease and implications for its chronotherapy with H2-receptor antagonist medication. In: *Ann Rev Chronopharmacol*, Vol 6. Reinberg A, Smolensky M, Labrecque G, eds. Oxford: Pergamon Press, pp. 113–135.

Moore RM, Zehmer RB. (1973) Listeriosis in the United States–1971. *J Infect Dis* 127(5): 610–611.

Moore Ede, MC. (1973) Circadian rhythms of drug effectiveness and toxicity. *Clin Pharmacol Therap* 14(6): 925–935.

Moore-Ede MC, Sulzman FM, Fuller CA. (1982) *The Clocks That Time Us. Physiology of the Circadian Timing System*. Cambridge: Harvard University Press, 448 pp.

Moore-Ede, M. (1986) Physiology of the circadian timing system: predictive versus reactive homeostasis. *Amer J Physiol* 19: R735–R752.

Moorthi KM, Hogan D, Lurbe E, Redon J, Batlle D. (2004) Nocturnal hypertension: will control of nighttime blood pressure prevent progression of diabetic renal disease? *Curr Hypertens Rep* 6(5): 393–399.

Morgan LM, Aspostolakou F, Wright J, Gama R. (1999) Diurnal variations in peripheral insulin resistance and plasma non-esterified fatty acid concentrations: a possible link? *Ann Clin Biochem* 36(Pt 4): 447–450.

Morgan T, Anderson A, Jones E. (1997) The effect on 24-hour blood pressure control of an angiotension converting enzyme inhibitor (perindopril) administered in the morning or at night. *J Hypertens* 15(2): 205–211.

Mormont MC, Berestka J, Mushiya T, Langevin T, Roemeling Rv, Rabatin J, Sothern R. Hrushesky W. Circadian dependency of vinblastine toxicity. (1986) In: *Ann Rev Chronopharmacol*, Vol 3. Reinberg A, Smolensky M, Labrecque G, eds. New York: Pergamon Press, pp. 187–190.

Mormont MC, von Roemeling R, Sothern RB, Berestka JS, Langevin TR, Wick M, Hrushesky WJM. (1988) Circadian rhythm and seasonal dependence in the toxicological response of mice to epirubicin. *Invest New Drugs* 6: 273–283.

Mormont C, Boughattas NA, Lévi F. (1989) Mechanisms of circadian rhythms in the toxicity and efficacy of anticancer drugs: relevance for the development of new analogues. In: *Chronopharmacology. Cellular and Biochemical Interactions.* Lemmer B, ed. New York: Marcel Dekker, Inc., pp. 395–437.

Mormont MC, Hecquet B, Bogdan A, Benavides M, Touitou Y, Lévi F. (1998) Noninvasive estimation of the circadian rhythm in serum cortisol in patients with ovarian or colorectal cancer. *Intl J Cancer* 78(4): 421–424.

Mormont MC, Waterhouse J, Bleuzen P, Giacchetti S, Jami A, Bogdan A, Lellouch J, Misset JL, Touitou Y, Lévi F. (2000) Marked 24-h rest/activity rhythms are associated with better quality of life, better response, and longer survival in patients with metastatic colorectal cancer and good performance status. *Clin Cancer Res* 6(8): 3038–3045.

Mormont MC, Langouet AM, Claustrat B, Bogdan A, Marion S, Waterhouse J, Touitou Y, Lévi F. (2002) Marker rhythms of circadian system function: a study of patients with metastatic colorectal cancer and good performance status. *Chronobiol Intl* 19(1): 141–155.

Mormont MC, Levi F. (2003) Cancer chronotherapy: principles, applications, and perspectives. *Cancer* 97(1): 155–169. Review.

Muller JE, Ludmer PL, Willich SN, Tofler GH, Aylmer G, Klangos I, Stone PH. (1987) Circadian variation in the frequency of sudden cardiac death. *Circulation* 75(1): 131–138.

Muller JE, Tofler GH. (1991) Circadian variation and cardiovascular disease. *N Engl J Med* 325(14): 1038–1039.

Muller JE. (1999) Circadian variation and triggering of acute coronary events. *Amer Heart J* 137(4 Pt 2): S1–S8. Review.

Muller D, Lampe F, Wegscheider K, Schultheiss HP, Behrens S. (2003) Annual distribution of ventricular tachycardias and ventricular fibrillation. *Amer Heart J* 146(6): 1061–1065.

Murch SJ, Simmons CB, Saxena PK. (1997) Melatonin in feverfew and other medicinal plants. *Lancet* 350(9091): 1598–1599.

Mustad V, Derr J, Reddy CC, Pearson TA, Kris-Etherton PM. (1996) Seasonal variation in parameters related to coronary heart disease risk in young men. *Atherosclerosis* 126(1): 117–129.

Nagayama H. (1999) Influences of biological rhythms on the effects of psychotropic drugs. *Psychosom Med* 61(5): 618–629 (Review).

Nelson W, Cornélissen G, Honkley D, Bingham C, Halberg F. (1983) Construction of rhythm-specified reference intervals and regions, with emphasis on 'hybrid' data, illustrated for plasma cortisol. *Chronobiologia* 10(2): 179–193.

Nelson RJ, Blom JMC. (1994) Photoperiodic effects on tumor development and immune function. *J Biol Rhythms* 9(3–4): 233–249.

Nelson RJ, Demas GE, Klein SL, Kriegsfeld LJ. (1995) The influence of season, photoperiod, and pineal melatonin on immune function. *J Pineal Res* 19(4): 149–165.

Newell GR, Lyncy HK, Gibeau JM, Skitz MR. (1985) Seasonal diagnosis of Hodgkin's disease among young adults. *JNCI* 74(1): 53–56.

Nicolau GY, Lakatua D, Sackett-Lundeen L, Haus E. (1984) Circadian and circannual rhythms of hormonal variables in elderly men and women. *Chronobiol Intl* 1(4): 301–319.

Nicolau GY, Haus E, Popescu M, Sackett-Lundeen L, Petrescu E. (1991) Circadian, weekly, and seasonal variations in cardiac mortality, blood pressure, and catecholamine excretion. *Chronobiol Intl* 8(2): 149–159.

O'Brien E, Sheridan J, O'Malley K. (1988) Dippers and non-dippers. *Lancet* 2(8607): 397.

O'Conner JF, Shelley EM, Stern LO. (1974) Behavioral rhythms related to the menstrual cycle. In: *Biorhythms and Human Reproduction*. Ferin M, Halberg F, Richart RM, Vande Wiele RL, eds. New York: John Wiley & Sons, Inc., pp. 309–324.

O'Donoghue C, Rubin M. (1977) Frequency of pain in post-cholecystectomy patients. In: *Proc XII Intl Conf Intl Soc Chronobiol*. Milan: Il Ponte, pp. 365–368.

Ogawa H, Yasue H, Oshima S, Okumura K, Matsuyama K, Obsta K. (1989) Circadian variation of plasma fibrinopeptide-A level in patients with variant angina. *Circulation* 80(6): 1617–1626.

Ollagnier M, Decousus H, Cherrah Y, Levi F, Mechkouri M, Queneau P, Reinberg A. (1987) Circadian changes in the pharmacokinetics of oral ketoprofen. *Clin Pharnacokinet* 12(5): 367–378.

Oren DA. (1991) Retinal melatonin and dopamine in seasonal affective disorder. *J Neural Transm Gen Sect* 83(1–2): 85–95 (Review).

Oren DA, Shannon NJ, Carpenter CJ, Rosenthal NE. (1991) Usage patterns of phototherapy in seasonal affective disorder. *Comprehen Psychiat* 32(2): 147–152.

Oren DA, Moul DE, Schwartz PJ, Brown C, Yamada EM, Rosenthal NE. (1994) Exposure to ambient light in patients with winter seasonal affective disorder. *Amer J Psychiat* 151(4): 591–593.

Ostfeld AM. (1963) The natural history and epidemiology of migraine and muscle contraction headache. *Neurology* 3(2): 11–15.

Otsuka K, Halberg F. (1994) Circadian profiles of blood pressure and heart rate of apparently healthy metropolitan Japanese. *Front Med Biol Eng* 6(2): 149–155.

Otsuka K, Cornelissen G, Halberg F, Oehlerts G. (1997) Excessive circadian amplitude of blood pressure increases risk of ischaemic stroke and nephropathy. *J Med Eng Technol* 21(1): 23–30.

Otsuka K, Cornelissen G, Shinagawa M, Halberg F. (1999) Blood pressure variability assessed by semiautomatic and ambulatorily functional devices for home use. *Clin Exp Hypertens* 21(5–6): 729–740.

Ozaki N, Ono Y, Ito A, Rosenthal NE. (1995) Prevalence of seasonal difficulties in mood and behavior among Japanese civil servants. *Amer J Psychiatry* 152(8): 1225–1227.

Otto ME, Svatikova A, Barretto RB, Santos S, Hoffmann M, Khandheria B, Somers V. (2004) Early morning attenuation of endothelial function in healthy humans. *Circulation* 109(21): 2507–2510.

Palinkas LA, Houseal M, Rosenthal NE. (1996) Subsyndromal seasonal affective disorder in Antarctica. *J Nerv Mental Dis* 184(9): 530–534.

Paloheimo J. (1961) Seasonal variations of serum lipids in healthy men. *Ann Med Exp Finn* 39(Suppl 8): 1–99.

Pang SF, Pang CS, Poon AM, Lee PD, Liu ZM, Shiu SY. (1998) Melatonin: a chemical photoperiodic signal with clinical significance in humans. *Chinese Med J (Engl)* 111(3): 197–203.

Panza JA, Epstein SE, Quyyumi AA. (1991) Circadian variation in vascular tone and its relation to alpha-sympathetic vasoconstrictor activity. *N Engl J Med* 325(14): 986–990.

Parry BL, Mahan AM, Mostofi N, Klauber MR, Lew GS, Gillin JC. (1993) Light therapy of late luteal phase dysphoric disorder: an extended study. *Amer J Psychiat* 150(9): 1417–1419.

Parry BL, Berga SL, Mostofi N, Klauber MR, Resnick A. (1997) Plasma melatonin circadian rhythms during the menstrual cycle and after light therapy in premenstrual dysphoric disorder and normal control subjects. *J Biol Rhythms* 12(1): 47–64.

Parry BL. (2002) Jet lag: minimizing it's effects with critically timed bright light and melatonin administration. *J Mol Microbiol Biotechnol* 4(5): 463–466 (Review).

Parulekar W, de Marsh RW, Wong R, Mendenhall W, Davey P, Zlotecki R, Berry S, Rout WR, Bjarnason GA. (2004) Phase I study of 5-fluorouracil and leucovorin by continuous infusion chronotherapy and pelvic radiotherapy in patients with locally advanced or recurrent rectal cancer. *Intl J Radiat Oncol Biol Phys* 58(5): 1487–1495.

Pasqualetti P, Colantonio D, Casale R, Acitelli P, Natali G. (1990) [The chronobiology of sudden cardiac death. The evidence for a circadian, circaseptimanal and circannual periodicity in its incidence.] [Italian]. *Minerva Med* 81(5): 391–398.

Pauley SM. (2004) Lighting for the human circadian clock: recent research indicates that lighting has become a public health issue. *Med Hypoth* 63(4): 588–596.

Peleg L, Goldman B, Ashkenazi IE. (1993) Effects of inter-individual and diurnal variations on the activity of HexA. Relevance to screening of Tay-Sachs disease carriers. In: *Biologie Prospective (Computes rendus du 8e Colloque de Pont-a-Mousson)*. Galteau M-M, Siest G, Henny J, eds. Montrouge: John Libbey Eurotext, pp. 309–312.

Perreau-Lenz S, Pevet P, Buijs RM, Kalsbeek A. (2004) The biological clock: the bodyguard of temporal homeostasis. *Chronobiol Intl* 21(1): 1–25 (Review).

Petit E, Milano G, Lévi F, Thyss A, Bailleul F, Schneider M. (1988) Circadian rhythm-varying plasma concentration of 5-fluorouracil during a five-day continuous venous infusion at a constant rate in cancer patients. *Cancer Res* 48(6): 1676–1679.

Petralito A, Gibiino S, Miano MF, Mangiafico RA, Cuffari MA, Fiore CE. (1982) Daily modifications of plasma fibrinogen, platelet aggregation, Howell's time, PTT, TT, and antithrombin III in normal subjects and in patients with vascular disease. *Chronobiologia* 9(2): 195–201.

Petrovsky N, Harrison LC. (1997) Diurnal rhythmicity of human cytokine production: a dynamic disequilibrium in T helper cell type 1/T helper cell type 2 balance? *J Immunol* 158(11): 5163–5168.

Petrovsky N, McNair P, Harrison LC. (1998) Diurnal rhythms of pro-inflammatory cytokines: regulation by plasma cortisol and therapeutic implications. *Cytokine* 10(4): 307–312.

Petrovsky N, Socha L, Silva D, Grossman AB, Metz C, Bucala R. (2003) Macrophage migration inhibitory factor exhibits a pronounced circadian rhythm relevant to its role as a glucocorticoid counter-regulator. *Immunol Cell Biol* 81(2): 137–143.

Piccoli B, Parazzoli S, Zaniboni A, Demartini G, Frashini F. (1991) [Non-visual effects of light mediated via the optical route: review of the literature and implications for occupational medicine.] [Italian]. *Medicina del Lavoro* 82(3): 213–232.

Pickering TG. (1990) The clinical significance of diurnal blood pressure variations. Dippers vs nondippers. *Circulation* 81(2): 700–702.

Pincus DJ, Szafler SJ, Ackerson LM, Martin RJ. (1997) Chronotherapy of asthma with inhaled steroids: the effect of dosage timing on drug efficacy. *J Allergy Clin Immunol* 95: 1172–1178.

Pincus DJ, Humeston TR, Martin RJ. (1997) Further studies on the chronotherapy of asthma with inhaled steroids: The effect of dosage timing on drug efficacy. *J Allergy Clin Immunol* 100: 771–774.

Pirke KM, Wuttke W, Schweiger U, eds. (1989) *The Menstrual Cycle and Its Disorders.* Berlin: Springer-Verlag, 192 pp.

Place VA. (1988) Testosterone replacement with a circadian pattern. In: *Ann Rev Chronopharmacol*, Vol 5. Reinberg A, Smolensky M, Labrecque G, eds. Oxford: Pergamon Press, pp. 93.

Place VA, Nichols KC. (1991) Transdermal delivery of testosterone with Testoderm to provide a normal circadian pattern of testosterone. *Ann NY Acad Sci* 618: 441–449.

Pointer JS. (1997) The diurnal variation of intraocular pressure in non-glaucomatous subjects: relevance in a clinical context. *Ophthalmic Physiol Opt* 17(6): 456–465.

Poirier L, Lefebvre J, Lacourciere Y. (1999) Chronotherapeutics: are there meaningful differences among antihypertensive drugs? *Curr Hypertens Rep* 1(4): 320–327 (Review).

Pollmacher T, Mullington J, Korth C, Schreiber W, Hermann D, Orth A, Galanos C, Holsboer F. (1996) Diurnal variations in the human host response to endotoxin. *J Infect Dis* 174(5): 1040–1045.

Pöllmann L, Harris PHP. (1978) Rhythmic changes in pain sensitivity in teeth. *Intl J Chronobiol* 5(3): 459–464.

Pöllmann L. (1982) Circannual variations in the frequency of some diseases of the oral mucosa. *J Interdiscipl Cycle Res* 13: 249–256.

Pöllmann L. (1984) Duality of pain demonstrated by the circadian variation in tooth sensitivity. In: *Chronobiology 1982–1983*. Haus E, Kabat H, eds. Basel: S. Karger, pp. 225–228.

Pöllmann L, Pöllmann B. (1988) Circadian variations of the efficiency of hepatitis B vaccination. In: *Ann Rev Chronopharmacol*, Vol 5. Reinberg A, Smolensky M, Labrecque G, eds. Oxford: Pergamon Press, pp. 45–48.

Pöppel E, Aschoff JC, Giedke H. (1970) Tageperiodische Veränderugen der reaktions Zeit der Wahlreaktion. *Zeit Exp Angew Psychol* 17(4): 537–552.

Portaluppi F, Montanari L, Massari M, Di Chiara V, Capanna M. (1991) Loss of nocturnal decline of blood pressure in hypertension due to chronic renal failure. *Amer J Hyperten* 4(1 Pt 1): 20–26.

Portaluppi F, Montanari L, Vergnani L, Tarroni G, Cavallini AR, Gilli P, Bagni B, degli Uberti EC. (1992) Loss of nocturnal increase in plasma concentration of atrial natriuretic peptide in hypertensive chronic renal failure. *Cardiology* 80(5–6): 312–323.

Procacci P, Della Corte M, Zoppi M, Maresca M. (1974) Rhythmic changes in the cutaneous pain threshold in man. A general review. *Chronobiologia* 1(1): 77–96.

Proc Conf Ferrara, Italy, Sep 10–12, 1995. (1996) Time-dependent structure and control of arterial blood pressure. *Ann NY Acad Sci* 783: 1–342.

Qureshi IA, Xiao RX, Yang BH, Zhang J, Xiang DW, Hui JL. (1999) Seasonal and diurnal variations of ocular pressure in ocular hypertensive subjects in Pakistan. *Singapore Med J* 40(5): 345–348.

Rabkin SW, Mathewson FA, Tate RB. (1980) Chronobiology of cardiac sudden death in man. *JAMA* 244(12): 1357–1358.

Rackley RJ, Meyer MC, Straughn AB. (1988) Circadian rhythm in theophylline disposition during a constant-rate intravenous infusion of aminophylline in the dog. *J Pharm Sci* 77(8): 658–661.

Ramirez-Lassepas M, Haus E, Lakatua DJ, Sackett L, Swoyer J. (1980) Seasonal (circannual) periodicity of spontaneous intracerebral hemorrhage in Minnesota. *Ann Neurol* 8(5): 539–541.

Ratajczak HV, Sothern RB Hrushesky WJM. (1988) Estrous influence on surgical cure of a mouse breast cancer. *J Exp Med* 168(1): 73–83.

Reddy AB, Field MD, Maywood ES, Hastings MH. (2002) Differential resynchronisation of circadian clock gene expression within the suprachiasmatic nuclei of mice subjected to experimental jet lag. *J Neurosci* 22(17): 7326–7330.

Redfern PH, Lemmer B, eds. (1997) *Physiology and Pharmacology of Biological Rhythms*. Berlin: Springer-Verlag, 668 pp.

Reinberg A, Ghata J. (1964) *Biological Rhythms*. New York: Walker, 138 pp.

Reinberg A. (1967) The hours of changing responsiveness or susceptibility. *Perspect Biol Med* 11(1): 111–128.

Reinberg A, Zagula-Mally Z, Ghata J, Halberg F. (1969) Circadian reactivity rhythm of human skin to house dust, penicillin, and histamine. *J Allergy* 44(5): 292–306.

Reinberg A, Gervais P, Halberg F, Gaultier M, Roynette N, Abulker Ch, Dupont J. (1973) Mortalité des adultes: rythmes circadiens et circannuels. *Nouv Presse Méd* 2: 289–294.

Reinberg A, Smolensky MH. (1974) Circatrigintan secondary rhythms related to the hormonal changes in the menstrual cycle: general considerations. In: *Biorhythms and Human Reproduction*. Ferin M, Halberg F, Richart RM, Vande Wiele RL, eds. New York: John Wiley & Sons, Inc., pp. 241–258.

Reinberg A, Lagoguey M. (1978) Annual endocrine rhythms in healthy young adult men: their implication in human biology and medicine. In: *Environmental Endocrinology*. Assenmacher I, Farner DS, eds. Berlin: Springer-Verlag, pp. 113–121.

Reinberg A, Smolensky M. (1982) Circadian changes in drug disposition in man. *Clin Pharmacokinet* 7(5): 401–420.

Reinberg A, Smolensky M. (1983) *Biological Rhythms and Medicine. Cellular, Metabolic, Physiopathologic, and Pharmacologic Aspects*. New York: Springer-Verlag, 305 pp.

Reinberg A, Koulbanis C, Soudant E, Nicolai A, Mechkouri M. (1990a) Circadian changes in the size of facial skin corneocytes of healthy women. In: *Ann Rev Chronopharmacol*, Vol 7. Reinberg A, Smolensky M, Labrecque G, eds. New York: Pergamon Press, pp. 331–334.

Reinberg A, Lévi F. (1990b) Dose-response relationships in chronopharmacology. In: *Ann Rev Chronopharmacol*, Vol 6. Reinberg A, Smolensky M, Labrecque G, eds. Oxford: Pergamon Press, pp. 25–46.

Reinberg A, Touitou Y, Botbol M, Gervais P. (1990c) Preservation of the adrenal function with oral morning dosing of corticoids in long term (3 to 11 years) treated corticodependent asthmatics. In: *Ann Rev Chronopharmacol*, Vol. 7. Reinberg A, Smolensky M, Labrecque G, eds. Oxford: Pergamon Press, pp. 327–330.

Reinberg AE, Ashkenazi IE. (1993) Interindividual differences in chronopharmacologic effects of drugs: a background for individualization of chronotherapy. *Chronobiol Intl* 10(6): 449–460 (Review).

Reinberg AE, Lewy H, Smolensky M. (2001) The birth of chronobiology: Julien Joseph Virey 1814. *Chronobiol Intl* 18(2): 173–186.

Reisin LH, Pancheva N, Berman M, Khalameizer V, Jafary J, Yosefy C, Blaer Y, Manevich I, Peled R, Scharf S. (2004) Circadian variation of the efficacy of thrombolytic therapy in acute myocardial infarction-isn't the time ripe for cardiovascular chronotherapy? *Angiology* 55(3): 257–263.

Reiter RJ. (1986) Normal patterns of melatonin levels in the pineal gland and body fluids of humans and experimental animals. *J Neural Transm* [Suppl] 21: 35–54.

Reiter RJ. (1991) Pineal melatonin: cell biology of its synthesis and its physiological interactions. *Endocr Rev* 12(2): 151–179.

Reiter RJ. (1998) Cytoprotective properties of melatonin: presumed association with oxidative damage and aging. Nutrition 14(9): 691–696 (Review).

Reiter RJ, Tan DX. (2002) Melatonin: an antioxidant in edible plants. *Ann NY Acad Sci* 957: 341–344 (Review).

Reiter RJ, Tan DX, Sainz RM, Mayo JC, Lopez-Burillo S. (2002) Melatonin: reducing the toxicity and increasing the efficacy of drugs. *J Pharm Pharmacol* 54(10): 1299–1321 (Review).

Reiter RJ. (2003) Melatonin: clinical relevance. *Best Pract Res Clin Endocrinol Metab* 17(2): 273–285 (Review).

Reiter RJ, Tan DX, Mayo JC, Sainz RM, Leon J, Czarnocki Z. (2003) Melatonin as an antioxidant: biochemical mechanisms and pathophysiological implications in humans. *Acta Biochim* Pol 50(4): 1129–1146 (Review).

Reiter RJ. (2004) Mechanisms of cancer inhibition by melatonin. *J Pineal Res* 37(3): 213–214 (Review).

Reiter RJ, Tan DX, Pappolla MA. (2004) Melatonin relieves the neural oxidative burden that contributes to dementias. *Ann NY Acad Sci* 1035: 179–196.

Reiter RJ, Tan DX, Leon J, Kilic U, Kilic E. (2005) When melatonin gets on your nerves: its beneficial actions in experimental models of stroke. *Exp Biol Med* (Maywood) 230(2): 104–117.

Rensing L, Meyer-Grahle U, Ruoff P. (2001) Biological timing and clock metaphor: oscillatory and hourglass mechanisms. *Chronobiol Intl* 18: 329–369.

Richter CP. (1965) *Biological Clocks in Medicine and Psychiatry*. Springfield, IL.: CC Thomas, 108 pp.

Rieck A, Damm F. (1975) Circadian variations in blood flow of the extremities at rest and during work (Abstract 49). *Pflügers Arch* 355(Suppl): R25.

Rietveld PEM, Rietveld WJ, Boon ME, Sothern RB, Hrushesky WJM. (1988) The incidence of cervical pathology in Dutch women. Seasonal rhythms in infections and neoplasia (Abstract). *J Interdiscipl Cycle Res* 19: 206.

Rietveld WJ, Boon ME, Meulman JJ. (1997) Seasonal fluctuations in the cervical smear detection rates for (pre)malignant changes and for infections. *Diagn Cytopathol* 17(6): 452–455.

Ritschel WA, Forusz H. (1994) Chronopharmacology: a review of drugs studied. *Methods Find Exp Clin Pharmacol* 16(1): 57–75 (Review).

Rivard GE, Infante-Rivard C, Hoyeux C, Champagne J. (1985) Maintenance chemotherapy for childhood acute lymphoblastic leukemia: better in the evening. *Lancet* 2: 1264–1266.

Rivard GE, Infante-Rivard C, Dresse MF, Leclerc JM, Champagne J. (1993) Circadian time-dependent response of childhood lymphoblastic leukemia to chemotherapy: a long-term follow-up study of survival. *Chronobiol Intl* 10(3): 201–204.

Rivera-Coll A, Fuentes-Arderiu X, Diez-Noguera A. (1993) Circadian rhythms of serum concentrations of 12 enzymes of clinical interest. *Chronobiol Intl* 10(3): 190–200.

Roberts N, Hogg D, Whitehouse GH, Dangerfield P. (1998) Quantitative analysis of diurnal variation in volume and water content of lumbar intervertebral discs. *Clin Anat* 11(1): 1–8.

Rocco MB, Barry J, Campbell S, Nabel E, Cook EF, Goldman L, Selwyn AP. (1987) Circadian variation of transient myocardial ischemia in patients with coronary artery disease. *Circulation* 75(2): 395–400.

Roenneberg T, Aschoff J. (1990a) Annual rhythm of human reproduction: I. Biology, sociology or both? *J Biol Rhythms* 5(3): 195–216.

Roenneberg T, Aschoff J. (1990b) Annual rhythm of human reproduction: II. Environmental correlations. *J Biol Rhythms* 5(3): 217–239.

Roenneberg T, Merrow M. (2003) The network of time: understanding the molecular circadian system. *Curr Biol* 13(5): R198–R207 (Review).

Roenneberg T, Daan S, Merrow M. (2003) The art of entrainment. *J Biol Rhythms* 18(3): 183–194.

Rosen LN, Targum SD, Terman M, Bryant MJ, Hoffman H, Kasper SF, Hamovit JR, Docherty JP, Welch B, Rosenthal NE. (1990) Prevalence of seasonal affective disorder at four latitudes. *Psychiatry Res* 31(2): 131–144.

Rosenblatt LS, Shifrine M, Hetherington NW, Paglierioni T, MacKenzie MR. (1982) A circannual rhythm in Rubella antibody titers. *J Interdiscipl Cycle Res* 13(1): 81–88.

Rosenthal NE, Sack DA, Gillin JC, Lewy AJ, Goodwin FK, Davenport Y, Mueller PS, Newsome DA, Wehr TA. (1984) Seasonal affective disorder. A description of the syndrome and preliminary findings with light therapy. *Arch Gen Psychiatry* 41(1): 72–80.

Rosenthal NE, Sack DA, Skwerer RG, Jacobsen FM, Wehr TA. (1988) Phototherapy for seasonal affective disorder. *J Biol Rhythms* 3(2): 101–120 (Review).

Rosenwasser AM, Wirz-Justice A. (1997) Circadian rhythms and depression: clinical and experimental models. In: *Physiology and Pharmacology of Biological Rhythms*. Redfern PH, Lemmer B, eds. Berlin: Springer-Verlag, pp. 457–486.

Rudic RD, McNamara P, Curtis AM, Boston RC, Panda S, Hogenesch JB, Fitzgerald GA. (2004) BMAL1 and CLOCK, two essential components of the circadian clock, are involved in glucose homeostasis. *PLoS Biol* 2(11): 1893–1899.

Rutenfranz J, Colquhoun P. (1979) Circadian rhythms in performance. *Scan J Work Environ Health* 5(3): 167–177.

Saarijarvi S, Lauerma H, Helenius H, Saarilehto S. (1999) Seasonal affective disorders among rural Finns and Lapps. *Acta Psychiatrica Scandinavic* 99(2): 95–101.

Sack RL, Lewy AJ, Blood ML, Stevenson J, Keith LD. (1991) Melatonin administration to blind people: phase advances and entrainment. *J Biol Rhythms* 6(3): 249–261.

Sack RL, Lewy AJ, Blood ML, Keith LD, Nakagawa H. (1992) Circadian rhythm abnormalities in totally blind people: incidence and clinical significance. *J Clin Endocrin Metab* 75(1): 127–134.

Sack RL, Lewy AJ. (2001) Circadian rhythm sleep disorders: lessons from the blind. *Sleep Med Rev* 5(3): 189–206.

Sadeh A, Hauri PJ, Kripke DF, Lavie P. (1995) The role of actigraphy in the evaluation of sleep disorders. *Sleep* 18(4): 288–302 (Review).

Sanchez de la Peña S. (1993) The feedsideward of cephalo-adrenal immune interactions. *Chronobiologia* 20: 1–52.

Sanchez CP. (2004) Chronotherapy of high-dose active vitamin D_3: is evening dosing preferable? *Pediatr Nephrol* 19(7): 722–723.

Sanders SW, Bishop AL, Moore JG. (1991) Intragastric pH and pharmacokinetics of intravenous ranitidine during sinusoidal and constant-rate infusions. *Chronobiol Intl* 8(4): 267–276.

Sasaki J, Kumagae G, Sata T, Ikeda M, Tsutsumi S, Arakawa K. (1983) Seasonal variation of serum high density lipoprotein cholesterol levels in men. *Atherosclerosis* 48(2): 167–172.

Sato G, Matsutani S, Maruyama H, Saisho H, Fukuzawa T, Mizumoto H, Burioka N. (1998) Chronobiological analysis of seasonal variations of variceal hemorrhage. *Hepatol* 28(3): 893–895.

Savides TJ, Messin S, Senger C, Kripke DF. (1986) Natural light exposure of young adults. *Physiol Behav* 38(4): 571–574.

Schernhammer ES, Laden F, Speizer FE, Willett WC, Hunter DJ, Kawachi I, Colditz GA. (2001) Rotating night shifts and risk of breast cancer in women participating in the nurses' health study. *JNCI* 93(20): 1563–1568.

Schernhammer E, Schulmeister K. (2004) Light at night and cancer risk. *Photochem Photobiol* 79(4): 316–318. Review.

Scheving LE. (1959) Mitotic activity in the human epidermis. *Anat Rec* 135: 7–20.

Scheving LE, Cardoso SS, Pauly JE, Halberg F, Haus E. (1974) Variation in susceptibility of mice to the carcinostatic agent arabinosyl cytosine. In: *Chronobiology*. Scheving LE, Halberg F, Pauly JE, eds. Tokyo: Igaku Shoin Ltd., pp. 213–217.

Scheving LE, Tsai TH, Pauly JE. (1986) Chronotoxicology and chronopharmacology with emphasis on carcinostatic agents. In: *Ann Rev Chronopharmacol*, Vol. 2. Reinberg A, Smolensky M, Labrecque G, eds. New York: Pergamon Press, pp. 177–197.

Schibler U, Ripperger J, Brown SA. (2003) Peripheral circadian oscillators in mammals: time and food. *J Biol Rhythms* 18(3): 250–260.

Schnur S. (1956) Mortality rates in acute myocardial infarction–IV. The seasonal variation in morbidity and mortality. *Ann Intern Med* 44(3): 476–481.

Schulz B, Ratzmann KP, Albrecht G, Bibergeil H. (1983) Diurnal rhythm of insulin sensitivity in subjects with normal and impaired glucose tolerance. *Exp Clin Endocrinol* 81(3): 263–272.

Sensi S, Capani F. (1976) Circadian rhythm of insulin-induced hypoglycemia in man. *J Clin Endocrinol Metab* 43(2): 462–465.

Seto TB, Mittleman MA, Davis RB, Taira DA, Kawachi I. (1998) Seasonal variation in coronary artery disease mortality in Hawaii: observational study. *BMJ* 316: 1946–1947.

Shadrin AS, Marinich IG, Taros LY. (1977) Experimental and epidemiological estimation of seasonal and climato-geographical features of non-specific resistance of the organism to influenza. *J Hyg Epidemiol Microbiol Immunol* 21(2): 155–161.

Shimanov M, Havat H, Chaitchik S, Brener J, Schachter P, Czerniak A. (2005) Combined systemic chronotherapy and hepatic artery infusion for the treatment of metastatic colorectal cancer confined to the liver. *Chemother* 51(2–3): 111–115.

Shinagawa M, Kubo Y, Otsuka K, Ohkawa S, Cornelissen G, Halberg F. (2001) Impact of circadian amplitude and chronotherapy: relevance to prevention and treatment of stroke. *Biomed Pharmacother* 55(Suppl 1): 125s–132s.

Shinkawa A, Veda K, Hasuo Y, Kiyohara Y, Fujishima M. (1990) Seasonal variation in stroke incidence in Hisayama, Japan. *Stroke* 21(9): 1262–1267.

Shirakawa T, Honma S, Honma K. (2001) Multiple oscillators in the suprachiasmatic nucleus. *Chronobiol Intl* 18: 371–387.

Shue JL, Sothern RB, Kanabrocki EL, Bremner FW, Nemchausky BM, Vesely DL, Feuers RJ, Scheving LE, Olwin JH. (1994) Consideration of temporal (circadian) relationships between intraocular pressure and several physiologic variables hypothesized to be involved with its regulation (Abstract OD-112). *Optom Vis Sci* 71(12s): 111–112.

Sica DA, Neutel JM, Weber MA, Manowitz N. (2004) The antihypertensive efficacy and safety of a chronotherapeutic formulation of propranolol in patients with hypertension. *J Clin Hypertens (Greenwich)* 6(5): 231–241.

Silverman RA, Stevenson L, Hastings HM. (2003) Age-related seasonal patterns of emergency department visits for acute asthma in an urban environment. *Ann Emerg Med* 42(4): 577–586.

Simonsen PE, Niemann L, Meyrowitsch DW. (1997) *Wuchereria bancrofti* in Tanzania: microfilarial periodicity and effect of blood sampling time on microfilarial intensities. *Trop Med Intl Health* 2(2): 153–158.

Simpson HW. (1987) Human thermometry in health and disease: the chronobiologist's perspective. In: *Chronobiotechnology and Chronobiological Engineering. NATO ASI Series.* Scheving LE, Halberg F, Ehret CF, eds. Dordrecht: Martinus Mijhoff Publication, pp. 141–188.

Singh N, Wagener MM, Gayowski T. (2001) Seasonal pattern of early mortality and infectious complications in liver transplant recipients. *Liver Transpl* 7(10): 884–889.

Sinha MK, Ohannesian JP, Heiman ML, Kriauciunas A, Stephens TW, Magosin S, Marco C, Caro JF. (1996) Nocturnal rise of leptin in lean, obese, and non-insulin-dependent diabetes mellitus subjects. *J Clin Invest* 97(5): 1344–1347.

Skene DJ, Deacon S, Arendt J. (1996) Use of melatonin in circadian rhythm disorders and following phase shifts. *Acta Neurobiol Exp (Wars)* 56(1): 359–362 (Review).

Skene DJ, Lockley SW, Arendt J. (1999) Use of melatonin in the treatment of phase shift and sleep disorders. *Adv Exp Med Biol* 467: 79–84 (Review).

Smaaland R, Laerum OD, Lote K, Sletvold O, Sothern RB, Bjerknes R. (1991) DNA synthesis in human bone marrow is circadian stage dependent. *Blood* 77: 2603–2611.

Smaaland R, Laerum OD, Sothern RB, Sletvold O, Bjerknes R, Lote K. (1992) Colony-forming units–Granulocyte/macrophage and DNA synthesis of human bone marrow are circadian stage-dependent and show covariation. *Blood* 79(9): 2281–2287.

Smaaland R, Laerum OD. (1992) Chronobiology of human bone marrow. In: *Biological Rhythms in Clinical and Laboratory Medicine.* Touitou Y, Haus E, eds. Berlin: Springer-Verlag, pp. 527–546.

Smaaland R, Lote K, Sothern RB, Laerum OD. (1993) DNA synthesis and ploidy in non-Hodgkin's lymphomas demonstrate intrapatient variation depending on circadian stage of cell sampling. *Cancer Res* 53(13): 3129–3138.

Smaaland R, Sothern RB. (1994) Circadian cytokinetics of murine and human bone marrow and human cancer. In: *Circadian Cancer Therapy.* Hrushesky WJM, ed. Boca Raton: CRC Press, pp. 119–163.

Smaaland R, Sothern RB, Lote K, Sandberg S, Aakvaag A, Laerum OD. (1995) Circadian phase relationships between peripheral blood variables and bone marrow proliferative activity in clinical health. *In Vivo* 9(4): 379–389.

Smith RE. (1969) Circadian variations in human thermoregulatory response. *J Appl Physiol* 26(5): 554–560.

Smith DHG, Neutel JM, Weber MA. (2001) A new chronotherapeutic oral drug absorption system for verapamil optimizes blood pressure control in the morning. *Amer J Hypertens* 14(1): 14–19.

Smolensky M, Halberg F, Sargent F. (1972) Chronobiology of the life sequence. In: *Advances in Climatic Physiology.* Ito S, Ogata K, Yohimura H, eds. Tokyo: Igaku Shoin Ltd., pp. 281–318.

Smolensky MH, Halberg F. (1977) Circadian rhythms in airway patency and lung volumes. In: *Chronobiology in Allergy and Immunology*. McGovern JP, Smolensky MH, Reinberg A, eds. Springfield: CC Thomas, pp. 117–138.

Smolensky MH. (1981) Chronobiologic factors related to the epidemiology of human reproduction. In: *Research on Fertility and Sterility*. Cortés-Prieto J, Campos da Paz A, Neves-e-Castro M, eds. Lancaster: MTP Press Ltd., pp. 157–181.

Smolensky MH. (1983) Aspects of human chronopathology. In: *Biological Rhythms and Medicine. Cellular, Metabolic, Physiopathologic, and Pharmacologic Aspects*. Reinberg A, Smolensky MH, eds. Springer-Verlag: New York, pp. 131–209.

Smolensky MH, Barnes PJ, Reinberg A, McGovern JP. (1986a) Chronobiology and asthma: I. Day–night differences in bronchial patency and dyspnea and circadian rhythm dependencies. *J Asthma* 23(6): 321–343.

Smolensky MH, Scott PH, Barnes PJ, Jonkman JHG. (1986b) The chronopharmacology and chronotherapy of asthma. In: *Ann Rev Chronopharmacol*, Vol 2. Reinberg A, Smolensky M, Labrecque G, eds. Oxford: Pergamon Press, pp. 229–273.

Smolensky MH, D'Alonzo GE. (1988) Biologic rhythms and medicine. *Amer J Med* 85: 34–46.

Smolensky MH. (1989) Chronopharmacology of theophylline and beta-sympathomimetics. In: *Chronopharmacology: Cellular and Biochemical Interactions*. Lemmer B, ed. New York: Marcel Dekker, pp. 65–113.

Smolensky MH, D'Alonzo, GE, Reinberg A. (1990) Current developments in the chronotherapy of nocturnal asthma. In: *Ann Rev Chronopharmacol*, Vol. 6. Reinberg A, Smolensky M, Labrecque G, eds. Oxford: Pergamon Press, pp. 81–112.

Smolensky MH, Reinberg A, Labrecque G. (1995) Twenty-four hour pattern in symptom intensity of viral and allergic rhinitis: treatment implications. *J Allergy Clin Immunol* 95: 1084–1096.

Smolensky MH, D'Alonzo GE. (1997) Progress in the chronotherapy of nocturnal asthma. In: *Physiology and Pharmacology of Biological Rhythms*. Redfern PH, Lemmer B, eds. Berlin: Springer-Verlag, pp. 205–249.

Smolensky MH. (1998) Knowledge and attitudes of American physicians and public about medical chronobiology and chronotherapeutics. Findings of two 1996 Gallup surveys. *Chronobiol Intl* 15(4): 377–394

Smolensky M, Lamberg L. (2000) *The Body Clock Guide to Better Heath: How to Use Your Body's Natural Clock to Fight Illness and Achieve Maximum Health*. New York: Henry Holt & Company, 428 pp.

Smolensky MH, Haus E. (2001) Circadian rhythms and clinical medicine with applications to hypertension. *Amer J Hypertens* 14(9 Pt 2): 280S–290S (Review).

Solomon GD. (1992) Circadian rhythms and migraine. *Cleve Clin J Med* 59(3): 326–329.

Sothern RB. (1974a) Chronobiologic serial section on 8876 oral temperatures collected during 4.5 years by presumably healthy man (age 20.5 at start of study). In: *Chronobiology*. Scheving LE, Halberg F, Pauly JE, eds. Tokyo: Igaku Shoin Ltd., pp. 245–248.

Sothern RB. (1974b) Low frequency rhythms in the beard growth of a man. In: *Chronobiology*. Scheving LE, Halberg F, Pauly JE, eds. Tokyo: Igaku Shoin Ltd., pp. 241–244.

Sothern RB, Leach C, Nelson WL, Halberg F, Rummel JA. (1974) Characteristics of urinary circadian rhythms in a young man evaluated on a monthly basis during the course of 21 months. In: *Chronobiological Aspects of Endocrinology*. Aschoff J, Halberg F, Ceresa F, eds. Stuttgart: FK Schattauer Verlag, pp. 71–80.

Sothern RB, Nelson WL, Halberg F. (1977a) A circadian rhythm in susceptibility of mice to the anti-tumor drug, adriamycin. In: *Proc XII Intl Conf Intl Soc Chronobiology.* Milano: Il Ponte, pp. 433–437.

Sothern RB, Rosen G, Nelson WL, Jovonovich JA, Wurscher TJ, Halberg F. (1977b) Circadian rhythm in tolerance of melphalan by mice. In: *Proc XII Intl Conf Intl Soc Chronobiology.* Milano: Il Ponte, pp. 443–450.

Sothern RB, Simpson HW, Leach C, Nelson WL, Halberg F. (1981a) Individually-assessable human circadian, circaseptan and circannual urinary rhythms (with temporal compacting and editing procedures). In: *Proc XIII Intl Conf Intl Soc Chronobiology.* Halberg F, Scheving LE, Powell EW, Hayes DK, eds. Milan: Il Ponte, pp. 363–371.

Sothern RB, Halberg F, Good RA, Simpson HW, Grage TB. (1981b) Difference in timing of circadian susceptibility rhythm in murine tolerance of chemically-related antimalignant antibiotics: adriamycin and daunomycin. In: *Chronopharmacology and Chronotherapeutics.* Walker CA, Winget CM, Soliman KFA, eds. Tallahassee: Florida A&M University Foundation, pp. 257–268.

Sothern RB, Halberg F, Hrushesky WJM. (1988a) Circadian stage not time of day characterizes doxorubicin susceptibility rhythm of mice in continuous light. In: *Ann Rev Chronopharm,* Vol. 5. Reinberg A, Smolensky M, Labrecque G, eds. New York: Pergamon Press, pp. 385–388.

Sothern RB, Haus R, Langevin TR, Popovich JA, Wick M, Roemeling Rv, Hrushesky WJM. Profound circadian stage dependence of mitomycin-C toxicity. (1988b) In: *Annual Rev Chronopharm,* Vol. 5. Reinberg A, Smolensky M, Labrecque G, eds. New York: Pergamon Press, pp. 389–392.

Sothern RB, Lévi F, Haus E, Halberg F, Hrushesky WJM. (1989) Control of a murine plasmacytoma with doxorubicin–cisplatin: dependence on circadian stage of treatment. *J Natl Cancer Inst* 8: 135–145.

Sothern RB, Rhame F, Suarez C, Fletcher C, Sackett-Lundeen L, Haus E, Hrushesky WJM. (1990) Oral temperature rhythmometry and substantial within-day variation in zidovudine levels following steady state dosing in human immunodeficiency virus (HIV) infection. In: *Chronobiology: Its Role in Clinical Medicine, General Biology and Agriculture, Part A, Proc XIXth Intl Conf Intl Soc Chronobiol,* Bethesda, MD, June 20–24, 1989. Hayes DK, Pauly JE, Reiter RJ, eds. New York: Wiley-Liss, Inc., pp. 67–76.

Sothern RB, Slover GPT, Morris RW. (1993) Circannual and menstrual rhythm characteristics in manic episodes and body temperature. *Biol Psychiatry* 33(3): 194–203.

Sothern RB. (1994) Circadian and circannual characteristics of blood pressure self-measured for 25 years by a clinically-healthy man. *Chronobiologia* 21(1–2): 7–20.

Sothern RB, Gruber SA. (1994) Further commentary: physiological parameters in laboratory animals and humans. *Pharm Res* 11(2): 349–350.

Sothern RB, Kanabrocki EL, Boles MA, Nemchausky BM, Olwin JH, Scheving LE. (1994) Marker rhythmometry: comparison of simultaneous circadian variations for cortisol & melatonin in 3 biological fluids of adult men (Abstract). In: *Biological Rhythms and Medications. Proc 6th Intl Conf Chronopharmacology & Chronotherapeutics.* Amelia Island, Florida, July 4–9, 1994: #XII–4.

Sothern RB. (1995) Time of day versus internal circadian timing references. *J Infus Chemother* 5(1): 24–30.

Sothern RB, Roitman-Johnson B, Kanabrocki EL, Yager JG, Roodell MM, Weatherbee JA, Young MRI, Nemchausky BM, Scheving LE. (1995a) Circadian characteristics of circulating interleukin-6 in adult men. *J Allergy Clin Immunol* 95(5): 1029–1035.

Sothern RB, Smaaland R, Moore JM. (1995b) Circannual rhythm in DNA synthesis (S-phase) in healthy human bone marrow and rectal mucosa. *FASEB J* 9(5): 397–403.

Sothern RB, Roitman-Johnson B. (2001) Biological rhythms and immune function. In: *Psychoneuroimmunology*, 3rd edn., Vol. 1. Ader R, Felten DL, Cohen N, eds. San Diego: Academic Press, pp. 445–479.

Sqalli A, Oustrin J, Houin G, Bugat R, Canal P, Carton M. (1989) Clinical chronopharmacokinetics of doxorubicin (DXR). In: *Ann Rev Chronopharmacol*, Vol. 5. Reinberg A, Smolensky M, Labrecque G, eds. Oxford: Pergamon, pp. 393–396.

Srivastava S, Sharma M. (1998) Seasonal affective disorder: report from India (latitude 26° 45'N). *J Affect Disorders* 49(2): 145–150.

Steinijans VW, Trautmann H, Sauter R, Staudinger H. (1990) Theophylline therapeutic drug monitoring in the case of a new sustained-release pellet formulation for once-daily evening administration. In: *Ann Rev Chronopharmacol*, Vol 7. Reinberg A, Smolensky M, Labrecque G, eds. Oxford: Pergamon Press, pp. 323–326.

Stevens RG, Davis S, Thomas DB, Anderson LE, Wilson BW. (1992) Electric power, pineal function, and the risk of breast cancer. *FASEB J* 6(3): 853–860 (Review).

Stevens RG, Davis S. (1996) The melatonin hypothesis: electric power and breast cancer. *Environ Health Perspect* 104(Suppl 1): 135–140 (Review).

Stevens RG. (2005) Circadian disruption and breast cancer. *Epidemiology* 16(2): 254–258.

Stokkan K-A, Reiter RJ. (1994) Melatonin rhythms in Arctic urban residents. *J Pineal Res* 16(1): 33–36.

Storms WW. (2004) Pharmacologic approaches to daytime and nighttime symptoms of allergic rhinitis. *J Allergy Clin Immunol* 114(5 Suppl): S146–S153.

Strempel H. (1977) Circadian cycles of epicritic and protopathic pain threshold. *J Interdiscipl Cycle Res* 8(3–4): 276–280.

Sumaya CV, Marx J, Ullis K. (1980) Genital infections with herpes simplex virus in a university student population. *Sex Transm Dis* 7(1): 16–20.

Svanes C, Sothern RB Sørbye H. (1998) Rhythmic patterns in incidence of peptic ulcer perforation over 5.5 decades in Norway. *Chronobiol Intl* 15(3): 241—264.

Swedo SE, Pleeter JD, Richter DM, Hoffman CL, Allen AJ, Hamburger SD, Turner EH, Yamada EM, Rosenthal NE. (1995) Rates of seasonal affective disorder in children and adolescents. *Amer J Psychiat* 152(7): 1016–1019.

Swerdlow AJ. (1985) Seasonality of presentation of cutaneous melanoma, squamous cell cancer and basal cell cancer in the Oxford Region. *Brit J Cancer* 52(6): 893–900.

Swoyer J, Haus E, Lakatua D, Sackett-Lundeen L, Thompson M. (1984) Chronobiology in the clinical laboratory. In: *Chronobiology 1982–1983*. Haus E, Kabat H, eds. Basel: S Karger, pp. 533–543.

Symonds C. (1956) A particular variety of headache. *Brain* 79(2): 217–232.

Tamarkin L, Ameida OF, Danforth DN Jr. (1985) Melatonin and malignant disease. *Ciba Found Symp* 117: 284–299.

Tampellini M, Bitossi R, Brizzi MP, Saini A, Tucci M, Alabiso I, Dogliotti L. (2004) Pharmacoeconomic comparison between chronochemotherapy and FOLFOX regimen in the treatment of patients with metastatic colorectal cancer: a cost-minimization study. *Tumori* 90(1): 44–49.

Tan DX, Manchester LC, Hardeland R, Lopez-Burillo S, Mayo JC, Sainz RM, Reiter RJ. (2003) Melatonin: a hormone, a tissue factor, an autocoid, a paracoid, and an antioxidant vitamin. *J Pineal Res* 34(1): 75–78 (Review).

Temte JL. (1989) Exploring environmental cycles in psychiatric patients. *Wis Med J* 88(1): 17–20.

Termorshuizen F, Hogewoning AA, Bouwes Bavinck JN, Goettsch WG, de Fijter JW, van Loveren H. (2003) Skin infections in renal transplant recipients and the relation with solar ultraviolet radiation. *Clin Transplant* 17(6): 522–527.

Thiberville L, Compagnon P, Moore N, Bastian G, Richard MO, Hellot MF, Vincent C, Kannass MM, Dominique S, Thuillez C, Nouvet G. (1994) Plasma 5-fluorouracil and alpha-fluoro-beta-alanin accumulation in lung cancer patients treated with continuous infusion of cisplatin and 5-fluorouracil. *Cancer Chemother Pharmacol* 35(1): 64–70.

Thomas CB, Holljes HWD, Eisenberg FF. (1961) Observations on seasonal variations in total serum cholesterol level among healthy young prisoners. *Ann Intern Med* 54: 413–430.

Thompson DR, Blandford RL, Sutton TW, Marchant PR. (1985) Time of onset of chest pain in acute myocardial infarction. *Intl J Cardiol* 7(2): 139–148.

Thompson DR, Sutton TW, Jowett NI, Pohl JE. (1991) Circadian variation in the frequency of onset of chest pain in acute myocardial infarction. *Br Heart J* 65(4): 177–178.

Toh KL, Jones CR, He Y, Eide EJ, Hinz WA, Virshup DM, Ptacek LJ, Fu YH. (2001) An h*Per2* phosphorylation site mutation in familial advanced sleep phase syndrome. *Science* 291(5506): 1040–1043.

Tokura H, Ohta T, Shimomoto M. (1979) Circadian change of sweating rate measured locally by the resistance method in man. *Experientia* 35(5): 615–616.

Touitou Y, Haus E, eds. (1994) *Biological Rhythms in Clinical and Laboratory Medicine*. Berlin: Springer-Verlag: 720 pp (see pp. 5, 27 for quotes).

Tsementzis SA, Kennet RP, Hitchcock ER, Gill JS, Beevers DG. (1991) Seasonal variation of cerebrovascular diseases. *Acta Neurochir (Wien)* 111(3–4): 80–83.

Tsinkalovsky O, Smaaland R, Rosenlund B, Sothern RB, Hirt A, Eiken HG, Steine S, Badiee A, Foss Abrahamsen J, Laerum OD. (2005) Circadian variations of clock gene expression in CD34+ progenitor cells in the human bone marrow (in preparation).

Tuchman M, Roemeling Rv, Lanning R, Sothern RB, Hrushesky WJM. (1988) Sources of variability of dihydropyrimidine dehydrogenase activity in human blood mononuclear cells. In: *Ann Rev Chronopharmacol*, Vol. 5. Reinberg A, Smolensky M, Labrecque G, eds. New York: Pergamon Press, pp. 399–402.

Ueda HR, Chen W, Minami Y, Honma S, Honma K, Iino M, Hashimoto S. (2004) Molecular-timetable methods for detection of body time and rhythm disorders from single-time-point genome-wide expression profiles. *Proc Natl Acad Sci USA* 101(31): 11227–11232.

Van Cauter E, Honinckx E. (1985) Pulsatility of pituitary hormones. In: *Ultradian Rhythms in Physiology and Behavior*. Schulz H, Lavie P, eds. Berlin: Springer-Verlag, pp. 41–60.

Van Cauter E, Desir D, Decoster C, Fery F, Balasse EO. (1989) Nocturnal decrease in glucose tolerance during constant glucose infusion. *J Clin Endocrinol Metab* 69(3): 604–611.

Van den Heiligenberg S, Déprés-Brummer P, Barbason H, Claustrat B, Reynes M, Lévi F. (1999) The tumor promoting effect of constant light exposure on diethylnitrosamine-induced hepatocarcinogenesis in rats. *Life Sci* 64(26): 2523–2534.

Vander A, Sherman J, Luciano D. (1998) *Human Physiology: The Mechanisms of Body Function*, 7th edn. Boston: McGraw-Hill, p. 558.

Veldhuis JD, Johnson ML, Iranmanesh A, Lizarralde G. (1992) Rhythmic and nonrhythmic modes of anterior pituitary hormone release in man. In: *Biological Rhythms in Clinical and Laboratory Medicine*. Touitou Y, Haus E, eds. Berlin: Springer-Verlag, pp. 277–291.

Verdecchia P, Schillaci G, Porcellati C. (1991) Dippers versus non-dippers. *J Hypertens* (Suppl) 9(8): S42–S44 (Review).

Verschoore M, Poncet M, Krebs B, Ortonne J-P. (1993) Circadian variations in the number of actively secreting sebaceous follicles and androgen circadian rhythms. *Chronobiol Intl* 10(5): 349–359.

Vesely DL, Kanabrocki EL, Sothern RB, Scheving LE, Tsai TH, Greco J, Bushnell DL, Kaplan E, Rumbyrt J, Sturtevant RP, Sturtevant FM, Bremner WF, Third JLHC, Hrushesky WJM, Olwin JH. (1990) The circadian rhythm of the N-terminus and C-terminus of the atrial natriuretic factor prohormone. *Chronobiol Intl* 7 (1): 51–57.

Vianna EO, Martin RJ. (1998) Bronchodilators and corticosteroids in the treatment of asthma. *Drugs Today (Barc)* 34(3): 203–223.

Vijayalaxmi, Thomas CR Jr, Reiter RJ, Herman TS. (2002) Melatonin: from basic research to cancer treatment clinics. *J Clin Oncol* 20(10): 2575–2601 (Review).

Vollman RF. (1974) Some conceptual and methodological problems in longitudinal studies on human reproduction. In: *Biorhythms and Human Reproduction*. Ferin M, Halberg F, Richart RM, Vande Wiele RL, eds. New York: John Wiley & Sons, Inc., pp. 161–170.

von Armin T, Höfling B, Schreiber M. (1985) Characteristics of episodes of ST elevation or ST depression during ambulatory monitoring in patients subsequently undergoing coronary angiography. *Br Heart J* 54(5): 484–488.

Voultsios A, Kennaway DJ, Dawson D. (1997) Salivary melatonin as a circadian phase marker: validation and comparison to plasma melatonin. *J Biol Rhythms* 12(5): 457–466.

Waldrop RD, Saydjari R, Rubin NH, Rayford PL, Townsend CM Jr, Thompson JC. (1989a) Photoperiod influences the growth of colon cancer in mice. *Life Sci* 45(8): 737–744.

Waldrop RD, Saydjani R, Rubin NH, Rayford PL, Townsend CM Jr, Thompson JC. (1989b) DNA synthetic activity in tumor-bearing mice. *Chronobiol Intl* 6(3): 237–243.

Waters WE, O'Connor PJ. (1971) Epidemiology of headache and migraine in women. *J Neurol Neurosurg Psychiat* 34(2): 148–153.

Waters DD, Miller DD, Bouchard A, Bosch X, Theroux P. (1984) Circadian variation in variant angina. *Amer J Cardiol* 54(1): 61–64.

Waterton JC, Solloway S, Foster JE, Keen MC, Gandy S, Middleton BJ, Maciewicz RA, Watt I, Dieppe PA, Taylor CJ. (2000) Diurnal variation in the femoral articular cartilage of the knee in young adult humans. *Magnetic Resonance Med* 43(1): 126–132.

Weber MA, Prisant LM, Black HR, Messerli FH. (2004) Treatment of elderly hypertensive patients with a delayed-release verapamil formulation in a community-based trial. *Amer J Geriatr Cardiol* 13(3): 131–136.

Wehr TA, Sack DA, Rosenthal NE. (1987) Seasonal affective disorder with summer depression and winter hypomania. *Amer J Psychiatry* 144(12): 1602–1603.

Wehr TA, Rosenthal NE. (1989) Seasonality and affective illness. *Amer J Psychiatry* 146(7): 829–839 (Review).

Wehr TA, Giesen HA, Schulz PM, Anderson JL, Joseph-Vanderpool JR, Kelly K, Kasper S, Rosenthal NE. (1991) Contrasts between symptoms of summer depression and winter depression. *J Affective Disorders* 23(4): 173–183.

Wehr TA, Giesen HA, Moul DE, Turner EH, Schwartz PJ. (1995) Suppression of men's responses to seasonal changes in daylength by modern artificial lighting. *Amer J Physiol* 269(1 Pt2): R173–R178.

Weiss KB. (1990) Seasonal trends in US asthma hospitalizations and mortality. *JAMA* 263(17): 2323–2328.

Weitzman ED, Czeisler CA, Moore-Ede MC. (1979) Sleep-wake, neuroendocrine and body temperature circadian rhythms under entrained and non-entrained (free-running) conditions in man. In: *Biological Rhythms and Their Central Mechanism*. Suda M, Hayaishi O, Nakagawa H, eds. Amsterdam: Elsevier/N-Holland BioMed Press, pp. 199–227.

Weydahl A, Sothern RB. (1997) Seasonal variation in glycemic response to exercise in the subarctic. *Biol Rhythm Res* 28: 42–55.

Wetterberg L, Arendt J, Paunier L, Sizonenko PC, Donselaar W van, Heyden T. (1976) Human serum melatonin changes during the menstrual cycle. *J Clin Endocrinol Metab* 42: 185–188.

White C, Smolensky MH, Sanders SW, Buchi KN, Moore JG. (1991) Day–night and individual differences in response to constant-rate ranitidine infusion. *Chronobiol Intl* 8(1): 56–66.

White WB, Anders RJ, MacIntyre JM, Black HR, Sica DA. (1995) Nocturnal dosing of a novel delivery system of verapamil for systemic hypertension. Verapamil Study Group. *Amer J Cardiol* 76(5): 375–380.

White WB. (2000) Ambulatory blood pressure monitoring: dippers compared with non-dippers. *Blood Press Monit* 5 (Suppl 1): S17–S23 (Review).

White WB, LaRocca GM. (2002) Chronopharmacology of cardiovascular therapy. *Blood Press Monit* 7(4): 199–207 (Review).

White WB. (2003) Relevance of blood pressure variation in the circadian onset of cardiovascular events. *J Hypertens* 21(Suppl 6): S9–S15 (Review).

Willich SN, Levy D, Rocco MB, Tofler GH, Stone PH, Muller JE. (1987) Circadian variation in the incidence of sudden cardiac death in the Framingham Heart Study population. *Amer J Cardiol* 60(10): 801–806.

Willich SN, Goldberg RJ, Maclure M, Perriello L, Muller JE. (1992) Increased onset of sudden cardiac death in the first three hours after awakening. *Amer J Cardiol* 70(1): 65–68.

Willich SN, Lowel H, Lewis M, Hormann A, Arntz HR, Keil U. (1994) Weekly variation of acute myocardial infarction. Increased Monday risk in the working population. *Circulation* 90(1): 87–93.

Woodhouse PR, Khaw KT, Plummer M, Foley A, Meade TW. (1994) Seasonal variations of plasma fibrinogen and factor VII activity in the elderly: winter infections and death from cardiovascular disease. *Lancet* 343(8895): 435–439.

Wu J. (1982) Neijing chronobiologic medical theories. *Chin Med J* 95(8): 569–578.

Wu MW, Li XM, Xian LJ, Levi F. (2004) Effects of meal timing on tumor progression in mice. *Life Sci* 75(10): 1181–1193.

Wyatt JK. (2004) Delayed sleep phase syndrome: pathophysiology and treatment options. *Sleep* 27(6): 1195–1203.

Yen SSC, Vandenberg G, Tsai CC, Siler T. (1974) Causal relationship between the hormonal variables in the menstrual cycle. In: *Biorhythms and Human Reproduction*. Ferin M, Halberg F, Richart RM, Vande Wiele RL, eds. New York: John Wiley & Sons, Inc., pp. 219–238.

Yetman RJ, Andrew-Casal M, Hermida RC, Dominguez BW, Portman RJ, Northrup H, Smolensky MH. (2002) Circadian pattern of blood pressure, heart rate, and double product in liver glycogen storage disease. *Chronobiol Intl* 19(4): 765–783.

Yosipovitch G, Sackett-Lundeen L, Goon A, Huak CY, Goh CL, Haus E. (2004) Circadian and ultradian (12 h) variations of skin blood flow and barrier function in non-irritated and irritated skin–effect of topical corticosteroids. *J Invest Dermatol* 122(3): 824–829.

Young MRI, Matthews JP, Kanabrocki EL, Sothern RB, Roitman-Johnson B, Scheving LE. (1995) Circadian rhythmometry of serum interleukin-2, interleukin-10, tumor necrosis factor-a, and granulocyte-macrophage colony-stimulating factor in men. *Chronobiol Intl* 12(1): 19–27.

Zimmet PZ, Wall JR, Rome R, Stimmler L, Jarrett RJ. (1974) Diurnal variation in glucose tolerance: associated changes in plasma insulin, growth hormone, and non-esterified fatty acids. *Br Med J* 1(906): 485–488.

12
Autorhythmometry

"That period of twenty-four hours, formed by the regular revolution of our earth, in which all its inhabitants partake, is particularly distinguished in the physical oeconomy of man. . . . It is, as it were the unity of our natural chronology."
— C.W. Hufeland (1762–1836), German physician

Introduction

Know thyself (Socrates) is a proverbial statement that adds meaning to those who are interested in learning about their own rhythms. In fact, the derivation of the word "autorhythmometry" is from the words for "self" (*autos*) and "to measure" (*metry*) placed before and after the word "rhythm," which explains the focus of this chapter on the self-measurement of one's own body rhythms (Figure 12.1). After learning about the presence of biological rhythms, most individuals become curious about their own body rhythms by asking themselves: *When am I the most alert? When do I perform and feel the best? Do my vital signs, such as heart rate, blood pressure, or body temperature, really change very much throughout the day? Do my rhythms show that I am a morning person, an evening person, or somewhere in-between, and what does that mean?*

Much of what we know about the status of our health or performance capabilities has been obtained from sporadic medical check-ups or possibly from participation in a group study, where somebody else conducted the sampling or monitoring. But is one value obtained at any single time of day or even only once a year always sufficient for accurate diagnostic purposes? For example, machines that measure blood pressure (BP) are available to the general public and can be found in various public places, including near pharmacies and booths in malls and elsewhere. Individuals may thus check their BP once in a while, but what does a single value mean, especially, since blood pressure can vary rhythmically by 30 mmHg or more over 24 h? While few individuals have had the time or opportunity to participate in a performance or medical-related study, most individuals can easily self-measure a number of variables on themselves and thereby learn about changes that naturally occur over the course of each day (and possibly over longer spans).

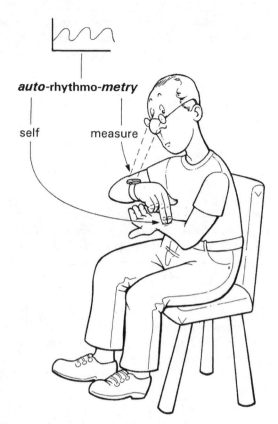

FIGURE 12.1. Schematic of derivation of the word "autorhythmometry" and a subject performing a self-measurement by counting his pulse.

Measuring Your Own Body Rhythms

The self-monitoring of physiological or psychological functions or performance and their statistical analysis for rhythms is called *autorhythmometry* (AR) (Halberg et al., 1972).

While a number of body functions can be measured throughout the day by individuals manually or with automatic instrumentation, oral temperature, blood pressure, heart rate, and possibly peak flow, are the most commonly-monitored variables. When individuals have taken readings of oral temperature every few hours throughout the day, a prominent daily variation is usually apparent. As mentioned in Chapter 11 on Clinical Medicine, normal limits change predictably throughout the 24 h. Thus, normal oral temperature can change by 2–3°F (1–2°C) throughout the day and may range from less than 97.0°F (36.1°C) near the end of sleep to 98.6°F (37.0°C) or more in the middle of the afternoon.

Similarly, blood pressure in a healthy (normotensive) individual may vary 20–50 mmHg or more from the lowest to the highest readings over a 24-h span (e.g., systolic blood pressure may be 110 mmHg in the morning and 140 mmHg in the evening, while the overall average may be 125 mmHg) (Halberg et al, 1984a). An individual's expected "normal" value for these variables, therefore, would be dependent upon the time of day and not some time-unvarying population-based number at any time of the day, such as 98.6°F (37.0°C)[1] for body temperature or 120 mmHg for systolic blood pressure.

What individuals have found for themselves through AR, as well as the diversity of those who have done so, is quite revealing.

School Children

Starting with youths, self-measurements over several days under usual conditions by 115 high-school students in Houston, TX, ages 15–18 years, documented daily rhythms in oral temperature, blood pressure, pulse, and in ratings of mood and vigor (Smolensky et al., 1981).[2] A similar project was also carried out over 72 h in over 400 students in three high schools in Little Rock, AR (Glasgow et al., 1982). The stability of circadian patterns in these variables was shown when 10 children in Minneapolis, MN, ages 9–13 years, self-measured most of the same variables plus adding speed and finger-counting for 3 weeks in the spring of 1 year and again for 3 weeks a year later (Rabatin et al., 1981).[3]

Circadian patterns were also found in BP in school children between the ages of 9–12 years who wore an ambulatory monitor that automatically recorded BP every 30 min during waking and hourly overnight for 24 h (Yetman et al., 1994). The authors concluded that when performing long-term tracking of BP in children, a longer sampling span was necessary to accurately characterize the individual 24-h pattern in the BP of active children, such as a minimum of 48 h of monitoring as has been suggested by others (Cugini et al., 1988).

---

[1] The value of 37.0°C (98.6°F) as the average temperature of healthy adults was originally derived from over one million *axillary* temperature readings from >25,000 patients by Carl Wunderlich in Germany in 1868 (Mackowiak et al., 1992a). Without referring to a biological rhythm, Wunderlich also noted that temperatures were lowest (36.2°C/97.2°F) between 02:00 and 08:00 and highest (37.5°C/99.5°F) between 16:00 and 21:00. However, the concept of 98.6°F as the most normal temperature persists to this day in both lay thinking and medical writing. This was documented in a survey of 268 physicians and medical students, where 75% indicated 98.6°F (37°C) as their definition of a normal body temperature (Mackowiak & Wasserman, 1995). Another 13% defined a narrow range around 98.6°F as indicating normal body temperature. In addition, only 4% specified a particular body site in their definition.

[2] Eleven of these students also collected their own urine samples for 24 h. They measured and recorded the volume and labeled each sample prior to storage in a refrigerator. Chemical analyses were performed by a technician and computer analysis found circadian rhythmicity in Na, K, Cl, and uric acid.

[3] These data resulted from a high-school science project on autorhythmometry that began when the first author of the paper was a student in the ninth grade and had schoolmates and siblings perform these self-measurements at 6 times each day during waking-only, including while at school and at home.

Adults

With regard to adults, BP results were also found to be more reliable when several hundred women were monitored for 48 h rather than for only 24 h (Hermida & Ayala, 2003). Home self-measurements of BP have been used to establish circadian patterns in normotensives (De Meyer & Vogelaere, 1990) and prior to selecting a temporal therapy for hypertensives (De Scalzi et al., 1986), to monitor the effectiveness of treatment for uncomplicated hypertension (Scarpelli et al., 1978; Engel et al., 1979, 1987; Mengden et al., 1992; Beliaev et al., 2002) and to follow morning hypertension in diabetics (Kamoi et al., 2002, 2003).

Performance

Results from some studies where individuals performed various tasks at different times of the day indicated that performance was better at certain times, with improvement associated with the diurnal rise in body temperature (Winget et al, 1985). This was the case for groups of 25–30 sailors, 17–33 years of age, who performed a number of highly practiced familiar skills or novel laboratory tests (Blake, 1967) and for a group of healthy athletes 19–23 years of age (21 women and 25 men) at a sports institute in Romania, who measured their "muscle power" (an index calculated from the maximal strength and speed of contraction of the right and left quadriceps), grip strength and manual dexterity (Lundeen et al., 1990). Sixteen children aged 7–9 years showed performance rhythms, mainly in speed to perform the tasks, with a peak in the afternoon when tested four times between 09:00 h and 19:30 h (Reinberg et al., 1988). There was even a circadian rhythm observed in the walking time every hour of a man who walked 1,000 miles in 1,000 h (1 mile per hour for 1,000 consecutive hours) to win a large monetary bet in 1809 (Crawford et al., 1990).

Ultradian Rhythms

In regard to higher frequency rhythms in performance, 90–100 min oscillations were observed when 8 subjects performed verbal and spatial matching tasks every 15 min for 8 h, with peaks between these two cognitive measures being out-of-phase (Klein & Armitage, 1979). A similar study of 24 subjects found an 80-min cycle in verbal matching and a 96-min cycle in spatial matching (Gordon et al., 1995). Cycles around 4 h were also found in these cognitive performance tasks (Klein & Armitage, 1979; Iskra-Golec, 2001).

Self-help Health Care

With more and more emphasis being placed on preventive health care, autorhythmometry, which can be essentially cost-free, provides information and self-awareness about one's own body functions in health and in sickness. In this regard, it has been suggested that self-measurement of body rhythms should be taught no later than high school (Scheving, 1990).

In relation to health education, self-measurements may be helpful in promoting positive health attitudes and practices, since individuals can become more knowledgeable about their physiological and psychological well-being by taking part in the study of their own health (Halberg et al., 1976). Thus, to a certain degree, AR is being used when an individual monitors his or her glucose levels to control diabetes, monitors body temperature for natural family planning or during fever, frequently measures their own blood pressure with or without treatment, or measures peak flow to follow the effects of asthma on lung function with or without treatment.

Monitoring Symptoms

In regard to lung function, self-measurements have been utilized in studying peak flow rhythms in (1) healthy and/or *asthmatic* children and adults (Hetzel & Clark, 1980; Henderson & Carswell, 1989; Albertini et al., 1989; Troyanov et al., 1994; Lebowitz et al., 1997), (2) in *treated asthmatics* (Reindl et al., 1969; Reinberg et al., 1983; Burioka et al., 2000, 2001), (3) in passive and active *smokers* (Casale et al., 1992, 1997), and (4) during *aging* (Halberg E et al., 1981).

In addition, individuals have performed self-measurements in order to monitor symptoms of a disease, including: (1) grip strength in *myasthenia gravis*, a neuromuscular disease (Simpson et al., 1971), (2) body temperature and other variables (blood pressure, heart rate, and urine) in *psychiatric illness* (Trapp et al., 1979; Slover et al., 1986; Sothern et al., 1993), (3) grip strength and finger joint sizes in *rheumatoid arthritis* (Kowanko et al., 1982), (4) pain in *osteoarthritis of the knee* (Bellamy et al., 1990), (5) pain, stiffness, and manual dexterity in *rheumatoid arthritis* and *hand osteoarthritis* (Bellamy et al., 1991, 2002), (6) pain, stiffness and fatigue in *fibromyalgia* (Bellamy et al., 2004), (7) pain intensity in *intractable pain* (Folkard et al., 1976), and (8) self-tonometry of intraocular pressure by *glaucoma* patients (Kothy et al., 2001).

Body Temperature

The circadian rhythm in body temperature (BT), which exhibits regular and predictable changes in relation to the sleep-wake schedule, has been studied perhaps more than any other rhythm (Reinberg & Smolensky, 1990). The BT rhythm is already noticeable by age 4–9 weeks and becomes more pronounced by age 2–3 months and thereafter, into adulthood (Hellbrügge, 1960).

One's own BT rhythm can be fairly easily, and accurately measured by autorhythmometry. If one were to take a reading of BT upon waking, just prior to retiring and every 2–4 h in-between and graph the values against the time each was measured, enough of the 24 h pattern would have been measured to allow the prominent circadian variation to be apparent to the naked eye. Indeed, this pattern is evident in Figure 12.2 that shows a circadian oral temperature pattern of a man

FIGURE 12.2. Historic literature from 1842: a circadian pattern is evident in the first recorded oral temperatures throughout the day by a healthy man (redrawn from Fig. 154 in Aschoff & Heise, 1972).

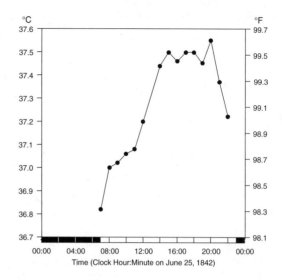

published in a thesis in 1842, which is possibly the first such curve ever recorded[4] (Aschoff & Heise, 1972).[5] While in good clinical health and under stable living conditions, this pattern repeats itself day after day (Figure 12.3) and if one wakes in the middle of the night to self-measure temperature, the reading will usually be quite low.

Internal Marker Rhythm

The phasing of the circadian rhythm in BT is very stable under standard light–dark conditions, yet will readjust its times of highs and lows after shifts in the light–dark synchronizer schedule of the environment (e.g., due to time zone

---

[4] While the temperature values were reported to be obtained orally, the range of values from 98.3–99.6°F is more consistent with levels found when measuring rectal temperatures (see additional figures in this chapter). The subject was reported to be in good health, so fever is also ruled out as elevating the observed temperature levels. Another early paper presented tables of actual values for oral temperature that was measured two or three times a day for 8 months by a healthy man, 55 years of age (Davy, 1845). Average temperature was highest upon arising (98.74°F), remained elevated during in the early afternoon (98.52°F), and was lowest at midnight before retiring (97.92°F). Over one 24-h span of 2 h measurements between awakening and going to bed, he recorded the highest value at 16:00 h (98.9°F) and the lowest value at 01:00 h (97.6°F). These studies in the 19th century were all the more remarkable since the thermometer used was 12.5 inches long, had a bulb about an inch long and half an inch wide that fit under the tongue, and could be read by the subject himself. Accurate readings were tedious and demanded much time and care (Davy, 1845).

[5] An excellent review of the history of human body temperature measurement, the development of thermometers and other measuring devices, including a "chronobra" that measures breast temperature (Simpson et al., 1982), and other early reports on diurnal variations in body temperature can be found in a chapter by Simpson et al., 1987.

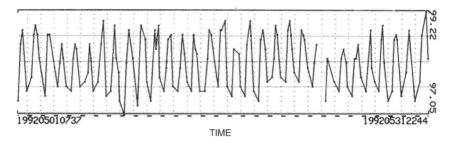

FIGURE 12.3. Circadian and other patterns seen in time plot of oral temperature self-measured during waking-only for 1 month by a healthy man (RBS, age 45 years). Dashed vertical lines = local midnight. Dark horizontal bars = approximate sleep times (Sothern, unpublished).

changes, shiftwork scheduling or daylight saving time). The regularity and stability of the temperature rhythm has led to its use as an *internal marker rhythm*, providing a meaningful phase reference in order to better understand the internal temporal organization of other rhythms (i.e., the relationships between the sequence of highs and lows among various variables), as opposed to using the external clock hour as a reference (Sothern, 1995).

For example, the peak in cortisol may normally be a certain number of hours before the peak in BT, while the peak in white blood cells may normally be several hours after the peak, and so on. In addition, core BT, along with plasma cortisol and plasma melatonin, which all show prominent circadian rhythms, are commonly used when studying phasing of the human circadian pacemaker (Klerman et al., 2002). Body temperature is, therefore, one of the main variables to be monitored to determine proper circadian synchronization of subjects in various medical and other studies.

Measurement Site

Body temperature measurement sites commonly include the mouth (oral), deep body (rectal, urine), and skin surface (axillary, ear). While rectal and oral temperatures have been simultaneously measured in some studies (Atkinson et al., 1993), oral temperature is often used, especially in field studies, where it is not feasible or necessary to obtain frequently-measured core temperature. This has been the case in studies of *children* (Rabatin et al., 1981; Smolensky et al., 1981; Bee & Webb, 1987), the *elderly* (Okamoto-Mizuno et al., 2000), *psychiatric patients* (Atkinson et al., 1975; Madjirova et al., 1995; Rao et al., 1995), persons with certain *infectious disease* conditions (e.g., HIV) (Sothern et al., 1990) or with *intestinal infections* (Mackowiak et al., 1992b), individuals on the job, such as *shift workers* (Reinberg et al., 1984; Fujiwara et al., 1992; Vangelova & Deyanov, 2000), *bus or truck drivers* (Pokorny et al, 1988; Stoynev & Minkova, 1997), *fire brigade control room operators* (Knauth et al., 1995), *ambulance personnel* (Motohashi & Takano, 1993), *air traffic controllers* (Stoynev & Minkova, 1998), *sailors* on board ship (Condon et al., 1984), or even *crew members* on long B–1 B bomber mission simulations (French et al., 1994).

In many studies, rectal temperature is considered the "gold standard" for monitoring a subject's BT rhythm, but as pointed out above, except for differences in overall levels, oral temperature can provide adequate estimates of the circadian amplitude and acrophase.[6] Thus, circadian profiles of oral temperature measured for up to a week will usually be sufficient to document a person's synchronization to his or her own sleep-wake schedule (Reinberg & Smolensky, 1983; Refinetti & Menaker, 1992), help to determine whether a person is a morning or evening "type" (Foret et al, 1982), as well as reveal any abnormalities and act as an indicator of illness (Davis & Lentz, 1989).

Axillary temperature has also been used as a marker rhythm when studying shift workers (Motohashi et al., 1987) and when studying rhythm characteristics of patients with major affective disorders (Bicakova-Rocher et al., 1996).[7] An overlooked source of reliable deep body temperature information is via the urine stream, where urine temperature has been shown to be closely correlated with rectal temperature (Fox et al., 1971, 1975; Brooke et al., 1973).

What and When is Normal?

While many medical textbooks and dictionaries still define 98.6°F (37.0°C) as the upper limit of "normal" (see Figure 11.2 in Chapter 11 on Clinical Medicine), others have reported that this upper limit: *"should be abandoned as a concept having any particular significance for the normal body temperature."* (Mackowiak et al., 1992a). These authors reported that 98.9°F (37.2°C) in the early morning and 99.9°F (37.7°C) later in the day should be regarded as the upper limit of the normal oral temperature in healthy adults. In addition, this report went on to state that individual variability and time of day need to be taken into account when interpreting someone's oral temperature values, rather than rely on a comparison to mean values derived from population studies.[8]

In addition to age, sex, activity, hormones, health status, and measurement site, there are systematic changes in normal body temperature due to time of day.

---

[6] Due to the more defined trough (bathyphase) of body temperature at night and the broader peak daytime peak, many investigators prefer to use rectal temperature that can be continuously recorded throughout the day and night, including while a person is sleeping. Even without statistical analysis, the trough can usually be fairly accurately located by the naked eye.

[7] It's interesting to note that oral temperature has been commonly used in the USA for everyday monitoring of body temperature, while rectal temperature is monitored in many clinical studies. In Europe, on the other hand, axillary temperature has been the site of choice for everyday monitoring of body temperature and oral temperature is used in many clinical studies.

[8] For example, an individual could monitor themselves while healthy and later use these values as their own reference values when ill.

Therefore, normal limits and circadian timing of the body temperature rhythm would ideally need to be defined for each individual by self- or automatic measurements of body temperature over an appropriate span of time. This is also the case for blood pressure, where self-assessment at home can almost immediately determine normotension and identify elevations at unusual times or under times of stress. Self-measurements can also provide positive feedback with regards to what is "normal" for individuals and encourage them to adhere to prescribed treatments or even certain exercise or weight loss regimens when viewing self-obtained evidence of a decrease in blood pressure. Repeated self-measurements of one's own vital signs thus have the potential to play an important role in preventive health care.

Blood Pressure

Rather than being random fluctuations or mere curiosities, it has been suggested that alterations from a person's own normal blood pressure values and/or in their times of highs and lows might serve as warning signs for early detection of illness or to follow a disease process that is or is not being treated (Levine et al., 1980; Halberg et al., 1984b). For example, trends and yearly rhythms have been shown in a time series covering 36 years in a man who self-measured blood pressure once a month beginning at the age 29 years and who subsequently developed hypertension at age 50 years and began taking hypotensive drugs at age 53 years following an ischemic brain attack (stroke) (Okajima et al., 2000).

Monitoring Hypertension

High blood pressure (hypertension) has been recognized as a widespread epidemic, with most people not knowing they have it. Since few people are aware of its danger, the World Health Organization has recommended home self-monitoring of blood pressure as an aid in the diagnosis and treatment of hypertension, since a blood pressure measured at a single time of day or on any one day may not be enough for accurate diagnostic purposes (WHO, 1988). This suggestion for home self-blood pressure monitoring (SBPM) has since been supported for its ease of performance, reliability, cost-effectiveness, and its value in research (Marolf et al., 1987; Mengden et al., 1994; Weisser et al., 1994; Vaisse et al., 2000; Parati et al., 2002). SBPM has also been shown to be useful with regards to identifying the "white-coat" effect wherein some individuals in the clinic but not at home show office hypertension, an early indication of hypertension (Middeke & Lemmer, 1996; Staessen et al., 2001; Pickering, 2002). In addition, home BP measurement in the morning and evening by nearly 5,000 treated hypertensives was significantly prognostic for a cardiovascular event, while office BP measurements were not, leading the authors to conclude that: *"BP should systematically be measured at home in patients receiving treatment for hypertension.*" (Bobrie et al., 2004).

Ambulatory Monitoring

It has been suggested that SBPM be used in conjunction with ambulatory BP monitoring (APBM), using home SBPM values via an interactive telemonitoring system for maintenance and follow-up of blood pressure after an initial diagnosis using ABPM (Parati et al., 2002; Pickering, 2002; Mengden et al., 2003). Normal (but time-unqualified) values for SBPM that correspond to office BP values have been established (Weisser et al., 2000), and SBPM is being encouraged for the elderly, diabetic hypertensives and pregnant women (Herpin et al., 2000; Broege et al., 2001).

Large studies have shown that up to 90% of patients are willing to self-measure their own BP, with results more reproducible than office BP and more representative of everyday life (Mengden et al., 2003). While SBPM can involve most individuals in the management of their own BP and is therefore recommended, a few individuals were not enthusiastic about SBPM and considered that it was a doctor's job to manage hypertension (Rickerby & Woodward, 2003).

Morningness–Eveningness

Some people like to wake up early each day, while others find this difficult. Conversely, some like to stay up late at night, while others "go to bed with the chickens." Differences in sleep preferences and the time of day for peaks or troughs in some physiologic variables or in best performance for some variables can depend upon an individual's circadian typology or *chronotype*, which indicates whether he/she is a "morning type" (sometimes called a lark, early bird or M-type) or an "evening type" (sometimes called an owl, night owl, or E-type).

Questionnaires

Several questionnaires are available that can categorize individuals on their morningness–eveningness orientation by their responses to a number of questions, including the *Morningness–Eveningness Questionnaire* (Horne & Ostberg, 1976), the *Diurnal-Type Scale* (Torsvall & Åkerstedt, 1980), the *Composite Scale of Morningness* (Smith et al., 1989), the *Circadian Type Questionnaire* (Folkard et al., 1979), and the *Early/Late Preferences Scale* (Smith et al., 2002). The latter two questionnaires each showed group consistency in the ability to assess morningness in spite of cross-cultural differences when studying more than 1800 individuals in six geographically different countries (USA, England, the Netherlands, Spain, Colombia, and India) (Smith et al., 2002). The recent *Munich ChronoType Questionnaire* (Roenneberg et al., 2003) considers work and free days separately and includes a self-assessed chronotype scale, while the *Circadian Type Inventory* uses only 11 items to determine a morningness score (Di Milia et al., 2004).[9]

Morningness vs. Life Factors

Rather than either/or, the morningness–eveningness score has been reported to be a continuum, with one survey of >2100 university students finding that 15.8% were M-types, 59.6% were intermediate types, and 24.5% were E-types (Adan & Natale, 2002). Thus, a more useful term to express morning versus evening orientation is simply *morningness* (Smith et al., 2002). Morning-evening scores have been shown to be associated with age (more M-types with increasing age) (Monk et al., 1991; Czeisler et al., 1992; Tankova et al., 1994; Carrier et al., 1997), sex (more M-types in females) (Vink et al., 2001; Adan & Natale, 2002), lifestyle regularity (M-types more regular in daily lifestyle) (Monk et al., 2004) and other life factors (e.g., work, school).

Endogenous Disposition

Results from some studies indicate that there actually may be an endogenous disposition based in genetics to being a lark or an owl (Katzenberg et al., 1998; Vink et al., 2001). Peaks in oral temperature occurred earlier in M-types (Horne & Ostberg, 1976) and in oral temperature, heart rate and adding speed in ten M-types compared with E-types self-measuring for 10–15 days (Gupta & Pati, 1994). Over two weeks of self-measurements, subjective mood peaked earlier in the day for M-types than E-types that were defined by the time of the peak in oral temperature (Kerkof, 1998). In studies where rectal temperature was automatically monitored, temperature peaked 2.2 h earlier (Kerkof & Van Dongan, 1996), up to 3 h earlier (Waterhouse et al., 2001) or 68 min earlier (Bailey & Heitkemper, 2001) in morning types than in evening types, suggesting that there is an endogenous component to the difference in temperature phase.

Body Temperature Phase

The lowest phase of core body temperature (BT) occurred earlier during sleep and started rising long before waking up in M-types, while the interval between low BT phase and waking up was much shorter in E-types (i.e., BT dropped more

---

[9] At a meeting in 1994, one of us (RBS) met a nursing researcher who reported that the following single question could categorize patients or subjects: "*Do you consider yourself an early person, a late person or have no preference?*"(Felver & Hoeksel, 1994). I have never been able to find a publication that validated this single question, but from personal experience, people who have taken the Horne-Ostberg Morningness-Eveningness questionnaire tend to agree with their chronotype score, indicating that they had a feeling about their degree of morningness, which the single question might identify. The *Munich ChronoType Questionnaire* also asks individuals to self-assess their chronotype (time-of-day-type) on a 0 to 6 scale (from early to late) (Roenneberg et al., 2003).

quickly during sleep and was rising again in M-types while it was still dropping in E-types) (Duffy et al., 1999). Among 68 young adults, morning types woke at a later internal circadian phase, even if E-types woke at a later external clock hour. During a constant routine, a shorter circadian period and advanced phase in core BT was associated with a higher morning-type score, indicating that diurnal preference and waking time may be grounded in the genetics of one's circadian physiology (the intrinsic circadian period) (Duffy et al., 2001).

Cognitive Tasks

When 36 students, aged 18–26 years, performed a search task and a cognitive task, and measured their oral temperature at different times of the day between 08:00 h and 23:00 h, circadian rhythms were found in each variable (Monk & Leng, 1986). As anticipated, slight differences in circadian patterns were found in oral temperature or performance speed (values were higher for M-types earlier in the day), but much more dramatic differences in cognitive ability were found: M-types performed better on the logical reasoning task shortly after awakening, while E-types performed best in the early afternoon. Morningness–eveningness differences have also been observed in the circadian rhythm in sleep propensity (Lavie & Segal, 1989), and sleep pressure during wakefulness (Hidalgo et al., 2003; Taillard et al., 2003).

Individual differences based upon chronotype thus have the potential to influence many areas, including the timing of circadian rhythms, sleep characteristics, sports performance, academic achievement, and adaptation to work schedules. Chronotype may also have an impact in scheduling patient responsiveness in hospitals, such as patient teaching, physical therapy, or getting out of bed for the first time after surgery.

When and How Long to Measure?

Students and others learning about biological rhythms are usually able to see many of their own circadian variations if they self-measure every 2–4 h for at least 2 or 3 days, but preferably for at least a week, then graph their data (Figure 12.4). Many external factors can mask the appearance of a circadian rhythm, resulting in measurements reflecting a combination of the pure rhythm (the endogenous signal) and error (noise) caused by ambient temperature, humidity, noises, illness, stress, lack of sleep, learning, and measurements at only a few times of the day and not others, etc. However, with a week or more of measurements, even if only throughout the waking span with no data obtained during sleep, the circadian pattern will usually protrude through all the "noise" in the data when looking at time-point averages.

Self-Measurements During Travel

Individuals have monitored their own body rhythms via self-measurements during school hours, during work, and to monitor phase-shifting after travel across

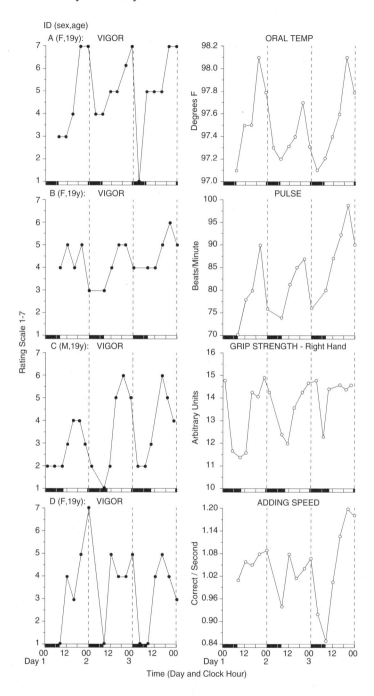

FIGURE 12.4. Some examples of circadian patterns in data obtained by four college students performing self-measurements over 3 days. Vigor rating (shown for each subject in *left column*) paired with another variable (in *right column*) measured by same student. Dark bars = time of sleep. Vertical dashed lines = midnight (Sothern & Koukkari, unpublished).

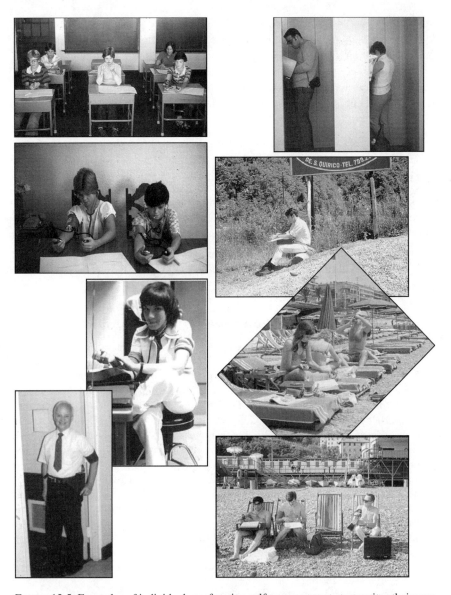

FIGURE 12.5. Examples of individuals performing self-measurements to monitor their own body rhythms during school projects (*upper left*), during work (*lower left*) and to monitor phase-shifting in pulse, temperature, peak flow, and blood pressure after travel across time zones to another country (*right side*). Photos by R. Sothern.

time zones (Figure 12.5). Even in difficult settings, individuals have found ways to collect rhythmic data on themselves via self-measurements. This was the case for four men who crossed the Greenland icecap on foot and on ski in the arctic summer even though the journey included some blizzard conditions and below freezing temperatures (Campbell et al., 1982, 1985). During five phases of this study, these men collected their saliva by spitting into small vials every 3 h for 1–3 days and kept the samples frozen in the snow for later chemical determination and documentation of circadian rhythms in many hormones, including the stress-related hormone cortisol. Explorers on a North Pole expedition also obtained body temperature rhythm data by measuring their own urine temperature throughout their journey in spite of ambient temperatures approaching −50°C (Simpson, 1970).

Self-Measurements During Isolation

One of the authors of this book (RBS) has been performing autorhythmometry on himself for more than 3 decades by measuring a dozen physiological and psychological variables an average of five times each day since May 1967 at the age of 20.5 years. This long self-measurement series began when the subject volunteered for a research project that involved studying his rhythms while in Germany in an underground isolation chamber away from all known time cues (described in Essay 12.1).

Essay 12.1. (by RBS): Self-Measurements in "Aschoff's Bunker"

My long self-measurement series of over 3.5 decades began after meeting a young German medical student (Dr. med. Jürgen Kriebel) sent by Prof. Jürgen Aschoff to Minnesota in the spring of 1967. After several weeks of adjustment to the 7-h time change from Germany to Minnesota, he was away from all time cues for about 3 weeks in a small isolation room in the Periodicity Analysis Laboratories of Prof. Franz Halberg at the University of Minnesota. During this time, he performed several self-measurements repeatedly throughout each waking span and collected urine samples upon urge around-the-clock. He explained to me that he was studying how the timing of his body rhythms would reset (e.g., phase-shift) due to changes in time zones following flights from east to west and west to east (from Germany to Minnesota and back to Germany again) in order to be in synch with local time. Perhaps more intriguing was the time he spent in isolation, both in an underground bunker in Germany and in the small room at the University of Minnesota, where, in the absence of time cues, his circadian rhythms free-ran with periods longer (i.e., 26.1 h) than precisely 24.0 h (Kriebel, 1974). Upon learning that I could help his project by doing the same study, but in reverse since I would begin in Minnesota, out of curiosity I agreed.

In May 1967, I began to self-measure oral temperature, pulse and time estimation several times each day, as well as keep a record of the times I went to bed and woke up. In June, I began to measure my blood pressure and in July, grip strength, peak flow, and mood and vigor ratings were added to the battery of tests I tried to do every 2 or 3 h during waking. In July, I also began to collect urine samples upon urge around-the-clock by first recording the time of each urine void, measuring the volume produced and then saving a 30-ml sample in a plastic bottle. These bottles were labeled and placed in a freezer for chemical analyses (I actually continued to collect urine samples for a total of 17 years, resulting in about 31,380 bottles that filled a lot of freezers!).

After obtaining baseline data over several months that established the circadian patterns of my body rhythms in Minnesota, I flew to Germany in August of that year and experienced my first jet lag due to a 7-h phase shift advance caused by the 7-h time-zone change, while continuing to self-measure all of the variables mentioned above. Another variable, the measurement of adding speed, was added to each measurement session while in Germany. The isolation chamber was located at the Institut für Verhaltensphysiologie (Institute for Entrained Physiology) under the direction of Prof. Jürgen Aschoff, one of the pioneers in biological rhythm research. This institute was located about an hour's drive to the southwest of Munich on a small site near Erling-Andechs, two villages each so small that they shared a post office (thus the hyphenated name).

The isolation chamber was called the "Bunker" and consisted of a concrete structure containing two separate chambers built into the side of a hill just below the main buildings of the institute. Each unit was totally soundproof and consisted of a large living room containing a table, chairs, cabinets and a bed, and two small side rooms: a kitchen and a bathroom. An entryway contained a refrigerator for new food to be placed in while I was asleep and for the urine samples I placed in a refrigerator there the previous "day" to be retrieved. There were doors at each end of the entryway, one from the outer hallway and one into the main isolation room and when one was open the other would be electronically locked. While a subject agreed to be in isolation, he or she actually could walk out and end the study at any time, so I could have escaped if necessary. Medical students usually volunteered for the isolation studies since it gave them an uninterrupted span to study and review their textbooks (plus free food for several weeks and a daily bottle of the famous Andechs beer!). Due to a conflict in the scheduling of the isolation chamber, I was not able to begin my stay in isolation in mid-August as planned. I therefore decided to remain in Germany and forego Fall quarter at the University (I was a 20-year-old student at the time) so that I could carry out the isolation study in early September.

After shopping for groceries and stocking the kitchen with things I knew how to prepare (unlike for other subjects, there would be no daily bottle of beer for me since it would interfere with some of the urinary variables to be determined), I entered the Bunker at around 2 PM in the afternoon on September 9, 1967 for what was to be a 20-day stay. On a countertop against one of the walls was a panel of buttons that were to be pressed whenever I did certain activities (Figure 12.6A), such as doing a self-measurement, collecting a urine sample, eating a meal, showering, exercising, going to bed, and waking up. A stripchart recorder outside of the isolation chamber would record each press of a button and later the actual times could be matched with the actual event. I was asked not to take any naps, but could sleep as much as I liked every "night." There was also a large round light on the wall that if lit would notify me that there was a message or items in the entryway for me to look at immediately, such as messages, additional food, questionnaires, mail, or reading materials. I also checked the entryway after awakening each day for additional food and supplies I requested. I later learned that the floors and the bed were wired with sensors to monitor my activity patterns both while awake and asleep, so in addition to collecting data on my activity automatically, the experimenters knew when it was safe to add or take things from the entryway without my knowing. The ceiling was totally covered with plastic panels containing fluorescent lights and these were constantly on, so I was in continuous light even when sleeping. Since the Bunker was totally soundproofed from outside noise, the only background sound came from a large

542   12. Autorhythmometry

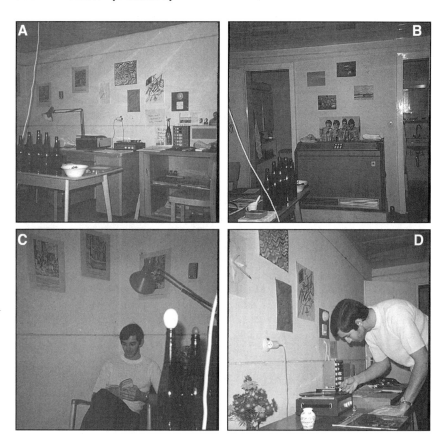

FIGURE 12.6. Room used to study human rhythms while in isolation from time cues. Scenes of the inside of one of the two isolation chambers (the "Bunker") at the institute of Prof. J. Aschoff in Erling-Andechs, Germany during temporal isolation: (*A*) Main room with panel of buttons to press to indicate certain activities, such as waking, sleeping, eating, doing a self-measurement, etc. A light on the wall, if lit, indicated that a message was in the entryway. The paper clock in A and D was made by the subject in order to have a sense of the passing of time; (*B*) View of the kitchen (*left*) and bathroom (*right*), overhead continuous lighting and air-conditioning unit; (*C*) Subject reading to pass the time. The white cord seen in *panels A, B*, and *C* is from the rectal temperature probe that was attached to the center of the ceiling and was worn by the subject throughout his 20 days in the isolation chamber; (*D*) Subject using the record player, which gave the only real clue as to the passing of time. *Note*: the watch on the subject's arm was not present during isolation: these photos were taken after the study was completed.

heating/cooling unit at the end of the main room that had a fan running continuously (Figure 12.6B).

In addition to my self-measurements, which I tried to do at least nine times during each waking span, and the collection of urine samples upon urge, I also had to wear a rectal thermometer, something that I had not known about and was at first reluctant to do.

Since microelectronics were not very small in those days, this consisted of a rather large probe that was attached to an electric cord that hung from the center of the ceiling in the main room (the cord is visible in Figures 12.6A, B, and C). Except for using the toilet and showering, the probe was to remain in place day and night and great care had to be taken to not catch the cord on a table edge or chair when moving about the room lest it catch and create an unpleasant tug on the inserted probe. In retrospect, the rectal temperature data proved to be very informative about the free-running body temperature rhythm, which showed the exact same free-running period of 25.0 h and amplitude of 0.38°C as the oral temperatures that I self-measured only while awake (Sothern, 1999).

The Bunker had no windows or other openings to the outside, no radio, no television, no telephone, or other means (this was before personal computers, the internet and cell phones, which would also be unavailable to any participants today) of determining what the local time was. One of the first things I left in the entryway for the researchers to retrieve was my wristwatch that they forget to take as they left the room on the first day (*note*: the photos in Figure 12.6C and D were taken after the study, hence a wristwatch is present). The first two days were perhaps the hardest in that the proposed 3-week stay seemed like a long time to be cut off from human contact. By the 3rd day I was deeply involved with trying to read a book a day (Figure 12.6C), writing letters and poetry, and collecting data. I think my emotions and creativity were heightened by the forced solitude and quiet, something that I now find is often missing in modern life. I did have a record player available (Figure 12.6D) and used it to play the two record albums I brought along (remember, these were pre-CD days!). One was 'Sgt. Pepper's Lonely Hearts Club Band' by the Beatles and I literally wore out the grooves on this LP.

Actually, I was also able to get a sense of time by playing the records, since the length of each song was listed on the album cover. For at least an hour or so I knew exactly how much time had passed. This was necessary because I found that I absolutely needed to have a sense of what time it was. This need was so strong that I soon devised a clock consisting of a clock face drawn on a piece of paper and two cut-out paper hands. This clock was held together by a pencil that I pushed through the hands and clock face and into a hole in the wall that I found (see Figure 12.6A and D). Each morning I set the hands at 8 AM as though that could be the time I got up and then throughout the day I moved the hour and minute hands by as much time as I thought it had taken me to finish an activity. For example, I guessed how much time it took to shower or do a self-measurement or that I had read a certain number of pages an hour, so I would move the clock hands accordingly. In addition to recording the sequential number of each measurement and pressing the appropriate event button on the event recorder, I would write down the time that I thought it was on my data page. Even with this crude attempt at keeping time, my days got longer since I was unknowingly staying up and sleeping a little later each day. For my last self-measurement in isolation, I wrote down 16:00 h on September 26 and it turned out that the actual time was 09:30 h on September 27—a difference of about 17.5 h!

Several hours after this measurement, I was trying to keep my eyes open to finish reading the last pages of a book when I noticed that the emergency wall light came on. I immediately went into the entryway and found a note telling me that the experiment was ending and could the researchers come in. I wrote back that I was very tired and that after finishing a book I was planning to self-measure and go to sleep, since it was evening—I asked if they could come back in the morning. They wrote back that it was already morning and they were coming in! Thus, time for me had slowed down where each day averaged about 25.0 h long and therefore I thought it was still the day before. To my knowledge I am still the only subject to have ever self-measured so many variables while being isolated in the Bunker.

In addition to studying free-running circadian rhythms in humans and other animals, such as birds, the laboratory in Germany also studied circannual rhythms. Upon learning this I decided to continue performing self-measurements 5–6 times each day and have done so for more than three decades after my Bunker experience (however, urine collections were stopped after 17 years) in order to obtain data not only on my circadian rhythms as I aged, but also on my own yearly and even longer rhythms.

Long Self-Measurement Series

Among other findings, the long self-measurement series of RBS has shown (1) a persistent circadian rhythm in oral temperature with a remarkable stability of the acrophase, except for expected phase advances and delays following time zone changes and phase-drifting in isolation (Sothern, 1974);[10] (2) the reproducibility of 24-h urinary rhythms over nearly 2 years (Sothern et al., 1974), as well as the detection and reproducibility of weekly and annual rhythms in several urinary variables (Sothern et al., 1981), including circannual aspects of catecholamine excretion (Sothern et al., 1984); and (3) the consistency of a circadian pattern in blood pressure over many years, in spite of some changes in the overall mean (Sothern & Halberg, 1986; Sothern, 1994). This can be seen when comparing the circadian patterns for systolic blood pressure when the subject was 20, 30, 40, and 50 years of age (Figure 12.7).

In addition to monitoring changes associated with age (Cornélissen et al., 1994), this long time series has the further value of allowing for analyses to look at low-frequency variations (e.g., annual and longer) in some of the variables (e.g., heart rate, blood pressure, and peak flow) and possible associations with cycles in the geophysical world (e.g., sunspots and other solar phenomena) (Cornélissen et al., 1996, 1998; Prigancová et al., 1997; Sothern et al., 1998, 2002). For example, data from the long time series of RBS have been used to document periods of about 6 months and slightly longer than the year (1.3 years) in blood pressure (Halberg et al., 2003a,b), while some associations have been found between heart rate and 7-day changes in solar activity (Cornélissen et al., 1996) and between the 10.5 years sunspot cycle and several of the self-measured variables, including blood pressure (Halberg et al., 2001, 2003b).

In another long time series by a woman with manic disorder who self-measured body temperature upon arising for 11 years in order to follow her monthly fertility cycle, an association was found between her episodes of mania and certain

---

[10] Self-measurements of oral temperature have also been used by others to document phase advances and delays after time zone changes to and from Europe (Haus, et al., 1981).

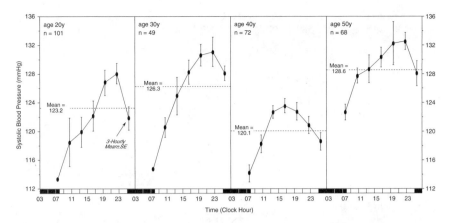

FIGURE 12.7. Persistent circadian patterns for self-measured systolic blood pressure in the same male (RBS) at age 20, 30, 40, and 50 years. Three-hourly averages ± SE were computed from data self-measured every 2–4 h during waking-only over the same 2 weeks (June 27–July 10) at the different ages (Sothern, unpublished).

times of the menstrual cycle (i.e., near menstruation), as well as with certain times of the year (winter and spring) (Sothern et al, 1993).

What can be Self-Measured?

While modern instrumentation allows for continuous around-the-clock monitoring of several physiologic variables, including body temperature (Simpson et al., 1987), blood pressure, heart rate, and activity (E. Halberg et al., 1987; Redmond & Hegge, 1987), self-measurement using inexpensive equipment often suffices to map the main features of a rhythm.

The most prominent and easily detected body rhythm may be that of temperature. In addition to oral temperature, blood pressure, and heart rate, other variables that have been self-measured include: respiration rate, grip strength, peak flow, and/or vital capacity of the lungs, time estimation, finger or toe joint circumferences, sleep-wake times, eye-hand skill, tapping tests, adding and short-term memory speeds and many other performance tests, ratings of mood, vigor, or symptoms scores (e.g., for pain, stiffness, and anxiety), and frequent collection of saliva, blood, or urine. Some of these tests are used during training sessions or rehabilitation, or to follow the effects of prescribed drugs. An example of a subject performing autorhythmometry can be seen in Figure 12.8.

Equipment

Most equipment necessary to perform autorhythmometry is relatively simple and readily available. Examples of some equipment used during self-measurement can be seen in Figure 12.8.

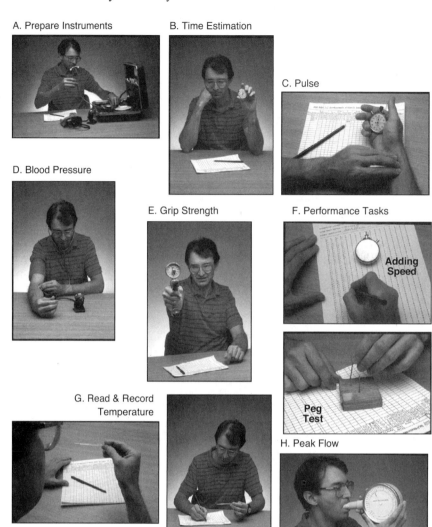

FIGURE 12.8. An individual performing some self-measurements of his body functions: (A) preparing instruments, (B) time estimation, (C) pulse, (D) systolic and diastolic blood pressure, (E) grip strength (physical performance), (F) performance tasks (adding speed, peg test), (G) read and record oral temperature, (H) peak expiratory flow (lung function).

Internal or External Body Temperature

Body temperature can be measured internally in the mouth (or the rectum, vagina, or urine outflow) or externally under the arm (axilla), in the ear, or on the skin, including the forehead or breast. However, according to the site of measurement, the overall level of temperature will vary (temperature will be higher internally

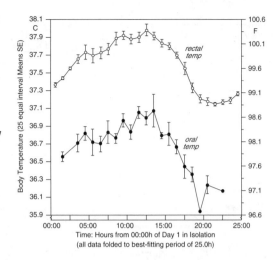

FIGURE 12.9. Body temperature sites compared: oral vs. rectal. Apart from differences in overall level (mesor), oral and rectal temperature can show the same circadian range of change (amplitude) and timing of highs and lows. Here, the free-running rhythm in body temperature during the last 7 days in isolation shows an identical period (25.0 h), amplitude (0.66°F vs. 0.68°F), and acrophase from both oral and rectal measurement sites for a healthy young man (RBS, age 20 years) (Sothern, unpublished).

and lower on the surface) (Erickson & Kirklin, 1993; Rabinowitz et al., 1996). In addition, the circadian pattern may differ (e.g., highest values obtained internally are generally found in the middle of the day, while highest temperatures on the surface of the skin are usually found at night due to heat loss as a function of internal conductance and ambient temperature) (cf. Aschoff et al., 1974).

Several figures have been prepared to illustrate some of the differences in temperature due to site of measurement. In Figure 12.9, the continuously-measured rectal temperature was higher overall than the self-measured oral temperature in a young man during the last 7 days of an isolation study (see Essay 12.1), yet both the circadian amplitude and phasing (and free-running period) were identical between the two sites of measurement (Sothern, 1999). Similarly, a comparison of oral temperature with urine outflow temperature self-measured by a healthy man over 6 days revealed nearly identical amplitudes and acrophases, but higher values for the core temperature as measured in the urine (Figure 12.10). Finally, ear temperature and oral temperature measured simultaneously by a healthy man five or six times a day for 2 months also revealed nearly identical acrophases, but lower temperature readings and a lower amplitude when measured in the ear (Figure 12.11).

Temperature Devices

Unbreakable electronic thermometers are now commonly used in place of mercury-containing thermometers, which were highly accurate but slower to use and more prone to breakage, resulting in exposure to the mercury, which is toxic and requires special disposal procedures. Electronic thermometers arrive quickly at an oral temperature value by monitoring the increasing slope of temperature and emitting a beep when this slope begins to taper off. Most people then remove the thermometer from their mouth and record the temperature value displayed.

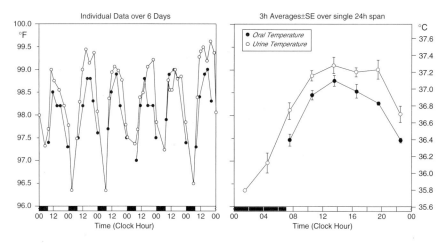

FIGURE 12.10. Body temperature sites compared: oral vs. urine outflow. Comparison of circadian variations in oral and core temperature (as measured in urine outflow) self-measured over 6 days by a healthy man (RBS, age 46 years). Dark bar = sleep. Mean oral temperature was significantly different than the mean for urine temperature (98.17°F vs. 98.40°F), while amplitudes (0.86°F vs. 1.08°F) and acrophases (15:34 h vs. 16:06 h) were not (Sothern, unpublished).

However, this value can be consistently lower by up to 0.5°F than that obtained with the mercury-containing thermometer left in place for 5 min (Figure 12.12, upper left panel). However, since temperature may continue to rise if the electronic thermometer is placed back in the mouth, the displayed value should be checked every minute for a few more minutes until it stops rising and a more accurate oral temperature value can be obtained.

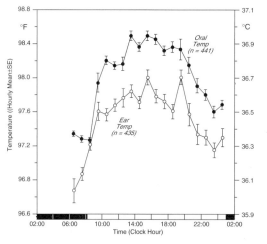

FIGURE 12.11. Body temperature sites compared: oral vs. ear. Ear temperature and oral temperature measured simultaneously 5/6 times a day by a healthy man (RBS, age 53 years) for two months revealed nearly identical acrophases (15:44 h vs. 16:09 h), but lower temperature readings (97.49 vs. 97.98°F) and a lower amplitude (0.50°F vs. 0.69°F) when measured in the ear (Sothern, unpublished).

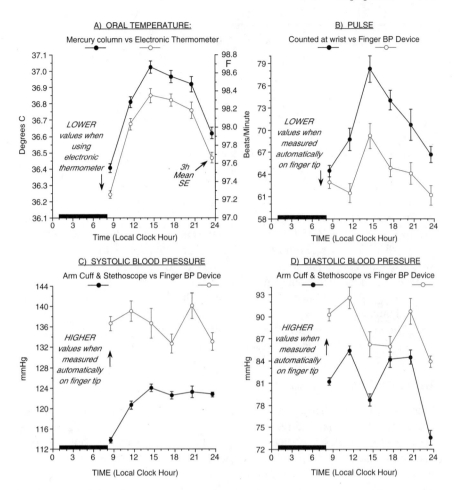

FIGURE 12.12. Different instruments used for autorhythmometry can obtain different values. Comparison of data obtained over 24 days by a healthy man (RBS, age 45 years): (*A*) oral temperature was *lower* when measured with an electronic thermometer as compared to a glass mercury column thermometer; (*B*) pulse was *lower* when measured electronically on the finger tip as compared with the manual count for a full minute at the wrist; (*C* and *D*) both systolic and diastolic blood pressures were *higher* when measured automatically on the finger tip as compared with measurements on the arm using an inflatable BP cuff and stethoscope (Sothern, unpublished).

A new generation of hand-held electronic devices can obtain a temperature from the inner ear within seconds, but as mentioned above (Figure 12.11), in good clinical health, ear temperature values are usually lower than oral temperature values and more erratic due to exposure to the environment. However, while ear thermometers are not particularly reliable in rhythm studies, they may quickly detect the presence of a fever. Finally, temperature-sensing capsules

Blood Pressure Devices

Blood pressure (BP) can be monitored using an inflatable cuff and a stethoscope (Figure 12.8D) or by using tabletop devices that digitally display systolic and diastolic BP, and usually heart rate, without the need for a stethoscope. Various home electronic monitors that measure BP on the upper arm, wrist, or finger have been compared with only small, clinically insignificant differences between them (Yarows & Brook, 2000). However, BP measured on the finger or at the wrist does not always correspond to BP monitored on the upper arm, leading to dubious reliability.[11] This is illustrated in Figure 12.12, where blood pressure (lower panels) and pulse (upper right panel) measured by an electronic monitor on the finger are compared with the same values measured manually on the arm. In this example, pulse measured by the finger device was lower than the actual manual count over a full minute, while both systolic and diastolic BP were significantly elevated over levels found when using a cuff and stethoscope. It is thus important to use the same monitor when repeatedly self-measuring and comparing BP.

Automatic Devices

Physiological monitoring around-the-clock can now be achieved in the home, work, and educational and recreational environments using automatic recording devices (cf. Miles & Broughton, 1990). Portable devices that can be carried on a belt or shoulder strap and connected to a cuff on the arm will monitor and record BP and heart rate at preset intervals, such as every 15 min or once an hour, throughout the day and night (even during sleep) for several days, resulting in time series such as the blood pressure data shown in Figure 12.13 that were collected for 7 days by a 23-year-old man. In spite of the short-term rising and falling of blood pressure values in response to this young man's activities, hourly averages computed from the 7 days of data show a clear circadian pattern, with higher values during the day and lower values during sleep (Figure 12.14).

---

[11] While BP is easier to measure on a finger than on the arm, arteries in the fingers show a greater responsiveness to changes in local conditions (Latman, 1992) and significant differences in systolic pressure have been noted between the fingers and position of the hand in relation to the chest (Grune et al., 1992), making finger BP insufficiently representative of general systemic BP. Similarly, wrist BP devices show larger variability in repeated BP measurements than arm cuff devices, possibly due to incomplete arterial occlusion at the wrist (Kikuya et al., 2002). Several international guidelines for home self-monitoring recommend measurement of BP using arm cuff devices, but not on the finger or wrist (Mengden et al., 2000; Imai, et al., 2003).

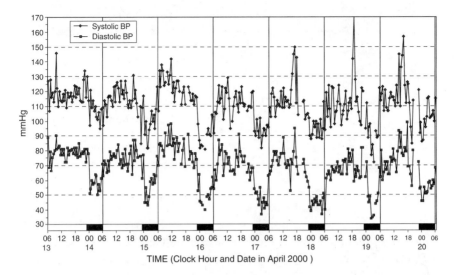

FIGURE 12.13. Circadian patterns in systolic and diastolic blood pressure measured by a portable automatic monitor every 30 min around-the-clock for 7 days in a healthy man (AMJ, age 23 years). In spite of the short-term rising and falling of blood pressure values in response to activities, day-night differences are evident (Sothern, unpublished).

Heart rate can be counted manually (Figure 12.8C), or can be monitored and recorded automatically using various portable electronic devices such as the BP device just mentioned, by a band around the wrist or chest or via a finger clip (Figure 12.15). Heart rate monitored automatically will not only reveal responses to exercise, but also show changes in heart rate that occur over the

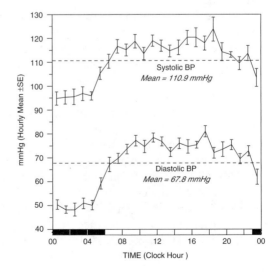

FIGURE 12.14. Overall circadian pattern in systolic and diastolic blood pressure of a healthy young man (AMJ, age 23 years). Hourly means ±SE were computed from the 320 readings shown in Figure 12.13 that were collected by a monitor that was programmed to take a reading every 30 min around-the-clock over 7 days (April 13–20, 2000) (Sothern, unpublished).

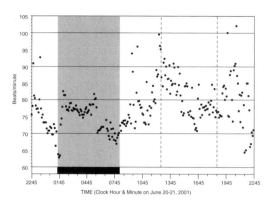

FIGURE 12.15. Heart rate monitored continuously around-the-clock. A variety of portable electronic monitors are available for this purpose. The data shown here are 5-min values averaged from 30-s data collected for 24 h by a healthy man (RBS, age 54 years) who used a fingertip clip attached to a portable pulse oximeter. Dark bar/shaded area = sleep (Sothern, unpublished).

entire 24-h pattern when the device is worn around-the-clock, as shown for four students in Figure 12.16. In addition to heart rate, other devices can be worn or carried around-the-clock in order to continuously monitor whole body activity, exposure to light and environmental temperature, as shown for an adult man in Figure 12.17.

Other Equipment

A stopwatch is useful for timing performance tasks, heart rate, and respiration. There are many kinds of stopwatches, from wind-up analog models to electronic digital ones, the latter often included on some wristwatches—all work adequately. When nothing else is available, some subjects have even quickly looked at the second hand on a wall clock at the beginning and end of their measurement and obtained a rough estimate of the time elapsed while performing a task.

Grip strength devices (Figure 12.8E), which are often used by persons with arthritis or during rehabilitation, are generally not readily available, but a simple device can be put together using a rubber bulb and a pressure gauge. Peak flow meters, which measure the peak amount of air that can be blown from the lungs during a sharp exhaled blast, are often used by persons with asthma or athletes in training and are available in standard size (Figure 12.8H) or in a mini-version, which is more affordable and transportable (Sothern et al., 2003).

Other performance devices can be manufactured to suit the particular variable being measured. For example, to measure eye-hand coordination, a darning needle holding a string knotted at the bottom end could be used to thread a specified number of beads as quickly as possible or a small square of wood with three holes drilled at the points of a triangle into its surface (Figure 12.8F) could be created in order to advance two nails one after the other from hole to hole around the triangle a specified number of times. Palmtop computers have also been programmed to test performance via vigilance, memory and/or reaction time tests (Varkevisser & Kerkhof, 2003).

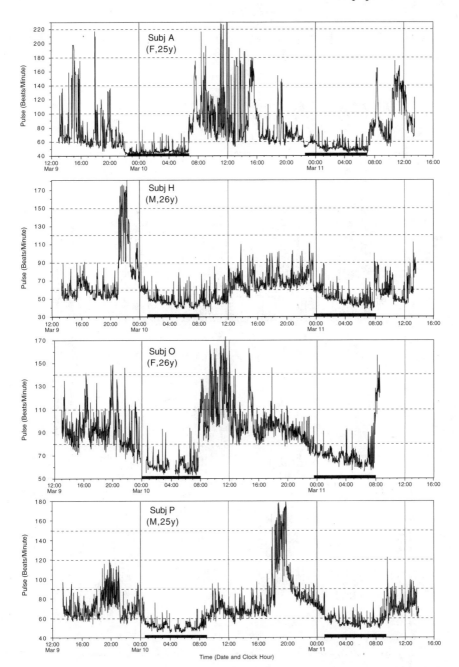

FIGURE 12.16. Circadian patterns in heart rate of four healthy young athletes. Each student wore a device consisting of a chest strap and wristwatch that automatically monitored heart rate every minute for 2 days. Elevated values occurred during spans of exercise, while lowest values were observed during sleep. Dark bars = times of sleep. Data courtesy of Prof. A. Weydahl, Finnmark College, Alta, Norway.

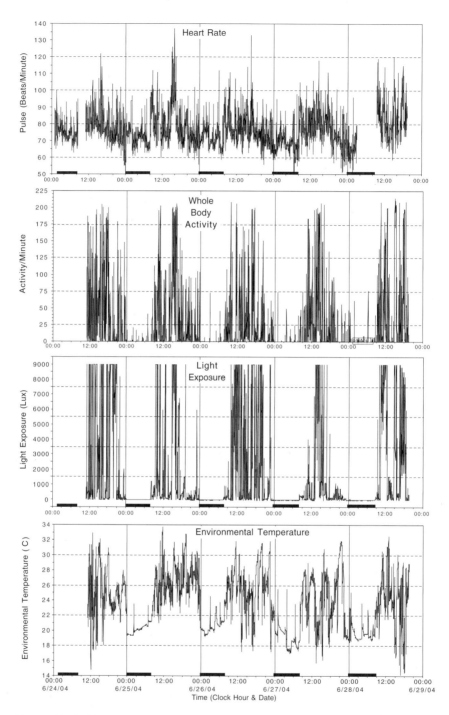

FIGURE 12.17. Electronic devices carried for several days by a healthy man (RBS, age 57 years) reveal 24-h patterns in heart rate, whole body activity, light exposure, and environmental temperature. Data collected in June during the time of the midnight sun in Alta, Norway. Equipment and data retrieval provided by Prof. A, Weydahl, Finnmark College, Alta, Norway.

Saliva, Urine, and Blood

Saliva can be collected in specially designed small tubes (called "salivettes") for chemical assay, while urine can be collected in a plastic graduated cylinder, often while directing the stream over the tip of a thermometer that has been hung inside the cylinder to get a very quick and reliable measure of core body temperature (Ehrenkranz, 1986a,b). From the whole urine sample, total volume, pH, and specific gravity can immediately be measured prior to storing a small amount of fluid in a plastic bottle for chemical determinations. While some people (diabetics) need to sample their own blood frequently for glucose assay, blood collection for white blood cell counts and/or other assays is best done with a trained professional and should not be done without the supervision of a physician.

Procedures for Self-Measurements

Detailed procedures for self-measurement of several physiological, psychological and urinary variables have been published (Halberg et al., 1972). In order to perform autorhythmometry, one will need to obtain reliable equipment, become skilled in using it, and learn how to accurately record data. This may mean practicing the task(s) several times before actually beginning a study. This applies to both those variables that are simply observed, such as body temperature and heart rate, and performance tasks that require conscious effort, such as time estimation or adding speed. In addition, one must also attempt to remain objective while doing self-measurements and not expect a certain answer at one time of the day or another. For performance tests, one should always try to obtain the fastest time or highest score.

Keeping Records

A recording sheet needs to be created on which to write the date, time and numerical values observed during each measurement session (see Items 1 and 2 in the Appendix for some sample pages that could be used). A "Comments" column should also be included in which to note events that might have influenced each reading (e.g., room temperature and humidity, illness, medications, food and drink, stress, etc.), as well as the times of going to bed and getting up. These comments can later be helpful when trying to interpret unusual values or trends in the data. A sample recording sheet is illustrated in Table 12.1.

Sampling Sequence

For conducting self-measurements, an individual should first try to sit down comfortably, at a table if possible, and away from noises and distractions. An overview of procedures and comments for monitoring oral temperature, mood,

TABLE 12.1. Example of self-measurement recording sheet.

Self-measurements of: ___(Subject's name here)___ Year:_____ Page:_____

Date	Time	Pulse	Mood Rating	Vigor Rating	Systolic BP	Diastolic BP	Oral Temperature	Comments
March 16	0800	70	4	1	112	68	97.2	Woke at 07:50 h
"	1210	79	3	4	121	75	98.3	Hungry
"	1615	83	4	3	124	77	98.6	Lunch at 12:30 h
"	1840	79	4	3	128	82	98.7	Hurrying
"	2330	64	5	2	126	80	97.9	Tired

vigor, pulse and blood pressure, and some performance tasks that can be obtained while seated is listed in Table 12.2, with more detailed, step-by-step instructions given in Item 3 of the Appendix. The sequence should be arranged to be both efficient and safe (e.g., do not move about with a thermometer in the mouth). Temperature is often obtained while other tasks are performed, and pulse and blood pressure are measured after sitting for some minutes. If ear-canal temperature is measured, it will be obtained within seconds and can be recorded immediately. If axillary temperature is measured, it may have to be taken and recorded after completing other tasks in order for the thermometer to remain in place long enough for an accurate reading without becoming dislodged by other self-measurement activities.

Looking at the Data

To monitor a rhythm, repeated measurements need to be taken over a sufficient duration of time (not just over a few hours or days) in order to observe and test for the periodicity of interest (ultradian, circadian, and infradian). But the problem as to what to do with so many numbers, rather than interpret a single value, can seem overwhelming to the untrained eye. This is especially pertinent when dozens of values are obtained by self-measurement or hundreds or even thousands of values have been obtained by automatic instrumentation for blood pressure, heart rate, or temperature.

Making Graphs

Values obtained by self-measurement can be graphed either by hand or computer to visually inspect the daily changes in the original data. A sample page for recording and graphing temperature, pulse, and blood pressure can be found in the Appendix (Item 2). Also, average values by an hour or more at different times of the day (or month or year) can be computed, as well as the highest and lowest values selected to find the extent of variation in readings over the observation span.

TABLE 12.2. Some tasks or tests that have been used when conducting autorhythmometry.

Task or Test	Procedure	Comments
Temperature	Follow manufacturer's instructions for site of measurement: oral, skin, ear, axilla, and other	Be aware of all thermometer safety issues prior to study and follow them
Mood	Rate mood on a scale from 1 to 7, where 4 is an average state, 1 is the lowest and 7 is the highest. Create adjectives to help use this rating scale (see Appendix Item 12.3).	This is a subjective rating, with 3 levels above and below a 4 that represents an individual's usual, average daily feeling of mental state
Vigor	Rate vigor on a scale from 1 to 7, where 4 is an average state, 1 is the lowest and 7 is the highest. Create adjectives to help use this rating scale (see Appendix Item 12.3).	This is a subjective rating, with 3 levels above and below a 4 that represents an individual's usual, average daily feeling of vitality
Time estimation	Start a stopwatch, estimate 1 min, stop the watch, and record actual time elapsed to nearest 0.1 min	Count to 60 silently by using 1,001, 1,002, 1,003, etc., or visualize a second-hand moving on a clock
Heart rate	Locate pulse at thumb-side of wrist and count the number of beats for one full minute	Pulse can also be found at the neck, by listening to the chest with a stethoscope or obtained electronically when using a blood pressure device
Blood pressure	Follow manufacturer's instructions and record the systolic and diastolic pressures	Training may be necessary to properly use the monitoring device(s). The average of two or three readings is usually more accurate than a single reading
Adding speed[a]	Use prepared sheet containing columns of 50 single-digit random numbers (see Appendix Item 12.4). Start stopwatch, quickly add consecutive overlapping numbers in one column, stop stopwatch, and record elapsed time	Check for pairs added incorrectly. Divide the number of correctly added pairs per se by time elapsed to get correct/s. A higher ratio = better performance
Short term memory[a]	Use prepared sheet containing rows of 10 sets of 7-digit random numbers (see Appendix Item 12.5). Start stopwatch, quickly view and memorize the first number set, cover and record numbers in space below, stop stopwatch, and record elapsed time	Check for numbers remembered incorrectly. Divide the number of correctly remembered single digits by time elapsed to get correct/s. A higher ratio = better performance

[a]Sample pages for these tests can be found in the Appendix as Items 12.4 and 12.5.

Testing for a Time-Effect

The data can also be entered into a computer for analysis by a program designed to (1) compute time-point and/or daytime vs. night-time means, (2) quantify rhythms for their statistical significance, (3) calculate the mean and amplitude of each rhythm, and (4) compute the average time when highest or lowest values occurred (see Chapter 13 on Analyzing Data).

Take-Home Message

Self-measurement of physiological and/or psychological variables show that nearly all functions undergo predictable circadian changes each day. This knowledge could help to define what is "normal" on an individual basis, not only overall, but also at different times of the day for someone in good health and/or during illness (e.g., high blood pressure, asthma, arthritis, fever, etc.).

References

Adan A, Natale V. (2002) Gender differences in morningness–eveningness preference. *Chronobiol Intl* 19(4): 709–720.

Albertini M, Politano S, Berard E, Boutte P, Mariani R. (1989) Variation in peak expiratory flow of normal and asymptomatic asthmatic children. *Pediatr Pulmonol* 7(3): 140–144.

Aschoff J, Heise A. (1972) Thermal conductance in man: its dependence on time of day and on ambient temperature. In: *Advances in Climatic Physiology*. Itoh A, Ogata K, Yoshimura H, eds. Tokyo: Igaku-Shoin Ltd., pp. 334–348.

Aschoff J, Biebach H, Heise A, Schmidt T. (1974) Day-night variation in heat balance. In: *Heat Loss from Animals and Man. Assessment and Control*. Monteith JL, Mount LE, eds. London: Butterworth, pp. 147–172.

Atkinson M, Kripke DF, Wolf SR. (1975) Autorhythmometry in manic-depressives. *Chronobiologia* 2(4): 325–335.

Atkinson G, Coldwells A, Reilly T, Waterhouse J. (1993) A comparison of circadian rhythms in work performance between physically active and inactive subjects. *Ergonomics* 36(1–3): 273–281.

Bailey SL, Heitkemper MM. (2001) Circadian rhythmicity of cortisol and body temperature: morningness–eveningness effects. *Chronobiol Intl* 18(2): 249–261.

Bee V, Webb WB. (1987) Temperature rhythms in native Alaskan and Caucasian children. *Biol Psychol* 24(2): 101–104.

Beliaev SD, Khetagurova LG, Zaslavskaia RM. (2002) [Assessment of time-dependent effects of long-acting altiazem in patients with hypertension stage II.] [Russian]. *Klin Med (Mosk)* 80(3): 62–66.

Bellamy N, Sothern RB, Campbell J. (1990) Rhythmic variations in pain perception in osteoarthritis of the knee. *J Rheumatol* 17(3): 364–372.

Bellamy N, Sothern RB, Campbell J, Buchanan WW. (1991) Circadian rhythm in pain, stiffness, and manual dexterity in rheumatoid arthritis: relation between discomfort and disability. *Annals Rheum Dis* 50(4): 243–248.

Bellamy N, Sothern R, Campbell J, Buchanan WW. (2002) Rhythmic variations in pain, stiffness and manual dexterity in hand osteoarthritis. *Annals Rheum Dis* 61(12): 1075–1080.

Bellamy N, Sothern R, Campbell J, Kaloni S. (2004) Aspects of diurnal rhythmicity in pain, stiffness and fatigue in fibromyalgia. *J Rheumatol* 31(2): 379–389.

Bicakova-Rocher A, Gorceix A, Reinberg A, Ashkenazi II, Ticher A. (1996) Temperature rhythm of patients with major affective disorders: reduced circadian period length. *Chronobiol Intl* 13(1): 47–57.

Blake MJF. (1967) Time of day on performance in a range of tasks. *Psychon Sci* 9(6): 349–350.

Bobrie G, Chatellier G, Genes N, Clerson P, Vaur L, Vaisse B, Menard J, Mallion JM. (2004) Cardiovascular prognosis of "masked hypertension" detected by blood pressure

self-measurement in elderly treated hypertensive patients. *JAMA* 291(11): 1342–1349.
Broege PA, James GD, Pickering TG. (2001) Management of hypertension in the elderly using home blood pressures. *Blood Press Monit* 6(3): 139–144.
Brooke OG, Collins JE, Fox RH, James S, Thornton C. (1973) Evaluation of a method for measuring urine temperature. *J Physiol* 231(2): 91P–93P.
Burioka N, Suyama H, Sako T, Shimizu E. (2000) Circadian rhythm in peak expiratory flow: alteration with nocturnal asthma and theophylline chronotherapy. *Chronobiol Intl* 17(4): 513–519.
Burioka N, Sako T, Tomita K, Miyata M, Suyama H, Igishi T, Shimizu E. (2001) Theophylline chronotherapy of nocturnal asthma using bathyphase of circadian rhythm in peak expiratory flow rate. *Biomed Pharmacother* 55(Suppl 1): 142s–146s.
Campbell IT, Walker RF, Riad-Fahmy D, Wilson DW, Griffiths K. (1982) Circadian rhythms of testosterone and cortisol in saliva: effects of activity-phase shifts and continuous daylight. *Chronobiologia* 9(4): 389–396.
Campbell IT, Wilson DW, Walker RF, Griffiths K. (1985) The use of salivary steroids to monitor circadian rhythmicity on expeditions in the arctic. *Chronobiol Intl* 2(1): 55–59.
Carrier J, Monk TH, Buysse DJ, Kupfer DJ. (1997) Sleep and morningness–eveningness in the "middle" years of life (20 y–50 y). *J Sleep Res* 6(4): 230–237.
Casale R, Natali G, Colantonio D, Pasqualetti P. (1992) Circadian rhythm of peak expiratory flow in children passively exposed and not exposed to cigarette smoke. *Thorax* 47(10): 801–803.
Casale R, Pasqualetti P. (1997) Cosinor analysis of circadian peak expiratory flow variability in normal subjects, passive smokers, heavy smokers, patients with chronic obstructive pulmonary disease and patients with interstitial lung disease. *Respiration* 64(4): 251–256.
Condon R, Knauth P, Klimmer F, Colquhoun P, Herrmann H, Rutenfranz J. (1984) Adjustment of the oral temperature rhythm to a fixed watchkeeping system on board ship. *Intl Arch Occup Environ Health* 54(2): 173–180.
Cornélissen G, Haus E, Halberg F. (1994) Chronobiologic blood pressure assessment from womb to tomb. In: *Biologic Rhythms in Clinical and Laboratory Medicine*. Touitou Y, Haus E, eds. Berlin: Springer-Verlag, pp. 428–452.
Cornélissen G, Halberg F, Wendt HW, Bingham C, Sothern RB, Haus E, Kleitman E, Kleitman N, Revilla MA, Revilla M Jr, Breus TK, Pimenov K, Grigoriev AE, Mitish MD, Yatsyk GV, Syutkina EV. (1996) Resonance of about-weekly rhythm in human heart rate with solar activity change. *Biologica (Bratisl)* 51(6): 749–756.
Cornelissen G, Halberg F, Sothern RB, Nikityuk BA, Alonso LG, Syutkina EV, Grafe A, Bingham Ch. (1998) Toward a chronoastrobiology: sunspot cycles and geomagnetism as well as sunshine may modulate human morphology. *Russian Morphological Newsletter* 3(4): 133–137.
Crawford JM, Radford PF, Simpson HW. (1990) Chronobiological analysis of 1000 miles walked in 1000 hours. In: *Chronobiology: Its Role in Clinical Medicine, General Biology and Agriculture, Part B. Progress in Clinical and Biological Research*, Vol. 341B. Hayes DK, Pauly JE, Reiter RJ, eds. New York: Wiley-Liss, pp. 291–297.
Cugini P, Girelli I, Latini M, Cogliati AA, Di Palma L, Battisti P, Felici W, Tucciarone L. (1988) Chronobiology of blood pressure in childhood: implications for tracking. *Chronobiologia* 15: 291–298.
Czeisler CA, Dumont M, Duffy JF, Steinberg JD, Richardson GS, Brown EN, Sánchez R, Ríos CD, Ronda JM. (1992) Association of sleep-wake habits in older people with changes in output of circadian pacemaker. *Lancet* 340(8825): 933–936.

Davy J. (1845) On the temperature of man. *Philosoph Trans Royal Soc London* 135: 319–333.

Davis C, Lentz MJ. (1989) Circadian rhythms: charting oral temperatures to spot abnormalities. *J Gerontol Nursing* 15(4): 34–39.

De Meyer F, Vogelaere P. (1990) Spectral resolution of cardio-circulatory variations in men measured by autorhythmometry over 2 years. *Int J Biometeorol* 34(2): 105–121.

De Scalzi M, de Leonardis V, Fabiano FS, Cinelli P. (1986) Circadian rhythms of arterial blood pressure. *Chronobiologia* 13(3): 239–244.

Di Milia L, Smith PA, Folkard S. (2004) Refining the psychometric properties of the circadian type inventory. *Personality Indiv Diffs* 36(8): 1953–1964.

Duffy JF, Dijk DJ, Hall EF, Czeisler CA. (1999) Relationship of endogenous circadian melatonin and temperature rhythms to self-reported preference for morning or evening activity in young and older people. *J Invest Med* 47(3): 141–150.

Duffy J, Rimmer DW, Czeisler CA. (2001) Association of intrinsic circadian period with morningness–eveningness, usual wake time, and circadian phase. *Behav Neurosci* 115(4): 895–899.

Ehrenkranz JR. (1986a) A new method for measuring body temperature. *N J Med* 83(2): 93–96.

Ehrenkranz JR. (1986b) Urine temperature and core temperature. *JAMA* 255(14): 1880–1881.

Engel R, Halberg F, Nelson W, Bartter FC. (1979) Rhythmometry gauges treatment of mesor-hypertension in the seventh and eight decades of life. *Intl J Chronobiol* 6(3): 163–178.

Engel R, Harvey J, Halberg F, Cornelissen G. (1987) Intermittent automatic chronobiologic monitoring complements daily self-measurements. *Chronobiologia* 14(1): 35–38.

Erickson RS, Kirklin SK. (1993) Comparison of ear-based, bladder, oral, and axillary methods for core temperature measurement. *Crit Care Med* 21(10): 1528–1534.

Felver L, Hoeksel R. (1994) Short assessment of morningness-eveningness: a tool for clinical practice and research. In: *Proc 6th Intl Conf Chronopharmacology & Chronotherapeutics*, July 5–9, 1994, Amelia Island, FL. Abstract XV-1.

Folkard S, Glynn CJ, Lloyd JW. (1976) Diurnal variation and individual differences in the perception of intractable pain. *J Psychosom Res* 20(4): 289–301.

Folkard S, Monk TH, Lobban MC. (1979) Towards a predictive test of adjustment to shift work. *Ergonomics* 22(1): 79–91.

Foret J, Benoit O, Royant-Parola S. (1982) Sleep schedules and peak times of oral temperature and alertness in morning and evening 'types.' *Ergonomics* 25(9): 821–827.

Fox RH, Woodward PM, Fry AJ, Collins JC, MacDonald IC. (1971) Diagnosis of accidental hypothermia of the elderly. *Lancet* 1(7696): 424–427.

Fox RH, Brooke OG, Collins JC, Bailey CS, Healey FB. (1975) Measurement of deep body temperature from the urine. *Clin Sci Mol Med* 48(1): 1–7.

French J, Bisson RU, Neville KJ, Mitcha J, Storm WF. (1994) Crew fatigue during simulated, long duration B-1B bomber missions. *Aviat Space Environ Med* 65(5 Suppl): A1–6.

Fujiwara S, Shinkai S, Kurokawa Y, Watanabe T. (1992) The acute effects of experimental short-term evening and night shifts on human circadian rhythm: the oral temperature, heart rate, serum cortisol and urinary catecholamine levels. *Intl Arch Occup Environ Health* 63(6): 409–418.

Glasgow DR, Scheving LE, Pauly JE, Bruce TA. (1982) Autorhythmometry – A new concept in biology and health education. *J Ark Med Soc* 79(2): 81–91.

Gordon HW, Stoffer DS, Lee PA. (1995) Ultradian rhythms in performance on tests of specialized cognitive function. *Intl J Neurosci* 83(3–4): 199–211.

Gregg I, Nunn AJ. (1973) Peak expiratory flow in normal subjects. *Br Med J* 3(5874): 282–284.

Grune S, Spuhler T, Edmonds D, Binswanger B, Weisser B, Mengden T, Vetter W. (1992) [Blood pressure measurement using the fingers.] [German]. *Schweiz Rundsch Med Prax* 81(5): 103–105.

Gupta S, Pati AK. (1994) Characteristics of circadian rhythm in six variables of morning active and evening active healthy human subjects. *Indian J Physiol Pharmacol* 38(2): 101–107.

Halberg E, Halberg J, Halberg Fcn, Sothern RB, Levine H, Halberg F. (1981) Familial and individualized longitudinal autorhythmometry for 5 to 12 years and human age affects. *J Gerontol* 36(1): 31–33.

Halberg E, Carandente F, Sothern RB, Halberg F. (1987) Chronobioengineering for human blood pressure. In: *Chronobiotechnology and Chronobiological Engineering. NATO ASI Series*. Scheving LE, Halberg F, Ehret CF, eds. Dordrecht: Martinus Mijhoff Publication, pp. 289–298.

Halberg F, Johnson EA, Nelson W, Runge W, Sothern RB. (1972) Autorhythmometry – procedures for physiologic self-measurement and their analysis. *The Physiology Teacher* 1: 1–11.

Halberg F, Lauro R, Carandente F. (1976) Autorhythmometry leads from single-sample medical check-ups toward a health science of time series. *Ric Clin Lab* 6(3): 207–250.

Halberg F, Drayer JM, Cornélissen G, Weber MA. (1984a) Cardiovascular reference database for recognizing circadian mesor- and amplitude-hypertension in apparently healthy men. *Chronobiologia* 11(3): 275–298.

Halberg F, Scheving LE, Lucas E, Cornélissen G, Sothern RB, Halberg E, Halberg J, Halberg Fcn, Carter J, Straub KD, Redmond DP. (1984b) Chronobiology of human blood pressure in the light of static (room-restricted) automatic monitoring. *Chronobiologia* 11(3): 217–247.

Halberg F, Cornélissen G, Watanabe Y, Otsuka K, Fiser B, Siegelova J, Mazankova V, Maggioni C, Sothern RB, Katinas GS, Syutkina EV, Burioka N, Schwartzkopff. (2001) Near 10-year and longer periods modulate circadians: intersecting anti-aging and chronoastrobiological research. *J Geront: Med Sci* 56A(5): M304–M342.

Halberg F, Cornélissen G, Schack B, Wendt HW, Minne H, Sothern RB, Watanabe Y, Katinas G, Otsuka K, Bakken EE. (2003a) Blood pressure self-surveillance for health also reflects 1.3-year Richardson solar wind variation: spin-off from chronomics. *Biomed Pharmacother* 57 (Suppl 1): 58s–76s.

Halberg F, Cornélissen G, Stoynev A, Ikonomov O, Katinas G, Sampson M, Wang Z, Wan C, Singh RB, Otsuka K, Sothern RB, Sothern SB, Sothern MI†, Syutkina EV, Masalov A, Perfetto F, Tarquini R, Maggioni C, Kumagai Y, Siegelova J, Fiser B, Homolka P, Dusek J, Uezono K, Watanabe Y, Wu J, Sonkowsky R, Schwartzkopff O, Hellbrügge T, Spector NH, Baciu I, Hriscu M, Bakken E. (2003B) Season's Appreciations 2002 and 2003. Imaging in time: the transyear (longer-than-the-calendar year) and the half-year. *Neuroendocrinol Lett* 24(6): 421-440.

Haus E, Sackett LL, Haus M, Swoyer J, Babb WK, Bixby EK. (1981) Cardiovascular and temperature adaptation to phase shift by intercontinental flights – longitudinal observations. In: *Advances in the Biosciences Vol 30: Night and Shift Work: Biological and Social Aspects*. Reinberg A, Vieux N, Andlauer P, eds. New York: Pergamon Press, pp. 375–390.

Hellbrügge T. (1960) The development of circadian rhythms in infants. In: *Biological Clocks. Cold Spring Harbor Symposia on Quantitative Biology*, Vol 25. New York: the Biological Laboratory, pp. 311–323.

Henderson AJ, Carswell F. (1989) Circadian rhythm of peak expiratory flow in asthmatic and normal children. *Thorax* 44(5): 410–414.

Hermida RC, Ayala DE. (2003) Sampling requirements for ambulatory blood pressure monitoring in the diagnosis of hypertension in pregnancy. *Hypertension* 42(4): 619–624.

Herpin D, Pickering T, Stergiou G, de Leeuw P, Germano G. (2000) Consensus Conference on Self-blood pressure measurement. Clinical applications and diagnosis. *Blood Press Monit* 5(2): 131–135 (Review).

Hetzel MR, Clark TJ. (1980) Comparison of normal and asthmatic circadian rhythms in peak expiratory flow rate. *Thorax* 35(10): 732–738.

Hidalgo MP, de Souza CM, Zanette CB, Nunes PV. (2003) Association of daytime sleepiness and the morningness/eveningness dimension in young adult subjects in Brazil. *Psychol Rep* 93(2): 427–434.

Horne JA, Ostberg O. (1976) A self-assessment questionnaire to determine morningness–eveningness in human circadian rhythms. *Intl J Chronobiol* 4(2): 97–110.

Imai Y, Otsuka K, Kawano Y, Shimada K, Hayashi H, Tochikubo O, Miyakawa M, Fukiyama K, Japanese Society of Hypertension. (2003) Japanese society of hypertension (JSH) guidelines for self-monitoring of blood pressure at home. *Hypertens Res* 26(10): 771–782.

Iskra-Golec I. (2001) Ultradian rhythms in processing speed of laterally exposed words and pictures. *J Hum Ergol (Tokyo)* 30(1–2): 241–244.

Kamoi K, Miyakoshi M, Soda S, Kaneko S, Nakagawa O. (2002) Usefulness of home blood pressure measurement in the morning in type 2 diabetic patients. *Diabetes Care* 25(12): 2218–2223.

Kamoi K, Imamura Y, Miyakoshi M, Kobayashi C. (2003) Usefulness of home blood pressure measurement in the morning in type 1 diabetic patients. *Diabetes Care* 26(8): 2473–2475.

Katzenberg D, Young T, Finn L, King DP, Takahashi JS, Mignot E. (1998) A CLOCK polymorphism associated with human diurnal preference. *Sleep* 21(6): 569–576.

Kerkhof GA, Van Dongen HPA. (1996) Morning-type and evening-type individuals differ in the phase position of their endogenous circadian oscillator. *Neurosci Lett* 218(3): 153–156.

Kerkhof GA. (1998) The 24-hour variation of mood differs between morning- and evening-type individuals. *Percept Motor Skills* 86(1): 264–266.

Kikuya M, Chonan K, Imai Y, Goto E, Ishii M; Research Group to Assess the Validity of Automated Blood Pressure Measurement Devices in Japan. (2002) Accuracy and reliability of wrist-cuff devices for self-measurement of blood pressure. *J Hypertens* 20(4): 629–638.

Klein R, Armitage R. (1979) Rhythms in human performance: 1 1/2-hour oscillations in cognitive style. *Science* 204(4399): 1326–1328.

Klerman EB. Gershengorn HB. Duffy JF. Kronauer RE. (2002) Comparisons of the variability of three markers of the human circadian pacemaker. *J Biol Rhythms* 17(2): 181–193.

Knauth P, Keller J, Schindele G, Totterdell P. (1995) A 14-h night-shift in the control room of a fire brigade. *Work Stress* 9(2–3): 176–186.

Kothy P, Vargha P, Hollo G. (2001) Ocuton-S self tonometry vs. Goldmann tonometry; a diurnal comparison study. *Acta Ophthalmol Scand* 79(3): 294–297.

Kowanko IC, Knapp MS, Pownall R, Swannell AJ. (1982) Domiciliary self-measurement in the rheumatoid arthritis and the demonstration of circadian rhythmicity. *Ann Rheum Dis* 41(5): 453–455.

Kriebel J. (1974) Changes in internal phase relationships during isolation. In: *Chronobiology*. Scheving LE, Halberg F, Pauly JE, eds. Tokyo: Igaku Shoin Ltd., pp. 451–459.

Latman NS. (1992) Evaluation of finger blood pressure monitoring instruments. *Biomed Instrum Technol* 26(1): 52–57.

Lavie P, Segal S. (1989) Twenty-four-hour structure of sleepiness in morning and evening persons investigated by ultrashort sleep-wake cycle. *Sleep* 12(6): 522–528.

Lebowitz MD, Krzyzanowski M, Quackenboss JJ, O'Rourke MK. (1997) Diurnal variation of PEF and its use in epidemiological studies. *Eur Respir J Suppl* 24: 49S–56S.

Levine H, Cornelissen G, Halberg F, Bingham C. (1980) Self-measurement, automatic rhythmometry, transmeridian flights and aging. In: *Chronobiology: Principles and Applications to Shifts in Schedules*. Scheving LE, Halberg F, eds. Alphen aan den Rijn: Sijthoff & Noordhoff, pp. 371–392.

Lundeen WA, Nicolau GY, Lakatua DJ, Sackett-Lundeen L, Petrescu E, Haus E. (1990) Circadian periodicity of performance in athletic students. In: *Chronobiology: Its Role in Clinical Medicine, General Biology and Agriculture, Part B. Prog Clin Biol Res*, Vol. 341B. Hayes DK, Pauly JE, Reiter RJ, eds. New York: Wiley-Liss, pp. 337–343.

Mackowiak PA, Wasserman SS, Levine MM. (1992a) A critical appraisal of 98.6°F, the upper limit of the normal body temperature, and other legacies of Carl Reinhold August Wunderlich. *JAMA* 268(12): 1578–1580.

Mackowiak PA, Wasserman SS, Levine MM. (1992b) An analysis of the quantitative relationship between oral temperature and severity of illness in experimental shigellosis. *J Infect Dis* 166(5): 1181–1184.

Mackowiak PA, Wasserman SS. (1995) Physicians' perceptions regarding body temperature in health and disease. *S Med J* 88(9): 934–939.

Madjirova NP, Tashev TG, Delchev NN, Bakalova RG. (1995) Interrelationship between cortisol levels in plasma and the circadian rhythm of temperature, pulse and blood pressure in depressed patients with good and disturbed sleep. *Intl J Psychophysiol* 20(3): 145–154.

Marolf AP, Hany S, Battig B, Vetter W. (1987) Comparison of casual, ambulatory and self-determined blood pressure measurement. *Nephron* 47(Suppl 1): 142–145.

Mengden T, Binswanger B, Weisser B, Vetter W. (1992) An evaluation of self-measured blood pressure in a study with a calcium-channel antagonist versus a beta-blocker. *Amer J Hypertens* 5(3): 154–160.

Mengden T, Weisser B, Vetter W. (1994) Ambulatory 24-hour blood pressure versus self-measured blood pressure in pharmacologic trials. *J Cardiovasc Pharmacol* 24(Suppl 2): S20–25 (Review).

Mengden T, Chamontin B, Phong Chau N, Luis Palma Gamiz J, Chanudet X. (2000) Use procedure for self-measurement of blood pressure. First International Consensus Conference on Self Blood Pressure Measurement. *Blood Press Monit* 5(2): 111–129.

Mengden T, Uen S, Baulmann J, Vetter H. (2003) Significance of blood pressure self-measurement as compared with office blood pressure measurement and ambulatory 24-hour blood pressure measurement in pharmacological studies. *Blood Press Monit* 8(4): 169–172.

Middeke M, Lemmer B. (1996) Office hypertension: abnormal blood pressure regulation and increased sympathetic activity compared with normotension. *Blood Press Monit* 1(5): 403–407.

Miles LE, Broughton RJ, eds. (1990) *Medical Monitoring in the Home and Work Environment*. New York: Raven Press, 336 pp.

Monk TH, Leng VC. (1986) Interactions between inter-individual and inter-task differences in the diurnal variation of human performance. *Chronobiol Intl* 3(3): 171–177.

Monk TH, Reynolds CF, Buysse DJ, Hoch CC, Jarrett DB, Jennings JR, Kupfer DJ. (1991) Circadian characteristics of healthy 80 year olds and their relationship to objectively recorded sleep. *J Gerontol* 46(5): M171–M175.

Monk TH, Buysse DJ, Potts JM, DeGrazia JM, Kupfer DJ. (2004) Morningness–eveningness and lifestyle regularity. *Chronobiol Intl* 21(3): 435–443.

Motohashi Y, Reinberg A, Levi F. (1987) Axillary temperature: a circadian marker rhythm for shift workers. *Ergonomics* 30(9): 1235–1247.

Motohashi Y, Takano T. (1993) Effects of 24-hour shift work with nighttime napping on circadian rhythm characteristics in ambulance personnel. *Chronobiol Intl* 10(6): 461–470.

Okajima Y, Togo M, Kitagawa G, Nishikawa S. (2000) Time series analysis of monthly body weight and blood pressures of one man from 29 to 65 years. *Amer J Human Biol* 12(4): 526–541.

Okamoto-Mizuno K, Yokoya T, Kudoh Y. (2000) Effects of activity of daily living and gender on circadian rhythms of the elderly in a nursing home. *J Physiol Anthropol Appl Human Sci* 19(1): 53–57.

Parati G, de Leeuw P, Illyes M, Julius S, Kuwajima I, Mallion JM, Ohtsuka K, Imai Y. (2002) 2001 Consensus Conference on Ambulatory Blood Pressure Monitoring participants. Blood pressure measurement in research. *Blood Press Monit* 7(1): 83–87 (Review).

Pickering T. (2002) Future developments in ambulatory blood pressure monitoring and self-blood pressure monitoring in clinical practice. *Blood Press Monit* 7(1): 21–25 (Review).

Pokorny ML, Blom DH, Opmeer CH. (1988) Effects of work and circadian rhythm on bus driver's oral temperature. *Chronobiol Intl* 5(4): 425–432.

Prigancová A, Strestík J, Sothern RB. (1997) Dynamics of some bioparameters in relation to external chronomodulation. In: *Proc 3rd Intl Workshop: Chronobiology and Its Roots in the Cosmos*, September 2–6, 1997. Mikulecky M, ed. Bardejov, Slovakia, pp. 153–161.

Rabatin JS, Sothern RB, Brunning RD, Goetz FC, Halberg F. (1981) Circadian rhythms in blood and self-measured variables of 10 children, 9–14 years of age. In: *Chronobiology, Proc XIII Intl Conf Intl Soc Chronobiol*. Halberg F, Scheving LE, Powell EW, Hayes DK, eds. Milan: Il Ponte, pp. 373–385.

Rabinowitz RP, Cookson ST, Wasserman SS, Mackowiak PA. (1996) Effects of anatomic site, oral stimulation, and body position on estimates of body temperature. *Arch Intern Med* 156(7): 777–780.

Rao ML, Strebel B, Halaris A, Gross G, Braunig P, Huber G, Marler M. (1995) Circadian rhythm of vital signs, norepinephrine, epinephrine, thyroid hormones and cortisol in schizophrenia. *Psychiatry Res* 57(1): 21–39.

Redmond DP, Hegge FW. (1987) The design of human activity monitors. In: *Chronobiotechnology and Chronobiological Engineering. NATO ASI Series*. Scheving LE, Halberg F, Ehret CF, eds. Dordrecht: Martinus Mijhoff Publication, pp. 202–215.

Refinetti R, Menaker M. (1992) The circadian rhythm of body temperature. *Physiol & Behav* 51(3): 613–637.

Reinberg A, Gervais P, Chaussade M, Fraboulet G, Duburque B. (1983) Circadian changes in effectiveness of corticosteroids in eight patients with allergic asthma. *J Allergy Clin Immunol* 71(4): 425–433.

Reinberg A, Smolensky M. (1983) Chronobiology and thermoregulation. *Pharmacol Ther* 22(3): 425–464 (Review).

Reinberg A, Andlauer P, De Prins J, Malbecq W, Vieux N, Bourdeleau P. (1984) Desynchronization of the oral temperature circadian rhythm and intolerance to shift work. *Nature* 308(5956): 272–274.

Reinberg A, Ugolini C, Motohashi Y, Dravigny C, Bicakova-Rocher A, Levi F. (1988) Diurnal rhythms in performance tests of school children with and without language disorders. *Chronobiol Intl* 5(3): 291–299.

Reinberg A, Smolensky M. (1990) Chronobiology and thermoregulation. In: *Thermoregulation: Physiology and Biochemistry*. Schönbaum E, Lomax P, eds. New York: Pergamon, pp. 61–100.

Reindl K, Falliers C, Halberg F, Chai H, Hillman D, Nelson W. (1969) Circadian acrophase in peak expiratory flow rate and urinary electrolyte excretion of asthmatic children: phase shifting of rhythms by prednisone given in different circadian system phases. *Rass Neurol Veg* 23(1): 5–26.

Rickerby J, Woodward J. (2003) Patients' experience and opinions of home blood pressure measurement. *J Hum Hypertens* 17(7): 495–503.

Roenneberg T, Wirz-Justice A, Merrow M. (2003) Life between clocks: daily temporal patterns of human chronotypes. *J Biol Rhythms* 18(1): 80–90.

Scarpelli PT, Romano S, Lamanna S, Buricchi L, Cai MG. (1978) Autorhythmometry in hypertension: some methodological aspects and clinical implications. *Chronobiologia* 5(4): 407–424.

Scheving LE. (1990) Autorhythmometry: a useful concept for teaching health education. In: *Medical Monitoring in the Home and Work Environment*. Miles LE, Broughton RJ, eds. New York: Raven Press, pp. 99–111.

Simpson HW. (1970) Urine temperature measurement in human circadian rhythm studies. *J Physiol* 212(2): 29P–30P.

Simpson HW, Kelsey C, Gatti RA, Good RA, Halberg F, Bohlen JG, Sothern RB, Delea CS, Haus E, Bartter FC. (1971) Autorhythmometry in Myasthenia Gravis. Detection of chronopathology and assessment of condition by rhythm-adjusted level of grip strength. *J Interdiscipl Cycle Res* 2: 397–416.

Simpson H, Wilson DW, Griffiths K, Bramen R, Holliday H. (1982) Thermorhythmometry of the breast. A review to 1981. In: *Biomedical Thermology*. Gautherie M, ed. New York: Alan R. Liss, pp. 133–154.

Simpson HW, Gruen W, Halberg E, Halberg F, Knauth P, Land DV, Moog R, Nougier JJ, Reinberg A, Smolensky M, Wilson D. (1987) Human thermometry in health and disease: the chronobiologist's perspective. In: *Chronobiotechnology and Chronobiological Engineering. NATO ASI Series*. Scheving LE, Halberg F, Ehret CF, eds. Dordrecht: Martinus Mijhoff Publication, pp. 141–188.

Slover G, Sothern RB, Scheving LE, Tsai TS, Halberg F. (1986) Urinary and self-measured circadian and circatrigintan rhythms before and after a human manic episode. In: *Proc 2nd Intl Conf Medico-Social Aspects of Chronobiology*, Florence, October 2, 1984. Halberg F, Reale L, Tarquini B, eds. Rome: Instituto Italiano di Medicina Sociale, pp. 427–444.

Smith CS, Reilly C, Midkiff K. (1989) Evaluations of three circadian rhythm questionnaires with suggestions for an improved measure of morningness. *J Applied Psychol* 74(5): 728–738.

Smith CS, Folkard S, Schmieder RA, Parra LF, Spelten E, Almiral H, Sen RN, Sahu S, Perez LM, Tisak J. (2002) Investigation of morning-evening orientation in six countries using the preferences scale. *Personality Indiv Diffs* 32(6): 949–968.

Smolensky MH, Ouellette J, Meyer M, Skeleton M, Iser L, McClellan D, Ouellette L, Martin D. (1981) The utilization of self-monitoring procedures to augment the high school health education curriculum. In: *Chronobiology, Proc XIII Intl Conf Intl Soc Chronobiol*. Halberg F, Scheving LE, Powell EW, Hayes DK, eds. Milan: Il Ponte, pp. 355–361.

Sothern RB. (1974) Chronobiologic serial section on 8876 oral temperatures collected during 4.5 years by presumably healthy man (age 20.5 at start of study). In: *Chronobiology*. Scheving LE, Halberg F, Pauly JE, eds. Tokyo: Igaku Shoin Ltd., pp. 245–248.

Sothern RB. Leach C, Nelson WL, Halberg F, Rummel JA. (1974) Characteristics of urinary circadian rhythms in a young man evaluated on a monthly basis during the course of 21 months. In: *Chronobiological Aspects of Endocrinology*. Aschoff J, Halberg F, Ceresa F, eds. Stuttgart: FK Schattauer Verlag, pp. 1–34.

Sothern RB, Simpson HW, Leach C, Nelson WL, Halberg F. (1981) Individually-assessable human circadian, circaseptan and circannual urinary rhythms (with temporal compacting and editing procedures). In: *Proc XIII Intl Conf Intl Soc Chronobiology, 1977*. Halberg F, Scheving LE, Powell EW, Hayes DK, eds. Milan: Il Ponte, pp. 363–371.

Sothern RB, Leach CS, Halberg F. (1984) Circannual aspects of urinary catecholamine excretion in a healthy man. In: *Chronobiology 1982–1983*. Haus E, Kabat H, eds. Basel: S. Karger, pp. 308–314.

Sothern RB, Halberg F. (1986) Circadian and infradian blood pressure rhythms of a man 20 to 37 years of age. In: *Proc 2nd Intl Conf Medico-Social Aspects of Chronobiology*, Florence, October 2, 1984. Halberg F, Reale L, Tarquini B, eds. Rome: Instituto Italiano di Medicina Sociale, pp. 395–416.

Sothern RB, Rhame F, Suarez C, Fletcher C, Sackett-Lundeen L, Haus E, Hrushesky WJM. (1990) Oral temperature rhythmometry and substantial within-day variation in zidovudine levels following steady state dosing in human immunodeficiency virus (HIV) infection. In: *Chronobiology: Its Role in Clinical Medicine, General Biology and Agriculture, Part A*. Hayes DK, Pauly JE, Reiter RJ, eds. New York: Wiley-Liss, Inc., pp. 67–76.

Sothern RB, Slover GPT, Morris RW. (1993) Circannual and menstrual rhythm characteristics in manic episodes and body temperature. *Biol Psychiat* 33(3): 194–203.

Sothern RB. (1994) Circadian and circannual characteristics of blood pressure self-measured for 25 years by a clinically-healthy man. *Chronobiologia* 21(1–2): 7–20.

Sothern RB. (1995) Time of day versus internal circadian timing references. *J Infus Chemother* 5(1): 24–30.

Sothern RB, Cornélissen G, Bingham C, Watanabe Y, Grafe A, Halberg F. (1998) Solar cycle stage: an important influence of physiology that must not be ignored (abstract). In: *Proc 3rd Intl Symp Chronobiol & Chronomed*, Kunming, China, October 7–12, 1998, p. 144.

Sothern RB. (1999) Oral temperature maps main features of circadian and other rhythms over 32 years (Abstract). *Chronobiol Intl* 16(suppl 1): 99.

Sothern RB, Burioka N, Cornélissen G, Engel P, Halberg Fcn, Siegelová J, Vlcek J, Halberg J, Halberg Franz. (2002) A circadecennian peak expiratory flow and the putative merit of self-measurement. *Scripta Medica (Brno)* 75(5): 261–265.

Sothern RB, Cornélissen G, Engel P, Fiser B, Siegelova J, Vlcek J, Dusek J, Halberg F. (2003) Chronomics: Instrumentation for monitoring of peak expiratory flow. *Scripta Medica (Brno)* 76(5): 313–316.

Staessen JA, Asmar R, De Buyzere M, Imai Y, Parati G, Shimada K, Stergiou G, Redon J, Verdecchia P. (2001) Participants of the 2001 Consensus Conference on Ambulatory Blood Pressure Monitoring. Task Force II: blood pressure measurement and cardiovascular outcome. *Blood Press Monit* 6(6): 355–370 (Review).

Stoynev AG, Minkova NK. (1997) Circadian rhythms of arterial pressure, heart rate and oral temperature of truck drivers. *Occup Med* 47(3): 151–154.

Stoynev AG, Minkova NK. (1998) Effect of forward rapidly rotating work on circadian rhythms of arterial pressure, heart rate and oral temperature in air traffic controllers. *Occup Med (Lond)* 48(2): 75–79.

Taillard J, Philip P, Coste O, Sagaspe, P, Bioulac B. (2003) The circadian and homeostatic modulation of sleep pressure during wakefulness differs between morning and evening chronotypes. *J Sleep Res* 12(4): 275–282.

Tankova I, Adan A, Buela-Casal G. (1994) Circadian typology and individual differences: A review. *Personality Indiv Diffs* 16(5): 671–684.

Torsvall L, Åkerstedt T. (1980) A diurnal type scale. Construction, consistency and validation in shift work. *Scand J Work Environ Health* 6(4): 283–290.

Trapp G, Eckert ED, Vestergaard P, Sothern RB, Halberg F. (1979) Psychophysiologic circadian rhythmometry on manic-depressive twins. *Chronobiologia* 6: 387–396.

Troyanov S, Ghezzo H, Cartier A, Malo JL. (1994) Comparison of circadian variations using FEV1 and peak expiratory flow rates among normal and asthmatic subjects. *Thorax* 49(8): 775–780.

Vaisse B, Genes N, Vaur L, Bobrie G, Clerson P, Mallion JM, Chatellier G. (2000) [The feasibility of at-home self-monitoring blood pressure in elderly hypertensive patients.] [French]. *Arch Mal Coeur Vaiss* 93(8): 963–967.

Vangelova KK, Deyanov CE. (2000) The effect of high ambient temperature on the adjustment of operators to fast rotating 12-hour shiftwork. *Rev Environ Health* 15(4): 373–379.

Varkevisser M, Kerkhof GA. (2003) 24-hour assessment of performance on a palmtop computer: validating a self-constructed test battery. *Chronobiol Intl* 20(1): 109–121.

Vink JM, Groot AS, Kerkhof GA, Boomsma DI. (2001) Genetic analysis of morningness and eveningness. *Chronobiol Intl* 18: 809–822.

Waterhouse J, Folkard S, Van Dongen H, Minors D, Owens D, Kerkhof G, Weinert D, Nevill A, Macdonald I, Sytnik N, Tucker P. (2001) Temperature profiles, and the effect of sleep on them, in relation to morningness–eveningness in healthy female subjects. *Chronobiol Intl* 18(2): 227–247.

Weisser B, Grune S, Burger R, Blickenstorfer H, Iseli J, Michelsen SH, Opravil R, Rageth S, Sturzenegger ER, Walker P, et al. (1994) The Dubendorf Study: a population-based investigation on normal values of blood pressure self-measurement. *J Hum Hypertens* 8(4): 227–231.

Weisser B, Mengden T, Dusing R, Vetter H, Vetter W. (2000) Normal values of blood pressure self-measurement in view of the 1999 World Health Organization-International Society of Hypertension guidelines. *Amer J Hyperten* 13(8): 940–943.

Winget CM, DeRoshia CW, Holley DC. (1985) Circadian rhythms and athletic performance. *Sports Exer* 17: 498–513.

World Hypertension League. (1988) Self-measurement of blood pressure. *Bull WHO* 66(2): 155–159.

Yarows SA, Brook RD. (2000) Measurement variation among 12 electronic home blood pressure monitors. *Amer J Hypertens* 13(3): 276–282.

Yetman RJ, West MS, Portman RJ. (1994) Chronobiologic evaluation of blood pressure in school children. *Chronobiol Intl* 11(2): 132–140.

Appendix 12.1

Recording Sheet for Self-Measurement of Oral Temperature (and other functions)

Subject Name:_____ Age:___yrs   Sex: M F   Page___

year: Date	Time	Oral Temp				Time* of Sleep	Comments**

*Please use 24h time (6 A.M. = 0600; 6 P.M. = 1800) or indicate A.M. or P.M.
**Under comments record any of the following since previous measurement: smoking (S), exercise (E), strenuous exercise (SE), Anxiety (A), Anger (A), Alcohol (specify), Coffee (Cf), Cola (Cl), Tea (T), medicine (indicate kind & dose), Meals (B=breakfast, L=lunch, D=dinner, Sn=snack), extremes in weather conditions (specify). Women should specify days of menstruation with an 'M.'
Please record times of going to bed and arising in sleep column.

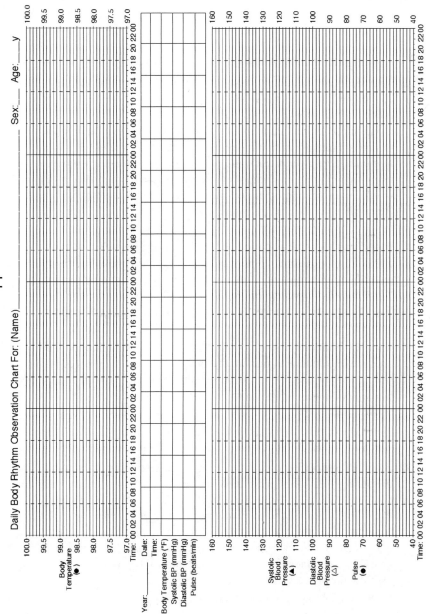

APPENDIX Item 12.3: Detailed Instructions for Performing Self-measurements

1. **Record Start Time:** Write exact local time (date, clock hour & minute) in columns for date and time. Use 24-hour time, where 6 AM = 0600 and 6 PM = 1800 or be sure to label each time with AM or PM. For computer analysis, these times will have to be entered only in 24h time. Note: write with something that won't smear if the recording sheet gets wet.

2. **Temperature:** Turn on thermometer, if electronic, or shake down mercury column to a reading below 96°F or 35°C. For oral temperature, place thermometer under tongue as far back as possible and hold firmly in place with mouth closed for at least 4 minutes (or until electronic thermometer beeps). *(When oral temperature is monitored it is the first variable that is initiated so that it is being determined while other variables are being measured, since it takes several minutes.)* Keep mouth closed while thermometer is in and leave in while performing other self-measurement tasks while remaining seated. *Instructions on the use of the thermometer and review of safety issues should be addressed prior to monitoring of temperature.*

3. **Mood and Vigor:** Rate your mood and vigor on a scale from 1 to 7, where 4 is a middle, average state, 1 is the lowest and 7 is the highest. You can create your own adjectives to go along with this scale, such as 1 = very depressed or tired, 2 = somewhat depressed or tired, 3 = slightly less than average, 4 = average, 5 = slightly above average, 6 = quite cheerful or active, 7 = super happy or active.

4. **Time Estimation:** Start a stopwatch with its face turned away from you and estimate the duration of one minute by some method, such as counting silently from 1 to 60 or visualizing the movement of a second hand on a clock. Stop the watch upon reaching 60 and record the actual seconds elapsed to the nearest 0.1 second.

5. **Heart Rate:** Find your pulse under the thumb-side of the wrist, start the stopwatch and count the number of beats for a full minute and record.

6. **Blood Pressure:** Be sure you are sitting and resting for a few minutes before measuring your blood pressure. Measure blood pressure and record both systolic (when beating starts) and diastolic (when beating disappears) pressures. *Prior instructions obtained from a health care professional may be helpful in learning to use the equipment and/or recognize the correct sounds.* If you are not sure of the reading you get, re-inflate the cuff and listen a second time. It is generally a good idea to measure and record BP two or even three times, then later use the average values.

7. **Performance Tasks:** Perform adding speed and/or short-term memory tasks using prepared sheets (see examples in Appendix Items 4 and 5). For adding speed, start the stopwatch, then quickly add consecutive overlapping pairs of numbers in a column of 51 random single-digit numbers, writing the sum to the right of each pair (50 sums total) (e.g., if the sequence in the column began

with 9, 1, 0, 8, 7, etc., the sums would be 10, 1, 8, 15, etc.). Stop the stopwatch and record the elapsed time. Check the sums for errors and calculate a ratio of the number of correct sums (50 possible) per second (total correct/elapsed time) and record. For random number memory, there should be 10 sets of 7-digit numbers in a horizontal row. Start the stopwatch, look at the first set of 7 numbers, quickly memorize them, then slide the stopwatch over the set to cover it and write the numbers in the space below. Look at the next set to the right, memorize, cover with the stopwatch, record and so on until the end of the line. Stop the stopwatch and record elapsed time. Check the numbers for accuracy and record the number of digits remembered or written down incorrectly, whether they are the wrong digit or in the wrong sequence. For example, if a sequence of 7 numbers was 3238765 and 3328765 was recorded, there are two errors, since the second and third digits were written in the wrong sequence. Calculate a ratio of the number of digits remembered correctly (70 possible) per second (correct/seconds elapsed) and record. Other performance tasks can include bead intubation or stringing or advancing two pins 50 times in a pegboard containing 3 holes (see Figure 12.8F).

8. **Record Temperature:** After finishing other measurements, remove thermometer from mouth and read the mercury column as accurately as possible (to the nearest tenth of a degree) or read the digital display. If time permits, the thermometer should be placed back in the mouth for a minute to see if the temperature still rises. If it goes up a bit, keep putting it back in the mouth for a minute at a time until the mercury (or digital display) stops rising and record the value. The digital thermometer often beeps when the rise in temperature slows down, but it may not have reached the correct physiologic temperature. Being in air conditioning or just coming inside during winter, talking a lot or drinking something cold can all artificially lower oral temperature, which will rise to its correct setting if the mouth is kept tightly closed for several more minutes. Clean and store thermometer in a safe place.

9. **Record Comments:** For all types of measurements, notes should be made in a "Comments" column of events and conditions that could have affected measurements. These comments might include medicines, ambient temperature or humidity, change in routine, exercise, food, anxiety, anger, cold or hot beverages, or meals. Women should specify days of menstruation. Also write down the times when you go to sleep and wake up.

Appendix 12.4

Random Number Adding Speed Test - Sheet 1

Instructions: use one column for each test time. Each column contains 51 random numbers. Start stopwatch, add consecutive overlapping pairs of numbers (col 1: 6+7=13, 7+5=12, etc; don't sum the column). Stop stopwatch after last pair addition, record seconds elapsed to 00.0, check for errors in addition, divide # correct pairs (50 possible) by seconds and record ratio of # correct per second.

#	col 1	col 2	col 3	col 4	col 5	col 6	col 7	col 8	col 9
1	6	8	1	8	0	8	5	0	7
2	7	2	9	9	2	3	3	9	0
3	5	4	3	5	1	5	8	3	1
4	5	5	4	7	7	1	5	4	3
5	7	7	5	5	7	3	6	4	3
6	6	7	9	0	3	5	6	3	7
7	5	0	3	6	7	6	8	2	7
8	9	7	7	5	7	7	7	1	9
9	9	7	2	0	9	9	7	3	7
10	1	3	2	7	3	3	2	6	8
11	3	8	0	7	2	9	4	2	3
12	3	1	1	1	7	0	0	6	0
13	2	8	1	0	8	5	0	5	1
14	1	5	5	1	1	8	7	2	2
15	1	9	4	3	1	9	4	9	2
16	0	7	7	4	3	7	6	1	8
17	4	0	1	4	4	8	6	3	0
18	1	7	7	4	9	6	4	8	2
19	6	0	1	8	1	8	8	2	0
20	7	6	0	1	4	6	6	3	7
21	6	3	4	5	8	5	7	3	5
22	0	2	6	0	1	8	7	3	1
23	7	3	3	9	5	6	5	6	8
24	9	1	1	3	1	9	0	6	0
25	7	3	0	9	4	2	8	7	1
26	6	4	5	1	2	3	7	2	1
27	8	4	8	3	3	4	5	0	7
28	0	0	7	3	9	9	3	7	6
29	3	8	2	3	0	8	5	7	3
30	3	6	1	3	1	2	3	0	9
31	0	8	8	2	6	8	1	9	7
32	1	3	6	0	6	2	3	1	4
33	5	5	6	2	7	4	7	3	5
34	4	6	8	3	8	6	9	8	1
35	9	5	8	2	8	8	5	0	9
36	7	6	5	1	4	9	5	8	1
37	7	5	5	2	2	4	3	6	3
38	7	8	8	3	0	8	0	7	8
39	8	5	2	3	3	0	1	7	1
40	6	1	7	2	2	2	4	9	3
41	5	0	8	8	4	8	7	2	3
42	8	1	6	8	6	0	1	7	8
43	7	4	7	5	6	1	0	5	7
44	7	7	7	1	9	3	3	3	6
45	9	4	3	5	5	2	4	8	3
46	6	6	4	3	2	5	1	4	9
47	2	4	1	0	6	8	6	0	2
48	8	0	8	9	0	6	0	5	0
49	0	3	5	7	8	8	4	4	9
50	7	1	1	1	0	4	4	8	1
51	1	1	5	0	4	4	8	1	1

Elapsed Time (secs): _____ _____ _____ _____ _____ _____ _____ _____ _____

# of errors: _____ _____ _____ _____ _____ _____ _____ _____ _____

correct: _____ _____ _____ _____ _____ _____ _____ _____ _____

#correct/sec: _____ _____ _____ _____ _____ _____ _____ _____ _____

Appendices 573

Appendix 12.4

Random Number Adding Speed Test - Sheet 2

Instructions: use one column for each test time. Each column contains 51 random numbers. Start stopwatch, add consecutive overlapping pairs of numbers (col 1: 6+7=13, 7+5=12, etc; don't sum the column). Stop stopwatch after last pair addition, record seconds elapsed to 00.0, check for errors in addition, divide # correct pairs (50 possible) by seconds and record ratio of # correct per second.

#	col 10	col 11	col 12	col 13	col 14	col 15	col 16	col 17	col 18
1	1	3	8	9	6	5	8	0	9
2	8	4	2	3	2	5	2	3	6
3	8	7	9	5	6	9	6	2	6
4	3	3	7	2	6	2	6	8	2
5	5	9	0	2	7	2	0	5	4
6	2	0	0	0	2	6	8	8	4
7	7	1	2	5	0	5	4	1	1
8	1	8	4	2	6	1	8	9	9
9	0	6	6	2	6	0	4	7	4
10	8	8	4	6	9	8	5	2	8
11	0	3	9	3	2	9	7	6	6
12	4	5	5	2	2	2	3	3	0
13	8	8	8	8	4	6	1	1	3
14	7	4	9	3	7	3	9	6	5
15	0	4	4	4	5	0	4	5	2
16	3	4	4	3	7	0	3	7	8
17	8	5	0	1	7	5	6	7	9
18	2	4	3	3	5	2	3	4	2
19	1	4	1	2	3	5	1	6	2
20	3	4	9	1	4	9	6	3	1
21	2	7	8	9	8	8	1	3	1
22	4	9	8	7	8	5	7	1	1
23	4	6	1	3	0	0	8	3	1
24	7	3	1	0	6	3	2	4	0
25	4	7	7	9	1	0	2	2	9
26	7	1	0	7	9	9	9	1	6
27	6	3	2	7	8	3	9	6	6
28	0	8	6	8	5	9	9	9	5
29	6	9	6	2	1	4	5	0	2
30	7	3	4	5	0	1	4	7	4
31	4	0	4	2	3	7	4	5	9
32	9	2	6	7	4	0	8	5	8
33	4	8	7	5	5	8	3	7	1
34	5	0	8	7	1	2	8	8	2
35	1	7	1	2	8	4	2	1	9
36	5	3	4	9	3	1	4	4	0
37	3	0	2	1	6	6	0	7	5
38	9	6	6	6	6	0	0	4	5
39	1	9	0	8	6	6	1	8	4
40	9	0	6	9	1	7	1	6	4
41	8	4	4	7	7	2	4	9	9
42	9	4	1	9	7	3	5	9	3
43	9	5	5	4	6	1	5	4	2
44	4	8	2	9	0	1	8	7	0
45	9	6	4	9	8	7	6	4	8
46	1	9	4	7	2	2	5	8	7
47	5	7	3	8	9	0	3	6	2
48	6	2	9	6	9	4	0	2	7
49	1	7	7	7	1	9	3	5	5
50	5	0	1	7	0	9	3	2	2
51	0	1	7	0	9	3	2	1	1

Elapsed Time (secs): ___ ___ ___ ___ ___ ___ ___ ___ ___

\# of errors: ___ ___ ___ ___ ___ ___ ___ ___ ___

\# correct: ___ ___ ___ ___ ___ ___ ___ ___ ___

\#correct/sec: ___ ___ ___ ___ ___ ___ ___ ___ ___

Appendix 12.5

Random Number Memory Test - Sheet 1

Instructions: use one column for each test time. Each column contains 10 sets of 7-digit random numbers. Start stopwatch, quickly look at and memorize first 7-number set, cover with stopwatch and write numbers from memory in box below the set. Repeat down the column and stop stopwatch after last (10th) set, record seconds elapsed, check for errors in transcription (wrong numbers or out of sequence), divide # correct (70 possible) by seconds and record ratio of # digits correctly remembered per second.

set	col 1	col 2	col 3	col 4	col 5	col 6
1	5446322	9675418	7588412	4262687	7915253	6590570
2	1538985	3435788	1677737	1605134	4456038	1885039
3	8594110	0631837	4623043	0824428	6832883	8241402
4	6114969	6211152	4290266	5949704	4693938	1128688
5	0521982	4753409	8100702	9715513	8354486	1065167
6	4141798	9861475	6808901	9840966	9162101	8771992
7	2835794	2485604	2041167	4547685	9189667	2065236
8	1778300	9688712	5821213	8930069	5575163	1080683
9	4095085	9080121	7057743	5005195	8515687	2988185
10	8299564	5516577	9452274	3175385	0752156	6616444

Elapsed Time (secs): _____ _____ _____ _____ _____ _____

\# of errors: _____ _____ _____ _____ _____ _____

\# correct: _____ _____ _____ _____ _____ _____

\#correct/sec: _____ _____ _____ _____ _____ _____

Appendix 12.5

Random Number Memory Test - Sheet 2

Instructions: use one column for each test time. Each column contains 10 sets of 7-digit random numbers. Start stopwatch, quickly look at and memorize first 7-number set, cover with stopwatch and write numbers from memory in box below the set. Repeat down the column and stop stopwatch after last (10th) set, record seconds elapsed, check for errors in transcription (wrong numbers or out of sequence), divide # correct (70 possible) by seconds and record ratio of # digits correctly remembered per second.

set	col 7	col 8	col 9	col 10	col 11	col 12
1	5565944	8431895	8565188	7725020	7936567	4736134
2	5336471	5855042	5719417	8363556	4224990	4569066
3	4992757	8020788	3385144	8336971	1385878	5042367
4	0724379	4613401	9041989	5862508	5892503	8929284
5	6787900	3969328	4029309	1570796	9251159	2341013
6	8446062	5723730	9576347	0490054	4661450	5984414
7	4489809	8995017	6510996	0415104	1624975	9879518
8	8062166	0646815	5074130	2110880	9153036	8608578
9	4281577	2496961	9163166	9549388	6280070	3739076
10	8366636	7165962	3131089	1223660	1008941	2842070

Elapsed Time (secs): _____

# of errors: _____

correct: _____

#correct/sec: _____

Appendix 12.5

Random Number Memory Test - Sheet 3

Instructions: use one column for each test time. Each column contains 10 sets of 7-digit random numbers. Start stopwatch, quickly look at and memorize first 7-number set, cover with stopwatch and write numbers from memory in box below the set. Repeat down the column and stop stopwatch after last (10th) set, record seconds elapsed, check for errors in transcription (wrong numbers or out of sequence), divide # correct (70 possible) by seconds and record ratio of # digits correctly remembered per second.

set	col 13	col 14	col 15	col 16	col 17	col 18
1	7230564	1740103	5653518	2908569	8667923	4205412
2	2146044	5445019	6490042	9632523	6033222	4474105
3	5938032	9421146	3956405	1626965	6311648	9131889
4	9471023	8121104	3045985	5286262	8569373	0117587
5	1015495	5843401	2306813	8450272	0254054	9138003
6	0690274	2079261	4617755	6488620	4873073	2042360
7	9305487	2593066	5907711	2114505	3976571	3922019
8	8262776	1165823	0226329	3998162	0243253	9619997
9	7604667	9142812	0066455	8474062	5261532	9669387
10	7964379	9836402	3910262	7825896	8136941	0599958

Elapsed Time (secs): _____

# of errors: _____

correct: _____

#correct/sec: _____

13
Chronobiometry: Analyzing for Rhythms

"Given for one instant an intelligence which could comprehend all the forces by which nature is animated and the respective positions of the beings which compose it, if moreover this intelligence were vast enough to submit these data to analysis, it would embrace in the same formula both the movements of the largest bodies in the universe and those of the lightest atom; to it nothing would be uncertain, and the future as the past would be present to its eyes."
— *Pierre Simon De Laplace (1749–1827),*
French mathematician, philosopher
"Theorie Analytique de Probabilites: Introduction,"
v. VII, Oeuvres (1812–1820)

Introduction

Studies that focus on biological rhythms, like those in other areas of biology, include the formulation and testing of hypotheses, as well as the statistical analysis of data representing results obtained from experiments and observations. Procedures by which results from studies of biological rhythms are analyzed range from the visual estimation of period, phase and amplitude, to the use of complex computer-assisted statistical programs based upon mathematical models.

Much of the early work on biological rhythms simply involved describing results that were presented in a table or a graph. Even pioneers in the field of biological rhythm research were sometimes reluctant to apply statistics to their data, as noted in the following statement by Frits Went (1903–1990), one of the outstanding plant physiologists of the 20th century: *"I am very little impressed by complicated and clever theoretical or mathematical constructions; in fact, I don't understand many of them. Nor can I follow or accept statistical analyses: if the facts don't speak for themselves clearly, no statistical treatment will make them palatable."* (Went, 1974).

Graphs of original data values can show rhythmic patterns that become apparent to the naked eye when plotted against time, with vertical lines indicating intervals of 24 h (Figure 13.1). Other graphs, especially activity or schedules of animals and humans, were and still are illustrated as double plots, wherein a

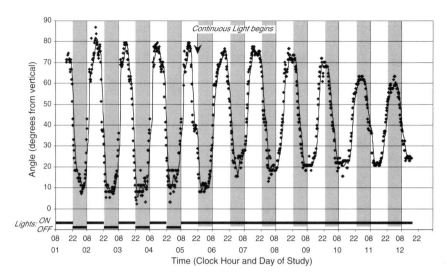

FIGURE 13.1. Simple chronogram showing a synchronized, high amplitude circadian leaf movement pattern for a Kentucky Wonder pole bean plant during 4.5 days in LD 14:10 h (*left*). The leaf movement rhythm free-runs when placed under constant light (LL, arrow) for a week. During LL, the 24-h period lengthens (as seen in the delay in the time of peak values from day to day) and the amplitude dampens (Sothern & Koukkari, unpublished).

second day is plotted both to the right and below the previous day in order to visualize stability or changes in the time of onset, duration and/or offset of the activity (Figure 13.2).

While it is sometimes adequate to view a graph of the measured data by eye and subjectively estimate a *period* and *amplitude* from patterns in the data, more precise values are often required. These may be obtained by using objective statistical procedures, such as periodogram techniques (for activity data) or by fitting a mathematical curve (cosine) to the data. In addition, when interpreting and comparing data, statistical analyses can help by indicating how well the observed data are described by these mathematical terms and when making comparisons of rhythm characteristics (e.g., amplitude or phase) between two or more variables or individuals. This is especially relevant in medical studies of clinical variables.

The objective of this chapter is to introduce some of the methods that have been and can be used to analyze data from rhythm studies.

Data Collection

To determine whether time has a systematic effect on a variable, it is necessary to collect data not only by making repeated measurements at different times, but also by scheduling them at appropriate intervals that may or may not be

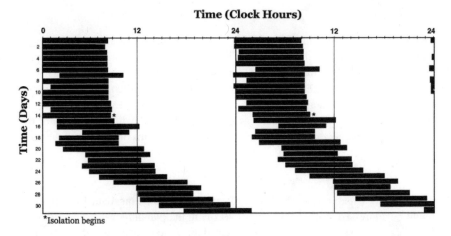

FIGURE 13.2. Double-plot of sleep–wake pattern of a healthy young man (RBS, age 20 years) before and during 19.5 days in an isolation chamber (see Essay 12.1). The delay in sleep onset from one day to the next during social isolation and lack of time cues indicates that the circadian sleep–wake cycle was "free-running" with a period longer than 24 h.

regular or equidistant (i.e., equally spaced) over several hours, days, weeks, months, or years.

Number of Timepoints

In order to analyze for a rhythm by curve-fitting procedures, there must be a minimum of three timepoints (e.g., 08:00, 16:00, and 04:00 h) covering the span of the period being studied, since values collected at merely two timepoints (e.g., 08:00 and 04:00 h) can only be connected by a straight line. However, the "Nyquist folding frequency" (cf. DePrins et al., 1986) implies that selection of a period for an analysis of a rhythm cannot go below an interval that is three times the sampling frequency. Thus, if data were collected every 4 h, the shortest frequency that could be tested (in addition to a 24-h period) would be 12 h (3×4 h $= 12$ h), but not 8 h or 6 h, etc., and if the data were collected every 8 h, the shortest period to be tested would be 24 h (3×8 h $= 24$ h), but not 12 h.

Ideally, data would be sampled such that there were more than three timepoints covering the period, since a cosine would fit a 3-timepoint series perfectly, with no possibility for testing its appropriateness as a model (e.g., goodness of fit, normality of residuals, etc.). However, data collection at only three timepoints is far from satisfactory in order to confidently describe a waveform or other characteristics of the rhythm under investigation. Therefore, for reasonably accurate rhythm detection and description, data collection at many more timepoints is essential.

How Long?

It is recommended that data collection be over a span that is at least twice the length of the period being studied (e.g., 48 h for a 24-h period, etc.) in order to show reproducibility of a pattern in the data.[1] This time series can then be used to quantitatively describe the data characteristics (number of data, mean, median, mode, minimum and maximum values, standard deviation, standard error, etc.) for the overall series and/or for separate timepoints, such as getting averages for different times of the day, days of the week, days of the menstrual cycle, or months or seasons of the year.

Sampling Often Enough

Methods and schedules selected for collecting and analyzing data from a study should be addressed when a research project is in the planning stage. For example, four timepoints may be all that can be collected due to technical, ethical or other difficulties, as in some invasive medical procedures (e.g., bone-marrow sampling) or due to time constraints, such as the need to sample in each season or only when a subject is awake. Sometimes if the interval between sampling is long, not only can one miss the detection of higher frequency oscillations, but due to *aliasing,* it could reveal a false periodicity. This is analogous to the stroboscopic effect that makes the wheels of a moving object appear to rotate in the opposite direction (see Figure 13.3).

Aliasing

If sampling density is sparse when measuring a higher frequency rhythm, a lower frequency rhythm may result as a phenomenon known as aliasing. For example, when designing a study to collect circadian data spaced 4-h apart around-the-clock, it may be necessary to sample each timepoint on different days due to lengthy procedures, such as bioassays or drug applications, or invasive procedures, such as blood sampling or bone-marrow aspirations. A schedule designed to collect a sample 4-h later each day (hence, every 28 h) or once a week, but 4-h later than the previous sample, would eventually sample the desired 6 circadian times around-the-clock. However, such a time series graphed out over the actual sampling span of several days might also appear to show a low-frequency oscillation that is only artifactual due to sampling interval aliasing and therefore not an actual longer rhythm.

Decision Making

Biological rhythms research often involves making decisions and statements (e.g., in publications) based upon comparisons of similarities or differences among groups, especially in reference to time. These decisions or statements should not be based upon an opinion from "eye-balling" the data, but rather upon an unbiased

---

[1] This requirement can be satisfied by sampling one individual over several days or, for population studies, several individuals over a single day.

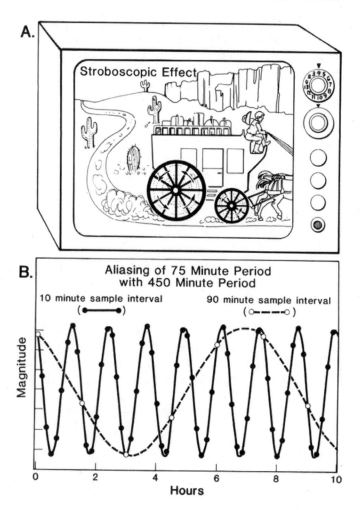

FIGURE 13.3. Example of aliasing due to inappropriate sampling schedule. Wheels on a vehicle sometimes appear to be going backwards as a result of a stroboscopic effect by light that is periodically interrupted at certain speeds of the wheel (A). When monitoring a rhythm (B), 40-min sampling of a variable with a 75-min ultradian rhythm can result in numbers that appear to have a 9-h rhythm (450 min).

statistical analysis of the data. For example, the mean (average) level of a substance, such as cortisol in a human blood sample or activity of a plant enzyme, may appear to be twice as high at one time of day than another in the same individual or group. However, the validity of this finding will depend upon a number of factors, such as sample sizes (i.e., the number of data) and variability among samples.

Thus, results from experiments or other types of studies, whether based upon visual observations of phenomena, surveys, or quantitative measurements, must be evaluated by the use of statistics to test for significance. The statistical method

TABLE 13.1. Sample data: heart rate (beats/min) measured every 4 h during waking for 4 days by a healthy male (RBS, age 20 years).

Time	Day 1	Day 2	Day 3	Day 4	Mean
08:30	82	87	89	90	87.00
12:30	94	90	92	101	94.25
16:30	92	93	76	76	84.25
20:30	81	80	64	84	77.25
00:30	73	76	73	77	74.75

may be as simple as comparing timepoint means by an appropriate test or by curve-fitting procedures in order to describe characteristics of a rhythmic pattern.

Data Preparation

Prior to data analysis, data should be graphed and, if necessary, edited or transformed.

Graphs and Visual Inspection

Regardless of the statistical method chosen, the first step in analyzing data for a rhythm should be a visual inspection of the values obtained. Organizing the data into a table (Table 13.1) and then plotting them into a graph (Figure 13.4) is

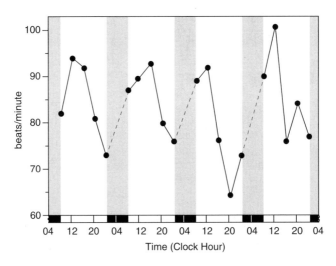

FIGURE 13.4. Graph of heart rate plotted against measurement times during waking-only over 4 days by a healthy young man (RBS, age 20 years) (see original data in Table 1). Main features of the circadian rhythm appear present in spite of the lack of data during sleep (dark bars) (Sothern, unpublished).

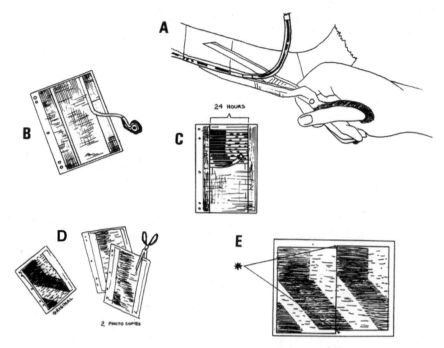

FIGURE 13.5. Example of creation of a double plot from activity recordings (A) by pasting successive 24-h strips both below (C) and to the right (E) of each other. Such double plots are used to show the time course (onset, duration, and offset) for such variables as whole body activity, sleeping times, etc., during synchronized and free-running environmental conditions, development of circadian sleep–wake patterns of infants or the presence of both 24.0-h solar and 24.8-h lunar activity rhythms in a crab.

almost always helpful. Activity data (e.g., from an animal running on a wheel in its cage, via telemetry in the wild, or from a human wearing a wrist activity monitor, etc.) can be organized into a double-plot scattergram (Figure 13.5). Alternatively, these usually dense data can be averaged or summed over a specified interval (e.g., hourly) and these values plotted against time.

A graph of either individual values or means on the vertical (y) axis against time on the horizontal (x) axis, called a *chronogram*, can be inspected for patterns, trends and/or outliers (e.g., extreme values) in the data. A view of the graph may indicate that the data need to be edited prior to rhythm analysis.

Editing or Transforming Data

Very high or low values may be erroneous due to an inaccuracy in the measurement technique or they may be outside of a normal distribution for some other known or unknown reason. If they are beyond 2 or 3 standard deviations from the mean, they may be objectively removed from the data set as mathematically determined extreme values (see Essay 13.1).

Essay 13.1 (by RBS): Standard Deviation and Error

A normal distribution of data values is symmetrical about the mean and looks like a bell-shaped curve. It is often called the *Gaussian distribution* in honor of the German mathematician, Carl Friedrich Gauss (1777–1855), who figured prominently in describing this statistic. The standard deviation (SD) is the average difference of each individual data value from the mean. In other words, the average of all deviations above and below the mean (ignoring the plus and minus signs) is *1 standard deviation*. In a normal distribution, 68% of all values lie within ±1 SD of the mean, 96% lie within ±2 SD and 99% lie within ±3 SD of the mean. Thus, individual values that lie outside 2 or 3 SD of the mean represent only 2% and 0.5% in each direction, respectively, and can sometimes be considered extreme outliers and candidates for removal from a database. It is noteworthy to point out that the often computed 95% 'confidence' limits represent an area that is ±1.96 SD, and not ±2 SD, of the mean.

The SD is used to describe the distribution of the individual data around the mean value, while the *standard error* (SE) is used to describe the error interval around a computed parameter, such as the mean, amplitude, or phase. The SE of the mean is obtained by dividing the SD by the square root of the total number of data points. For example, if an SD of 12 was computed from a sample size of 100 values, then the SE would be 1.2 (12 divided by 10, the square root of 100). The inferential statistics involved in the construction of confidence intervals and significance testing are based on standard errors.

Some studies obtain values that consistently increase or decrease over the course of the measurement span (i.e., show a trend) due to the study design or other reasons (e.g., in the absence of food or water, during illness or aging, etc.). It may, therefore, be necessary to use a standard statistics program to test for the presence of a linear or polynomial trend. This procedure involves the least-squares fit (see below) of an increasing or decreasing straight line to the data. If the total variance in the data is significantly less about the trend line than about the mean, then there is a trend in the data. A significant trend can be subtracted from the original values and the original mean value added back to each of the residual values to obtain a "detrended" data set prior to rhythm analysis (Figure 13.6).

There is also the possibility that a trend in a short time series can be part of a longer rhythm. For example, data collected over several days may show a circadian pattern with an increasing or decreasing trend in the daily means, but there may also be an underlying monthly rhythm that was not completely monitored and the increasing or decreasing values represent subspans of this longer rhythm (e.g., a menstrual or annual cycle).

Normalizing Data

Before combining data from several subjects for a grouped rhythm analysis, data are often *normalized* or *standardized* to reduce variability caused by differences

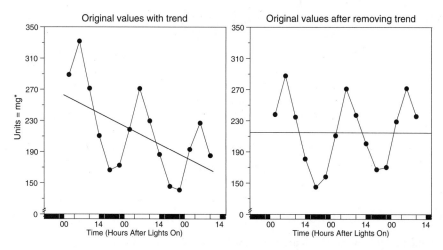

FIGURE 13.6. Example of a time-series before (left) and after (*right*) detrending. Data shown represent 4-hourly change in weight (mg) due to transpiration in a plant (*Arabidopsis*, genotype = Wild-type) over 2.5 days under LD14:10 schedule. Dark bar = lights off span. *P*-value for trend = 0.020. If significant, the trend can be statistically removed and the original arithmetic mean added back to the residual values after detrending to produce data shown in the graph on the right. Mean for both graphs = 216.1 mg/4 h (cf. Sothern et al, 2004).

in levels. One of the most common procedures used to reduce inter-subject variability is to transform each individual's values to percent of his/her overall mean or express each value as a ratio or percent of his/her highest value. A set of data can also be normalized in various ways (e.g., log values or square roots) if they are not normally distributed (e.g., if the range from lowest to highest values is very large).

Partitioning Data Spans

Also to be decided is if the entire data span should be analyzed as a whole or be partitioned into experimentally determined subspans. This is necessary if behavior and/or variables before, during or after a procedure (e.g., a treatment) or after a change in environmental conditions (e.g., a change in the light–dark times, summer vs. winter, stages of the menstrual cycle, etc.) are to be tested and compared. For example, data collected during sleep could be partitioned to look for rhythms in the first, middle and last parts of the span or data could be partitioned to look at rhythm characteristics before and after flying across time zones. Rhythms may also have been altered due to aging, fatigue, or stress due to the length of an experiment, and long time-series may need to be partitioned accordingly for analyses of separate subspans.

Statistical Detection of Time Effects

If single or mean values at different test-times appear to differ or there is a recurring pattern that appears obvious to the naked eye, it may be hypothesized that a meaningful circadian or other variation is present and that this variation represents an underlying rhythm. However, this pattern can only be described as a time-dependent variation until it has been statistically validated as being rhythmic.

Objective statistical methods[2] for time-effect involve testing for differences among data obtained at either two timepoints, or three or more timepoints. A flow chart of steps used in the most common time-series analyses is shown in Figure 13.7.

Using Two Timepoints

When only two test-times are being evaluated (e.g., sampling in mid-L vs. mid-D, morning vs. evening, winter vs. summer), they are usually compared by a *t-test*, which tests for a significant difference between the *means*, or by *chi-square*, which tests for a significant difference in the *proportions* (incidence), such as 50 (74%) positive results out of 68 total tests vs. 17 (22%) positive results out of 70 total tests (see Figure 13.7—left side). If these statistical comparisons detect a difference at a probability level (p-value) of 0.05 (5%) or less (i.e., the number of times out of a hundred that this result could have occurred by chance), then the results are considered significantly different and, therefore, this difference may indicate an effect of time.

It can be hypothesized that this significant difference reflects an underlying rhythm, but the complete characteristics of the rhythm itself have not been described, since only two timepoints were measured. In the example just cited, the proportions (74% vs. 22%) are different at the $p<0.0001$ level, indicating that there is a significant difference in positive responses between the two test-times.[3]

Using Three or More Timepoints

When three or more test-times are available, additional statistical procedures may be applied to the data. The time-series may be tested for time-effect by chi-square if comparing proportions (as above) or by an *analysis of variance (ANOVA)*,

---

[2] Unfamiliar mathematical terms used in this section, such as normal distribution, standard deviation (SD), standard error (SE) (see Essay 13.1), p-value, chi-square, t-test, ANOVA, or least-squares method, are common terms that are defined in books on statistics.

[3] The data shown on the left middle section of Figure 13.7 actually represent the number of mice that survived a potentially lethal dose of an anti-cancer drug. Each healthy mouse received the same dose adjusted for body weight, but at different times. When mice were tested at six different times around-the-clock, a circadian rhythm in susceptibility was found, as shown in Figure 13.7—right side (Sothern et al., 1977).

Statistical Detection of Time Effects 587

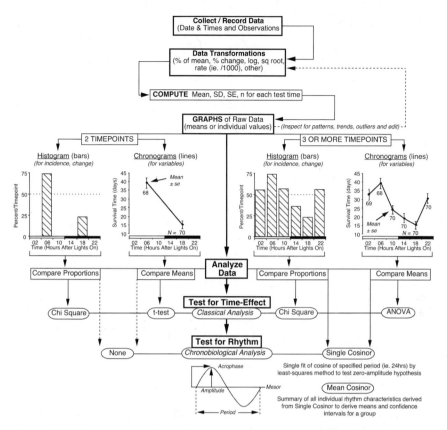

FIGURE 13.7. There are several steps in time-series analysis, beginning with recording the time and value of an observation. Prior to analysis, data may need to be detrended or normalized (e.g., by re-expressing values as percent of individual series mean, square roots, rates, etc.) when combining values from different subjects. Means and error estimates can be computed overall and by timepoint and graphs prepared. How data can be analyzed for a rhythm will depend upon the number of timepoints that are measured. When only two timepoints are measured (*left panel*), proportions (e.g., dead vs. alive, yes vs. no) can be compared by chi-square test, while means can be compared by *t*-test. When three or more timepoints are available (*right panel*), proportions can still be compared by chi-square, while means can be tested for time-effect by analysis of variance (ANOVA). In addition, rhythm characteristics of period, mesor (rhythm-adjusted mean), amplitude, and acrophase can be derived by the least-squares fit of a cosine (single-cosinor method) to all data (Sothern, 1992).

which tests for significant differences among three or more individual timepoint means (as opposed to the 2-group *t*-test mentioned above) (Figure 13.7—right side). ANOVA, Chi-square, *t*-test, and trend analyses are called *parametric* statistical procedures. When three or preferably more timepoints are available, the time series can also be tested for the presence of a rhythm by the fit of a *cosine* model, a *non-parametric* procedure.

Statistical Detection of Rhythms

Evaluating time-series for rhythm(s) is a critical task in all fields that study biological rhythms. Times series analysis helps in the objective, unbiased description of features of data that are collected over time (Bingham, 1978). Several computerized methods are available to analyze times-series. Each has advantages and disadvantages depending on the type and schedule of data collection.

Analyzing Time-Series by Standard Methods

Behavioral or activity data (e.g., wheel running, perch hopping or other movements, sleep–wake records) that are usually yes-no or on-off data that can be coded as zeros and ones (or ones and twos) can be analyzed for the amplitude and period of a rhythm by standard "classical" regression methods of spectral analysis for signal detection (i.e., a high amplitude), such as *autocorrelation*, *Fourier transformation,* and *Periodogram* (cf. Dörrscheidt & Beck, 1975; Bingham, 1978; Morgan & Minors, 1995; Hassnaoui et al., 1998). These procedures can be applied to other types of dense, continuously recorded data, as well (e.g., leaf movements, wrist activity, beat-to-beat heart rate, and core temperature). These procedures can provide a satisfactory statistical appraisal of the presence of a rhythm by estimating its period and amplitude. Several of these standard methods are available in computer programs for the PC (cf. Domoslawski et al., 1991; Plautz et al., 1997,[4] among others), while a *chi-square Periodogram* (Sokolove & Bushell, 1978) has been adapted to use moving windows (time-intervals) of data to detect changes in the period due to experimental conditions or erratic behavioral responses (Hassnaoui et al., 1998).

Limitations of Standard Methods

Classical methods generally require equally spaced data over integral (whole) multiples of the period investigated. To use one of these methods to look for a circadian rhythm, the data must be collected at equal intervals (e.g., at 5 or 10 min, or 1 or 2 h, etc.) over complete whole days (e.g., 1 day, 2 days, 3 days, etc.), and not portions of days (i.e., 1.2 days, 2.5 days, 3.8 days, etc.). There can be no extra values (e.g., a few measurements were obtained 1-h apart, while the remainder were obtained 3-h apart) and no "holes" or gaps in a time series (e.g. no missing

---

[4] FFT-NLLS (Fast Fourier Transform-nonlinear least squares) rhythm analysis software written by Dr. M. Straume of the NSF Center for Biological Timing is described and reviewed in 2004 (Straume, 2004). It interfaces with Microsoft Excel spreadsheet and acquisition packages commonly used in plant and insect circadian studies (e.g., gene expression research, bioluminescence, and leaf movements) where large data sets are automatically recorded at frequent intervals, if not continuously. The software, which estimates period and phase, can be downloaded for free from several web sites, including: http://www.scripps.edu/cb/kay/shareware/ and http://template.bio.warwick.ac.uk/staff/amillar/PEBrown/BRASS/BrassPage.htm.

FIGURE 13.8. Single Cosinor formula: mathematical model used to analyze time series of data and estimate rhythm parameters. This model determines the best-fitting cosine by the method of least-squares (cf., Nelson et al., 1979).

$$y_i = M + A \cos(\omega \tau_i + \emptyset) + e_i$$

- y_i = fitted value of the function at time τ_i
- M = the mean level (mesor)
- A = amplitude of the cosine (1/2 peak-trough range)
- ω = angular frequency of the cosine (= $2\pi i/\tau$)
- τ = period of the fitted cosine
- $\emptyset$ = phase angle of the peak of the cosine (acrophase)
- e_i = residual error at τ_i

timepoints), which is often the case in a biologic time series.[5] For these classic analyses, a missed data point or points need(s) is often estimated and inserted by computing the mean of the value preceding and the value following the missing value(s).

Examples of dense time series include monitoring of activity, heart rate, body temperature, and/or light exposure, where values are obtained every few seconds and dropouts are common. An "interpolated" value is then inserted at the missing timepoint and, while satisfying a statistical requirement, may or may not be a reflection of the true value that might have been obtained had an actual measurement been made at the missing time. This may not be important to the analytical outcome for activity data, but may be crucial and decisive in determining true rhythm characteristics for other biologic variables. In-depth reviews of the many procedures and other considerations in data acquisition and analysis are available (Reinberg & Smolensky, 1983; DePrins et al., 1986; Minors & Waterhouse, 1989; Refinetti, 1993; Morgan & Minors, 1995).[6]

Analyzing Time-Series by Curve Fitting

One of the most commonly used quantitative methods for analysis of periodic phenomena, especially in the area of medicine (e.g., Reinberg & Smolensky, 1983; Arendt et al., 1989), is the *Cosinor* technique, involving the statistical fitting of a cosine to a time-series of data by a procedure known as least-squares linear regression (Halberg et al, 1972). This mathematical model[7] (Figure 13.8) is very robust in the sense that it can analyze a time-series in which the data are not necessarily equidistant or sampled over integral multiples of the period in question (Figure 13.9).

---

[5] The periodogram can be used to correctly evaluate a period in data containing gaps, but the amplitude will depend on the number and distribution of gaps in the time-series (Dörrscheidt & Beck, 1975).

[6] Some additional statistical articles of interest are listed in the reference list of this chapter.

[7] See Chapter 5 on Models for a discussion of various mathematical models, such as limit cycles.

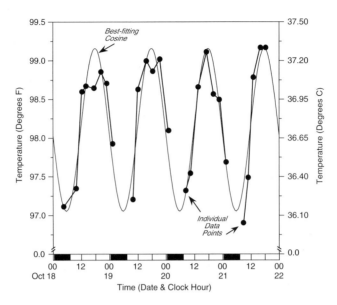

FIGURE 13.9. Example of how a 24-h cosine can be fit to data collected at unequal intervals over a non-integral number of days. In this example, core-body (urine) temperature was measured at times of void upon urge by a healthy man (RBS, age 31 years) for 3.75 days. Even though the data collection schedule was unbalanced (e.g., timepoints are unequally spaced, often with no values during sleep [dark bars] and do not cover 4 full days), the 24-h cosine is a good fit to the data and describes the circadian waveform adequately (Sothern, unpublished).

While there are some reservations regarding the application of the cosinor technique, especially when a time-series is not perfectly sinusoidal or contains multiple peaks (Wever, 1979; Enright, 1981, 1989; Klemfuss & Clopton, 1993), there are numerous benefits, especially when interpreting rhythmic patterns and characteristics (Mikulecky, 1991; Bourdon et al., 1995; Fernandez & Hermida, 1998). Some cosinor programs available for the PC and Macintosh computers include: *Cosina* by Monk & Fort, 1983; *Pharmfit* by Mattes et al 1991; *Chronolab* by Mojón et al, 1992; and *Circadia* Software (Behavioral Cybernetics, Cambridge, MA). Cosinor software is also available as part of a set of analysis programs (Refinetti, 2000); time-series analysis and forecasting systems (www.euroestech.net/indexuk.php), or available as a web-based application (*Ritmom* by Alberola et al., 1999). In addition, a single-cosinor technique has been designed for use in a commercially available spreadsheet (Microsoft Excel®) (Bourdon et al., 1995). There are also many more references with regards to analyzing data for rhythms, some of which are listed at the end of this chapter under "Additional References."

The Least-Squares Technique

When measurements of a biological variable are repeated several times, the exact same number is usually not obtained each time. When describing the data collected, it is therefore necessary to summarize all the values collected, usually as an arithmetic average (mean) value, with an estimate of how much scatter or deviation there is above and below this number. In effect, a mean value represents a flat horizontal line determined to have an equal amount of deviation of individual data points above it as below it in a set of data. Thus, if each value is subtracted from the mean, the sum of the deviations of the individual data points above the mean is equal to the sum of the deviations below the mean.

In order to calculate the total variance in a data set, the deviations of individual values are summed after first squaring each one in order to remove the negative sign from those values below the mean (*note*: if this wasn't done, the sum of all the positive and negative deviations from the mean would be zero). However, squaring each value also gives more weight to outliers. The result is a *sum of the squares* and this value represents the *total variance* in the data.

Any mathematical model fit to the data by a *least-squares method*, such as an increasing or decreasing trend line (linear or binomial) or a cosine, is done in an effort to see if the model can significantly reduce the total variance about the arithmetic mean in order to obtain a smaller (or least) sum of the squares (hence, "least squares") and thereby use that model to better describe a pattern in the data (e.g., an increasing or decreasing trend, or an oscillation) when it appears to be different from the flat horizontal line represented by the mean.

The Best-Fitting Curve

The cosinor software program analyzes a time-series using a *least-squares* technique based on a cosine wave of variable *period* (the length of the curve), level (the middle value of the curve, called the *mesor*, above and below which variability of the data are equal), *amplitude* (extent of the rise and fall of the curve around the mean) and *phase* (the location in time of the peak and trough of the curve).

If the period is set to a defined value (e.g., 24.0 h), the program varies the phase, amplitude, and level to find and obtain the best fit of the cosine with that period length (e.g., the best-fitting curve will be that which results in the smallest sum of the squares).[8] Alternatively, the program may test a range of periods (e.g., a so-called least-squares window or band of periods, such as 20–28 h with 0.1-h increments between trial periods), wherein the program fits cosines of each length separately. The curve with the lowest residual sum of squares (i.e., the "least squares") (which usually coincides with the highest amplitude) is then selected as representing the frequency of the underlying periodicity (rhythm) in the time series.

The difference between the total variability about the arithmetic mean and the remaining variability around the fitted cosine can be expressed as percent, is called *percent rhythm*, and is comparable to the R^2 value[9] obtained in regression

---

[8] The method for determining the best fit of any straight line is to sum the squares of all the distances of each data point from every possible straight line. The line with the smallest "sum of squares" is the best-fitting line. The cosinor method is identical, but instead of a straight line, the model fits a cosine function.

[9] $R^2 =$ the proportion of the variance of Y attributed to its linear regression on X.

tests, such as correlations. Thus, a cosine which results in a residual sum of the squares value that is only a third (33%) of the total variability computed around the mean (which represents a horizontal flat line) is said to account for 66% of the total variability in the time series (i.e., 100% minus 33% = 66% rhythm).

Statistical Significance

However, even though a curve might fit the data better than a flat line, it still may not do so with statistical significance. An oscillation is considered a statistically validated rhythm when the *p-value* (probability of outcome) is ≤0.05. A *p*-value of 0.05 or less is generally accepted as the level at which a finding is considered statistically significant and, therefore, representing a true difference between groups being tested, rather than a random artifact. A *p*-value represents the probability (e.g., the number of times out of a 100) that a given result would happen by chance if the same experiment or procedure were repeated 100 times (e.g., a *p*-value of 0.03 means that the finding is likely to happen only 3 out of a 100 times by chance and, therefore, may be considered more likely to be a factual result).

A *p*-value is computed mathematically based upon the amount of data and a comparison of the total variance around the arithmetic mean versus the residual variance around the fitted cosine. In cosinor analysis, this is called a *zero-amplitude test*, and if significant, there is less total variance around the cosine than there is around a flat line (i.e., the mean). A curved line (a cosine) with a significant (*p*-value ≤0.05) amplitude will better describe the variations in the data than does the flat line representing the mean and its error estimates (SD or SE) alone. However, by definition, this mathematical model is symmetrical, even though the data being analyzed may not be, and its peak or trough may be estimated statistically to be between actual measurements or at a time when no data were collected. Nevertheless, detection of a statistically significant cosine fit with a given period implies a regular recurring pattern, and therefore non-randomness, in a time- series, with values oscillating at that frequency, as opposed to complete randomness about a homeostatic mean in a time-series.

Complex Waveforms

A 24-h rhythm may not be as smooth as a single-component (24-h) cosine, but may consist of a major peak and one or more minor peaks or a broad peak in the data due to ultradian variations and/or unequal proportions of the light–dark cycle. This sort of time series can often be more accurately analyzed and described by the least-squares fit of a multiple-component cosine model (Tong et al., 1977; Fernández & Hermida, 1998).

A simultaneous fit of two or more components that are each statistically significant[10] (e.g., 24 h with 12 h and/or 8 h and/or 7 days, etc., in the cosine model) can be applied to the data to see if a composite cosine will account for more of the

---

[10] Note: *p*-values are determined for each trial period in a multiple-component cosine model and only those with a $p \leq 0.05$ are included in the final model.

overall variability (e.g., reduce the sum of the squares) and follow more closely the actual waveform (ups and downs) of the data (see example in Figure 13.10). Thus, the cosinor technique is an extremely versatile, easy to use statistical method, which can describe virtually any *waveform* in a time-series by using an appropriate number of components (i.e., 24 h, 12 h, and 8 h for a circadian waveform) in the final cosine model (Tong et al., 1977; Cornélissen et al., 1990; De Prins & Hecquet, 1994; Fernandez & Hermida, 1998).[11]

Rhythm Parameter Comparisons

The cosinor technique is useful in objectively quantifying individual rhythm characteristics, which can later be summarized for groups of individuals. Once rhythm characteristics for individual series have been obtained (these are called "first-order statistics"), they may be used to compare series or groups ("second-order statistics"), such as comparing amplitudes and/or acrophases between control and experimental groups or before and after a treatment in an individual or for a group.

Several publications present the methods for estimating and comparing single-component (e.g., 24-h) rhythm parameters (e.g., amplitude, acrophase, and mesor) between or among individuals or groups (Halberg et al., 1967, 1972; Nelson et al, 1979; Bingham et al, 1982; DePrins et al., 1986; Faure et al, 1990; Fuentes-Arderiu & Rivera-Coll, 1990; Fernandez & Hermida, 1998). Statistical parametric and nonparametric techniques have also been developed to compare the parameters of multiple-component cosine models (Fernández et al., 2004). Thus, for a 24-h and 12-h composite cosine model, the mesor, amplitude and/or orthophase for the overall composite waveform, as well as the amplitude and acrophase for each fitted single component (e.g., 24 h, 12 h) can be compared between two or more series. These statistical procedures can also be used to compute confidence intervals for the global amplitude, orthophase, and bathyphase of a multiple-component model. Such confidence intervals are usually not provided by programs that fit a multiple-component cosine, but are useful when comparing results between individuals or populations.

Lack of Rhythm Detection

Most variables that are measured in living systems do show some circadian and other oscillations in the sense that their numbers do not remain constant, but rather rise and fall throughout the day. However, the detection of a significant rhythm will depend on several factors, including adequate sampling over an adequate number of cycles or using enough subjects.

For example, measuring something at only three or four different times during waking, but never at night, may not provide enough information to detect a significant circadian rhythm, since only a small portion of the overall 24-h span was monitored.

---

[11] Others have tried a periodic spline model when the pattern is not symmetric, e.g., when the peak and trough are not 12-h apart or additional peaks or troughs are not harmonic to the main period (Wang & Brown, 1995).

594   13. Chronobiometry: Analyzing for Rhythms

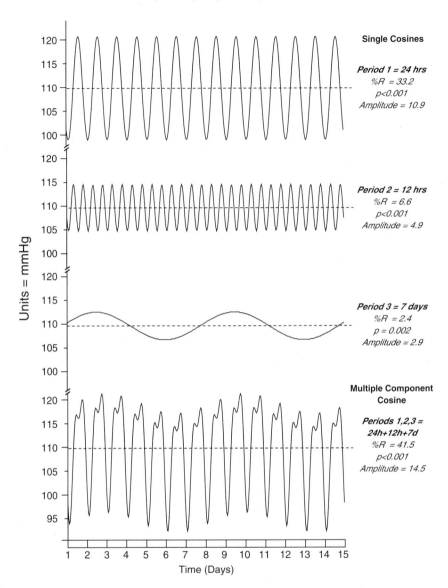

FIGURE 13.10. Construction of a multiple-component cosine model by the simultaneous least-squares fit of two or more periods each determined to be significant by single cosine fitting. In this example, different cosines were fit to systolic blood pressure monitored automatically every 30 min around-the-clock for 2 weeks (4/12–25/2000; $n = 532$ values) by a healthy man (AMJ, age 23 years) and three periods were statistically significant: 24 h, 12 h, and 7 days (Sothern, unpublished; cf. Figure 12.13). The waveform for each cosine is shown separately and as a composite model. For descriptive characteristics of each cosine: %R = percent of total variability reduced by fitted cosine model; *p*-value from zero-amplitude test; Amplitude = distance from middle of cosine (= mesor [dashed line]) to peak or trough of cosine model. *Note*: due to asymmetrical waveform, amplitude for a multiple-component cosine represents half the peak-trough difference and not distance from mesor.

Also, the gap time between successive measurements may not allow for an accurate estimation of the true period, due to statistical considerations (e.g., too few data over the period being tested, such as measuring only every 6 h). It may be necessary to increase the sample size by collecting more data at more frequent timepoints, collecting over a longer time span or collecting from more individuals to enlarge the group.

In some instances, when it is not possible to quantify a rhythm mathematically, a waveform that is reproducible from 1 day to the next or among several individuals or that will change in a predictable way following manipulation of one or more synchronizers (lighting, feeding) may be evidence of a rhythm (Minors & Waterhouse, 1989). Also, a portion of a rhythm is sometimes monitored as evidence of the overall circadian rhythm. For example, many investigators measure blood or saliva for melatonin for only an hour or so before sleep onset and then only for several more hours during sleep in order to determine the location of the melatonin peak, which occurs shortly after falling asleep (Lewy & Sack, 1989; Lewy et al., 1999). Since a prominent circadian rhythm in melatonin has been established (Arendt, 1989), with high values during darkness and very low values during the light span, sampling a portion of the circadian span (in this case, the first half of the dark span) is often adequate to make conclusions about the circadian rhythm in melatonin.

Descriptive Rhythm Parameters

Several parameters that describe a rhythm's characteristics can be derived from the best-fitting curve. These include: the *period* (duration of one complete cycle); the *mesor* (the mid-value of the cosine curve representing a rhythm-adjusted mean),[12] the *amplitude* (half the distance from the peak to the trough of the fitted cosine indicating the predictable rise and fall around the mesor); and the *acrophase* (the time of the peak of the fitted curve, representing the average time of high values in the data). The time of the lowest point of a curve representing a rhythm is called the *bathyphase*. Depending on the variable, either the peak or the trough may be more useful when discussing its rhythm.

The objective, statistically determined mesor, amplitude, and acrophase, as opposed to a subjective, "eye-balled" evaluation, can be extremely useful in describing the characteristics of a rhythm (see Chapter 2 of Characteristics of Rhythms for additional discussion). Group rhythm characteristics can be obtained either by fitting a cosine to all data from all individuals as a single series[13] or by summarizing all individual amplitudes and acrophases from members of the group

---

[12] The value for the mesor is the same as the arithmetic mean when data are equally distributed over the observation span. However, the mesor can reflect a truer overall mean than the arithmetic mean if more data were obtained at the peak or trough, which would then pull the average higher or lower, respectively, simply due to unequal sampling. Some choose to use all capital letters for MESOR to signify it as an acronym for Midline Estimated Statistic of Rhythm (Halberg et al., 1977).

[13] Normalizing data by converting each series to percent of individual mean before combining all data will negate differences in overall levels (mesors) and generally result in better rhythm detection.

or groups by a statistical procedure known as the *population-mean cosinor* (cf. Nelson et al., 1979; Bingham et al., 1982). The single or population-mean numerical parameters are also useful when comparing two or more rhythms in an individual or between groups (e.g., when comparing the timing [acrophase] between two variables, such as cortisol and melatonin, or when comparing the timing of a variable before and after a change in conditions).

The Cosinor Illustrated

The use of the Cosinor method in rhythm detection and description is illustrated in Figure 13.11. Panel A shows a scatter of values graphed without considering time of measurement. The variability in the measurements may result from random or periodic fluctuations. When these same data are plotted against time of measurement (Panel B), there appears to be a time-related change underlying the differences. Because of unequidistant sampling, with more values collected in the afternoon and none during the night, the arithmetic mean is closer to the higher values (Panel C) and may not be representative of the actual overall 24-h mean. Note that even with the lack of night-time data, a rhythmic pattern seems evident.

The fit of a 24-h cosine by the least-squares (cosinor) method reveals that a best-fitting curved line follows, and therefore better describes, the actual pattern in the data more closely than a flat line (the mean) (Panel D). The mid-value of the fitted curve, the mesor, can be used to represent a rhythm-adjusted mean that is not biased by dense sampling at some times and not at others (Panel E). As stated previously, the mean and the mesor will be the same if data are collected at equal intervals over the period being investigated. Other characteristics of the rhythm include the acrophase, the bathyphase, and the amplitude, A, (Panel F).

Example of a Cosinor Program

The ChronoLab™ software, written for the Macintosh computer (Mojón et al., 1992), is an easy to use, readily available cosinor program used to test data for a rhythm.[14] It provides a way to fit curves to time-coded data and also to obtain a print-out of rhythm results with or without an accompanying graph. One of the authors of this book (RBS) has used this software since its adaptation from a mainframe computer and many of the figures throughout this chapter and book were first generated by this program package.

The ChronoLab program is capable of doing many different sorts of analyses and manipulations of the data that may be necessary, such as reading in more than

---

[14] The ChronoLab program is available free of charge over the internet by contacting its authors at: http://www.tsc.uvigo.es/BIO/bioing.html (click on "chronobiological signal processing"). An instructor may have this program available on a server computer as part of a course, or the authors of this book can be contacted, since we use the program extensively. While this program is fairly easy to use, the manual accompanying the Chrono-Lab program will be useful for further details, as will contact with someone familiar with the various additional aspects of the program not mentioned here.

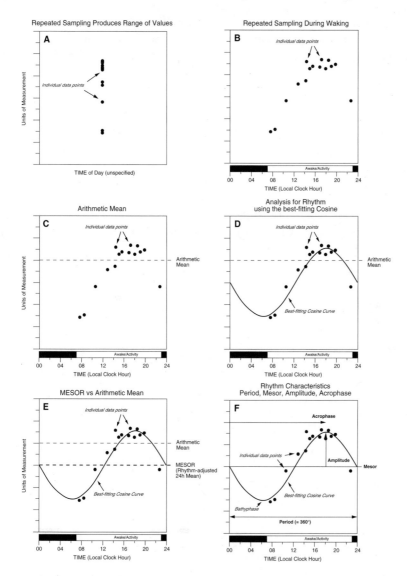

FIGURE 13.11. Characteristics of rhythmic data when time is not considered can be described as a mean and range of values (Panel *A*). These same data plotted against the times they were measured may reveal a systematic (e.g., rhythmic) pattern (Panels *B* and *C*). A cosine curve may follow more closely the pattern in the data than does the plot of a line representing the arithmetic mean of all data (Panel *D*). The arithmetic mean can be artificially raised due to more frequent sampling during one time of the day, but can be more correctly estimated by computation of the mesor, the middle of the cosine (Panel *E*). When time of obtaining each value is considered, the data can be described in additional terms from a best-fitting cosine, including when the highest values occurred (acrophase), what the middle value of the rhythm was (mesor) and how far above and below this value the points extended (amplitude) (Panel *F*) (Sothern, 1992).

one file at a time, eliminating extreme outliers, transforming data, fitting a trend plus cosine, fitting cosines with multiple components (e.g., 24 h, 12 h, and 8 h), summarizing rhythm parameters across groups of subjects (population-mean cosinor), comparing rhythm parameters between single series or groups, and analyzing sub-sequences in longer time-series based on the moving-window principle (serial section analysis). Serial section results can also be plotted, as well as polar plots generated that display an ellipse representing a joint 95% confidence interval for the amplitude and phase of a rhythm.

Graphs of the data can be generated in a variety of ways, including showing the data and the best-fitting cosine over the entire time span of monitoring (Figure 13.12A) or averaging data to show time-point means over one average cycle (Figure 13.12B). An example of adding significant ultradian periods to the 24-h cosine model in order to better describe the actual circadian pattern is shown in Figure 13.13 for systolic blood pressure measured automatically for one week by a female student.

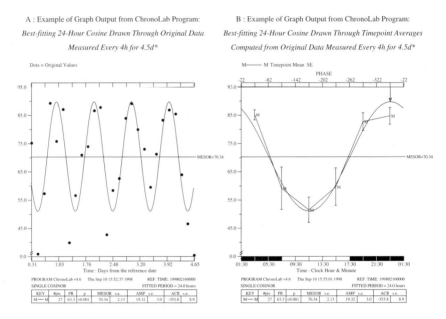

FIGURE 13.12. Examples of some of the graphic displays that can be produced by the ChronoLab Program. A best-fitting cosine can be drawn through the original data (left) or timepoint means (right). Data represent 4-hourly feeding activity (percent filtering) of an aquatic insect (*Brachycentrus*) monitored over 4.5 days (original data of Koukkari & Sothern, 1998). Dark bar = daily dark span. Note that the rhythm characteristics (percent rhythm (PR), mesor, amplitude, acrophase, and *p*-value) are the same for both curves, since the same data were analyzed: only the display of the data is different (means vs. original points). In addition, adjustments to each graph that may be necessary (e.g., connecting the original data dots, labeling the *y*-axis for units, relabeling the *x*-axis for appropriate time units, adding titles, etc.) can be done in a graphics program.

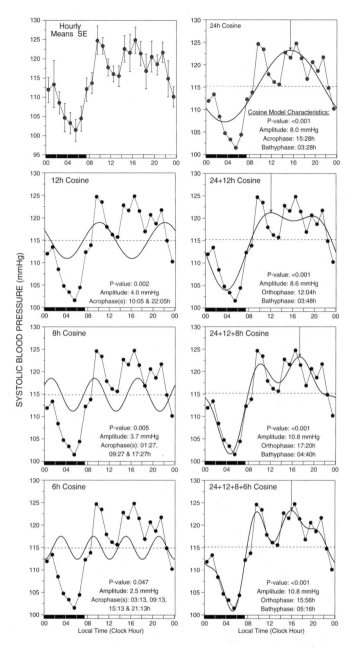

FIGURE 13.13. Example of multiple-component cosine model fit to data. Additional significant periods can be added to a single-cosinor model to often better describe characteristics of a circadian rhythmic pattern, here demonstrated for systolic blood pressure monitored automatically over 7 days by a healthy student (AO, female, age 20 years, October 28–November 3, 2000, N of data = 219). Acrophase or orthophase = high point of cosine (arrows); bathyphase = low point of cosine (Sothern, unpublished).

Take-Home Message

Evaluating a biological time-series for a rhythm by an appropriate statistical technique to establish statistical validity is essential in the study of biological rhythms. Fortunately, software packages are available for this purpose. Data collected over time can be analyzed to determine the length of any oscillating periodicity in the time-series and to confirm whether it is statistically significant (i.e., whether the rhythm is significantly different from a flat line or random variations). If a cosine is fit to the data, its characteristics (period, amplitude, mesor [mean], and phase) can be useful in describing a rhythm and when comparing one rhythm with another, such as when rhythm characteristics are being compared (1) in different variables, (2) in the same variable in two or more individuals, or (3) in the same individual on repeated occasions.

References

Alberola C, Revilla M, Mazariegos R. (1999) Web-base chronobiological analysis. *Biol Rhythm Res* 30(5): 477–496.

Arendt J. (1989) Melatonin and the pineal gland. In: *Biological Rhythms in Clinical Practice*. Arendt J, Minors DS, Waterhouse JM, eds. London: Wright, pp. 184–206.

Bingham C. (1978) Time series. In: *International Encyclopedia of Statistics*, Vol 2. Kruskal WH, Tanur JM, eds. New York: the Free Press, pp. 1167–1189.

Bingham C, Arbogast B, Cornélissen-Guillaume G, Lee JK, Halberg F. (1982) Inferential statistical methods for estimating and comparing cosinor parameters. *Chronobiologia* 9: 397–439.

Bourdon L, Buguet A, Cucherat M, Radomski MW. (1995) Use of a spreadsheet program for circadian analysis of biological/physiological data. *Aviat Space Environ Med* 66(8): 787–791.

Cornélissen G, Bakken E, Delmore P, Orth-Gomér K, Åkerstedt T, Carandente O, Carandente F, Halberg F. (1990) From various kinds of heart rate variability to chronocardiology. *Amer J Cardiol* 66: 863–867.

DePrins J, Cornélissen G, Malbecq W. (1986) Statistical procedures in chronobiology and chronopharmacology. In: *Annual Review of Chronopharmacology*. Vol 2. Reinberg A, Smolensky M, Labrecque G, eds. New York: Pergamon Press, pp. 27–141.

DePrins J, Hecquet B. (1994) Data processing in chronobiological series. In: *Biologic Rhythms in Clinical and Laboratory Medicine*, Touitou Y, Haus E, eds. Berlin: Springer-Verlag, pp. 90–113.

Dörrscheidt GJ, Beck L. (1975) Advanced methods for evaluating characteristics parameters (τ, α, ϕ) of circadian rhythms. *J Math Biol* 2: 107–121.

Domoslawksi J, Zajac W, Lewandowski MH, Hess G. (1991) Chronos–experimental data processing package for chronobiology. *J Interdiscipl Cycle Res* 22(2): 108–109.

Enright JTE. (1981) Data analysis. In: *Biological Rhythms. Handbook of Behavioral Neurobiology*, Vol 4. Aschoff J, ed. New York: Plenum Press, pp. 21–39.

Enright JT. (1989) The parallactic view, statistical testing and circular reasoning. *J Biol Rhythms* 4(2): 295–304.

Faure A, Nemoz C, Claustrat B. (1990) A graphical and statistical technique for investigation of time series in chronobiology according to cosinor procedure. *Comput Biol Med* 20: 319–329.

Fernández JR, Hermida RC. (1998) Inferential statistical method for analysis of nonsinusoidal hybrid time series with unequidistant observations. *Chronobiol Intl* 15(2): 191–204.

Fernández JR, Mojón A, Hermida RC. (2004) Comparison of parameters from rhythmometric models with multiple components on hybrid data. *Chronobiol Intl* 21(3): 469–484.

Fueuntes-Arderiu X, Rivera-Coll A. (1993) Impact of circadian rhythms on the interpretation of stat measurements. *Clin Chim Acta* 214: 113–118.

Halberg F, Tong YL, Johnson EA. (1967) Circadian system phase–an aspect of temporal morphology; procedures and illustrative examples. In: *The Cellular Aspects of Biorhythms*. von Mayersbach H, ed. Berlin: Springer-Verlag, pp. 20–48.

Halberg F, Johnson EA, Nelson W, Runge W, Sothern RB. (1972) Autorhythmometry–procedures for physiologic self-measurement and their analysis. *The Physiology Teacher* 1: 1–11.

Hassnaoui M, Pupier R, Attia J, Blanc M, Beauchaud M, Buisson B. (1998) Some tools to analyze changes of rhythms in biological time series. *Biol Rhythm Res* 29(4): 353–366.

Klemfuss H, Clopton PL. (1994) Seeking Tau: a comparison of six methods. *J Interdiscipl Cycle Res* 24(1): 1–16.

Koukkari WL, Parks TW, Sothern RB. (1999) Individual circadian rhythms in filtering behavior of Trichoptera during synchronized and constant lighting conditions (abstract). *NABS* 16(1): 218.

Lewy AJ, Sack RL. (1989) The dim light melatonin onset (DLMO) as a marker for circadian phase position. *Chronobiol Intl* 6: 93–102.

Lewy AJ, Cutler NL, Sack RL. (1999) The endogenous melatonin profile as a marker for circadian phase position. *J Biol Rhythms* 14(3): 227–236.

Mattes A, Witte K, Hohmann W, Lemmer B. (1991) Pharmfit–a nonlinear fitting program for pharmacology. *Chronobiol Intl* 8: 460–476.

Mikulecky M. (1991) Enright's criticism of cosinor analysis: the solution? *J Interdiscipl Cycle Res* 22(2): 157.

Minors DS, Waterhouse JM. (1989) Analysis of biological times series. In: *Biological Rhythms in Clinical Practice*. Arendt J, Minors DS, Waterhouse JM, eds. London: Butterworth & Co., pp. 272–293.

Mojón A, Fernández JR, Hermida R. (1992) Chronolab: an interactive software package for chronobiologic time series analysis written for the Macintosh computer. *Chronobiol Intl* 9: 403–412.

Monk TH, Fort A. (1984) "Cosina": a cosine curve fitting program suitable for small computers. *Intl J Chronobiol* 8: 193–224.

Morgan E, Minors DS, eds. (1995) Special Issue: methods in biological time-series analysis. *Biol Rhythm Res* 26: 124–252.

Nelson W, Tong YL, Lee JK, Halberg F. (1979) Methods for cosinor rhythmometry. *Chronobiologia* 6: 305–323.

Plautz JD, Straume M, Stanewsky R, Jamison CF, Brandes C, Dowse HB, Hall JC, Kay SA. (1997) Quantitative analysis of *Drosophila period* gene transcription in living animals. *J Biol Rhythms* 12(3): 204–217. (Erratum in: J Biol Rhythms 1999; 14(1): 77.)

Refinetti R. (1993) Laboratory instrumentation and computing: comparison of six methods for the determination of the period of circadian rhythms. *Physiol Behav* 54(5): 869–875.

Refinetti R. (2000) *Circadian Physiology*. Boca Raton: CRC Press, 184 pp. (*Note*: programs for analyses for time-effects are also available for PCs from this author at: http://www.circadian.org/softwar.html).

Reinberg A, Smolensky MH. (1983) Investigative methodology for chronobiology. In: *Biological Rhythms and Medicine. Cellular, Metabolic, Physiopathologic and Pharmacologic Aspects.* Reinberg A, Smolensky MH, eds. New York: Springer-Verlag, pp. 23–46.

Sokolove PG, Bushell WN. (1978) The chi-square periodogram: its utility for analysis of circadian rhythms. *J Theor Biol* 72: 131–160.

Sothern RB, Nelson WL, Halberg F. (1977) A circadian rhythm in susceptibility of mice to the anti-tumor drug, adriamycin. In: *Proc XII Intl Conf Intl. Soc Chronobiology*, Washington, DC, Aug 10–13, 1975. Milano: Il Ponte, pp. 433–438.

Sothern RB. (1992) *Studies on Biologically Rhythmic Natural Modifiers of Cancer Treatment in Mice and Man—Chronotherapeutic Concepts and Applications* (PhD thesis). Dept of Pathology, University of Bergen, Norway.

Sothern RB, Tseng TS, Orcutt S, Olszewski N, Koukkari WL. (2002) *GIGANTEA* and *SPINDLY* genes linked to the clock pathway that controls circadian characteristics of transpiration in *Arabidopsis. Chronobiol Intl* 19(6): 1005–1022.

Straume M. (2004) DNA microarray time series analysis: automated statistical assessment of circadian rhythms in gene expression patterning. *Methods Enzymol* 383: 149–166 (Review).

Tong YL, Nelson WL, Sothern RB, Halberg F. (1977) Estimation of the orthophase–timing of the high values on a non-sinusoidal rhythm – illustrated by the best timing for experimental cancer chronotherapy. In: *Proc XII Intl Conf Intl Soc Chronobiol*, Washington, DC, Aug 10–13, 1975. Milan: Il Ponte, pp. 765–769.

Wang Y, Brown MB. (1996) A flexible model for human circadian rhythms. *Biometrics* 52: 588–596.

Went FW. (1974) Reflections and speculations. *Ann Rev Plant Physiol* 25: 1–26.

Wever RA. (1979) *The Circadian System of Man. Results on Experiments Under Temporal Isolation.* New York: Springer-Verlag, 276 pp.

Additional References of Interest

Bloomfield E. (1976) *Fourier Analysis of Time Series: An Introduction.* New York: John Wiley & Sons.

Cugini P, DiPalma L. (1994) Cosint analysis: a procedure for estimating biological rhythms as integral function by measuring the area under their best-fitting waveform profile. *Biol Rhythm Res* 25: 15–36.

Del Pozo F, Jimenez J, De Feudis F. (1979) Method for the acquisition and analysis of chronopharmacological data. In: *Chronopharmacology.* Reinberg A, Halberg F, eds. Oxford: Pergamon Press, pp. 185–192.

Hickey DS, Kirkland JL, Lucas SB, Lye M. (1984) Analysis of circadian rhythms by fitting a least squares sine curve. *Comp Biol Med* 14: 217–233.

Martin W. (1981) Estimation of parameters of circadian rhythms, if measurements cannot be performed around the clock. The case of a band limited signal plus noise. In: *Night and Shiftwork: Biological and Social Aspects.* Reinberg A, Vieux N, Andlauer P, eds. Oxford: Pergamon Press, pp. 37–44.

Sollberger A. (1970) Problems in the statistical analysis of short periodic time series. *J Interdiscipl Cycle Res* 1: 49–88.

Wever RA. (1965) A mathematical model for circadian rhythms. In: *Circadian Clocks.* Aschoff J, ed. Amsterdam: North-Holland Publication, pp. 47–63.

van Cauter E. (1974) Methods for the analysis of multifrequency biological time series. *J Interdiscipl Cycle Res* 5: 131–148.

Author Index

Aakvaag A, 367, 519
Abbaticchio G, 284
Abdel-Monem MM, 369
Abecia JA, 375
Abella ML, 18
Abercrombie GF, 235
Abodeely M, 203
Abrahamsen JF, 205, 367, 491, 523
Abrami G, 231
Abulker C, 18
Acebo C, 416
Achiron A, 416
Acitelli P, 513
Ackerson LM, 514
Adam R, 499, 500, 507, 508
Adams DB, 284, 506
Adams R, 506
Adan A, 558, 567
Adanan CR, 369
Adenis A, 496
Ader R, 522
Adkisson PL, 333, 368
Adriani A, 103
Aeschbach D, 64
Afolayan SB, 367
Agoramoorthy G, 288
Ahlbom A, 498
Ahlgren A, 415, 501
Ahmad M, 103, 132
Ahmed S, 134
Aida K, 370
Aidoo A, 417
Akers RM, 371
Åkerstedt T-L, 510, 567, 600
Akslen L, 491

Alabadi D, 194
Alabiso I, 522
Alafaci E, 499
Alberola C, 600
Albers HE, 369
Albertini M, 558
Albertson TM, 198
Albrecht G, 518
Albrecht PG, 335, 501
Albrecht U, 204
Albright DL, 132, 284
Albulker Ch, 515
Alheit J, 333
Ali MA, 333, 375
Alila-Johanasson A, 367
Alingh Prins AJ, 204
Allan JS, 417
Allan TM, 497
Allaouchiche B, 497
Allard HA, 334
Allen AJ, 522
Allen MJ, 371
Allison KC, 420, 424
Alm K, 367
Almendral A, 290
Almiral H, 565
Alonso I, 503
Alonso LG, 559
Amasino RM, 198, 337
Ameida MC, 492
Ameida OF, 522
Amer Vet Med Assoc, 367
Anand A, 291
Anders RJ, 525
Anderson A, 511

603

Anderson CR, 374
Anderson CS, 498
Anderson GE, 371
Anderson JD, 338
Anderson JL, 524
Anderson LE, 57, 63, 491, 522
Anderson RN, 334
Anderson-Bernadas C, 333
Andersson CR, 105, 199
Andersson H, 200
Andersson M, 367
Andlauer P, 565
Andre M, 493
Andreotti FA, 491
Andrew-Casal M, 525
Andrews RV, 57
Andronov AA, 194
Anisimov VN, 493
Anjo A, 500
Anon, 285
Antoch, 194, 199, 500
Antunes CM, 368, 499
Antunes-Rodriguez J, 492
Anwar YA, 491
Aoki K, 371
Aoki S, 199
Apfelbaum M, 418
Arai T, 201
Arakawa K, 518
Araki CT, 367
Araki H, 491
Aral SO, 292
Aranda A, 416
Araujo JF, 367
Arbogast B, 600
Aréchiga H, 231
Arendt J, 57, 415, 417, 422, 489, 491, 501, 509, 519, 525, 600
Armitage R, 57, 562
Armstrong SM, 424
Arnedt JT, 415
Arntz HR, 492, 525
Aronson BD, 194, 285
Arrese JE, 503
Arrhenius S, 231
Arthur JM, 333
Ascari E, 499
Aschoff J, 9, 15, 57, 135, 136, 291, 333, 416, 492, 517, 558

Aschoff JC, 514
Aschoff JWL, 59
Aserinsky E, 15
Ash Mr, 285
Ashburner M, 194
Ashkenazi I, 291
Ashkenazi IE, 423, 499, 513, 515
Ashkenazi II, 558
Ashmore LJ, 194
Asmar R, 566
Aspostolakou F, 511
Assad C, 61
Assenmacher I, 367
Assenza A, 373
Astre C, 494
Astronomical Almanac, 416
Athanassenas G, 133
Atkins CE, 367
Atkinson G, 416, 417, 424, 510, 532, 558
Atkinson M, 558
Attar-Levy D, 492
Attia J, 59, 601
Augar D, 423
Augustinc G, 289, 370, 505
Auliciems A, 333
Auvil-Novak SE, 421, 492
Avissar S, 136, 424
Avivi L, 291
Ayala DE, 503, 562
Aylmer G, 511
Aymard N, 18
Ayo JO, 367

Babb WK, 561
Babu P, 200
Bachmann F, 500
Baciu I, 561
Bäckström M, 374
Backwell PRY, 231
Badiee A, 205, 523
Badot PM, 61
Badwe RA, 492
Bagni B, 514
Bailey CS, 560
Bailey J, 204
Bailey SL, 558
Bailleul F, 507, 508, 513
Baird DD, 286, 292
Bakalova RG, 563

Baker BS, 291
Baker FC, 285
Baker IA, 504
Baker SM, 416
Bakken E, 561, 600
Bakken EE, 59, 495, 561
Bakst MR, 367
Balakrishnan C, 497
Balasse EO, 523
Baldissera SF, 492
Baldwin IT, 336
Ball EE, 231
Ball SK, 292
Ballario P, 194, 201
Balsalobre A, 57, 194
Balzer I, 15, 416, 419
Banerjee D, 194
Bangsberg D, 373
Banks-Schlegel S, 420, 509
Barak Y, 416
Barak-Glantz I, 136, 424
Baras M, 502
Barbason H, 523
Barbosa J, 501
Barens PJ, 520
Barett K, 419
Bargiello TA, 194
Barnes DM, 495
Barnett JE, 57, 80, 103
Barnhart A, 510
Barnhart K, 285
Barnum CP, 104
Barnwell FH, 231
Barr K, 416
Barr W, 231
Barrett PR, 419
Barretto RB, 513
Barrozo RB, 367
Barry DM, 368
Barry J, 517
Bartal JP, 423
Bartko JJ, 134, 506
Bartoli V, 492
Bartter FC, 289, 492, 501, 508, 560, 565
Bastian G, 523
Bastow RM, 198
Batejat D, 492
Bateman JRM, 492
Batlle D, 511

Battig B, 563
Battisti P, 559
Baturin DA, 493
Baudet MF, 494
Bauer VK, 134
Baulk SK, 419
Baulmann J, 563
Baxter C, 416, 423
Baxter W, 195
Baylies MK, 194
Bazin H, 369, 371, 421
Bazin R, 291
Bean JT, 505
Beard CM, 492
Beard KH, 236
Beare AN, 421
Beauchaud M, 59, 601
Beaumont M, 492
Beaute-Lafitte A, 494, 507
Beck L, 600
Bee V, 558
Beersma DG, 135, 510
Beevers DG, 523
Bégout Anras M-L, 416
Behrens S, 511
Behrmann G, 419
Bélanger PM, 492, 506
Beliaev SD, 558
Beling I, 103
Bellamy N, 492, 493, 558
Bell-Pedersen D, 62, 194, 202
Benavides M, 507, 508, 511
Ben-David Y, 194, 285, 367, 493
Beniashvili DS, 493
Benjamin S, 493
Bennett DA, 498
Bennett J, 103
Benoit O, 560
Benraad TJ, 291
Bentley AE, 492
Bentley KS, 289, 508
Benton D, 424
Benton LA, 367
Benzer S, 60, 200, 289
Beppu T, 202
Berard E, 558
Berestka J, 511
Berg H, 501
Berga SL, 289, 508, 513

Berge C, 371, 420
Berger PJ, 285, 290, 291
Berger S, 105
Bergin ME, 231
Bergquist JC, 103
Berlin R, 373
Berman M, 516
Bernard D, 291
Berners DJ, 333
Bernert RA, 508
Berrettini W, 133
Berry S, 513
Berson DM, 57, 59, 61, 63, 136
Bertheault-Cvitkovic F, 499, 500
Berthold P, 301, 333
Bertolucci C, 373
Bertram L, 285
Bertrand P, 233
Bertrand WE, 425
Betz A, 194
Beuscart R, 291
Bewsher C, 509
Bhattacharjee C, 232
Bibergeil H, 518
Bicakova-Rocher A, 423, 558, 565
Biebach H, 492, 558
Biersner RJ, 421
Bigoni M, 420
Billy BD, 421
Bingham C, 103, 104, 289, 512, 559, 563, 566, 600
Binkley S, 15, 291, 367
Binkley SA, 356
Binswanger B, 561, 563
Bioulac B, 567
Bird T, 505
Bird TJ, 288
Birketvedt G, 420
Birkhauser M, 290
Bisdee JT, 493
Bishop AL, 517
Bishop J, 418
Bismuth H, 499, 507, 508
Bissery MC, 500
Bisson RU, 560
Biti G, 492
Bitman J, 15, 333, 367, 371
Bitossi R, 522
Bittman EL, 371

Bittner JJ, 104, 335, 501
Bixby EK, 561
Bjarnason G, 496
Bjarnason GA, 194, 285, 367, 493, 504, 513
Bjerknes R, 291, 519
Bjerner B, 416
Bjorndal KA, 339
Black A, 367
Black HR, 524, 525
Blacky AR, 504
Blaer Y, 516
Blake G, 57
Blake MJF, 558
Blake W, 416,
Blanc M, 59, 601
Blandford RL, 523
Blask DE, 58, 493, 496
Blau J, 194, 195, 199, 203
Blazsek I, 508
Blei AT, 495
Bleuzen P, 511
Blickenstorfer H, 567
Block KI, 493
Blom DHJ, 422, 564
Blom JMC, 459, 460, 512
Blomquist CH, 285
Blood ML, 134, 517
Bloom M, 416
Bloomfield E, 602
Blosser TH, 369
Blüher M, 493
Bluming AZ, 504
Blumsohn A, 506
Blunt W, 103
Boari B, 420, 499, 509
Bobrie G, 558, 567
Bochnik HJ, 493
Boden BP, 103
Boeing WJ, 103
Bogaty P, 493
Bogdan A, 511
Bohlen JG, 565
Bohn G, 232
Bohn TM, 501
Boiteux A, 198
Boivin DB, 416, 493, 495
Boles MA, 521
Bolliet V, 416

Bonadona JFDe, 497
Bondjers G, 498
Bonner JJ, 194
Bonnet B, 17, 61, 290
Bonvon MV, 17
Bonzon M, 196
Boomsma DI, 567
Boon ME, 516
Boorstin DJ, 103
Borch RF, 372
Borer K, 502
Borges FNS, 418
Borgese MB, 205
Born J, 493
Borner MM, 493
Borsarini A, 416
Borthwick HA, 135
Bosch X, 524
Boselli E, 497
Bossone E, 509
Boston RC, 517
Botbol M, 515
Bouchard A, 524
Boudette NE, 333
Boughattas NA, 508, 511
Bouhuys AL, 135, 510
Boujard T, 416
Boulos Z, 57
Bounias M, 234
Bourdallé-Badie C, 493
Bourdeleau P, 565
Bourdon L, 600
Bourdon R, 18
Boutte P, 558
Bouwes Bavinck JN, 523
Bowcock AM, 204
Boyazoglu PA, 368
Brace RA, 367
Brackney EL, 335, 501
Bradfield CA, 195
Bradley P, 232
Brainard GC, 416, 420, 493, 496
Brambl R, 202
Bramen R, 602
Brandes C, 601
Brandt WH, 285
Braunig P, 564
Braunroth E, 137, 236
Breasted JH, 103

Bremner FW, 290, 510, 518
Bremner WF, 493, 494, 524
Bremnern F, 289, 371, 505
Brener J, 518
Brennan PJ, 494
Bressolle F, 494
Breur GJ, 371
Breus T, 495
Breus TK, 103, 559
Brewer RL, 288
Brewerton T, 133
Brewerton TD, 132
Brezinski DA, 494
Bridges EJ, 494
Brienza S, 499
Briggs TS, 195
Briggs WR, 203, 234
Brink P, 285
Brito AC, 368, 499
Brizzi MP, 522
Brock B, 63
Brodie KH, 286, 497
Brody S, 195, 200
Broege PA, 559
Brommage R, 354, 370
Bronson FH, 367
Brook RD, 567
Brooke OG, 559, 560
Broughton RJ, 564
Brower L, 333
Brown AC, 494
Brown BW, 501
Brown C, 135, 512
Brown EN, 559
Brown F, 195
Brown FA Jr, 103, 232
Brown IR, 195
Brown MB, 602
Brown MH, 289, 508
Brown RL, 57
Brown SA, 57, 204, 518
Brown WR, 285
Browne RC, 416
Bruce TA, 560
Bruce VA, 57, 62, 105, 202
Bruggemann T, 492
Bruguerolle B, 15, 493, 494, 495, 498
Brunner M, 201
Brunning R, 418

Brunning RD, 564
Brush MG, 494
Bryant MJ, 136, 517
Bryson RW, 285
Bryssine B, 497
Bubenik GA, 132, 136
Bucala R, 514
Buchanan BW, 333
Buchanan WW, 492, 493, 558
Buchi KN, 494, 525
Buchsbaum AM, 195
Buck J, 416
Buckland RW, 333
Buckley J, 501
Buckley NA, 232
Buckley P, 424
Buela-Casal G, 567
Bugat R, 522
Buguet A, 600
Buhr ED, 65, 375
Buijs RM, 424, 513
Buisson B, 59, 601
Bulgaris C, 133
Bullough JD, 418
Bunger MK, 195
Bünning E, 15, 57, 58, 59, 195, 232, 285
Burau K, 422
Burger R, 567
Burgess HJ, 494
Burgin RE, 425
Burgoon PW, 60, 416
Burgoyne RD, 195
Buricchi L, 565
Burioka N, 530, 559, 561, 566
Burki F, 507
Burns ER, 422, 501
Burns TA, 367
Burt AD, 284
Burton JL, 285
Busch LA, 494
Bush DE, 292
Bush R, 510
Bushell WN, 602
Bushnell DL, 289, 370, 505, 524
Buss D, 423
Busse H-G, 195
Busse W, 510
Butel JS, 508
Butler LJ, 509

Butte M, 495
Buxtorf JC, 494
Buysse DJ, 421, 559, 564
Byers SW, 368
Byrne 13, 416

Cabera J, 495
Cabibbo A, 194
Cadotte L, 421
Cadotte LM, 369
Cagnacci A, 285
Cagnoni M, 501
Cai MG, 565
Caillard V, 494, 507
Cajochen C, 417, 420
Calaban M, 205
Caldas MC, 367
Callard D, 416, 494
Calo G, 499
Calvo C, 503
Camagna A, 501
Cambar J, 373, 493
Cambie G, 494, 507
Cambrosio A, 58
Camello PJ, 368
Campbell CJ, 285
Campbell IT, 559
Campbell J, 492, 493, 558
Campbell S, 517
Campbell SS, 132, 495
Campos da Paz A, 292, 520
Canal P, 522
Canaple L, 495
Cane MA, 333
Caniato R, 416
Cano JP, 498
Canon C, 507
Cantaluppi A, 16
Cantiani M-G, 236
Cao S, 201
Caola G, 373
Capani F, 424, 518
Capanna M, 514
Capildeo R, 501
Caplan SR, 197
Capodaglio PF, 496
Cappato R, 499
Cappelletti EM, 416
Caputo KP, 292

Carandente F, 59, 501, 561, 600
Carandente O, 600
Carbone F, 419
Cardemil L, 292
Cardno P, 500
Cardoso S, 16,
Cardoso SS, 495, 502, 518
Carlidagger P, 420
Carlson GL, 235
Carlson L, 335, 336
Caro JF, 424, 519
Carpenter CJ, 135, 512
Carpenter JE, 337
Carré IA, 195, 201, 204
Carrier J, 416, 421, 559
Carrillo E, 368
Carskadon MA, 416, 421
Carswell F, 562
Carter J, 561
Cartier A, 567
Carton M, 522
Carvalho S, 424
Casale G, 495
Casale R, 513, 559
Casciano DA, 202, 368, 417
Case AM, 495
Cashmore AR, 103, 132
Caspari EW, 333
Caspers H, 232
Casselman AL, 334
Cassone VM, 417, 495
Castrillon DH, 291
Catchpole A, 333
Cauchy RT, 58, 493, 496
Caussanel JP, 507
Cavallini AR, 514
Cawley BM, 335, 369
Ceccotti P, 496
Celec P, 285
Cella LK, 417, 424
Centola GM, 286
Cermak J, 333
Cermakian N, 495
Chabaud A, 494, 507
Chabre M, 195
Chadee DD, 232
Chai H, 565
Chaitchik S, 518
Chamontin B, 563

Champagne J, 516
Chan CT, 509
Chance B, 15, 195
Chang AM, 194, 199
Chang HS, 198
Chang JG, 495
Chanudet X, 563
Chaouat D, 18
Charrmont AJ, 423
Chassard D, 495, 497
Chatellier G, 558, 567
Chatterjee B, 497
Chaudary MA, 492
Chauffard F, 492
Chauffournier JM, 289
Chaussade M, 564
Chaves I, 105
Chavez EA, 201
Chay TR, 195
Chaykin S, 286, 495
Chemineau P, 368
Chen J-P, 58
Chen S-CG, 339
Chen ST, 495
Chen W, 523
Chen Y, 204
Chen YR, 369
Chenault B, 508
Chenault CB, 289
Cheng P, 195, 198
Chernavskii DS, 195
Cherrah Y, 512
Chertow GM, 506
Cherukuri S, 500
Chevalier V, 496
Chevelle C, 508
Chiba Y, 15
Chipponi J, 496
Chodos D, 497
Chollet P, 496, 500, 508
Chonan K, 562
Choo KB, 495
Cho-Park PF, 194, 493
Chou H-F, 368
Chou J, 197
Chou TC, 58
Chow CS, 232
Christensen JM, 499
Christy JH, 231

Chu C, 495
Chu N-M, 509
Chu P, 495
Chua NH, 200, 201
Chueh PJ, 202
Cieslak D, 505
Cinelli P, 560
Cirillo VP, 196
Citri Y, 206
Claessens N, 503
Clair NP, 286
Clark DH, 372
Clark NL, 367
Clarke SW, 492
Clarke TJ, 562
Claustrat B, 286, 498, 511, 523, 600
Clayton RK, 132
Clench J, 18
Clendenin C, 195
Clerson P, 558, 567
Clopton PL, 601
Coburn JW, 369
Cochran WW, 333
Coen G, 496
Coenen G, 132
Coffeen T, 233
Cogliati AA, 559
Cohen MC, 495
Cohen P, 495
Cohn C, 4 417
Colantonio D, 513, 559
Colditz GA, 518
Coldwells A, 558
Cole CL, 333, 368
Coleman RA, 290
Coleman RJ, 374
Coleman RM, 417
Collins JC, 560
Collins JE, 559
Cologne Light Symposium, 495
Colot H, 196
Colquhoun P, 517, 559
Colquhoun WP, 417 495
Colson FH, 103
Cometti G, 418
Compagnon P, 523
Condon R, 559
Connolly MS, 422
Connors ES, 494

Conroy RT, 417
Constantinou CE, 374
Contor CR, 333
Cook EF, 517
Cooke HG, 286
Cooke S, 103
Cooke WJ, 422
Cookson ST, 564
Cooper LS, 495
Cooreman W, 509
Copeland EM, 509
Copeland J, 421
Coppola E, 509
Corden S, 204
Cordes P, 500
Córdoba J, 495
Corker CS, 286, 287
Cornélissen G, 16, 59, 103, 104, 287, 333, 334, 368, 495, 496, 501, 502, 512, 518, 559, 560, 561, 563, 566, 600
Cornélissen-Guillaume G, 600
Cortellessa MC, 509
Cosgriff JP, 135
Cosgrove JW, 195
Costa-Neto JB, 367
Coste O, 492, 567
Cosyns P, 509
Coté CG, 200
Coto-Montes A, 198
Cotte JP, 373
Couderchet M, 333
Coudert B, 496, 500
Coupland G, 204
Courboulay V, 373
Cournoyer C, 289, 370, 505
Court JM, 368
Covelo M, 503
Cowan WM, 59, 370
Coyle PV, 496
Crafts R, 286
Craviotto C, 496
Crawford JM, 559
Crawford VL, 496
Crea F, 507
Creinin MD, 286
Crespi V, 498
Cressey D, 418, 419
Crews D, 286
Crews ST, 177, 195, 286

Crockford G, 500
Crosby TJ, 505
Cross KW, 498
Crosthwaite SK, 194, 195
Croze M, 497
Cucherat M, 600
Cuffari MA, 513
Cugini P, 449, 496, 559, 602
Cumming BG, 132, 286
Cummings FW, 195
Cummings P, 420
Cunliffe WJ, 285
Curé H, 496
Curtis AM, 517
Cussler K, 369
Cusulos M, 502
Cutkomp L, 335, 369
Cutkomp LK, 15, 16, 58, 334
Cutler NL, 134, 601
Cutler WB, 103, 232, 286, 290
Cutolo M, 496
Cyran SA, 195
Czarnocki Z, 423, 516
Czeisler CA, 416, 417, 418, 419, 420, 424, 494, 524, 559, 560
Czekala N, 375
Czekala NM, 286
Czerniak A, 518

D'Alonzo GE, 494, 496, 520
D'Attino RM, 500
D'Hondt P, 509
Daan S, 203, 517
Dabbs JM Jr, 286, 496
Dahl KD, 286
Dahlbom M, 367
Dahlitz M, 510
Dalton KD, 496
Damiola F, 202
Damm F, 516
Danforth DN Jr, 522
Dangerfield P, 517
Danilenko KV, 417
Darwin C, 58
Darwin F, 58
Dauchy R, 493
Dauncey MJ, 370
Davenne D, 416, 418, 424, 494, 496
Davenport Y, 136, 517

Davey P, 513
Davidenko JM, 195
David-Gray Z, 58, 133
Davidson TL, 368
Davies B, 496
Davies P, 135, 500
Davis C, 560
Davis FC, 62, 97, 373, 417
Davis RB, 18
Davis RH, 286
Davis RL, 499
Davis S, 63, 496, 522
Davis SJ, 198
Davy J, 560
Dawes C, 368, 496
Dawson AH, 232
Dawson D, 524
Dawson S, 289, 371, 494, 505
Dawut L, 137
Day PR, 287
de Boer JM, 137
De Buyzere M, 566
De Candolle AP, 103
de Castro JM, 232
De Feudis F, 602
de Fijter JW, 523
de Fini, 284
De France O, 16
De Gier J, 198
De Greef JA, 18, 63, 105, 235, 292
de Leeuw P, 562, 564
de Leonardis V, 560
De Mairan J, 103
de Marsh RW, 509, 513
De Meyer F, 509, 560
De Moraes CM, 333
de Nicola P, 495
De Prins J, 565, 600
De Scalzi M, 560
De Solla Price DJ, 103
de Souza CM, 562
de Toma D, 499
de Waard ER, 602
Deacon S, 415, 519
Deacon SJ, 417
Deardon DJ, 235
DeBartolo M, 505
Debon R, 497
Dechaud H, 286

Deck J, 291, 422
Decker M, 132, 284
Decoster C, 523
DeCoursey PJ, 16, 231, 232
DeCousus H, 512
Decousus HA, 497
Deed JR, 372
Deedwania PC, 499
degli Uberti EC, 514
DeGrazia JM, 564
DeGrip W, 58
DeGrip WJ, 62, 133
Deitzer GF, 205
Deka AC, 497, 501
Del Pozo F, 602
Delahaye-Brown AM, 197
Delchev NN, 563
Delea C, 501, 508
Delea CS, 289, 492, 565
Delgadillo JA, 368
Della Corte M, 290, 514
Delmore P, 501, 600
Delongchamp RR, 368
Delyukov A, 501
Demakis JG, 505
Demartini G, 422, 514
Demas GE, 512
Demey-Ponsard R, 507
Denault D, 196
Denault DL, 196
Denis J, 492
Denis JB, 498
Denny R, 336
Denome SA, 201
Deprés-Brummer P, 523
DePrins J, 600
DeRoshia CW, 425, 567
Derr J, 511
Deshler J, 199
Desir D, 523
DeVecchi A, 16
Devlin PF, 136
Devoto L, 285
Dewan EM, 286, 497
Dewhurst K, 286
Deyanov CE, 567
Di Chiara V, 514
Di Marzo A, 496
Di Milia L, 560

Di Palma L, 496, 559
di Paola ED, 496
Diddams SA, 103
Dieppe PA, 524
Diez-Noguera A, 516
Dijk DJ, 419, 560
DiPalma L, 602
DiSanto A, 497
Distante A, 509
Dittrich H, 504
Ditty JL, 196
Doalto L, 498
Dobra KW, 58, 196
Docherty JP, 136, 517
Dodson S, 334
Doering CH, 286, 497
Dogliotti L, 497, 508, 522
Doherty S, 137
Doireau P, 492
Dolci D, 499
Dominguez BW, 525
Dominguez MJ, 503
Dominique S, 523
Domoslawksi J, 600
Donahoo WT, 422
Donahue M, 103
Donati L, 501
Done J, 339
Donkin EF, 368
Donselaar W, 525
Doran DL, 334
Dorfman P, 15
Dorigo B, 492
Dörrscheidt GJ, 600
Dotan A, 291
Douglas AS, 497
Douglas RH, 59
Douylliez C, 15
Dove WF, 199, 205
Dowell SF, 497
Down A, 423
Dowse HB, 60, 601
Dowsett KF, 368
Doyle MR, 198
Doyle N, 493
Dragovic Z, 201
Drake DJ, 16, 334
Drance SM, 497
Drash A, 368

Dravigny C, 565
Dray F, 289
Drayer JIM, 501
Drayer JM, 561
Dresse MF, 516
Driver HS, 285
Drugan A, 418
Drummond RO, 368
Drust B, 417
Du F, 204
du Preez ER, 368
Duarte G, 368
Dubbels R, 497
Dubey DP, 369, 497
Dubinsky C, 506
Duboule D, 202
Duburque B, 564
Duffour J, 494
Duffy J, 560
Duffy JF, 416, 559, 560, 562
Duffy PH, 417
Duflo F, 497
Duhamel Du Monceau HL, 103
Duke GE, 368
Duke SH, 16, 17, 289, 334, 335, 336
Dumont M, 133, 503, 559
Dunberg A, 338
Duncan WC Jr, 64, 497
Dunea G, 232
Dunlap J, 196
Dunlap JC, 62, 64, 104, 194, 195, 196, 197, 200, 201, 202, 285, 336
Dunn FA, 57
Dunnigan MG, 497
Dunson D, 292
Dunson DB, 286
Dupont J, 18, 515
Dupont M, 494
Durgan M, 17, 337
Dusek J, 501, 561, 566
Dusing R, 567
Dutta TK, 497
Dvornyk V, 103
Dyrenfurth I, 286, 497

Eagle KA, 509
Earl CR, 419
Eastell R, 506
Eastman CI, 494

Eastwood MR, 132, 133, 497
Ebata T, 498
Eberly S, 286
Ebling FJ, 58, 286
Echizenya M, 200, 506
Eckert ED, 567
Eder J, 420
Edery I, 201, 204
Edmonds D, 561
Edmunds GF Jr, 286
Edmunds LN Jr, 59, 196, 291, 374
Edwards E, 417, 424
Edwards SW, 17, 60,
Eesa N, 16, 334, 368
Efron B, 424
Ehlers C, 497
Ehrenkranz JR, 560
Ehrenkranz JRL, 133
Ehret CF, 16, 58, 196, 286, 368, 417
Ehret DL, 334
Ehrly AM, 498
Eichele G, 204
Eichelmann C, 58
Eichler VB, 61, 372
Eichner ER, 417
Eide EJ, 523
Eigenberg RA, 369
Eiken HG, 205, 523
Eisenberg FF, 523
Eisenberg S, 502
Ekbom KA, 498
Ekelund LG, 500
Ekström P, 368
el Sanadi N, 492
Elfving DC, 287, 334
Elimam A, 417
Elizar A, 416
Ellinwood WE, 373
Elliot J, 61, 135
Elliot JA, 136, 204
Elliot JM, 334
Elliot WJ, 498
Elveback LR, 492
Emens JS, 508
Emerson M, 496
Emery P, 103, 196
Engel P, 560
Engel R, 287, 566
Engeli M, 287, 419, 501

Engelmann W, 58, 63, 196, 199
Engler R, 504
Englund CE, 414, 421
Enright JT, 58, 232
Enright JTE, 600
Epstein JT, 513
Erickson RS, 560
Eriguchi M, 498
Erikson L, 367
Eriksson ME, 196
Erkert HG, 232, 288
Erlich H, 104
Ernst R, 368
Ernst RA, 368
Erny P, 493
Erren TC, 417, 498
Erritzoe J, 372
Erwin J, 334
Erzen J, 235
Eshel G, 333
Eskin A, 61
Espagnet S, 373
Estabrook RW, 15, 196
Evans G, 375
Evans JW, 16, 334, 368
Evans LT, 58, 133, 137, 286
Evanson OA, 368
Exley D, 286, 287

Fabiano FS, 560
Fabiano N, 505
Fabsitz R, 418
Fager G, 498
Faggiuolo R, 500, 508
Fahrenkrug J, 59
Faiman C, 287
Fairhurst S, 417
Fakouhi TD, 499
Falcón J, 375
Falgairette G, 287
Falliers C, 565
Faloona F, 104
Fan C-M, 195
Farber MS, 374
Farner DS, 367
Farré EM, 196
Farsund T, 504
Faure A, 600
Fayomi A, 367

Fazio F, 373
Fehm HL, 493
Feigin VL, 498
Feijo JA, 288
Feinleib M, 418
Feissner JM, 509
Feldman JF, 58, 196
Feldman SR, 502
Felici W, 559
Felver L, 560
Fentiman I, 498
Fentiman IS, 492, 495, 498, 510
Fenton S, 509
Ferin M, 286
Fernandes G, 502
Fernandez JR, 498, 503, 601
Ferrario VF, 498
Fersini C, 499, 509
Fery F, 523
Fetzer JL, 336
Feuer G, 506
Feuers RJ, 202, 368, 417, 518
Fewell JE, 368
Feychting M, 498
Fidanza F, 287
Field MD, 515
Figala J, 64, 334, 339
Figueiro MG, 418
Filer RB, 287
Filippini R, 416
Filipski E, 498, 500
Fincham JRS, 256, 287
Fingerman M, 232, 233, 334
Fink H, 497
Finn L, 562
Fiore CE, 513
Fireman FA, 368, 499
Firth RG, 423
Fischer D, 58
Fischer FM, 418
Fischer-Barnicol D, 506
Fiser B, 501, 561, 566
Fishbein SJ, 371
Fisher C, 287
Fisher KC, 17
Fisher LB, 287
Fisher RS, 424
Fiske S, 58
Fitzgerald GA, 517

Fleck A, 291
Fleegler FM, 368
Fleischer AB Jr, 502
Fleming DM, 498
Fleming KA, 601
Fletcher C, 521, 566
Fleury JJ, 367
Flisinska-Bojanowska A, 371
Floyd RD, 337
Foa A, 373
Focan C, 496, 498, 499, 500, 508
Focan-Henrard D, 498
Foldes A, 373
Foley A, 525
Folk GE, 57
Folkard S, 418, 424, 499, 510, 560, 565, 567
Follett BK, 202
Fontana S, 496
Fontes G, 368, 499
Forbes EB, 334
Forcada F, 375
Forchhammer MC, 334
Forel A, 104
Foret J, 560
Fornasari PM, 499
Forpomés O, 373
Forsyth JJ, 418
Fort A, 601
Forusz H, 516
Foss Abrahamsen J, 205, 523
Foster DL, 371
Foster JE, 524
Foster R, 58, 133
Foster RG, 58, 59, 61, 62, 133, 134, 196, 373
Foster WC, 425
Foulley JL, 507
Fox AW, 499
Fox BA, 201
Fox CA, 287
Fox RH, 559, 560
Fraboulet G, 564
Franchi L, 201
Francis CD, 197
Francoeur D, 291, 422
Frangos E, 133
Frank KD, 334
Franke H-D, 233

Franken P, 60
Franken RJ, 204
Fraser C, 287, 500
Frashini F, 422, 514
Freedman MS, 58, 133
Freedman RR, 287
Freedman SH, 499
Freeman L, 197, 287
Freifelder D, 104
Freitag WO, 417
French J, 492, 560
Friedler RM, 369
Friedman E, 286
Friedrich JW, 334, 509
Friend DJC, 334
Frishman WH, 499
Froberg JE, 499
Froehlich A, 61
Froehlich AC, 196, 197
Fromentin JM, 334
Fromm RE Jr, 418
Frosch S, 205
Froy O, 334
Fry AJ, 560
Fu L, 499
Fu YH, 523
Fuentes-Arderiu X, 516, 601
Fuhrberg B, 419
Fuhrman M, 233
Fujimaki Y, 499
Fujimori T, 197
Fujimoto K, 198
Fujishima M, 518
Fujiwara S, 560
Fukiyama K, 562
Fukuhara C, 205
Fukuzawa T, 518
Fuller CA, 61, 104, 201, 337, 345, 369, 371, 510
Furman I, 285
Furuya M, 200
Fuster V, 492
Futscher B, 505
Fuzeau-Braesch S, 338, 373
Fyfe T, 497

Gachihi GS, 497
Gaddy J, 510
Gagnon FA, 507

Gagnon R, 507
Gaillard AW, 423
Gaist PA, 134, 506
Galanos C, 514
Gale HG, 234
Gallagher SS, 419
Gallepp G, 58
Gallerani M, 420, 499, 509
Galliot M, 18
Galston AW, 203
Gama R, 511
Gamage K, 370
Gamba G, 499
Gamus D, 499
Gandy S, 524
Garceau N, 197
Garcia CR, 103, 290
Garcia L, 503
Garcia Sainz M, 501
Garcia-Fernandez JM, 133
Garde AH, 499
Gardner HF, 199
Gardner KH, 198
Gardner PD, 285, 291
Garlick PJ, 493
Garner WW, 133, 334
Garnett GP, 287, 500
Garufi C, 508
Gassett JW, 373
Gastiaburu J, 508
Gatika SM, 499
Gatti RA, 565
Gatty R, 418
Gaul GB, 500
Gaultier C, 499
Gaultier M, 515
Gautherie M, 565
Gauthier A, 418, 424, 494, 496
Gavaghan D, 89
Gayowski T, 519
Gegout-Pottie P, 369
Gehrels T, 233
Gehring W, 104
Gehrke CW, 369
Gekakis N, 197
Gelfand EW, 499
Geliebter A, 418
Genes N, 558, 567
Geng R, 199

Georg B, 59
George JE, 368
Gerisch G, 197
Germano G, 562
Gerner E, 416
Gerrits RJ, 369
Gershengorn HB, 562
Gervais A, 18
Gervais C, 16
Gervais P, 16, 18, 515, 564
Ghata J, 18, 291, 423, 515
Ghezzo H, 567
Ghosh AK, 15
Giacchetti S, 499, 500, 511
Giacomoni M, 287
Giacovazzo M, 496
Giagulli VA, 284
Gianotti L, 496
Gibeau JM, 233, 512
Gibiino S, 513
Gibson RN, 500
Giedke H, 15, 514
Giese AC, 16, 233, 235
Giesen HA, 524
Gilad R, 416
Giles R, 453
Gill J, 369, 371
Gill JS, 500, 523
Gill SK, 500
Gillanders SW, 197
Gillet P, 369
Gillett CE, 495
Gilli P, 514
Gillin JC, 136, 513, 517
Gilpin EA, 504
Ginsburg H, 494, 507
Ginther OJ, 369, 371
Giorgino R, 284
Giovannini C, 496
Girard F, 499
Girelli I, 559
Gittelson B, 418
Giudice E, 373
Giwercman A, 287
Gjessing L, 291
Gjessing R, 287
Glasgow DR, 560
Glass L, 197, 199
Glasser S, 499

Glasser SP, 500
Glezen WP, 507
Glickman G, 416
Gliessman P, 373
Glossop NR, 195
Glova GJ, 338
Glover TD, 368
Gluck ME, 418
Glynn C, 500
Glynn CJ, 499, 560
Gobel H, 500
Godfrey R, 424
Goebel A, 497
Goehring L, 334
Goel N, 418
Goettsch WG, 523
Goetz F, 419, 501
Goetz FC, 418, 564
Goft M, 338
Goh CL, 525
Gold AR, 284
Goldberg RJ, 525
Goldbeter A, 197
Golden SS, 105, 196, 197, 199, 200, 204
Goldman B, 513
Goldman L, 517
Goldsmith GW, 335
Goldstein I, 287
Goldstick O, 418
Gomeni R, 494
Gomes WR, 335, 369
Gooch VD, 197, 287
Good RA, 369, 521, 565
Goodman RL, 371
Goodwin BC, 197
Goodwin FK, 136, 508, 517
Goodwin SF, 291
Gooley JJ, 58
Goon A, 525
Gorbacheva VY, 488, 500
Gorceix A, 558
Gordon C, 369
Gordon HW, 561
Gordon WR, 58
Gordon DJ, 500
Gordon PC, 500
Gordy J, 505
Gorgo Y, 501
Gorl M, 201

Gorman MR, 197
Goss RJ, 16, 133
Goto E, 562
Goto M, 64
Gotter AL, 334
Goudevenos JA, 500
Goudsmit EM, 494
Graber RR, 133
Graeber RC, 418
Grafe A, 559, 566
Grage TB, 521
Gragnani A, 204
Graham CA, 287, 418
Graham D, 509
Graham L, 505
Granda TG, 498, 500
Grandi E, 499
Grassly NC, 287, 500
Grasso F, 373
Gratton L, 499
Greco J, 289, 370, 505, 524
Green CB, 500
Green J, 418
Green MS, 506
Greenberg B, 418
Greenberg G, 494
Greene AV, 202
Greep RO, 372
Greeson JM, 416
Gregg I, 506, 561
Gregory W, 498
Gregory WM, 492, 510
Griffith JS, 333, 351
Griffith MK, 369
Griffiths K, 501, 559, 565
Grigoriev AE, 103, 559
Grill JT, 419
Grimaudo V, 500
Grisaru N, 136, 424
Grobbelaar N, 197
Groh KR, 368
Gronfier C, 418
Groot AS, 567
Gross G, 564
Gross J, 287
Gross JD, 197, 410
Grossman AB, 514
Gruber SA, 374, 504, 521
Gruen W, 565

Gruia G, 499
Grune S, 561, 567
Gruner R, 61
Grunwald GK, 425
Gruska M, 500
Grutsch JF, 508
Guberan E, 500
Gubin D, 495, 501
Gudkov AV, 500
Guerin MV, 372
Guevara A, 286
Guillaume FM, 16, 58, 335
Guilleminault C, 421, 424
Guillon D, 233
Guillon P, 233
Guingamp C, 369
Guinot C, 507
Gulline HF, 339
Gullion GW, 16
Gully RJ, 335, 501
Gunaydin B, 504
Gundel A, 419
Gunsolus JL, 17, 337
Günther R, 289, 506
Gupta BD, 497, 501
Gupta S, 561
Guthmann H, 233
Guyer B, 419
Guyot R, 497
Gwinner E, 16, 133, 333, 335
Gykkenborg J, 287

Haberman S, 501
Hack LM, 501
Hackett GR, 288
Hadfield C, 203
Haest CWM, 198
Hafenrichter AL, 335
Hafs HD, 373
Hagen E, 333
Hahn GL, 369, 372
Hahn PM, 510
Hahn RA, 501
Haikonen PM, 419
Haim RA, 59
Hajdukovic RM, 510
Halaban R, 59, 335
Halaris A, 564
Halawani ME, 372

Halberg E, 104, 105, 419, 500, 5015, 561, 565
Halberg F, 15, 16, 17, 18, 58, 59, 62, 103, 104, 105, 233, 235, 236, 287, 288, 289, 291, 334, 335, 336, 368, 369, 370, 371, 372, 374, 418, 419, 421, 424, 492, 495, 496, 497, 499, 500, 501, 502, 503, 504, 505, 506, 508, 510, 512, 515, 518, 519, 520, 521, 524, 559, 560, 561, 563, 564, 565, 566, 567 600, 601, 602
Halberg Fcn, 16, 233, 501, 502, 561, 566
Halberg J, 59, 104, 335, 369, 501, 502, 561, 566
Halderson JL, 339
Halhuber M, 501
Hall A, 196, 198, 205
Hall EF, 560
Hall J, 287
Hall JC, 17, 60, 103, 196, 198, 200, 203, 206, 288, 289, 291, 601
Hallek M, 291, 510
Halliday IM, 496
Halliday KJ, 198
Hallin TA, 337
Hamblen M, 206
Hamburg DA, 13, 286, 497
Hamburger C, 287, 419, 501
Hamburger SD, 522
Hameed S, 375
Hamilton JB, 288
Hamm H, 375
Hammann S, 236
Hamner KC, 133, 135, 198, 288, 292, 335
Hamovit JR, 136, 517
Han B, 198
Hanano S, 196, 198
Hancox JG, 502
Handler AM, 59
Haney JF, 335
Hanifin JP, 416
Hankins MW, 59, 133
Hannan PJ, 500
Hannibal J, 59
Hanny BW, 337
Hansen K, 493
Hansen AM, 499
Hansen V, 133
Hanson CA, 425
Hanson FE, 419

Hany S, 563
Hara M, 288, 370, 502
Harari G, 506
Hardeland R, 15, 198, 416, 419, 522
Harder JD, 288
Hardin PE, 195, 198, 201, 288
Hardin TA, 132, 133
Hardtland-Rowe R, 233
Hare EH, 133
Harkness RA, 502
Harkness RS, 288
Harland WA, 497
Harlap S, 502
Harley CB, 198
Harmer SL, 194, 196, 198
Harmon FG, 196
Harms E, 198
Harris PHP, 514
Harris W, 419
Harrison LC, 513
Hart RW, 417
Hartenbower DL, 369
Harter L, 288
Hartinger J, 369
Hartmann WK, 59
Harvey J, 560
Harvill EK, 333
Hashimoto S, 504, 523
Hasler BP, 508
Hassan AA, 369
Hassnaoui M, 59, 601
Hastings HM, 519
Hastings JW, 59, 63, 64, 202, 205
Hastings MH, 59, 105, 200, 498, 502, 515
Hastings W, 195
Hasuo Y, 518
Hattan DG, 417
Hattar S, 59, 61
Hattori A, 288, 502
Hauert J, 500
Hauri PJ, 517
Haus E, 16, 103, 104, 288, 289, 339, 369, 370, 371, 375, 418, 419, 420, 421, 423, 497, 501, 502, 505, 506, 512, 515, 518, 520, 521, 522, 523, 525, 559, 561, 563, 565, 566
Haus M, 371, 420
Haus R, 521
Hauty GT, 425

Hawk E, 369, 371, 372, 374
Hawking F, 370, 503
Hay M, 373
Hayasaki M, 370
Hayashi H, 562
Hayes DK, 335, 369
Hayes RM, 370
Haywood KM, 419
Hazel JR, 198
Hazlerigg D, 200
He Q, 198
He Y, 523
Healey FB, 560
Heath AC, 509
Heath HW, 374
Hébert M, 133, 423, 503
Hecht BP, 236
Hecht GM, 509
Hecquet B, 498, 511, 600
Hegge FW, 564
Heide OM, 133
Heiman ML, 519
Heintzen C, 62, 195, 202
Heise A, 492, 558
Heitkemper MM, 558
Hejl Z, 503
Hekimi S, 200
Heldmaier G, 375
Helenius H, 517
Hellbrügge T, 561, 562
Heller HC, 60, 291
Hellot MF, 523
Helms K, 335
Helms PJ, 497
Helson VA, 334
Henderson AJ, 562
Hendricks SB, 133, 135, 335
Hendrickson AE, 59, 370
Hendrix DL, 335
Henning H, 504
Henry F, 503
Henson CA, 16, 335
Hentschel B, 493
Hepler PK, 288
Herget M, 506
Hering DW, 16, 104
Herman TS, 524
Hermand E, 291
Hermann D, 514

Hermida Dominguez RC, 503
Hermida R, 601
Hermida RC, 63, 288, 498, 503, 504, 525, 562, 601
Herpin D, 562
Herrick JR, 288
Herrmann H, 559
Herxheimer A, 504
Herzog ED, 203
Hess B, 195, 198
Hess G, 600
Hetherington NW, 374, 517
Hetzel MR, 562
Heyden T, 525
Hibberd V, 198
Hickey DS, 602
Hicks KA, 198
Hidalgo MP, 562
Hildebrandt G, 287, 419
Hiles LG, 375
Hill B, 419
Hill CB, 137
Hill DR, 104
Hill DW, 419
Hill JO, 422, 425
Hilleren M, 419
Hillman D, 287, 419, 501, 565
Hillman WS, 59, 134, 198, 288, 335
Hills R, 336
Hindersson P, 59
Hines MN, 233
Hinz WA, 523
Hinze E, 17, 289
Hirai M, 506
Hirata K, 370
Hirose M, 205
Hirsch EE, 419
Hirt A, 205, 523
His B, 284
His BP, 420
Hisa T, 498
Hitchcock ER, 523
Hiusman HO, 137
Hjalmarson A, 504
Ho LC, 334
Hoag E, 499
Hoban TM, 420
Hobbs JD, 289
Hoch CC, 564

Hockey GRJ, 418
Hodoglugil U, 504
Hoeksel R, 560
Hoenen MM, 61
Hoffman CL, 522
Hoffman H, 136
Hoffman M, 517
Hoffmann K, 59
Hoffmann M, 513
Hoffmann RF, 57
Hoffmann WE, 371
Höfling B, 524
Hofmann F, 59
Hogan D, 511
Hogben LT, 233
Hogenesch JB, 195, 198, 235, 373, 517
Hogewoning AA, 523
Hogg D, 517
Hohmann W, 601
Holaday JW, 16
Holdaway-Clarke TL, 288
Holland GJ, 70
Holley DC, 425, 567
Holliday H, 565
Hollingdal M, 63
Holljes HWD, 523
Hollo G, 562
Holm A, 416
Holmes E, 289, 371, 505
Holmgren M, 198
Holowachuk EW, 493
Holsboer F, 514
Holt CS, 336
Holt Jr JP, 285
Holzmann M, 510
Homolka P, 501, 561
Hong HK, 65, 375
Honinckx E, 523
Honkley D, 512
Honma K, 198, 504, 518, 523
Honma S, 198, 504, 518, 523
Honrado G, 419
Hopkins P, 288, 370
Hoppensteadt FC, 16, 199
Hormann A, 525
Horn G, 104
Horne J, 419
Horne JA, 419, 562
Hornecker JP, 233

Hostmark J, 504
Hotchkiss CE, 370
Hotz Vitaterna M, 137
Hou MF, 495
Houin G, 522
Houseal M, 513
Hoyeux C, 516
Hoyle MN, 58, 196
Hrafnkelsson J, 422
Hriscu M, 561
Hrushesky W, 62, 511
Hrushesky WJ, 194, 285, 372, 508
Hrushesky WJM, 62, 285, 387, 493, 494, 504, 505, 511, 515, 516, 521, 523, 524, 566
Hsi B, 292
Hsi BP, 132
Hsu DS, 137
Hsueh AJ, 286
Huak CY, 525
Huang P, 499
Huang TC, 197
Huber G, 564
Huber SC, 335
Hudson D, 58
Huggins GR, 103, 290
Hughes DA, 233
Huguet G, 419
Hui JL, 515
Hume P, 501
Humeston TR, 18, 514
Humphrey J, 336
Humphreys D, 135
Hunt PD, 324, 419
Hunter DJ, 518
Hunter-Ensor M, 202, 204
Huntington JB, 371
Huq S, 497
Hurtt ME, 289, 508
Huston JP, 367
Hwang DS, 16, 502
Hyde J, 500

Iacobelli S, 508
Iannarelli M, 509
Ibach B, 506
Ibrahim CA, 17
Igishi T, 559
Iglesias A, 139

Iglesias M, 503
Iigo M, 288, 370, 502
Iijima S, 202
Iino M, 523
Ikeda M, 199, 518
Ikemoto H, 201
Ikonomov O, 561
Illnernova H, 420
Illyes M, 564
Imai Y, 562, 564, 566
Imamura Y, 562
Imeson JD, 504
Infante-Rivard C, 516
Ingraham JL, 204
Ingram DL, 370
Inonog S, 201
Inouye SI, 63
Inouye ST, 60, 375
Iranmanesh A, 523
Iseli J, 567
Iser L, 566
Ishii M, 562
Ishiura M, 60, 199, 200
Iskra-Golec I, 562
Ismail AA, 287
Ito A, 512
Ito S, 202
Itoh A, 558
Itoh M, 288, 502
Itzhaki M, 499, 508
Iwasaki H, 106, 199, 205
Iwasaki K, 60, 104, 288
Iwasaki S, 498

Jablonski W, 291
Jacklet JW, 199
Jackson DP, 420
Jackson FR, 194
Jackson JM, 134
Jackson S, 134
Jacob S, 288
Jacobs DR, 500
Jacobsen FM, 134, 136, 504, 517
Jacotot B, 494
Jacquier AC, 206
Jafary J, 516
Jain V, 509
Jalife J, 195
James FO, 194, 493

James GD, 559
James K, 509
James S, 559
James SP, 504
James WPT, 493
Jami A, 511
Jamison CF, 601
Janes JT, 422
Janssen E, 287
Janzen DH, 16, 60
Jarillo JA, 132
Jarjour NN, 504
Jarosz D, 425
Jarrett DB, 564
Jarrett RJ, 500, 504, 525
Jasmin C, 499, 508
Jaubert JC, 497
Jefferts SR, 103
Jellyman DJ, 233
Jenik J, 333
Jenkin G, 288, 370
Jennings JR, 564
Jeon M, 199
Jewelewicz R, 286, 497
Jewett ME, 418, 420
Jiang G, 62
Jilge B, 370
Jimenez J, 602
Jin X, 200
Jinyi W, 104
Jobst K, 288, 292
Johansson AM, 495
Johansson BB, 504
Johnson AL, 368
Johnson CH, 60, 104, 105, 106, 199, 200, 206
Johnson CK, 60
Johnson E, 501
Johnson EA, 561, 601
Johnson J, 201
Johnson KA, 194, 285
Johnson L, 370
Johnson M, 499
Johnson MA, 17, 60, 336
Johnson ML, 52
Johnsson A, 16, 60, 199
Johnstone MT, 505
Jones CB, 422
Jones CR 523

Jones E, 511
Jones RS, 506
Jonkman JHG, 520
Jordan D, 286
Jordan R, 285, 367, 493
Jordan RS, 194
Jores A, 17, 505
Joseph-Vanderpool JR, 524
Jost D, 506
Joulia JM, 494
Jovonovich JA, 521
Jowett NI, 523
Joyce MC, 335, 369
Joyce PR, 135
Juhl CB, 63
Julius S, 564
Jung G, 498
Junges W, 17, 289
Juric X, 510

Kadman-Zahavi A, 134, 233
Kadono H, 336, 370
Kaff A, 500
Kahn S, 289, 370, 505
Kaiser HE, 200, 501
Kaiser IH, 505
Kakizawa T, 495
Kalb C, 288
Kallner B, 423
Kalmus H, 17
Kaloni S, 558
Kalsbeek A, 513
Kam LWG, 367
Kamide R, 498
Kamoi K, 562
Kampa EM, 103
Kanabrocki EL, 288, 289, 290, 370, 494, 501, 505, 510, 518, 521, 524, 525
Kanabrocki JA, 505
Kanai S, 104
Kaneko M, 103, 196
Kaneko S, 562
Kannass MM, 523
Kaplan D, 199
Kaplan E, 289, 370, 505, 524
Karacan I, 289
Karasek M, 505
Kark JD, 502
Karlsen P, 285

Karlsson HG, 199
Karnik JA, 506
Karsch FJ, 371
Kasal CA, 371
Kasper S, 133, 134, 506, 524
Kasper SF, 136
Kass DA, 371
Katayama M, 204
Katayama T, 61
Katinas G, 501, 502
Katinas GS, 59, 495, 561
Kato Y, 198
Katoh T, 200, 506
Katsanou N, 133
Katzenberg D, 562
Kauffman S, 205
Kaufmann MR, 287, 334
Kautsky N, 204
Kauzmann W, 202
Kavaliers M, 419
Kavanagh JJ, 492
Kawachi I, 518
Kawamoto T, 198
Kawamura T, 506
Kawano Y, 562
Kawasaki T, 496, 501
Kay SA, 63, 106, 136, 194, 196, 198, 199, 201, 204, 235, 601
Kay T, 425
Kazantsev A, 137
Keating P, 58
Keatts H, 135
Keen H, 504
Keen MC, 524
Keene KS, 506
Keener JP, 199
Keiding N, 287
Keil U, 525
Keitel WA, 508
Keith LD, 517
Kellendonk C, 57
Keller J, 562
Keller JB, 16
Kellicott WE, 289
Kelly K, 524
Kelly TL, 419
Kelsey C, 565
Kennaway DJ, 375, 419, 506, 524
Kennedy BJ, 372

Kennedy BW, 335, 336
Kennedy KS, 421
Kennet RP, 523
Kerenyi NA, 506
Kerfoot WB, 233
Kerkhof G, 562, 567
Kerkhof GA, 137, 562
Kern R-I, 420
Kerner von Marilaun A, 104, 336
Kerr CM, 423
Kerr HD, 506
Kerr IG, 493
Kessler A, 336
Kettlewell PS, 17, 336
Keutmann HT, 374
Kevei E, 205
Keverline S, 286
Key JL, 200
Khaikin SE, 194
Khalameizer V, 516
Khalil TM, 420
Khalsa SB, 420
Khandheria B, 513
Kharour HH, 235
Khaw KT, 525
Khetagurova LG, 558
Kiddy CA, 371
Kihlstrom JE, 289, 420
Kikuno R, 104
Kikuya M, 562
Kilic E, 516
Kilic U, 516
Kim WY, 199, 201
Kim YS, 502
King DP, 194, 197, 199, 204, 205, 562
King RW, 60, 133, 134, 289
King AI, 336
Kingsmill SF, 337
King VM, 498
Kippert F, 60
Kirkham KE, 287
Kirkland JL, 602
Kirklin SK, 560
Kita M, 202
Kitagawa G, 564
Kitayama Y, 199
Kivimae S, 198
Kiyohara Y, 518
Kiyosawa K, 60

Klangos I, 511
Klante G, 375
Klauber MR, 513
Klebs G, 134
Klein H, 333
Klein R, 562
Klein SL, 512
Kleitman E, 103, 559
Kleitman N, 15, 17, 103, 233, 420, 599
Klem ML, 425
Klemfuss H, 601
Klenke E, 497
Klepfish A, 423
Klerman EB, 562
Klimmer F, 559
Kloppenborg PW, 291
Kloss B, 199, 200, 203
Klotter K, 200
Klotz JH, 336
Kluft C, 491
Knapp E, 289, 506
Knapp M, 563
Knauth P, 559, 562, 565
Kneesel S, 374
Knight M, 60, 104
Knudsen SM, 59
Ko CH, 65, 375
Kobayoshi C, 562
Kobilanski C, 60, 371
Koch HJ, 506
Koehler PG, 371
Kogan J, 509
Koiwaya Y, 491
Kolakowski LF Jr, 204
Kolár J, 420
Kolka MA, 60
Kolodziej-Maciejewka H, 505
Komosa M, 371
Kompanowska-Jezierska E, 369
Kondo H, 506
Kondo T, 60, 104, 105, 199, 200, 204, 205
Kondratov RV, 500
Konopka RJ, 59, 60, 200, 289
Kooistra LH, 371
Kopal Z, 233
Kornhauser JM, 199, 205
Korte J, 506
Korth C, 514
Koskinen E, 367

Kothy P, 562
Koukkari WL, 16, 50, 58, 60, 63, 200, 289, 290, 333, 334, 335, 336, 337, 339, 601, 602
Koulbanis C, 515
Kowanko IC, 563
Kozma-Bognár L, 205
Kraemer HC, 286, 497
Kraenzlin M, 372
Krahn DD, 132
Kramer PM, 497
Krauchi K, 417
Krause JA, 493, 496
Krebs B, 524
Kreitzman L, 196
Kreps JA, 198
Kretschmer H, 105
Kreutz F, 498
Kriauciunas A, 519
Kriebel J, 563
Krieger A, 103
Kriegsfeld LJ, 512
Kripke DF, 17, 289, 506, 508, 517, 518, 558
Kris-Etherton PM, 511
Kristal-Boneh E, 506
Kronauer RE, 416, 417, 418, 562
Kruithof EKO, 500
Krzanowski M, 371
Krzywanek H, 372
Krzyzanowski M, 563
Kubo Y, 518
Kucera H, 420
Kucera J, 333
Kudels M, 285
Kudoh Y, 564
Kühl J, 291
Kuhl JF, 502
Kühl JFW, 16, 289, 502, 506
Kulbs D, 369
Kumagae G, 518
Kumagai Y, 561
Kumar D, 506
Kumar V, 134
Kumazawa S, 201
Kume K, 105, 200
Kunkel JG, 288
Kuno N, 200
Kunstlinger F, 499, 508

Kunz SE, 368
Kuo K, 369
Kuo SJ, 495
Kupfer DJ, 421, 559, 564
Kurokawa Y, 560
Kurucz CN, 420
Kusanagi H, 200, 506
Kutsuna S, 60, 199, 200
Kuwabara N, 371
Kuwajima I, 564
Kuzel M, 501
Kuznetsov YA, 204
Kvolvik SK, 337
Kwan JW, 492
Kwiatkowski F, 496
Kwok A, 194
Kyriacou CP, 17, 60, 105, 200, 289

L'Azou B, 373
L'Hermite M, 290
La Cecilia O, 420
Laakso ML, 367
Labrecque G, 17, 493, 520
Lacourciere Y, 514
Laden F, 518
Ladlow JF, 371
Laerum OD, 205, 291, 367, 491, 504, 519, 523
Lafaye J-Y, 416
LaFontaine G, 336
Lagarde D, 416, 492, 494
Lagardére J-P, 416
Lago AD, 334
Lagoguey M, 289, 291, 501, 515
Lakatos P, 506
Lakatua D, 288, 370, 501, 502, 512, 522
Lakatua DJ, 371, 420, 501, 515
Lakin-Thomas PL, 194, 197, 200, 287
Lakowski B, 171, 200
Lamanna S, 565
Lambe M, 420
Lamberg L, 63, 235, 292, 520
Lambert PW, 233
Lampe F, 511
Lampl Y, 416
Lampo S, 505
Lance WR, 373
Land DV, 565
Landau I, 494, 507

Landsdorp PM, 201
Lang A, 134
Lang HJ, 233
Lange OL, 338
Langevin T, 371, 511
Langevin TR, 503, 511, 521
Langlois PH, 420, 507
Langouet AM, 511
Lanning R, 523
Lansac J, 233
Lanza GA, 507, 509
Lanzinger I, 60, 371
Laporte A, 423
Larimer JL, 234
Larkin JE, 60
LaRocca GM, 525
Larregain-Fournier D, 500
Larue J, 424
Lasley BL, 285
Last KS, 105
Lataif L, 495
Lathe R, 104
Latini M, 559
Latman NS, 563
Latreille J, 507
Laudet V, 495
Lauerma H, 517
Lauro R, 501, 561
Lauwers GY, 509
Lavery CE, 495
Lavie P, 17, 369, 517, 563
Law SP, 233
Lawley HJ, 103, 290
Lawlor DW, 337
Lazar MA, 206
Lazarus JM, 506
Lazzari CR, 367
Le Bail N, 500
Le Fur I, 507
Le M, 203
Le Rol A, 500
Le Saunier F, 508
Leach C, 520, 521, 566
Leach CS, 507, 566
Leakey JE, 417
Lebowitz MD, 120
Lebre P, 494
Leclerc JM, 516
Lecolazet R, 233

Lednicky JA, 508
Ledsome JR, 507
Lee C, 499
Lee CC, 105, 204
Lee CF, 507
Lee JK, 289, 335, 369, 501, 502, 506, 600, 601
Lee K, 104, 196, 200
Lee PA, 561
Lee PD, 513
Lee RE, 507
Lee T, 63
Lee TM, 418
Leech DM, 103, 104
Lees AD, 134, 200
Lefcourt A, 15, 333, 367
Lefcourt AM, 371
Lefebvre J, 514
Lefler BJ, 508
Legan SJ, 371
Legre G, 494
Leidenberger F, 505
Leikin JB, 422
Lein A, 289
Lellouch J, 511
Lem J, 59
Lemmer B, 420, 497, 507, 515, 563, 601
Leng VC, 510, 564
Lenn NJ, 61, 372
Lenton EA, 290
Lentz MJ, 560
Leon J, 516
Leonard L, 289
Leone G, 496
Lepers JP, 507
Leppla NC, 371
Leroy-Martin B, 291
Lessoway V, 507
Lester D, 134, 507
Letizia C, 496
Letourneau Y, 500
Lévi F, 18, 372, 423, 425, 496, 497, 498, 499, 500, 507, 508, 510, 511, 512, 513, 515, 521, 523, 525, 565
Levine H, 282, 289, 418, 508, 561, 563
Levine MM, 509, 563
Levine RJ, 289, 508
Levine RL, 418
Levine SM, 137

Levnaic S, 500
Levy D, 18, 525
Lew GS, 513
Lew NL, 506
Lew RA, 422
Lewandowski MH, 600
Lewczuk B, 336, 372
Lewis H, 336
Lewis M, 525
Lewis RD, 234
Lewis SM, 417
Lewy AJ, 134, 136, 420, 508, 517, 601
Lewy H, 516
Leyendecker G, 289
Li Q, 194, 285, 367, 493
Li X, 498
Li XM, 425, 525
Liao HW, 59
Lieber AL, 233
Liesegang A, 372
Lim JD, 337
Lim P, 286
Lin CY, 200
Lin HY, 197
Lin MC, 195, 289, 508
Lin SY, 194
Lincoln D, 285, 367, 493
Lincoln DW, 194
Lincoln GA, 134, 200, 372
Lindberg FH, 423
Linden H, 200, 201
Lindsley G, 60
Lindstrom SA, 373
Ling PD, 508
Lipson SF, 372
Lis CG, 508
Little MA, 508
Liu DW, 60
Liu J, 499
Liu X, 498
Liu Y, 195, 197, 198, 201, 236
Liu Z, 508
Liu ZM, 513
Lizarralde G, 523
Llerena LA, 286
Llory JF, 500
Lloyd AL, 201
Lloyd CW, 286
Lloyd D, 17, 60, 61, 201

Lloyd JW, 499, 560
Lobban MC, 370, 560
Lobotsky J, 286
Locatelli I, 501
Lockley SW, 420, 501, 509, 519
Lockwood APM, 235
Lockwood GW, 337
Lofts B, 134
Loher W, 136
Lohrer H, 420
Loken MK, 502
Long EM, 133
Lopez JE, 503
Lopez S, 507
Lopez-Burillo S, 516, 522
Lopez-Molina L, 202
Loraine JA, 287
Lörcher L, 40, 61
Loros J, 201
Loros JJ, 62, 104, 194, 195, 196, 197, 200, 201, 202, 285, 336
Lote KK, 291, 519
Louzada FM, 418
Love DN, 287
Love RJ, 375
Lovett-Douse JW, 17
Low KG, 509
Lowe ME, 334
Lowel H, 525
Lowrey PL, 61, 65, 199, 205, 372, 375, 509
Loy SF, 416
Lu J, 58
Lucas E, 561
Lucas RJ, 58, 59, 61, 133, 134
Lucas SB, 602
Luce G, 17
Luce GG, 420
Lucente M, 507, 509
Lucia P, 496
Luciano D, 523
Ludmer PL, 511
Ludwig H, 17, 289
Luis Palma Gamiz J, 563
Lukas KE, 375
Luke D, 424
Lumsden PJ, 134, 201
Lund E, 133
Lundeen WA, 563

Lurbe E, 511
Lydic R, 369
Lye M, 602
Lyncy HK, 512
Lysenko TD, 337

Macaulay AS, 372
Macchi M, 57
MacDonald I, 510, 567
Macdonald I, 567
MacDonald IC, 560
Macdonald M, 493
Macdonald MJ, 367
Macháchová I, 420
Machi MM, 57
Maciewicz RA, 524
Macino G, 194, 200, 201
MacIntyre JM, 525
MacKenzie MR, 517
Mackey MC, 197
Mackowiak PA, 509, 563, 564
MacLean AW, 415
Maclure M, 525
Madden PA, 509
Madjirova NP, 563
Madrid JA, 368
Maes M, 509
Maggioni C, 495, 501, 561
Magosin S, 519
Magrelli A, 194
Mahan AM, 513
Maharaj V, 337
Mahoney MM, 289
Maier CT, 337
Majercak J, 201, 204
Major RH, 104
Makis AC, 500
Malbecq W, 565, 600
Malik S, 134
Mallion JM, 558, 564, 567
Malo JL, 567
Malpaux B, 368
Manchester LC, 522
Mandel J, 501
Mandoli DF, 234
Manevich I, 516
Manfredini R, 420, 499, 509
Mangiafico RA, 513
Manolfi M, 507

Manoukian AS, 201
Manowitz N, 519
Månson JC, 290, 509
Manzoli U, 509
Maplaux B, 368
Maple TL, 375
Marbell A, 509
Marbury MO, 337
Marcacci L, 57
Marchant PR, 523
Marco C, 519
Marcovitch S, 134
Marcus C, 417
Maresca M, 290, 514
Mariani R, 558
Marinich IG, 518
Marion S, 511
Markey SP, 508
Markhart AH III, 290, 337
Marks G, 505
Marks GE, 505
Marks JS, 505
Marler M, 564
Marolf AP, 563
Maron BJ, 509
Marques M, 61, 337
Marques N, 104, 367
Marsh RW, 509
Marshak RE, 333
Marshall HM, 420
Marshall J, 17,
Marshall TJ, 423
Martens UM, 201
Martin A, 418
Martin C, 494
Martin D, 566
Martin DF, 285
Martin GB, 372
Martin L, 234
Martin MJ, 424
Martin P, 510
Martin RJ, 18, 420, 509, 514, 524
Martin W, 602
Martinek S, 201
Martinez HM, 16
MartinNG, 509
Martins SA, 195
Martinson KB, 17, 337
Martrenchar A, 373

Maruyama H, 518
Marx J, 292, 522
Más P, 201
Masaki S, 134, 135
Masalov A, 561
Maseri A, 507
Massari M, 514
Massry SG, 369
Masumoto K, 63
Mathé G, 507, 508
Mather FB, 368
Mathew RM, 289
Mathewson FA, 515
Matsubara N, 202
Matsuoka M, 501
Matsutani S, 518
Matsuyama K, 512
Mattes A, 601
Matthews CD, 288, 370, 372
Matthews JP, 525
Matthews REF, 337
Matthews VJ, 374
Mauer AM, 290, 509
Mauget R, 136
Maw MG, 234
Maxia N, 285
Maxson SJ, 337
Mayer W, 61
Mayersbach Hv, 501
Mayhugh MA, 417
Mayo JC, 423, 516
Maywood ES, 59, 105, 200, 502, 515, 516
Mazankova V, 561
Mazariegos R, 600
Mazy V, 498
Mbugua JM, 499
McCabe CM, 339
McCall MA, 61
McCann M, 496
McClellan D, 566
McClintock MK, 288, 290, 292, 420, 421, 423, 424
McClung CR, 201, 203
McCormick BB, 509
McCormick JB, 289, 371, 494, 505, 509
McCoy NL, 286
McDonald JD, 205
McGovern JP, 507, 520
McGrew WC, 287, 418

McIntyre GA, 335
McMeekin RR, 425
McMichael BL, 337
McNair P, 513
McNamara P, 517
McNeese MC, 422
McWatters HG, 198
Meade TW, 504, 525
Meares R, 509
Mecacci L, 423
Mechkouri M, 291, 423, 507, 512, 515
Mehes K, 288, 292
Mehta RH, 509
Mehta SK, 508
Mehta TS, 234
Meier AH, 63
Meijer JH, 61
Meinert JC, 368
Meissl H, 368
Melchart D, 510
Melchior P, 234
Melin D, 17, 61
Mellenberger RW, 374
Menaker M, 58, 61, 62, 64, 65, 135, 136, 137, 203, 204, 371, 372, 373, 375, 417, 564
Menard J, 558
Mendel G, 290
Mendel VE, 372
Mendelsohn FAO, 509
Mendenhall W, 513
Mendenhall WM, 509
Mengden T, 561, 563, 567
Mercier J, 17, 61
Mermall H, 289, 290, 370, 505, 510
Merrow M, 194, 201, 203, 336, 517, 565
Mersch PP, 135, 510
Mescher MC, 333
Messerli FH, 524
Messin S, 518
Messmore HL, 505
Mesure M, 338
Methew RM, 508
Metz C, 514
Metzger G, 507
Meulman JJ, 516
Meyer HJ, 370
Meyer M, 566, 515
Meyer U, 203

Meyer WJ, 289, 508
Meyer-Grahle U, 203, 516
Meylor JS, 425
Meyn LA, 286
Meyrowitsch DW, 519
Miall WE, 494
Miano MF, 513
Michael TP, 201
Michaels SD, 337
Michel D, 236
Michelinidagger V, 420
Michelsen SH, 567
Michelson AA, 234
Middeke M, 563
Middendorp HM, 135, 510
Middleton BJ, 524
Midkiff K, 565
Migitaka H, 288, 502
Mignot E, 562
Mikulecky M, 234, 235, 601
Milano G, 513
Miles LE, 510, 564
Milgrom H, 510
Milgrom-Friedman J, 509
Millam JR, 368
Millar AJ, 196, 198, 199, 201, 205
Miller AJ, 201
Miller CD, 234
Miller CS, 337
Miller DD, 524
Miller EA, 199
Miller J, 500
Miller KV, 373
Miller R, 59, 337
Miller S, 420
Miller SL, 104
Millet B, 17, 61, 290
Mills JN, 510
Milnamow M, 204
Minami Y, 523
Minis DH, 135
Minkova NK, 567
Minn Motor Veh Crash Facts, 421
Minne H, 495, 561
Minoli SA, 367
Minors D, 418, 419, 424, 510, 567
Minors DS, 57, 61, 421, 510, 600, 601
Minorsky N, 201
Minton JE, 369

Mirick DK, 496
Mishima K, 200, 506
Misset JL, 499, 500, 507, 508, 511
Mitcha J, 560
Mitchell D, 285
Mitchell MD, 288, 370
Mitish MD, 103, 559
Mitler MM, 510
Mitsui A, 201
Mittleman M, 505
Mittleman MA, 495, 518
Miyakawa M, 562
Miyakosi M, 562
Miyamoto Y, 137
Miyata M, 559
Mizumoto H, 518
Mizuno T, 197, 202
Modan B, 495
Moeneclaey N, 498
Mohotti AJ, 337
Mohr E, 372
Mohr KL, 289, 508
Mohr PE, 510
Mohsenzadeh S, 203
Moiseff A, 421
Mojón A, 63, 503, 601
Molle M, 493
Moller AP, 372
Møller SG, 200
Monk TH, 416, 421, 510, 559, 560, 564, 601
Monnerjahn S, 203
Montalbetti N, 499
Montanari L, 514
Moog R, 565
Moore JG, 285, 494, 510, 517, 525
Moore JM, 522
Moore MV, 337
Moore N, 523
Moore RM, 510
Moore RY, 61, 204, 372
Moore TR, 367
Moore-Ede M, 17
Moore-Ede MC, 61, 104, 201, 337, 369, 371, 372, 375, 417, 421, 510, 511, 524
Moorthi KM, 511
Moran J, 368
Moran SM, 195
Moran VC, 236,

Moreira EF, 62
Moreira LF, 372
Moreno CRC, 418
Morgan AH, 290
Morgan E, 234, 601
Morgan LM, 511
Morgan LW, 202
Morgan T, 511
Mori T, 106, 206
Morizot F, 507
Morler H, 105
Mormede P, 373
Mormont C, 510
Mormont MC, 498, 511
Morré DJ, 202
Morré DM, 202
Morris JE, 57, 491
Morris NM, 290, 292
Morris RW, 292, 422, 521, 566
Morris T, 496
Morris TR, 372
Mortimer D, 290
Mosca CL, 425
Moser I, 58, 232
Mostofi N, 513
Motohashi Y, 564, 565
Motta LD, 492
Moul DE, 135, 512, 524
Mount LE, 370
Mowrowicz I, 507
Mrosovsky N, 59, 337, 419
Mudd CA, 289, 371, 505
Mueller PS, 136, 517
Mulder RT, 135
Muller D, 511
Muller JE, 18, 494, 495, 505, 511, 525
Muller MN, 372
Mullington J, 514
Mullis K, 104
Mumford L, 421
Muñoz M, 58, 133
Munt PW, 415
Murakami N, 61
Murano G, 496
Murch SJ, 511
Murphy PJ, 132, 495
Murray D, 61
Mushiya T, 511
Müssle L, 15, 285

Mustad V, 511
Musumoto KH, 375
Myers DH, 135
Myers MP, 197, 202, 204

Nabel E, 517
Nagayama H, 512
Nagy F, 205
Naidoo N, 202
Naitoh P, 417, 421
Nakagaki U, 491
Nakagawa H, 517
Nakagawa O, 562
Nakajima M, 105, 205
Nakamichi M, 202
Nakamura M, 491
Nakamura RM, 367
Nakao M, 504
Nalishiti V, 419
Nanda KK, 135
Nankin HR, 290
Naqvi RH, 291
Nasu T, 61
Natale V, 558
Natali G, 513, 559
Natelson BH, 16
Natesan AK, 417, 495
Nath P, 424
National Sleep Foundation, 421
Naylor E, 234, 236
Neels H, 509
Negus NC, 285, 290, 291
Nelson DE, 62
Nelson R, 63
Nelson RJ, 203, 512
Nelson W, 62, 369, 372, 421, 501, 512, 560, 561, 601
Nelson WL, 369, 421, 501, 520, 521, 566
Nemchausky BA, 505
Nemchausky BM, 289, 290, 370, 494, 505, 510, 518, 521
Nemoz C, 600
Neri DF, 419
Nesbit ME, 497
Netter P, 369
Neugebauer O, 104
Neuhaus-Steinmetz U, 15
Neuman D, 234
Neutel JM, 519

Nevill A, 424, 567
Neville KJ, 560
Nevo E, 103
Newell GR, 512
Newsome DA, 136, 508, 517
Ngam S, 367
Nguyen HB, 197
Nguyen-Clausen A, 421
Ni R-X, 374
Nicholls N, 338
Nicholls P, 375
Nichols KC, 514
Nicod P, 504
Nicolai A, 515
Nicolau G, 502
Nicolau GY, 288, 370, 502, 512, 563
Nicolson GL, 204
Nielsen NC, 287
Niemann L, 519
Nienaber JA, 369
Niezgodzki G, 496
Niimura M, 498
Nikaido SS, 104
Nikityuk BA, 559
Nishi R, 61
Nishikawa S, 564
Nishiyama M, 202
Nisimura T, 62
Niwa H, 136
Nixon EH, 290, 337
Njuguna MN, 499
Njus D, 202
Noden PA, 373
Noeske TA, 63
Noeske-Hallin TA, 337
Nokin J, 290
Nold G, 420
Noll SL, 372
Nomura M, 199
Nonaka Y, 498
Norberg J, 204
Norman RL, 373
Norrving B, 504
North C, 202
North JD, 104
Northrup H, 525
Norton D, 287
Noshiro M, 198
Nougier JJ, 565

Nouvet G, 523
Novak RD, 421, 492
Nowakowska-Jankiewicz B, 505
Noyman Y, 423
Numata H, 62
Nunes PV, 562
Nunez AA, 63
Nunn AJ, 561
Nylund RE, 339

O'Brien C, 424
O'Brien E, 512
O'Brien M, 417, 423
O'Conner JF, 512
O'Connor PJ, 524
O'Donoghue C, 512
O'Hara PD, 231
O'Malley K, 512
O'Reardon JP, 420
O'Rourke MK, 563
Oakley NW, 504
Oates CW, 103
Oberhauser KS, 334, 337
Obsta K, 512
Odén P-C, 338
Oehlerts G, 512
Ogawa H, 512
Ogden LJE, 337
Oh WJ, 65, 375
Ohannesian JP, 519
Ohayon MM, 421
Ohkawa S, 518
Ohno M, 506
Ohno Y, 506
Ohta T, 523
Ohtani-Kaneko R, 288, 370, 502
Ohtsuka K, 564
Oka T, 63
Okajima J, 370
Okajima Y, 564
Okamoto N, 506
Okamoto-Mizuno K, 564
Okamura H, 63, 202, 205, 375
Økland M, 194
Okumura K, 512
Okusami AE, 339
Okuyama H, 202
Oladele SB, 367
Olive PJ, 105

Ollagnier M, 512
Olofsson SO, 498
Olwin J, 505
Olwin JH, 289, 371, 494, 505, 518, 521, 524
Omae T, 501
Ono Y, 512
Onyeocha FA, 338, 373
Opmeer CH, 564
Oppe C, 506
Opravil R, 567
Oren DA, 512
Orev E, 136
Orians GH, 291
Orr D, 200
Orth A, 514
Orth-Gomér K, 600
Ortonne J-P, 524
Osborn DA, 373
Oshima S, 512
Ostatnikova D, 285
Ostberg O, 562
Osterlund B, 498
Ostfeld AM, 512
Ostwald D, 233
Osugi D, 506
Otis JS, 362, 372
Otsuka K, 59, 495, 501, 512, 518, 561, 562
Otto ME, 513
Ouellette I, 566
Ouellette J, 566
Oustrin J, 522
Ouyang Y, 105, 106
Overland L, 290
Overmeer WPJ, 137
Owens D, 510, 567
Owings D, 233
Oxender WD, 373
Ozaki N, 512
Ozawa R, 205

Page TL, 234
Paglierioni T, 517
Paine MW, 417
Painter DP, 338
Palamarchuk EK, 195
Palinkas LA, 513
Palmer JD, 195, 234, 235, 290
Palmiotto JO, 236

Paloheimo J, 513
Pancheva N, 516
Panda S, 198, 235, 373, 517
Pandian PS, 373
Pandula E, 506
Pang CS, 513
Pang SF, 419, 459
Panza JA, 513
Papadimitriou ED, 500
Papathanasiou A, 500
Pappas K, 500
Pappolla MA, 516
Paquet J, 133, 503
Paquet N, 423
Parati G, 564, 566
Parazzoli S, 422, 514
Pareschi PL, 499
Parikh V, 204
Park YH, 232
Parker MW, 135
Parker NC, 337
Parkes D, 510
Parkhurst AM, 369
Parks DK, 422
Parks TW, 60, 601
Parra LF, 565
Parry BL, 289, 508, 513
Parsons RH, 418
Parulekar W, 513
Parvathy Rajan K, 235
Pasceri V, 507
Pasotti C, 495
Pasqualetti P, 513, 559
Patel NK, 495
Paterson AM, 373
Pati AK, 561
Patterson BD, 336
Patterson RS, 371
Patti G, 507
Pauchet F, 18
Paul RJ, 105
Pauley SM, 422, 513
Pauly JA, 18
Pauly JE, 16, 202, 289, 368, 371, 422, 502, 505, 518, 560
Paunier L, 525
Pavia D, 492
Pavlidis T, 62, 202
Pawlikowski M, 505

Payne SR, 235
Payne WD, 17
Peacey MJ, 197
Peacocke J, 132, 497
Pearcey SM, 232
Pearse JS, 16, 233, 235
Pearse VB, 16, 233, 235
Pearson TA, 511
Peeters D, 509
Peiper D, 134, 233
Pekeris CL, 235
Peled R, 516
Peleg L, 513
Pelicano H, 499
Pellegrino AM, 496
Peltier MR, 373
Pen F, 234
Penberthy DR, 506
Penev P, 495
Peng R, 508
Pengelley ET, 17
Pepe PE, 418
Perez LM, 565
Perez Vega E, 501
Perez-Polo JR, 371
Perfetto F, 495, 561
Perkins DD, 286
Perpoint B, 496, 500, 508
Perpoint BM, 497
Perreau-Lenz S, 513
Perriello L, 525
Persinger MA, 422
Pertsov AV, 195
Peter AM, 497
Peters A, 338
Peters JC, 422
Petersen A, 338
Peterson EL, 202
Petit C, 137
Petit E, 513
Petralito A, 513
Petrangeli CM, 496
Petrescu E, 502, 512, 563
Petri G, 103
Petrovsky N, 513, 514
Pevet P, 513
Pezet D, 496
Pfeifer GW, 290
Phanvijhitsiri K, 106

Philip P, 567
Philippe L, 369
Philippens KM, 373, 422
Philippu A, 60, 371
Phong Chau N, 563
Piccione G, 373, 374
Piccoli B, 422, 514
Pickering T, 564
Pickering TG, 468, 514, 559
Piekarski C, 417, 498
Pierard C, 492
Piérard GE, 503
Piérard-Franchimont C, 503
Pierce SM, 337
Pierratos A, 509
Pierson DL, 508
Pimenov K, 103, 559
Pincus DJ, 18, 514
Pine ES, 202
Pinguet F, 494
Pinto LH, 194, 199, 205
Piovan A, 416
Pipkin JL, 417
Pirke KM, 514
Pirrotta V, 203
Pittendrigh CS, 57, 59, 62, 105, 135, 202
Pizzorni C, 496
Place VA, 514
Planque B, 334
Plautz JD, 601
Pleasants ME, 372
Pleeter JD, 522
Plesofsky-Vig N, 202
Pletcher J, 202
Plummer M, 525
Podnieks I, 17
Poeggeler B, 198
Pohjola-Sintonen S, 494
Pohl JE, 523
Poincaré H, 202
Pointer JS, 514
Poirier L, 514
Pokorny ML, 564
Pokorny MLI, 422
Pol F, 373
Polezhaev AA, 195
Politano S, 558
Politoff L, 290
Pollak CP, 418

Pollmacher T, 514
Pöllmann B, 495, 514
Pöllmann L, 495, 514
Poncet M, 524
Pons M, 373
Ponticelli C, 16
Poole CJ, 424
Poolsanguan B, 235
Poon AM, 513
Poon SSS, 201
Popescu M, 512
Popovich JA, 521
Pöppel E, 15, 514
Porcellati C, 523
Porksen N, 63
Portaluppi F, 507, 509, 514
Portman RJ, 525, 567
Post E, 334
Poston DG, 508
Potter VR, 58
Potts JM, 564
Pourquier P, 493
Powell BL, 235
Pownall R, 563
Pozo MJ, 368
Prat M, 15
Pregueiro A, 196
Pregueiro AM, 62, 202
Preitner N, 202
Prendergast BJ, 203
Presser HB, 18, 62, 235, 290, 422
Preti A, 135, 290
Preti G, 290
Price D, 500
Price JL, 199, 203
Price S, 338
Price-Lloyd N, 62, 202
Prigancová A, 564
Prinsen E, 420
Prisant LM, 524
Proc Conf Ferrara, 514
Procacci P, 290, 514
Proffit WR, 507
Prolo LM, 203
Proschan MA, 509
Proux S, 423
Provencio I, 58, 62, 63, 133, 136
Przybylska-Gornowicz B, 336, 372
Psilolignos P, 133

Ptacek LJ, 523
Pupier R, 59, 601
Purchase HG, 369
Puricelli L, 416
Purves WK, 291
Putterill J, 204
Putz Z, 285
Pye EK, 62, 195, 203

Quackenboss JJ, 563
Quadagno DM, 291, 422
Quadri SF, 290, 510
Quail PH, 137
Queneau P, 512
Queneau PM, 497
Quentin E, 15
Quigley BM, 422
Quill H, 105
Qureshi IA, 515
Quyyami AA, 513

Rabarison P, 507
Rabatin J, 511
Rabatin JS, 564
Rabinowitz L, 373
Rabinowitz RP, 564
Rabkin RP, 133
Rabkin SW, 515
Rackley J, 477, 515
Radcliffe SW, 195
Radford RJ, 559
Radke A, 501
Radomski MW, 600
Rafnasson V, 422
Rageth S, 567
Raghavan GV, 372
Rahman S, 497
Rahman WA, 369
Rahmani R, 498
Rajaratnam SM, 422
Ralph MR, 62, 203, 373
Ralston SL, 367
Ramcharan S, 499
Ramirez-Lassepas M, 515
Ramsay N, 204
Ramsay R, 424
Rani S, 134
Rao KP, 203
Rao ML, 564

Raschka C, 506
Rashid MZ, 369
Rassoul F, 493
Ratajczak HV, 62, 515
Rathinavel S, 338
Ratzmann KP, 518
Rauscher WC, 195
Rautenbach GH, 368
Ravecca D, 495
Rayford PL, 64, 524
Raymond R, 505
Raymond RC, 422
Raynal DM, 510
Rea MS, 418
Rebuzzi AG, 509
Recht LD, 422
Reddy AB, 59, 502, 515
Reddy CC, 511
Reddy KL, 195
Reddy P, 174, 183, 203
Redfern P, 424
Redfern PH, 515
Redman PH, 494
Redmond DP, 561, 564
Redon J, 511, 566
Refinetti R, 62, 373, 374, 564, 601
Regal PJ, 422
Regensberg C, 508
Reichardt HM, 57
Reid BL, 336
Reid MS, 336
Reid RL, 495
Reilly C, 565
Reilly KB, 367
Reilly T, 291, 416, 417, 418, 422, 423, 424, 558
Reinberg A, 16, 18, 105, 261, 266, 289, 291, 338, 371, 419, 423, 497, 499, 501, 507, 508, 512, 515, 520, 564, 565, 589, 602
Reinberg AE, 291, 423, 506, 516
Reinberg MA, 18
Reinberg O, 423
Reindl K, 565
Reiners R, 338
Reisin LH, 481, 516
Reiter C, 500
Reiter RJ, 288, 417, 423, 497, 498, 502, 516, 522, 524

Renner M, 105
Rensing L, 203, 516
Renstrom RA, 203
Reppert SM, 65, 105, 200, 204, 334, 374, 375
Resnick A, 513
Retamal P, 135
Reucher E, 419
Reutter R, 372
Revilla M Jr, 103, 559
Revilla M, 600
Revilla MA, 103, 559
Reyner L, 419
Reyner LA, 419
Reynes M, 499, 523
Reynolds CF, 564
Reynolds III CF, 421
Rhame F, 521, 566
Rhodus N, 501
Riad-Fahmy D, 559
Ribak J, 423
Ricci L, 499
Rich I, 508
Rich TA, 506
Richard JL, 494
Richard MO, 523
Richards MA, 492, 510
Richards M, 498
Richardson DC, 371
Richardson GS, 471, 559
Richert RM, 18, 62
Richter CP, 62, 374, 516
Richter DM, 522
Richter V, 493
Rickerby J, 565
Ricou F, 504
Riddiford LM, 64
Riebman JB, 367
Rieck A, 516
Riemersma JBJ, 423
Rienstein S, 291
Rietveld PEM, 516
Rietveld WJ, 63, 374, 516
Righini AS, 367
Rikin A, 338
Riley C, 505
Rimmer DW, 560
Rimmington GM, 338
Rinaldi S, 203, 204

Rinehart JS, 235
Ring A, 416
Ringelberg J, 338
Ringelstein JG, 505
Ringwald P, 507
Rinzel J, 199
Riond JL, 372
Ríos D, 417
Ripperger J, 204, 518
Riss P, 420
Risteli J, 372
Ritschel WA, 516
Rivard GE, 516
Rivera-Coll A, 516, 601
Robert S, 493
Robert-Granie C, 507
Roberts JK, 200
Roberts ML, 495
Roberts N, 517
Roberts R, 61, 135
Roberts WC, 509
Robertson RN, 203
Robinson G, 373
Robinson JE, 371
Robinson PR, 57
Robos A, 133
Robyn C, 290
Rocco MB, 18, 517, 525
Rocha EB, 368, 499
Rodahl A, 423
Rodgers A, 498
Rodgers CW, 423
Rodiek AV, 375
Rodriguez IR, 62
Roemeling Rv, 504, 511, 521, 523
Roenneberg T, 135, 136, 201, 203, 291, 517, 565
Rogers KD, 368
Rohmert W, 419
Rohtla KM, 495
Roijackers RM, 204
Roitman B, 501
Roitman G, 136
Roitman-Johnson B, 505, 521, 522, 525
Rollag MD, 62, 63, 136, 416
Romano A, 499
Romano S, 565
Rome R, 525
Romit A, 496

Ronda JM, 417, 559
Rongstad OJ, 338
Roodell MM, 521
Rook A, 291
Rooney CM, 508
Rosbash M, 103, 104, 196, 198, 203, 206, 288
Rose FC, 501
Rosen G, 419, 521
Rosen GD, 418
Rosen LN, 136, 517
Rosenberg A, 369
Rosenblatt LS, 374, 517
Rosenbloom AL, 368
Rosenlund B, 205, 523
Rosenthal J, 335, 369
Rosenthal NE, 132, 133, 134, 504, 506, 509, 512, 513, 517, 522, 524
Rosenwasser AM, 517
Rosenzweig NS, 202
Ross AM, 498
Ross CW, 136, 203, 235, 338, 423
Ross J Jr, 504
Ross MW, 289
Rossi B, 423
Rotenberg L, 418
Rothenfluh A, 199, 200, 203
Rothenfluh-Hilfiker A, 204
Rothschild R, 494
Rougvie AE, 199
Round FE, 234
Rout WR, 513
Rovensky J, 234
Rowan W, 136, 338
Royant-Parola S, 560
Royles P, 419
Roynette N, 515
Rubens RD, 492
Rubin M, 512
Rubin ML, 202
Rubin NH, 64, 285, 494, 524
Rubnitz ME, 505
Ruckebusch Y, 506
Rudic RD, 517
Rudran R, 288
Ruff F, 18
Ruhoff MS, 493
Rumbyrt J, 289, 371, 505, 524
Rummel JA, 508, 520, 566

Runge W, 561, 601
Ruoff P, 203, 516
Rush AJ, 57
Rutenfranz J, 419, 517, 559
Rutila JE, 203
Ryan M, 494
Ryan RJ, 287
Rydell J, 339
Ryman D, 419
Ryner LC, 291
Ryo H, 104

Sääf J, 374
Saarijarvi S, 517
Saarilehto S, 517
Sack DA, 134, 136, 504, 517, 524
Sack RL, 134, 420, 508, 517, 601
Sackett L, 501, 515
Sackett LL, 561
Sackett-Lundeen L, 288, 289, 370, 371, 375, 420, 502, 505, 512, 521, 522, 525, 563, 566
Sackett-Lundeen LL, 420
Sadava D, 291
Sadeh A, 517
Saez L, 197, 198, 199, 200
Sagar PM, 338
Sagaspe P, 567
Sage LC, 338
Sahu S, 565
Saiki R, 104
Saini A, 522
Saino N, 372
Saint Pol P, 291
Sainz RM, 423, 501, 516, 522
Saisho H, 518
Sakaki Y, 205
Sako T, 559
Salandri A, 496
Salgado LE, 61
Salido GM, 368
Salis PJ, 289
Salisbury FB, 136, 203, 235, 423
Salmi R, 499, 509
Salmon M, 338
Salomé PA, 201, 203
Salomonsz R, 195
Samach A, 204
Sampson M, 561

Sancar A, 137
Sanchez CP, 517
Sanchez de la Peña S, 63, 501, 517
Sandberg G, 338,
Sandberg R, 417, 559
Sandberg S, 367, 519
Sandeen M, 232
Sanders AF, 423
Sanders EH, 285, 291
Sanders SA, 287
Sanders SW, 517, 525
Sangoram AM, 194, 199, 204
Santoro G, 284
Santos S, 513
Saper CB, 58
Sargent F, 18, 519
Sargent ML, 197, 203
Sarova-Pinhas I, 416
Sasaki J, 518
Sasaki T, 423
Sasser LB, 57, 491
Sasseville A, 423
Sassi ML, 372
Sassolas G, 286
Sata T, 518
Satav T, 418
Sato E, 197, 202
Sato F, 198
Sato G, 518
Satoh K, 200, 506
Satter RL, 199, 203
Sauer L, 493
Sauer LA, 58, 493, 496
Saunders DS, 136, 197, 202
Sauter R, 522
Savides TJ, 518
Saxena BB, 289
Saxena PK, 511
Saydjari R, 64, 524
Sayre JW, 424
Scally AJ, 232
Scammell TE, 58
Scantelbury M, 59
Scarpelli PT, 565
Scatena FN, 236
Scavo D, 496
Schachter P, 518
Schack B, 561
Schaefer KE, 423

Schaffer E, 501
Schäffer R, 204
Schams D, 132
Schank JC, 423
Scharf S, 104, 516
Scharpe S, 509
Scheffer FAJL, 424
Scheffer M, 198, 203, 204
Scherer I, 61
Schernhammer E, 518
Schernhammer ES, 518
Scheving L, 62
Scheving LE, 16, 18, 202, 288, 289,
    290, 291, 368, 369, 370, 371, 372, 373,
    374, 421, 422, 493, 495, 497, 501, 502,
    505, 510, 518, 521, 524, 525, 560, 561,
    565
Schibler U, 57, 202, 204, 518
Schillaci G, 523
Schilman PE, 367
Schindele G, 562
Schiwara HW, 497
Schloot W, 497
Schmidle A, 63
Schmidt B, 41, 64
Schmidt CW Jr, 424
Schmidt F, 374
Schmidt T, 492, 558
Schmieder RA, 565
Schmitz O, 204
Schmoor C, 201
Schnakenberg E, 497
Schneider M, 513
Schneider N, 103
Schnur S, 518
Schoeller DA, 417, 424
Schoeman HS, 368
Schoknecht PA, 367
Schopf JW, 105
Schrader LE, 334
Schreiber C, 492
Schreiber G, 136, 424
Schreiber M, 524
Schreiber W, 514
Schrempf M, 58
Schulmeister K, 518
Schultheiss HP, 492, 511
Schultz TF, 204
Schulz B, 518

Schulz PM, 524
Schuman L, 501
Schuster J, 63
Schuster JL, 338
Schutz G, 57
Schwartz BD, 425
Schwartz MD, 63
Schwartz PJ, 135, 512, 524
Schwartz WJ, 36, 422
Schwartzkopff O, 59, 502, 561
Schweiger HG, 105, 204
Schweiger M, 204
Schweiger U, 514
Schwemmle B, 338
Schwiebert E, 338
Science Magazine Editors, 204
Scott BIH, 339
Scott PIH, 520
Scully NJ, 135
Segal S, 563
Sehgal A, 194, 197, 202, 204, 205
Seifer R, 416
Seim G, 334
Seki K, 371
Sel'kov E, 194
Selby CP, 137
Sellers A, 369
Selwyn AP, 517
Sempere AJ, 136
Sen RN, 565
Seng JE, 417
Senger C, 518
Sensi S, 424, 518
Senske B, 418
Seriolo B, 496
Sesboue B, 424
Sestito A, 507
Seto TB, 518
Sévigny J, 423
Sexton TJ, 374
Sha LR, 235
Shadrin AS, 518
Shafer AI, 494
Shaffer JW, 424
Shafor R, 510
Shahrem MR, 369
Shaikh AA, 291
Shaikh SA, 291
Shanahan TL, 419

Shanas U, 59
Shankaraiah K, 16, 233
Shannon NJ, 135, 512
Shapses SA, 367
Sharma M, 522
Sharp DC, 373
Sharp J, 374
Shaw MK, 204
Shaw P, 419
Shearman LP, 65, 105, 200, 204, 375
Shelley EM, 512
Shelstad J, 289, 370, 505
Shelton JN, 374
Shepard RC, 506
Sheridan J, 512
Sheridan SC, 502
Sherman J, 523
Shi M, 196
Shideler SE, 285
Shifrine M, 374, 517
Shigeyoshi Y, 205
Shimada K, 562, 566
Shimizu E, 559
Shimizu T, 200, 506
Shimomoto M, 523
Shimomura K, 65, 375
Shimonov M, 518
Shinagawa M, 512, 518
Shinkai S, 560
Shinkawa A, 518
Shinoda H, 506
Shinohara K, 63
Shinomura T, 200
Shiotsuka RN, 16, 502
Shirakawa T, 518
Shiramizu K, 370
Shirazi P, 494, 505
Shiu SY, 513
Shogren B, 336
Short RV, 291, 371
Short SR, 201
Shubeita HE, 291, 422
Shue JL, 518
Shuster S, 285
Sica DA, 519, 525
Sideris DA, 500
Sidote D, 201, 204
Siegelová J, 501, 561, 566
Siegmund R, 506

Siepka SM, 65, 375
Siler T, 525
Silva D, 514
Silva I, 503
Silverman RA, 519
Simmons CB, 511
Simmons MJ, 292
Simmons MS, 424
Simon MC, 195
Simon N, 494
Simonon MA, 369
Simonsen PE, 519
Simpson H, 565
Simpson HW, 291, 500, 501, 519, 521, 559, 565, 566
Sinclair RM, 235
Singer SJ, 204
Singh N, 519
Singh RB, 561
Sinha MK, 424, 519
Sisson DV, 372
Sizonenko PC, 525
Skakkebaek NE, 287
Skeleton M, 566
Skene DJ, 501, 509, 519
Skitz MR, 512
Skopik SD, 136
Skovgaard LT, 499
Skwerer RG, 136, 517
Slack CR, 339
Slack FJ, 194
Slama O, 424
Sletvold O, 291, 519
Slome D, 233
Slotowitz HI, 424
Slover G, 565
Slover GPT, 292, 521, 566
Smaaland R, 205, 291, 367, 374, 491, 519, 522, 523
Smale L, 63, 289
Smals AG, 291
Smiley AM, 18
Smirnova OC, 132
Smith AM, 135
Smith CS, 565
Smith DHG, 519
Smith EO, 508
Smith GP, 418
Smith ID, 418

Smith JC, 419
Smith KD, 291, 292
Smith L, 424
Smith M, 232
Smith MM, 500
Smith P, 492
Smith PA, 560
Smith RE, 519
Smith RS, 424
Smith SC Jr, 504
Smithies O, 137
Smith-Sivertsen T, 133
Smolensky M, 63, 105, 235, 288, 292, 503, 507, 509, 515, 516, 519, 565
Smolensky MH, 16, 18, 132, 284, 291, 292, 420, 422, 423, 492, 494, 507, 510, 515, 520, 525, 566, 602
Snedeker PW, 505
Snustad DP, 292
So WV, 103, 196, 203
Socha L, 514
Soda S, 562
Sok M, 235
Sokolove PG, 136, 602
Soliman MRI, 425
Sollberger A, 602
Solloway S, 524
Solomon GD, 520
Solyanik GI, 195
Somero GN, 204
Somers DE, 63, 136, 199, 201, 204
Sone M, 493
Song EJ, 194
Song KH, 370
Song W, 202
Song XH, 235
Soni B, 58, 133
Sonkowsky R, 561
Sonkowsky RV, 502
Soong L, 369
Sorbetti-Guerri F, 236
Sørbye H, 18, 522
Sothern MI, 561
Sothern R, 419, 493, 505, 511, 558
Sothern RB, 16, 17, 18, 60, 62, 63, 103, 105, 194, 205, 233, 235, 285, 287, 289, 290, 291, 292, 336, 337, 339, 367, 369, 370, 371, 374, 418, 424, 491, 492, 493, 494, 495, 496, 497, 500, 501, 502, 504,

505, 508, 510, 511, 515, 516, 518, 519, 520, 521, 522, 523, 524, 525, 559, 561, 564, 565, 566, 567, 601, 602
Sothern SB, 561
Soudant E, 515
Souissi N, 424
Soulen TK, 200
Soussan A, 507
Southern MM, 196
Soutoul JH, 233
Souverijn JH, 137
Soveri T, 367
Spector NH, 561
Speirs L, 416
Speizer FE, 518
Spelten E, 565
Spiegelhalter F, 375
Spieler RE, 63, 337, 339, 367
Spies HG, 373
Spira A, 292
Spiteri NJ, 204
Sprinkle RL, 423
Spruyt E, 18, 63, 105, 235, 292
Spuhler T, 561
Sqalli A, 522
Srere P, 196
Sriram S, 105, 200
Srivastava S, 522
St John JB, 338
Staessen JA, 566
Stahle H, 370
Staknis D, 197
Standford L, 61
Stanewsky R, 204, 601
Stanisiewski EP, 374
Stark JC, 339
Stauble TN, 495
Staudinger H, 496, 522
Steeves TD, 199
Steeves TDL, 204
Stein Y, 502
Steinberg JD, 559
Steinberger E, 291, 292
Steine S, 205, 523
Steinijans V, 496
Steinijans VW, 522
Steliu J, 420
Stenseth NC, 334
Stephan FK, 63, 375

Stephens AN, 501
Stephens TW, 519
Stephenson LA, 60
Stergiou G, 562, 566
Stern K, 292, 424
Stern LO, 512
Stern PH, 506
Stern R, 492
Stetson MH, 136, 204
Stevens RG, 57, 63, 424, 491, 496, 522
Stevenson J, 517
Stevenson L, 519
Steward GM, 504
Stewart I, 204
Stiasny S, 133
Stice S, 417
Stieglitz A, 375
Stimmler L, 525
Stoffer DS, 561
Stoinski TS, 375
Stokkan K-A, 522
Stolen PD, 339
Stone P, 499
Stone PH, 18, 511, 525
Stoney P, 418, 419
Stoney PH, 500
Storigatz SH, 417
Storm WF, 560
Storms WW, 522
Storti RV, 195
Storz UC, 105
Stout RW, 496
Stoynev A, 561
Stoynev AG, 567
Straub KD, 561
Straughn AB, 515
Straume M, 198, 601, 602
Strayer CA, 201
Strebel B, 564
Strempel H, 522
Strestík J, 564
Stroud B, 15, 333, 367
Strubbe JH, 204
Strubel H, 61
Strumpf IJ, 424
Stull CL, 375
Stunkard AJ, 420, 424
Sturtevant F, 289, 371

Sturtevant FM, 524
Sturtevant RP, 289, 371, 505, 524
Sturzenegger ER, 567
Suarez C, 521, 566
Sucher E, 499
Suen D-F, 339
Sujino M, 63, 375
Suller MT, 61
Sulli A, 496
Sullivan W, 235
Sullivan WN, 335, 369
Sulon J, 507
Sulzman FM, 61, 104, 201, 202, 337, 369, 371, 372, 375, 510
Sumaya CV, 292, 522
Summala H, 419
Sun ZS, 204, 493
Sundararaj BI, 18, 235, 236, 424
Sundararajan KS, 338
Sunderland R, 498
Sundwall A, 374
Sung S, 198
Suri V, 203
Susak LE, 507
Susalka SJ, 374
Sutovsky M, 375
Sutovsky P, 375
Suttle MA, 337
Sutton TW, 523
Suyama H, 559
Suzuki T, 136, 288, 502
Svanes C, 18, 522
Svatikova A, 513
Svensson AM, 339
Swannell AJ, 563
Swedo SE, 522
Sweeney B, 205
Sweeney BM, 63, 64, 205
Sweeney O, 496
Swensson A, 416
Swerdlow AJ, 522
Swoyer J, 375, 502, 515, 522, 561
Symonds C, 522
Symonds RL, 136
Sytnik N, 510, 567
Syutkina EV, 103, 502, 559, 561
Szabo S, 204
Szafler SJ, 514
Szecsey A, 506

Tabata M, 136, 370
Taillard J, 567
Taira DA, 518
Taiz L, 136
Takagi Y, 198
Takahashi A, 201
Takahashi JS, 60, 61, 62, 64, 65, 137, 194, 195, 197, 199, 203, 204, 205, 372, 375, 500, 509, 562
Takamura T, 135
Takano T, 564
Takao M, 57, 59, 61
Takeda M, 136
Takeda Y, 498
Takimoto A, 292
Talbott JH, 105
Tall J, 423
Talora C, 194
Tamakoshi A, 506
Tamarkin L, 522
Tamburi S, 509
Tampellini M, 497, 522
Tan DX, 423, 516, 522
Tanabe A, 199
Tanaka H, 60
Tanaka M, 199
Tanfer K, 292
Taniguchi Y, 199
Tankova I, 567
Taponen J, 367
Targum SD, 136, 292, 517
Tarjan G, 506
Taros LY, 518
Tarquini B, 501
Tarquini R, 495, 561
Tarroni G, 514
Tashev TG, 563
Tashkin DP, 424
Tast A, 375
Tate RB, 515
Tatti B, 501
Tauber E, 105
Tavadia HB, 501
Taylor BJ, 291
Taylor BL, 205
Taylor CJ, 524
Taylor N, 374
Taylor NJ, 374
Taylor SR, 285

Taylor WR, 62
Tedeschi E, 492
Téhard B, 423
Tei H, 205
Telsfer S, 375
Templeton AA, 290
Temte JL, 522
Tenenhaus M, 507
Tepper B, 369
Terazawa T, 506
Terlain B, 369
Terman JS, 137
Terman M, 57, 136, 137, 417, 517
Termorshuizen F, 523
Terrill CE, 369
Tester JR, 64, 334, 338, 339
Thapan K, 509
The Prairie Astronomy Club, 424
Theologides A, 497
Theroux P, 524
Thiberville L, 523
Third JLHC, 289, 371, 493, 494, 505, 524
Thomas B, 134, 137
Thomas CB Jr, 524
Thomas CB, 523
Thomas DB, 63, 522
Thomas JH, 60, 104, 288
Thomas KH, 134
Thommen G, 424
Thompson DR, 523
Thompson DT, 337
Thompson JC, 64, 524
Thompson M, 522
Thompson S, 59
Thompson SG, 494
Thorburn GD, 288, 370
Thornby JL, 289
Thorndike L, 105
Thornton C, 559
Thresher RJ, 137
Thuillez C, 523
Thyss A, 513
Ticher A, 558
Tikasingh ES, 232
Tisak J, 565
Tjoa WS, 292
To K-Y, 339
Tobin EM, 205
Tochikubo O, 562

Todd HE, 285
Todo T, 104
Tofler G, 505
Tofler GH, 18, 494, 511, 525
Togo M, 564
Toh H, 104
Toh KL, 523
Tokura H, 523
Tolbert MG, 338
Tomita J, 105, 205
Tomita K, 559
Tøndering C, 77, 105
Tong YL, 601, 602
Torres L, 103
Torsvall L, 567
Tosini G, 372, 375
Tóth R, 205
Totterdell P, 562
Touboul JP, 507
Touitou Y, 104, 288, 291, 339, 419, 502, 507, 511, 515, 523
Tournois J, 137, 292, 339
Townsend CM Jr, 64, 524
Toyama J, 506
Tranchot J, 373
Trapp G, 567
Trautmann H, 522
Traver JR, 286
Travis LB, 368
Tredici G, 498
Trevan DJ, 197
Trewavas A, 60, 104
Trinder J, 424
Tritsch G, 501
Troen P, 290
Tronche F, 57
Trost DC, 500
Troyanov S, 567
Trucco E, 196
Truman JW, 64
Tsai CC, 525
Tsai TH, 202, 289, 368, 370, 422, 505, 518, 524
Tsai TS, 565
Tschachler E, 507
Tsementzis SA, 523
Tsinkalovsky O, 205, 523
Tsinoremas NF, 60, 200
Tsitourides S, 133

Tsuchida T, 506
Tsutsumi S, 518
Tubiana-Mathieu N, 496
Tucci M, 522
Tucciarone L, 559
Tuchman M, 523
Tucker HA, 374
Tucker P, 510, 567
Tucker W, 205
Tulinius H, 422
Tumlinson JH, 333
Turek FW, 60, 64, 199, 205
Turner CM, 333
Turner EH, 522, 524
Turner MP, 61
Turner RM, 375
Turturro A, 417
Tuunainen E, 367
Tyler NJ, 134
Tyson JJ, 205
Tzahor Z, 136, 424
Tzischinsky O, 416

Ucieda R, 503
Udry JR, 290, 292
Uebelmesser ER, 64
Ueda HR, 523
Uemoto S, 137, 340
Uen S, 563
Ueno K, 501
Uezono K, 501, 561
Uglow RF, 235
Ugolini C, 565
Ullis K, 292, 522
Underwood H, 61, 64, 135, 137, 205
Uozumi T, 202
Upcroft JA, 339
Urey HC, 104
US Dept Transportation, 424
Usami E, 336, 370

Vahed S, 493, 494
Vaisse B, 558, 567
Valachova A, 234
Van Beers P, 492
Van Cauter E, 417, 523
van Deenen LLM, 198
Van den Heiligenberg S, 523
Van Den Heuvel CJ, 495
van den Hoofdakker RH, 135, 510
van der Horst GT, 63, 105, 375
van der Mark J, 205
van der Pol B, 205
Van Dongen H, 567
Van Dongen HP, 137
Van Dongen HPA, 562
van Dongen W, 420, 424
van Doornen LJP, 424
Van Helden A, 105
Van Heyden T, 525
Van Hoecke J, 416, 418, 494
van Leeuwen P, 422
van Loveren H, 523
Van Nes EH, 204
van Onckelen H, 420
van Wüllen M, 15
van Zon AQ, 620
Vande Wiele RL, 286, 497
Vandenberg G, 525
Vander A, 436, 523
Vangelova KK, 567
Vannetzel JM, 507, 508
Vargha P, 562
Varkevisser M, 567
Vasal S, 18, 235, 236
Vasta M, 501
Vaughan GM, 58, 496
Vaur L, 558, 567
Vauthey J-N, 509
Veatch R, 505
Veda K, 518
Vedral D, 18
Veerman A, 137
Vejvoda M, 419
Vekemans M, 290
Veldhuis JD, 523
Vendola G, 284
Verbelen J-P, 18, 63, 105, 235, 292
Verdecchia P, 523, 566
Vergnani L, 514
Verhoeven KJF, 338
Verkerk R, 509
Verschoore WB, 431, 439, 524
Vesely DL, 289, 370, 505, 518, 524
Vestergaard P, 567
Vetter H, 563, 567
Vetter W, 561, 563, 567
Vianna EO, 524

Vieira EF, 285
Vierstra RD, 137
Vieux N, 423, 565
Vijayalaxmi, 524
Vilchez RA, 508
Villella A, 291
Vincent C, 523
Vincent WJ, 416
Vince-Prue D, 137
Vink JM, 567
Vinogradova O, 103
Vinsjevik M, 203
Virshup DM, 523
Vitaterna MH, 194, 199, 205
Vitiello MV, 421
Vitt AA, 194
Vittorioso P, 194
Vlcek J, 566
Voda AM, 132, 284
Voegelin MR, 492
Vogelaere P, 560
Vogt DJ, 236
Vogt KA, 236
Voit M, 103
Volkers P, 369
Vollman RF, 524
Volpe A, 285
von Armin T, 524
Von Gaertner T, 137, 236
von Goetz C, 416
von Mayersbach H, 373, 422, 501
von Roemeling R, 511
von Schantz M, 58, 133
Voracek M, 500
Voultsios A, 428, 506, 524
Vrignaud P, 500

Waalen J, 64
Wagener MM, 519
Wager-Smith K, 202
Wagner DR, 198
Wagner E, 205
Wagner H, 510
Wagoner N, 59, 370
Wakai K, 506
Waldrop RD, 64, 524
Walford J, 500
Walker P, 567
Walker RF, 559

Wall D, 501
Wall JR, 525
Wallach LA, 497, 501
Waller-Scanlon L, 417
Walsh HM, 337
Walsh TJ, 422
Walter S, 500
Walter SD, 133
Wan C, 561
Waner JI, 285
Wang DY, 492, 510
Wang L, 198
Wang X, 198
Wang Y, 602
Wang Z, 561
Wang ZY, 205
Wanner M, 372
Ward FAB, 105
Warde SB, 200, 336
Warner R, 334
Warren M, 286, 422, 497
Warwick EJ, 369
Wasserman L, 64
Wasserman SA, 291
Wasserman SS, 509, 563, 564
Watanabe T, 560
Watanabe Y, 495, 501, 502, 561, 566
Waterhouse J, 337, 417, 424, 504, 510,
    511, 536, 558, 567, 595
Waterhouse JM, 57, 61, 421, 510,
    601
Waters DD, 493, 524
Waters TF, 336, 339
Waters WE, 524
Waterton JC, 524
Watt I, 524
Wauters A, 509
Wax Y, 495
Weatherbee JA, 521
Weaver DR, 65, 105, 200, 204, 375
Webb H, 419
Webb HM, 232, 424
Webb WB, 558
Weber JM, 417
Weber MA, 501, 519, 524, 561
Weghaupt K, 420
Wegscheider K, 511
Wehr TA, 64, 132, 133, 134, 136, 497,
    504, 506, 508, 517, 524

Weinberg CR, 286, 292
Weinert D, 567
Weir FW, 420
Weiss KB, 524
Weisser B, 561, 563, 567
Weissman A, 418
Weitz CJ, 197
Weitzman ED, 424, 510, 524
Welch B, 136, 517
Wendt H, 419
Wendt HW, 103, 501, 559, 561
Went FW, 64, 135, 205, 336, 339, 602
Wergin WPP, 338
Wesley CS, 199, 202
West MS, 567
Wetterberg L, 292, 374, 505, 525
Wever R, 15
Wever RA, 64, 416, 425, 602
Weydahl A, 292, 525
Whaley FS, 500
Wharton JT, 492
Wheeler DA, 203
White C, 525
White M, 420
White RF, 507
White WB, 491, 525
White ZS, 508
Whitehouse GH, 517
Whitrow GJ, 105, 205
Whitton JL, 497
Whyte IM, 232
Wick M, 511, 521
Widner H, 504
Wiklund O, 498
Wilcox AJ, 286, 292
Wilde GJ, 415
Wildervanck C, 423
Wildgruber C, 416
Wilkins MB, 64
Willett WC, 518
Williams BG, 234, 236
Williams GH, 494
Williams JA, 205
Williams MK, 500
Williams P, 136
Williams PH, 137
Williams RL, 289
Williams SB, 196, 199, 204
Williamson CE, 103, 104

Willich SN, 18, 492, 494, 511, 525
Willrich LM, 421
Wilsbacher LD, 194, 195, 197, 199
Wilson BJ, 232
Wilson BW, 63, 522
Wilson C, 501
Wilson D, 495, 501, 565
Wilson DW, 559, 585
Wilson FD, 374
Wilson MA, 510
Wiltschko W, 335
Winfree AT, 59, 205, 206
Wing PC, 507
Wing PR, 425
Wingate D, 506
Winget CM, 368, 425, 567
Winkler M, 500
Winstead DK, 425
Wirz-Justice A, 417, 517, 565
Wishart GJ, 367
Witherington BE, 338, 339, 340
Wititsuwannakul R, 340
Witte K, 601
Woelfle MA, 106
Wojtczak-Jarosozwa J, 425
Wolcott JH, 425
Wolf SR, 518
Wolfe M, 510
Wolfson A, 64
Wolfson AR, 416
Wollnik F, 60, 64
Wong R, 513
Wood DL, 15, 333, 367, 371
Wood P, 508
Wood PA, 194, 285, 367, 493
Wood WML, 340
Woodhouse PR, 525
Woods SL, 494
Woodward DO, 203
Woodward J, 565
Woodward PM, 560
Woodward S, 287
Working PK, 289
World Hypertension League, 567
Worms MJ, 370
Wright EM Jr, 292
Wright J, 509, 511
Wright KP Jr, 418
Wu A, 194, 493

Wu CH, 287
Wu J, 561, 525
Wu JY, 504
Wu LY, 202
Wu MW, 425, 525
Wulff K, 506
Wurscher TJ, 521
Wuttke W, 514
Wyatt HR, 422, 425
Wyatt JK, 525

Xian LJ, 425, 525
Xiang LI, 515
Xiao RX, 515
Xiong H, 194, 493
Xu NT, 235
Xu X, 200
Xu Y, 106, 206

Yager JG, 521
Yamada EM, 135, 512, 522
Yamaguchi S, 63, 375
Yamamoto K, 202, 288, 502
Yamamoto Y, 233
Yamashino T, 197, 202
Yamashita O, 235
Yamauchi H, 373
Yamazaki S, 64, 65, 375
Yanagie H, 498
Yang BH, 515
Yang Y, 195, 198
Yanovsky MJ, 194
Yanowitch RE, 425
Yardim S, 504
Yarows SA, 567
Yasue H, 512
Yates FE, 367
Yatsyk GV, 103, 559
Yau KW, 59, 61
Ychou M, 494
Yeh KT, 495
Yellin AM, 425
Yen SSC, 525
Yetman RJ, 422, 525, 567
Yin HC, 64
Yin L, 206
Yokoya T, 564
Yonovsky MJ, 196
Yoo KC, 137, 340

Yoo OJ, 65, 375
Yoo SH, 65, 375
Yosefy C, 516
Yoshimura Y, 374
Yosipovitch G, 525
You M, 508
Young BS, 506
Young MRI, 525, 521
Young MW, 106, 194, 195, 197, 198, 199,
    200, 201, 202, 203, 204, 206
Young T, 562
Youthed GJ, 236
Yu Q, 206
Yunis EJ, 502

Zachmann A, 375
Zagula-Mally Z, 18, 515
Zajac W, 600
Zakany J, 202
Zaltzman J, 509
Zanette CB, 562
Zani A, 423
Zaniboni A, 422, 514
Zanotti C, 420
Zaslavskaia RM, 558
Zaslavskaya R, 502
Zee P, 495
Zeevaart JAD, 137
Zehmer RB, 510
Zehring WA, 203
Zeiger E, 136, 292
Zemek R, 234
Zengil H, 504
Zerubavel E, 106
Zhang J, 515
Zhang LP, 235
Zhang R, 500
Zhang Y, 235
Zhao Y, 194, 199, 204
Zheng B, 105
Zhu T, 198
Zhulin IB, 177, 205
Zidani R, 18, 499, 500, 508
Zidek V, 333
Zieher SJ, 505
Zieleniewski J, 505
Zimmerman NH, 375
Zimmet PZ, 525
Zinn JG, 106

Zinneman H, 369
Zinneman HH, 421
Zlotecki R, 513
Zoppi M, 290, 514
Zorn A, 290
Zsifkovits S, 288, 292
Zubidad Ael S, 59

Zuch J, 287
Zuchenko O, 204
Zucker I, 63, 375
Zulley J, 419
Zuniga O, 375
Zürcher E, 236
Zylka MJ, 65, 200, 204, 375

Subject Index

abscisic acid, 311
absorption spectrum, 120
Acanthamoeba, 9, 41
accidents, 11, 228, 383, 394–396
adaptive advantages, 98, 385
acrophase, 20, 120, 352, 595–597
acrophase charts, 351–352, 432, 445, 454
action spectrum, 120
adrenal, 34, 36, 277, 278, 467, 482
aftereffects, 38, 48
agents of stress, 304–309, 327
agriculture, 293
Albizzia (plant), 165, 166, 311
alcohol, 42, 228
aliasing, 20, 580–581
amino acids, 38, 93, 164, 168
amplitude, 20, 29–32, 42–43, 193, 209, 214–215, 308, 329–330, 346, 467–469, 591–593, 595–597
AM & PM, 74, 86
analyzing for rhythms, 577
anesthesia, 7, 480
Antheraea (Chinese oak silk moth), 34–35
aquatic insects, 318–322, 330
Arabidopsis circadian clock, 187–189
arthritis, 4, 442, 481, 530
Aschoff's Rule, 40
asexual reproduction, 237, 255–257
asthma, 442, 452, 454, 468, 477, 480–482, 530
athletes, 390, 392
atmosphere, 95, 99, 152
ATP (adenosine triphosphate), 156–157, 168
autorhythmometry, 526–576

bamboo (*Phyllostachys*), 8, 10, 26, 260
bean (*Phaseolus*), 9, 10, 22, 25–26, 29, 40–41, 50–51, 228, 248, 252, 305, 308, 578
beard growth, 275–276, 280–283
bees, 87, 89, 228, 310, 380
Belousov-Zhabotinsky reaction, 155
Biloxi soybean, 113, 125, 260, 294, 296
biochemical & metabolic models, 155–156
biological clocks, 28, 34, 75, 85, 87, 89, 90, 92, 94, 98, 101, 102, 139, 141–142, 167, 171, 174–175, 192, 300, 328, 427, 488
biological oscillations (spectrum), 22
birding, 317–318, 332
birds, 26, 36, 39–40, 44, 111–112, 116, 118, 123, 126–127, 300–302, 311, 332, 356, 369, 361–362, 380, 409, 456
birth (human), 6, 8, 129, 228, 261, 271, 442, 445
birth control, 268–271
blood, 121, 273, 349, 351, 358, 406, 428–429, 440, 445, 449–450, 467
blood cells, 187, 279, 346, 351, 355, 434, 441, 445, 463–466, 532
blood pressure, 264, 428, 441, 445, 468, 479–481, 527, 534, 539, 544, 549–550
body temperature, 1, 9, 22, 41, 44, 48, 264, 269, 312, 342, 390, 399, 401, 427, 437, 453, 488, 530
body weight, 272, 274, 404, 408
bone, 334, 354, 432
bone marrow, 87, 187, 246, 432, 464, 471, 473, 484
breast cancer, 410, 447, 455, 461–462, 477, 484, 486–487
Bünning's hypothesis, 125

649

Subject Index

caddisflies (Trichoptera), 319–322
cAMP, 153–156. 156, 158, 437
calendars, 72, 79
cancer, 49, 245, 402, 410–412, 414, 443, 447, 455, 459–462, 467, 471–490, 586
carbon dioxide (CO_2), 43, 95, 244, 279, 293, 311, 339
cardiovascular disease, 8, 442, 444–445, 448–451, 480
cat (*Felix*), 25, 343–345, 348–349, 358, 386, 456
cattle, 343, 348
cell division, 100, 142, 189, 240, 248, 244–245
cellular clock, 12, 440, 486–488
cellulose, 9, 229
central clock, 96–94, 101, 173
chaos, 153
chemicals (effects), 4, 7, 38, 42, 44–45, 217, 238, 259, 268, 304, 360, 384
chicken (*Gallus*), 343–345, 349–350, 356
Chlamydomonas, 30, 100
cholesterol, 344, 351, 404, 436, 448, 450–451
chromosome, 30–32, 142, 167, 171, 183, 186, 239, 249
chronobiology, 12–13, 19, 29
chronodiagnosis, 360, 462–469, 490
chronon concept, 167
chronopharmacology, 478, 486, 491
chronophytopathology, 308
chronotherapy, 360, 475–486, 488, 490
chronotolerance, 304
chronotoxicity, 470–475
chronotype, 391–392, 535–537
circadian clock, 28, 32–38, 41, 44–45, 90, 96–98, 101–102, 173–192, 303–304, 356, 403, 409, 486–488
circadian rhythm, 23–25
circadian system, 44, 100, 173, 192–193, 311, 391
circadian-infradian rule, 248
circadian time (CT), 20, 86, 257
circalunar, 207, 217, 278
circalunidian, 207, 229, 230
circannual, 22, 26, 126, 222, 301, 358–361, 363, 445–453, 478,
circaseptan, 10, 22, 26, 79, 228–229, 449

circatidal, 207, 216–217, 230
circatrigintan, 22, 26, 42, 262, 453–454
circumnutation, 9, 49, 56, 165, 308
civil twilight, 108, 110
clepsydra, 74, 77
clinical medicine, 426–491
clock genes (general), 28, 93–94, 96, 98, 171–172, 174, 176–192, 304
clock genes (human), 249–250, 431, 440, 486–488
clock mutants, 30–33, 36, 99, 167–173, 488
clock proteins, 102, 176–192
clocks, 1, 32, 66, 69, 72, 75, 79, 88, 92, 140, 167, 172, 179
cockelbur (*Xanthium*), 118, 120
color change, 110–111, 224–226, 297
constant environmental conditions, 20, 38, 40, 48, 216
constant fixed dosing, 476–478
constant temperature, 38, 45, 91, 216
corticosteroids, 7, 467, 478
cortisol, 9, 44, 312, 343, 350, 410, 429, 433, 467
cosine curve, 29, 247, 579, 587, 589, 590, 594, 599
cosinor programs, 590
courtship & mating, 9, 30, 32, 218, 220, 222, 224, 258
crab, 217–218, 220, 222, 224
critical daylength or nightlength, 114–115
cryptochromes, 100, 121, 122, 172, 177, 188, 209
cyanobacteria circadian clock, 97, 189–192
cyclic environment, 48, 49, 261, 293

damping, 20–21, 43
Darwin's footsteps, 49
day (time), 26–29, 38, 68–69, 74, 77–79
daylight saving time, 85
DD (constant dark), 20, 38, 40
death, 4, 6, 8, 48, 305, 332, 402, 442, 448–449, 470
deer, 10–11, 128, 238, 396
dental, 432, 442–443, 446, 448
Desmodium (telegraph plant), 9, 42
rhythm detection by senses, 2
Dictyostelium (cellular slime mold), 153–154, 158

Subject Index   651

diel rhythm, 22–23, 322, 330
diet, 259, 397, 401, 407, 450
digestive system, 436
disease (animal), 341, 357–359
disease (human), 8, 261–262, 441–451, 459–460, 464, 467–468, 487
disease (plant), 308
diurnal (day-active), 2, 4, 23, 36, 40
DLMO (Dim Light Melatonin Onset), 489, 595
DNA, 27, 93–98, 99–100, 167–170, 239, 240–242, 366
Dog Star (Sirius), 73
dog, 343–344, 348, 351, 354
drift (in streams), 311, 332, 330
Drosophila circadian clock, 181–184
drugs, 7, 410, 426, 460, 470–486

earth, 26, 67, 82, 98
earth tides, 214–216
eclosion, 25, 30–33, 183, 258, 318
ecosystem, 322, 327–328
ECG (electrocardiogram), 5
EEG (electroencephalogram), 6, 9, 25
endogenous, 21–25, 28, 44, 48, 90, 123, 125–127, 130, 141, 216–217, 224, 229–231, 301–302, 309, 343, 346, 355, 357, 360, 360, 380, 389, 394, 446, 448, 373, 488, 536–537
entrainment, 20, 37, 40, 44, 118, 121–122, 125, 142, 178, 181, 209, 217, 304, 410, 427
enzymes, 94, 100, 156, 158, 168, 177, 311, 344, 356
estrus, 42, 266, 343, 361, 363–365, 455
Euglena, 9, 248
European starling (*Sturnus*), 40, 126
evolution, 92–102
exogenous components, 23–25, 44, 90, 141, 216, 298, 401
external coincidence model, 125, 143–144
external timing hypothesis, 141

fatty acids, 159, 161, 163–164
feedback loop, 94, 102, 174–178
fish, 10, 45, 111, 118, 122, 151, 217, 219–220, 222, 296, 301, 310–313, 318–322, 330, 344, 357, 361, 380, 463
flies and fleas, 359–360

flower clock, 75, 87–88
flower, 117, 241–243, 250–251, 260–261, 295, 315–316
fly-fishing, 318–322
forward genetics, 171
Fourier transform, 588
fox, 297, 323–324
free-running, 7, 20–21, 24, 38–40
frequency multiplication/demultiplication, 39
fruit fly (*Drosophila*), 30–33, 41, 142, 173, 178, 181–184, 249, 258

Galileo Galilei, 69, 75, 90, 92
garden warbler (*Sylvia*), 301–302
gating (or gate), 140–144
genetics, 29–32
germination of seeds, 10, 117, 142, 261
G-E-T effect, 248
global workplace, 376, 388
glossary, 20
glycolytic oscillations, 157–158
goat, 259, 344–345, 351, 365
gonads (development), 126, 222, 259, 277, 312
Goodwin oscillator, 148
Greenwich Mean Time (GMT), 81–82

HALO (hours after lights on), 86, 472
hamster (*Mesocricetus*), 4, 30, 36–37, 39, 44, 111, 123–124, 174, 184, 227, 345, 354, 384, 397,
hands of the clock, 93, 144–145, 155
harmonic oscillator, 146, 148
headache, 443, 448–449, 452, 454
heart (cardiac conduction system), 5, 89
Helianthus tuberosus (Jerusalem artichoke), 111, 118
Hellenistic, 74, 77–78
herbicides, 47, 304–306
hibernation, 10, 25, 111
hobbies, 317–322
homeostasis, 433, 470, 474
hormones, 261–264, 270–273, 277–278, 404, 409, 428–429, 433, 456
horse, 9, 111, 312, 343–345, 351, 354, 359, 361, 364
hour (time), 74, 77
hourglass, 75, 123, 141–143

652   Subject Index

hours of changing resistance, 304, 470
human body systems, 427, 430–441
humidity, 44, 47, 193
hypertension, 448, 468–469, 479–181, 489, 529, 534–535
Hyalophora cecropia (giant silk moth), 33
hypothalamus, 34–37, 140, 262, 355, 433, 456, 457

imbibition of seeds, 10, 26, 228–229, 261
immune system, 434, 459–460
IAA (indole-3-acetic acid), 311
infradian rhythms, 7, 10, 22, 25–26, 32, 42, 46, 126, 248–249, 273–284, 344, 351, 354, 384, 414, 449
input pathway, 93, 173, 192–193, 304
insecticides, 7, 305, 307
integrate and fire model, 143
intercourse, 261, 269
internal coincidence model, 125, 144
International Date Line, 82
intertidal zones, 207, 210–211, 217–218
ions, 56, 155–156, 159, 162, 165

Japanese morning glory (*Pharbitis nil*), 126, 260
Japanese radish (*Raphanus sativus*), 128–129, 187, 299
jet lag, 38, 85, 382, 392–394, 397, 457, 460, 487–488

latitude, 90, 107, 110, 116, 129–130, 389, 445, 453
LD (light-dark), 20, 44
leaf movements, 9, 15, 40, 50, 54, 75, 90, 166, 187
learning activities, 50–56, 570
least-squares method, 590
Lemna (duckweed), 39, 296
light, 22, 40, 44, 99–101, 112, 118–122, 130–132, 177–178, 207–209, 410, 456–458, 552, 554
light pollution, 328–332, 408–411, 456–462
limit cycle, 147–152
lipids, 159, 160, 163
litigation, 396
lizard, 111, 343, 363
LL (constant light), 20, 30, 40, 48–49
longitude, 75, 76, 80–82, 84–85, 90, 214

longitude clock, 80
Lorenz attractor, 151–152
Lotka-Volterra equations, 148
luciferase, 170, 180, 187, 190
lunar cycles, 74, 76, 207, 212, 216, 222, 226–229, 265–266, 378

mammalian circadian clock, 184–187
marker rhythm, 428, 488–489, 531–533
Mars, 29, 399
Maryland Mammoth tobacco, 113, 294–295
masking, 20, 47, 48, 226, 346–347, 355, 488–489
mathematical models, 145–153
mayflies (Ephemeroptera), 319–322
meal-timing, 313, 346, 402–408, 488
mechanical models, 141–145
mechanisms, 139–141
Medical Chronobiology Aging Project (MCAP), 428–429, 441, 465–466
meiosis, 239, 240–244, 248, 251, 257, 283
melanopsin, 36–38, 121–122, 178, 356
melatonin, 43, 259–260, 271, 312, 344, 351, 354–359, 361–363, 365–366, 393, 397, 399, 401, 409–411, 428–429, 433, 448, 454, 456–462, 488–489, 532
membrane models, 158–167
menstrual cycle, 32, 42, 46, 73, 227, 262, 384, 414, 456–455, 458, 544
mesor, 29, 468, 591, 595, 597
microfilaria, 357–359, 467
migration, 2, 6, 10, 26, 101, 116, 126, 300
military time, 86
Mimosa (sensitive plant), 91
mitosis, 143, 240–241, 244–245, 248, 464
molecular models, 167–192
monarch butterfly (*Danaus plexippus*), 302–304
monkey, 9, 210, 227, 266, 343–345, 351, 353–354, 358, 380, 404
month (time), 71–77
mood, 131, 268–270, 278, 279, 401, 403, 453, 536, 557
moon, 26–29, 66–67, 71–74, 76, 79, 95, 112, 207–215, 223–224, 226–229, 329–330, 476
moon phases and humans, 227–228

Morningness-Eveningness (chronotype), 386, 391–392, 535–537
moths, 33, 115, 238, 305, 308, 331
mouse (*Mus*), 7, 36–37, 39–40, 42, 48, 184–186, 246–248, 266, 305, 345–347, 351, 402–403, 455, 460, 471–473, 482–484, 487–488

NADH, 9, 47, 155, 157, 164
National Institute of Standards and Technology, 67, 84
Nanda-Hamner protocol, 123
natural resources, 293
nematode (*Caenorhabditis elegans*), 25, 89, 171
Neurospora circadian clock, 179–181
Night Blooming Jessamine (*Cestrum nocturnum*), 260
night dippers, 468
nocturnal (night-active), 4, 23, 36, 40, 48, 86, 210, 218, 228, 246–247, 259, 272, 302, 311, 322, 324–325, 330–331, 343, 355, 386, 396, 403, 461, 464, 467–468, 473, 481–482, 484
nucleotides, 93, 95, 127, 155, 157–158, 167–170
Nyquist folding frequency, 579

onion (*Allium cepa*), 114, 244–245
opsin, 37, 119, 361
oral mucosa, 186, 246, 249–250, 431, 440, 446
oral temperature, 45, 388, 391, 393, 398, 437–438, 527–528, 530–533, 536–538, 547–549
oscillator, 123, 138–146, 148–158, 162–164, 171, 174, 176, 179, 199–193, 217, 229–231
Oscillatoria, 96

pain, 4, 269, 273, 275, 277, 384, 392, 412, 428, 432, 442–443, 454, 476–477, 480, 490
palolo worm (*Eunice*), 2, 223
parasites, 357
peak, 14, 20, 29, 591–593, 595, 597
penile tumescence, 272
performance, 7, 12, 379, 383, 386–387, 389–396, 399, 427, 529, 537–538, 552

period, 20–22, 30, 38–42, 591, 595, 597
periodogram, 578, 588–589
pest management, 304–309
phase, 14, 20–21, 29–30, 36, 44–47, 146, 536, 591
phase-response curve (PRC), 46, 125
phase-shifting, 44–46, 186, 387, 410, 414, 540
pheromones, 238, 268, 384
phospholipids, 159–163, 166
phosphorylation, 95, 157, 168, 176, 183–184, 189, 191–192
photomorphogenesis, 99, 177, 296
photoperiodic response types, 113–115
photoperiodism, 109, 111–113
photophil, 125, 144
photoreceptors &/or pigments, 93, 99, 100, 112, 117–119, 122
photosynthesis, 99–100, 189
phototaxis, 99, 101, 227
phylogenetic tree, 96–97
phytochromes, 112, 119, 121–122, 177
pig, 341, 344–346, 351, 355, 362, 365
pineal gland, 34–35, 119, 121, 140, 312, 356–357, 361, 409–410, 456–457, 459, 462
polypeptides, 38, 93, 168–170
post-lunch dip, 395–396
predator-prey limit cycle, 147, 149–150
primary circadian clock, 32–38, 192–193, 355–357
Prime Meridian, 75, 82
proteins, 93–97, 102, 156, 160, 162–179
pseudoscience (biorhythms), 367, 377, 412, 414, 418, 420, 422
pulse (heart rate), 2, 89, 92, 264, 390, 410, 428, 435, 441, 527, 538–539, 544, 549–550, 552–554
pulselogium, 75, 90, 92
Punnett Square, 240–241, 244, 249

Q_{10}, 41, 50, 163, 223, 300

rat, 48, 246–247, 345, 355, 402–403, 482–484
recording physical & biological time, 85–87
REM/non-REM, 9, 272, 386
relaxation oscillator, 143

reproduction, 220, 237, 250, 255, 259–261, 360, 363,
retinal ganglion cells (RGCs), 37–38
retinohypothalamic tract (RHT), 37–38
rhodopsin, 37, 119, 122
rhythm characteristics, 20–21, 356, 426, 467–469, 595, 597
rhythm (defined), 1, 20–21
rhythm domains, 6, 9, 22, 25
rhythm method, 269
RNA, 36, 93–98, 167–170, 249–250, 304, 487, 489
ruffed grouse (*Bonasa*), 8, 10, 323, 326
running wheels, 36, 184–185

SAD (Seasonal Affective Disorder), 131–132, 452–453, 467
saliva, 278, 393, 428–429, 435–436, 445, 488
SCN (suprachiasmatic nucleus), 34–38, 45, 140, 178, 186, 356, 363, 433, 448, 457, 462, 487–488
scotophil, 125, 144
second (time), 9, 75, 77, 82–83
self-measurement, 12, 391, 526–558
semen, 284, 365, 439, 445
semidian (12 h) rhythm, 126, 245, 248, 354, 399–401
sexuality, 237–268, 259–261
sex-related rhythms in men, 272–284
sheep (*Ovis*), 312, 343–345, 361–363
shiftwork, 11, 38, 377, 379, 385–388, 461–462, 487–488, 533
sidereal day, 69
sidereal month, 71, 208, 214
skeleton photoperiods, 123, 362, 364
skin, 7, 37, 118, 159, 186, 246, 249, 411, 428, 431, 440, 446, 474, 532, 546–547, 557
slave oscillator or clock, 83, 142
sleep, 9, 11, 25, 44, 131, 272, 377–380, 385–392, 394–399, 401, 404, 428, 452–453, 457, 460–461, 467–468, 481, 487, 489, 535–537, 579
social isolation, 39, 280, 401, 540–544, 547, 579
social synchronization, 44, 46, 268, 380–385, 397, 401
society, 376–415

solstice, 70–71, 76, 113
sound, 2, 22, 44, 75, 77, 80, 305, 353, 471
spatiotemporal systems, 153, 155
sperm, 223, 237, 251, 366, 445
S-phase (cell cycle), 87, 246, 249, 432, 445, 447, 460, 464, 473, 484–485
splitting, 39
sports-related performance, 389–393, 529, 537
squirrel, 9, 10, 48, 223, 323–324, 327, 345
stem diameter, 229
stomata, 165, 244, 311
subjective day, 125, 186
subjective night, 56, 125
submarine personnel, 399–401
suicide, 131, 228, 452
susceptibility, 11, 302, 309, 403, 447, 470–472, 474, 478, 487
synchronizer, 20, 33, 36, 44, 67, 90, 93, 126, 130, 139, 173, 178, 193, 209, 217, 223, 229, 301, 346, 356, 380, 397, 531
synchrony, 1, 11, 26, 86, 98, 197, 201, 209, 215–216, 220, 264, 268, 287, 290, 379–380, 384, 389, 418, 420–421

telemetry, 322–327, 349, 549–550
temperature (environmental), 41–42, 178–179, 297–300
temperature compensation, 41–42, 159, 163–164, 217, 300
thermoperiodism, 297–298
tidal rhythms, 22, 26, 210–214, 216–221, 224–226, 231
tides, 207, 210, 212, 214, 227
time zones, 84
time, 26–29, 73–75, 473
tomato (*Solanum lycopersicon*), 48, 114, 298, 300, 311
transients, 38, 44
transpiration, 9, 25, 187
travel, 44, 80, 126, 302, 392–393, 396, 399
twilight, 44, 74, 108, 325, 326–329, 380

ulcer, 436, 442, 446–448, 478
ultradian rhythm, 7, 9, 22, 25–26, 30, 32, 41, 44, 52–54, 89, 92, 98, 125–126, 153–155, 158, 164, 248–249, 255, 258, 272, 312, 343, 348–349, 354, 380, 468, 529, 581, 598

Subject Index   655

ultraviolet light (UV), 22, 94, 99–101, 122, 178, 304, 332
urine, 275, 278–279, 349, 351–355, 428–430, 437, 441, 445, 488, 533, 540, 546, 547–548, 555

van der Pol equations, 148
vehicle accidents, 381–383, 394–396
velvetleaf (*Abutilon*), 2, 7, 47, 56
vernalization, 128–129, 298–300
veterinary medicine, 341–366
virus, 261, 308–309, 446–447, 464, 477

water (H_2O), 42, 44, 74, 77, 95, 99, 162, 193
waveform, 8, 42, 48, 467, 592–595
weather patterns, 313–314, 446
week (time), 78–79
warbler (bird), 10, 127, 301–302

year (time), 69–77
yeast, 9, 25, 157–158

zeitgeber, 20, 24, 44, 86
zeitgeber time (ZT), 86
zero-amplitude test, 352
zugunruhe, 126–127